Introduction to Technical Mathematics
Fourth Edition

Allyn J. Washington
Mario F. Triola

Dutchess Community College
Poughkeepsie, New York

ADDISON-WESLEY PUBLISHING COMPANY

Reading, Massachusetts • Menlo Park, California • New York • Don Mills, Ontario • Wokingham, England
Amsterdam • Bonn • Sydney • Singapore • Tokyo • Madrid • San Juan • Milan • Paris

Other books by
Allyn J. Washington

Basic Technical Mathematics, Fourth Edition
Basic Technical Mathematics with Calculus, Fourth Edition
Basic Technical Mathematics with Calculus, Metric, Fourth Edition
Technical Calculus with Analytic Geometry, Second Edition
Arithmetic and Beginning Algebra
Essentials of Basic Mathematics, Third Edition

Mario F. Triola

Elementary Statistics, Third Edition
Mathematics and the Modern World, Second Edition

Sponsoring Editor: Sally Elliott
Production Editor: Greg Hubit
Design: Nancy Benedict
Cover photo: © Mark Greenberg / Visions 1986

Library of Congress Cataloging in Publication Data

Washington, Allyn J.
  Introduction to technical mathematics.

  Includes index.
  1. Mathematics—1961–   I. Triola, Mario F.
II. Title.
QA39.2.W374   1988   513'.14   87-30657
ISBN 0-8053-9538-5

11 12 13 14 15 16 17 18 19 20 DO 959493

Preface

Introduction to Technical Mathematics is designed for students in technical programs at colleges and technical institutes. It is intended for those who require a basic knowledge of mathematics for use in their particular programs and professions. The general topics included are an arithmetic review, algebra, geometry, and trigonometry. The level of presentation is suitable for those whose backgrounds lack algebra or geometry or require a review of the basic topics. The concepts discussed in this book make it suitable for a course that can be used as preparation for more advanced work in mathematics or as a terminal course. This text has been designed primarily for formal courses, but it could be easily used by individuals wishing to follow a self-study program.

The examples and exercises include many applications from different fields of technology, such as electronics, mechanics, machine design, civil engineering, forestry, architecture, automotive engineering, physics, chemistry, computer science, and computer operations. These applications are intended to illustrate the use of mathematics. The technical material itself is not developed and no prior technical knowledge is required.

The content of this book is developed in an informal and intuitive manner. All important terms are carefully introduced, and basic concepts are developed in order to give the student an understanding of the material and how it relates to other topics. More detail has been included for those concepts which are generally found to be more difficult. The emphasis throughout the text is on the development of the mathematics necessary for technical work.

New Features of This Edition

- There is a new chapter (Chapter 20) dealing with complex numbers.
- There is a new section (14-4) that includes the use of determinants in solving systems of equations.
- There is a new section (14-5) describing algebraic solutions for systems of three equations with three unknowns.
- There is a new section (8-4) that covers factoring of the sum and difference of two perfect cubes.
- The sections in Chapter 7 have been rearranged, with a new section created.
- In Chapter 12, the first two sections have been combined into one, with greater emphasis placed on converting between logarithmic and exponential forms.
- This edition incorporates the use of a second color that enhances the design.
- A special margin symbol ▶ and diagrammed comments are used to identify and clarify the more difficult concepts.
- Each chapter begins with separate introductory remarks and an introductory problem. All of these introductory problems are related to the theme of aviation.
- The important formulas from the geometry chapters are summarized inside the back cover.
- At the end of each chapter, all chapter formulas are listed together for easy reference.
- There are now more than 6500 exercises, including approximately 1600 new exercises. That is an increase of 42%. Each exercise set is numbered according to the section it follows. Exercises are grouped so that there is generally an even-numbered exercise equivalent to each odd-numbered exercise.
- There are more than 700 examples, with 20% of them new. The examples and exercises include applications from the latest technological developments.
- The number of figures has been substantially increased so that there are now about 575 figures. This is an increase of 63% over the last edition.

Additional Features

- Stated problems appear in most sections. The solving of such problems makes the student more accustomed to them. The simplest forms of stated problems are found in the earlier chapters, and Chapter 5 devotes an entire section to the analysis of such problems.

- There is an appendix (Appendix B) that discusses the fundamentals of the BASIC programming language. That appendix includes fifteen programs, along with sample runs, many of which correspond directly to examples in the text.

- The examples are used to introduce concepts as well as to clarify and illustrate material already presented.

- Coverage of the use of calculators is extensive. Calculator usage is discussed throughout the text in appropriate sections, and Appendix A is a reference for most of the commonly used calculator keys.

- The order of coverage may be changed in several places without loss of continuity. Also, certain sections may be omitted without loss of continuity. Any omissions or changes of order will, of course, depend on the type of course and the completeness required.

- Review exercises are included at the end of each chapter. These may be used as a source of additional problems or for review assignments.

- The answers to all odd-numbered exercises are given at the back of the book. Included are answers to graphical problems and other types that are often left out of other textbooks.

- The instructor's manual for *Introduction to Technical Mathematics, 4th Ed.*, contains general comments and suggestions on the use of the material in each chapter and the answers to the exercises (both even and odd). This information could be helpful in preparing and designing the course to meet the needs of the students.

Acknowledgments

The authors wish to acknowledge the many helpful comments and suggestions provided for the preparation of this book. In particular, we wish to thank the following reviewers who supplied many valuable suggestions: Dale Boye, Schoolcraft College; Douglas Cook, Owens Technical College; Charles McSurely, Nashville State Technical Institute; Barbara Miller, Lorain County Community College; Robert Opel, Waukesha County Technical Institute; Linda A. Plastas, Montgomery College; Jack D. Scrivener, Clackamas Community College; William Webb, University of Akron; Frank Weeks, Mt. Hood Community College; and Jean Wilberg, SUNY Morrisville. We also appreciate the suggestions made by users of earlier editions. We thank Gloria Langer for her work in checking answers to the exercises. The assistance, cooperation, and diligent efforts of the Benjamin/Cummings staff are also greatly appreciated.

A.J.W.
M.F.T.

Contents

THIRTEEN Graphs 387

FOURTEEN Simultaneous Linear Equations 426

FIFTEEN Additional Topics from Geometry 466

SIXTEEN Trigonometry of Right Triangles 515

Arithmetic

We live in an era of continued rapid technological growth. This technology provides us with dramatic results that affect and enhance all of our lives. Computers, the space program, and nuclear energy are but three of the major growth areas that constantly demand technological improvements. We now routinely use integrated circuits in calculators, televisions, cars, microwave ovens, and video cassette recorders. Lasers are routinely used in surveying and surgery. We have aircraft that are faster, more maneuverable, and able to carry heavier payloads with greater fuel efficiency. Our homes are more energy efficient. Our leisure activities are enriched by stereo color televisions, video cassette recorders, computer video games, and high-quality sound systems.

All of these developments depend on technical knowledge founded in mathematics. Mathematics is a necessary and critical component of contemporary technology. There is, of course, beauty in mathematics itself. But from a more practical viewpoint, an understanding of basic mathematics opens the door to the many different fields of technology. In this text we begin the development of the mathematics required for entry into these scientific and technical fields. This text includes examples and exercises from such areas as electrical and mechanical technology, civil engineering, solar engineering, physics, and computer science; however, successful solution of applied problems will *not* require a knowledge of the relevant field of application. In addition to the mathematics itself, this text is also designed to increase your familiarity with the metric system and the use of calculators.

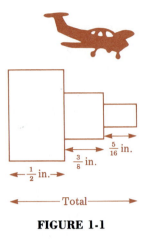

FIGURE 1-1

This text includes a wide variety of different exercises and examples that illustrate applications of the related mathematical concepts. In this chapter, for example, the applications refer to such topics as oil drilling, fuel-oil consumption, cassette tapes, satellites, electrical resistors, computers, gears, surveying, solar energy, chemistry, weather, architecture, and aviation. At the beginning of each chapter, we will present a problem that is related to aviation. Although these opening problems all relate to aviation, they involve a wide variety of different mathematical concepts. These problems will be solved in their respective chapters. Among the topics in Chapter 1 is the addition of fractions. If an aircraft firewall is $\frac{1}{2}$ in. thick and that thickness is increased by a $\frac{3}{8}$-in. washer and a $\frac{5}{16}$-in. bolt, what is the combined total thickness? See Fig. 1-1. This is only one of the many types of problems included in Chapter 1. This particular problem will be solved later in the chapter.

1-1 Addition and Subtraction of Whole Numbers

Much of the material in this chapter may be review. It is important to be skilled in this fundamental material so that the following chapters can be studied. Finding the solutions to most applied problems in science and technology involves arithmetical computations. To perform these computations, we use the basic arithmetical operations of addition, subtraction, multiplication, and division. These operations are essential to performing computations, and they are fundamental to the development of the various branches of mathematics itself. It is important that these operations be performed accurately and with reasonable speed. Although we assume that you are familiar with these operations, we shall include a brief discussion here for review and reference.

The most fundamental use of numbers is that of counting. *The numbers used for counting, the* **natural numbers,** *are represented by the symbols 1, 2, 3, 4, and so on. When we include* **zero,** *we have the* **whole numbers,** *which are represented by 0, 1, 2, 3, and so on.*

Any of the whole numbers can be written with the use of the 10 symbols 0, 1, 2, 3, 4, 5, 6, 7, 8, 9 if the actual *position* of a symbol in a given number is properly noted. This important feature used in writing numbers— that of *placing each symbol in a specified position in order to give it a particular meaning—is referred to as* **positional notation.**

EXAMPLE A   In the number 3252, read as "three thousand two hundred fifty-two," the left 2 represents the number of hundreds and the right 2 represents the number of ones, because of their respective positions. Even though the symbol is the same, its position gives it a different value. Also, the 3 represents the number of thousands and the 5 represents the number of tens, because of their respective positions.

In the number 325, the 2 represents the number of tens and the 5 represents the number of ones. In the number 352, the 5 represents the number of tens and the 2 represents the number of ones. Even though the same symbols are used, the different positions result in different values. □

The process of finding the total number of objects in two different groups of these objects, without actually counting the objects, is **addition.** *The numbers being added are the* **addends,** *or* **terms,** *and the result is the* **sum.** In performing an addition we must be careful to *add only like quantities,* such as feet to feet or quarts to quarts. If we are adding or subtracting pure numbers without units like gallons or miles, we can add or subtract in the obvious way. But if the numbers are associated with units like gallons or miles, we cannot add or subtract unless the units are the same. This means that we cannot add 2 qt to 3 gal without first expressing both quantities in the same units, such as quarts. We cannot add 2 qt to 3 mi since there is no way to make the units agree.

EXAMPLE B If one container has a capacity of 4 qt and another has a capacity of 2 qt, we find the combined capacity of the two containers by adding

$$4 \text{ qt} + 2 \text{ qt} = 6 \text{ qt}$$ □

In Example B we used qt as the symbol for quart. When such units of measurement are designated by their appropriate symbols, and these symbols might be unfamiliar, they will be defined when used in this chapter. It is, however, assumed that you are familiar with the following units and their symbols:

Unit	Symbol	Unit	Symbol
inch	in.	gallon	gal
foot	ft	minute	min
yard	yd	pound	lb
mile	mi	quart	qt
		ounce	oz

A discussion of units and their symbols is found in the first section of Chapter 2.

When we are adding two whole numbers, we must take into account positional notation and add only those numbers with the same positional value. In this way we are adding like quantities.

EXAMPLE C   When we add 46 and 29, we are saying

$$4 \text{ tens} + 6 \text{ ones}$$
$$\underline{2 \text{ tens} + 9 \text{ ones}}$$
$$6 \text{ tens} + 15 \text{ ones}$$

Since 15 ones = 1 ten and 5 ones, we then have

$$6 \text{ tens} + 1 \text{ ten} + 5 \text{ ones} = 7 \text{ tens} + 5 \text{ ones}$$

We usually perform this addition as follows.

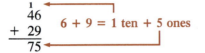

where the 1 shows the number of tens "carried" from the ones column to the tens column. □

You should know the basic sums through 9 + 9 = 18. If you are at all unsure of any of these, write them out so that you can review them. Being able to perform addition accurately and with reasonable speed comes from knowing the basic sums well; this takes practice.

It is also wise to form the habit of checking your work. Several methods of checking addition are available. A simple and effective method is to add the columns in the direction opposite to that used in finding the sum originally.

EXAMPLE D   If we find the sum of the indicated numbers by adding the columns downward, we can check the results by adding again, this time upward.

$$
\begin{array}{cc}
\overset{2\,2}{327} & \overset{2\,2}{327} \\
582 & 582 \\
695 \quad \text{add} & 695 \quad \text{check} \\
\underline{419} & \underline{419} \\
2023 & 2023
\end{array}
$$

□

The process of **carrying** *a number from one column to the next is necessary whenever the sum of the digits in a column exceeds 9.* A very similar situation occurs when we add distances expressed in more than one measuring unit. Consider the following example.

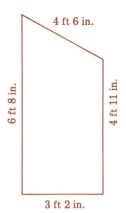

4 ft 6 in.

6 ft 8 in.

4 ft 11 in.

3 ft 2 in.

```
 6 ft   8 in.
 4 ft   6 in.
 4 ft  11 in.
 3 ft   2 in.
17 ft  27 in.
```

FIGURE 1-2

EXAMPLE E

The **perimeter** *of a plane geometric figure is the total distance around it.* Suppose a solar home has a window with a shape described by Figure 1-2. To find the perimeter of this window, we must add the indicated lengths. The result is 17 ft 27 in., but a more useful result is obtained if we use the fact that 27 in. = 2 ft 3 in. We can then state that the perimeter is 19 ft 3 in. In this addition process, we essentially "carried" 2 from the inch column to the foot column.   ☐

We must often determine how much greater one number is than another. This leads to the operation of **subtraction,** the inverse of addition. *Subtraction consists of reducing the number from which the subtraction is being made (the* **minuend***) by the number being subtracted (the* **subtrahend***). The result is called the* **difference.**

EXAMPLE F   If we wish to subtract 29 from 73, we find that for the number of ones involved, we are to reduce 3 by 9. When we consider natural numbers, we cannot perform this operation. However, if we "borrow" 10 ones from the tens of 73, the subtraction amounts to subtracting 2 tens + 9 ones from 6 tens + 13 ones. We can show the subtraction as

$$
\begin{array}{r} 7 \text{ tens} + 3 \text{ ones} \\ -\ 2 \text{ tens} + 9 \text{ ones} \end{array}
\quad \text{or} \quad
\begin{array}{r} 6 \text{ tens} + 13 \text{ ones} \\ -\ 2 \text{ tens} +\ 9 \text{ ones} \end{array}
$$

Using the second form, we see that the result is 4 tens + 4 ones, or 44. The usual form of showing the subtraction is

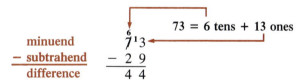

```
                    73 = 6 tens + 13 ones
    minuend       7¹3
 −  subtrahend   − 2 9
    difference     4 4
```

Here the small 1 shows the number of tens borrowed, and the small 6 shows the remaining tens after the borrowing. Borrowing in subtraction is essentially the opposite of carrying in addition.

We can check subtraction by adding the difference and the subtrahend. The sum should be the minuend. This follows directly from the meaning of subtraction. The check of the preceding subtraction is as follows:

```
    29        subtrahend
 + 44       + difference
   73         minuend
```
                                                                          ☐

In many subtraction problems it is necessary to borrow more than once. It might be necessary to borrow from the tens and then again from the hundreds or thousands. The next example illustrates this type of subtraction.

EXAMPLE G The subtraction $8203 - 4659$ is shown as follows:

$$\begin{array}{r} {}^{7}\!\!8{}^{1}\!2{}^{9}\!{}^{1}\!0{}^{1}\!3 \\ -\ 4\ 6\ 5\ 9 \\ \hline 3\ 5\ 4\ 4 \end{array}$$ ← $8203 = 7$ thousands $+ 11$ hundreds $+ 9$ tens $+ 13$ ones

Here we see that it was necessary to borrow from the tens, although there were initially no tens to borrow from. We had to borrow first from the hundreds and then from the tens. Finally, it was necessary to borrow from the thousands to complete the subtraction in the hundreds column. □

Although the additions and subtractions discussed in this section can be easily and quickly performed on a calculator, you must understand these basic arithmetic operations and be able to perform them mentally. A good understanding of all the basic operations is important to comprehending algebra and other topics we will develop. Only after you understand these operations and can perform them accurately should you consider using a calculator. Many gross errors have been made with calculators because they have been used improperly. We discuss the use of calculators in Section 1-11.

Exercises 1-1

In Exercises 1 through 12, add the given numbers. Be sure to check your work.

1. 36	2. 45	3. 627	4. 433	5. 446	6. 809
29	89	83	612	915	826
87	37	524	109	992	278
				67	548

7. 8028	8. 7695	9. 3873	10. 989	11. 30,964	12. 87,657
4756	4803	9295	3216	9,877	93,984
4803	986	4082	4807	92,286	57,609
3823	7375	399	736	5,547	8,726
		7646	9297	965	92,875

In Exercises 13 through 24, perform the indicated subtractions. Check your work.

13. 8704	14. 5162	15. 873	16. 921	17. 8305	18. 2006
− 3102	− 2041	− 292	− 224	− 7356	− 1197

19. 36,047	20. 32,105	21. 40,165	22. 10,906	23. 290,078	24. 872,110
− 26,249	− 22,116	− 9,586	− 9,928	− 194,396	− 682,324

In Exercises 25 through 32, perform the indicated operations. Be sure to express answers with the proper units.

25. 5 ft 8 in. + 2 ft 6 in.   **26.** 6 ft 10 in. + 5 ft. 8 in.   **27.** 8 lb 13 oz + 9 lb 3 oz

28. 2 lb + 20 oz

29. 7 ft 11 in. − 2 ft 6 in.

30. 6 ft 4 in. − 3 ft 5 in.

31. 9 lb 2 oz − 4 lb 7 oz

32. 3 lb − 25 oz

In Exercises 33 through 40, find the perimeters of the given figures. Be sure to express answers with the proper units.

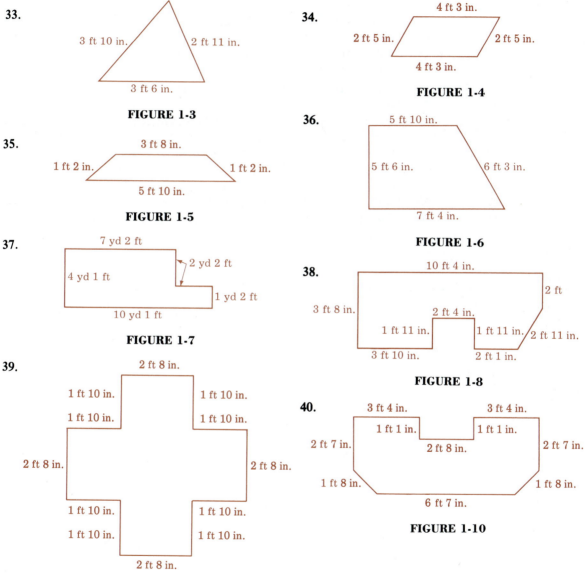

33.

3 ft 10 in.   2 ft 11 in.

3 ft 6 in.

FIGURE 1-3

34.

4 ft 3 in.

2 ft 5 in.   2 ft 5 in.

4 ft 3 in.

FIGURE 1-4

35.

3 ft 8 in.

1 ft 2 in.   1 ft 2 in.

5 ft 10 in.

FIGURE 1-5

36.

5 ft 10 in.

5 ft 6 in.   6 ft 3 in.

7 ft 4 in.

FIGURE 1-6

37.

7 yd 2 ft

2 yd 2 ft

4 yd 1 ft

1 yd 2 ft

10 yd 1 ft

FIGURE 1-7

38.

10 ft 4 in.

2 ft

3 ft 8 in.

2 ft 4 in.

1 ft 11 in.   1 ft 11 in.   2 ft 11 in.

3 ft 10 in.   2 ft 1 in.

FIGURE 1-8

39.

2 ft 8 in.

1 ft 10 in.   1 ft 10 in.

1 ft 10 in.   1 ft 10 in.

2 ft 8 in.   2 ft 8 in.

1 ft 10 in.   1 ft 10 in.

1 ft 10 in.   1 ft 10 in.

2 ft 8 in.

FIGURE 1-9

40.

3 ft 4 in.   3 ft 4 in.

1 ft 1 in.   1 ft 1 in.

2 ft 7 in.   2 ft 8 in.   2 ft 7 in.

1 ft 8 in.   1 ft 8 in.

6 ft 7 in.

FIGURE 1-10

In Exercises 41 through 48, find the difference in length of the longest and the shortest sides in the indicated figure.

41. Figure 1-3

42. Figure 1-4

43. Figure 1-5

44. Figure 1-6

45. Figure 1-7

46. Figure 1-8

47. Figure 1-9

48. Figure 1-10

In Exercises 49 through 64, solve the given problems and check your work. Where appropriate, the following units of measurement are designated by the given symbols:

Unit	Symbol	Unit	Symbol
second	s	ohm	Ω
hour	h	gram	g
square mile	mi^2	kilogram	kg

49. An oil well drilling rig goes down 312 ft on the first day, 126 ft the second day, and 92 ft the third day. How deep is the well at the end of the third day?

50. A furnace burns 327 gal of oil in November, 415 gal in December, and 398 gal in January. How much oil is burned during these three months?

51. In buying electrical testing equipment, a technician spends $789, $624, and $278 for three instruments. How much did she spend?

52. Three songs on a cassette tape require 2 min 40 s, 3 min 15 s, and 3 min 42 s. What is the total time required for these three songs?

53. The times for each of three successive orbits of a certain satellite circling the earth are 1 h 36 min 26 s, 1 h 34 min 47 s, and 1 h 30 min 17 s. How long do these three orbits take?

54. Three parcels weigh 2 lb 9 oz, 4 lb 11 oz, and 3 lb 14 oz. What is the sum of their weights?

55. In June, 345 gal of water evaporate from a pond and in July, 623 gal evaporate. How much more water evaporates in July than in June?

56. An electrical resistor has a resistance of 175 Ω. Because of overheating, the resistance changes to 241 Ω. By how much does the resistance change?

57. During a certain time interval, a projectile rises from a height of 256 ft to a height of 412 ft. What is the change in altitude during that time interval?

58. A chemist has 1064 g of a certain sample before drying and 846 g after drying. What is the loss in grams during the drying process?

59. How long is it from sunrise to sunset on a day in which the sun rises at 7:35 A.M. and sets at 6:18 P.M.?

60. One toxic waste container holds 14 gal 1 qt and another holds 8 gal 3 qt. How much more does the larger container hold than the smaller one?

61. Three objects together weigh 503 kg. Two of these objects weigh 87 kg and 228 kg. How much does the third object weigh?

62. A person purchases several computer components for a total of $203. Two of the items cost $87 and $79. What is the cost of the other items?

63. A satellite is used to photograph three land areas totalling 56,410 mi^2. If two of the included regions have areas of 26,340 mi^2 and 13,820 mi^2, what is the area of the remaining region?

64. A small airplane weighs 3872 lb when fully loaded. What is the total weight after fuel weighing 84 lb is used up and 120 lb of baggage is removed?

1-2 Multiplication and Division of Whole Numbers

Multiplication *is a short-cut method for doing repeated additions of the same number. The number being multiplied is called the* **multiplicand,** *the number of times it is taken is the* **multiplier,** *and the result is the* **product.** The multiplicand and multiplier are also called **factors.** The basic notations used to denote multiplication are ×, ·, and parentheses.

EXAMPLE A   By the expression 3 × 5 we mean

$$3 \times 5 = 5 + 5 + 5 = 15$$

Here 3 is the multiplier, 5 is the multiplicand, and 15 is the product. The product can also be expressed as

$$3 \cdot 5 = 15 \quad \text{or} \quad (3)(5) = 15 \qquad \square$$

To perform multiplication accurately and with reasonable speed, it is necessary, as with addition, to know the basic products through 9 × 9 = 81 without hesitation. If you have any doubt about your knowledge of these products, review them and practice until they have been mastered.

The process of multiplication has certain basic properties. One of these, known as the **commutative law,** states that *the order of multiplication does not matter.* Another of these, the **associative law,** deals with the multiplication of more than two factors. It states that *the grouping of the numbers being multiplied (***factors***) does not matter.* Since multiplication is basically a process of addition, these properties also hold for addition.

EXAMPLE B   The commutative and associative laws are illustrated below.

	commutative law	associative law
addition	3 + 5 = 5 + 3	(3 + 5) + 7 = 3 + (5 + 7)
multiplication	2 × 6 = 6 × 2	(2 × 6) × 4 = 2 × (6 × 4)

both products   both products
are 12   are 48 $\square$

Another important property of numbers that involves multiplication and addition is known as the **distributive law.** This law states that *if the sum of two numbers is multiplied by another given number, each is multiplied by the given number and the products are added to find the final result.* This may sound complicated, so examine the following example closely.

EXAMPLE C　　Applying the distributive law to the product $(4)(3 + 6)$, we have

$$(4)(3 + 6) = (4)(3) + (4)(6) = 12 + 24 = 36$$

We see that this gives the same result as the product $(4)(9) = 36$, since $3 + 6 = 9$.　　□

　　　Many of us can execute the usual multiplication process even though we might not totally understand the reasons for this procedure. In the next example we first illustrate the multiplication procedure and we then use the distributive law to explain why it works.

EXAMPLE D　　When multiplying 27 by 3, the usual procedure is shown as

$$\begin{array}{r} \overset{2}{2}7 \\ \times\ \ 3 \\ \hline 81 \end{array}$$　　$3 \times 7 = 2$ tens $+ 1$ one

We multiply 7 by 3 and obtain 21, placing the 1 under the 3 and carrying the 2 into the tens column. The 2 is then added to the product $3 \times 2 = 6$ to obtain the digit 8 which appears in the product 81.

　　　We could have multiplied 27 by 3 by using the distributive law as follows:

$$\begin{aligned} 3(27) &= 3(20 + 7) \\ &= 3(20) + 3(7) \\ &= 60 + 21 = 60 + (20 + 1) \\ &= (60 + 20) + 1 = 81 \end{aligned}$$

□

EXAMPLE E　　To find the product of 26 and 124, we would proceed as follows:

$$\begin{array}{r} 124 \\ \times\ 26 \\ \hline 744 \\ 248\ \ \ \ \\ \hline 3224 \end{array}$$　　displaced one place to left

If we now consider the product as

$$\begin{aligned} (124)(26) &= (124)(20 + 6) \\ &= (124)(20) + (124)(6) = 2480 + 744 = 3224 \end{aligned}$$

we see that the first line (744) obtained in the multiplication process is also one of the products found by using the distributive law. We then note that the 248 found in the multiplication process is equivalent to the 2480 product found with the distributive law. The final zero is not written, but it is represented since the 248 is displaced one position to the left.　　□

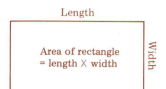

Area of rectangle
= length × width

FIGURE 1-11

There are many applications of multiplication, some of which are demonstrated in the exercises. One of the most basic applications is in determining the **area** *of a* **rectangle**, *which is defined to be the product of the rectangle's* **length** *and* **width** (see Figure 1-11).

EXAMPLE F   A rectangular plate is used as a ramp for the handicapped. If its length is 4 ft and its width is 6 ft, as shown in Fig. 1-12, its area is given by

Area = (4 ft) × (6 ft) = 24 ft²

Here ft² is the symbol for square foot.   ☐

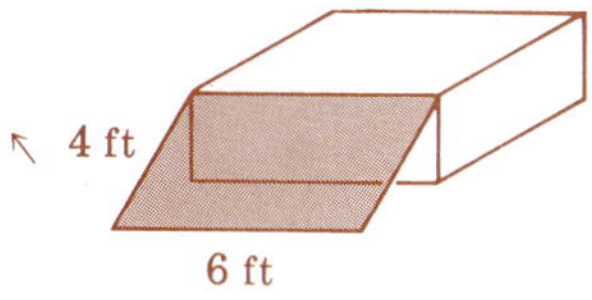

4 ft

6 ft

FIGURE 1-12

Just as subtraction is the inverse process of addition, division is the inverse process of multiplication. Since multiplication can be thought of as a process of repeated addition, we can think of division as a process of repeated subtraction. If we subtract 2 from 10 five times, for example, the result is zero. This means that 10 divided by 2 is 5. Here 10 is the **dividend,** 2 is the **divisor,** and 5 is the **quotient.** A common notation for this is 10 ÷ 2 = 5, where ÷ indicates division.

If the division "comes out even," the product of the quotient and the divisor will be equal to the dividend. This gives us a way to check that the division has been done correctly. It is assumed that you know how to perform the division process, but the following example is given for review and also as an explanation of why the division process works.

EXAMPLE G   Suppose that we want to divide 3288 by 24. We set up the problem as follows:

$$
\begin{array}{r}
137 \\
24\overline{)3288} \\
2400 \\
\hline
888 \\
720 \\
\hline
168 \\
168 \\
\end{array}
$$

quotient
dividend
24 × 100

24 × 30

24 × 7

divisor

We can now see that

3288 = 24 × 137 = (24 × 100) + (24 × 30) + (24 × 7)
= 24(100 + 30 + 7)

$$
\begin{array}{r}
137 \\
24\overline{)3288} \\
24 \\
\hline
88 \\
72 \\
\hline
168 \\
168 \\
\end{array}
$$

24 × 1

24 × 3

24 × 7

The meanings of the products 2400, 720, and 168 can be seen in this example. Usually, the extra zeros are not written, and only a sufficient number of digits are "brought down." Normally, the division would appear as shown at the left.   ☐

In many divisions, the divisor will not divide exactly into the dividend. In these cases, there is a **remainder** in the answer. The following example illustrates division with a remainder.

EXAMPLE H   Divide 5286 by 25.

$$
\begin{array}{r}
211 \\
25\overline{)5286} \\
50 \\
\hline
28 \\
25 \\
\hline
36 \\
25 \\
\hline
11
\end{array}
$$

25 × 200 = 5000, shown as 25 × 2 = 50

25 × 10 = 250, shown as 25 × 1 = 25

25 × 1 = 25

remainder

Thus 5286 = (25 × 211) + 11. Since 11 is smaller than the divisor 25, the division process is discontinued and 11 becomes the remainder. □

There are many real applications of division. The following example illustrates one such application, and others are found in the exercises.

EXAMPLE I

$$
\begin{array}{r}
305 \\
27\overline{)8235} \\
81 \\
\hline
135 \\
135 \\
\hline
\end{array}
$$

A certain computer component costs $27. If a shipment of these components is valued at $8235, how many components are in the shipment?

Since we have the total value of the shipment and the value of one component, we may find the number of components by dividing $8235 by $27, as shown at the left. Two digits are brought down since 27 will not divide into 13. We see from the division that there are 305 components in the shipment. □

There is only one number that can never be used as a divisor, and that number is zero. In Chapter 3 we will explain why we can never divide by zero.

Exercises 1-2

In Exercises 1 through 8, perform the indicated multiplications.

1. 23 × 458

2. 27 × 835

3. (218)(6032)

4. (256)(1024)

5. 1024
     × 1024

6. 4108
     × 3897

7. 61547
     × 3849

8. 78793
     × 5698

In Exercises 9 through 16, perform the indicated divisions.

9. 3$\overline{)732}$

10. 81$\overline{)3159}$

11. 54$\overline{)17496}$

12. 32$\overline{)16256}$

13. 65536 ÷ 32

14. 62387 ÷ 28

15. 608271 ÷ 307

16. 918885 ÷ 725

In Exercises 17 through 20, verify the associative law of multiplication by first multiplying the numbers in parentheses and then completing the multiplication on each side.

17. $(17 \times 38) \times 74 = 17 \times (38 \times 74)$

18. $16 \times (312 \times 42) = (16 \times 312) \times 42$

19. $(326 \times 45) \times 217 = 326 \times (45 \times 217)$

20. $52 \times (36 \times 132) = (52 \times 36) \times 132$

In Exercises 21 through 24, perform the indicated multiplications by using the distributive law. Check your answer by adding the numbers within the parentheses first and then complete the multiplication.

21. $15(3 + 7)$ **22.** $14(8 + 12)$ **23.** $628(29 + 86)$ **24.** $4159(387 + 832)$

Exercises 25 through 32 give the lengths and widths of certain rectangles. Determine their areas. (In Exercise 29, cm is the symbol for centimeter.)

25. Length = 20 mi, width = 8 mi

26. Length = 44 ft, width = 24 ft

27. Length = 17 in., width = 14 in.

28. Length = 682 ft, width = 273 ft

29. Length = 296 cm, width = 35 cm

30. Length = 543 yd, width = 274 ft

31. Length = 18 in., width = 2 ft

32. Length = 34 in., width = 3 ft

In Exercises 33 through 48, solve the given problems. Where appropriate, the following units of measurement are designated by the given symbols: miles per hour—mi/h, hour—h, revolution—r, square foot—ft^2, square meter—m^2, meter—m, millimeter—mm, square millimeter—mm^2, square inch—in^2.

33. A jet plane averages a speed of 595 mi/h for 17 h. How far does it travel?

34. A certain gear makes 78 r/min (revolutions per minute). How many revolutions does it make in an hour?

35. What is the cost of carpeting a rectangular room 8 yd by 5 yd if the total cost is $32 per square yard? See Figure 1-13.

36. Suppose that the area of a rectangle is 61,884 ft^2 and the length is 573 ft. Find the width. See Figure 1-14.

37. A surveyor is partitioning a parcel of land into 1-acre rectangular plots. One acre is 43,560 ft^2. If one such plot is 132 ft wide, what is its length?

38. Suppose that the area of a rectangle is 56,056 m^2 and the width is 98 m. What is the length?

39. If a car traveled 234 mi on 13 gal of gasoline, what was its gas consumption in miles per gallon?

40. If a class of 27 students allots $945 for computer usage, what allotment does each individual student have?

41. Oil enters a pipeline and travels at the rate of 8 mi/h. If the oil reaches the exit point in 5 h, how long is the pipeline?

42. A satellite travels 27,000 mi in 24 h. What is its average speed?

43. A land developer divides a tract of land into 253 equally valued parcels and one lesser parcel valued at $430. If the tract is valued at $1,440,000, what is the value of each parcel?

44. Find the area of the piece of land shown in Figure 1-15.

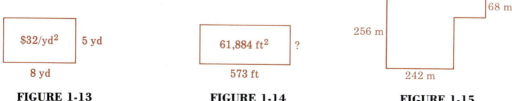

FIGURE 1-13   FIGURE 1-14   FIGURE 1-15

45. Shearing pins for a special machine cost $16 each. If the cost of a shipment of these pins is $792, which includes $24 in shipping charges, how many pins were in the shipment?

46. A flywheel on an aircraft engine makes 2300 r/min. How many revolutions does it make in 4 h of flight?

47. It is estimated that a home will require solar panels of at least 120 ft² in area. Will two identical solar panels meet this need if they are rectangular with length 5 ft and width 10 ft?

48. The floor in a laboratory has an area of 96,000 in.². How many tiles are needed to cover this floor if the tiles are squares with sides of 9 in.? See Figure 1-16.

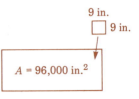

FIGURE 1-16

1-3 Fractions

Mathematics can be applied to many situations in which only whole numbers are necessary. However, there are many other cases in which parts of a quantity or less than the total of a group must be used. An industrial plant might have a total energy cost that is the sum of fuel and electricity expenses. If we consider only the fuel cost, we are considering only a fractional part of the total energy cost.

Considerations such as these lead us to fractions and the basic operations with them. In general, *a* **fraction** *is the indicated division of one number by another*. It is also possible to interpret this definition as a certain number of equal parts of a given unit or given group.

EXAMPLE A The fraction $\frac{5}{8}$ is the indicated division of 5 by 8. In this fraction, 5 is the **numerator** and 8 is the **denominator**.

$$\frac{5 \longleftarrow \text{ numerator}}{8 \longleftarrow \text{ denominator}}$$

It is also possible to interpret the fraction $\frac{5}{8}$ as referring to 5 of 8 equal parts of a whole (see Figure 1-17). Here the line is the given unit, of which 5 parts of 8 are being considered. That is, the line segment *AB* is $\frac{5}{8}$ of the line segment *AC*.

FIGURE 1-17

If, in a group of eight batteries, five are 6-V batteries, we could say that $\frac{5}{8}$ of the group are 6-V batteries. Here V is the symbol for volt.   □

If the numerator of a fraction is numerically less than the denominator, the fraction is called a **proper fraction.** *If the numerator equals or is numerically greater than the denominator, the fraction is called an* **improper fraction.** Since an improper fraction in which the numerator is greater than the denominator represents a number numerically greater than 1, it is often convenient to use a **mixed number**—*a whole number plus a proper fraction*—to represent the same number. Consider the following example.

EXAMPLE B   The fraction $\frac{4}{9}$ is a proper fraction, whereas $\frac{9}{4}$ is an improper fraction. It is

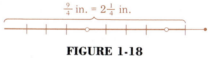

possible to interpret $\frac{9}{4}$ in terms of a number of equal parts of a given unit. Consider a line which has been divided in $\frac{1}{4}$-in. intervals (see Figure 1-18). The fraction $\frac{9}{4}$ could then be interpreted as representing 9 of the $\frac{1}{4}$-in. units. The line segment is also seen to be $2\frac{1}{4}$ in. long. Here $2\frac{1}{4}$ is a mixed number meaning $2 + \frac{1}{4}$.

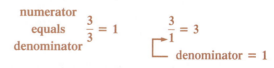

FIGURE 1-18

An improper fraction in which the numerator equals the denominator is equal to the whole number 1. Also, an improper fraction with a denominator of 1 equals the number in the numerator. Each of these can be seen to be valid if we think of the fraction in terms of division.

EXAMPLE C   The fraction $\frac{3}{3}$ equals 1, since $3 \div 3 = 1$. In the same way, $\frac{7}{7} = 1$ and $\frac{73}{73} = 1$. This is true since any number (except zero) divided by itself is 1.

numerator equals denominator $\dfrac{3}{3} = 1$ $\dfrac{3}{1} = 3$ denominator = 1

The fraction $\frac{3}{1}$ equals 3, since $3 \div 1 = 3$. In the same way, $\frac{7}{1} = 7$ and $\frac{73}{1} = 73$. This is true since any number divided by 1 equals that number. □

To convert an improper fraction to a mixed number, divide the numerator by the denominator. The number of times the denominator divides evenly is the whole-number part of the mixed number. The remainder obtained is the numerator of the fraction part of the mixed number, and the denominator of this fraction is the same as the denominator of the improper fraction.

EXAMPLE D   To convert $\frac{9}{4}$ to a mixed number, we divide 9 by 4. The result is 2 with a remainder of 1. Thus $\frac{9}{4} = 2\frac{1}{4}$.

Converting $\frac{73}{14}$ to a mixed number, we divide 73 by 14, obtaining 5 with a remainder of 3. Thus $\frac{73}{14} = 5\frac{3}{14}$ as illustrated below.

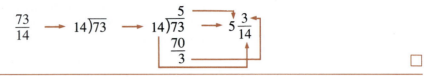

To convert a mixed number to an improper fraction, multiply the whole number of the mixed number by the denominator of the fraction. Add this result to the numerator of the fraction. Place this result over the denominator of the fraction.

EXAMPLE E   The selling price of a stock is listed as $2\frac{1}{4}$ (dollars). Write $2\frac{1}{4}$ as an improper fraction.

To convert $2\frac{1}{4}$ to an improper fraction, we multiply 2 by 4, obtaining 8. This is added to 1, obtaining 9, which is now the numerator of the improper fraction. Thus

$$2\frac{1}{4} = \frac{(2 \times 4) + 1}{4} = \frac{8 + 1}{4} = \frac{9}{4}$$

Exercises 1-3

In Exercises 1 through 4, write fractions for the indicated selected parts.

1. Five equal parts from among nine such parts.   **2.** Four equal parts from among eleven such parts.

3. One part from among seven such equal parts.   **4.** Eight equal parts from among eight such parts.

In Exercises 5 through 8, write fractions equal to each indicated quotient.

5. $7 \div 13; 3 \div 16$ **6.** $1 \div 3; 11 \div 17$ **7.** $9 \div 8; 1 \div 12$ **8.** $23 \div 4; 2 \div 35$

In Exercises 9 through 12, write the whole number which equals the given fraction.

9. $\frac{6}{6}; \frac{6}{1}$ **10.** $\frac{19}{19}; \frac{19}{1}$ **11.** $\frac{32}{1}; \frac{32}{32}$ **12.** $\frac{503}{1}; \frac{503}{503}$

In Exercises 13 through 20, convert the given improper fractions to mixed numbers.

13. $\frac{5}{3}$ **14.** $\frac{18}{5}$ **15.** $\frac{64}{13}$ **16.** $\frac{55}{32}$

17. $\frac{278}{75}$ **18.** $\frac{315}{32}$ **19.** $\frac{1329}{25}$ **20.** $\frac{4376}{118}$

In Exercises 21 through 28, convert the given mixed numbers to improper fractions.

21. $3\dfrac{2}{5}$ 22. $6\dfrac{3}{4}$ 23. $9\dfrac{7}{8}$ 24. $12\dfrac{3}{5}$

25. $17\dfrac{2}{13}$ 26. $34\dfrac{3}{8}$ 27. $105\dfrac{3}{4}$ 28. $235\dfrac{16}{25}$

In Exercises 29 through 40, solve the given problems.

29. A case of oil contains 24 cans. If 7 cans are removed, write a fraction for the part of the case that remains.

30. Write a fraction which indicates the cars with automatic transmission from a group of cars of which seven have automatic transmission and five have standard transmission.

31. Write a fraction which indicates the 60-W electric light bulbs from a container which has ten 60-W bulbs and thirteen 100-W bulbs. (W is the symbol for watt.)

32. The break-even point for a certain production crew is found to be 48 units per hour. If 55 units are produced each hour, what fraction of the break-even point is the current production level?

33. A standard sheet of typing paper is $\frac{17}{2}$ in. wide. Express this width as a mixed number.

34. A cylinder holds $\frac{39}{4}$ oz of fuel. Express this capacity as a mixed number.

35. The radiator of a car holds $\frac{15}{4}$ gal. Express this capacity as a mixed number.

36. A floppy disk has a diameter of $5\frac{1}{4}$ in. Write this value as an improper fraction.

37. The snowfall from a storm was $3\frac{7}{10}$ in. deep. Express the snowfall as an improper fraction.

38. A person took $5\frac{17}{100}$ s to solve a problem with a calculator. Express this time as an improper fraction. (The symbol s denotes second.)

39. A fuel gauge gives the indication shown in Figure 1-19. Write a fraction which indicates the part of the tank that contains fuel.

40. A scale gives the indication shown in Figure 1-20. If the 1 represents 1 kg, write a fraction for the part of a kilogram which is indicated. (The symbol kg denotes kilogram.)

FIGURE 1-19

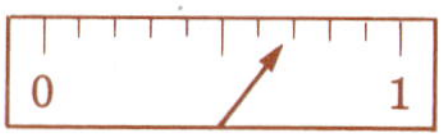

FIGURE 1-20

1-4 Equivalent Fractions

We have considered the basic meanings of various types of fractions. Now we shall determine how a fraction may be equivalent to other fractions. Determining equality of two or more fractions is very important in working with fractions and performing the basic operations on them.

Consider the area of rectangle A as shown in Figure 1-21. It has been divided into four equal parts, one of which is shaded. Thus the fraction $\frac{1}{4}$ can be used to represent the shaded part. Rectangle B is the same as rectangle A, but it has been divided again by a line through the middle. Now the shaded area consists of 2 of 8 equal parts, and the fraction $\frac{2}{8}$ can be used to represent the shaded area. However, the shaded areas of the two rectangles are the same, which means that $\frac{1}{4} = \frac{2}{8}$. Since the two fractions represent the same fractional part, they must be equal.

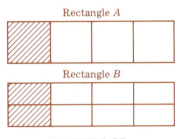

Rectangle A

Rectangle B

FIGURE 1-21

The fraction $\frac{2}{8}$ could be obtained by doubling both the numerator and the denominator of the fraction $\frac{1}{4}$. Also, we may obtain the fraction $\frac{1}{4}$ by dividing both the numerator and the denominator of the fraction $\frac{2}{8}$ by 2. These two fractions are known as **equivalent fractions**. In general: *Two fractions are equivalent (equal) if the numerator and the denominator of one of the fractions can be multiplied, or divided, by the same number (not zero) to obtain the other fraction.*

EXAMPLE A   The fraction $\frac{18}{24}$ can be obtained by multiplying the numerator and the denominator of the fraction $\frac{3}{4}$ by 6. That is,

$$\frac{3}{4} = \frac{3 \times 6}{4 \times 6} = \frac{18}{24}$$

Therefore $\frac{3}{4}$ and $\frac{18}{24}$ are equivalent fractions.

If we multiply the numerator and the denominator of $\frac{3}{4}$ by 3, we obtain the fraction $\frac{9}{12}$. Thus $\frac{3}{4}$ and $\frac{9}{12}$ are equivalent, which means that $\frac{3}{4} = \frac{9}{12}$. Other fractions equivalent to $\frac{3}{4}$ may be obtained by multiplying the numerator and the denominator by other nonzero numbers.   ☐

EXAMPLE B   Following are equivalent fractions which have been obtained by performing the indicated operation on the numerator and the denominator of the left-hand fraction.

$$\frac{2}{4} = \frac{4}{6} \quad \text{multiplication by 2} \qquad \frac{2}{3} = \frac{14}{21} \quad \text{multiplication by 7}$$

$$\frac{5}{8} = \frac{20}{32} \quad \text{multiplication by 4} \qquad \frac{5}{8} = \frac{45}{72} \quad \text{multiplication by 9}$$

$$\frac{28}{100} = \frac{7}{25} \quad \text{division by 4} \qquad \frac{12}{54} \quad \frac{2}{9} \quad \text{division by 6}$$

$$\frac{121}{154} = \frac{11}{14} \quad \text{division by 11} \qquad \frac{156}{84} = \frac{13}{7} \quad \text{division by 12} \qquad \square$$

In changing one fraction to an equivalent fraction, we say that *the fraction has been converted to* **higher terms** *if the resulting numerator and denominator are larger. A fraction is said to be* **reduced** *if it is changed to a fraction in which the numerator and denominator are smaller.*

One of the most important operations performed on a fraction is reducing it to **lowest terms** or **simplest form.** *This means that the resulting numerator and denominator are not both evenly divisible by any whole number other than 1.*

EXAMPLE C   The fraction $\frac{10}{24}$ is not in lowest terms, since both the numerator and the denominator are divisible by 2. To reduce $\frac{10}{24}$ to lowest terms we divide the numerator and the denominator by 2 and obtain $\frac{5}{12}$. Since 5 and 12 are not both divisible by any whole number other than 1, the lowest term of $\frac{10}{24}$ is $\frac{5}{12}$. Therefore $\frac{10}{24} = \frac{5}{12}$. $\square$

EXAMPLE D   The fraction $\frac{42}{54}$ is not in simplest form. Dividing both 42 and 54 by 2, we obtain the following:

$$\frac{42}{54} = \frac{42 \div 2}{54 \div 2} = \frac{21}{27}$$

▶   However, we note that 21 and 27 are both evenly divisible by 3. Although $\frac{21}{27}$ is a reduced form of $\frac{42}{54}$, *it is not the simplest form.* Dividing 21 and 27 by 3, we get

$$\frac{21}{27} = \frac{21 \div 3}{27 \div 3} = \frac{7}{9}$$

Now 7 and 9 are not both evenly divisible by any whole number other than 1. Thus $\frac{7}{9}$ is the simplest form of $\frac{42}{54}$. (If we had originally noted that 42 and 54 were both divisible by 6, we would have obtained $\frac{7}{9}$ directly.) $\square$

Reducing an improper fraction to lowest terms is done in precisely the same way as with a proper fraction. The final form should be left as an improper fraction unless specifically noted otherwise.

When reducing a fraction to lowest terms, we may find that the largest number that will divide into both the numerator and the denominator is not obvious, as illustrated in Example D.

For purposes of reducing fractions with large numerators and denominators, as well as performing the basic operations on fractions, it is convenient to determine the various factors of a whole number. We recall that each of the numbers being multiplied together is a **factor** of a given product. Here the whole number under consideration is the product in question. Thus *the process is that of* **factoring** *whole numbers, which means we shall be determining those numbers which, when multiplied together, give the whole number.*

EXAMPLE E

factors

$$10 = 2 \times 5$$

The whole number 10 is equal to 10×1. In general, any number can be considered to be the product of itself and 1. Thus 10 and 1 are factors of 10. However, we also note that $10 = 2 \times 5$, which means that 2 and 5 are also factors of 10. Therefore the number 10 has 1, 2, 5, and 10 as factors—each of these whole numbers can be multiplied by another whole number to obtain 10. □

Those factors which are the most important are called **prime numbers.** *A prime number is a whole number that is evenly divisible only by itself and 1.* Neither 0 nor 1 is considered to be a prime number.

EXAMPLE F

2 is prime, since it is divisible evenly only by 2 and 1.

3 is prime, since it is divisible evenly only by 3 and 1.

4 is not prime, since $4 = 2 \times 2$. That is, 4 is divisible by a whole number other than itself or 1.

5 is prime, since it is divisible evenly only by 5 and 1.

6 is not prime, since $6 = 2 \times 3$.

Other prime numbers are 7, 11, 13, 17, and 19.

Other numbers which are not prime are 8 ($8 = 2 \times 2 \times 2$), 9 ($9 = 3 \times 3$), 10 ($10 = 5 \times 2$), 12 ($12 = 2 \times 2 \times 3$), 14 ($14 = 2 \times 7$), 15 ($15 = 3 \times 5$), 16 ($16 = 2 \times 2 \times 2 \times 2$), and 18 ($18 = 2 \times 3 \times 3$). □

Since the most useful factorization of a whole number consists of prime factors, *we shall factor all whole numbers until only prime factors are present. When the original number has been factored so that all of the factors are prime numbers, we say that the number has been* **factored completely**.

If the whole number being factored is the product of two other numbers, we should express it as that product. Then we should examine each factor to determine whether or not it is prime. If we have no specific clues as to the factors, we simply start dividing by the prime numbers. In doing so, however, we note that even numbers have a factor of 2 and numbers ending in 0 or 5 have a factor of 5.

EXAMPLE G

prime
factors

$$18 = 2 \times 3 \times 3$$

In factoring 18, we may note that $18 = 2 \times 9$. However, 9 is not prime, so we further factor 9 as 3×3. Therefore $18 = 2 \times 3 \times 3$. Since all factors are now prime, the factorization is complete. We also note that the prime factor 3 appears twice.

In factoring 34, we note that 34 is even so that 2 must be a factor. This leads to the factorization $34 = 2 \times 17$, which is the complete factorization, since 17 is prime.

In factoring 55, we note that 55 ends in 5 so that 5 must be a factor. We can express 55 as 5×11 and, since 5 and 11 are both prime, we say that 55 has been factored completely.

In factoring 91, we divide by the prime numbers. The first prime number to divide evenly is 7 so that $91 = 7 \times 13$, which is the complete factorization because 7 and 13 are both prime.   □

The actual order in which factors are found is not important. Often several variations in the factorization of a number are possible. Remember: *What matters is that **the final factorization is complete with only prime factors**.*

To reduce a fraction by use of factoring, first factor both the numerator and the denominator into prime factors. At this point any factors which are common to both the numerator and the denominator can be determined by inspection. Any such factors can then be divided out.

EXAMPLE H

To reduce the fraction $\frac{65}{78}$, first factor 65 and 78 into prime factors. This gives

$$\frac{65}{78} = \frac{5 \times 13}{2 \times 3 \times 13} \quad \text{prime factors}$$

We note that the factor 13 appears in both the numerator and the denominator so that both may be divided by 13. Dividing both numerator and denominator by 13 we get

$$\frac{65}{78} = \frac{65 \div 13}{78 \div 13} = \frac{5}{6}$$ □

In Example H we factored both the numerator and the denominator, noted the common factor, and then divided by the common factor. However, it is not necessary to perform the actual division directly. Any factor common to both the numerator and the denominator can simply be crossed out to get the same result as dividing. *This process is often referred to as* **cancellation.** *It must be remembered, however, that* **cancellation is actually a process of** dividing **both the numerator and the denominator by the common factor.**

After we have factored the numerator and the denominator, it often happens that more than one common factor appears in each. When this happens, every factor common to both the numerator and the denominator should be canceled.

EXAMPLE I   In reducing the fraction $\frac{21}{70}$, we first factor both 21 and 70 to get

$$\frac{21}{70} = \frac{3 \times 7}{2 \times 5 \times 7} \quad \text{prime factors}$$

Note that the factor 7 appears in both the numerator and the denominator. We can do the division by crossing out the 7's. Performing this cancellation, we get

$$\frac{21}{70} = \frac{3 \times \cancel{7}^{1}}{2 \times 5 \times \cancel{7}_{1}} = \frac{3}{10} \quad \text{Cancel common prime factors}$$

We must remember that we really divided by 7, leaving quotients of 3 and 10.

In reducing the fraction $\frac{140}{56}$, we get the following:

$$\frac{140}{56} = \frac{\cancel{2}^{1} \times \cancel{2}^{1} \times 5 \times \cancel{7}^{1}}{\cancel{2}_{1} \times \cancel{2}_{1} \times 2 \times \cancel{7}_{1}} = \frac{5}{2} \qquad \square$$

When we reduce a fraction and get a result that is an improper fraction, we usually leave it in that form unless there are specific instructions or reasons for doing otherwise.

If all prime factors of the numerator are canceled, we must remember that *cancellation amounts to division.* This implies that a 1 remains in the numerator as a result of the division. This is true since any number divided by itself is 1, and we are dividing the number in the numerator by itself.

EXAMPLE J   In reducing $\frac{15}{75}$, we factor the numerator and the denominator as follows:

$$\frac{15}{75} = \frac{3 \times 5}{3 \times 5 \times 5}$$

In canceling, note that both the 3 and the 5 of the numerator are to be canceled. Since this amounts to dividing the numerator by 15, we get

$$\frac{15}{75} = \frac{\overset{1}{\cancel{3}} \times \overset{1}{\cancel{5}}}{\underset{1}{\cancel{3}} \times \underset{1}{\cancel{5}} \times 5} = \frac{1}{5} \longleftarrow \text{ All prime factors canceled} \qquad \square$$

EXAMPLE K   In reducing $\frac{72}{6}$ we factor the numerator and denominator to get

$$\frac{72}{6} = \frac{2 \cdot 2 \cdot 2 \cdot \cancel{3} \cdot 3}{\cancel{2} \cdot \cancel{3}} = \frac{12}{1} = 12 \longleftarrow \text{ All prime factors canceled}$$

Instead of expressing the result as $\frac{12}{1}$, we write it simply as 12. $\square$

Exercises 1-4

In Exercises 1 through 12, determine fractions equivalent to the given fractions by performing the indicated operation on the numerator and the denominator.

1. $\frac{3}{7}$ (multiply by 2)   2. $\frac{5}{9}$ (multiply by 3)   3. $\frac{16}{20}$ (divide by 4)

4. $\frac{15}{125}$ (divide by 5)   5. $\frac{4}{13}$ (multiply by 6)   6. $\frac{8}{15}$ (multiply by 11)

7. $\frac{60}{156}$ (divide by 12)   8. $\frac{140}{42}$ (divide by 7)   9. $\frac{13}{25}$ (multiply by 7)

10. $\frac{17}{15}$ (multiply by 12)   11. $\frac{1024}{64}$ (divide by 32)   12. $\frac{289}{340}$ (divide by 17)

In Exercises 13 through 24, reduce each of the given fractions to lowest terms.

13. $\frac{4}{8}$ 14. $\frac{12}{18}$ 15. $\frac{15}{10}$ 16. $\frac{21}{6}$ 17. $\frac{20}{25}$ 18. $\frac{21}{28}$

19. $\frac{27}{45}$ 20. $\frac{16}{40}$ 21. $\frac{24}{60}$ 22. $\frac{55}{66}$ 23. $\frac{30}{75}$ 24. $\frac{9}{78}$

In Exercises 25 through 36, find the value of the missing numbers.

25. $\frac{1}{3} = \frac{?}{12}$ 26. $\frac{5}{8} = \frac{?}{16}$ 27. $\frac{2}{3} = \frac{10}{?}$ 28. $\frac{7}{8} = \frac{21}{?}$ 29. $\frac{12}{64} = \frac{?}{16}$ 30. $\frac{27}{36} = \frac{?}{16}$

31. $\frac{?}{6} = \frac{2}{3}$ 32. $\frac{?}{18} = \frac{4}{9}$ 33. $\frac{3}{?} = \frac{6}{80}$ 34. $\frac{5}{?} = \frac{30}{18}$ 35. $\frac{13}{?} = \frac{52}{64}$ 36. $\frac{7}{?} = \frac{21}{192}$

In Exercises 37 through 48, factor the given numbers into their prime factors.

37. 20   38. 28   39. 16   40. 32   41. 36   42. 44

43. 48   44. 52   45. 57   46. 84   47. 105   48. 132

In Exercises 49 through 60, reduce each fraction to lowest terms by factoring the numerator and the denominator into prime factors and then canceling any common factors.

49. $\dfrac{30}{35}$ 50. $\dfrac{28}{63}$ 51. $\dfrac{24}{30}$ 52. $\dfrac{45}{75}$ 53. $\dfrac{56}{24}$ 54. $\dfrac{99}{27}$

55. $\dfrac{52}{78}$ 56. $\dfrac{70}{84}$ 57. $\dfrac{17}{68}$ 58. $\dfrac{19}{57}$ 59. $\dfrac{63}{105}$ 60. $\dfrac{78}{117}$

In Exercises 61 through 72, solve the given problems. Reduce all fractions to lowest terms.

61. How would a $\frac{12}{16}$-in. wrench be labeled if all wrenches in a set are labeled with fractions reduced to lowest terms?

62. A $\frac{3}{8}$-in. hole must be drilled for a bolt so that a voltage regulator can be mounted in a truck. If the drill bits are measured in 64ths of an inch, which bit should be used?

63. One road has 32 homes and 14 of them use solar energy to heat water. What fraction of the homes on this road heat their hot water with solar energy?

64. Sixty digital watches are set at 12:00. These watches are checked exactly 12 h later and 53 of them indicate 12:00. What fraction of the watches displayed the correct time 12 h after they were set? (The symbol h denotes hour.)

65. An electric current is split between two wires. If 6 A pass through the first wire and 14 A pass through the second wire, what fraction of the total current passes through the first wire? (A is the symbol for ampere.)

66. Two firms share a computer. If, in a given hour, one firm uses the computer for 32 min and the other firm uses it the remainder of the hour, what fraction of the cost of the computer should the first firm pay for that hour?

67. A chemist mixes 18 mL of sulfuric acid with 27 mL of water. What fraction of the mixture is pure acid? (The symbol mL denotes milliliter.)

68. The production of a certain component costs $13 for labor and $6 for material. What fraction of the total cost is the cost of labor?

69. A small business acquires a microcomputer at a cost of $4500. Two years later this same microcomputer is sold for $3600. What fraction of the original cost is lost?

70. An object falling under the influence of gravity falls 1296 ft during the first 9 s and 304 ft during the tenth second. What fraction of the distance fallen during the first 10 s occurs during the tenth second? (The symbol s denotes second.)

71. A large oil holding tank is filled by a large pipe and two smaller pipes. The large pipe supplies 64 gal/min while each of the smaller pipes supplies 30 gal/min. If the pipes are all used until the tank is full, what fraction of the oil is supplied by the large pipe? See Figure 1–22.

72. In the Gateway Arch in St. Louis, each of the two bases could not be allowed to be off by more than $\frac{1}{64}$ in. If an error was determined to be $\frac{5}{256}$ in., is it acceptable? Explain.

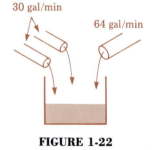

30 gal/min

64 gal/min

FIGURE 1-22

1-5 Addition and Subtraction of Fractions

In Section 1-1 we stressed that only like quantities can be added or subtracted. This principle is true regardless of the nature of the numbers. We can add 3 oz and 4 oz to get 7 oz, but we cannot add 3 oz and 4 mi.

We can think of a fraction as a number of equal parts of a given unit or group so that each of these parts or units is equal. Consequently, *fractions in which the denominators are equal are called* **like fractions.**

EXAMPLE A The fractions $\frac{3}{8}$, $\frac{5}{8}$, and $\frac{11}{8}$ are like fractions because all have denominators of 8. With 3 eighths, 5 eighths, and 11 eighths, we have like quantities since they are all eighths. The fractions $\frac{1}{3}$ and $\frac{2}{6}$ are not like fractions since the denominators are different. They are equivalent fractions, but they are not like fractions. □

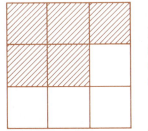

To develop the procedure for adding fractions, consider the squares in Figure 1-23. Add the three small shaded squares of the top row (which represent $\frac{3}{9}$ of the large square) to the small shaded squares of the middle row (which represent $\frac{2}{9}$ of the large square). The sum is five small squares, or $\frac{5}{9}$ of the large square. This suggests that

$$\frac{3}{9} + \frac{2}{9} = \frac{5}{9}$$

FIGURE 1-23

Following this reasoning, *we add like fractions by placing the sum of the numerators over the denominator which is common to them. In the same way, when we subtract fractions we place the difference of numerators over the common denominator.* These procedures are illustrated in the following example.

EXAMPLE B

sum of numerators

1. $\dfrac{2}{9} + \dfrac{5}{9} = \dfrac{2+5}{9} = \dfrac{7}{9}$ ← common denominator

2. $\dfrac{5}{11} - \dfrac{3}{11} = \dfrac{5-3}{11} = \dfrac{2}{11}$

3. $\dfrac{3}{8} + \dfrac{1}{8} = \dfrac{3+1}{8} = \dfrac{4}{8} = \dfrac{1}{2}$

4. $2\dfrac{1}{7} + \dfrac{3}{7} = \dfrac{15}{7} + \dfrac{3}{7} = \dfrac{15+3}{7} = \dfrac{18}{7}$

5. $\dfrac{7}{12} + \dfrac{1}{12} - \dfrac{5}{12} = \dfrac{7+1-5}{12} = \dfrac{8-5}{12} = \dfrac{3}{12} = \dfrac{1}{4}$

In (4) the mixed number is first changed to an improper fraction; then the addition is performed. Although this is not the only procedure that can be followed, it is standard. In (5), where more than two fractions are being combined, the addition is performed first and then the subtraction. (Until signed numbers are considered in Chapter 3, this procedure should be followed.) Note that the results in (3) and (5) are reduced to simplest form. □

▶ *If the fractions being added or subtracted do not have the same denominators, it is necessary to convert them so that all denominators are equal.* This is consistent with the requirement that we can add or subtract only like quantities.

EXAMPLE C   To perform the addition $\frac{1}{3} + \frac{2}{9}$, we must first change the fraction $\frac{1}{3}$ to its equivalent form $\frac{3}{9}$. When this is done we have

$$\frac{1}{3} + \frac{2}{9} = \frac{3}{9} + \frac{2}{9} = \frac{3+2}{9} = \frac{5}{9}$$

denominators must be equal in order to add

▶ *Originally, in adding $\frac{1}{3}$ and $\frac{2}{9}$, we cannot add the numerators 1 and 2,* for this would be combining the unlike quantities of thirds and ninths improperly. For proper addition of fractions, the denominators must be the same. ☐

Although any proper common denominator can be used for adding fractions, there are distinct advantages if this denominator is the least possible value. The following example illustrates this.

EXAMPLE D   If we wish to add $\frac{5}{12}$ and $\frac{7}{8}$, we might note that both fractions can be converted to fractions with a common denominator of 96. Therefore, converting these fractions, we have

$$\frac{5}{12} = \frac{40}{96} \quad \text{and} \quad \frac{7}{8} = \frac{84}{96}$$

Adding $\frac{5}{12}$ and $\frac{7}{8}$ is equivalent to adding $\frac{40}{96}$ and $\frac{84}{96}$, or

$$\frac{5}{12} + \frac{7}{8} = \frac{40}{96} + \frac{84}{96} = \frac{124}{96}$$

This final fraction, $\frac{124}{96}$, should be reduced to its simplest form. This is done by dividing both numerator and denominator by 4.

$$\frac{5}{12} + \frac{7}{8} = \frac{40}{96} + \frac{84}{96} = \frac{124}{96} = \frac{31}{24}$$

If we had first recognized that a denominator of 24 could be used, the addition would be

$$\frac{5}{12} + \frac{7}{8} = \frac{10}{24} + \frac{21}{24} = \frac{31}{24}$$

In this case, the conversions are simpler and the final result is already in its simplest form. In other cases, the result might not be in simplest form, but this use of a smaller denominator does simplify the process. ☐

In Example D we saw that either 96 or 24 could be used as a common denominator. Actually, many other possibilities such as 48, 72, and 120 also exist. However, 24 is the smallest of all of the possibilities. It is the **lowest common denominator** for these fractions. In general: *The lowest* *common denominator of a set of fractions is the smallest number that is evenly divisible by all denominators of the set.* The advantages of using the lowest common denominator are illustrated in Example D. The conversions to higher terms are simpler, as is the simplification of the result. Therefore, when adding or subtracting fractions, we shall always use the lowest common denominator. We shall now see how the lowest common denominator is determined.

In many cases the lowest common denominator can be determined by inspection. For example, if we are adding the fractions $\frac{1}{2}$ and $\frac{3}{8}$, we can easily determine that the lowest common denominator is 8. When adding $\frac{1}{2}$ and $\frac{2}{3}$, the lowest common denominator is clearly 6. However, if we cannot readily determine the lowest common denominator by observation, we need a systematic method that we can follow.

When we discussed equivalent fractions, we saw how whole numbers are factored into their prime factors. We can also determine the lowest common denominator of a set of fractions by factoring the denominators into their prime factors. Therefore: *After each denominator has been factored into its prime factors, **the lowest common denominator is the product of all the different prime factors, each taken the greatest number of times it appears in any one of the denominators.***

The procedure for finding the lowest common denominator can be summarized as follows.

1. List all denominators.
2. Next to each denominator, write the number as a product of its prime factors.
3. The lowest common denominator is found by multiplying the prime factors from step 2. For any prime factor that is repeated within one or more denominators, repeat it as a factor in the lowest common denominator. For each denominator, find the number of times that the prime factor occurs, then select the largest of these numbers of occurrences. That is the number of times the factor is repeated in the lowest common denominator.

EXAMPLE E When we want to find the sum $\frac{5}{12} + \frac{7}{8}$, we must find the prime factors of 12 and 8. Since $12 = 2 \times 2 \times 3$ and $8 = 2 \times 2 \times 2$, we see that the only different prime factors are 2 and 3. *Since 2 occurs twice in 12 and three times in 8, it is taken three times in the least common denominator. Since 3 occurs once in 12 and does not occur in 8, it is taken once.* The lowest common denominator is

$$2 \times 2 \times 2 \times 3 = 24$$

This means that $\qquad \dfrac{5}{12} + \dfrac{7}{8} = \dfrac{10}{24} + \dfrac{21}{24} = \dfrac{31}{24}$

Convert to fractions with lowest common demoninator

Compare this result with Example D. $\qquad\qquad\qquad\qquad\qquad \square$

EXAMPLE F   Add: $\frac{5}{18} + \frac{3}{14}$

First we determine the prime factors of 18 and 14 as follows.

$$18 = 2 \times 3 \times 3, \qquad 14 = 2 \times 7$$

▶ We now see that the prime factors for the lowest common denominator are 2, 3, and 7. The greatest number of times each appears is *once for 2 (once each in 18 and 14),* twice for 3 (in 18), and once for 7 (in 14). The lowest common denominator is

$$2 \times 3 \times 3 \times 7 = 126$$

Now converting to fractions with 126 as the denominator and adding, we have

$$\frac{5}{18} + \frac{3}{14} = \frac{35}{126} + \frac{27}{126} = \frac{62}{126} = \frac{31}{63}$$

▶ Here we see that *the initial result of $\frac{62}{126}$ is not in lowest terms and must be reduced* by cancelling the common factor of 2.   ☐

EXAMPLE G   Combine: $\frac{2}{15} + \frac{11}{27} - \frac{7}{50}$.

First we determine the prime factors of 15, 27, and 50 as follows.

$$15 = 3 \times 5, \quad 27 = 3 \times 3 \times 3, \quad 50 = 2 \times 5 \times 5$$

We now observe that the prime factors for the lowest common denominator are 2, 3, and 5. The greatest number of times each appears is once for 2 (in 50), three times for 3 (in 27), and twice for 5 (in 50). The lowest common denominator can now be expressed as

$$2 \times 3 \times 3 \times 3 \times 5 \times 5 = 1350$$

We can now proceed to combine the fractions as follows.

$$\frac{2}{15} + \frac{11}{27} - \frac{7}{50} = \frac{2(90)}{15(90)} + \frac{11(50)}{27(50)} - \frac{7(27)}{50(27)}$$

$$= \frac{180}{1350} + \frac{550}{1350} - \frac{189}{1350} = \frac{180 + 550 - 189}{1350}$$

$$= \frac{730 - 189}{1350} = \frac{541}{1350}$$

▶ We know that *the only prime factors of 1350 are 2, 3, and 5. Since none of these divides evenly into 541, the result is in simplest form.*   ☐

Sometimes a whole number is involved in addition or subtraction with fractions. In such cases we express the whole number as an improper fraction with the proper common denominator of the fractions. (Note that this procedure is the same as the procedure for changing a mixed number to an improper fraction.)

EXAMPLE H Combine: $3 - \frac{16}{21} + \frac{4}{9}$.

We first determine the lowest common denominator of the fractions. Factoring the denominators, we have $21 = 3 \times 7$ and $9 = 3 \times 3$. The lowest common denominator is $3 \times 3 \times 7 = 63$. Now, since $3 = \frac{3}{1}$, we can change 3 into an improper fraction with a denominator of 63 by multiplying both numerator and denominator by 63. We get

$$3 - \frac{16}{21} + \frac{4}{9} = \frac{3(63)}{1(63)} - \frac{16(3)}{21(3)} + \frac{4(7)}{9(7)}$$

$$= \frac{189}{63} - \frac{48}{63} + \frac{28}{63} = \frac{189 - 48 + 28}{63}$$

$$= \frac{169}{63}$$

The result is in simplest form. ☐

EXAMPLE I One section of an aircraft firewall is $\frac{1}{2}$ in. thick. That thickness is increased by a $\frac{3}{8}$-in. washer and a $\frac{5}{16}$-in. bolt. See Figure 1-24. What is the total combined thickness?

Since we must add the three fractions, we begin by determining the lowest common denominator. Factoring the denominators, we get the three results 2, $2 \times 2 \times 2$, and $2 \times 2 \times 2 \times 2$, so the lowest common denominator is $2 \times 2 \times 2 \times 2 = 16$. We then change each fraction to an equivalent fraction with a denominator of 16.

$$\frac{1}{2} + \frac{3}{8} + \frac{5}{16} = \frac{1(8)}{2(8)} + \frac{3(2)}{8(2)} + \frac{5}{16}$$

$$= \frac{8}{16} + \frac{6}{16} + \frac{5}{16} = \frac{19}{16} \text{ in.}$$

The total combined thickness is $\frac{19}{16}$ in. (or $1\frac{3}{16}$ in.). ☐

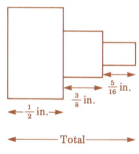

FIGURE 1-24

Exercises 1-5

In Exercises 1 through 8, find the lowest common denominator, assuming that the given numbers are denominators of fractions to be added or subtracted.

1. 2, 4 **2.** 2, 3 **3.** 6, 8 **4.** 6, 9

5. 8, 12, 18 **6.** 6, 10, 14 **7.** 10, 12, 25 **8.** 22, 24, 33

In Exercises 9 through 36, perform the indicated additions or subtractions, expressing each result in simplest form. In Exercises 23 through 36, use factoring to determine the lowest common denominator.

9. $\frac{1}{5} + \frac{3}{5}$ **10.** $\frac{2}{11} + \frac{5}{11}$ **11.** $\frac{5}{7} - \frac{3}{7}$ **12.** $\frac{4}{5} - \frac{1}{5}$

13. $\frac{1}{2} + \frac{1}{4}$ **14.** $\frac{1}{5} + \frac{2}{15}$ **15.** $\frac{23}{24} - \frac{5}{6}$ **16.** $\frac{2}{3} - \frac{1}{6}$

17. $\dfrac{1}{3} + \dfrac{3}{4}$

18. $\dfrac{3}{5} + \dfrac{5}{6}$

19. $\dfrac{1}{2} - \dfrac{2}{5}$

20. $\dfrac{2}{7} - \dfrac{1}{9}$

21. $5 - \dfrac{8}{3}$

22. $3 - \dfrac{11}{4}$

23. $\dfrac{11}{20} + \dfrac{5}{8}$

24. $\dfrac{2}{9} + \dfrac{5}{21}$

25. $\dfrac{5}{6} - \dfrac{3}{26}$

26. $\dfrac{19}{28} - \dfrac{5}{24}$

27. $3\dfrac{5}{6} - \dfrac{3}{8}$

28. $2\dfrac{4}{9} + \dfrac{5}{12}$

29. $\dfrac{4}{7} + \dfrac{1}{3} - \dfrac{3}{14}$

30. $\dfrac{1}{2} + \dfrac{5}{6} - \dfrac{9}{22}$

31. $\dfrac{3}{8} + \dfrac{7}{12} + \dfrac{1}{18}$

32. $3\dfrac{2}{9} + \dfrac{7}{15} - \dfrac{4}{25}$

33. $\dfrac{1}{3} + 1\dfrac{9}{14} - \dfrac{5}{21}$

34. $\dfrac{26}{27} - \dfrac{7}{18} + \dfrac{1}{10}$

35. $\dfrac{3}{4} + \dfrac{13}{30} - \dfrac{7}{12} + 2$

36. $5 - \dfrac{4}{25} + \dfrac{4}{5} + \dfrac{3}{35}$

In Exercises 37 through 48, solve the given problems.

37. One section of an aircraft firewall is $\frac{1}{4}$ in. thick. That thickness is increased by a $\frac{1}{16}$-in. washer and a $\frac{3}{16}$-in. bolt. What is the total combined thickness?

38. A flat machine plate has four sides whose lengths are $\frac{2}{3}$ ft, $\frac{1}{6}$ ft, $\frac{3}{4}$ ft, and $\frac{5}{12}$ ft. Find the perimeter. See Figure 1-25.

39. In an electric circuit, three resistors are connected in series. The total resistance is found by adding the individual resistances. What is the sum of $3\frac{3}{4}$ Ω, $2\frac{1}{2}$ Ω, and $3\frac{1}{4}$ Ω? (The symbol Ω denotes ohm.)

40. An engine coolant mixture consists of $3\frac{3}{4}$ gal of ethylene glycol and $4\frac{1}{5}$ gal of water. What is the total amount of coolant?

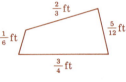

41. A ramp for the disabled is to be bordered with a special reflective tape. What length of tape is needed if the ramp is a rectangular plate $4\frac{5}{12}$ ft wide and $8\frac{11}{12}$ ft long?

FIGURE 1-25

42. A computer printer cable is $8\frac{2}{3}$ ft long. If a $3\frac{1}{12}$-ft section is cut off, what is the length of the remaining piece?

43. A weather station records precipitation of $\frac{1}{4}$ in., $1\frac{1}{12}$ in., and $\frac{3}{8}$ in. on three successive days. What is the total amount of precipitation?

44. A technician experiments with the conducting properties of an alloy. If the alloy is $\frac{2}{3}$ copper, $\frac{1}{5}$ gold, and the remaining part is silver, what fraction is silver?

45. A gauge indicates that a fuel tank is $\frac{1}{4}$ full. If $\frac{5}{8}$ tank of fuel is added and $\frac{2}{3}$ tank is later consumed, what fraction of a tank of fuel remains?

46. A reservoir is $\frac{3}{4}$ full at the start of summer. During the summer, consumers use $\frac{2}{9}$ of the total reservoir capacity. Another $\frac{1}{18}$ is lost through evaporation and leakage, but rain adds $\frac{1}{12}$ of the total capacity. How full is the reservoir now?

47. A shaft begins with a thickness of $1\frac{3}{4}$ in. The top surface is covered with a $\frac{1}{32}$-in. protective coating. The bottom is ground down so that $\frac{3}{48}$ in. is removed. What is the new thickness of the shaft?

48. An architect specifies a wall with support studs that are $1\frac{7}{8}$ in. thick. The inside of the wall is covered with $\frac{1}{2}$ in. thick sheetrock and $\frac{1}{4}$ in. paneling. The outside of the wall is covered with textured wood that is $\frac{7}{16}$ in. thick. What is the total thickness of the wall?

1-6 Multiplication and Division of Fractions

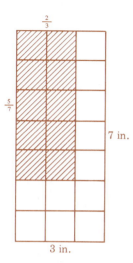

$\frac{2}{3}$

$\frac{5}{7}$

7 in.

3 in.

FIGURE 1-26

To develop the procedure for multiplying one fraction by another, we shall consider the area of a rectangle. See Figure 1-26. If the rectangle is 7 in. long and 3 in. wide, the area is 21 in.^{2}. Let us now mark both the length and the width at 1-in. intervals and divide the area into 21 equal parts, as shown. Each square has an area of 1 in.^{2}. If we now find the area of a section which is 5 in. long and 2 in. wide, it is 10 in.^{2}. This is equivalent to finding the area of a smaller rectangle whose length is $\frac{5}{7}$ of the length of the original rectangle and whose width is $\frac{2}{3}$ of the width of the original rectangle. We note that the resulting area is $\frac{10}{21}$ of the area of the original rectangle. Since we find area by multiplying length by width, we have

$$\frac{5}{7} \times \frac{2}{3} = \frac{10}{21}$$

This example suggests that multiplication is accomplished by multiplying the numerators and multiplying the denominators. In general: *The product of two fractions is the fraction whose numerator is the product of the numerators and whose denominator is the product of the denominators.*

EXAMPLE A

$$\frac{5}{7} \times \frac{2}{3} = \frac{5 \times 2}{7 \times 3} = \frac{10}{21} \longleftarrow \text{product of numerators}$$
$$\phantom{\frac{5}{7} \times \frac{2}{3} = \frac{5 \times 2}{7 \times 3} = \frac{10}{21}} \longleftarrow \text{product of denominators}$$

$$\frac{2}{5} \times \frac{8}{9} = \frac{2 \times 8}{5 \times 9} = \frac{16}{45}$$

$$\frac{9}{4} \times \frac{7}{2} = \frac{9 \times 7}{4 \times 2} = \frac{63}{8}$$

$$\frac{3}{14} \times \frac{15}{4} = \frac{3 \times 15}{14 \times 4} = \frac{45}{56} \qquad \square$$

If the resulting fraction is not in simplest form, it should be reduced to this form. We can do this by multiplying the numerators and denominators and reducing the resulting fraction, but this approach can lead to much more arithmetic than is necessary. Since a fraction is reduced to its lowest terms by dividing both numerator and denominator by the same number, we can first divide any factor present in both the numerator and denominator before performing the multiplication. In the following example, compare the two methods used for finding the product.

EXAMPLE B   In finding the product $\frac{24}{7} \times \frac{17}{32}$, we can multiply directly to get

$$\frac{24}{7} \times \frac{17}{32} = \frac{408}{224}$$

We can now reduce this result by dividing both numerator and denominator by 8, giving

$$\frac{24}{7} \times \frac{17}{32} = \frac{408}{224} = \frac{51}{28}$$

However, if we only indicate the multiplication as

$$\frac{24}{7} \times \frac{17}{32} = \frac{24 \times 17}{7 \times 32}$$

we note that both 24 and 32 are divisible by 8. If we perform this division before we multiply the factors in the numerator, we have

$$\frac{24}{7} \times \frac{17}{32} = \frac{24 \times 17}{7 \times 32} = \frac{3 \times 17}{7 \times 4} = \frac{51}{28}$$

▶ We obtain the same result, but the arithmetic operations are simpler and the numbers involved are smaller. Consequently, we should *divide out any factors which are common to both numerator and denominator* before *we actually multiply numerators and denominators.*   ☐

EXAMPLE C

$$\frac{4}{5} \times \frac{3}{8} = \frac{4 \times 3}{5 \times 8} = \frac{1 \times 3}{5 \times 2} = \frac{3}{10}$$ divide 4 and 8 by 4

$$\frac{18}{25} \times \frac{4}{27} = \frac{18 \times 4}{25 \times 27} = \frac{2 \times 4}{25 \times 3} = \frac{8}{75}$$ divide 18 and 27 by 9

$$\frac{16}{15} \times \frac{5}{12} = \frac{16 \times 5}{15 \times 12} = \frac{4 \times 1}{3 \times 3} = \frac{4}{9}$$ divide 16 and 12 by 4 and divide 5 and 15 by 5

$$\frac{30}{7} \times \frac{28}{33} = \frac{30 \times 28}{7 \times 33} = \frac{10 \times 4}{1 \times 11} = \frac{40}{11}$$ divide 30 and 33 by 3 and divide 28 and 7 by 7   ☐

▶ In many cases a factor which is common to both the numerator and the denominator of the resulting fraction is fairly obvious after the multiplication has been indicated. If such a factor is not evident, *the resulting numerator and denominator can be factored completely to determine common factors.*

EXAMPLE D   Specifications require that a rectangular plate be $\frac{9}{14}$ in. wide and $\frac{49}{15}$ in. long. Its area (in square inches) can be found by multiplying those two fractions. Find the product of the two fractions.

Indicating the product, we have

$$\frac{9}{14} \times \frac{49}{15} = \frac{9 \times 49}{14 \times 15}$$

In the resulting fraction the common factors may not be immediately evident. Assuming this to be the case, we may factor the numerator and the denominator. This leads to

$$\frac{9}{14} \times \frac{49}{15} = \frac{9 \times 49}{14 \times 15} = \frac{(3 \times 3) \times (7 \times 7)}{(2 \times 7) \times (3 \times 5)} = \frac{3 \times 3 \times 7 \times 7}{2 \times 7 \times 3 \times 5}$$

$$= \frac{3 \times 1 \times 1 \times 7}{2 \times 1 \times 1 \times 5} = \frac{21}{10}$$

After the factoring, the common factors of 3 and 7 are obvious and are divided out. The resulting area is $\frac{21}{10}$ in.^{2}.

*To multiply a whole number by a fraction, we can write the whole number as a fraction with 1 as the denominator. The result is the same if the whole number is multiplied by the numerator of the fraction. If a mixed number is to be multiplied by another number, **the mixed number must first be converted to an improper fraction.***

EXAMPLE E

$$4 \times \frac{3}{7} = \frac{4}{1} \times \frac{3}{7} = \frac{4 \times 3}{1 \times 7} = \frac{12}{7}$$

$$9 \times \frac{5}{12} = \frac{9 \times 5}{12} = \frac{3 \times 5}{4} = \frac{15}{4}$$

$$2\frac{1}{3} \times \frac{3}{5} = \frac{7}{3} \times \frac{3}{5} = \frac{7 \times 3}{3 \times 5} = \frac{7 \times 1}{1 \times 5} = \frac{7}{5}$$

$$6 \times 3\frac{2}{5} = 6 \times \frac{17}{5} = \frac{6 \times 17}{5} = \frac{102}{5}$$

If more than two fractions are to be multiplied, the resulting numerator is the product of the numerators and the resulting denominator is the product of the denominators.

EXAMPLE F

$$\frac{2}{5} \times \frac{3}{7} \times \frac{4}{11} = \frac{2 \times 3 \times 4}{5 \times 7 \times 11} = \frac{24}{385}$$

$$\frac{3}{7} \times \frac{4}{8} \times \frac{5}{13} = \frac{3 \times 4 \times 5}{7 \times 8 \times 13} = \frac{3 \times 1 \times 5}{7 \times 2 \times 13} = \frac{15}{182}$$

$$7 \times \frac{5}{28} \times \frac{2}{3} = \frac{7 \times 5 \times 2}{28 \times 3} = \frac{1 \times 5 \times 1}{2 \times 3} = \frac{5}{6}$$

$$\frac{25}{12} \times \frac{8}{15} \times \frac{3}{10} = \frac{25 \times 8 \times 3}{12 \times 15 \times 10} = \frac{5 \times 2 \times 3}{3 \times 3 \times 10}$$

$$= \frac{1 \times 1 \times 1}{3 \times 1 \times 1} = \frac{1}{3} \qquad \square$$

We will now illustrate the usual procedure used when dividing by a fraction. *To divide a number by a fraction, invert the divisor and multiply,* as in Example G.

EXAMPLE G

$$\frac{\frac{3}{7}}{\frac{4}{9}} = \frac{3}{7} \times \frac{9}{4} = \frac{3 \times 9}{7 \times 4} = \frac{27}{28}$$
↱ invert and multiply

$$\frac{8}{5} \div \frac{4}{15} = \frac{8}{5} \times \frac{15}{4} = \frac{8 \times 15}{5 \times 4} = \frac{2 \times 3}{1 \times 1} = 6$$
↱ invert and multiply

$$\frac{\frac{6}{2}}{\frac{2}{7}} = 6 \times \frac{7}{2} = \frac{6 \times 7}{2} = \frac{3 \times 7}{1} = 21$$

$$3 \div \frac{4}{7} = 3 \times \frac{7}{4} = \frac{3 \times 7}{4} = \frac{21}{4}$$

$$\frac{\frac{2}{5}}{\frac{4}{1}} = \frac{\frac{2}{5}}{4} = \frac{2}{5} \times \frac{1}{4} = \frac{2 \times 1}{5 \times 4} = \frac{1 \times 1}{5 \times 2} = \frac{1}{10}$$

$$\frac{2}{3} \div 6 = \frac{2}{3} \div \frac{6}{1} = \frac{2}{3} \times \frac{1}{6} = \frac{2 \times 1}{3 \times 6} = \frac{1 \times 1}{3 \times 3} = \frac{1}{9} \qquad \square$$

The method used in Example G can be justified by the fact that we can express a quotient like $\frac{3}{7} \div \frac{4}{9}$ as a fraction with a numerator of $\frac{3}{7}$ and a denominator of $\frac{4}{9}$. We could then proceed as follows.

$$\frac{\dfrac{3}{7}}{\dfrac{4}{9}} = \frac{\dfrac{3}{7} \times \dfrac{9}{4}}{\dfrac{4}{9} \times \dfrac{9}{4}} = \frac{\dfrac{3 \times 9}{7 \times 4}}{\dfrac{4 \times 9}{9 \times 4}} = \frac{\dfrac{27}{28}}{\dfrac{1 \times 1}{1 \times 1}} = \frac{\dfrac{27}{28}}{1} = \frac{27}{28} \quad \begin{array}{l} \text{final product} \\ \text{is } \dfrac{3}{7} \times \dfrac{9}{4} = \dfrac{27}{28} \end{array}$$

multiply numerator

and denominator by $\frac{9}{4}$

We have just seen that when dividing by a fraction it is necessary to invert the divisor. In general, *the* **reciprocal** *of a number is 1 divided by that number*. In following this definition we find that *the reciprocal of a fraction is the fraction with the numerator and denominator inverted*.

EXAMPLE H   The reciprocal of 8 is $\frac{1}{8}$. The reciprocal of $\frac{1}{3}$ is

$$\frac{1}{\dfrac{1}{3}} = 1 \times \frac{3}{1} = 3$$

The reciprocal of $\frac{7}{12}$ is

$$\frac{1}{\dfrac{7}{12}} = 1 \times \frac{12}{7} = \frac{12}{7} \longleftarrow \begin{array}{l} \text{numerator and denominator} \\ \text{of } \dfrac{7}{12} \text{ inverted} \end{array}$$

Finally, we consider the case in which the numerator or denominator, or both, of a fraction is itself the sum or difference of fractions. Such a fraction is called a **complex fraction**. *To simplify a complex fraction, we must first perform the additions or subtractions.*

EXAMPLE I   perform addition first

$$\frac{\dfrac{1}{2} + \dfrac{3}{4}}{8} = \frac{\dfrac{2}{4} + \dfrac{3}{4}}{8} = \frac{\dfrac{5}{4}}{8} = \frac{5}{4} \times \frac{1}{8} = \frac{5}{32}$$

then perform division

EXAMPLE J

perform addition and subtraction first

$$\frac{\dfrac{1}{5}+\dfrac{7}{10}}{\dfrac{2}{6}-\dfrac{1}{8}}=\frac{\dfrac{2}{10}+\dfrac{7}{10}}{\dfrac{8}{24}-\dfrac{3}{24}}=\frac{\dfrac{2+7}{10}}{\dfrac{8-3}{24}}=\frac{\dfrac{9}{10}}{\dfrac{5}{24}}=\frac{9}{10}\times\frac{24}{5}$$

$$=\frac{9\times 12}{5\times 5}=\frac{108}{25}$$

then perform division

□

Exercises 1-6

In Exercises 1 through 36, perform the indicated multiplications and divisions.

1. $\dfrac{2}{7}\times\dfrac{3}{11}$

2. $\dfrac{7}{8}\times\dfrac{3}{5}$

3. $3\times\dfrac{2}{5}$

4. $\dfrac{2}{7}\times 1\dfrac{2}{3}$

5. $\dfrac{7}{8}\div\dfrac{5}{6}$

6. $\dfrac{7}{4}\div\dfrac{2}{5}$

7. $\dfrac{5}{9}\div 3$

8. $2\dfrac{1}{2}\div\dfrac{8}{3}$

9. $\dfrac{8}{9}\times\dfrac{5}{16}$

10. $\dfrac{5}{12}\times\dfrac{3}{7}$

11. $2\dfrac{1}{3}\times\dfrac{9}{14}$

12. $\dfrac{22}{25}\times\dfrac{15}{33}$

13. $\dfrac{8}{15}\div\dfrac{12}{35}$

14. $\dfrac{21}{44}\div\dfrac{28}{33}$

15. $\dfrac{8}{17}\div 4$

16. $\dfrac{39}{35}\div\dfrac{13}{21}$

17. $\dfrac{3}{5}\times\dfrac{15}{7}\times\dfrac{14}{9}$

18. $\dfrac{2}{7}\times\dfrac{19}{4}\times\dfrac{21}{38}$

19. $\left(\dfrac{3}{4}\times\dfrac{28}{27}\right)\div\dfrac{35}{6}$

20. $\left(\dfrac{2}{3}\div\dfrac{7}{9}\right)\times\dfrac{14}{15}$

21. $\left(\dfrac{6}{11}\div\dfrac{12}{13}\right)\div\dfrac{26}{121}$

22. $\dfrac{18}{25}\div\left(\dfrac{17}{5}\div\dfrac{34}{15}\right)$

23. $\dfrac{9}{16}\times\left(\dfrac{1}{2}+\dfrac{1}{4}\right)$

24. $\left(\dfrac{7}{15}-\dfrac{3}{10}\right)\times\dfrac{12}{7}$

25. $6\div\left(\dfrac{1}{14}+\dfrac{3}{4}\right)$

26. $\left(\dfrac{9}{16}-\dfrac{1}{6}\right)\div 4\dfrac{3}{4}$

27. $\dfrac{\dfrac{1}{2}+\dfrac{1}{3}}{\dfrac{2}{3}}$

28. $\dfrac{\dfrac{15}{14}}{\dfrac{7}{8}-\dfrac{1}{4}}$

29. $\dfrac{\dfrac{1}{3}+\dfrac{5}{6}}{\dfrac{5}{12}-\dfrac{1}{4}}$

30. $\dfrac{\dfrac{7}{18}-\dfrac{2}{9}}{2-\dfrac{4}{15}}$

31. $\dfrac{\dfrac{4}{21}+\dfrac{5}{9}}{\dfrac{3}{4}+\dfrac{9}{14}}$

32. $\dfrac{\dfrac{2}{6}+\dfrac{7}{15}}{\dfrac{7}{10}-\dfrac{1}{4}}$

33. $\dfrac{\dfrac{3}{4}+\dfrac{1}{5}}{\dfrac{9}{10}-\dfrac{5}{6}}$

34. $\dfrac{\dfrac{13}{64}-\dfrac{1}{8}}{\dfrac{13}{16}-\dfrac{2}{5}}$

35. $\dfrac{\dfrac{26}{11}-2}{\dfrac{3}{5}+\dfrac{2}{3}}$

36. $\dfrac{\dfrac{25}{8}-\dfrac{7}{3}}{4-\dfrac{3}{8}}$

In Exercises 37 through 44, find the reciprocals of the given numbers.

37. 5; 13

38. 2; 9

39. $\dfrac{1}{2}$; $\dfrac{1}{5}$

40. $\dfrac{2}{9}$; $\dfrac{3}{7}$

41. $5\dfrac{1}{3}$; $3\dfrac{1}{2}$

42. $3\dfrac{1}{4}$; $9\dfrac{1}{32}$

43. $5\dfrac{1}{16}$; $7\dfrac{1}{10}$

44. $12\dfrac{1}{10}$; $15\dfrac{1}{3}$

In Exercises 45 through 56, solve the problems. (Where appropriate, h is the symbol for hour.)

45. The area of a rectangular sheet of typing paper is $93\frac{1}{2}$ in.^{2}. If the paper is 11 in. long, find its width.

46. The surface of a rectangular microprocessor chip is $\frac{11}{16}$ in. by $\frac{12}{33}$ in. What is its area? See Figure 1-27.

47. A machinist makes 16 shearing pins in 3 h. How many will be made in $2\frac{1}{4}$ h?

48. If 30 carbon-zinc flashlight batteries are connected in series, the total voltage is found by adding the individual voltages. What is the sum of the voltages of 30 batteries, each of which is $1\frac{1}{2}$ V? (V is the symbol for volt.)

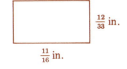

FIGURE 1-27

49. If a civil engineer can survey $2\frac{1}{4}$ acres in $6\frac{1}{2}$ h, what is his rate of surveying?

50. Cinder blocks are $7\frac{5}{8}$ in. high and each block is topped with an additional $\frac{1}{2}$ in. of mortar. How many blocks (with mortar) are needed for a column that is 130 in. tall?

51. A car's cooling system contains $14\frac{2}{3}$ qt, of which $\frac{3}{5}$ is alcohol. How many quarts of alcohol are in the system?

52. The acceleration due to gravity on Mars is about $\frac{2}{5}$ of that on the earth. If the acceleration due to gravity on earth is $32\frac{1}{5}$ ft/s^2, what is it on Mars? (The symbol ft/s^2 denotes feet per second per second.)

53. One inch equals about $2\frac{1}{2}$ cm. How many inches are there in 1 cm? (The symbol cm denotes centimeter.)

54. In a given amount of a certain alloy, there are 9 kg of iron. If the mass of the iron is $\frac{3}{8}$ of the mass of the alloy, what is the total mass of the alloy? (The symbol kg denotes kilogram.)

55. A dropped object travels $\frac{32}{165}$ mi in $\frac{1}{450}$ h. What is its average speed?

56. Two meshed gears have 20 teeth and 50 teeth, respectively. How many turns will the smaller gear make while the larger gear makes $3\frac{1}{2}$ turns?

1-7 Decimals

In discussing the basic operations on fractions in the last few sections, we have seen how to express and combine numbers that represent parts of quantities. In many applied situations, fractions are quite useful. However, there are many times in scientific work when a different way of expressing parts of quantities is more convenient. Measurements such as meter readings and distances are often expressed in terms of whole numbers and **decimal** parts. Computers and calculators display or print numbers in decimal form. Also, fractions are not normally used with the metric (SI) system which will be discussed later.

Fractions whose denominators are 10, 100, 1000, and so forth are called **decimal fractions.** For example, $\frac{7}{10}$ and $\frac{193}{10000}$ are decimal fractions. Making further use of positional notation, as introduced in Section 1-1, we place a **decimal point** to the right of the units digit and let the first position to the right stand for the number of tenths, the digit in the second position to the right stand for the number of hundredths, and so on. *Numbers written in this form are called* **decimals.**

EXAMPLE A The meaning of the decimal 6352.1879 is illustrated in Figure 1-28. We may also show the meaning of this decimal number as

$$6(1000) + 3(100) + 5(10) + 2(1) + \frac{1}{10} + \frac{8}{100} + \frac{7}{1000} + \frac{9}{10{,}000}$$

In this form we can easily see the relation between decimal and decimal fraction. □

thousands	hundreds	tens	units		tenths	hundredths	thousandths	ten-thousandths
6	3	5	2	·	1	8	7	9

FIGURE 1-28

Since a fraction is the indicated division of one number by another, *we can change a fraction to an equivalent decimal by division.* We place a decimal point and additional zeros to the right of the units position of the numerator and then perform the division.

EXAMPLE B A stock has a listed value of $\frac{5}{8}$ (dollars). Express that value in decimal form.

To change $\frac{5}{8}$ into an equivalent decimal, we perform the division shown at the left. Therefore $\frac{5}{8} = 0.625$. (It is common practice to place a zero to the left of the decimal point in a decimal less than 1. It clarifies that the decimal point is properly positioned.) □

```
  0.625
8)5.000
  4 8
    20
    16
     40
     40
```

It often happens that in converting a fraction to a decimal, the division does not come out even, regardless of the number of places to the right of the decimal point. When this occurs, the most useful decimal form is one that is *approximated* by **rounding off** the result of the division. (We take up the subject of approximate numbers in detail in Section 2-3.) *When we round off a decimal to a required accuracy, we want the decimal which is the closest approximation to the original but with the specified number of decimal positions.*

EXAMPLE C Since $\frac{3}{7} = 0.428\ldots$ (the three dots indicate that the division can be continued), we can round off the result to one decimal place (the nearest tenth) as $\frac{3}{7} = 0.4$, since 0.4 is the decimal closest to $\frac{3}{7}$, with the tenth position as the last position which is written.

Rounded off to two decimal places (the nearest hundredth), $0.428\ldots = 0.43$. Note that when we rounded off the number to hundredths, we increased the hundredths position by one. □

These examples suggest that when rounding off a decimal to a specified number of positions after the decimal point, the following rules apply:

1. *If the digit to the right of the round-off place is 5 or more, increase the digit in the round-off place by 1. If the digit to the right of the*

round-off place is 4 or less, do not change the digit in the round-off place.

2. *Delete all digits to the right of the round-off digit.*

This procedure is illustrated in the following example.

EXAMPLE D

hundredths
(4 or less) delete

$$0.862 = 0.86$$ (to two decimal places or hundredths)

hundredths

$$0.867 = 0.87$$ (to two decimal places or hundredths)

(5 or more) delete
and add 1 to 6

$$0.09326 = 0.093$$ (to three decimal places or thousandths)

(4 or less) delete
and delete the 6

$$0.09326 = 0.0933$$ (to four decimal places or ten thousandths)

(5 or more) delete
and add 1 to 2 ☐

Increasing the digit in the round-off place by 1 may cause a change in some of the digits to its left. For example, if 0.0598 is rounded to three decimal places, the result is 0.060.

To change a decimal to a fractional form, we write it in its decimal fraction form and then reduce to lowest terms. Consider the illustrations in the following example.

EXAMPLE E

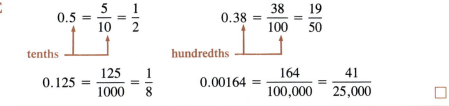

$$0.5 = \frac{5}{10} = \frac{1}{2} \qquad\qquad 0.38 = \frac{38}{100} = \frac{19}{50}$$

tenths   hundredths

$$0.125 = \frac{125}{1000} = \frac{1}{8} \qquad 0.00164 = \frac{164}{100,000} = \frac{41}{25,000}$$ ☐

When we consider the addition and subtraction of decimals, we see that the principles are the same as those for the whole numbers. Since only like quantities can be added or subtracted, we must then add tenths to tenths, hundredths to hundredths, and so on. Therefore, when we add and subtract decimals, we align the decimal points and perform the addition or subtraction as we did with the whole numbers.

EXAMPLE F

$$326.49$$
$$98.362$$
$$+ \; 5937.8$$
$$\overline{6362.652}$$

$$7862.472$$
$$- \;\; 794.56$$
$$\overline{7067.912}$$

☐

To develop the method for multiplying one decimal by another, we express each decimal in its decimal fraction form and then perform the multiplication. An observation of the results leads to the standard method. Consider the following example.

EXAMPLE G   To multiply 0.053 by 3.4, we express the multiplication as

$$(0.053)(3.4) = \frac{53}{1000} \cdot \frac{34}{10} = \frac{1802}{10,000} = 0.1802$$

We note that there are three decimal places (to the right of the decimal point) in the first of the numbers being multiplied, one decimal place in the second number, and four in the final result. Also, the numerator in the product is the product of 53 and 34 and is not affected by the denominators. In general, the positioning of the decimal point in the numbers being multiplied depends only on the denominators.

☐

Considering the results of Example G, when we multiply one decimal by another, *the number of decimal places (to the right of the decimal point) in the product is the sum of the number of decimal places in the numbers being multiplied*. It is not necessary to line up decimal points as we do in addition and subtraction. The numbers are multiplied as if they were natural numbers, and the decimal point is then properly positioned in the product.

EXAMPLE H

5.307	**3 places**
× 2.63	**2 places**
15921	
31842	
10614	
13.95741	**5 places**

0.004
× 0.02
0008
0000
0000
0.00008

☐

As in the other basic operations, the division of one decimal by another is very similar to the division of whole numbers. *Before the division is actually performed, however, the decimal point in the divisor is moved to the right a sufficient number of places to make the divisor a whole number. The decimal point in the dividend is then moved the same number of places.*

(This is equivalent to multiplying the numerator and denominator of the fraction form of the division by the same number.) The decimal point in the quotient is directly above that of the dividend.

EXAMPLE I   A circuit has a voltage level of 74.362 V and its resistance is 3.26 Ω. The current (in amperes) can be found by dividing 74.362 by 3.26. Find the quotient.

We perform the division as follows.

$$
\begin{array}{r}
22.8 \\
3.26.\overline{)74.36.2} \\
\end{array}
$$

← decimal point moved two places to right

$$
\begin{array}{r}
65\ 2 \\
\hline
9\ 16 \\
6\ 52 \\
\hline
2\ 642 \\
2\ 608 \\
\hline
34
\end{array}
$$

The result is expressed as 22.8 A.   □

Exercises 1-7

In Exercises 1 through 4, write the given decimals as a sum of 1's, 10's, 100's, etc., and decimal fractions, as in Example A.

 1. 47.3   **2.** 29.26   **3.** 429.486   **4.** 5230.3727

In Exercises 5 through 12, round off the given decimals as indicated.

 5. 27.32 (tenths)   **6.** 404.878 (hundredths)   **7.** 57.544 (two decimal places)

 8. 6.3833 (three decimal places)   **9.** 8.0327 (hundredths)   **10.** 0.4063 (hundredths)

 11. 17.3846 (tenths)   **12.** 86.30241 (thousandths)

In Exercises 13 through 20, change the given fractions into equivalent decimals. Where necessary round off the result to the accuracy indicated.

 13. $\dfrac{2}{5}$ **14.** $\dfrac{7}{16}$ **15.** $\dfrac{4}{19}$ (hundredths)   **16.** $\dfrac{11}{23}$ (hundredths)

 17. $\dfrac{47}{27}$ (tenths)   **18.** $\dfrac{882}{67}$ (tenths)   **19.** $\dfrac{362}{725}$ (thousandths)   **20.** $\dfrac{29}{426}$ (thousandths)

In Exercises 21 through 28, change the given decimals into equivalent fractions in simplest form.

 21. 0.8   **22.** 0.002   **23.** 0.45   **24.** 0.075

 25. 5.34   **26.** 17.6   **27.** 0.0252   **28.** 0.0084

In Exercises 29 through 44, perform the indicated operations.

29. 3.26 + 18.941 + 9.094

30. 18.87 + 8.9 + 182.825

31. 18.046 + 1924.3 + 874.91

32. 0.046 + 19.35 + 186.6942

33. 18.623 − 9.86

34. 2.057 − 1.388

35. 0.03106 − 0.00478

36. 0.8694 − 0.0996

37. (2.36)(5.932)

38. (37.4)(8.207)

39. (0.27)(36.6)

40. (0.0805)(1.006)

41. 5.6(3.72 + 18.6)

42. 0.042(3.072 + 92.23)

43. 6.75(0.107 − 0.089)

44. 0.375(4.70 − 2.92)

In Exercises 45 through 52, perform the indicated divisions, rounding off as indicated.

45. 32.6 ÷ 2.6   (tenths)

46. 37.9 ÷ 41.6   (tenths)

47. 192 ÷ 0.65   (tenths)

48. 132 ÷ 2.35   (tenths)

49. $\dfrac{96.288}{18.5}$ (hundredths)

50. $\dfrac{96.7}{0.826}$ (hundredths)

51. $\dfrac{6.238}{13.5}$ (hundredths)

52. $\dfrac{82.75}{103.6}$ (hundredths)

In Exercises 53 through 68, solve the given problems. (Where appropriate, h is the symbol for hour.)

53. A power supply line to a house is carrying 117.6 V. One day later, a reading of 118.2 V is obtained. By what amount did the voltage change? (V is the symbol for volt.)

54. The diameter of a tree is measured as 2.94 ft. One year later the same diameter is measured as 3.11 ft. By what amount did the diameter increase?

55. If an employee is given $416.34 for expenses but spends only $238.85, how much is left?

56. A spacer washer is 0.078 in. thick. If 0.012 in. is shaved off, what is the resulting thickness?

57. Three electrical resistors have resistances of 13.2 Ω, 6.9 Ω, and 8.4 Ω, respectively. What is the sum of these resistances? (The symbol Ω denotes ohm.)

58. Machine circuitry is mounted on a rectangular plate which is 13.56 cm long and 7.84 cm wide. What is the perimeter? (The symbol cm denotes centimeter.) See Figure 1-29.

7.84 cm

13.56 cm

FIGURE 1-29

59. If a tank has 318.62 L of fuel and 82.74 L are removed, how much remains? (The symbol L denotes liter.)

60. One end of a machine part is $5\frac{7}{16}$ cm wide and the other end is 4.982 cm wide. How much wider is the first end?

61. While using a calculator, a surveyor obtains an indicated length of 8.25 ft. First express that value as a mixed number; then express that distance in feet and inches.

62. A satellite circling the earth travels at 16,500 mi/h. If one orbit takes 1.58 h, how far does it travel in one orbit?

63. A machine is rented for $16.80 per hour for the first 8 h. The cost is $1\frac{1}{2}$ times the regular rate for any use past the 8 h limit. If the machine is used for $14\frac{1}{2}$ h, what is the total cost?

64. High speeds are often compared to the speed of sound. To determine the Mach number for a speed, we divide the given speed by the speed of sound (742 mi/h). Determine, to the nearest tenth, the Mach number of a jet that is traveling at 1340 mi/h.

65. A microcomputer is purchased for $1522.50 (including taxes). It is paid for in 12 monthly payments of $144.64. How much are the finance charges?

66. A surveyor uses an altimeter to measure heights, and obtains readings of 3.25 m, 12.67 m, and 4.71 m. To find the total true height, each reading must first be multiplied by a correction factor of 0.992 and then the results are added. Find that corrected total and round off the result to hundredths. (The symbol m designates meter.)

67. If 1 m = 39.37 in., how many meters are there in 8 ft? (Round off the result to hundredths.)

68. The relative density of an object can be defined as its weight in air divided by its loss of weight when weighed in water. A gold-aluminum alloy weighs 4.15 lb in air and 3.66 lb in water. What is its relative density? (Round off the result to tenths.)

1-8 Percent

Prior to this section we have used fractions and decimals for representing parts of a unit or quantity. We have seen that decimals have advantages: We use the same basic procedures for their basic operations as whole numbers, and they are more easily compared in size than fractions. In this section we consider the concept of **percent**.

The word "percent" means "per hundred" or hundredths, so *percent represents a decimal fraction with a denominator of 100.* The familiar % symbol is used to denote percent.

EXAMPLE A   3% means $\frac{3}{100}$ or 3(0.01) = 0.03. Also:

$$25\% = \frac{25}{100} = 0.25, \quad 300\% = \frac{300}{100} = 3.00, \quad 0.4\% = \frac{0.4}{100} = 0.004$$

Percent is very convenient, since we need consider only hundredths. With fractions we might use halves, fifths, tenths, and so on, and comparing fractions is difficult unless the denominators are the same. With decimals we use tenths, hundredths, thousandths, and so on. Comparisons of percents are easy, since only one denominator is used.

EXAMPLE B   Suppose you are told that $\frac{3}{20}$ of brand A fuses are defective and $\frac{4}{25}$ of brand B fuses are defective. You would first have to determine that $\frac{3}{20} = \frac{15}{100}$ and $\frac{4}{25} = \frac{16}{100}$ before you could compare. However, if you are told that 15% of brand A fuses are defective and 16% of brand B fuses are defective, the comparison is easy.

Using the meaning of percent allows us to convert percents to decimals, decimals to percents, percents to fractions, and fractions to percents. The following two examples illustrate these conversions.

EXAMPLE C

$$5\% = \frac{5}{100} = 0.05 \qquad \text{(percent to decimal)}$$

$$132\% = \frac{132}{100} = 1.32 \qquad \text{(percent to decimal)}$$

$$0.2\% = \frac{0.2}{100} = 0.002 \qquad \text{(percent to decimal)}$$

two places to left

$$0.45 = \frac{45}{100} = 45\% \qquad \text{(decimal to percent)}$$

$$0.3 = \frac{3}{10} = \frac{30}{100} = 30\% \qquad \text{(decimal to percent)}$$

$$0.826 = \frac{826}{1000} = \frac{82.6}{100} = 82.6\% \qquad \text{(decimal to percent)}$$

two places to right

Note: To convert a percent to a decimal or a decimal to a percent, the following rules apply.

1. *To change a percent to a decimal, move the decimal point two places to the left and omit the % sign.*

2. *To change a decimal to a percent, move the decimal point two places to the right and attach the % sign.*

EXAMPLE D

$$20\% = \frac{20}{100} = \frac{1}{5} \qquad \text{(percent to fraction)}$$

$$0.6\% = \frac{0.6}{100} = \frac{6}{1000} = \frac{3}{500} \qquad \text{(percent to fraction)}$$

$$\frac{3}{8} = 0.375 = 37.5\% \qquad \text{(fraction to percent)}$$

two places to right

$$\frac{15}{12} = 1.25 = 125\% \qquad \text{(fraction to percent)}$$

two places to right

From Example D we see that *to change a percent to its equivalent fractional form we first write the percent in its decimal fraction form*. This is done by writing the number of percent over 100. *This fraction is then reduced to lowest terms. To change a fraction to a percent, we first change the fraction to its decimal form by dividing the numerator by the denominator. The resulting decimal is then changed to a percent and rounded off if necessary.*

There are many applications of percent. Bank interest, income taxes, sales taxes, sales records of corporations, unemployment records, numerous statistics in sports, efficiency ratings of machines, compositions of alloys—these are only a few of the many uses of percent. The following example shows percent considered directly from its meaning. Other examples follow.

EXAMPLE E A manufacturing process has an 80% yield, meaning that 80% of the items produced are acceptable. This does not necessarily mean that there are exactly 100 items produced. If there are 200 items, 160 are acceptable since 80% of 200 = 0.80 × 200 = 160. If there are 40 items, 32 are acceptable since 80% of 40 = 0.80 × 40 = 32. ☐

In Example E we found the number of acceptable items by multiplying the percent in decimal form by the number of total items. In general: *The number of which we are finding the percent is the* **base,** *the percent expressed as a decimal is the* **rate,** *and the product of the rate and the base is the* **percentage.** Thus

$$\boxed{\textbf{Percentage} = \textbf{rate} \times \textbf{base}} \qquad \textbf{(1-1)}$$

This formula is known as the **percentage formula.** In using it we must be careful to identify the base and percentage properly. We must also remember that the rate is the percent expressed as a decimal.

EXAMPLE F A state has a sales tax of $5\frac{1}{2}\%$. What is the sales tax on a word-processing device costing $4000?

Here we must recognize that we are to find 5.5% of $4000. This means that 0.055 is the rate, $4000 is the base, and the tax is the percentage to be determined. Thus

$$\text{percentage} = \text{rate} \times \text{base}$$
$$\text{Tax} = 0.055 \times \$4000 = \$220$$

Therefore the sales tax on $4000 is $220. ☐

Many situations arise in which the percentage and base are known and the rate is to be found. Considering the case in which the base is 50, the rate is 0.40 (40%), and the percentage is 20, we have 0.40 × 50 = 20. If we now divide 20 (the percentage) by 50 (the base), we obtain 0.40 (the rate). In general,

$$\textbf{Rate} = \frac{\text{percentage}}{\text{base}} \qquad\qquad \textbf{(1-2)}$$

is an alternative way of writing the percentage formula when it is the rate we seek.

In these percentage, rate, and base problems, it is helpful to remember that the base is the total amount, the rate corresponds to the percent, and the percentage is the amount found when you take a percent of the base.

Percentage = rate × base	*20 = 0.40 × 50*
Base: Total amount	50
Percent: Rate	0.40
Percentage: Amount found by multiplying the rate and base	20

EXAMPLE G A manufacturer bought a machine for $28,000 and later sold it for $12,000. By what percent did the machine decrease in value?
▶ *The machine dropped in value by $16,000 and this amount represents the percentage. Since the value dropped from the original amount of $28,000, that is the base.* We now calculate

$$\text{Rate} = \frac{\overset{\text{percentage}}{16,000}}{\underset{\text{base}}{28,000}} = \underset{\text{rate}}{0.571} \qquad \text{(rounded off)}$$

We conclude that the machine decreased 57.1% in value.   ☐

Another problem arises when the percentage and rate are known and the base is to be determined. Consider again the case 0.40 × 50 = 20, where 0.40 is the rate, 50 is the base, and 20 is the percentage. If we divide 20 (the percentage) by 0.40 (the rate), we obtain 50 (the base). In general,

$$\textbf{Base} = \frac{\text{percentage}}{\text{rate}} \qquad\qquad \textbf{(1-3)}$$

is a third way we may express the percentage formula when it is the base that is to be found.

EXAMPLE H A fuel mixture is 40% alcohol. One shipment of the mixture contains 300 gal of alcohol. How many gallons are in the shipment?

▶ *The percentage is 300 gal, since there are 300 gal of alcohol in the mixture. The rate is 0.40, since 40% of the mixture is alcohol. Thus, the base is to be found.* This means

$$\text{Total number of gallons} = \frac{\overset{\text{percentage}}{\underset{\text{rate}}{300}}}{\underset{\text{rate}}{0.40}} = \underset{\text{base}}{750} \text{ gal}$$

Thus, there are 750 gal in the shipment. ☐

Exercises 1-8

In Exercises 1 through 8, change the given percent to equivalent decimals.

1. 8%   **2.** 78%   **3.** 236%   **4.** 482%

5. 0.3%   **6.** 0.082%   **7.** 5.6%   **8.** 10.3%

In Exercises 9 through 16, change the given decimals to percent.

9. 0.27   **10.** 0.09   **11.** 3.21   **12.** 21.6

13. 0.0064   **14.** 0.0007   **15.** 7   **16.** 8.3

In Exercises 17 through 24, change the given percent to equivalent fractions.

17. 30%   **18.** 48%   **19.** 2.5%   **20.** 0.8%

21. 120%   **22.** 0.036%   **23.** 0.57%   **24.** 0.14%

In Exercises 25 through 32, change the given fractions to percent. Round off to the nearest tenth of a percent where necessary.

25. $\dfrac{3}{5}$ **26.** $\dfrac{7}{20}$ **27.** $\dfrac{4}{7}$ **28.** $\dfrac{16}{11}$

29. $\dfrac{8}{35}$ **30.** $\dfrac{18}{29}$ **31.** $\dfrac{8}{3}$ **32.** $\dfrac{9}{7}$

In Exercises 33 through 64, solve the given problems.

33. Find 20% of 65.   **34.** Find 2.6% of 230.   **35.** What is 0.52% of 1020?

36. What is 126% of 300?   **37.** What percent of 72 is 18?   **38.** What percent of 250 is 5?

39. 3.6 is what percent of 48?   **40.** 0.14 is what percent of 3.5?   **41.** 25 is 50% of what number?

42. 3.6 is 25% of what number?

43. If 1.75% of a number is 7, what is the number?

44. If 226% of a number is 3.7, what is the number? (Round off to tenths.)

45. A software package retails for $48, but it is sold at a 15% discount. What is the discounted price?

46. The sales tax in a certain state is 6%. What is the tax in this state on tools with a total price of $378?

47. A company pays taxes of $27,200 on earnings of $85,000. What percent of earnings is paid in taxes?

48. An incandescent light bulb converts 4% of its input electrical energy into light. The remaining energy is given off as waste heat. Of 52,400 J of energy supplied, how much is actually *wasted* by such a bulb? (J is the symbol for joule.)

49. A machine rents for $16.80 per hour. If that cost is increased by 5%, what is the new cost per hour?

50. A factory uses a solar system to heat 42% of the hot water it needs. In one day, there is a need for 6500 gal of hot water. How many gallons were heated by the solar system?

51. A microcomputer costing $3995 is discounted by 15% for educational institutions. What does a college pay for one such unit?

52. The flue gas of a domestic oil burner is 84.7% nitrogen when measured by volume. What is the volume of nitrogen in 872,000 cm^3 of flue gas? (The symbol cm^3 denotes cubic centimeter.)

53. An alloy weighs a total of 640 lb, and it includes 224 lb of zinc. What percent of the alloy is zinc?

54. If $730 is deposited in a company credit union account, and $65.70 interest is earned for a one-year term, what is the annual rate of interest (in percent)?

55. If 2890 g of ocean water is found to contain 9.1 g of salt, what percent of ocean water is salt? (The symbol g denotes gram.)

56. A company manager sets a production goal of 850 tires for one particular day, but only 793 tires were produced. What percent of the goal was met?

57. A particular light bulb is labeled 60 watts, but actual tests show that the electric power level is really 58.2 W. What percent of the advertised wattage was actually achieved by this bulb? (W is the symbol for watt.)

58. One microcomputer has 16,384 bytes of memory. If a particular program requires 2460 of those bytes, what percent is left for other uses?

59. A sales representative earns $200 per week plus 12% commission on all sales he makes. What are his earnings for a week in which he sells $2600 worth of merchandise?

60. If a sales tax of 6% amounts to $34.80, what is the original cost of the goods purchased?

61. The efficiency of a motor is defined as power output divided by power input, usually expressed in percent. What is the efficiency of an electric motor whose power input is 850 W and whose power output is 561 W?

62. The effective value of an alternating current is 70.7% of its maximum value. What is the maximum value (to the nearest ampere) of a current whose effective value is 8 A? (A is the symbol for ampere.)

63. In one hour, a solar cell can convert 110.6 Btu of solar energy to 4.95 Btu of electrical energy. The efficiency of the conversion is the energy output divided by the energy input, expressed in percent. Find the efficiency. (The symbol Btu denotes British thermal unit.)

64. The minimum distance from the earth of a satellite is about 46% of its maximum distance from the earth. If its minimum distance is 1020 mi, what is its maximum distance (to the nearest mile)?

1-9 Powers and Roots

In many situations we encounter a number which is to be multiplied by itself several times. Rather than writing the number over and over repeatedly, we can use a notation in which *we write the number once and write the number of times it is a factor as an* **exponent.** The exponent is a small number to the right and slightly above the number. *The expression is usually referred to as a* **power** *of the number; the number itself is called the* **base.**

EXAMPLE A

Instead of writing 3×3, we write 3^2. Here the 2 is the exponent and the 3 is the base. The expression is read as "3 to the second power" or "3 squared."

Rather than writing $7 \times 7 \times 7$, we write 7^3. Here the 3 is the exponent and the 7 is the base. The expression is read as "7 to the third power" or "7 cubed."

The product $5 \times 5 \times 5 \times 5 \times 5 \times 5$ is written as 5^6. Here 6 is the exponent and 5 is the base. The expression is read as "the sixth power of 5" or "5 to the sixth power." ☐

2 factors of 3

$3 \times 3 = 3^2$

3 factors of 7

$7 \times 7 \times 7 = 7^3$

If a number is raised to the first power—that is, if it appears only once in a product—the exponent 1 is not usually written. Also, if we are given a number written in terms of an exponent, we can write the number in ordinary notation by performing the indicated multiplication. Consider the illustrations in the following example.

EXAMPLE B

$5^1 = 5$ (normally the 1 would not appear)

$4^5 = 4 \times 4 \times 4 \times 4 \times 4 = 1024$

$(2^3)(3^2) = (2 \times 2 \times 2)(3 \times 3) = (8) \times (9) = 72$ ☐

We can write another variation of the expanded form of a number using powers of 10. Considering the meanings of the various digit positions of a number (1's, 10's, 100's, 1000's, and so on), we see that we can express them in terms of powers of 10. Actually the word "decimal" means "based on the number 10," and we see that the base used in the expanded form of a number is 10. For this reason, 10 is the base of our number system.

EXAMPLE C

The number 3625 can be written as

$(3 \times 1000) + (6 \times 100) + (2 \times 10) + (5 \times 1)$

Since $1000 = 10 \times 10 \times 10 = 10^3$ and $100 = 10 \times 10 = 10^2$, we write

$3625 = (3 \times 10^3) + (6 \times 10^2) + (2 \times 10) + (5 \times 1)$

We can see that the number of times a given power of 10 (the base) is counted is determined by the value and position of a given numeral. ☐

EXAMPLE D   Any decimal may be written with powers of 10. However, for the digits to the right of the decimal point, the powers of 10 are in the denominator. Since

$$5.26 = (5 \times 1) + \left(2 \times \frac{1}{10}\right) + \left(6 \times \frac{1}{100}\right)$$

we may write

$$5.26 = (5 \times 1) + \left(2 \times \frac{1}{10}\right) + \left(6 \times \frac{1}{10^2}\right)$$ □

Now that we have discussed the meaning of a power of a number, we consider the concept of the roots of a number. Essentially, finding the root of a number is the reverse of raising a number to a power. *The* **square root** *of a given number is that number which when squared equals the given number*. We place the **radical sign** $\sqrt{}$ over the given number to indicate its square root.

EXAMPLE E   The square root of 9 is 3, since $3^2 = 9$. We write this as $\sqrt{9} = 3$.
The square root of 64 is 8, since $8^2 = 64$. We write this as $\sqrt{64} = 8$.
$\sqrt{25}$ (square root of 25) = 5, since $5^2 = 25$.
$\sqrt{144}$ (square root of 144) = 12, since $12^2 = 144$.   □

EXAMPLE F   The time (in seconds) it takes an object to fall 576 ft is found by evaluating

$$\sqrt{\frac{576}{16}}$$

After first dividing 576 by 16 to get a quotient of 36, we proceed to evaluate $\sqrt{36}$.

$$\sqrt{36} = 6 \text{ since } 6^2 = 36.$$

The time is 6 s.   □

Square roots are extremely important in basic mathematics. In addition to square roots, other roots are also used in many applications. We will now briefly consider some of the other roots.

The **cube root** *of a given number is that number which when cubed equals the given number. The* **fourth root** *of a given number is that number which when raised to the fourth power equals the given number*. The fifth root, sixth root, and so on are defined in a similar manner. The notation **if no number appears, it means square root** $\sqrt{}$ $\sqrt[3]{}$ is used for a cube root, $\sqrt[4]{}$ is used for the fourth root, and so on.

EXAMPLE G The cube root of 27 is 3, since $3^3 = 3 \times 3 \times 3 = 27$. This cube root is written as $\sqrt[3]{27} = 3$.

The fourth root of 625 is 5, since $5^4 = 5 \times 5 \times 5 \times 5 = 625$. This fourth root is written as $\sqrt[4]{625} = 5$.

$\sqrt[4]{81}$ (fourth root of 81) $= 3$, since $3^4 = 81$.

$\sqrt[3]{125}$ (cube root of 125) $= 5$, since $5^3 = 125$.   □

So far we have determined only powers and roots of whole numbers. We may also determine powers and roots of any decimal. In doing so, we must carefully note the position of the decimal point.

EXAMPLE H Squaring 0.6, we have $0.6^2 = 0.6 \times 0.6 = 0.36$.

In the same way, $0.6^3 = 0.6 \times 0.6 \times 0.6 = 0.216$.

Also, $2.7^2 = 7.29$ and $2.73^2 = 7.4529$.

$\sqrt{0.36}$ (square root of 0.36) $= 0.6$, since $0.6^2 = 0.36$.

$\sqrt{1.44} = 1.2$, since $1.2^2 = 1.44$ □

In this section we have introduced the meanings of the power of a number and the root of a number. In the examples and the following exercises, expressions are evaluated by use of the definitions and all values are exact. We may find the exact value of a power of a number by definition in all cases. However, the roots of most numbers do not result in exact values. Therefore, in Section 1-10 we consider methods of determining square roots of numbers so that we can deal with such cases.

Exercises 1-9

In Exercises 1 through 8, rewrite the given expressions by using the appropriate base and exponent. (Don't evaluate.)

1. $8 \times 8 \times 8$ **2.** $4 \times 4 \times 4 \times 4 \times 4$ **3.** $2 \times 2 \times 2 \times 2$
4. $5 \times 5 \times 5$ **5.** $3 \times 3 \times 3 \times 3 \times 3$ **6.** $12 \times 12 \times 12$
7. $10 \times 10 \times 10 \times 10 \times 10$ **8.** $6 \times 6 \times 6 \times 6 \times 6 \times 6 \times 6 \times 6$

In Exercises 9 through 16, write the given expressions as products. (Don't evaluate.)

9. 8^2 **10.** 2^5 **11.** 3^6 **12.** 4^3 **13.** 7^8 **14.** 10^4 **15.** 5^6 **16.** 6^5

In Exercises 17 through 20, write the given decimals in the expanded form using powers of 10, as in Examples C and D.

17. 8543   **18.** 3527   **19.** 5.739   **20.** 56.872

In Exercises 21 through 48, evaluate the given expressions.

21. 3^5 **22.** 2^7 **23.** 4^3 **24.** 6^4
25. 0.3^2 **26.** 0.12^2 **27.** 3.5^3 **28.** 0.08^3

29. $\sqrt{16}$	30. $\sqrt{81}$	31. $\sqrt{121}$	32. $\sqrt{400}$
33. $\sqrt[3]{64}$	34. $\sqrt[4]{16}$	35. $\sqrt[5]{32}$	36. $\sqrt[5]{243}$
37. $\sqrt{0.16}$	38. $\sqrt{0.81}$	39. $\sqrt{0.09}$	40. $\sqrt{1.21}$
41. $(3^3)(2^2)$	42. $(5^2)(4^3)$	43. $(5^3)(6^2)$	44. $(7^3)(10^4)$
45. $(\sqrt{121})(3^4)$	46. $(\sqrt{144})(4^3)$	47. $(0.7^2)(\sqrt{1.69})$	48. $(3.5^2)(\sqrt{0.04})$

In Exercises 49 through 60, solve the given problems.

49. The amount of electrical energy required to heat one cup of coffee is about $7 \times (10^4)$ J. Write this number in ordinary notation. (J is the symbol for joule.)

50. In UHF television broadcasting, the signal may have a frequency of $2 \times (10^9)$ cycles per second. Write this number in ordinary notation.

51. In determining the speed (in radians per second) of a cantilevered shaft, we must evaluate $\sqrt{625}$. Find that value.

52. The number of memory units in many microcomputers is 2^{18}. Write this number in ordinary notation.

53. Find the side of a square whose area is 900 ft². (The side equals the square root of the area.) (The symbol ft² denotes square foot. Note the use of the exponent with the unit symbol.) See Figure 1-30.

54. The time (in seconds) required for an object to fall 225 ft due to gravity can be found by evaluating

$$\frac{\sqrt{225}}{4}$$

Find the time required.

$A = 900$ ft²   ?

?

FIGURE 1-30

55. Find the impedance (in ohms) of a certain electrical circuit by evaluating $\sqrt{900}$.

56. A surveyor determines a particular horizontal distance (in meters) by evaluating $\sqrt{256}$. Find the distance.

57. A certain computer can perform $6 \times (10^8)$ operations in 1 s. Write this number in ordinary notation. (The symbol s denotes second.)

58. In calculating the diameter (in inches) of a vent pipe for a commercial oil burner, a technician must evaluate $\sqrt{484}$. Find that value.

59. An industrial plant uses security codes consisting of two letters followed by three digits. The number of different possible codes is $(26^2)(10^3)$. Evaluate that number.

60. Suppose you save 1¢ on the first day of January, 2¢ on the second day, 4¢ on the third, and you continue to double your deposit each day. The number of cents required for the last day of January is 2^{30}. What is the value of 2^{30} cents?

1-10 Evaluating Square Roots

A major objective of this section is to gain valuable experience in working with tables. We will soon see that, using calculators, it is very easy to obtain square roots of numbers, but we will also use a table for evaluating square roots. Tables of mathematical and physical values are often used in science and technology, and our work in this section is designed to better acquaint the reader with the use of those tables.

At the end of the previous section, we noted that the square root of most numbers does not come out exactly. In this section we develop specific methods by which square roots may be determined.

There are a number of different methods of finding the square root of a number. In Section 1-11 we show how calculators are used for a variety of computations, including the process of finding square roots of numbers. To find the square root of a number with a calculator, simply enter the number and then press the square root key.

EXAMPLE A   To find the square root of 3.24, first enter that number in the calculator and then press the square root key.

3.24 $\boxed{\sqrt{x}}$

The calculator should display the result of 1.8.   □

Square roots of numbers are found most often through the use of a calculator. Another way to evaluate square roots is to use a table like Table 1 in Appendix D, which is located near the end of the book. Even if you plan to use a calculator for all square root calculations, you should learn how to use Table 1 since it is a very good preparation and practice for using other types of tables as well. Since scientific and technical applications frequently involve tables of all sorts, it is wise to become adept at using them.

In Table 1 at the end of this book, numbers from 1.0 to 9.9 are shown with their squares and square roots. The square roots are expressed to the nearest thousandth. This table may be used to find the square or square root of any number shown, as well as any power of 10 times these numbers. Although our primary interest in this section is with square roots, we shall also deal briefly with squares since they are shown in the table.

EXAMPLE B   To find the square of 2.8, we locate 2.8 in the column labeled N. Then the value of the square is located opposite 2.8 in the column labeled N^2. Thus $2.8^2 = 7.84$.

To find the square root of 2.8, we again locate 2.8 in the column labeled N. The value of $\sqrt{2.8}$ is then found opposite 2.8 in the column labeled $\sqrt{N}$. We get $\sqrt{2.8} = 1.673$.

Since $28 = 2.8 \times 10$, we can find the value of $\sqrt{28}$ opposite 2.8 in the column labeled $\sqrt{10N}$. This column is used for the square roots of numbers 10 times the value of N. We get $\sqrt{28} = 5.292$. In the same way,

$7.3^2 = 53.29$ **7.3 under N, 53.29 under N²**

$\sqrt{7.3} = 2.702$ **7.3 under N, 2.702 under √N**

$\sqrt{73} = 8.544$

Again, since $73 = 7.3 \times 10$, the value of $\sqrt{73}$ is found in the $\sqrt{10N}$ column opposite 7.3.   □

We can use Table 1 to determine the square root of any power of 10 times the listed numbers. However, unless the number is listed directly, the decimal point of the square root will be placed differently from that shown. Nevertheless, the digits of the square root can be found from the table.

We can correctly locate the decimal point by using the two possible sets of digits in making estimates or guesses about the value of the square root. We can isolate the correct estimate or guess by multiplying each estimate or guess by itself to determine the one product which is close to the original number.

EXAMPLE C A square parking lot has an area of 6500 ft^2. The length of a side (in feet) can be found by evaluating $\sqrt{6500}$. Use Table 1 to determine the value of $\sqrt{6500}$.

From Table 1 we see that when $N = 6.5$, $\sqrt{N}$ and $\sqrt{10N}$ correspond to 2.550 and 8.062. These values suggest that initial guesses for the value of $\sqrt{6500}$ would be 25.50, 255.0, 80.62, and 806.2. If we consider each of these guesses and multiply each one by itself, we will see that only 80.62×80.62 will yield a product close to the original number of 6500. We therefore conclude that

$$\sqrt{6500} = 80.62$$

The length of one side is 80.62 ft. Some of the guesswork can be eliminated if we recognize that $80 \times 80 = 6400$. In any case, we can eliminate much arithmetic work by squaring approximations for each of the guesses. □

EXAMPLE D Use Table 1 to determine the value of $\sqrt{380}$.

Referring to Table 1, we see that with $N = 3.8$, $\sqrt{N}$ and $\sqrt{10N}$ correspond to 1.949 and 6.164. Guesses for the value of $\sqrt{380}$ include 19.49, 194.9, and 61.64, but only 19.49×19.49 has a value close to 380 so that we get

$$\sqrt{380} = 19.49$$

Here we can approximate 19.49 by 20, and note that $20 \times 20 = 400$, which is reasonably close to 380. □

EXAMPLE E Use Table 1 to determine the value of $\sqrt{0.0085}$.

With $N = 8.5$, Table 1 can be used to get $\sqrt{N} = 2.915$ and $\sqrt{10N} = 9.220$. Guesses for the value of $\sqrt{0.0085}$ might consist of 0.02915, 0.002915, 0.9220, 0.09220, and 0.009220. We check each guess by multiplying its approximation by itself. Since

$$0.09 \times 0.09 = 0.0081$$

where 0.09 is the approximation for 0.09220, we conclude that

$$\sqrt{0.0085} = 0.09220$$ □

It is possible to use Table 1 without having to guess at the result. The procedure is discussed in Section 10-4.

There are exact procedures for manually calculating square roots without using tables or calculators, but the widespread availability of calculators, computers, and tables has greatly reduced the importance of these procedures. Since such procedures are complicated and rarely used, we will not discuss them here.

Exercises 1-10

In Exercises 1 through 20, evaluate the given expression by using Table 1.

1. 2.1^2
2. 5.4^2
3. 3.9^2
4. 8.1^2
5. $\sqrt{2.1}$
6. $\sqrt{5.4}$
7. $\sqrt{3.9}$
8. $\sqrt{8.1}$
9. $\sqrt{32}$
10. $\sqrt{75}$
11. $\sqrt{55}$
12. $\sqrt{91}$
13. $\sqrt{190}$
14. $\sqrt{760}$
15. $\sqrt{4300}$
16. $\sqrt{96,000}$
17. $\sqrt{0.044}$
18. $\sqrt{0.71}$
19. $\sqrt{0.0065}$
20. $\sqrt{0.00013}$

In Exercises 21 through 32, solve the given problems.

21. In doing calculations required for a heat loss formula, we must find the square root of the wind speed, as measured in meters per second. Evaluate $\sqrt{18.0}$ to the nearest hundredth.

22. A steel disk is used to seal a blowout port on a truck engine. Its diameter (in inches) is found by evaluating $2\sqrt{1.30}$. Find the diameter to the nearest hundredth of an inch.

23. Near the surface of the moon, the time (in seconds) it takes an object to fall 12 ft is about

$$\sqrt{\frac{12}{2.7}}$$

Find that time to the nearest tenth.

24. A surveyor determines the horizontal distance (in feet) between two points by evaluating $\sqrt{58,000}$. Find that distance to the nearest tenth.

25. The period (in seconds) of a pendulum whose length is 5 ft is found by evaluating $(1.6)(\sqrt{2.5})$. By evaluating the square root, find the period (to hundredths).

26. In certain circumstances, the frequency of vibration in hertz (Hz), where one hertz equals one cycle per second, in an electric circuit can be found by evaluating

$$\frac{1}{(6.3)(\sqrt{0.000026})}$$

Find the frequency to the nearest tenth.

27. Under certain conditions, the pressure (in kilopascals—symbol kPa) of 40 cm³ (cubic centimeters) of a gas can be calculated to be $(0.3)(\sqrt{40})^3$. Calculate the pressure to the nearest tenth.

28. The velocity (in feet per second—symbol ft/s) of a freely falling object is approximately eight times the square root of the distance (in feet) it has fallen. Find the velocity (to the nearest unit) of an object which has fallen 620 ft.

29. The longest straight-line segment which can be drawn on a 3 in. by 5 in. index card is $\sqrt{34}$ in. Evaluate that length to the nearest tenth.

30. A quality control engineer determines that the variation among a collection of measurements can be described by the number

$$\frac{\sqrt{1600}}{\sqrt{20}\sqrt{19}}$$

Find that value to the nearest tenth.

31. One gallon of a special thermal protective paint will cover an area of 150 ft². If the entire gallon is used to paint a square area, find (to the nearest tenth) the length of a side of that square. (The symbol ft² denotes square foot.)

32. A zoning ordinance requires a particular store to have a parking lot with an area of at least 800 m². If this parking lot is designed to be square, find the length of a side to the nearest tenth. (The symbol m² denotes square meter.) See Figure 1-31.

FIGURE 1-31

1-11 Arithmetic with Calculators

In addition to Appendix A, you will find references to calculators throughout this book. A calculator will be very helpful for this course, and its use will simplify many of the calculations. However, it should be stressed that an understanding of mathematical principles cannot be replaced by the use of a calculator. Without an understanding of the associated mathematics, the calculator is effectively worthless.

In this section we show how to use the calculator for the fundamental arithmetic operations already presented in this chapter. We will assume that you have a scientific calculator which uses **algebraic logic**. (Algebraic logic is also called the **algebraic operating system,** or AOS.) The type of logic your calculator uses can be quickly checked by pressing the following sequence of keys.

$$2 \boxed{+} 3 \boxed{\times} 4 \boxed{=}$$

If the result is 20, then your calculator uses **chain logic**. If the result is 14, then your calculator uses algebraic logic, which is generally more suitable for scientific and technical applications. Such calculators are designed to automatically perform the operations according to a specific **hierarchy.** *The hierarchy of algebraic logic requires that multiplications and divisions be performed before additions and subtractions.* With the preceding sequence of keys, algebraic logic requires that the multiplication of 3 × 4 be completed before the operation of addition is executed. With chain logic, the operations are simply executed from left to right in order of occurrence. *It is extremely important to know the type of logic your calculator uses.*

Throughout the remainder of this text, *we will assume that the calculator uses algebraic logic;* this is the case with most scientific calculators.

EXAMPLE A   The sequence of keys

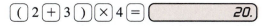

$9 \boxed{+} 2 \boxed{\times} 4 \boxed{-} 12 \boxed{\div} 3 \boxed{=} \boxed{\qquad 13.} \longleftarrow$ **final display**

gives us a result of 13. The multiplication and division are executed first so that the expression first becomes $9 + 8 - 4$, which in turn becomes 13. □

Many scientific calculators incorporate the parentheses keys which can be used to override the hierarchy of operations. Let us assume that in the expression $2 + 3 \times 4$ we really want the sum of 2 and 3 to be multiplied by 4. We can use parentheses to group the $2 + 3$ component by using the following sequence of keys.

$\boxed{(} 2 \boxed{+} 3 \boxed{)} \boxed{\times} 4 \boxed{=} \boxed{\qquad 20.}$

When you enclose an expression within parentheses, you force the calculator to evaluate that part first.

If the parentheses keys are not available, we could first add the 2 and 3, press the equals key, and then multiply by 4 as follows.

$2 \boxed{+} 3 \boxed{=} \boxed{\times} 4 \boxed{=} \boxed{\qquad 20.}$

When you press the equals key, the calculator will complete all of the operations which have been entered. The result of the last expression will be 20.

We will often encounter mathematical expressions which include parentheses, as in $2(3 + 4)$. The value of $2(3 + 4)$ is 14; the operation of multiplication is implied. But a calculator entry of

$2 \boxed{(} 3 \boxed{+} 4 \boxed{)} \boxed{=}$

usually will not yield the correct answer of 14 since the implied operation of multiplication is not recognized by most calculators. Instead, we must explicitly specify the operation of multiplication as follows.

$2 \boxed{\times} \boxed{(} 3 \boxed{+} 4 \boxed{)} \boxed{=} \boxed{\qquad 14.}$

EXAMPLE B   The correct calculator sequence to be used for evaluating $(2 + 3)(4 + 5)$ is

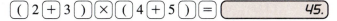

$\boxed{(} 2 \boxed{+} 3 \boxed{)} \boxed{\times} \boxed{(} 4 \boxed{+} 5 \boxed{)} \boxed{=} \boxed{\qquad 45.}$

In this way we obtain the result of 45.   □

In Sections 1-3 through 1-6 we discussed fractions, and in Section 1-7 we considered decimals. With calculators we essentially ignore fractions and work with equivalent decimal representations instead.

EXAMPLE C   To evaluate $\frac{5}{12} + \frac{7}{8}$ using a calculator, we use the entries

$$5 \;\boxed{\div}\; 12 \;\boxed{+}\; 7 \;\boxed{\div}\; 8 \;\boxed{=}\; \boxed{\;\textit{1.2916667}\;}$$

and obtain a result of 1.2916667. With most calculators, we will not get a display of $\frac{31}{24}$, which is the correct answer. The calculator display of 1.2916667 is a rounded-off result which is *approximately* equal to the correct answer of $\frac{31}{24}$. ☐

EXAMPLE D   A blueprint specifies that a hole be drilled with a diameter of $5\frac{13}{16}$ in. Use a calculator to write that number in decimal form.

To enter $5\frac{13}{16}$ in the calculator, press this sequence of keys:

$$5 \;\boxed{+}\; 13 \;\boxed{\div}\; 16 \;\boxed{=}\; \boxed{\;\textit{5.8125}\;}$$

This tells us that $5\frac{13}{16} = 5.8125$. ☐

When we write a mixed number such as $5\frac{3}{14}$, we imply that the 5 and the $\frac{3}{14}$ are added. Remember, a mixed number was defined to be a whole number *plus* a proper fraction. Again, the implied operation of addition is not recognized by most calculators, so we must explicitly enter the operation as in the following example.

EXAMPLE E   To find the sum of $5\frac{3}{14} + 6\frac{7}{8}$, press the following keys.

$$5 \;\boxed{+}\; 3 \;\boxed{\div}\; 14 \;\boxed{+}\; 6 \;\boxed{+}\; 7 \;\boxed{\div}\; 8 \;\boxed{=}\; \boxed{\;\textit{12.089286}\;}$$

When we make this entry, we express $5\frac{3}{14} + 6\frac{7}{8}$ as $5 + \frac{3}{14} + 6 + \frac{7}{8}$, which is an equivalent form that makes the implied additions explicit. The final display tells us that $5\frac{3}{14} + 6\frac{7}{8} = 12.089286$. ☐

It is calculations like the one given in Example E that illustrate the usefulness of calculators. To manually find the given sum, we would have to first determine the lowest common denominator, then convert both mixed numbers to improper fractions having that lowest common denominator, and then add. However, the manual addition will yield the exact value while most calculators will yield a rounded-off decimal form of the exact sum. (There are some calculators that perform operations with fractions. The results are fractions reduced to lowest terms.)

In Section 1-6 we saw that the reciprocal of a number is 1 divided by that number. The reciprocal key is usually labeled $1/x$ and it is used after the number itself has been entered.

EXAMPLE F   To find the reciprocal of 8, enter

$$8 \; \boxed{1/x} \quad \boxed{\qquad 0.125 \qquad}$$

The final display shows that the decimal equivalent of $\frac{1}{8}$ is 0.125.   ☐

In Section 1-8 we discussed percent. Many calculators incorporate a percent key which, when pressed, automatically converts a percent to the equivalent decimal. The percent key simply causes an entered number to be divided by 100.

EXAMPLE G   To convert 62% to a decimal, enter

$$62 \; \boxed{\%} \quad \boxed{\qquad 0.62 \qquad}$$

This means that 62% = 0.62.   ☐

Sections 1-9 and 1-10 involved powers and roots. These operations are extremely important in scientific and technical calculations, and any good scientific calculator will have keys for these operations. The square root key is used to find the square root of an entered number, as illustrated by Example H and in Section 1-10.

EXAMPLE H   To find the square root of 5.29, first enter that number and then press the square root key.

$$5.29 \; \boxed{\sqrt{x}} \quad \boxed{\qquad 2.3 \qquad}$$

Thus, $\sqrt{5.29} = 2.3$.   ☐

To raise a number to a power, use the power key which is usually labeled as either x^y or y^x.

EXAMPLE I   To calculate the value of 5^4, enter

$$5 \; \boxed{x^y} \; 4 \; \boxed{=} \quad \boxed{\qquad 625. \qquad}$$

This tells us that $5^4 = 625$.   ☐

Additional calculator techniques will be introduced in appropriate sections throughout this text.

Exercises 1-11

In Exercises 1 through 16, first determine the expected result without using a calculator and then check your answer with a scientific calculator.

1. $1.03 + 2.04$
2. $8.7 - 6.3$
3. $9.88 - 3.44$
4. $20.35 + 10.40$

5. 8×7
6. $36 \div 4$
7. $18 \div 3$
8. 9×6

9. $\sqrt{9}$
10. $\sqrt{16}$
11. $\sqrt{64}$
12. $\sqrt{49}$

13. 2^3
14. 3^4
15. 2^5
16. 5^3

In Exercises 17 through 36, use a scientific calculator to evaluate the given expression.

17. $\dfrac{2}{3} + \dfrac{3}{4}$
18. $\dfrac{3}{8} - \dfrac{5}{18}$
19. $\dfrac{7}{9} - \dfrac{5}{19}$
20. $\dfrac{5}{64} + \dfrac{11}{64}$

21. $5\dfrac{1}{4} + 6\dfrac{2}{3}$
22. $3\dfrac{1}{8} + 7\dfrac{2}{11}$
23. $2\dfrac{6}{7} + 3\dfrac{3}{4}$
24. $8\dfrac{1}{2} + 9\dfrac{1}{4}$

25. $\left(\dfrac{11}{3}\right)\left(\dfrac{6}{5}\right)$
26. $\dfrac{23}{21} \div \dfrac{7}{12}$
27. $\dfrac{3}{4} \div \dfrac{5}{6}$
28. $\left(\dfrac{3}{4}\right)\left(\dfrac{5}{6}\right)$

29. $(10 + 2)(6 + 4)$
30. $6(8 + 9)$
31. $12(2 + 9)$
32. $(8 - 3)(10 - 2)$

33. $\sqrt{19.135}$
34. $\sqrt{0.001}$
35. $\sqrt{0.0001}$
36. $\sqrt{0.88}$

In Exercises 37 through 40, use the reciprocal key on a calculator to find decimal equivalents of the given numbers.

37. $\dfrac{1}{3}$
38. $\dfrac{1}{12}$
39. $\dfrac{1}{45}$
40. $\dfrac{1}{16}$

In Exercises 41 through 44, use the percent key on a calculator to find the decimal equivalents of the given percent.

41. 34%
42. 256.3%
43. 345.6%
44. 8%

In Exercises 45 through 52, use a scientific calculator to solve the given problems.

45. The volume (in cubic inches) of a cylinder head can be determined by evaluating $3.14 \times 1.52^2 \times 5.48$. Calculate the volume to the nearest hundredth.

46. In analyzing product failures, a quality control specialist must evaluate

$$\sqrt{\dfrac{1362}{35}}$$

Find that value to the nearest hundredth.

47. Reactance (in ohms) in an electric circuit can be found by evaluating

$$\dfrac{1}{2 \times 3.14 \times 398 \times 0.187}$$

Find that value and round to the nearest thousandth.

48. An architect can establish the distance (in feet) between two electrical outlets by evaluating

$$\sqrt{18.2^2 + 7.5^2}$$

Find that distance to the nearest tenth.

49. In analyzing the rate of growth of bacteria, a researcher must evaluate 2.718^6. Find that value to the nearest tenth.

50. One arm of an industrial robot is designed so that the number of positions it can take is given by $64 \times 2^6 \times 5^4$. Find that number.

51. A sample of an alloy contains 23.4 g of silver. If the mass of this silver is $\frac{3}{16}$ of the total mass of the sample, find the total mass of the sample. (The symbol g denotes gram.)

52. During a test of radioactive decay of the isotope barium 140, a sample of 72.6 g is reduced to a mass of 63.9 g. By what percent did the sample of barium decrease?

Chapter 1 Formulas

$$\text{Percentage} = \text{rate} \times \text{base} \tag{1-1}$$

$$\text{Rate} = \frac{\text{percentage}}{\text{base}} \tag{1-2}$$

$$\text{Base} = \frac{\text{percentage}}{\text{rate}} \tag{1-3}$$

1-12 Review Exercises for Chapter 1

In Exercises 1 through 44, perform the indicated operations.

1. $3126 + 328 + 9876$

2. $98{,}076 + 8992 + 3964$

3. $8764 - 5985$

4. $19{,}264 - 9397$

5. $8.12 + 19.092 + 93.9$

6. $986.42 + 93.7 + 8.966$

7. $2706.46 - 829.5$

8. $2.9064 - 0.918$

9. 476×9172

10. $8076 \times 79{,}064$

11. $14{,}980 \div 35$

12. $511{,}098 \div 129$

13. 3.93×18.4

14. 0.0362×19.41

15. $0.04536 \div 3.24$

16. $8.0698 \div 0.0257$

17. $3.96(0.042 + 9.33)$

18. $92.6(18.4 + 7.82)$

19. $\frac{3}{13} + \frac{2}{13}$

20. $\frac{1}{2} + \frac{1}{8}$

21. $2\frac{1}{5} - \frac{3}{10}$

22. $\frac{5}{12} - \frac{3}{14}$

23. $\frac{13}{30} + \frac{7}{12} + \frac{1}{6}$

24. $\frac{16}{63} + \frac{5}{14} - \frac{1}{21}$

25. $\frac{7}{50} + \frac{6}{35} + \frac{19}{42}$

26. $\frac{1}{27} + \frac{11}{45} + \frac{7}{18}$

27. $\left(\frac{11}{2}\right)\left(\frac{4}{5}\right)$

28. $\left(\frac{4}{9}\right)\left(\frac{15}{48}\right)$

29. $\left(\frac{18}{77} \times \frac{14}{27}\right) \times \frac{121}{8}$

30. $\left(3\frac{1}{6} \times 1\frac{1}{5}\right) \times \frac{125}{38}$

31. $\frac{32}{21} \div \frac{8}{7}$

32. $\frac{24}{17} \div \frac{12}{7}$

33. $\frac{5 \times 19 \times 31}{3 \times 7 \times 11} \div \frac{19 \times 31}{7 \times 43}$

34. $\frac{2 \times 9 \times 20}{3 \times 49 \times 63} \div \frac{8 \times 35}{14 \times 27}$

35. $\sqrt{49}$

36. $\sqrt{144}$

37. $\sqrt{0.0016}$

38. $\sqrt{0.0144}$

39. $\sqrt[3]{1000}$

40. $\sqrt[6]{64}$

41. $3(4^2)$

42. $(2^3)(5^2)$

43. $3^4 - 2^5$

44. $3(6^2 + 4^4)$

In Exercises 45 through 48, change the given mixed numbers to improper fractions and change the given improper fractions to mixed numbers.

45. $3\frac{3}{7}$; $\frac{17}{5}$

46. $7\frac{3}{8}$; $\frac{21}{19}$

47. $32\frac{1}{8}$; $\frac{123}{7}$

48. $56\frac{3}{4}$; $\frac{172}{17}$

In Exercises 49 through 52, reduce each of the given fractions to lowest terms.

49. $\dfrac{18}{30}$ **50.** $\dfrac{15}{27}$ **51.** $\dfrac{26}{65}$ **52.** $\dfrac{60}{75}$

In Exercises 53 through 56, find the reciprocals of the given numbers.

53. $\dfrac{2}{9}$; $3\dfrac{1}{7}$ **54.** $\dfrac{4}{3}$; $8\dfrac{2}{5}$ **55.** $1\dfrac{1}{9}$; $\dfrac{34}{43}$ **56.** $4\dfrac{4}{9}$; $\dfrac{7}{17}$

In Exercises 57 through 60, change the given fractions to equivalent decimals and change the given decimals into equivalent fractions.

57. $\dfrac{9}{32}$ **58.** $\dfrac{121}{400}$ **59.** 0.56   **60.** 3.155

In Exercises 61 through 64, change the given percents to equivalent decimals and to equivalent fractions.

61. 82%   **62.** 250%   **63.** 0.55%   **64.** 0.0225%

In Exercises 65 through 68, change the given decimals and fractions to percents.

65. 0.934   **66.** 87.28   **67.** $\dfrac{2}{25}$ **68.** $\dfrac{15}{4}$

In Exercises 69 through 76, determine the value of the given expression from Table 1 at the end of the book.

69. $\sqrt{4.4}$ **70.** $\sqrt{8.7}$ **71.** $(3.8)^2$ **72.** $(9.4)^2$
73. $\sqrt{43}$ **74.** $\sqrt{77}$ **75.** $\sqrt{16{,}000}$ **76.** $\sqrt{0.88}$

In Exercises 77 through 80, determine the square roots by using a calculator. Round off results to the number of decimal places indicated in parentheses.

77. $\sqrt{0.538756}$ (three)   **78.** $\sqrt{948.64}$ (one)   **79.** $\sqrt{9.372}$ (two)   **80.** $\sqrt{8724}$ (one)

In Exercises 81 through 84, use a scientific calculator to determine the value of the given expression. Round off results to the number of decimal places indicated in parentheses.

81. $3\dfrac{3}{4} + \dfrac{5}{18}$ (three)   **82.** 5.1^6 (one)   **83.** $2\dfrac{3}{7} + \sqrt{10}$ (two)   **84.** $2.1(3.6 + 7.5)$ (one)

In Exercises 85 through 108, solve the given problems. Where appropriate, the following units of measurement are designated by the given symbols: meter—m, square foot—ft², miles per hour—mi/h, hour—h, ohm—Ω, second—s, volt—V, degrees Fahrenheit—°F, degrees Celsius—°C, kelvin—K, watt—W.

85. Find the perimeter of the triangle (three sides) whose sides are 3.68 in., 8.21 in., and 6.09 in. See Figure 1-32.

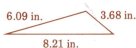

86. Find the perimeter of a rectangle of length 18.2 m and width 9.3 m.

FIGURE 1-32

87. A barometer reads 30.21 in. before a storm. It then drops 0.67 in. What is the later reading?

88. If a solution costs 87.9¢ per gallon, how much does it cost to fill a tank that holds 12.6 gal?

89. Given that the area of a rectangle is 70.305 ft² and its width is 3.27 ft, find its length. See Figure 1-33.

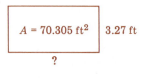

90. A satellite circling the earth travels at 17,300 mi/h. If one orbit takes 1.55 h, how far does it travel in one orbit?

91. The total resistance of 18 equal electrical resistors is 370.8 Ω. What is the resistance of each?

FIGURE 1-33

92. Several years ago a manufacturer acquired a computer capable of performing a simple operation in 0.0000002 s. A new version of a similar computer can now do a simple operation in 0.0000000006 s. How many times faster is the new model?

93. Both sides of a metal plate $\frac{3}{16}$ in. thick are coated with a film $\frac{1}{64}$ in. thick. What is the resulting thickness of the plate?

94. Two-fifths of a piece of wire $11\frac{1}{4}$ ft long is cut off. What is the length of the remaining piece?

95. A bottle contains $3\frac{3}{4}$ qt of acid which is $\frac{6}{25}$ sulfuric acid. If $\frac{2}{3}$ of the acid is poured from the bottle, how much sulfuric acid is left in the bottle?

96. In a house plan, a living room 24 ft 6 in. long, a hall 4 ft 10 in. wide, and a bedroom 15 ft 8 in. long are shown across the front of the house. The interior walls are 5 in. thick and the exterior walls are 8 in. thick. What is the width of the front of the house? See Figure 1-34.

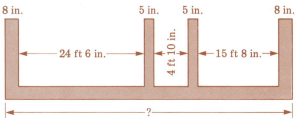

97. A surveyor measures a distance and obtains a value of 84.23 m. This value must be multiplied by 0.998 to correct for the sag in the measuring tape. What is the corrected length? (Round off result to nearest hundredth of a meter.)

FIGURE 1-34

98. The approximate length in feet of a certain pulley belt is found by making the following calculation: $2(4.27) + 3.14(0.83 + 1.42)$. Find the length of the pulley belt to the nearest hundredth of a foot.

99. Approximately 0.71% of uranium is U-235 (the type used for atomic energy). How much U-235 is contained in 50 lb of uranium ore which is 2.13% uranium? (Round off result to nearest thousandth of a pound.)

100. A manufacturer makes an article for $18.50. He sells it to a dealer, making a profit of 40% of the cost. The dealer sells it, making a 50% profit of his cost. What is the price charged by the dealer?

101. The price of a component used in making a helium-neon laser is $60. The supplier of this part reduces the price by 15%. This discounted price is later increased by 15%. What is the current price?

102. Two months is 6.67% of the guaranteed life of an automobile battery. What is the guaranteed life of the battery to the nearest month?

103. Seventy-five pounds of salt water contains 4% salt. If 45 lb of fresh water is added to this solution, what is the percentage of salt in the final solution?

104. Under certain conditions the voltage in a given electric circuit is found by calculating $\sqrt{(16.1)(8.86)}$. Find the voltage to the nearest tenth of a volt.

105. The impedance (effective resistance) (in ohms) of a certain electric circuit is found by calculating $\sqrt{(5.68)^2 + (10.07 - 2.42)^2}$. Calculate the impedance of this circuit to the nearest tenth of an ohm.

106. The Fahrenheit degree is $\frac{5}{9}$ of the Celsius degree. If the Celsius temperature rises by 40°, by how much does the Fahrenheit temperature rise?

107. A metal bar is cut into six pieces, each $3\frac{3}{8}$ in. long. Each cut wastes $\frac{1}{16}$ in. Determine the length of the original bar.

108. The thermodynamic temperature (in kelvins) of the filament of a light bulb equals approximately 1000 times the fourth root of the wattage. Find the Celsius temperature (°C: 273 less than the thermodynamic temperature) of the filament of a 16-W bulb.

CHAPTER TWO

Units of Measurement and Approximate Numbers

In this chapter we will learn about the metric system and how to convert measurements between that system and the U.S. Customary System. We will also study procedures for changing measurements within a system, as in reducing centimeters to millimeters. We will study the importance of significant digits in calculations, and we will learn the correct procedures for rounding numbers. All of these topics have great practical importance for scientists and technicians.

We will also learn to solve many different types of problems. As one example, suppose that a pilot estimates that she must take on 12 gal of avgas (aviation grade gasoline). If she is in Canada, she must convert that amount to liters. How many liters should she request? We will solve this and many other such problems in this chapter.

2-1 Units of Measurement: The Metric System

Most scientific and technical calculations involve numbers that represent a measurement or count of a specific physical quantity. *Such numbers are called* **denominate numbers;** *associated with these denominate numbers are* **units of measurement.** A measured distance of 3 ft, for example, has the denominate number 3 associated with the foot as the unit of measurement. In this case, the foot is a unit of measurement in the U.S. Customary System. (The U.S. Customary System was also known as the British system, but the metric system is now used in Great Britain.) In the United States we use both the metric and U.S. Customary Systems of measurement. Every other major industrial country uses the metric system, but the United States is in a very long state of transition. Some of the major U.S. industrial companies have already converted to the metric system, and most others will do so soon. We must be familiar with both systems, since they are often used. Important advantages of the metric system are:

1. There is a definite need for the world to have one standard system of units. Industries not using the worldwide standard will be at a disadvantage.

2. In the metric system, *units of different magnitudes vary by a power of 10, and therefore only a shift of a decimal point is necessary when changing units*. No similar arrangement exists in the U.S. Customary System.

3. When measuring some physical quantity, only one metric unit will apply. Length, for example, is measured in terms of meters. The U.S. Customary System measures lengths in several different units, including inches, feet, yards, miles, and rods.

To better see the advantages of the metric system, consider the following example.

EXAMPLE A To convert 3 mi to inches in the U.S. Customary System, we get

$$3 \text{ mi} = 3 \times 5280 \text{ ft} = 3 \times 5280 \times 12 \text{ in.} = 190{,}080 \text{ in.}$$

To convert 3 km to centimeters in the metric system, we get

$$3 \text{ km} = 3 \times 1000 \text{ m} = 3 \times 1000 \times 100 \text{ cm} = 300{,}000 \text{ cm}$$

Although both of these conversions are similar, the required arithmetic is much simpler in the metric system. □

The system now referred to as the metric system is actually the **International System of Units,** or the **SI** system. The metric system uses seven **base units** which are used to measure these quantities: (1) length, (2) mass, (3) time, (4) electric current, (5) temperature, (6) amount of substance, and (7) luminous intensity. Other units, referred to as **derived units,** are given in terms of the units of these quantities.

EXAMPLE B The unit for electric charge, the *coulomb*, is defined as one ampere second (the unit of electric current times the unit of time). Therefore electric charge is defined in terms of the base units of electric current and time. □

The coulomb is a derived unit with a special name, but some derived units are not given special names. Speed, for example, might be in terms of meters per second, but there is no special name given to the units of speed.

In the U.S. Customary System, the base unit of length is the **foot** while the **meter** is the base unit of length in the metric system. Both systems use the **second** as the base unit of time. A basic problem arises with the base units for mass and force. The weight of an object is the *force* with which it is attracted to the earth, but the *mass* of an object is a measure of the amount of material in it. Mass and weight are therefore different quantities, and it is not strictly correct to convert **pounds** (force) to **kilograms** (mass). However, the kilogram is commonly used for weight even though it is actually a unit of mass. (The metric unit of force, the **newton,** is also used for weight.)

As for the other base units, the SI system defines the **ampere** as the base unit of electric current, and this unit can also be used in the U.S. Customary System. As for temperature, **degrees Fahrenheit** are used with the U.S. Customary System and **degrees Celsius** are used with the metric system. (Actually, the **kelvin** is defined as the base unit of thermodynamic temperature, although for temperature intervals, one kelvin equals one degree Celsius.) In the SI system, the base unit for the amount of a substance is the **mole** and the base unit of luminous intensity is the **candela.**

For convenience, when designating a unit of measurement, we use a **symbol** rather than the complete name of the unit.

EXAMPLE C   In denoting feet, the commonly used symbol is ft. Inches are denoted by in. and miles by mi.   ☐

Since the metric system has units of different magnitude varying by powers of 10, the arithmetic is simpler and requires only a shift of the decimal point. No consistent arrangement exists in the U.S. Customary System, so we must contend with a variety of multiplying factors such as 12, 16, 32, 36, and 5280. Also, the metric system uses a specific set of **prefixes** which denote the particular multiple being used. Table 2-1 shows some of the common prefixes with their symbols and meanings. A complete table is given as Table 6 in Appendix D.

EXAMPLE D   The symbol for the meter is m. A kilometer therefore equals one thousand meters. This is shown as

$$1 \text{ km} = 10^3 \text{ m} \quad \text{or} \quad 1 \text{ km} = 1000 \text{ m}$$

represents 1000   represents meter   ☐

When changing a quantity from one metric prefix to another, we can use Figure 2-1 to simplify the process of shifting the decimal point. This is illustrated in the following example.

TABLE 2-1 Metric Prefixes   *[handwritten: 1,000,000,000,000 = 10^{12} Trillion]*

[handwritten: Tera   T]

Prefix	Symbol	Factor by which unit is multiplied	Common meaning
[handwritten: GIGA]	*[handwritten: G]*	*[handwritten: 1,000,000,000 = 10^9]*	*[handwritten: billion]*
mega	M	$1{,}000{,}000 = 10^6$	million
kilo	k	$1000 = 10^3$	thousand
centi	c	$0.01 = \dfrac{1}{10^2}$	hundredth
milli	m	$0.001 = \dfrac{1}{10^3}$	thousandth
micro	μ	$0.000001 = \dfrac{1}{10^6}$	millionth
nano	n	$0.000000001 = \dfrac{1}{10^9}$	billionth
pico	p	$0.000000000001 = \dfrac{1}{10^{12}}$	trillionth

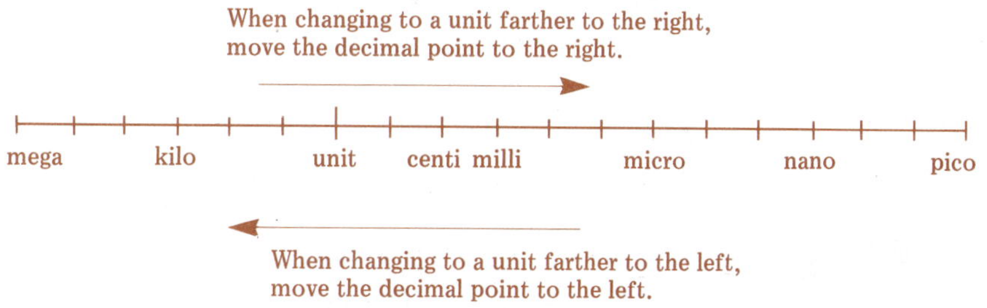

When changing to a unit farther to the right, move the decimal point to the right.

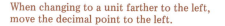

mega   kilo   unit   centi milli   micro   nano   pico

When changing to a unit farther to the left, move the decimal point to the left.

FIGURE 2-1

EXAMPLE E   Figure 2-1 indicates that a change from kilograms to centigrams involves moving the decimal point to the right (toward centi) five places. Therefore 0.0345 kilograms is equal to 3450 centigrams, which is expressed as 0.0345 kg = 3450 cg.

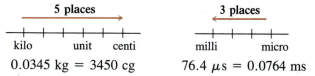

5 places	3 places
kilo unit centi	milli micro

0.0345 kg = 3450 cg   76.4 μs = 0.0764 ms

A change from microseconds to milliseconds requires that the decimal point be moved three places to the left. As a result, 76.4 microseconds becomes 0.0764 milliseconds, or 76.4 μs = 0.0764 ms. Such changes are examples of *reduction*, which will be considered in Section 2-2.

TABLE 2-2 Metric and U.S. Customary Units

Quantity	Metric			U.S. Customary		
	Unit	Symbol		Unit	Symbol	
Length	meter	m		foot	ft	
	centimeter	cm	1 cm = 0.01 m	yard	yd	1 yd = 3 ft
	millimeter	mm	1 mm = 0.001 m	inch	in.	1 in. = $\frac{1}{12}$ ft
	kilometer	km	1 km = 1000 m	mile	mi	1 mi = 5280 ft
Mass (SI units) or Force (U.S. Customary units)	kilogram	kg		pound	lb	
	gram	g	1 g = 0.001 kg	ounce	oz	1 oz = $\frac{1}{16}$ lb
	milligram	mg	1 mg = 0.001 g	short ton	t	1 t = 2000 lb
Capacity	liter	L		quart	qt	
	milliliter	mL	1 mL = 0.001 L	pint	pt	1 pt = $\frac{1}{2}$ qt
	kiloliter	kL	1 kL = 1000 L	gallon	gal	1 gal = 4 qt
Temperature	degrees Celsius	°C		degrees Fahrenheit	°F	
Time (both systems)	second	s				
	minute	min	1 min = 60 s			
	hour	h	1 h = 3600 s			
Electrical units (both systems)	ampere	A	current			
	volt	V	potential (voltage)			
	ohm	Ω	resistance			
	coulomb	C	charge			
	watt	W	power			

Note: The SI symbol for liter is l, but since this symbol may be confused with the numeral 1, the symbol L is recognized for use in the U.S. and Canada. Also, ℓ is recognized in several countries.

Table 2-2 lists some of the units most commonly used in the metric system, U.S. Customary System, or both. Also shown are the symbols and some of the basic relationships. A more complete table is shown in Table 5 in Appendix D.

The metric prefixes are used with the various units of the metric system as in Example E. Example F further illustrates the use of units, prefixes, and symbols.

EXAMPLE F

Unit	Symbol	Meaning
centiliter	cL	1 cL = 0.01 L

represents 0.01   represents liter

millisecond	ms	1 ms = 0.001 s

represents 0.001   represents second

microvolt	μV	1 μV = 0.000001 V
megacoulomb	MC	1 MC = 1,000,000 C

As mentioned earlier, the derived units for other quantities are expressible in the base units. For example, the units of area, volume, and speed are expressed in terms of the base units. The units for energy and pressure have special names, but they may be expressed in terms of the base units.

EXAMPLE G We have already noted that we find the area of a rectangle by multiplying one length by another. Considering the meaning of exponents given in Section 1-9, we can think of area as being measured in units of length $\times$ length = (length)^2. Therefore area is measured in square feet (ft^2), square miles (mi^2), square meters (m^2), and so on.

To find the volume of a rectangular solid, we multiply the length by the width by the depth. This means that volume is measured in units of length $\times$ length $\times$ length = (length)^3. Thus we measure volume in cubic feet (ft^3), or cubic centimeters (cm^3), and so on.

Density, which is weight (or mass) per unit volume, is measured in lb/ft^3 or kg/m^3, for example.

If we have the product of two different units, we show that the units are to be multiplied by placing a raised dot between the symbols. To show the unit kilogram meter (simply a space between unit names indicates product), we use kg·m.

In designating units of area or volume, we use square or cubic units and we will use exponents in the designation. As in Example G, we use m^2 rather than sq m and in.^{3} rather than cu in.

If a unit is used in the SI system or in both systems, we use the SI symbol for the unit. As an example, we use s rather than sec for seconds. Standard U.S. Customary symbols are used for units in that system only. However, in writing numbers we use the notation which is still prevalent. For long numbers, we use commas to separate groupings of three. (In the SI system spaces, not commas, are used.) For example, we write twenty-five thousand as 25,000 (not as 25 000, which is the SI style). It is expected, however, that the SI style will eventually become universal.

In this section we have considered the meaning and use of units of measurement, along with an introduction to the metric system. In this discussion we considered the base units as well as derived units directly associated with the base units. In Section 2-2 we shall consider the problem of changing, or converting, one set of units to another.

Exercises 2-1

In Exercises 1 through 8, give the symbol and meaning of the given unit.

1. milliampere *mA*
2. micrometer *μm*
3. kilovolt *kV*
4. megawatt
5. kilowatt *kw*
6. nanosecond
7. megaliter *ML*
8. centigram

In Exercises 9 through 16, give the name and meaning for the units whose symbols are given. (See Example F.)

9. MV *Megavolt*
10. kV
11. μs *microsecond*
12. kW
13. cV *centi volt*
14. mΩ
15. nA *nanoAmpere*
16. ps

In Exercises 17 through 20, use only base units to give the proper symbols (a) in the metric system and (b) in the U.S. Customary System for the units of the given quantities.

17. area *sq·ft*
18. volume
19. speed (distance per unit of time)
20. acceleration (distance per squared unit of time)

In Exercises 21 through 32, rewrite the given quantity so that all numbers are between 1 and 10. Use the proper metric symbol for the given unit. For example, 0.03 meter would be expressed as 3 cm.

21. 4 seconds
22. 3 hours
23. 8 meters
24. 2 kilometers
25. 0.04 second
26. 0.07 liter
27. 0.000003 second
28. 0.06 meter
29. 0.08 gram
30. 0.005 meter
31. 4000 grams
32. 9000 watts

In Exercises 33 through 44, give the proper symbols of the units of the indicated quantities in terms of the units given.

33. The sides of a rectangle are given in millimeters. What is the symbol for the units of area of the rectangle?

34. The edges of a box are given in inches. What is the symbol for the units of volume of the box?

35. The driver of a car determines the number of miles traveled and the number of gallons of gasoline consumed. What is the symbol for the gasoline consumption of the car?

36. The distance a rocket travels is measured in kilometers, and the time of travel is measured in seconds. What is the symbol for the units of the speed of the rocket?

37. The unit of frequency is the *hertz* (Hz), which has units of the reciprocal of seconds. What is the symbol for the hertz in terms of seconds?

38. The unit of electric capacitance is the *farad* (F), which has units of seconds per ohm. What is the symbol for the farad in terms of seconds and ohms?

39. Pressure may be expressed as the number of pounds per square foot. What is the symbol for the units of pressure?

40. The U.S. Customary unit for power is the *horsepower* (hp), which has units of foot pounds per second. What is the symbol for the units of the horsepower in terms of feet, pounds, and seconds?

41. The metric unit for force is the *newton* (N), which has units of kilogram meters per second squared (only seconds are squared). What is the symbol for the newton in terms of kilograms, meters, and seconds?

42. The metric unit for energy is the *joule* (J), which has units of a newton meter. What is the symbol for the joule in terms of newtons and meters?

43. The unit of electric charge, the *coulomb* (C), has units of an ampere second. What is the symbol of the coulomb in terms of amperes and seconds?

44. The unit for electric potential (voltage) is the *volt* (V), which has units of a joule per ampere second. What is the symbol for the volt in terms of joules, amperes, and seconds?

2-2 Reduction and Conversion of Units

Recall that denominate numbers represent measurements or counts, and they are associated with units of measurement. When we are working with denominate numbers, it is sometimes necessary to change from one set of units to another. *A change within a system is called a* **reduction.** *A change from one system to another is called a* **conversion.**

EXAMPLE A   To find the cost of floor covering for a room, we must know the area of the room, and this is generally given in square feet. Floor-covering cost is often given in cost per square yard so that it is necessary to reduce square feet to square yards. Here we are changing from square feet to square yards, which are both U.S. Customary units. This is called a reduction because we are changing units within the same system.

Distances in Europe are given in kilometers. To determine equivalent distances in miles, we must convert kilometers to miles. Here we are changing from metric units to U.S. Customary units. This is called a conversion because we are changing units in one system to those in another.   □

For purposes of changing units, Table 2-3 gives some basic reduction and conversion factors. Other reduction factors are given in Table 2-2.

TABLE 2-3   Reduction and Conversion Factors

Reduction factors

$144 \text{ in.}^2 = 1 \text{ ft}^2$	(all are exact)	$100 \text{ mm}^2 = 1 \text{ cm}^2$
$9 \text{ ft}^2 = 1 \text{ yd}^2$		$10{,}000 \text{ cm}^2 = 1 \text{ m}^2$
$1728 \text{ in.}^3 = 1 \text{ ft}^3$		$1000 \text{ mm}^3 = 1 \text{ cm}^3$
$27 \text{ ft}^3 = 1 \text{ yd}^3$		$1000 \text{ cm}^3 = 1 \text{ L}$

Conversion factors

1 in. = 2.54 cm (exactly)		$1 \text{ ft}^3 = 28.32 \text{ L}$
1 m = 39.37 in.	(all others are approx.)	$1 \text{ pt} = 473.2 \text{ cm}^3$
1 mi = 1.609 km		1 L = 1.057 qt
1 lb = 453.6 g		
1 kg = 2.205 lb		

To change a given number of one set of units to another set of units, *we perform multiplications and divisions with the units themselves.* These computations are essentially the same as the ones we did when we were working with fractions. The units themselves may be divided out.

EXAMPLE B If we have a number representing feet per second to be multiplied by another number representing seconds per minute, as far as the units are concerned we have

$$\frac{ft}{s} \times \frac{s}{min} = \frac{ft \times s}{s \times min} = \frac{ft}{min}$$

This means that the final result would be in feet per minute. □

See Appendix B for a related computer program.

In changing a number of one set of units to another set of units, we use reduction and conversion factors from Tables 2-2 and 2-3 and apply the principle illustrated in Example B for operating with the units themselves. The convenient way to use the values in the tables is in the form of fractions. *Since the given values are equal to each other, their quotient is 1.* For example, since 1 in. = 2.54 cm,

$$\frac{1 \text{ in.}}{2.54 \text{ cm}} = 1 \qquad \text{or} \qquad \frac{2.54 \text{ cm}}{1 \text{ in.}} = 1$$

since each represents the division of a certain length by itself. Multiplying a given quantity by 1 does not change its value. The following examples illustrate reduction and conversion of units.

EXAMPLE C Convert a distance of 2.800 mi to kilometers.
Since 1 mi = 1.609 km, the fraction

$$\frac{1.609 \text{ km}}{1 \text{ mi}}$$

is equal to 1 and we can therefore multiply 2.800 mi by that fraction to get

$$2.800 \text{ mi} \left(\frac{1.609 \text{ km}}{1 \text{ mi}} \right) = 4.505 \text{ km}$$

In this example we know to use the fraction 1.609 km/1 mi since we want the number of kilometers per mile. Also, we see that *this fraction works since the mi units cancel.* If we used the fraction 1 mi/1.609 km and tried to set up the conversion as

$$2.800 \text{ mi} \left(\frac{1 \text{ mi}}{1.609 \text{ km}} \right)$$

we see that no cancellation of units occurs. This shows that 1 mi/1.609 km is the wrong form to use. □

EXAMPLE D   Reduce 378 ft³ to cubic yards.

Multiplying the given quantity by 1 yd³/27 ft³ (which is equal to 1), we get

$$378 \text{ ft}^3 \left(\frac{1 \text{ yd}^3}{27 \text{ ft}^3} \right) = 14 \text{ yd}^3$$

□

EXAMPLE E   Reduce 30 mi/h to feet per second.

$$30 \text{ mi/h} = \left(30 \frac{\text{mi}}{\text{h}} \right) \left(\frac{5280 \text{ ft}}{1 \text{ mi}} \right) \left(\frac{1 \text{ h}}{60 \text{ min}} \right) \left(\frac{1 \text{ min}}{60 \text{ s}} \right)$$

$$= \frac{(30)(5280) \text{ ft}}{(60)(60) \text{ s}} = 44 \text{ ft/s}$$

Note that the only units remaining are those that were required.   □

EXAMPLE F   Reduce 575 g/cm³ to kilograms per cubic meter.

$$575 \frac{\text{g}}{\text{cm}^3} = \left(575 \frac{\text{g}}{\text{cm}^3} \right) \left(\frac{100 \text{ cm}}{1 \text{ m}} \right)^3 \left(\frac{1 \text{ kg}}{1000 \text{ g}} \right)$$

$$= \left(575 \frac{\text{g}}{\text{cm}^3} \right) \left(\frac{1000000 \text{ cm}^3}{1 \text{ m}^3} \right) \left(\frac{1 \text{ kg}}{1000 \text{ g}} \right)$$

$$= 575 \times 1000 \frac{\text{kg}}{\text{m}^3} = 575{,}000 \frac{\text{kg}}{\text{m}^3}$$

Note that the fraction 100 cm/1 m was cubed so that the cm³ units could then be divided out.   □

EXAMPLE G   Convert 62.80 lb/in.² to kilograms per square meter.

$$62.80 \text{ lb/in.}^2 = \left(62.80 \frac{\text{lb}}{\text{in.}^2} \right) \left(\frac{1 \text{ kg}}{2.205 \text{ lb}} \right) \left(\frac{1 \text{ in.}}{2.54 \text{ cm}} \right)^2 \left(\frac{100 \text{ cm}}{1 \text{ m}} \right)^2$$

$$= \left(62.80 \frac{\text{lb}}{\text{in.}^2} \right) \left(\frac{1 \text{ kg}}{2.205 \text{ lb}} \right) \left(\frac{1 \text{ in.}^2}{2.54^2 \text{ cm}^2} \right) \left(\frac{10000 \text{ cm}^2}{1 \text{ m}^2} \right)$$

$$= \frac{(62.80)(10000)}{(2.205)(2.54^2)} \frac{\text{kg}}{\text{m}^2} = 44{,}150 \text{ kg/m}^2$$

□

Many scientific calculators have keys that allow automatic conversion from one system to another. The most common conversions are gallons to liters, pounds to kilograms, and inches to centimeters or millimeters. The corresponding conversions can also be made from the metric system to the U.S. Customary system. (That is, liters to gallons, kilograms to pounds, and so on.) Scientific calculators vary in the way these conversions are done, so it is best to consult the manual which comes with your particular calculator. However, the most common conversion procedure involves keys which serve dual purposes. Here is a typical procedure: First enter the number to be converted, then press the function key (often labeled F or FCN or 2nd), and then press the dual-purpose key which corresponds to the desired conversion.

EXAMPLE H   A pilot wants to take on 12 gal of avgas (aviation grade gasoline), but she is in Canada and must convert that amount to liters. On one particular calculator, 12 gal is converted to liters by pressing

gal-L

12 [F] [5]   (45.424941)

The display will show the number of liters in 12 gal. The key shown to the extreme right is an example of a dual purpose key. This key is usually used to enter the number 5, but it can also be used to convert gallons to liters. Note that this key's secondary purpose is used only when it is preceded by the function key.

Using the calculator, we find that 12 gal is equivalent to 45.424941 L. Actually, this is somewhat misleading since an eight-digit result is not appropriate here. In Section 2-3 we will discuss accuracy and precision, and we will see how such results should be presented. For now we might simply note that 12 gal is equivalent to 45 L.   ☐

Exercises 2-2

In Exercises 1 through 16, use only the values given in Tables 2-2 and 2-3. Where applicable, round off to the indicated place value.

1. How many millimeters are there in 1 cm?

2. How many centigrams are there in 1 kg?

3. How many inches are there in 1 mi?

4. How many ounces are there in 1 t?

5. Reduce 2 ft² to square inches.

6. Reduce 50 cm³ to cubic millimeters.

7. Convert 12 in. to centimeters.

8. Convert 8 kg to pounds. (Round off to tenths.)

9. Reduce 55 gal to quarts.

10. Reduce 60 kL to milliliters.

11. Convert 25 qt to liters. (Round off to tenths.)

12. Convert 27 cm to inches. (Round off to tenths.)

13. Reduce 5.2 m² to square centimeters.

14. Reduce 0.205 L to cubic centimeters.

15. Convert 256.3 L to cubic feet. (Round off to hundredths.)

16. Convert 967 in.³ to liters. (Round off to tenths.)

In Exercises 17 through 24, use a scientific calculator to make the conversions. Round off to the indicated place value.

17. Convert 3.26 qt to liters (round off to hundredths).

18. Convert 27.6 mi to kilometers (round off to tenths).

19. Convert 152.4 lb to kilograms (round off to hundredths).

20. Convert 13.256 m to inches (round off to hundredths).

21. Convert 3.76 kg to pounds (round off to hundredths).

22. Convert 8.762 L to gallons (round off to thousandths).

23. Convert 6.000 ft to millimeters (round off to units).

24. Convert 22.1 gal to liters (round off to tenths).

In Exercises 25 through 40, solve the given problems. Round off to the indicated place value.

25. A road requires 180,000 lb of gravel fill. How many tons is this?

26. An airplane can climb to a maximum altitude of 22,460 ft. What is the altitude in miles? (Round to hundredths.)

27. While driving in France, a sales representative encounters a speed limit sign indicating 80 km/h. Convert that speed limit to mi/h. (Round off to units.)

28. A foreign car's weight is given as 1764 kg. Convert that value to pounds. (Round off to tens.)

29. The manual for a car specifies that the crankcase oil capacity is 4 qt. Convert that capacity to liters. (Round off to tenths.)

30. A spark plug socket has an inside diameter of $\frac{13}{16}$ in. Convert that distance to millimeters. (Round off to units.)

31. Near the earth's surface, the acceleration due to gravity is about 980 cm/s². Convert this to feet per second squared. (Round off to tenths.)

32. The speed of sound is about 770 mi/h. Change this speed to feet per second. (Round off to units.)

33. The density of water is about 62.4 lb/ft³. Convert this to kilograms per cubic meter (round off to units).

34. The average density of the earth is about 5.52 g/cm³. Express this in kilograms per cubic meter (round off to units).

35. The moon travels about 1,500,000 mi in about 28 days in one rotation about the earth. Express its velocity in feet per second (round off to tens).

36. The amount of water that can be pumped from a well is found to be 72 gal/min. Convert that rate to liters per second. (Round off to hundredths.)

37. A unit used to measure the flow of water is cubic foot per minute. Use a calculator to convert 1 ft³/min to liters per second (round off to hundredths).

38. The metric unit for force is the *newton* (N), where 1.00 N = 0.225 lb. Use a calculator to convert 250 ft·lb into newton meters (round off to units).

39. A commonly used unit of heat is the kilocalorie (kcal), whereas the SI unit of heat is the joule (J), where 1.000 kcal = 4185 J. Use a calculator to convert 225 kJ to kilocalories. (Round off to tenths.)

40. Acceleration can be expressed in units of ft/s² or m/s², where 1.000 m/s² = 3.281 ft/s². Use a calculator to convert 32.0 ft/s² to m/s².

2-3 Approximate Numbers and Significant Digits

The last example in Section 2-2 showed that a calculator could be used to convert 12 gal to liters. It was noted that the result of 45.424941 L is misleading since an eight-digit result is not warranted. The number of digits in the result determines its accuracy. In this section we consider the proper use of both accuracy and precision, which will be defined soon.

Some numbers are exact while others are approximate. When we perform calculations on numbers, we must consider the accuracy of these numbers, since they affect the accuracy of the results obtained. *Most numbers involved in technical and scientific work are approximate, since they represent some measurement. However, certain other numbers are exact, having been arrived at through some definition or counting process*. We can decide if a number is approximate or exact if we know how the number was determined.

EXAMPLE A If a surveyor measures the distance between two benchmarks as 156.2 ft, we know that the 156.2 is approximate. A more precise measuring device may cause us to determine the length as 156.18 ft. However, regardless of the method of measurement used, we can never determine this length exactly.

If a voltage shown on a voltmeter is read as 116 V, the 116 is approximate. A more precise voltmeter may show the voltage as 115.7 V. However, this voltage cannot be determined exactly. ☐

EXAMPLE B If a computer counts the number of students majoring in science or a technology and prints this number as 768, this 768 is exact. We know the number of students was not 767 or 769. Since 768 was determined through a counting process, it is exact.

When we say that 60 s = 1 min, the 60 is exact, since this is a definition. By this definition there are exactly 60 s in 1 min. As another example illustrating how a number can be exact through definition, the inch is now officially defined so that 1 in. = 2.54 cm exactly. ☐

When we are writing approximate numbers, we must often include some zeros so that the decimal point will be properly located, as in 0.003 or 5600. However, except for these zeros, all other digits are considered to be **significant digits.** When we make certain computations with approximate numbers, we must know the number of significant digits. The following example illustrates how we determine this.

EXAMPLE C   All numbers in this example are assumed to be approximate.

34.7 has three significant digits.

8900 has two significant digits. We assume that *the two zeros are place-holders* (unless we have specific knowledge to the contrary). (Note: The overbar symbol (‾) is sometimes used to show that trailing zeros are significant digits. For example, 89,0$\overline{0}$0 has four significant digits.)

706.1 has four significant digits. The zero is not used for the location of the decimal point. It shows specifically the number of tens in the number.

▶   5.90 has three significant digits. The *zero is not necessary as a place-holder and should not be written unless it is significant.*   □

Examine Table 2-4, which gives additional examples of approximate numbers along with the proper number of significant digits.

TABLE 2-4

Approximate number	Number of significant digits	Comment
87,000	2	The zeros are placeholders and are not counted as significant digits.
408,000	3	The zero farthest to the left is not a placeholder but the other three zeros are placeholders and are not counted as significant digits.
4.0005	5	The zeros are not placeholders; they do count as significant digits.
0.004	1	All zeros are placeholders and do not count as significant digits.
4.000	4	The zeros are not *required* as placeholders and they do count as significant digits.
0.000503	3	The three digits farthest to the right are significant, but the four zeros farthest to the left are placeholders and they do not count as significant digits.

Note that all nonzero digits are significant. Zeros, other than those used as placeholders for proper positioning of the decimal point, are also significant.

In computations involving approximate numbers, the position of the decimal point as well as the number of significant digits is important. *The precision of a number refers directly to the decimal position of the last significant digit.* That is, precision refers to the value that the rightmost significant digit measures. *The* **accuracy** *of a number refers to the number of significant digits in the number.* See Figure 2-2.

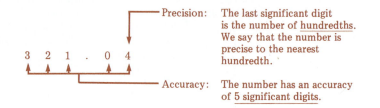

FIGURE 2-2

EXAMPLE D Suppose that you are measuring an electric current with two ammeters. One ammeter reads 0.031 A and the other reads 0.0312 A. The second reading is more precise—the last significant digit is the number of ten-thousandths, whereas the first reading is expressed only to thousandths. The second reading is also more accurate, since it has three significant digits rather than two.

A machine part is measured to be 2.5 cm long. It is coated with a film 0.025 cm thick. The thickness of the film has been measured to a greater precision, although the two measurements have the same accuracy: two significant digits.

A segment of a newly completed highway is 9270 ft long. The concrete surface is 0.8 ft thick. Of these two numbers, 9270 is more accurate, since it contains three significant digits, and 0.8 is more precise, since it is expressed to tenths.   □

We can formalize the process for determining accuracy as follows: To determine the accuracy of a number, count the number of significant digits by following the rule for the appropriate case, as follows.

Case 1. The number has no digits to the right of the decimal point:

Count the digits from left to right, but stop with the last nonzero digit.

3 0 8 0 0   Accuracy: 3 significant digits

Case 2. The number has at least one digit to the right of the decimal point:

Begin with the leftmost nonzero digit and count all digits to the right.

0 . 0 0 2 0 3 0   Accuracy: 4 significant digits

The last significant digit of an approximate number is not completely accurate. It has usually been determined by estimation or rounding off. However, we do know that it is in error by at most one-half a unit in its place value.

EXAMPLE E When we measure the distance in Example A to be 156.2 ft, we are saying that the length is at least 156.15 ft and no longer than 156.25 ft. Any value between these two, rounded off to tenths, would be expressed as 156.2 ft.

In converting the fraction $\frac{2}{3}$ to the decimal form 0.667, we are saying that the value is between 0.6665 and 0.6675. □

In Chapter 1, we introduced briefly the process of **rounding off** *a number. The principle of rounding off is to write the closest approximation, with the last significant digit in a specified position, or with a specified number of significant digits.* We shall now formalize the process of rounding off as follows:

If we want a certain number of significant digits, we examine the digit in the next place to the right. If this digit is less than 5, we accept the digit in the last place. If the next digit is 5 or greater, we increase the digit in the last place by 1, and this resulting digit becomes the final significant digit of the approximation. If necessary, we use zeros to replace other digits in order to locate the decimal point properly. Except when the next digit is a 5, and no other nonzero digits are discarded, we have the closest possible approximation with the desired number of significant digits.

EXAMPLE F 70360 rounded off to three significant digits is 70400.

third sig. digit   5 or greater
70360
Add 1 to the 3 and replace 6 with 0

70430 rounded off to three significant digits is 70400.
187.35 rounded off to four significant digits is 187.4.
71500 rounded off to two significant digits is 72000. □

EXAMPLE G 39.72 rounded off to two significant digits is 40.

When 1 is added to the second significant digit, 9, we get 10, and ***the 1 must be carried to the tens.*** Also, since there are no significant digits in the result to the right of the decimal point, ***the 7 and 2 are not replaced with zeros.*** □

Some computers and calculators round off results while others use a *process of* **truncation**, *in which the digits beyond a certain place are discarded.* For example, 3.17482 truncated to thousandths is 3.174. For our purposes in this text, when working with approximate numbers we shall use only rounding off. A quick and easy calculator test can be run by entering 2 ÷ 3 = . If the last digit is 7, the calculator rounds off results. If the last digit is 6, the calculator truncates results.

Exercises 2-3

In Exercises 1 through 8, determine whether the numbers given are exact or approximate.

1. An engine has 6 cylinders.

2. The speed of sound is 770 mi/h.

3. A copy machine weighs 67 lb.

4. A gear has 36 teeth.

5. The melting point of gold is 1063°C.

6. The 25 math students had a mean test grade of 79.3.

7. A semiconductor 1 cm by 1 mm is priced at $2.80.

8. A calculator has 40 keys and its battery lasts for 987 h of use.

In Exercises 9 through 16, determine the number of significant digits in the given approximate numbers.

9. 563;   4029

10. 46.8;   5200

11. 3799;   2001

12. 0.0025;   0.6237

13. 5.80;   5.08

14. 7000;   7000.0

15. 10060;   403020

16. 9.000;   0.009

In Exercises 17 through 24, determine which of the two numbers in each pair of approximate numbers is (a) more precise and (b) more accurate.

17. 3.764;   2.81

18. 0.041;   7.673

19. 30.8;   0.01

20. 70,370;   50,400

21. 0.1;   78.0

22. 7040;   37.1

23. 7000;   0.004

24. 50.060;   8.914

In Exercises 25 through 36, round off the given approximate numbers (a) to three significant digits and (b) to two significant digits.

25. 5.713

26. 53.72

27. 6.934

28. 27.81

29. 4096

30. 287.4

31. 46792

32. 32768

33. 501.46

34. 7435

35. 0.21505

36. 0.6350

In Exercises 37 through 48, solve the given problems.

37. A surveyor measured the road frontage of a parcel of land and obtained a distance of 128.3 ft. Based on this result, find the lowest possible value and the highest possible value.

38. A micrometer caliper is used to measure the diameter of a bolt and a reading of 0.768 in. is obtained. Based on this measurement, find the smallest possible diameter and the largest possible diameter.

39. An automobile manufacturer claims that the gasoline tank on a certain car holds approximately 82 L. What are the very least and the very greatest capacities?

40. The maximum weight an industrial robot can lift is measured as 46 lb. Based on this measurement, what is the lowest possible value and the highest possible value of this maximum weight?

41. A chemist has a container of 164.0 mL of sulfuric acid. Convert this volume to quarts and use the same degree of accuracy.

42. A design specification calls for a washer that is 0.0625 in. thick. Convert this thickness to millimeters and use the same degree of accuracy.

43. For the 100-yd length of a football field, assume that all three digits are significant. Convert this length to meters and use the same degree of accuracy.

44. An architect specifies that a support beam must be 7.315 m long. Convert that length to feet and use the same degree of accuracy.

45. Use your calculator to enter $2 \div 3$. How many significant digits are in the calculator display?

46. Use your calculator to evaluate $7 \div 9$. If the last digit displayed is 7, your calculator truncates results. If the last digit is 8, your calculator rounds off. Determine whether your calculator truncates or rounds off.

47. Use your calculator to enter 24.000. This number has an accuracy of five significant digits. Now press the key labeled $=$. What happens to the accuracy?

48. What is the accuracy of the largest number your calculator will display? What is the precision of the smallest number your calculator will display?

2-4 Arithmetic Operations with Approximate Numbers

With the widespread use of calculators and computers, results are often displayed with an improper degree of precision or accuracy. Even without calculators or computers it is often easy to make the error of giving a false indication of the accuracy or precision of a result.

EXAMPLE A   A pipe used in a solar-heating system is made in two sections. The first is measured to be 16.3 ft long and the second is measured to be 0.927 ft long. A plumber wants to know what the total length will be when the two sections are put together.

At first, it appears we might simply add the numbers to obtain the necessary result:

$$\begin{array}{r} 16.3 \ \ \text{ft} \\ \underline{0.927 \ \text{ft}} \\ 17.227 \ \text{ft} \end{array}$$

However, the first length is precise only to tenths, and the digit in this position was obtained by rounding off. It might have been as small as 16.25 ft or as large as 16.35 ft. If we consider only the precision of this first number, the total length might be as small as 17.177 ft or as large as 17.277 ft. These two values agree when rounded off to two significant digits (17). They vary by 0.1 when rounded off to tenths (17.2 and 17.3). When rounded to hundredths, they do not agree at all, since the third significant digit is different (17.18 and 17.28). Therefore there is no agreement at all in the digits after the third when these two numbers are rounded off to a precision beyond tenths. This is reasonable, since the first length is not expressed beyond tenths. The second number does not change the precision of the result, even though it is expressed to thousandths. Therefore we may conclude that *the result must be rounded off at least to tenths*, the precision of the first number.   □

EXAMPLE B   We can find the area of a rectangular storage lot by multiplying the length, 207.54 m, by the width, 81.4 m. Performing the multiplication, we find the area to be

$$(207.54 \text{ m})(81.4 \text{ m}) = 16{,}893.756 \text{ m}^2$$

However, we know that this length and width were found by measurement and that the least each could be is 207.535 m and 81.35 m. Multiplying these values, we find the minimum possible area to be

$$(207.535 \text{ m})(81.35 \text{ m}) = 16{,}882.97225 \text{ m}^2$$

The maximum possible value for the area is

$$(207.545 \text{ m})(81.45 \text{ m}) = 16{,}904.54025 \text{ m}^2$$

We now note that the least possible and greatest possible values of the area agree when rounded off to three significant digits (16,900 m²). There is no agreement in digits beyond this if the two values are rounded off to a greater accuracy. We conclude that *the accuracy of the result is good to three significant digits or certainly no more than four*. We also note that the width was accurate to three significant digits and the length to five significant digits.   □

The following rules are based on reasoning similar to that in Examples A and B. We shall use these rules when we perform the basic arithmetic operations on approximate numbers.

▶   1. *When approximate numbers are* **added** *or* **subtracted,** *the result is expressed with the precision of the least precise number.*

▶   2. *When approximate numbers are* **multiplied** *or* **divided,** *the result is expressed with the accuracy of the least accurate number.*

▶   3. *When the* **root** *of an approximate number is found,* *the accuracy of the result is the same as the accuracy of the original number.*

4. *Before and during the calculation, all numbers except the least precise or least accurate may be rounded off to one place beyond that of the least precise or least accurate. This procedure is helpful when a calculator is not used. It need not be followed when a calculator is used.*

Rules 1 to 3 are summarized in Figure 2-3. The last of these rules is designed to make the calculation as easy as possible, since carrying the additional digits is meaningless in the intermediate steps. If a calculator is used, this would be an unnecessary use of time and might lead to error if the numbers were reentered incorrectly. The following examples illustrate the use of these four rules.

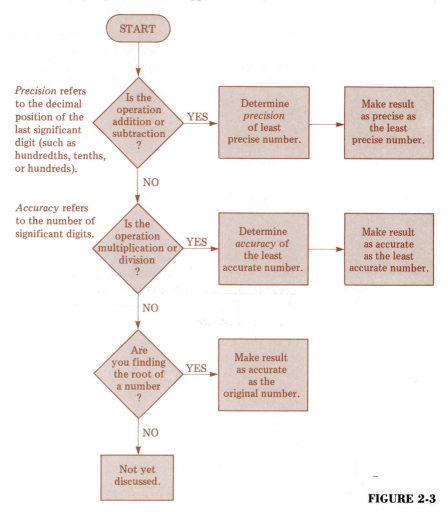

Precision refers to the decimal position of the last significant digit (such as hundredths, tenths, or hundreds).

Accuracy refers to the number of significant digits.

FIGURE 2-3

EXAMPLE C Add the approximate numbers 73.2, 8.0627, 93.57, 66.296.

The least precise of these numbers is 73.2. Therefore before performing the addition we may round off the other numbers to hundredths. If we are using a calculator, this need not be done. In either case, after the addition we round off the result to tenths. This leads to:

Without calculator	*With calculator*	
73.2	73.2	← least precise (tenths)
8.06	8.0627	
93.57	93.57	
66.30	66.296	
241.13 ←	241.1287 ←	answer must be rounded to tenths

The sum is expressed with the precision of the least precise number (73.2), so *we round off the result to tenths to get a final result of 241.1.* □

EXAMPLE D   Divide 292.6 by 3.4, where each is an approximate number.

Since the divisor is accurate only to two significant digits, the final result is accurate to two significant digits. Therefore we may round off the dividend to three significant digits and divide until we have three significant digits in the quotient. The result will then be rounded off to two significant digits.

$$
\begin{array}{r}
86.1 \\
34\overline{)2930.0} \\
272 \\
\hline
210 \\
204 \\
\hline
60 \\
34 \\
\hline
\end{array}
$$

From the division shown above, we conclude that the result is 86. If a calculator (which displays eight digits) is used, we obtain $292.6 \div 3.4 = 86.058824$. Again *we round off the result to 86.*   □

EXAMPLE E   If we use a calculator when we subtract 36.1 from 727.842, we get

$$
\begin{array}{l}
727.842 \\
\underline{36.1} \quad \longleftarrow \text{ least precise (tenths)} \\
691.742 \longleftarrow \text{ answer must be rounded to tenths}
\end{array}
$$

Therefore *the result is 691.7.* Since we have subtracted, we make the result as precise as the least precise number, and 36.1 is expressed only to the precision of tenths.

When we find the product of 2.4832 and 30.5, we get

$$(2.483)(30.5) = 75.7315 \text{ (without calculator)}$$

or

$$(2.4832)(30.5) = 75.7376 \text{ (with calculator)}$$

Therefore *the final result is 75.7,* since we have multiplied and 30.5 has the accuracy of three significant digits.

When we use a calculator to find the square root of 3.7, we get $\sqrt{3.7} = 1.9235384$. *We round off the final result to 1.9,* since 3.7 has two significant digits.   □

If an exact number is included in a calculation, there is no limit to the number of decimal positions it may take on. The accuracy of the result is limited only by the approximate numbers involved.

EXAMPLE F Considering the exact number 144 and the approximate number 2.7, we express the result to tenths if these numbers are added or subtracted. If these numbers are to be multiplied or divided, we express the result to two significant digits:

precision of 2.7

$$144 + 2.7 = 146.7 \qquad 144 - 2.7 = 141.3$$
$$144 \times 2.7 = 390 \qquad 144 \div 2.7 = 53$$

accuracy of 2.7 □

We should now use the rules for rounding off approximate numbers when we make reductions or conversions of the type discussed in Section 2-2.

EXAMPLE G Reduce 12.34 ft to inches.

$$12.34 \text{ ft} = (12.34 \text{ ft.})\left(\frac{12 \text{ in.}}{1 \text{ ft.}}\right)$$
$$= 148.08 \text{ in.}$$

The result should have only four significant digits (the accuracy of 12.34), so the correct result is 148.1 in. We did not round off to two significant digits because the number 12 is exact (1 ft = 12 in. exactly). □

EXAMPLE H Convert 12.34 ft to centimeters.

$$12.34 \text{ ft} = 12.34 \text{ ft} \left(\frac{12 \text{ in.}}{1 \text{ ft}}\right)\left(\frac{2.54 \text{ cm}}{1 \text{ in.}}\right)$$
$$= 376.1232 \text{ cm}$$

This is rounded off to 376.1 cm. Here the 12 and the 2.54 are both exact, so the answer is rounded off to four significant digits to correspond to the accuracy of the original number. □

EXAMPLE I Convert 2.5835 mi to kilometers.

$$2.5835 \text{ mi} = (2.5835 \text{ mi})\left(\frac{1.609 \text{ km}}{1 \text{ mi}}\right)$$
$$= 4.1568515 \text{ km}$$

In this case, the 1.609 is not exact so the product of 2.5835 and 1.609 should be rounded off to four significant digits, the accuracy of 1.609, which is the least accurate number. Therefore the correct result is 4.157 km. □

EXAMPLE J   Evaluate the following expression in which all numbers are approximate.

$$(3.14)(93.125) + \frac{436.15}{25.21}$$

We first do the multiplication, and we use the *accuracy* of the least accurate number. However, we carry an extra digit at intermediate stages of a calculation. We get $(3.14)(93.125) = 292.4$, which is rounded to four significant digits (three-digit accuracy plus one extra digit). The remaining operations are illustrated below.

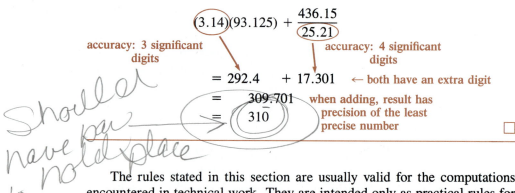

$$(3.14)(93.125) + \frac{436.15}{25.21}$$

accuracy: 3 significant digits   accuracy: 4 significant digits

$$= 292.4 \quad + 17.301 \quad \leftarrow \text{both have an extra digit}$$

$$= \quad 309.701 \quad \text{when adding, result has precision of the least precise number}$$

$$= \quad 310$$

Shoulld nave par to hold place

The rules stated in this section are usually valid for the computations encountered in technical work. They are intended only as practical rules for working with approximate numbers. It was recognized in Examples A and B that the last significant digit obtained by these rules is subject to error. Therefore it is possible that the most accurate result is not obtained by their use, although this is not often the case. Results obtained with a calculator may differ slightly from those obtained with the use of rule 4.

Exercises 2-4

In Exercises 1 through 4, add the given approximate numbers.

1. 2.5
36.37
0.173

2. 81
219.36
4.748

3. 0.84
0.046
0.729
0.1654

4. 0.047
49.3
2.0548
312.49

In Exercises 5 through 8, subtract the given approximate numbers.

5. 802.35
48.4

6. 29.257
23.4

7. 7.35
4.7637

8. 5.04853
0.71

In Exercises 9 through 12, multiply the given approximate numbers.

9. (47.5)(2.38)   **10.** (38.3)(0.105)   **11.** (0.53)(805.6)   **12.** (0.004050)(5038.61)

In Exercises 13 through 16, divide the given approximate numbers.

13. $803 \div 7.2$ **14.** $0.3795 \div 506$ **15.** $\dfrac{63750}{81}$ **16.** $\dfrac{94.338}{0.4105}$

In Exercises 17 through 20, find the square roots of the given approximate numbers.

17. $\sqrt{48}$ **18.** $\sqrt{8.2}$ **19.** $\sqrt{87.4}$ **20.** $\sqrt{0.0483}$

In Exercises 21 through 24, evaluate the given expression. All numbers are approximate.

21. $3.862 + 14.7 - 8.3276$ **22.** $(3.2)(0.386) + 6.842$

23. $\dfrac{8.60}{0.46} + (0.9623)(3.86)$ **24.** $9.6 - 0.1962(7.30)$

In Exercises 25 through 28, perform the indicated operations. The first number given is approximate and the second number is exact.

25. $3.62 + 14$ **26.** $17.382 - 2.5$ **27.** $(0.3142)(60)$ **28.** $8.62 \div 1728$

In Exercises 29 through 40, use a scientific calculator to perform the indicated operations. All numbers are approximate, and results should be rounded off to the proper degree of precision or accuracy.

29. $3.5 + 4.67 + 5.001$ **30.** $13.987 - 6.66666$ **31.** $8.4 - 0.333333$

32. $17.3 - 8.45 + 9.001$ **33.** $(6.022169)(3.874)$ **34.** $739.4 \div 8.73$

35. $6.25 \div 9.8876$ **36.** $(156.03)(22.7)$ **37.** $\sqrt{3872}$

38. $\sqrt{0.0091}$ **39.** $\sqrt{0.135}$ **40.** $\sqrt{49000}$

In Exercises 41 through 48, solve the given problems and round off the results to the proper degree of precision or accuracy.

41. Three adjacent lots have road frontages measured to be 75.4 ft, 39.66 ft, and 81 ft, respectively. What is the total length of the frontages of these three lots?

42. A force of 16.7 lb and another force of 10.258 lb are acting in the same direction on an object. Find the sum of these two forces.

43. If 1K of computer memory consists of exactly 1024 bytes, how many bytes are there in a 256K computer? (The number 256 is also exact.)

44. In an electric circuit, the power (in watts) is found by multiplying the current (in amperes) and the voltage (in volts). Find the power developed in a circuit with a current of 0.0225 A and a voltage of 1.657 V.

45. We can find the average velocity of a dropped object by dividing the distance traveled by the time the object falls. Find the average velocity for an object that falls 259 ft in 4.028 s.

46. Three sales representatives earn commissions of $512.81, $410.35, and $809.25, respectively. Find the mean of these amounts by first adding them and then dividing the total by 3 (an exact number).

47. A square piece of land has an area of 23,700 m². What is the length of one side?

48. We can find the radius (in inches) of a floppy disk by evaluating $\sqrt{86.59 \div 3.14159}$. Find the radius.

Some of Exercises 49 through 56 will require the use of the reduction and conversion tables (Tables 2-2 and 2-3). Of those that are listed, the reduction factors are exact and the conversion factors are approximate (except that 1 in. = 2.54 cm is exact).

49. Reduce 1.03 mi to feet.   **50.** Reduce 65.37 ft to inches.

51. Convert 15.00 gal to liters.   **52.** Convert 26 mi to kilometers.

53. We can find the voltage in an electric circuit by adding 15.2 Ω, 5.64 Ω, and 101.23 Ω and then multiplying the sum by 3.55 A. Find the voltage.

54. The density of a certain type of iron is 7110 kg/m³. The density of a type of tin is 448 lb/ft³. Which is greater? By how many kilograms per cubic meter?

55. A flash of lightning strikes an object 3.25 mi distant from a person. The thunder is heard 15 s later. The person then calculates the speed of sound and reports it to be 1144 ft/s. What is wrong with the conclusion?

56. In determining the intensity of sound, an acoustical engineer must evaluate $(19.7392)(0.213)(36.0^2)(1.2^2)$. All factors are approximate. Use a calculator to find the product and then round off the result to the proper degree of precision or accuracy.

2-5 Review Exercises for Chapter 2

In Exercises 1 through 4, determine the number of significant digits in the given approximate numbers.

1. 6508;   70.43   **2.** 0.06;   0.60   **3.** 6070;   6007   **4.** 10.00;   0.002030

In Exercises 5 through 8, determine which of each pair of approximate numbers is (a) more precise and (b) more accurate.

5. 7.32;   73.2   **6.** 8000;   80.0   **7.** 6.49;   207.31   **8.** 98.568;   0.0021

In Exercises 9 through 16, round off each of the given approximate numbers (a) to three significant digits and (b) to two significant digits.

9. 98.46	**10.** 2.734	**11.** 60540	**12.** 219500
13. 672.8	**14.** 69005	**15.** 0.7000	**16.** 4935

In Exercises 17 through 32, perform the indicated operations on the approximate numbers.

17.
$$\begin{array}{r} 26.3 \\ 412.07 \\ +\ \ 0.349 \\ \hline \end{array}$$

18.
$$\begin{array}{r} 6.8072 \\ 14.4 \\ +\ \ 8.626 \\ \hline \end{array}$$

19.
$$\begin{array}{r} 19.8062 \\ -\ \ 8.92 \\ \hline \end{array}$$

20.
$$\begin{array}{r} 806 \\ -\ \ 4.92 \\ \hline \end{array}$$

21. $(3.96)(0.030)$ **22.** $(9.52)(4000)$ **23.** $4.924 \div 86$ **24.** $6.80 \div 0.0327$

25. $\sqrt{47}$ **26.** $\sqrt{3.04}$ **27.** $\sqrt{12} + \dfrac{5.87}{1.42}$

28. $(0.3920)(14.65) - 2.96$

29. $\sqrt{988.5}$ **30.** $\sqrt{2736} \div 3.14$ **31.** $\dfrac{67.3}{\sqrt{5.4}}$ **32.** $\sqrt{58.00} \div \sqrt{673}$

In Exercises 33 through 36, give the meaning of each metric unit. Also, in Exercises 33 and 34, give the name of the unit; in Exercises 35 and 36, give the proper symbol.

33. μg   **34.** cA   **35.** kilosecond   **36.** megavolt

In Exercises 37 through 48, the given numbers are approximate. Make the indicated reductions and conversions.

37. Reduce 385 mm^3 to cubic centimeters.   **38.** Reduce 0.475 m^2 to square centimeters.

39. Convert 5.2 in. to centimeters.   **40.** Convert 46.5 in. to meters.

41. Reduce 4.452 gal to pints.   **42.** Reduce 18.5 km to centimeters.

43. Convert 27 ft^3 to liters.   **44.** Convert 3.206 km to miles.

45. Reduce 0.43 ft^3 to cubic inches.   **46.** Reduce 28.3 in.^{2} to square feet.

47. Convert 2.45 mi/h to meters per second.   **48.** Convert 52 ft/min to centimeters per hour.

In Exercises 49 through 52, use a scientific calculator to make the indicated reductions or conversions. All numbers are approximate.

49. Reduce 14.73 lb/in.^{2} to tons per square foot.

50. Reduce 12.3 g/cm^2 to kilograms per square meter.

51. Convert 18.03 ft^3/s to cubic centimeters per minute.

52. Convert 17.08 ft^3/s to cubic meters per second.

In Exercises 53 through 68, solve the given problems.

53. A machinist mills a template and measures the thickness to be 0.186 in. Based on that measurement, find the lowest value and the highest value of the thickness.

54. A carbon-zinc D size flashlight battery has a voltage measured as 1.5 V. Based on that measurement, find the lowest value and the highest value of the voltage.

55. A company opens an escrow account with a deposit of $8600. If the annual interest rate is 6.25%, find the amount of simple interest earned in the first year.

56. A jet has an airspeed (speed relative to the air) of 543 mi/h. It is slowed by flying directly into a headwind of 60 mi/h. What is the ground speed of the jet?

57. A chemist combines three solids with masses of 2.841 g, 3.729 g, and 15.27 g. What is the total of these three masses?

58. A wire 4.39 m long is cut into 12 equal sections. What is the length of each section?

59. One meter of steel will increase in length by 0.000012 m for each 1°C increase in temperature. What is the increase in length of a steel girder 38 m long if the temperature increases by 45°C?

60. A board measures 5.25 in. wide after shrinking 12% while drying. What was the original width of the board?

61. A computer performs 232 calculations in 0.7245 s. What is the rate of calculations per second?

62. If a voltage level of 110.3 V is cut back by 5%, what is the reduced voltage level?

63. If the density of gasoline is 5.6 lb/gal, find its density in kilograms per liter.

64. An FAA flight service station reports that the winds aloft at 10,000 ft are at a speed of 45 knots. Find that speed in miles per hour. Assume that a knot is exactly one nautical mile per hour and that one nautical mile is about 6080 ft.

65. Cross-sectional areas of wires are often measured in square mils, where 1 mil = 0.001 in. exactly. Find the cross-sectional area (in square inches) of a wire with an area of 287 mil^2.

66. A technician records the time required for a robot's arm to swing from the extreme right to the extreme left. A time of 4.3 s is recorded. The time for the return swing is then recorded as 4.75 s. What is the difference between the two times?

67. A student wrote in her laboratory report that the time for a full swing of a pendulum was 2 s and the time of the next full swing was 2.2 s. Her conclusion was that the times differed by 0.2 s. What is the error in the conclusion (if the data are correct)?

68. If acceleration has units of velocity divided by units of time, what are the units of acceleration in the fundamental units of the metric system?

CHAPTER THREE

Signed Numbers

FIGURE 3-1

Prior to this chapter we have not tried to subtract a number from a smaller number, as in $33 - 58$. If we use only the numbers considered so far, we can't find solutions to such problems. Yet many real applications do require that we solve problems of this type, so it now becomes necessary to introduce a new set of numbers which we will refer to as **negative numbers.** Negative numbers are used in a variety of situations. Measurements of temperature routinely include negative numbers, as shown in the thermometer of Figure 3-1. Altitudes are usually measured with sea level as a zero reference point so that elevations below sea level are represented by negative numbers. In monetary transactions, losses or debts are typically represented as negative numbers. In electronics, voltage dropping below a given reference level is measured in negative volts.

In this chapter we study signed numbers and the principles of arithmetic that apply to them. We will learn to solve a variety of different applied problems. For example, while flying an instrument approach to an airport, a pilot at 4000 ft descends at the rate of 500 ft/min for 3.0 min and then climbs at the rate of 800 ft/min for 5.0 min. Using signed numbers, we will solve the problem of determining the resulting altitude.

3-1 Signed Numbers

A negative number is a number less than zero. The symbols for addition $(+)$ and subtraction $(-)$ are also used to indicate positive and negative numbers, respectively. For example, +5 (positive 5) is a positive number and −5 (negative 5) is a negative number. Since positive numbers correspond directly to the numbers we have been using (we made no attempt to associate a sign with them, only using + and − to designate addition and subtraction), the plus sign may be omitted before a positive number if there is no danger of confusion. However, we shall never omit the negative sign before a negative number.

natural numbers (or positive integers):
1, 2, 3, . . .
negative integers:
−1, −2, −3, . . .

integers: . . . −3, −2, −1, 0, 1, 2, 3, . . .

Another name given to the natural numbers is **positive integers.** *The negative counterparts of the positive integers are called* **negative integers.** *If we combine the positive integers, zero (which is neither positive nor negative), and the negative integers, we get the set of all* **integers.** In the following sections, we will sometimes refer to the integers.

EXAMPLE A

The numbers 7 (which equals +7) and +7 are positive integers. The number −7 is a negative integer. The number 0 is an integer, but it is neither positive nor negative. The number $\frac{2}{3}$ is not an integer, since it is not a natural number or the negative of a natural number. □

In this discussion, we see that *the plus and the minus signs are used in two senses. One is to indicate the operation of addition or subtraction; the other is to designate a positive or negative number.* Consider the following example.

EXAMPLE B

signed numbers

(+3) + (+6)

indicates addition

signed numbers

(+3) − (−6)

indicates subtraction

The expression 3 + 6 means "add the number 6 to the number 3." Here the plus sign indicates the operation of addition, and the numbers are **unsigned numbers,** *which are positive numbers.*

The expression (+3) + (+6) means "add the number +6 to the number +3." Here the middle plus sign indicates addition, whereas the other plus signs denote signed numbers. Of course, the result would be the same as 3 + 6. See diagram at left.

The expression 3 − 6 means "subtract the number 6 from the number 3." Here the minus sign indicates subtraction, and the numbers are unsigned.

The expression (+3) − (+6) means "subtract the number +6 from the number +3." Note that the positive signs only designate signed numbers.

The expression (+3) − (−6) means "subtract the number −6 from the number +3." Here the first minus sign indicates subtraction, whereas the second designates the negative number. See diagram at left.

The method of performing these operations is discussed in Section 3-2. □

Most calculators allow the entry of negative numbers. A common format is to enter the value of the number first and then press the *change sign* key which is usually labeled as $+/-$ or CHS.

EXAMPLE C To enter the negative number designated by -3, enter 3 followed by the change sign key as follows:

$3\ \boxed{+/-}$

(Although it is rare, some calculators do allow you to press the change sign key first.) If numbers are entered in a calculator without the change sign key, it is assumed that they are unsigned numbers and are therefore positive. After entering the keys indicated here, you should note the way your calculator displays the negative sign.   □

In addition to serving as a means for entering negative numbers, the change sign key also suggests an operation of finding the **negative of a number.** *To find the negative of a number, change the sign of that number.*

EXAMPLE D The negative of 64 is -64.
The negative of -37 is $+37$ or 37.
The negative of 0 is 0.   □

Positive and negative numbers can be illustrated on a scale as in Figure 3-2. On a horizontal line we choose a point which we call the **origin,** *and here we locate the integer 0. We then locate positive numbers to the right of 0 and negative numbers to the left of 0.* As we see, every positive number has a corresponding opposite negative number.

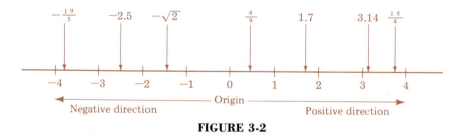

FIGURE 3-2

When numbers are marked positive or negative, they are called **signed numbers** (or *directed numbers*) *to indicate their opposite qualities, as shown on the scale.* Since positive numbers are directed to the right of the origin, it follows that numbers increase from left to right. Therefore: *Any number* *on the scale is smaller than any number located farther to the right. The symbols > and < mean* **"greater than"** *and* **"less than,"** *respectively.* Their use is shown in the following example.

EXAMPLE E

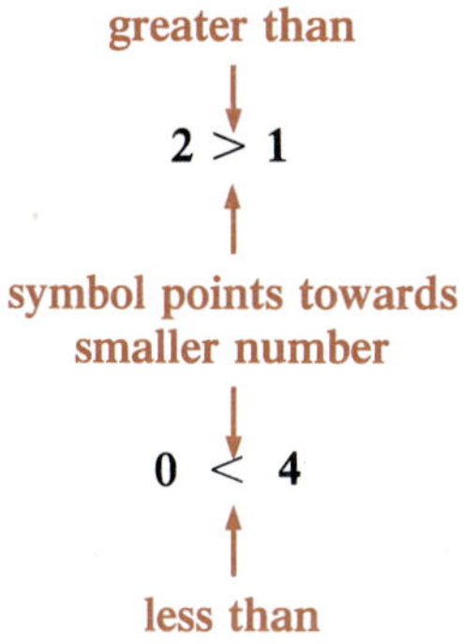

greater than

$\downarrow$

2 > 1

$\uparrow$

symbol points towards
smaller number

$\downarrow$

0 < 4

$\uparrow$

less than

The expression $2 > 1$ means "the number 2 is greater than the number 1." It also means that on the scale in Figure 3-2 we would find 2 to the right of 1. See diagram at left.

The expression $0 < 4$ means "0 is less than 4." This means that 0 is to the left of 4. See diagram at left.

The expression $-3 < 0$ means "-3 is less than 0." The number -3 lies to the left of 0 on the scale.

The expression $-3 > -5$ means "-3 is greater than -5." The number -3 is to the right of the number -5 on the scale.

In general, larger numbers are farther to the right while smaller numbers are farther to the left. ☐

Although the positive direction is conveniently taken to the right of the origin on the number scale, we may select any direction as the positive direction so long as we take the opposite direction to be negative. Consider the illustrations in the following example.

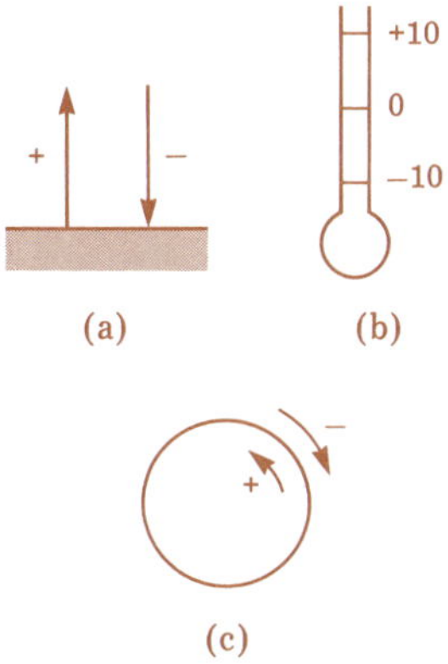

EXAMPLE F

When we are dealing with the motion of an object moving vertically with respect to the surface of the earth, if the positive direction is up we then take the negative direction to be down. The zero position can be chosen arbitrarily; often it is ground level. See Figure 3-3a.

If a temperature above zero (an arbitrarily chosen reference level) is called positive, a temperature below zero would be called negative. See Figure 3-3b.

For an object moving in a circular path, if a counterclockwise movement is called positive, a clockwise movement would be called negative. See Figure 3-3c. ☐

FIGURE 3-3

The value of a number without its sign is called its **absolute value.** That is, if we disregard the signs of $+5$ and -5, the value would be the same. This is equivalent to saying that the distances from the origin to the points $+5$ and -5 on the number scale are equal. (See Figure 3-4). The absolute value of a number is indicated by the symbol $| \;\; |$. The number $+5$ does not equal the number -5, but $|+5| = |-5|$. Absolute values can be used in performing arithmetic operations with signed numbers. Consider the following example.

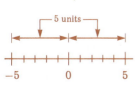

FIGURE 3-4

EXAMPLE G   The absolute value of $+8$ is 8. The absolute value of -2 is 2. The absolute value of $-\frac{3}{2}$ is $\frac{3}{2}$. We may write these equalities as $|+8| = 8$, $|-2| = 2$, and $|-\frac{3}{2}| = \frac{3}{2}$. Also,

$$|-7.3| = 7.3 \qquad\qquad |9.2| = 9.2$$
$$|-22.6| = 22.6 \qquad\qquad |524| = 524$$

$\square$

Exercises 3-1

In Exercises 1 through 8, state the meanings of the given expressions as in Example B.

1. $6 + 3$ **2.** $6 - 3$ **3.** $(+2) + (7)$ **4.** $(+9) - (+5)$

5. $(-8) + (-2)$ **6.** $(+4) - (-1)$ **7.** $(-3) - (+6)$ **8.** $(-1) - (-7)$

In Exercises 9 through 16, locate the approximate positions of the given numbers on a number scale such as that in Figure 3-2.

9. 5; -5 **10.** -0.5; 0.5 **11.** -2.3; $+3.2$ **12.** $+1.43$; -3.14

13. $+\sqrt{3}$; $-\sqrt{2}$ **14.** $-\dfrac{17}{6}$; $+\dfrac{3}{5}$ **15.** $+\dfrac{13}{22}$; $-\dfrac{22}{13}$ **16.** $-\dfrac{13}{4}$; $+\dfrac{28}{5}$

In Exercises 17 through 20, determine which of the given numbers is the largest.

17. 2; -9; 0 **18.** -3; -6; 0 **19.** -3; -1; -5 **20.** -7; -8; -9

In Exercises 21 through 32, insert the proper sign ($>$ or $<$ or $=$) between the given numbers.

21. $6 \quad 2$ **22.** $8 \quad -3$ **23.** $0 \quad 4$ **24.** $-3 \quad 0$

25. $-3 \quad -7$ **26.** $-9 \quad -8$ **27.** $-7 \quad -5$ **28.** $-1 > -5$

29. $\sqrt{5} \quad 2.2$ **30.** $\sqrt{10} \quad -3.2$ **31.** $|6| = |-6|$ **32.** $|-3| \quad -|-3|$

In Exercises 33 through 36, find the absolute value of each of the given numbers.

33. $+6$; -6 **34.** -5; 5 **35.** $-\dfrac{6}{7}$; $\dfrac{8}{5}$ **36.** 2.4; -0.1

In Exercises 37 through 52, certain applications of signed numbers are indicated.

37. In writing a computer program for a bank, the programmer represents a \$50 deposit as $+50$. What signed number would represent a \$30 withdrawal?

38. The Great Pyramid in Egypt was built about 2600 B.C. Assuming that years designated A.D. are considered positive, represent this date as a signed number.

39. In an electronic device, the voltage at a certain point is 0.2 V above that of the reference and is designated as $+0.2$ V. How is the voltage 0.5 V below the reference designated?

40. If a surveyor represents the altitude of a point 12 m above sea level by $+12$, what signed number represents an altitude that is 8 m below sea level?

41. The amounts (in millions of dollars) of indebtedness of two corporations are represented by -3 and -6. Which corporation has the smaller debt?

42. If the number of years between the current year and the year 2000 is represented by a positive number, what number represents the number of years between the current year and 1995?

43. Manufacturing output is increased by 28 units per day, which is represented by +28. What number represents a decrease of 37 units per day?

44. The image of an object formed by a lens can be formed either to the right or to the left of the lens, depending on the position of the object. If an image is formed 10 cm to the left and its distance is designated as −10 cm, what would be the designation of an image formed 4.5 cm to the right?

45. Use the proper sign (> or <) to show which temperature, −30°C or −5°C, is higher.

46. Use the proper sign (> or <) to show which elevation is higher: −80 m or −60 m.

47. Use the proper sign (> or <) to show which voltage reading is higher: −2 V or −5 V.

48. A quality control specialist analyzes production data by calculating correlation coefficients. Use the proper sign (> or <) to show which coefficient is the larger number: −0.94 or −0.58.

49. After clearing your calculator, first press the change sign key, then press the 5 key followed by the equal key. What is the display? Does your calculator allow you to use the change sign key *before* the number itself is entered?

50. In entering −5 on your calculator, the change sign key is pressed once. What happens if the change sign key is pressed twice instead of once?

51. For calculators that do not have a change sign key, how might you get a display of −3?

52. The calculator change sign key should be used for negative numbers and not the operation of subtraction. What is the result if you try to evaluate 8 − 5 by using the change sign key as follows?

$$8 \boxed{+/-} 5 \boxed{=}$$

3-2 Addition and Subtraction of Signed Numbers

Section 3-1 introduced signed numbers and ways to express them. This section presents the method for performing the operations of addition and subtraction with signed numbers, and the following sections consider multiplication and division of signed numbers. We want our procedures for the addition and subtraction of signed numbers to conform to the way things generally work in the real world. We have already established that positive numbers are equal to the equivalent unsigned numbers. As a result, *the sum of positive numbers is found the same way we find the sum of unsigned numbers*. This first principle is illustrated in the following example.

EXAMPLE A   Since 4 + 5 = 9, we have (+4) + (+5) = +9. Thus the sum of +4 and +5 is +9.

In the same way, since 3 + 12 = 15, we have (+3) + (+12) = +15. □

Let us now consider the addition of two negative numbers. If we regard a negative number as representing a loss, we can think of adding one negative number to another as adding one loss to another. When we add one loss to another, our result is a still greater loss. *This suggests that when we add two negative numbers, we add the absolute values and designate the sum as negative*.

EXAMPLE B From the preceding considerations, we see that $(-4) + (-5) = -(4 + 5) = -9$. Here we have added the absolute value of -4 and the absolute value of -5 in the expression $(4 + 5)$.

In the same way, $(-3) + (-12) = -(3 + 12) = -15$. □

When considering the addition of two numbers of *unlike* signs, we may think of combining a gain with a loss. If the gain is greater than the loss, we have a net gain. If the loss is greater than the gain, we have a net loss. This suggests that *the number of smaller absolute value is subtracted from the one of greater absolute value, with the result taking on the sign of the number of greater absolute value.*

EXAMPLE C
$$(+5) + (-4) = +(5 - 4) = +1$$
$$(-5) + (+4) = -(5 - 4) = -1$$
$$(-3) + (+12) = +(12 - 3) = +9$$
$$(+3) + (-12) = -(12 - 3) = -9$$ □

The process is summarized as follows.
To add signed numbers with *like signs:*

1. Add the absolute values of the numbers.

2. The sum has the sign common to the original numbers.

To add signed numbers with *unlike signs:*

1. Find the difference between the absolute values of the numbers.

2. The sum has the sign of the number with the larger absolute value.

EXAMPLE D
$$(+8) + (+2) = +(8 + 2) = +10$$

like signs   common   add absolute values
              sign

$$(-8) + (-2) = -(8 + 2) = -10$$
$$(+8) + (-2) = +(8 - 2) = +6$$

unlike signs   sign of $+8$ subtract absolute values
               sign of -8

$$(-8) + (+2) = -(8 - 2) = -6$$ □

In subtracting signed numbers, we use the same principle for subtraction that we used in Chapter 1. That is, we find a number (the difference) which, when added to the number being subtracted (the subtrahend), will give the other number (the minuend).

Consider the illustrations in the following example.

EXAMPLE E

usually written
as $7 - 3 = 4$

$(+7) - (+3) = +4$	since	$(+4) + (+3) = +7$
$(+3) - (+7) = -4$	since	$(-4) + (+7) = +3$
$(-7) - (+3) = -10$	since	$(-10) + (+3) = -7$
$(+3) - (-7) = +10$	since	$(+10) + (-7) = +3$
$(-7) - (-3) = -4$	since	$(-4) + (-3) = -7$
$(-3) - (-7) = +4$	since	$(+4) + (-7) = -3$

If we examine the results of Example E, we note that the same result would be obtained in each case if we changed the operation to addition and changed the sign of each of the numbers being subtracted.

Therefore, the process of subtracting a signed number from another signed number is summarized as follows:

1. Change the sign of the number being subtracted.
2. Change the subtraction symbol to the addition symbol.
3. Proceed as in the addition process.

EXAMPLE F

subtraction   addition

change sign

$(+7) - (+3) = +4$	and	$(+7) + (-3) = +4$
$(+3) - (+7) = -4$	and	$(+3) + (-7) = -4$
$(-7) - (+3) = -10$	and	$(-7) + (-3) = -10$
$(+3) - (-7) = +10$	and	$(+3) + (+7) = +10$
$(-7) - (-3) = -4$	and	$(-7) + (+3) = -4$
$(-3) - (-7) = +4$	and	$(-3) + (+7) = +4$

There are several important conclusions to be drawn from the illustrations of Example F. The first is that:

To subtract one signed number from another, we may change the sign of the subtrahend (the number after the subtraction symbol) and then add it to the minuend, using the rules for the addition of signed numbers.

Also, from the first two illustrations, we see that:

Subtracting a positive number is equivalent to adding a negative number of the same absolute value.

Finally, the last three illustrations point out that:

Subtracting a negative number is equivalent to adding a positive number of the same absolute value.

This last point may be thought of as removing (subtracting) a loss (a negative number), which results in a gain.

EXAMPLE G

$$(+8) - (+5) = (+8) + (-5) = +3$$
$$(+5) - (+8) = (+5) + (-8) = -3$$
$$(+8) - (-5) = (+8) + (+5) = +13$$
$$(+5) - (-8) = (+5) + (+8) = +13$$

Thus we see that $-(+5)$ and $+(-5)$ are the same and $-(-5)$ and $+(+5)$ are the same. □

In general, when working with signed numbers we find it preferable and convenient to:

Change the operations on negative numbers so that the result may be obtained by the addition and subtraction of positive numbers, written as unsigned numbers.

The steps to do this are as follows:

1. Change negative numbers to positive numbers and change the operation on these numbers.
2. Change the positive numbers to unsigned numbers.
3. Perform the operations as on positive numbers.

Consider the illustrations in the following example.

EXAMPLE H

change positive numbers to unsigned numbers

$$(+5) + (+2) = 5 + 2 = 7$$
$$(+5) - (+2) = 5 - 2 = 3$$
$$(+2) - (+5) = 2 - 5 = -3 \qquad \text{result is a negative number}$$

change negative number
to positive number

$$(+5) + (-2) = (+5) - (+2) = 5 - 2 = 3$$

change operation

$$(-5) + (+2) = -(+5) + (+2) = -5 + 2 = -3$$
$$(-5) - (-2) = -(+5) + (+2) = -5 + 2 = -3$$

When there is no sign before a signed number, we assume that it may be treated as if a + sign preceded it. That is, we treat (-5) as $+(-5)$ and treat $(+5)$ as $+(+5)$. □

When you become more familiar with these operations, you will be able to do them without actually writing in the step where a negative number is changed to a positive number. This can be shown immediately with signed numbers. When this is done we must remember that *adding a negative number is equivalent to subtracting an unsigned number and subtracting a negative number is equivalent to adding an unsigned number.*

EXAMPLE I

adding a negative number

$$(+7) + (-3) = 7 - 3 = 4$$

subtracting a negative number

$$(+7) - (-3) = 7 + 3 = 10$$
$$(-7) - (-3) = -7 + 3 = -4$$
$$(-7) + (-3) = -7 - 3 = -10$$
$$(-4) + (-2) - (-7) = -4 - 2 + 7 = -6 + 7 = 1$$
$$(-5) - (-2) - (+9) = -5 + 2 - 9 = -3 - 9 = -12$$

The following examples illustrate the applied use of addition and subtraction of signed numbers.

EXAMPLE J   An oil well which is 30 m below ground level is drilled 40 m deeper. The total depth (in meters) is represented as

$$(-30) + (-40) = -70$$

EXAMPLE K   The temperature of an engine coolant is $-8°C$. If the temperature increases by $12°C$, its new value is represented (in degrees Celsius) by

$$(-8) + (+12) = +4$$

When adding or subtracting signed numbers with a calculator, we may use the change sign key to enter negative numbers. In Example L, we assume that the calculator requires that a negative number be entered by pressing the change sign key *after* the relevant value has been entered. We also assume that a positive number is entered without use of the change sign key. That is, a number like $+5$ is entered by simply pressing the 5 key.

EXAMPLE L   To compute $(+5) - (-2)$ with a calculator, enter

$$5\;\boxed{-}\;2\;\boxed{+/-}\;\boxed{=}$$

to obtain the result 7.

To compute $(-5) - (-2)$ with a calculator, enter

$$5\;\boxed{+/-}\;\boxed{-}\;2\;\boxed{+/-}\;\boxed{=}$$

to obtain the result -3.

To compute $(-5) + (+2)$ with a calculator, enter

$$5\;\boxed{+/-}\;\boxed{+}\;2\;\boxed{=}$$

to obtain the result -3.   ☐

Exercises 3-2

In Exercises 1 through 32, perform the indicated operations on the given signed numbers.

1. $(+2) + (+9)$ 2. $(+3) + (+8) = 11$ 3. $(-6) + (-9)$ 4. $(-6) + (-5)$ -11
5. $(-3) + (+6)$ 6. $(-1) + (+8)$ 7. $(+2) + (-10)$ 8. $(+3) + (-11)$ -8
9. $(+12) - (+5)$ 10. $(+14) - (+6)$ 11. $(+7) - (+10) = -3$ 12. $(+2) - (+15)$
13. $(-6) - (+4)$ 14. $(-3) - (+8)$ -11 15. $(+7) - (-9)$ 16. $(+14) - (-2)$ 16
17. $(-6) - (-7)$ 18. $(-9) - (-5)$ -4 19. $(+1) + (-5) + (-2)$ 20. $(+6) + (+5) + (-3)$
21. $(+2) + (-8) - (+2)$ 22. $(-4) - (+8) + (-9)$
23. $(+5) - (-3) - (+7)$ 24. $(-7) - (-1) - (+6)$
25. $(-7) - (-15) + (-2)$ 26. $(-6) + (-13) - (-11)$
27. $(-9) - (-5) + (-8) - (+5)$ 28. $(-7) - (-5) + (-4) - (+6)$
29. $(+3) - (-7) - (+9) + (-3)$ 30. $(+6) + (-11) - (-5) - (+8)$ -8
31. $(-6) - (+9) - (-12) + (-4) - (-1)$ 32. $(-5) + (+6) - (+3) - (-14) + (-3)$

In Exercises 33 through 40, set up the given problems in terms of the addition or subtraction of signed numbers and then perform the operations. (The operation of addition or subtraction that should be used in setting up the problem is indicated in each case.)

33. The temperature of a fluid is $-10°C$. (a) If the temperature increases (use addition) by $5°C$, what is the resulting temperature? (b) If the temperature decreases (use subtraction) by $5°C$, what is the resulting temperature?

34. A woman has a debt of $17. Another person owes her $11. What is the sum (use addition) of these "assets" to her?

35. In programming an industrial robot, consider a 37-cm upward movement of an arm positive while a downward movement of 26 cm is negative. What is the distance between these high and low points? (Use subtraction.)

36. A fluid boils at $120°C$ and freezes at $-30°C$. How much higher is the boiling point than the freezing point? (Use subtraction.)

37. The current in an electric circuit changes from 12 A to −5 A. How much higher is 12 A than −5 A? (Use subtraction.)

38. In testing an aircraft altimeter, technicians use air pressure to simulate an altitude of 6000 ft, and an error of −150 ft is noted. What is the altimeter reading at this simulated altitude? (Use addition.)

39. In four successive years, the population of a certain city changes as follows: increases by 5000; increases by 2000; decreases by 3000; increases by 1000. Find the sum (use addition) of these changes.

40. What rule of English grammar is equivalent to subtracting a negative number in mathematics?

In Exercises 41 through 48, use a scientific calculator to perform the indicated operations on the given signed numbers.

41. $(-2) + (-5)$

42. $(-8) + (+3)$

43. $(+7) - (-2)$

44. $(+3) - (-8)$

45. $(-4) - (-3) + (-2)$

46. $(+8) - (-5) + (-1)$

47. $(-2) - (-5) - (-4)$

48. $(+12) - (-20) - (-15) + (-3)$

3-3 Multiplication of Signed Numbers

In developing the rules for the multiplication of signed numbers, we want to preserve the definitions already presented. In our work with the natural numbers, we have already defined multiplication to be a process of repeated addition. We now extend this same concept to signed numbers so that the multiplication of signed numbers is the process of repeated addition, where equal numbers are added together a specified number of times.

Recall from Chapter 1 that *the numbers to be multiplied are called* **factors** *and the result is the* **product.** Also, according to the commutative law, the product of two factors is not changed if the factors are interchanged. With these ideas in mind, consider the illustrations of the following example.

EXAMPLE A Since $3 \times 4 = 12$ and since $3 = +3$, $4 = +4$, and $12 = +12$, we have $(+3) \times (+4) = +12$.

Since $3 \times (-4)$ means $(-4) + (-4) + (-4)$, since $(-4) + (-4) + (-4) = -12$, and since $3 = +3$, we have $(+3) \times (-4) = -12$.

Also, since multiplication is commutative, we have $(-4) \times (+3) = (+3) \times (-4) = -12$. □

We see from Example A that the product of two positive numbers is positive. Also, the product of a positive number and a negative number is negative. We get this general principle:

Changing the sign of one factor changes the sign of the product.

Applying this principle, we may determine the product of two negative numbers.

EXAMPLE B   From Example A, we have seen that

$$(-4) \times (+3) = -12$$

Applying the principle that changing the sign of one factor changes the sign of the product, we change the factor $+3$ to -3 and the result changes from -12 to $+12$. This leads to

$$(-4) \times (-3) = +12$$

In general,

The product of two negative numbers is positive.

This last illustration may be thought of as subtracting -3 four times or $-(-3) - (-3) - (-3) - (-3) = +12$. We may also think of this as negating four \$3 debts, which would result in a gain of \$12. Note that this example is consistent with the fact that subtracting a negative number is equivalent to adding a positive number of the same absolute value. ☐

From the preceding discussion, we can state the following results:

    1. If two factors have like signs, their product is positive.

    2. If two factors have unlike signs, their product is negative.

If there are three or more factors, we can use the associative and commutative laws and multiply two factors at a time, in any order, observing each time the sign of the product. We can then conclude that:

    1. The product is positive if all factors are positive or there is an even number of negative factors.

    2. The product is negative if there is an odd number of negative factors.

If any factor is zero, it follows that the product is zero because zero times any number is zero.

EXAMPLE C

like signs   unlike signs

$$(+4) \times (+7) = +28 \qquad (+4) \times (-7) = -28$$
$$(-4) \times (-7) = +28 \qquad (-4) \times (+7) = -28$$

$(+2) \times (+3) \times (+5) = +30$	no negative factor
$(-2) \times (+3) \times (+5) = -30$	one negative factor
$(-2) \times (-3) \times (+5) = +30$	two negative factors
$(-2) \times (-3) \times (-5) = -30$	three negative factors
$(-1) \times (-2) \times (-3) \times (-5) = +30$	four negative factors
$(-1) \times (0) \times (-3) \times (-5) = 0$	zero is a factor ☐

We saw in Section 1-9 that if a given number appears as a factor in a product more than once, it is customary to use exponents. We use exponents with signed numbers in the same way we did with unsigned numbers in arithmetic. Therefore: *The exponent indicates the number of times a given number is a factor.*

EXAMPLE D Instead of writing $(+5) \times (+5)$, we could write $(+5)^2$. Here 2 is the exponent and $+5$ is the base.

Rather than writing $(-2)(-2)(-2)(-2)(-2)(-2)$, we write $(-2)^6$. Here 6 is the exponent and -2 is the base. □

We must note the use of parentheses in raising a signed number to a power. The meaning of $(-2)^6$ is indicated above. However, -2^6 means the negative of 2^6. *Since parentheses are not placed around -2, only the 2 is being raised to the sixth power and we get $-2^6 = -(2^6)$. Evaluating $(-2)^6$ shows it has a value of 64 while -2^6 has a value of -64.*

From our earlier discussion of the multiplication of signed numbers, we can see that a positive number raised to a power will have a positive result. However, *a negative number raised to a power will have a positive result if the power is an even number and a negative result if the power is odd.* This situation is due to the fact that the exponent indicates the number of factors of the number. We must also *be careful to note whether it is a signed number being raised to a power or the negative of an unsigned number being raised to a power.*

EXAMPLE E

$(+3)^2 = +9$ since $(+3)^2 = (+3) \times (+3)$

$(+4)^3 = +64$ since $(+4)^3 = (+4) \times (+4) \times (+4)$

$(-3)^2 = +9$ since $(-3)^2 = (-3) \times (-3)$

$-3^2 = -9$ since $-3^2 = -(3^2) = -(3 \times 3)$

$(-3)^3 = -27$ since $(-3)^3 = (-3) \times (-3) \times (-3)$

$-3^3 = -27$ since $-3^3 = -(3^3) = -(3 \times 3 \times 3)$

$(-2)^4 = +16$ since $(-2)^4 = (-2) \times (-2) \times (-2) \times (-2)$

$-2^4 = -16$ since $-2^4 = -(2^4) = -(2 \times 2 \times 2 \times 2)$

Also,

$(-1)^{20} = +1$ but $-1^{20} = -1$

parentheses show that -1 is raised to power

and

negative of the power of 1

$(-1)^{21} = -1$ and $-1^{21} = -1$ □

In any product where the factors are signed numbers, the multiplication of these signed numbers may be indicated by writing them, in their parentheses, adjacent to each other. This eliminates the need for using $\times$ or $\cdot$ as signs of multiplication.

EXAMPLE F   We may write $(-3) \times (+4)$ as $(-3)(+4)$ because we realize that this form indicates the multiplication of -3 and $+4$.
    Therefore $(-2)(-3)(+4) = +24$, since
$(-2)(-3)(+4) = (-2) \times (-3) \times (+4)$.   ☐

To use a calculator to find products of signed numbers, we normally enter the product using the $\boxed{\times}$ key, and we must also use the change sign key after values that are supposed to be negative. When we use the $\boxed{x^y}$ key on a calculator, an error display occurs on many calculators when the base is negative. In such cases it is best first to *determine the correct sign of the result and then to use the calculator to determine the numerical value with a positive base.*

EXAMPLE G   To calculate $(+6)(-4)$ enter $6 \boxed{\times} 4 \boxed{+/-} \boxed{=}$ to obtain -24.
    To calculate $(-2)^5$ we first determine that the result is negative and then enter $2 \boxed{x^y} 5 \boxed{=}$. The result is therefore -32. If we attempt to directly calculate $(-2)^5$ by entering

$$2 \boxed{+/-} \boxed{x^y} 5 \boxed{=}$$

 many calculators will display an error.   ☐

EXAMPLE H   While flying an instrument approach to an airport, a pilot at 4000 ft descends at the rate of 500 ft/min for 3.0 min and then climbs at the rate of 800 ft/min for 5.0 min.
    To determine the resulting altitude, we note that the altitude decreases in a descent and increases in a climb. We therefore represent lost altitude with a negative number and gained altitude with a positive number. Combining the changes, we get

start   descend   climb

$$(+4000) + (-500)(+3.0) + (+800)(+5.0)$$

dist. = rate $\times$ time

$$= 4000 - 1500 + 4000$$
$$= 6500 \text{ ft}$$

The plane will be at an altitude of 6500 ft.   ☐

Exercises 3-3

In Exercises 1 through 28, find the indicated products.

1. $(+8) \times (+10)$
2. $(+12) \times (+5)$
3. $(+7)(-9)$
4. $(-4)(+11)$
5. $(-7) \times (+12)$
6. $(+15) \times (-4)$ -60
7. $(-2) \times (-15)$
8. $(-9) \times (-8)$
9. $(+4) \times (-2) \times (+7)$
10. $(-3) \times (+6) \times (+5)$
11. $(-10)(-3)(+8)$
12. $(-12)(+4)(-10)$
13. $(+2) \times (-5) \times (-9) \times (+4)$
14. $(-7) \times (-6) \times (+10) \times (-1)$ $+420$
15. $(+8)(0)(-4)(-1)$
16. $(-1)(-3)(-5)(-2)$
17. $(+3)(-1)(+4)(-7)(-2)$ -420
18. $(-5)(+3)(-2)(0)(-9)$ 0
19. $(+6)^2; \quad (-6)^2$
20. $(+5)^3; \quad (-5)^3$ -125
21. $(-7)^2; \quad -7^2$
22. $-9^2; \quad (-9)^2$ 81
23. $(-4)^5$
24. $(-4)^6$
25. $(-1)^{16}; \quad -1^{16}$
26. $(-1)^{19}; \quad -1^{19}$ -1
27. $(-2)^7; \quad -2^7$
28. $-2^8; \quad (-2)^8$

In Exercises 29 through 36, set up the given problems in terms of the multiplication of signed numbers and then perform the operations.

29. Metal expands when heated and contracts when cooled. The length of a particular metal rod changes by 2 mm for each 1°C change of temperature. What is the total change in length of the rod as the temperature changes from 0°C to −25°C?

30. The price of a stock falls $3 each day for four consecutive days. Find the net change in the price of the stock during this period.

31. At a given point in its flight, an airplane had been descending at the rate of 1000 ft/min for 7 min. If the plane is now at an altitude of 0 ft, what was its altitude 7 min earlier?

32. The voltage in a certain electric circuit is decreasing at the rate of 2 mV/s. At a given time the voltage is 220 mV. What was the voltage 10 s earlier?

33. A hydraulic jack is used to support a heavy load. If the load is reduced by 36 lb/min and is currently 562 lb, what was the load 12 min earlier?

34. A flare is shot up from the top of a tower. Distances above the flare gun are positive while those below it are negative. After 4 s the distance (in feet) is found by evaluating $(+120)(4) + (-16)(4)^2$. Find this distance.

35. An engineer is designing the slope of a roadside so that it drops 2 ft in height for every 3 ft of horizontal distance. At one point there is a drop of 18 ft below the starting point. How much of a drop is there for a point on the slope with a horizontal distance 12 ft closer to the starting point?

36. Identify the pattern that develops when you evaluate $(-1)^2$, $(-1)^3$, $(-1)^4$, $(-1)^5$, $(-1)^6$, and so on.

In Exercises 37 through 44, use a scientific calculator to evaluate the given expressions.

37. $(-3) \times (-6)$
38. $(+4)(-2)$
39. $(-7)(+8)$
40. $(-5)(-4)$
41. $(-2)^3$
42. $-(+2)^3$
43. $-(+3)^4$
44. $(-3)^4$

3-4 Division of Signed Numbers

After having discussed addition, subtraction, and multiplication, we now consider the division of positive and negative numbers. We want to continue to make our definitions consistent with those already given. We have seen that division is the process of determining how many times one number is contained in another. Here, as in Chapter 1, *the number being divided is called the* **dividend,** *the number being divided into the dividend is the* **divisor,** *and the result is the* **quotient.**

    As we noted in Chapter 1, division is the inverse process of multiplication, since the product of the quotient and the divisor equals the dividend. We can now establish the principles used in the division of signed numbers from those we have already established for the multiplication of signed numbers. The basis for these principles is seen in Example A.

EXAMPLE A

$$(+15) \div (+5) = +3 \quad \text{since} \quad (+5) \times (+3) = +15$$
$$(+15) \div (-5) = -3 \quad \text{since} \quad (-5) \times (-3) = +15$$
$$(-15) \div (+5) = -3 \quad \text{since} \quad (+5) \times (-3) = -15$$
$$(-15) \div (-5) = +3 \quad \text{since} \quad (-5) \times (+3) = -15 \qquad \square$$

We see from the illustrations of Example A that:

    The quotient of two numbers of like signs is positive and the quotient of two numbers of unlike signs is negative.

    We should note that the division of two signed numbers follows the same principles as the multiplication of two signed numbers.

    The division of one signed number by another is often expressed in the form of a fraction, as illustrated in the following example.

EXAMPLE B

$(+21) \div (-3)$ in fractional form is $\dfrac{+21}{-3}$ so that

$$\overset{\text{unlike}}{\underset{\text{signs}}{}}\; \frac{+21}{-3} = -7 \qquad \overset{\text{unlike}}{\underset{\text{signs}}{}}\; \frac{-21}{+3} = -7 \qquad \overset{\text{like}}{\underset{\text{signs}}{}}\; \frac{-21}{-3} = +7 \qquad \square$$

    If either the numerator or the denominator (or both) of a fraction is the product of two or more signed numbers, this product may be found and then the value of the fraction can be determined by the appropriate division. However, since the same basic principles apply for the multiplication and the division of signed numbers, we need only determine the number of negative factors of the numerator and denominator combined. *If the total number of negative factors is odd, the quotient is negative. Otherwise, it is positive.* The reduction of the fraction to lowest terms then follows as with any arithmetic fraction. *Also, factors common to both the numerator and the denominator can be divided out.*

EXAMPLE C

two negative factors

$$\frac{(-2)(+12)}{-4} = +\frac{(2)(12)}{4} = \frac{(1)(12)}{2} = \frac{6}{1} = 6$$

The sign of the result is positive because the original fraction has two negative factors (-2 and -4) in the numerator and denominator. The resulting fraction is then simplified by dividing out two factors of 2 from both the numerator and the denominator. Since the resulting denominator is 1, it does not have to be written.

In a similar manner, we have

three negative factors

$$\frac{(-3)(+6)(-5)}{(-1)(+10)} = -\frac{(3)(6)(5)}{(1)(10)} = -\frac{(3)(6)(1)}{(1)(2)} = -\frac{(3)(3)(1)}{(1)(1)} = -9$$

The result is negative since the original fraction had three negative factors. The fraction was then simplified by dividing out a factor of 5 and a factor of 2. ☐

EXAMPLE D

Six partners form a company that suffers a total loss of $7200. Express the loss for each partner as a single value.

It is common to express gains as positive amounts and losses as negative amounts so that the loss for each partner is

$$\frac{-\$7200}{6} = -\$1200$$ ☐

Zero is the one and only number that cannot be used as a divisor. Since division is the inverse of multiplication, we can say that $0 \div$ (any number) $= 0$ because (any number) $\times 0 = 0$. This means that zero divided by any number (except zero) is equal to zero. However, if zero is a divisor as in (any number) $\div 0$, we cannot determine the quotient which will equal the *number* when multiplied by zero. (Any quotient that may be tried gives a product of zero when multiplied by zero and does not give the *number*.) In the case of $0 \div 0$, any positive or negative number can be a quotient, because $0 \times$ (*any number*) $= 0$. This result is said to be indeterminate.

Since *division by zero* yields either no solution or an indeterminate solution:

Division by zero is excluded at all times.

We must remember, however, that all other operations with zero are perfectly valid.

EXAMPLE E

1. $\frac{0}{4} = 0$, since $0 \times 4 = 0$.

2. $\frac{-5}{0}$ is excluded since (no number) $\times 0 = -5$.

3. $3 + 0 = 3$ and $0 - (-3) = 0 + 3 = 3$.

▶ Note that *division* by zero *is the excluded operation.* □

When we introduced subtraction where the result was not positive, we in turn introduced negative numbers. This was necessary since the results could not be expressed as positive numbers. When we divide one integer by another, it often happens that the results are not integers. *The name* **rational number** *is given to any number which can be expressed as the division of one integer by another (not zero).* For example, $\frac{2}{3}$, $\frac{8}{4}$, and $\frac{1097}{431}$ are rational numbers. We shall later learn of numbers which cannot be expressed in this way (although we have already met a few: $\sqrt{2}$, for example).

Division of signed numbers with a calculator follows the same pattern established in the preceding sections. The following example illustrates typical cases.

EXAMPLE F   To evaluate $(+18) \div (+3)$ on a calculator, enter

18 $\boxed{\div}$ 3 $\boxed{=}$ to get the result 6

To evaluate $(-18) \div (-3)$, enter

18 $\boxed{+/-}$ $\boxed{\div}$ 3 $\boxed{+/-}$ $\boxed{=}$ to get 6

To evaluate $(+18) \div (-3)$, enter

18 $\boxed{\div}$ 3 $\boxed{+/-}$ $\boxed{=}$ to get -6 □

Exercises 3-4

In Exercises 1 through 20, evaluate the given expressions by performing the indicated operations.

1. $(+24) \div (+6)$ **2.** $(+32) \div (+4)$ **3.** $(+27) \div (-3)$ **4.** $(-48) \div (+16)$

5. $\frac{-22}{+2}$ **6.** $\frac{+20}{-4}$ -5 **7.** $\frac{-60}{-4}$ **8.** $\frac{-52}{-13}$ 4

9. $0 \div (-6)$ ϕ **10.** $0 \div (+3)$ ϕ **11.** $\frac{-16}{0}$ *undef*   **12.** $\frac{+14}{0}$ *undef*

13. $\frac{(+8)(-4)}{+2}$ **14.** $\frac{(-6)(+3)}{-9}$ **15.** $\frac{(-5)(+15)}{(+25)(-1)}$ **16.** $\frac{(-36)(+12)}{(-4)(-27)}$ -432 $+108$

-4

17. $\dfrac{(-3)(-24)}{(-2)(-12)}$ **18.** $\dfrac{(-6)(+4)}{(+3)(+8)}$ **19.** $\dfrac{(+10)(-6)(-14)}{(-21)(+8)}$ **20.** $\dfrac{(-22)(+8)(+18)}{(+6)(-44)}$

In Exercises 21 through 28, set up the given problems in terms of the division of signed numbers. Then perform the operations.

21. In business, debits are often expressed as negative numbers. If a debit of $3380 is divided equally among 26 partners, express the debit for each one.

22. A fluid leaks from a tank at the rate of 3 gal each day. How long does it take to lose 75 gal from leakage?

23. A parachutist descends 1000 ft at a constant rate for 4 min. What is his change in altitude each minute?

24. An integrated circuit voltage of 0.6 V decreases by 0.3 V/s. How long does it take to reach a voltage of -0.9 V?

25. The temperature of a coolant used in a nuclear reactor is dropping by 2°C each day. At this rate, how long would it take for the temperature to drop by 10°C?

26. For a certain type of lens, if the object is 4 cm to the left of the lens and the image is 12 cm and also to the left, these distances are expressed as $+4$ cm and -12 cm. The magnification is the negative of the image distance divided by the object distance. If the magnification is positive, the image is erect with respect to the object. Otherwise it is inverted. Determine the magnification and whether or not the image is erect.

27. In designing an instrument approach to an airport, an FAA specialist requires that airplanes descend from 3000 ft at the rate of 500 ft/min for $3\frac{1}{2}$ min. For a missed approach, the plane must then climb 2000 ft/min for 2 min. If a plane makes an approach from 3000 ft and then climbs for a missed approach, what is its altitude?

28. An energy management program results in additional expenses of $2400 each year for 4 years and savings of $3000 each year for 8 years. What is the average annual net gain or loss during the 12-year period?

In Exercises 29 through 36, use a scientific calculator to evaluate the given expressions.

29. $(+14482) \div (-26)$ **30.** $(-3293) \div (89)$ **31.** $\dfrac{(+26)(-12)}{(+13)(-3)}$ **32.** $(-63) \div (-3)$

33. $(+63) \div (-3)$ **34.** $\dfrac{(+2)(+3)(+4)}{(-2)(-6)}$ **35.** $\dfrac{(-8)(-2)(0)}{(-7)(+3)}$ **36.** $\dfrac{12}{0}$ (note display)

3-5 Order of Operations

The preceding sections of this chapter involved the basic operations with signed numbers. Section 3-2 covered addition and subtraction, Section 3-3 covered multiplication and powers, and Section 3-4 covered division of signed numbers. In this section, we will consider expressions involving combinations of those operations. In evaluating such expressions, it is extremely important to know that *there is an order in which the operations are performed*. If we consider the expression $(+4)+(-3)(-4) - (+2^3) \div 4$, the correct result is 14, but incorrect answers of -3, -6, 5, and -54 can be easily obtained if we disregard the order in which the operations must be performed. Here is the order:

1. Evaluate any expressions enclosed within parentheses or other symbols of grouping.
2. Evaluate all powers.
3. Evaluate all multiplications and divisions as they occur in order from left to right.
4. Evaluate all additions and subtractions as they occur in order from left to right.

EXAMPLE A

To correctly evaluate $(-12) \div (+3) - (-2)(+5)$, *we must first perform the division and then the multiplication. **We do not perform the subtraction until these other operations are completed.*** After dividing and multiplying we get

$$(-4) - (-10)$$

We can now do the subtraction by changing to unsigned numbers as follows:

$$(-4) + (+10) = -4 + 10 = 6 \qquad \square$$

In Example A there were no powers or expressions enclosed within parentheses so the order of operations simply required that the division and multiplication preceded the subtraction. The next example includes powers.

EXAMPLE B

To correctly evaluate $(+2) + (+2)^3 - (+6)^2 \div (+3)$, we note that the expression involves addition, subtraction, two powers, and division. *The order of operations indicates that the powers be evaluated first*, so we get

$$(+2) + (+8) - (+36) \div (+3)$$

The operation of division is the next priority; dividing leads to

$$(+2) + (+8) - (+12)$$

Finally, we perform the addition and subtraction by converting to unsigned numbers as follows:

$$(+2) + (+8) - (+12) = 2 + 8 - 12 = -2 \qquad \square$$

EXAMPLE C

In the evaluation of $[(+2) - (+5)] + (+3)^2(+4)$ *our order of operations requires that we begin with the expression enclosed within* **brackets.** We evaluate $(+2) - (+5)$ to get $2 - 5 = -3$, and the problem is now reduced to

$$(-3) + (+3)^2(+4)$$

After expressions enclosed within grouping symbols, the next priority involves powers, so the expression is further simplified to

$$(-3) + (+9)(+4)$$

The third priority here is the multiplication which, when performed, gives

$$(-3) + (+36)$$

Finally, we add to get $-3 + 36 = 33$. Note that this example has parentheses nested within other grouping symbols (brackets). When this occurs, we should begin with the innermost set of parentheses and then proceed to work outward. ☐

The greatest number of errors in evaluating expressions probably occurs when dealing with minus signs. Some of these common difficulties are illustrated in the following examples.

EXAMPLE D In the expression

$$-(-7) - (-2)^4 - (-6)(+2)$$

▶ *there are sometimes difficulties with the $-(-7)$ since there is nothing before it.* We can look at this in either one of two ways. First, we could simply place a zero before it, and we see that it amounts to the subtraction of -7. The other way is to treat the minus sign in front as -1. It then becomes a factor in a multiplication. We therefore get

$$-(-7) = 0 - (-7) \qquad \text{or} \qquad -(-7) = (-1)(-7)$$

▶ Another difficulty arises when it is noticed that there is a minus sign before $(-2)^4$. *There is a tendency to make $-(-2)^4$ equal to $+(+2)^4$, but this is not true.* Remember, we must evaluate the $(-2)^4$ first, and it equals $+16$. Thus we have the subtraction of 16, not the addition of 16 as $+(+2)^4$ would suggest. We therefore get:

$$-(-7) - (-2)^4 - (-6)(+2) = +7 - (+16) - (-12)$$
$$= 7 - 16 + 12$$
$$= 3 \qquad ☐$$

We have already seen that parentheses group numbers and those numbers must be considered first. In addition to numbers enclosed within parentheses or other grouping symbols, we must also give top priority to numbers which are grouped as parts of a fraction as in the next example. In effect, the fraction bar serves as a grouping symbol.

EXAMPLE E In evaluating the expression

$$\frac{5}{2-3} + \frac{(-4)^2}{-2}$$

we see that we must first subtract 3 from 2 in the first denominator before we can perform the division. This does not violate our basic order of operations since the $2 - 3$ is grouped very specifically by being the denominator of the fraction. It is a step that must be made prior to performing the basic division.

▶ As for $\frac{(-4)^2}{-2}$, *there is often a tendency to make this* $\frac{(+4)^2}{+2}$, *but again this is not true.* We must remember the meaning of $(-4)^2$ as $(-4)(-4)$. Therefore, we should evaluate the power before we try to evaluate the fraction. Thus,

$$\frac{5}{2-3} + \frac{(-4)^2}{-2} = \frac{5}{-1} + \frac{(-4)^2}{-2} = \frac{5}{-1} + \frac{+16}{-2} = -5 + (-8)$$
$$= -5 - 8$$
$$= -13 \qquad \square$$

Example F further illustrates the order of operations which has been outlined in this section.

EXAMPLE F

$$\overset{\text{multiplication}}{(+7) - (-3)(+4) - \frac{(-6)(+1)}{-2}} = (+7) - (-12) - (+3)$$

multiplication and division

$$= 7 + 12 - 3$$
$$= 16$$

$$\overset{\text{power}}{3(+7) - (-5)^2 - \frac{12}{(-2)(-1)}} = 3(+7) - (+25) - \frac{12}{(-2)(-1)}$$

multiplication | multiplication and division

$$= (+21) - (+25) - (+6)$$
$$= 21 - 25 - 6$$
$$= -10$$

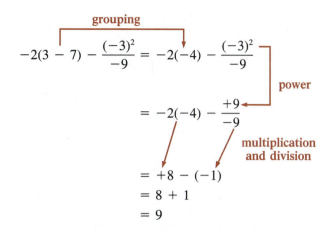

We note in the second illustration that $3(+7)$ was treated as $(+3)(+7)$, and in the third illustration $-2(-4)$ was treated as $(-2)(-4)$. Also, note in the third illustration that we performed the subtraction within the parentheses before multiplying. □

EXAMPLE G During a certain chemical reaction, the temperature of a mixture starts at 20°C, then drops 5°C/h for 3 h, and then rises 4°C/h for 7 h. Representing decreasing temperatures by negative numbers and increasing temperatures by positive numbers, we can express the resulting temperature as

$$20°C + (-5°C/h)(3\ h) + (4°C/h)(7\ h)$$

start   temperature drop   temperature rise

$$= 20°C - 15°C + 28°C = 33°C$$ □

Different calculators have different priorities in the way the operations are performed. In Section 1-11 it was stated that some calculators simply execute the operations in the order in which they occur. Fortunately, most scientific calculators are designed to follow the order of operations described in this section. If you want to test your calculator to see if it follows the correct order of operations, evaluate

$$(+4) + (-3)(-4) - (+2^3) \div 4$$

A typical scientific calculator will evaluate this expression with the following sequence of keys:

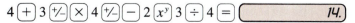

The result of 14 indicates that the calculator is using the *algebraic operating system* (AOS), which conforms to our order of operations. In any event, you should clearly understand how your calculator handles the operations if you plan to use it effectively.

Computer programmers know that different programming languages incorporate different orders for executing operations. The language BASIC, for example, does use the same order of operations described in this section, whereas a language called APL executes operations from right to left in the order they occur. Whatever system is used, it is imperative that the order of operations be known.

Exercises 3-5

In Exercises 1 through 28, evaluate the given expressions.

1. $(+4) - (-8)(+2)$

2. $(-6) + (-3)(-9)$

3. $(-9) + \dfrac{-10}{+2}$

4. $(-1) + \dfrac{+18}{-9}$

5. $(-1)(+7) + (-6)(+7)$

6. $(-5)(+8) - \dfrac{+15}{-3}$

7. $\dfrac{-22}{+2} - (+5)(-2)$

8. $(+3)(-10) + (-12)(+5)$

9. $(-8)(+9) - \dfrac{-9}{+3}$

10. $\dfrac{-60}{+4} + (-1)(-9)$

11. $\dfrac{-18}{+2} - \dfrac{+24}{-6}$

12. $\dfrac{+40}{-5} - \dfrac{-90}{+15}$

13. $(-7)(+15) - (-1)^2$

14. $(-6)(+4) - (-2)^2$

15. $-(+4) - (-4)^2$

16. $-(-9) + (-5)^3$

17. $-2(-9) - (-1)^4 + \dfrac{-18}{+6}$

18. $-3(+7) + (-3)^3 + \dfrac{-48}{+4}$

19. $5(-7) + \dfrac{-16}{+8} - (-2)^6$

20. $(+15) - (-2)^4 + (-6)(-5)$

21. $(+7)(-1) - \dfrac{+8}{-4} - 3(-8)$

22. $(-6)(-4) - \dfrac{(-6)(-1)}{-2} + 2(-4)$

23. $(+8)(0) - \dfrac{(+7)(-1)}{+1} - (-2)^2$

24. $\dfrac{-10}{-2} - \dfrac{(0)(-6)}{-2} - (-4)^2$

25. $\dfrac{7 + 1}{2} - (-2) - 2(-3)^2$

26. $(6 + 1)(-1) - 7(-1)^6 - \dfrac{-15}{-5}$

27. $\dfrac{18}{1 - 10} + \dfrac{(-4)^2}{-2} - (-8)$

28. $\dfrac{(-8)^2}{5 - 9} - (-1)^2 + 4(-9)$

In Exercises 29 through 36, set up the given problems in terms of operations on signed numbers and then solve. Unless otherwise indicated, all numbers are approximate.

29. A plane descends 1000 ft/min for 4 min and then ascends 500 ft/min for 6 min. Find its net change in altitude during this time.

30. In a certain city, the temperature rises 2°F per hour for 5 h, and then falls 3°F per hour for 10 h. Find the net change in temperature during this time.

31. The change in altitude of an object falling under gravity is determined by finding the product of -9.75 and the square of the time of the fall, and then dividing by 2. Determine the change in altitude (in meters) of an object that has fallen for 4.00 s. (The number 2 is exact.)

32. In converting a certain temperature from the Celsius scale to the Fahrenheit scale, a meteorologist must evaluate

$$\frac{(9)(27.0)}{5} + 32$$

What is the result? (The numbers 9, 5, and 32 are exact.)

33. When drilling for oil, a geologist calculates the amount of water which must be used to lighten mud. The amount of water (in gallons) is

$$\frac{102(18.05 - 15.02)}{15.02 - 8.33}$$

Find that amount.

34. A solar engineer needs to compute the total roof area of an industrial plant with an irregular shape. The area (in square feet) can be found from

$$(15)(20) + (32)^2 + (32 - 6)(10)$$

Find the area.

35. A holding tank containing 2000 gal of oil is drained at the rate of 120 gal/h for 6.0 h and then filled at the rate of 300 gal/h for 4.0 h. How much oil is in the tank now?

36. An electric current of 3.25 mA decreases at the rate of 0.15 mA/s for 12 s and then increases by 0.05 mA/s for 6.0 s. What is the resulting current?

In Exercises 37 through 44, use a scientific calculator to evaluate the given expressions.

37. $(+5)(+3) - (+2) + (+7) + (+8) \div (-2)$

38. $(-2)^3 - (+3)(+8) + (+5)(+10)$

39. $(+5) - (+3)(+6) + (-2)^2$

40. $\frac{(+15)}{10 - 7} - \frac{-2}{-1}$

41. $\frac{(-3)^4}{8 - 2} - \frac{36}{-4}$

42. $\frac{(+5)^4}{6 - 8} + \frac{9 - 4}{(+3)^2}$

43. $\frac{2 - 5}{6 - 18} - \frac{(+4)^3 - (+3)^4}{53 - 70}$

44. $(-2)^6 - (+3)^6 - \frac{(-2)^5 - (+2)^3}{(-2)(-2)(+5)}$

3-6 Review Exercises for Chapter 3

In Exercises 1 through 4, insert the proper sign ($>$ or $<$) between the given numbers.

1. $-3 \quad +1$

2. $-2 \quad -9$

3. $|-2| \quad -3$

4. $-|-4| \quad 3$

In Exercises 5 and 6, locate the approximate positions of the given numbers on a number scale.

5. $+2.5; \quad -\frac{5}{4}$

6. $\frac{3}{7}; \quad -\frac{2}{9}$

In Exercises 7 and 8, find the absolute value of each of the given numbers.

7. $\frac{2}{3}; \quad -\frac{1}{4}$

8. $-15; \quad \frac{7}{4}$

In Exercises 9 through 40, evaluate the given expressions by performing the indicated operations.

9. $(+4) + (-6)$
10. $(-6) + (-2)$
11. $(-5) - (+8)$
12. $(-3) - (-7)$

13. $(-9)(+7)$
14. $(-8)(-12)$
15. $(-63) \div (-7)$
16. $(-72) \div (+9)$

17. $(-3) - (-9) + (+4)$
18. $(+10) + (-4) - (-7)$

19. $(+9) - (+2) - (-12)$
20. $(-6) - (+5) + (-8)$

21. $(-2)(-6)(+8)$
22. $(-1)(+4)(+7)$
23. $(+5)(+3)(-2)(+2)$
24. $(-10)(+6)(-2)(-3)$

25. $\dfrac{(-5)(-6)}{+3}$
26. $\dfrac{(-8)(+7)}{-14}$
27. $\dfrac{(-18)(+4)}{(+6)(+2)}$
28. $\dfrac{(+8)(-9)}{(-4)(-3)}$

29. $(-4)^4$
30. $(-7)^3$
31. $\dfrac{(-1)(-5)(+45)}{(+9)(+5)(+1)}$
32. $\dfrac{(-8)(-16)(+3)}{(-2)(+6)(-4)}$

33. $(-2)(+4) - \dfrac{(-6)}{(+2)} - (-5)$

34. $-(-2)(-5) + \dfrac{(-9)}{(+3)} - \dfrac{(-16)}{(+4)}$

35. $\dfrac{(-2)(-3)}{-6} + (-4)(+2) - \dfrac{-8}{+4}$

36. $(-5) - (-2)(-5) + \dfrac{+4}{-2}$

37. $-(-7) - (-6)(+2) - (-3)^2$

38. $(-4) - \dfrac{-8}{+2} - (-2)^4$

39. $-\dfrac{(-2)(-3)}{-6} + (-4)(+2) - 3^2$

40. $(-5) + \dfrac{-14}{-2} - 2^4 - (-1)^2$

In Exercises 41 through 56, solve the given problems.

41. A car starts with 6 qt of oil. If it consumes 3 qt of oil and 2 qt are added, how much oil is left?

42. An astronomer writes a program and represents the year 2000 A.D. as +2000. How would she represent the year that is 5000 years before 2000 A.D.?

43. The lowest floor of a house is 4 ft below ground level, and the rooftop is 22 ft above ground level. Find the distance from the lowest floor to the rooftop.

44. If 3 s after the launching of a spacecraft is shown as +3 s, how should the instant 10 s *before* blastoff be shown? (When watching this type of event on television, note the use of signed numbers.)

45. The boiling point of mercury is about 357°C, and its melting point is about −39°C. What is the difference between the boiling and melting points of mercury?

46. Death Valley is 276 ft below sea level and Mt. Whitney is 14,502 ft above sea level. What is the vertical distance from the bottom of Death Valley to the top of Mt. Whitney?

47. In a golf tournament, a golfer is 2 strokes over par in the first round, 1 stroke under par in the second round, 3 strokes under par in the third round, and 2 strokes under par for the fourth round. What is his net score relative to par for the tournament? (If you watch these events on television, you will see signed numbers used in calculating scores.)

48. In three successive months in a certain area, the cost of living decreases 0.1%, increases 0.3%, and decreases 0.4%. What is the net change in the cost of living for these three months?

49. An employee has an expense account of $800. If she spends $45 each week for 8 weeks, what will her balance be?

50. A machine shop owner is paying for a lathe purchased with a 3-year installment loan. If $240 has been paid each month for the past 36 months, how much was paid during the first 30 months? If the lathe sold for $5995, what is the total amount of interest that will be paid?

51. After reaching the bottom of a steep incline, a skier slows down at a rate of 0.7 m/s each second. What is the net change in her speed 10 s later?

52. A cylinder is used to compress water. The pressure (in lb/in.^{2}) required to compress 600 in.^{3} of water by 3.0 in.^{3} can be found by evaluating

$$\frac{(-330000)(-3.0)}{600}$$

Find that pressure.

53. The current (in amperes) in a certain electric circuit is found by evaluating

$$\frac{(+5.96) - (-0.795)(8.04)}{8.04 + 3.85}$$

Find the current.

54. In analyzing the weather, a meteorologist notes that on one day the temperature started at $-8°C$. The temperature dropped 2°C each day for 4 days and it then rose 3°C each day for 3 days. What was the temperature at the end of the time period?

55. A nurse records changes in a patient's temperature and finds an increase of 1.2°F, an increase of 0.4°F, and three consecutive decreases of 0.8°F. Express the net change in temperature.

56. A computer has 65,536 bytes of memory available. Data entries use 14,072 bytes, and a program uses 25,360 bytes. If 5080 bytes of the data are erased, how many bytes are now available?

In Exercises 57 through 64, use a scientific calculator to evaluate the given expressions.

57. $(-65)(-92)$

58. $(+14)(-37)$

59. $\dfrac{(-32)(-63)}{(-18)(+56)}$

60. $\dfrac{(+16)(-85)}{(-68)(-5)}$

61. $(-2) - (+3)(+4) + (+3)^2$

62. $(+8) + (-2)(-6) - (+2)^4$

63. $\dfrac{(-18)(+24)}{(+36)(-3)} - \dfrac{(+52)(+4)}{(-13)(-8)}$

64. $\dfrac{-(+54) - (-5)^4}{130 - 5} - \dfrac{180 - 16}{37 + 4}$

CHAPTER FOUR

Introduction to Algebra

See Appendix B for a computer program that converts temperature readings from Fahrenheit degrees to Celsius degrees.

Many employers require at least a basic knowledge of algebra, since it helps us to solve many problems that would otherwise remain unsolved. One important feature of algebra is its use of letters to represent unknown quantities. Consider the following two statements which describe the relationship between the Celsius and Fahrenheit temperature scales:

Verbal form: The Celsius temperature equals five-ninths of the Fahrenheit temperature minus thirty-two degrees.

Algebraic form: $C = \frac{5}{9}(F - 32°)$

In addition to being unclear or ambiguous, the verbal statement does not express the relationship with the clarity of the algebraic statement. Any calculations can be done with much greater ease if the algebraic expression is used instead of the verbal expression. Also, we shall see that the algebraic form enables us to solve some problems that could never be solved with the verbal form.

In this chapter we explore the meaning of algebra and some of the important terminology associated with it. We also examine the basic operations, although they will be more fully developed in later chapters.

In addition to the general theory of algebra, this chapter also includes applied problems. For example, since aviation weather reports give some temperatures in the Fahrenheit scale and others in the Celsius scale, a pilot may need to convert a reported temperature of 10°F to a temperature on the Celsius scale. Using the above algebraic form of the equation relating Celsius and Fahrenheit temperatures, we will show (in Section 4-2) how this can be done.

4-1 Literal Numbers and Formulas

Algebra is a generalization of arithmetic which uses letters to represent numbers. The various other symbols, such as the plus sign and equals sign, are used to represent the operations and relations among them.

If we wished to find the cost of a piece of sheet metal, we would multiply the area, say in square feet, by the cost of the metal per square foot. If the piece is rectangular, we find the area by multiplying the length by the width. Consequently, the cost equals the length (in feet) times the width (in feet) times the cost of the sheet metal (per square foot). Rather than writing out such statements as the one just given, we can write

$$C = l \cdot w \cdot c$$

where it is specified and understood that C is the total cost, l is the length, w is the width, and c is the cost per unit area of the metal. Note that C and c represent different quantities.

The algebraic equation $C = l \cdot w \cdot c$ is an example of a **formula.** The letters C, l, w, c are **literal symbols** (letters that represent numbers). These letters are called **variables** since they are literal symbols that can represent different values. Formulas sometimes contain **constants,** which are literal symbols representing values that do not vary. For example, $d = \frac{1}{2} g t^2$ might be a formula with variables d and t and the constant g. This means that d and t can assume different values, but g would be a fixed constant value.

A major advantage of a formula is its general use in many different cases. The formula $C = l \cdot w \cdot c$ can be applied to an infinite number of different situations. If we want to find the cost of a rectangular piece of sheet metal with specific dimensions, we need only **substitute** the appropriate numbers into the formula.

EXAMPLE A   If a rectangular piece of sheet metal is 5 ft long and 4 ft wide and costs $3/ft^2, we have

$$l = 5 \text{ ft} \qquad w = 4 \text{ ft} \qquad c = \$3/\text{ft}^2$$

The formula $C = l \cdot w \cdot c$ becomes, in this case,

substitute

$$C = 5 \cdot 4 \cdot 3 = \$60$$

It would cost $60 to purchase that piece. For another piece with different dimensions and cost we would still substitute into the same formula, for it is a general expression that can be used in many different cases.   □

In developing the language of algebra for the operations on literal symbols (*the letters which represent numbers*), we use certain procedures and terms in writing algebraic expressions. These procedures are adopted because they make statements and notation concise and convenient. As various topics are developed, we shall introduce many of these. Since they are used throughout mathematics, it is important that we understand the precise meaning of each term and procedure.

At this point we shall discuss basic notation and terminology involving **factors** (*quantities being multiplied*). If numbers represented by symbols are to be multiplied, the expression is written without the signs of multiplication. The symbols are simply placed adjacent to each other. *One of these may be a numerical constant, in which case it is normally written first and is called the* **numerical coefficient** *of the expression.*

EXAMPLE B   Instead of writing $C = l \cdot w \cdot c$ or $C = l \times w \times c$, the expression for the cost of the sheet metal given earlier would be written as

$$C = lwc \qquad lwc = l \cdot w \cdot c = l \times w \times c$$

If every case of soda in a shipment contains 12 bottles, then the total numbers of bottles b is

$$b = 12c \qquad 12c = 12 \cdot c$$

where c is the number of cases. Here 12 is the numerical coefficient of c.   □

We can use exponents with literal symbols the same way we did with numbers. We must remember that the exponent is written only next to the quantity which is being raised to the indicated power.

EXAMPLE C   Instead of writing aa, meaning a times a, we write a^2. We write b^3 rather than bbb. The product $xxxxx$ is written as x^5.

The expression ab^3 means a times the cube of b. **The symbol a is not to be cubed; only the symbol b is to be cubed.**

The expression $5x^2y^3$ means five times the square of x times the cube of y.

If we wish to group letters and raise the entire group to a power, we use parentheses: $(axy)^2$ means axy times axy.   □

In a given product, the quantities being multiplied are called **factors** *of the product, just as they are in arithmetic. In general, the quantities multiplying a factor are the* **coefficient** *of that factor.*

EXAMPLE D

In the expression lwc, the l, w, and c are factors. The coefficient of c is lw and the coefficient of lw is c.

In the expression $3ab^2$, the 3, a, and b^2 are factors. The coefficient of b^2 is $3a$ and the coefficient of ab^2 is 3.

factors
↓ ↓ ↓
$3\ a\ b^2$
— numerical coefficient
— coefficient of ab^2

In the expression $3ab^2$, we know that $b^2 = bb$. We may list the factors of this expression without the use of exponents as 3, a, b, b. ☐

Wherever mathematics may be used, we may use symbols to represent the quantities involved. A verbal statement must be translated into an algebraic expression so that it may be used in algebra. The following example illustrates several verbal statements and their equivalent algebraic formulas.

EXAMPLE E

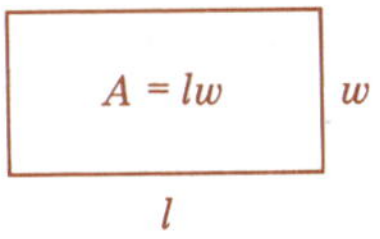

$A = lw$ w

l

FIGURE 4-1

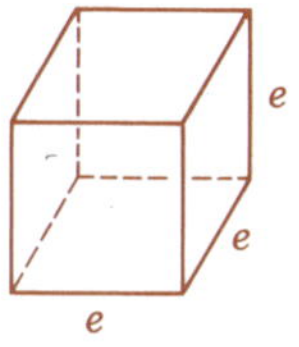

e

e

e

FIGURE 4-2

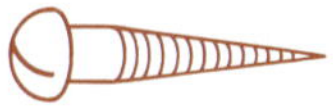

FIGURE 4-3

Verbal statement	Formula	Meaning of literal symbols
1. The area of a rectangle equals the length times the width.	$A = lw$	A is the area. l is the length. w is the width. (See Figure 4-1.)
2. The volume of a cube equals the cube of the length of one edge.	$V = e^3$	V is the volume. e is the length of an edge. (See Figure 4-2.)
3. The pitch of a screw thread times the number of threads per inch equals 1.	$pN = 1$	p is the pitch. N is the number of threads/in. (See Figure 4-3.)
4. The distance an object travels equals the average speed multiplied by the time of travel.	$d = rt$	d is the distance. r is the average speed. t is the elapsed time.
5. The simple interest earned on a principal equals the principal times the rate of interest times the time the money is invested.	$I = Prt$	I is the interest. P is the principal. r is the rate of interest. t is the time.
6. The voltage across a resistor in an electric circuit equals the current in the circuit times the resistance.	$V = IR$	V is the voltage. I is the current. R is the resistance.

☐

These formulas are valid for the given conditions. As mentioned before, if we wish to determine the result of using any of these formulas for specific values, we substitute these values for the letters in the formula and then calculate the result. The following example illustrates this in the case of two of these formulas.

EXAMPLE F

The volume of a cube 3.0 in. on an edge is found by substituting 3.0 for e in the formula $V = e^3$. We get

$$V = (3.0)^3 = 27 \text{ in.}^3$$

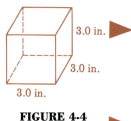

3.0 in.

3.0 in.

3.0 in.

FIGURE 4-4

If the cube had been 3.0 cm on an edge, the volume would have been 27 cm³. *We must be careful to attach the proper units to the result.*

If the current in a certain electric circuit is 3.0 A, the voltage across a 6.0-Ω resistor in the circuit is found by substituting 3.0 for I and 6.0 for R in the formula $V = IR$. We get

$$V = (3.0)(6.0) = 18 \text{ V}$$

Note that the letter V has a different meaning in each of these cases. In the formula $V = IR$, the literal symbol V represents voltage. But the symbol V in 18 V represents the symbol for volts, a unit of measure. Here we have a unit symbol which is the same as the literal symbol in a formula, and we must be careful to avoid confusion between these two uses. ☐

When selecting letters to be used as literal symbols, it is often helpful to choose letters that suggest the quantity being represented. The letter V is commonly used for voltage or volume. Also, t frequently represents time, d and s are often used for distance, r for a rate, and so on. It is also common to use the letters x, y, or z to represent unknown variable quantities, while the letters a, b, c, and k are often used to represent unknown constant quantities.

Exercises 4-1

In Exercises 1 through 8, the listed numbers are factors of a product. Determine the product of these listed factors.

1. a, b　　　　**2.** x, y, z　　　　**3.** x, x　　　　**4.** a, a, a, a

5. $2, w, w$　　　**6.** $6, a, a, c$　　　**7.** a, a, a, b, b　　　**8.** $3, 2, a, a, b, c, c, c$

In Exercises 9 through 16, identify the individual factors, without exponents, of the given algebraic expressions. (List them in the same way as shown for Exercises 1 through 8.)

9. bc　　　　**10.** $2ax$　　　　**11.** $7pqr$　　　　**12.** $3xyz$

13. i^2R　　　**14.** $17a^2b$　　　**15.** abc^3　　　**16.** $\pi r^2 h$

In Exercises 17 through 24, identify the coefficient of the factor that is listed second by examining the expression listed first.

17. $6x$; x **18.** $3s^2$; s^2 **19.** $2\pi r$; r **20.** $8a^2b$; b

21. $4\pi emr$; mr **22.** $qBLD$; D **23.** mr^2w^2; r^2 **24.** $36a^2cd$; a^2d

In Exercises 25 through 32, use the literal symbols listed at the end of each problem to translate the given statements into algebraic formulas.

25. A first number x equals four times a second number y. (x, y)

26. A first number equals the square of a second number. (x, y)

27. In a given length, the number of millimeters equals 10 times the number of centimeters. (m, c)

28. The total surface area A of a cube equals six times the square of the length of one of its edges e. (A, e)

29. The volume, in gallons, of a rectangular container equals 7.48 times the length times the width times the depth (these dimensions are measured in feet). (V, l, w, d) See Figure 4-5.

30. The heat H developed (per second) in a resistor in an electric circuit equals the product of the resistance R and the square of the current i in the circuit. (H, R, i)

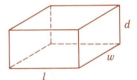

FIGURE 4-5

31. The distance d an object falls due to gravity equals $\frac{1}{2}$ times the acceleration g due to gravity times the square of the time of fall t. (d, g, t)

32. The rate of emission R of energy per unit area of the filament of an electric light bulb equals about 0.00002 times the fourth power of the thermodynamic temperature T of the filament. (R, T)

In Exercises 33 through 40, give the required formula.

33. The number N of feet in a distance of x miles.

34. The area A of a square of side s.

35. The number N of square feet in a rectangular area x yd by y yd.

36. The cost C of renting a masonry drill for t hours if the rental fee is \$3/h.

37. The number of bolts N in a shipment if there are n boxes of bolts and each box contains 24 bolts.

38. The number N of bits in a computer with x bytes if each byte consists of 8 bits.

39. The cost C of putting an edge strip around a square piece of wood of side s, if the strip cost c ¢/ft.

40. The total area A of N square tiles if each tile has sides of length s cm.

In Exercises 41 through 52, evaluate the required formula for the given values.

41. A missile travels at the rate of 6000 mi/h. Find the distance it will travel in 15 min. (See Example E and note the units given.)

42. Find the simple interest on \$500 at an interest rate of 8% for 3 years. (See Example E.)

43. Find the voltage in a circuit in which the current is 0.075 A and the resistance is 20 Ω. (See Example E.)

44. Find the volume of a cube for which the edge is 4.3 cm. (See Example E.)

45. A holding tank for industrial waste is a rectangular container 4.0 ft by 8.0 ft by 3.0 ft. Find the capacity of this tank in gallons. (See Exercise 29.)

46. Find the heat (in joules) developed each second in a 12-Ω resistor if the current in the circuit is 2.0 A. (See Exercise 30.)

$$H = Ri^2$$
$$= (12)(2)^2$$
$$= 12 \cdot 4 = 48 \text{ joules (J)}$$

47. In Exercise 31, find d if $g = 32.2$ ft/s² and $t = 2.25$ s.   **48.** In Exercise 28, find A if $e = 1.25$ m.

49. In Exercise 33, find N if $x = 2.5$.   **50.** In Exercise 35, find N if $x = 3.00$ and $y = 4.00$.

51. In Exercise 37, find N if $n = 144$.   **52.** In Exercise 40, find A if $N = 320$ and $s = 20.0$.

In Exercises 53 through 60, use a scientific calculator to find the indicated quantity.

53. In Exercise 29, find V if $l = 4.56$ m, $w = 2.77$ m, and $d = 2.09$ m.

54. In Exercise 28, find A if $e = 2.375$ m.

55. In Exercise 31, find d if $g = 32.2$ ft/s² and $t = 3.27$ s.

56. In Exercise 30, find the heat H (in joules) if the resistance is 32.2 Ω and the current is 3.45 A.

57. If an edge strip costs exactly 85 ¢/ft, find the cost of putting an edge strip around a square piece of wood of side 12.0 ft. (See Exercise 39.)

58. If a megabyte is a million bytes, how many bits are in a 20-megabyte computer? (See Exercise 38.)

59. Find the area of a square in square meters if $s = 67.35$ m.

60. Find the volume of a cube in cubic centimeters if $e = 102.4$ cm.

4-2 Basic Algebraic Expressions

In the last section we discussed literal symbols and we used them in some simple formulas. The discussion, examples, and exercises of that section used only the multiplication of literal symbols. We must often add, subtract, multiply, divide, and perform other operations on literal symbols. This section presents an introduction to these operations. Here we establish the need for such operations, and we also introduce some new terms necessary for later topics.

Many algebraic expressions require the addition or subtraction of algebraic quantities. See the following example.

EXAMPLE A   The conversion from the Celsius temperature scale to the Kelvin temperature scale (used in thermodynamics) is given by the formula

$$K = C + 273$$

The perimeter p of a triangle with sides a, b, and c is given by the formula

$$p = a + b + c$$

The net profit P from a sale of articles from which the income is I and which originally cost C is

$$P = I - C$$

The total surface area A of a rectangular box (the total area of all six faces) of dimensions l, w, and h is

$$A = 2lw + 2lh + 2wh \qquad \text{(See Figure 4-6.)} \qquad \square$$

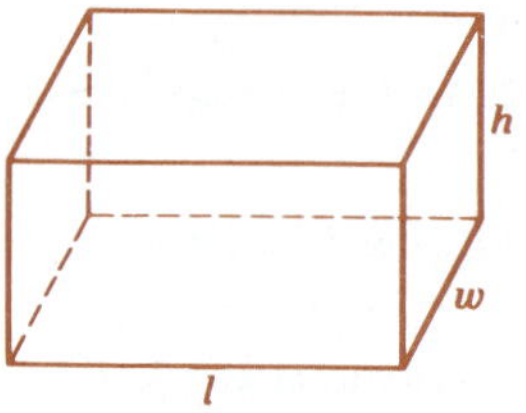

FIGURE 4-6

In a sum or difference, the quantities added or subtracted are called **terms.** *If the literal part of one term is the same as the literal part of another term, then the terms are called* **like terms** *or* **similar terms.** Recall that literal symbols are letters which represent numbers. Like terms or similar terms therefore have the same literal parts, as in $3xy$, $-xy$, and $4.2xy$. The following example illustrates like terms.

EXAMPLE B The algebraic expression $x + 3xy - 2x$ contains three terms: x, $3xy$, and $-2x$. The terms x and $-2x$ are like terms, since the literal parts are the same. The $3xy$ term is not like the x and $-2x$ terms since the *factor y* makes the literal part different.

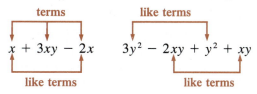

The expression $3y^2 - 2xy + y^2 + xy$ has four terms: $3y^2$, $-2xy$, y^2, and xy. The terms $3y^2$ and y^2 are like terms, and the terms $-2xy$ and xy are like terms.

The algebraic expression $3ab^2 + 4ab + 7b^2a$ has $3ab^2$ and $7b^2a$ as like terms. The term $4ab$ is not like $3ab^2$, but $7b^2a$ is like $3ab^2$, since $b^2a = ab^2$. ☐

Some algebraic expressions have only one term while others have two, some have three, and so on. Specific names have been given to some of these expressions so that they can be referred to with ease.

A **monomial** *is an expression with one term. A* **binomial** *is an expression with two terms. A* **trinomial** *is an expression with three terms. A* **multinomial** *is an expression with more than one term.* Thus binomials and trinomials can also be considered as multinomials.

EXAMPLE C The following table gives several examples of monomials, binomials, trinomials, and multinomials.

Monomials	Binomials	Multinomials	
		Trinomials	Other multinomials
a^2x	$a^2x + y$	$a^2x + y + x$	$a^2x + y + x + 3a$
$3abc$	$3abc + bc$	$3abc + bc - ab$	$3abc + bc - ab + a - 2b$
a	$a + c^2$	$a + c^2 + a^2$	$a + c^2 + a^2 - 2b + a^3 - c$

☐

We now consider multiplication. Recall that in Section 4-1 we discussed products such as *lwc* and $3x^2y$. However, if we want to multiply two algebraic expressions, at least one of which is itself the sum of terms, or the difference of terms, we must be able to indicate that the sum or difference is to be considered as one of the factors. *For this purpose, we use parentheses to group the terms together.* The following example illustrates this use of parentheses.

EXAMPLE D   A supplier of electronics components charges *p* dollars for a certain transformer. If it cost *c* dollars to make, the profit on the sale of a transformer is $p - c$ dollars. If *n* transformers are sold, the total profit *P* is

$$P = n(\overset{\displaystyle \longleftarrow p - c \text{ grouped by parentheses}}{p - c})$$

▶ *We do not write this as $np - c$;* this arrangement would indicate that *n* and *p* are to be multiplied and *c* is to be subtracted from the product *np*.

To indicate that the sum of two numbers *a* and *b* is to be multiplied by the difference between two other numbers *c* and *d*, we write

$$(a + b)(c - d) \quad \text{grouped by parentheses}$$

This expression is a *one-term* expression, although the *factors $a + b$* and $c - d$ each have two terms.   □

In algebra, division of one number by another is most commonly designated as a fraction. That is, $a \div b$ would be written as

$$\frac{a}{b}$$

or occasionally as a/b. Here *a* is the numerator and *b* is the denominator. The following example illustrates that division is necessary for the algebraic solution of certain problems.

EXAMPLE E   The frequency *f* of a radio wave equals its velocity *v* divided by its wavelength *l*. This is written as

$$f = \frac{v}{l}$$

In a particular type of electric circuitry, the combined resistance *C* of two resistors *r* and *R* is found to be

$$C = \frac{rR}{r + R}$$

Here we note that the numerator is the product of *r* and *R* and the denominator is the sum of *r* and *R*.   □

Another important type of algebraic expression involves the square root of a number. In Chapter 1 we defined the square root of a number to be one of two equal factors of the number. Algebraically this can be shown as

$$\sqrt{n} = a \quad \text{if} \quad a^2 = n \quad \text{and } a \text{ is nonnegative} \qquad (4\text{-}1)$$

(At this time we are considering only nonnegative numbers for n and a.) Here again we see the use of a concise algebraic expression.

Numerous technical applications in algebra involve the use of square roots. The following example illustrates one of them.

EXAMPLE F   One application of square roots is the determination of the time it takes an object to fall. The approximate time (in seconds) that it takes for an object to fall due to the influence of gravity equals the square root of the quotient of the distance (in feet) fallen divided by 16. The formula for this is

$$t = \sqrt{\frac{d}{16}} \qquad \qquad \square$$

As we noted in Section 4-1, an algebraic expression can be evaluated for a specific set of values by substituting the proper values. The following example illustrates the evaluation of some expressions given in earlier examples.

EXAMPLE G   In Example A, if $C = 20°$, the temperature in the Kelvin scale can be found as follows:

$$K = C + 273 \qquad \text{general expression}$$
$$= 20 + 273 \qquad \text{substitute 20 for } C$$
$$= 293 \qquad \text{result for specific case of } C = 20°C$$

In Example D, if $p = \$210$, $c = \$120$, and $n = 6$, the profit on these sales, $P = n(p - c)$, becomes

$$P = 6(210 - 120) = 6(90) = \$540$$

In Example F, if an object falls 144 ft, the time of fall,

$$t = \sqrt{\frac{d}{16}}$$

becomes

$$t = \sqrt{\frac{144}{16}} = \sqrt{9.0} = 3.0 \text{ s} \qquad \qquad \square$$

EXAMPLE H   The formula $C = \frac{5}{9}(F - 32°)$ relates temperatures on the Celsius and Fahrenheit scales. Since aviation weather reports give some temperatures in the Fahrenheit scale and others in the Celsius scale, a pilot may need to convert a reported temperature of 10°F to the Celsius scale. Using the above formula we get

$$C = \frac{5}{9}(10° - 32°) = \frac{5}{9}(-22°) = -12°$$

That is, 10°F is equivalent to $-12°$C (rounded to the nearest degree).   □

Exercises 4-2

In Exercises 1 through 4, identify the terms of the given algebraic expressions.

1. $x^2 + 4xy - 7x$ **2.** $a + 2ab - \dfrac{a}{b}$ **3.** $12 - 5xy + 7x - \dfrac{x}{8}$ **4.** $3(x + y) - 6a + 3x$

In Exercises 5 through 12, identify the like terms in the given expressions.

5. $3x - 7y + 2x$ **6.** $9a - 2b + 5b$ **7.** $3 + x + 5x - 3y$ **8.** $a + ab + 2a - 3b$

9. $5m^2 - 8mn + m^2n - mn + \dfrac{m}{n}$ **10.** $6R - bR + 3R^2 - 5bR$

11. $6(x - y) + (x + y) - 3(x - y)$ **12.** $5(a - b) - 3a + 7b + (a - b)$

In Exercises 13 through 16, indicate the multiplication of the given factors. (Only indicate the multiplication—do not multiply.)

13. $6, a, a - x$ **14.** $3, x^2, a - x^2$ **15.** $x^2, a - x, a + x$ **16.** $3, a^2, a^2 + x^2, a^2 - x^2$

In Exercises 17 through 20, express the indicated divisions as fractions.

17. $2 \div 5a$ **18.** $x^2 \div a$ **19.** $6 \div (a - b)$ **20.** $(3x^2 - 2x + 5) \div (x + 2)$

In Exercises 21 through 24, evaluate the formulas by using the given values.

21. $K = C + 273$, for $C = 55$ **22.** $P = n(p - c)$, for $n = 8$, $p = \$250$, $c = \$140$

23. $t = \sqrt{\dfrac{d}{16}}$, for $d = 64$ **24.** $C = \dfrac{5}{9}(F - 32°)$, for $F = 68°$

In Exercises 25 through 36, write the required formula.

25. A section x ft long is cut from a 50.0-ft piece of cable. Express the length L of the remaining cable in terms of x.

26. A rectangular plate has length l and width w. Write a formula for its perimeter, which is the sum of the lengths of all four sides.

27. A shipping carton is a rectangular box with square ends. If the square ends have sides of length x and the overall length of the box is l, write a formula for the total area A (of the six sides) of the surface in terms of x and l. See Figure 4-7.

FIGURE 4-7

28. A company deposits an amount of principal p in an account that pays simple interest at the rate r. The total amount A in the account after a period of time t can be found by adding the principal to the interest earned. Find the formula for A in terms of p, r, and t.

29. The voltage V across an electric circuit equals the current I times the resistance in the circuit. If the resistance in a certain circuit is the sum of resistances R and r, write the formula for the voltage.

30. The midrange m of a collection of numbers is one-half the sum of the smallest number s and the largest number l. Express this statement as a formula.

31. The total surface area A of a right circular cylinder equals the product of $2\pi r$ and the sum of r and h, where r is the radius and h the height of the cylinder. Express this statement as a formula. See Figure 4-8.

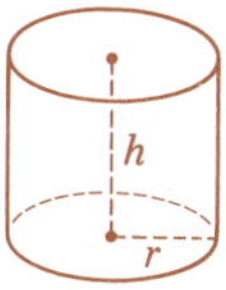

32. The value V of a machine depreciates so that its value after t years is its original value p divided by the sum of t and 1. Express this statement as a formula.

FIGURE 4-8

33. The arithmetic mean A of n numbers is the sum of these numbers divided by n. Express a formula for the arithmetic mean of the numbers a, b, c, d, and e.

34. The efficiency E of an engine is defined as the difference of the heat input I and heat output P (subtract P from I) divided by the heat input. Express this statement as a formula.

35. The time T for one complete oscillation of a pendulum equals approximately 6.28 times the square root of the quotient of the length l of the pendulum and the acceleration g due to gravity. Find the resulting formula.

36. According to the Pythagorean theorem, the length c of the hypotenuse of a right triangle equals the square root of the sum of the squares of the other two sides, a and b. Express the formula for the Pythagorean theorem. See Figure 4-9.

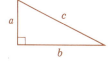

In Exercises 37 through 44, evaluate the required formula for the given values.

FIGURE 4-9

37. In Exercise 25, find the length of the remaining piece if 27.4 ft is cut off.

38. In Exercise 26, find the perimeter if the length and width are 29.2 cm and 37.9 cm, respectively.

39. In Exercise 27, find the total surface area if $x = 2.91$ ft and $l = 5.23$ ft.

40. In Exercise 29, find the voltage if $I = 0.00427$ A, $R = 82.6\ \Omega$, and $r = 1.08\ \Omega$.

41. In Exercise 33, find the arithmetic mean of the numbers $a = 2$, $b = 5$, $c = 3$, $d = 17$, and $e = 8$.

42. What is the value of a \$4000 machine after it has depreciated for 3 years? (Use the information from Exercise 32.)

43. In Exercise 35, find the time T (in seconds) if $l = 5.26$ ft and $g = 32.2$ ft/s².

44. In Exercise 34, find the efficiency if the heat input is 21,500 J (joules) and the heat output is 7600 J.

In Exercises 45 through 52, use a scientific calculator to evaluate the required formula for the given values.

45. In Exercise 27, find the total area if $x = 1.238$ m and $l = 1.348$ m.

46. What is the value of a \$5297 machine after it has depreciated for 4.08 years? (Use the information from Exercise 32.)

47. In Exercise 29, find the voltage if $I = 0.03672$ A, $R = 82.64\ \Omega$, and $r = 2.359\ \Omega$.

48. A quality control engineer tests five radios used for aircraft navigation. The voltage levels at which they failed are 18.21 V, 21.33 V, 17.67 V, 15.48 V, and 20.29 V. Find the arithmetic mean of these five values. (See Exercise 33.)

49. In Exercise 35, find the time T (in seconds) if $l = 4.087$ ft and $g = 32.21$ ft/s².

50. In Exercise 34, find the efficiency if the heat input is 4372 J and the heat output is 1305 J.

51. In Exercise 27, find the total area in *square meters* if $x = 2.357$ ft and $l = 4.928$ ft.

52. Find the length c in Exercise 36 if $a = 1.237$ m and $b = 2.344$ m.

4-3 **Introduction to Algebraic Operations**

In this section we consider the addition and subtraction of algebraic terms and then we demonstrate certain operations involving multiplication and division. These operations are necessary in the development of the following sections. However, a more complete and detailed discussion of algebraic operations is needed for later chapters, so we again consider this material in Chapter 7.

If a multinomial contains like as well as unlike terms, we may **simplify** *it by combining the like terms*. The like terms are combined by adding their numerical coefficients. The result is expressed as a sum of unlike terms. The following example illustrates the simplification of a multinomial.

EXAMPLE A   In the multinomial

$$3a^2 - 2ab + a^2 - b + 5ab$$

the $3a^2$ and a^2 are like terms and the $-2ab$ and $5ab$ are like terms. Adding $3a^2$ and a^2, we obtain $4a^2$. **(The coefficient of a^2 is 1, which is not written.)** When we subtract $2ab$ from $5ab$, we obtain $3ab$. Thus the given multinomial simplifies to

$$4a^2 + 3ab - b$$

Since there are no like terms in this result, no further simplification is possible and we now know that

$$3a^2 - 2ab + a^2 - b + 5ab = 4a^2 + 3ab - b \qquad \square$$

When we perform addition, we are making use of the basic **axiom** (*an accepted but unproved statement*) that the order of addition of terms does not matter. This axiom, known as the **commutative law of addition,** was mentioned in Chapter 1 in connection with the addition of numbers. Since algebraic terms represent numbers, it also holds in algebra. In general, it can be stated for numbers a and b as

$$a + b = b + a \qquad\qquad (4\text{-}2)$$

The following example illustrates the way we actually use this axiom to simplify a multinomial.

EXAMPLE B   The multinomial

$$x^2 + 5x - 3 + 4x + 5$$

is equivalent to

$$x^2 + 5x + 4x - 3 + 5$$

if we interchange the third and fourth terms according to the commutative law. We then combine the like terms $5x$ and $4x$, obtaining

$$x^2 + 9x + 2$$

In general, we do not have to rewrite such expressions in order to add. However, the fact that we can add like terms is due to this axiom. Thus

$$x^2 + 5x - 3 + 4x + 5 = x^2 + 9x + 2$$

Let us now consider multiplication, restricting our consideration at this time to the multiplying of certain multinomials by monomials. To perform this operation, we use the **distributive law**, which was introduced in Chapter 1. In algebraic form, the distributive law for numbers a, b, and c is

$$a(b + c) = ab + ac \tag{4-3}$$

Illustrations of the use of the distributive law are shown in the following example.

EXAMPLE C

$$3(x + y) = 3x + 3y$$

$$x(3 + 2y) = x(3) + x(2y) = 3x + 2xy$$
$$2x(x - y) = 2xx + 2x(-y) = 2x^2 + (-2xy) = 2x^2 - 2xy$$
$$3ab(2a - 5bc) = (3ab)(2a) + (3ab)(-5bc)$$
$$= 6aab - 15abbc = 6a^2b - 15ab^2c$$

Note that if the sign within the parentheses is a minus sign, the multiplication is equivalent to multiplying by a negative number.

Following is an example of the simplification of an expression involving the distributive law and the combining of like terms.

EXAMPLE D   Simplify: $3s + 2(s + 2t) - s(7 - 2t)$

First removing the parentheses by using the distributive law and then combining like terms, we have the following solution:

$$3s + 2(s + 2t) - s(7 - 2t) = 3s + 2s + 2(2t) - s(7) - s(-2t)$$
$$= 3s + 2s + 4t - 7s + 2st$$
$$= -2s + 4t + 2st$$

The like terms we combined were $3s$, $2s$, and $-7s$. The $4t$ term and the $2st$ term are not similar to the other terms.   □

In multiplying these expressions, we have also been using the **commutative law of multiplication,** *which states that the order of multiplication does not matter.* This was introduced in Chapter 1 and can be shown algebraically as

$$ab = ba \tag{4-4}$$

Other basic axioms we have been using in these operations are the **associative laws of addition and multiplication.** *These state that the order of grouping terms does not matter.* Algebraically they are shown as

$$a + (b + c) = (a + b) + c \tag{4-5}$$

and

$$a(bc) = (ab)c \tag{4-6}$$

In Example B the associative law of addition is used; in Example C the associative law of multiplication is used. Both are used in Example D.

In the division of algebraic expressions, the principle to be followed is the same as in the simplification of an arithmetic fraction:

Factors common to both numerator and denominator can be divided out.

We followed this principle in simplifying fractions in Chapter 1 and expressions involving signed numbers in Chapter 3. In Example E, division involving algebraic factors is illustrated.

EXAMPLE E   In simplifying the algebraic fraction

$$\frac{3xy}{ax}$$

we note that both the numerator and the denominator have a factor of x. Thus, dividing out this common factor of x, we have

$$\frac{3xy}{ax} = \frac{3\cancel{x}y}{a\cancel{x}} = \frac{3y}{a}$$

In dividing out the factor of x, we are stating that $x/x = 1$, and we know that multiplying the remaining factors by 1 does not change the resulting product.

In the following illustration, the resulting *denominator* is 1 and is not included in the final result.

$$\frac{3xy}{x} = \frac{3\cancel{x}y}{\cancel{x}} = \frac{3y}{1} = 3y$$

However, if the resulting *numerator* is 1, it must be retained, as in the following illustration:

$$\frac{x}{ax} = \frac{\cancel{x}}{a\cancel{x}} = \frac{1}{a}$$

When you become more familiar with this operation, you will not have to write in the step where the indicated cancellations are shown. □

In Example F, additional illustrations of dividing out factors common to both the numerator and the denominator of a fraction are shown.

EXAMPLE F

$$\frac{a}{10abc} = \frac{1}{10bc} \qquad \text{common factor of } a$$

$$\frac{4x}{6xyz} = \frac{2}{3yz} \qquad \text{common factor of } 2x$$

$$\frac{3x^2y}{x} = \frac{3xxy}{x} = \frac{3xy}{1} = 3xy \qquad \text{common factor of } x$$

$$\frac{5a^2bc}{10ab} = \frac{5aabc}{2(5)ab} = \frac{ac}{2} \qquad \text{common factor of } 5ab$$

$$\frac{3(x+y)^2}{(x+y)} = 3(x+y) \qquad \text{common factor of } (x+y) \qquad □$$

We conclude this discussion with an example that combines the various operations presented in the section.

EXAMPLE G   Three circuits are connected in series so that the total voltage is found by adding the three component voltages. Find the total voltage (in volts) by evaluating

$$3a + \frac{2ab}{b} + 5a(3 - 2b)$$

for $a = -5$ and $b = 7$.

It would be possible to substitute immediately in order to evaluate the expression. However, it almost always makes the evaluation simpler if *the basic algebraic operations are performed first*. Thus

$$3a + \frac{2ab}{b} + 5a(3 - 2b) = 3a + 2a + 15a - 10ab$$

$$= 20a - 10ab$$

Evaluating for $a = -5$ and $b = 7$, we get

$$20a - 10ab = 20(-5) - 10(-5)(7)$$

$$= -100 + 350$$

$$= 250$$

The total voltage is 250 **V**.

Exercises 4-3

In Exercises 1 through 36, simplify the given expressions by performing any indicated multiplications and divisions and then combining like terms.

1. $x + y + 5x$ **2.** $3x + 5xy - x$ **3.** $3a - 5b^2 + 4a + b^2$

4. $2x - y + 5x - y$ **5.** $2s + 3t - s + 4s$ **6.** $-5m + 2n - m + 8m$

7. $5(x + y)$ **8.** $6(s + t)$ **9.** $3(2a - b)$

10. $7(x - 3y)$ **11.** $\dfrac{7xy}{2y}$ **12.** $\dfrac{4abc}{5b}$

13. $\dfrac{x}{2x^2y}$ **14.** $\dfrac{3a}{6ab^2}$ **15.** $3(a - 5b) + 4a$

16. $4(x + 2y) - 6x$ **17.** $2(3x - y) + 3(2x + y)$ **18.** $4a(3 - b) - b(a - 4)$

19. $4R + 5(R + 3S) - S(3 - 5R)$ **20.** $6xy(3x - 5yz) + 12xyz$

21. $-2(x - y) - 3(x - y) - 4xy$ **22.** $-5(ab + bc) - 4(ac - ab)$

23. $\dfrac{ax}{a} + 3x$ **24.** $\dfrac{2y}{2} - 4y$ **25.** $\dfrac{2rs^2}{s} - 3r(5 - s)$ **26.** $\dfrac{3a^2b}{ab} - 2a(3 - b)$

27. $\dfrac{6abc^2}{2c} + c(ab - c) + b(ac - 1)$ **28.** $\dfrac{12mx^3}{3x} - \dfrac{2km^2x^2}{km} - 3mx(2 - x)$

29. $2(ax - 2) + \dfrac{9a^2x}{3a} + \dfrac{x}{x}$ **30.** $5(3a^2 - 2) - \dfrac{15a^3}{3a} + \dfrac{a^2}{a^2}$

31. $\dfrac{16a(x-y)}{4a} - \dfrac{3x^2 y}{3xy} + 2(x-3y)$

32. $\dfrac{24x^2(b+c)}{6x} + \dfrac{2b^2 x}{2b} - \dfrac{50c^3 x^2}{25c^2 x}$

33. $\dfrac{18x(x-y)}{9x} + \dfrac{6x(x-y)}{x-y} + \dfrac{ab^2}{ab}$

34. $\dfrac{12a(a-b)^2}{6(a-b)} + 5b(a+c) - 7ab$

35. $\dfrac{22x^2(x+y)}{11x} - \dfrac{4ax^2}{4a} + \dfrac{28x(x+y)}{7}$

36. $\dfrac{15a(x+y)^2}{5(x+y)^2} - \dfrac{9ax^2 y}{3ax} + \dfrac{3a^2}{a} - \dfrac{12x^2 y^3}{3xy^2}$

In Exercises 37 through 44, express the appropriate expressions in simplest form.

37. One transmitter antenna is measured to be $(x-a)$ ft long and another is measured to be $(x+2a)$ yd long. What is the sum of their lengths in feet?

38. Two newspapers print $5x+1000$ and $4x-2000$ papers, respectively, each day. What is the sum of their daily printings?

39. Find an expression for the total cost of $2a+1$ resistors costing 5¢ each and $a+b$ resistors costing 10¢ each.

40. One car goes 30 km/h for t hours, and a second car goes 40 km/h for $t+2$ hours. Find the expression for the sum of the distances traveled by the two cars.

41. One rectangular plate has dimensions x and y, another $x+2$ and y, and a third 3 and $y+5$. What is the expression for the sum of the areas? See Figure 4-10.

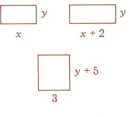

42. While making support beams, a carpenter cuts x ft from each of six 20-ft boards, y ft from each of four 15-ft boards, and x ft from a y-ft board. Write an expression for the total length of the remaining pieces.

FIGURE 4-10

43. A computer processes $12axy$ accounts and its average rate is $4x$ accounts per minute. How long did it take to process these accounts?

44. When $5(x-y)$ resistors are connected in series, they produce a total resistance of $125(x-y)^2$ ohms. What is the arithmetic mean of the resistance levels? (See Exercise 33 of Section 4-2.)

In Exercises 45 through 56, evaluate the required expression for the given values. Perform all algebraic simplifications before evaluating.

45. $a(x+3) + a(a-2)$, for $a=5$ and $x=2$.

46. $4x(x+2y) - x(x-y)$, for $x=6$ and $y=4$.

47. $\dfrac{12axy}{4a} + 3(y+8)$, for $x=2$ and $y=6$.

48. $\dfrac{18abc^2}{6abc} + a(ac-b) + \dfrac{5a^2 bc}{a}$, for $a=3$, $b=4$, and $c=6$.

49. In Exercise 39, let $a=5$ and let $b=7$. Find the total cost of the resistors.

50. In Exercise 40, let $t=2$ and find the sum of the distances traveled by the two cars.

51. In Exercise 41, let $x=12.3$ cm and let $y=14.9$ cm. Find the sum of the areas of the three plates.

52. In Exercise 44, let $x=8$ and let $y=2$. Find the arithmetic mean of the resistance levels.

53. $\dfrac{3a^2(b+2)}{a} + 7a(b+5) - 9ab$, for $a=18$ and $b=6$

54. $\dfrac{x^2 y(a+y)}{xy} + 7x(y-2a)$, for $a=8$, $x=13$, and $y=29$

55. $\dfrac{x^2(a-b)^2}{(a-b)} - a(x^2 - y) + x(x + y^2)$, for $a = 12$, $b = 9$, $x = 14$, and $y = 8$.

56. $\dfrac{18x^2y^2}{3xy} - \dfrac{5(x-y)^2}{(x-y)} + \dfrac{20x^2(x+y)}{4x^2}$, for $x = 3$ and $y = 7$.

In Exercises 57 through 60, first simplify algebraically and then use a scientific calculator to evaluate the expression for the given values.

57. $P(1 + rt) - P$, for $P = 68{,}720$, $r = 0.0825$, and $t = 1.75$.

58. $\frac{1}{2}(6.350 - mv^2) + mv^2 - 4.930$, for $m = 2.734$ and $v = 9.766$.

59. $x(x + y) + y(x - y) + \dfrac{x(x-y)^2}{x-y}$, for $x = 3.759$ and $y = 8.337$.

60. $\dfrac{18ax^2y}{3xy} - 3a(x - y)$, for $a = 1.173$, $x = 2.308$, and $y = 8.775$.

Chapter 4 Formulas

$$\sqrt{n} = a \quad \text{if} \quad a^2 = n \quad \text{and} \quad a \text{ is nonnegative.} \tag{4-1}$$

$$a + b = b + a \tag{4-2}$$

$$a(b + c) = ab + ac \tag{4-3}$$

$$ab = ba \tag{4-4}$$

$$a + (b + c) = (a + b) + c \tag{4-5}$$

$$a(bc) = (ab)c \tag{4-6}$$

4-4 Review Exercises for Chapter 4

In Exercises 1 through 4, identify the like terms in the given expressions.

1. $6ax - 7a + 5a + 8x$ **2.** $3ax^2 + 4ax + 5a^2x - 6ax^2$

3. $5(a - b) + 6ab - 7(a - b)$ **4.** $8x^2 - 2x^2 + 5x - 2$

In Exercises 5 through 32, simplify the given expressions.

5. $a + 3b + 6a - 7b$ **6.** $6a^2b - a^2b + 5ab - 2a^2b$ **7.** $6x(2a - 3b)$

8. $-3a(8 - 5b)$ **9.** $3(x + y) - 2y$ **10.** $x(2x + 6y) - x^2$

11. $\dfrac{a^2c}{a} - 2ac + c$ **12.** $5xy - \dfrac{xy^2}{y} - y^2$ **13.** $5(a + b) - 3(a + 2b)$

14. $a(a - x) + a(3a + 2x)$

15. $3(a + 2b) + 2b + b(3 - a)$

16. $x(2x - y) + y(2x + 1)$

17. $\dfrac{ax^2}{x} + a(x + 2)$

18. $\dfrac{by}{b} - (7 - y)$

19. $\dfrac{abx}{b} - \dfrac{a^2x}{a} + a(2 - x)$

20. $\dfrac{a^2c}{c} + \dfrac{a^3c}{ac} + 2a(a + c)$

21. $xy - 2x(y - 1) + 3x$

22. $4ac - a(2c - 3) - 5a$

23. $ab + 2(a - b) + a(2b - 1)$

24. $x(y + 1) + 6(x - y) - y(x - 2)$

25. $\dfrac{a}{a} + 2(a + 1) + a(3 + x)$

26. $\dfrac{2ax}{a} + \dfrac{x^2}{x} + \dfrac{3b^2x}{b^2} - 8x$

27. $\dfrac{6x(a + b)}{x} + 3(2a - b) + \dfrac{ab}{b}$

28. $\dfrac{c^2x^2}{cx} + c(2x - 3) + \dfrac{2cx}{x}$

29. $\dfrac{36a(a + b)}{12a} + \dfrac{2a(x + y)}{x + y} + \dfrac{b^2}{b}$

30. $\dfrac{abc^2}{b} + c(ac - b) + a(3c^2 + b)$

31. $\dfrac{9x(a - y)^2}{3(a - y)} - 2y(x - a) + ax$

32. $m(2a - n) + \dfrac{mn^3}{n^2} - a(2m - a)$

In Exercises 33 through 36, in the given expression identify the coefficient of the factor that is shown in parentheses.

33. $6a^2x$; (x)

34. $3abc^2$; (c^2)

35. $12abx$; (ab)

36. $15mn^2p$; (mn^2)

In Exercises 37 through 44, evaluate the given expression for the indicated values. Perform all algebraic operations before substituting.

37. $x(2x - 1) + 8x^2$ for $x = -2$

38. $2y(y - 3) - y^2$ for $y = -3$

39. $3x^2 - 5y + 2x^2 + 8y$ for $x = 4$ and $y = 6$

40. $18xy + 6x^2 + x(2y + x)$ for $x = 5$ and $y = 11$

41. $2(x - 1) + 8x^2 + x(8 - 7x)$ for $x = -4$

42. $15(x + 2a) + 6a(2 + x) + 8ax$ for $a = 7$ and $x = 9$

43. $\dfrac{a^2bc^3}{a} + c^2(2abc + 5)$ for $a = 6$, $b = 7$, and $c = 12$

44. $\dfrac{a^2(x - y)^2}{(x - y)} - x(a^2 + y) + y(a^2 + 5x)$ for $x = 8$ and $y = -6$

In Exercises 45 through 60, determine the appropriate formula in simplest form.

45. A number x is equal to 7 less than its square.

46. The sum of two numbers x and y equals twice their product.

47. In electricity, the capacitance C of a capacitor is equal to the charge Q on either plate, divided by the voltage V across the capacitor.

48. The temperature F in degrees Fahrenheit is equal to 32 more than the product of $\frac{9}{5}$ and the temperature C in degrees Celsius.

49. The three sides of an equilateral triangle are equal. Find the formula for the perimeter p of an equilateral triangle of side s. See Figure 4-11.

50. What is the formula for the sum S of the areas of a square of side x and a rectangle with sides x and y? See Figure 4-12.

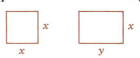

FIGURE 4-12

FIGURE 4-11

51. The increase I in population of a certain city was 1000 times the square of the number of years t after a census was taken. Express this statement as a formula.

52. In the theory of relativity, Einstein found that the equivalent energy E of a mass equals the product of the mass m and the square of the speed of light c. Write this as a formula.

53. A company has a cash surplus which is invested in two separate accounts. If x dollars are invested at rate r and $3x$ dollars are invested at rate $r + 0.02$, find a formula for the total interest I earned by the two accounts.

54. A firm pays 20 employees w dollars per week and 10 employees $w - 10$ dollars per week. Find an expression for the total wages T paid these employees each week.

55. A machine mount requires n bolts costing 5¢ each, $n + 2$ bolts costing 10¢ each, and $n + 5$ bolts costing 25¢ each. Find a formula for the total cost C (in cents) of these bolts.

56. A certain wire contracts when it is cooled. If the temperature drops from a to b, this wire's length decreases by an amount equal to the difference in temperatures divided by 1000. Find a formula for the decrease L in the length.

57. One printer prints x characters per second for t seconds. A second printer prints $2x$ characters per second for $t + 10$ seconds, while a third printer prints $5x + 8$ characters per second for $6t$ seconds. Find a formula for the total number N of characters printed by the three printers.

58. Three liquids are mixed to form a cleaning solution. One tank provides x gallons per hour for t hours. A second tank provides $2x$ gallons per hour for $t + 2$ hours, while a third tank provides $3x$ gallons per hour for $t - 3$ hours. Find a formula for the total amount A provided by all three tanks.

59. The electric current in a circuit equals the voltage divided by the resistance. The voltage in one circuit is expressed as $2x(t - 1)$ and the voltage in a second circuit is $6(8 - 3t)$. What is the sum I of the currents in these circuits, given that the resistances are x and 3, respectively?

60. A computer is rented for c dollars per day during the first week, and $2c$ dollars per day after the first week. Find a formula for the total cost C if the computer is rented for t days and t is greater than 7.

In Exercises 61 through 68, evaluate the indicated expressions.

61. In Exercise 49, find p given that $s = 7.0$ in.

62. In Exercise 50, find S given that $x = 45.7$ ft and $y = 76.3$ ft.

63. In Exercise 53, find the total annual interest I, given that $x = 40,000$ and $r = 0.06$.

64. In Exercise 53, find the total annual interest I, given that $x = 22,500$ and $r = 0.055$.

65. In Exercise 55, find C, given that (a) $n = 8$ and (b) $n = 12$.

66. In Exercise 57, find N, given that $x = 20$ and $t = 180$ s.

67. In Exercise 57, find N if $x = 18$ and t is the number of seconds in one hour.

68. In Exercise 60, let $c = 15$ and find the rental cost for (a) 9 days and (b) 30 days.

In Exercises 69 through 72, use a calculator to solve the given problems. All whole numbers are exact and all others are approximate.

69. Evaluate $\dfrac{11.2x^2y}{3x^2} + 4.3y - \dfrac{1.5ay^2}{ay}$ for $y = 7.9326$.

70. Evaluate the expression from Exercise 40 for $x = 152.7$ and $y = 348.3$

71. Evaluate the expression from Exercise 41 for $x = -19.734$.

72. Evaluate the expression from Exercise 44 for $x = 13.35$ and $y \doteq -32.48$.

Simple Equations and Inequalities

An **equation** *is a mathematical statement that two algebraic expressions are equal.* Equations can include a wide variety of different algebraic expressions, but this chapter includes equations in which only one literal symbol is the unknown. *By* **solving** *an equation, we mean finding a value of the unknown which, upon substitution, makes the two sides of the equation equal.* An equation is solved if the unknown is alone on one side of the equation while the other side consists of a single number or expression. We begin with simple equations, and we then proceed to consider formulas, inequalities, ratios, proportions, and variations.

Among the applied problems included in this chapter is the following. In the study of forces acting on the wing of an aircraft, the equation $M = R(L - x)$ is used. For a certain analysis it is necessary to solve for L. Later in the chapter we will show how this is done.

5-1 Solving a Simple Equation

We have stated that an equation is a mathematical statement that two algebraic expressions are equal. We begin this section with an example of an equation, and we then proceed to study methods of solving equations.

EXAMPLE A $3x - 4 = 2x - 1$ is an equation. In this equation it is understood that the variable x represents the unknown. If we substitute $x = 3$ into each side of the equation, we obtain $5 = 5$. Thus $x = 3$ is the solution of the equation. This is often stated as "3 **satisfies** the equation." If we try any other value of x, we will find that the two sides of the equation are not equal. If we substitute zero for x, for example, we get $-4 = -1$ so that $x = 0$ is not a solution of the equation.   □

It is the objective of this chapter to show methods which can be used to solve equations such as the one in Example A. At this time we will work only with equations with unknowns raised to the first power. The methods of solution are based on the basic operations discussed in Chapter 4. Later chapters show how other types of equations are solved. This chapter also includes brief coverage of simple inequalities.

A large part of algebra is devoted to finding solutions to various equations. In particular, technical and scientific work involves equations of all kinds, and their solutions are needed for finding or confirming scientific information. The following example illustrates some statements about numbers that lead to equations.

EXAMPLE B   In each case we shall let x be the unknown number.

Statement	*Equation*
1. A number decreased by seven equals twelve.	$x - 7 = 12$
2. Nine added to a number equals three.	$x + 9 = 3$
3. A number divided by two equals seven.	$\dfrac{x}{2} = 7$
4. Five times a number equals twenty.	$5x = 20$
5. Three less than twice a number equals one more than the number.	$2x - 3 = 1 + x$
6. Sixteen less three times a number equals two added to four times the number.	$16 - 3x = 2 + 4x$

□

Since an equation has both sides equal for the appropriate value of x, we have these basic properties of equations:

1. If the same number is added to each side, the two sides are still equal.
2. If the same number is subtracted from each side, the two sides remain equal to each other.
3. If the two sides are multiplied by the same nonzero number, they remain equal.
4. If the two sides are divided by the same number (provided it is not zero), they remain equal.

By performing these operations, we may isolate x on one side of the equation and the other numerical quantities on the other side. The following examples illustrate the method used to solve the equations of Example B.

EXAMPLE C Solve $x - 7 = 12$.

We note that when we add 7 to the left side, only x will remain. By adding 7 to both sides, we obtain

$$x - 7 + 7 = 12 + 7$$
$$x = 19$$

because $-7 + 7 = 0$ and $12 + 7 = 19$. The solution is $x = 19$. ☐

It is always wise to *check* the solution by following these simple steps:

1. *In the original equation,* substitute the solution for the unknown.
2. Simplify both sides of the equation by performing the indicated operations.
3. If both sides of the equation simplify to the same number, then the solution is correct. (If both sides simplify to different numbers, the solution is wrong.)

The solution of $x = 19$ in Example C can be checked by following the preceding steps. Upon substitution of 19 for x in the original equation we get $19 - 7 = 12$, which simplifies to $12 = 12$ so that the solution of $x = 19$ is correct.

EXAMPLE D Solve $x + 9 = 3$.

When we subtract 9 from each side, only x will remain on the left side. Performing this operation, we have

$$x + 9 - 9 = 3 - 9$$
$$x = -6$$

Substitution into the original equation gives $3 = 3$ so that the solution $x = -6$ checks. ☐

EXAMPLE E   **1.** Solve $\frac{x}{2} = 7$.

If we multiply the left side by 2, it becomes $2x/2$, or x. Thus, by multiplying both sides by 2, we obtain the solution

$$2\left(\frac{x}{2}\right) = 2(7)$$
$$x = 14$$

Substitution of the solution $x = 14$ into the original equation gives $7 = 7$, which means that it checks.

2. Solve $5x = 20$.

If we divide both sides by 5, the left side then becomes x and the solution is

$$\frac{5x}{5} = \frac{20}{5}$$
$$x = 4$$

Substitution of the solution $x = 4$ in the equation gives $20 = 20$.   □

Each of the equations in the above examples was solved by one of the basic operations. However, many equations require that we perform several operations to obtain the solution. In solving such equations it is usually a good strategy to follow these steps:

1. Eliminate any fractions by multiplying both sides of the equation by the lowest common denominator.

2. Use multiplication to remove any grouping symbols.

3. For each side of the equation, combine any like terms.

4. Use addition or subtraction to isolate the unknown on one side of the equation.

5. If the unknown is multiplied by a number, remove that number by division.

6. Check the solution by substituting it for the unknown in the original equation. (See the preceding list of steps.)

EXAMPLE F   Solve $2x - 3 = 1 + x$.

By combining terms containing x on one side and the other terms on the other side, we may isolate x and thereby find the solution. The solution proceeds as follows:

$$2x - 3 = 1 + x \qquad \text{original equation}$$
$$2x - 3 - x = 1 + x - x \qquad \text{subtract } x \text{ from each side}$$
$$x - 3 = 1 \qquad \text{combine like terms}$$
$$x - 3 + 3 = 1 + 3 \qquad \text{add 3 to each side}$$
$$x = 4 \qquad \text{simplify}$$

Since the last equation gives the desired value of x directly, the solution is complete.

▶ Check: Substituting *in the original equation* we obtain

$$2(4) - 3 = 1 + 4, \quad \text{or} \quad 5 = 5 \qquad \square$$

EXAMPLE G   In trying to find the center of gravity of a beam, an engineer needs to solve the equation $16 - 3x = 2 + 4x$. Solve for x.

The solution proceeds as follows:

$$16 - 3x = 2 + 4x \qquad \text{original equation}$$
$$16 - 3x + 3x = 2 + 4x + 3x \qquad \text{add } 3x \text{ to each side}$$
$$16 = 2 + 7x \qquad \text{combine like terms}$$
$$16 - 2 = 2 + 7x - 2 \qquad \text{subtract 2 from each side}$$
$$14 = 7x \qquad \text{combine like terms}$$
$$\frac{14}{7} = \frac{7x}{7} \qquad \text{divide each side by 7}$$
$$2 = x \quad \text{or} \quad x = 2 \qquad \text{simplify}$$

Check: Substituting 2 for x in the original equation, we obtain $10 = 10$. □

When you are more familiar with solving equations like these, you can do the step of adding or subtracting a term, or dividing by a factor, by inspection. See the following examples.

EXAMPLE H   Solve $3(2x + 1) = 2x - 9$.

We may proceed as follows:

$$3(2x + 1) = 2x - 9 \qquad \text{original equation}$$
$$6x + 3 = 2x - 9 \qquad \text{perform multiplication to remove parentheses}$$
$$4x + 3 = -9 \qquad \text{subtract } 2x \text{ from each side (by inspection)}$$
$$4x = -12 \qquad \text{subtract 3 from each side (by inspection)}$$
$$x = -3 \qquad \text{divide each side by 4 (by inspection)}$$

Check: Substituting in the original equation, we obtain

$$3(2(-3) + 1) = 2(-3) - 9$$
$$3(-6 + 1) = -6 - 9 \quad \text{or} \quad -15 = -15 \qquad \square$$

EXAMPLE I In solving

$$\frac{x}{3} - \frac{1}{2} = \frac{x}{2} - \frac{5}{2}$$

for x, it would be wise to first multiply each side by the common denominator 6 so that fractions are no longer involved. In so doing, we must be sure to multiply each term by 6. We get

$2x - 3 = 3x - 15$	multiply each side by 6
$-3 = x - 15$	subtract $2x$ from each side
$12 = x$ or $x = 12$	add 15 to each side

Substituting 12 for x in the original equation, we obtain $\frac{7}{2} = \frac{7}{2}$, so that the solution of $x = 12$ is verified.   ☐

Earlier, we listed steps for a solution strategy, but the order of those steps can vary. No specific order is required and most equations can be solved by any one of several different procedures. In the following example, the solution does not follow the order suggested by the solution strategy listed earlier, but it does yield the correct result.

EXAMPLE J In solving

$$500(2x - 3) = 200(x - 2) + 500$$

for x, we can work with smaller numbers by first dividing both sides by 100 to get

$$5(2x - 3) = 2(x - 2) + 5$$

We can now proceed with the order suggested by the earlier solution strategy.

$10x - 15 = 2x - 4 + 5$	remove parentheses
$10x - 15 = 2x + 1$	combine
$8x - 15 = 1$	subtract $2x$ from each side
$8x = 16$	add 15 to each side
$x = 2$	divide each side by 8

We check the solution by substituting 2 for x in the original equation. Since the result simplifes to $500 = 500$, the solution is verified.   ☐

Now we come to two final points about equations: First, *some equations are valid for* <u>all</u> *values of x. Such equations are referred to as* **identities.** *The equations we have been solving, those which are true only for specific values of the unknown, are often called* **conditional equations.** *Second, some equations are not valid for any values of x, since no values of the un-*

known will satisfy them. *Such equations are called* **contradictions.** *The following example illustrates an identity and a contradiction.*

EXAMPLE K

1. $2x + 2 = 2(x + 1)$ is true for all values of x. Proceeding with the solution, we would arrive at $0 = 0$, which is true for any value of x. This equation is an *identity*.

2. $x + 1 = x + 2$ is an equation, although in attempting to solve it we subtract x from both sides to get $1 = 2$, which *is not true for any value of x*. This equation has no solution and is a *contradiction*. ☐

Exercises 5-1

In Exercises 1 through 32, solve the given equations for x. Check the solution of each equation by substituting the value found in the equation.

1. $x + 3 = 5$

2. $x - 4 = 7$

3. $x + 5 = 8$

4. $x + 6 = 15$

5. $2x = 14$

6. $3x = 21$

7. $\dfrac{x}{7} = 5$

8. $\dfrac{x}{3} = 8$

9. $\dfrac{x}{6} = -3$

10. $\dfrac{x}{5} = -8$

11. $x + 6 = 2$

12. $x - 1 = -7$

13. $2x - 5 = 13$

14. $3x + 20 = 5$

15. $4x + 11 = 3$

16. $3x - 2 = 16$

17. $3 + 6x = 24 - x$

18. $5 - x = 8x - 13$

19. $3x - 6 = x - 18$

20. $14 - 2x = 5x + 7$

21. $2(x - 1) = x - 3$

22. $3(x + 2) = 2x + 11$

23. $\dfrac{x}{2} = x - 4$

24. $\dfrac{x}{3} = 12 - x$

25. $\dfrac{x}{3} + \dfrac{x}{4} = \dfrac{7}{12}$

26. $\dfrac{x}{2} + 2 = \dfrac{x}{5} + 5$

27. $3(x - 1) + x = 2(x - 1)$

28. $3(2 - 3x) = 7(x + 1) - 33$

29. $6x - 1 = 3(x - 2) + 6$

30. $4(2x + 1) = 3(x - 3) - 7$

31. $0.2(3 - x) - 0.3(x - 2) = 0.1x$

32. $0.2(3x - 4) = -0.2(3 - 4x) + 0.7$

In Exercises 33 through 44, set up the appropriate equation from the given statement and then solve.

33. Twice a number is 18.

34. Three more than a number is 11.

35. Half a number is 16.

36. A number less six equals seven.

37. Five more than a number is eight.

38. Three times a number is thirty-three.

39. A number divided by four is seventeen.

40. Six more than twice a number equals four times the number.

41. Three less than five times a number is the number less eleven. $5x - 3 = x - 11$

42. Half a number equals twice the difference of the number and twelve.

43. Five less four times a number equals one-fifth of the number.

44. Seven less six times a number equals the number less fourteen.

In Exercises 45 through 48 answer the given questions.

45. Distinguish between $x + 5 = x + 7$ and $x + 5 = x + 5$ as to the type of solution.

46. Distinguish between $x + 1 = x - 5$ and $x + 1 = 5 - x$ as to the type of solution.

47. Which of the following are identities?
  (a) $2(x + 1) + 1 = 3 + 2x$ (b) $3(x + 1) = 1 + 3x$ (c) $5x = 4(x + 1) - x$
  (d) $2x - 3 = 3(x - 1) - x$ (e) $4x - 5 = 4(x - 5)$

48. Which of the following equations are contradictions?
  (a) $x - 7 = x - 8$ (b) $1 - x = 1 + x$ (c) $2x + 2 = 2(x + 1)$
  (d) $3x - 6 = 3(x - 6)$ (e) $2x + 4 = 5 + 2x$

In Exercises 49 through 52, use a scientific calculator to solve the given equations for x. (Assume all numbers are approximate.)

49. $3.76 = 2.34x + 1.64$ **50.** $\dfrac{x}{0.813} = -2.798$

51. $1.26x - 3.65 = 2.75 - 3.14x$ **52.** $-2.54x = -3.12 + 7.28$

5-2 Simple Formulas and Literal Equations

In Section 5-1 we considered only equations containing the unknown x and other specific numbers. All the formulas included in Chapter 4 are actually equations, since they express equality of algebraic expressions. Most of these formulas contain more than one literal symbol. In this section we extend the methods of solving equations to formulas and other equations containing more than one literal symbol, as in the following example.

EXAMPLE A   In converting temperatures from the Celsius scale to the Kelvin scale, the equation

$$K = C + 273$$

is used. Since it is sometimes necessary to convert temperatures from the Kelvin scale to the Celsius scale, we want an expression for C. Solve for C.

Since the objective is to isolate C, we simply subtract 273 from each side of the equation to get

$$K - 273 = C \text{ or } C = K - 273$$

which is the required solution.   □

In Example A our result for C includes the literal symbol K instead of a single specific number, as in Section 4-1. Any equation containing literal symbols other than the unknown, or required symbol, is solved in the same general manner. In each case the result will include literal symbols and not only a specific number. Also, *many of the operations will be in terms of literal symbols*. Consider the following examples.

EXAMPLE B   In the equation $ay + b = 2c$, solve for b.

Since the object is to isolate b, we subtract ay from each side of the equation. This results in

$$b = 2c - ay$$

which is the required solution. □

EXAMPLE C   In the study of forces on the wing of an aircraft, the equation $M = R(L - x)$ is used. Solve for L.

$$M = R(L - x) \qquad \text{original equation}$$
$$M = RL - Rx \qquad \text{use distributive law}$$
$$M + Rx = RL \qquad \text{add } Rx \text{ to each side}$$
$$\frac{M + Rx}{R} = L \qquad \text{divide each side by } R$$

or

$$L = \frac{M + Rx}{R}$$

▶ This last equation gives the required solution. Note that **R cannot be divided out of the numerator and the denominator since it is not a factor of the entire numerator.** □

EXAMPLE D   In analyzing the transfer of heat through a wall, a solar engineer must solve the equation

$$q = \frac{KA(B - C)}{L}$$

for B. The solution proceeds as follows.

$$q = \frac{KA(B - C)}{L} \qquad \text{original equation}$$
$$qL = KA(B - C) \qquad \text{multiply each side by } L$$
$$qL = KAB - KAC \qquad \text{multiply to remove parentheses}$$
$$qL + KAC = KAB \qquad \text{add } KAC \text{ to each side}$$
$$\frac{qL + KAC}{KA} = B \qquad \text{divide each side by } KA$$

or

$$B = \frac{qL + KAC}{KA}$$

The last formula is the desired result. Note that KA cannot be divided out of the numerator and denominator since it is not a factor of the entire numerator. □

In many problems it is necessary to refer to two or more values of a quantity. For example, we may wish to represent the resistances in several different integrated circuits, representing resistance by the letter R. Instead of choosing a different letter for each resistance, we may use **subscripts** on the letter R for this purpose. Thus R_1 could be the resistance of the first circuit, R_2 the resistance of the second circuit, and so on. It must be remembered that R_1 *and* R_2 *are different literal numbers* just as x and y are different. Example E shows the solution of a literal equation involving subscripts.

EXAMPLE E Solve the equation $as_1 + cs_2 = 3a(s_1 + a)$ for s_1.
The solution proceeds as follows:

$$as_1 + cs_2 = 3a(s_1 + a)$$ original equation
$$as_1 + cs_2 = 3as_1 + 3a^2$$ remove parentheses
$$cs_2 = 2as_1 + 3a^2$$ subtract as_1 from each side
$$cs_2 - 3a^2 = 2as_1$$ subtract $3a^2$ from each side
$$\frac{cs_2 - 3a^2}{2a} = s_1$$ divide each side by $2a$

or

$$s_1 = \frac{cs_2 - 3a^2}{2a}$$

This last equation gives the required solution. ☐

EXAMPLE F In doing calculations with pulleys, the equation

$$L = 3.14(r_1 + r_2) + 2d$$

must be solved for r_1. The solution proceeds as follows:

$$L = 3.14(r_1 + r_2) + 2d$$ original equation
$$L = 3.14r_1 + 3.14r_2 + 2d$$ remove parentheses
$$L - 3.14r_2 - 2d = 3.14r_1$$ subtract $3.14r_2$ and $2d$ from each side
$$\frac{L - 3.14r_2 - 2d}{3.14} = r_1$$ divide each side by 3.14

or

$$r_1 = \frac{L - 3.14r_2 - 2d}{3.14}$$

The solution is shown in the last equation. ☐

It often happens that problems are presented in verbal form and they must be restated in an algebraic form before the solution is possible. In Section 5-4, we will consider general methods for converting verbal statements into equations, but the next example illustrates how a formula may be set up from a statement and then solved for one of its symbols.

EXAMPLE G One missile travels at a speed of v_1 mi/h for 3 h and another missile goes v_2 mi/h for $3 + t$ hours, and the total distance they travel is d. Solve the resulting formula for t.

If the first missile goes v_1 mi/h for 3 h, then it goes a distance of $3v_1$ mi. Similarly, the distance the second missile travels is $v_2(3 + t)$, so we have the total distance d as

$$d = 3v_1 + v_2(3 + t)$$

Solving this formula for t, we get the following:

$$d = 3v_1 + 3v_2 + v_2t \qquad \text{remove parentheses}$$

$$d - 3v_1 - 3v_2 = v_2t \qquad \text{subtract } 3v_1 \text{ and } 3v_2 \text{ from each side}$$

$$t = \frac{d - 3v_1 - 3v_2}{v_2} \qquad \text{divide each side by } v_2 \text{ and then switch sides}$$

The last formula is the desired result. ☐

Exercises 5-2

The formulas in Exercises 1 through 28 are used in the technical areas listed at the right. Solve for the indicated literal number.

1. $N = r(A - s)$, for r (engineering: stress)

2. $D = 2R(C - P)$, for C (business: depreciation)

3. $R_1L_2 = R_2L_1$, for R_2 (electricity)

4. $S = \dfrac{A - B}{A}$, for B (electricity: motors)

5. $v_2 = v_1 + at$, for v_1 (physics: motion)

6. $C = a + bx$, for a (economics: cost analysis)

7. $PV = RT$, for T (chemistry: gas law)

8. $E = IR$, for I (electricity)

9. $I = \dfrac{5300\ CE}{d^2}$, for C (atomic physics)

10. $E = \dfrac{mv^2}{2}$, for m (physics: energy)

11. $l = \dfrac{yd}{mR}$, for R (optics)

12. $P = \dfrac{N + 2}{D_0}$, for N (mechanics: gears)

13. $A = 180 - (B + C)$, for B (geometry)

14. $T_d = 3(T_2 - T_1)$, for T_1 (drilling for oil)

15. $R = \dfrac{CVL}{M}$, for L (water evaporation)

16. $r = \dfrac{g_2 - g_1}{L}$, for g_2 (surveying)

17. $D_p = \dfrac{MD_m}{P}$, for P (aerial photography)

18. $F = \frac{9}{5}C + 32$, for C (temperature conversion)

19. $L = 3.14(r_1 + r_2) + 2d$, for r_2 (pulleys)

20. $Q_1 = P(Q_2 - Q_1)$, for Q_2 (refrigeration)

21. $L = L_0(1 + at)$, for a (temperature expansion)

22. $a = V(k - PV)$, for k (biology)

23. $p - p_a = dg(y_2 - y_1)$, for y_2 (pressure gauges)

24. $F = A_2 - A_1 + P(V_2 - V_1)$, for V_2 (chemistry)

25. $A = \dfrac{n_1p_1 + n_2p_2}{n_1 + n_2}$, for p_1 (economics)

26. $Q = \dfrac{kAT(t_2 - t_1)}{d}$, for t_2 (heat conduction)

27. $f = \dfrac{f_s u}{u + v_s}$, for v_s (sound)

28. $P = \dfrac{V_1(V_2 - V_1)}{gJ}$, for J (jet engine power)

In Exercises 29 through 44, solve for the indicated literal number.

29. $a = bc + d$, for d

30. $x - 2y = 3t$, for x

31. $ax + 3y = f$, for x

32. $3ay + b = 7q$, for y

33. $2a(x + y) = 3y$, for x

34. $3x(a - b) = 2x$, for a

35. $\dfrac{a}{2} = b + 2$, for a

36. $\dfrac{y}{x} = 2 - a$, for y

37. $x_1 = x_2 + a(3 + b)$, for b

38. $s_1 + s_2 = 2(a - b)$, for a

39. $R_3 = \dfrac{R_1 + R_2}{2}$, for R_2

40. $m + n = \dfrac{2n + 1}{3}$, for n

41. $7a(y + z) = 3(y + 2)$, for z

42. $3x(x + y) = 2(3 - x)$, for y

43. $3(x + a) + a(x + y) = 4x$, for y

44. $2a(a - x) = 3a(a - b) + 2ax$, for b

In Exercises 45 through 52, set up the required formula and solve for the indicated letter.

45. The mean A of three numbers a, b, and c equals their sum divided by 3. Solve for c.

46. Computer output is obtained with three printers. The first printer can do x lines in one minute, the second printer can do $x + 100$ lines in one minute, and the third printer can do $x + 300$ lines per minute. A total of T lines is printed in one minute. Solve for x.

47. The current I in a circuit with a resistor R and a battery with voltage E and internal resistance r equals E divided by the sum of r and R. Solve for R.

48. A vending machine contains x nickels, $x + 5$ dimes, and $x - 6$ quarters. If the total value of these coins is y cents, solve for x.

49. If x dollars are invested at a rate of r_1, and $x + 1000$ dollars are invested at a rate of r_2, the total interest for one year is I. Solve for r_2.

50. One computer does C calculations per second for $t + 8$ seconds. A second computer does $C + 100$ calculations per second for t seconds. Together they do a total of N calculations. Solve for C.

51. A microwave transmitter can handle x telephone connections while seven separate cables can handle y connections each. This combined system can handle C connections. Solve for y.

52. A fire protection unit uses two water pumps to drain a flooded basement. One pump can remove A gal/h while the second pump can drain B gal/h. The first pump is run for t hours and the second pump is run for $t - 2$ hours so that K gallons are removed. Solve for B.

5-3 Simple Inequalities

Skip

In Section 5-1, we defined an equation to be a mathematical statement that two algebraic expressions are equal. In the same manner, *an* **inequality** *is a mathematical statement that one algebraic expression is greater than, or less than, another algebraic expression.*

Equation-solving plays a major role in mathematics and its applications. However, there are also occasions that require the solutions of inequalities. Some of these are shown through the examples and exercises of this section.

In Chapter 3, we introduced the signs of inequality $>$ and $<$. We now review those symbols and introduce two new symbols.

1. The symbol $<$ means "less than" and $a < b$ is interpreted as "a is less than b." For example, $2 < 5$.

2. The symbol $>$ means "greater than" and $a > b$ is interpreted as "a is greater than b." For example, $8 > 3$.

3. The symbol $\leq$ means "less than or equal to." The expression $a \leq b$ is interpreted as "a is less than or equal to b." For example, $3 \leq 7$ and $3 \leq 3$ are both true.

4. The symbol $\geq$ means "greater than or equal to." The expression $a \geq b$ is interpreted as "a is greater than or equal to b." For example, $9 \geq 6$ and $6 \geq 6$ are both true.

These signs define the **sense** *of the inequality, which refers to the direction in which the inequality sign is pointed. The two sides of the inequality are known as* **members** *of the inequality.*

same

EXAMPLE A

The inequalities $x + 3 < 6$ and $x + 9 < 12$ have the *same sense* because both inequality symbols are in the same direction.

The inequalities $x + 4 > 1$ and $3 - x < 2$ have *opposite senses* since the inequality symbols are in opposite directions. □

The **solution** *of an inequality consists of all values of the variable that satisfy the inequality.* That is, when any such value is substituted into the inequality, it becomes a correct statement mathematically. For most of the inequalities we consider here, the solution will consist of an unlimited number of values bounded by a specific real number.

EXAMPLE B

In the inequality $x + 4 > 1$, if we substitute any number greater than -3 for x, we find that the inequality is satisfied. That is, if we substitute -2, 0, 4, π, $-\sqrt{2}$, and so on for x, the inequality is satisfied. Thus the solution consists of *all* real numbers which are greater than -3. No value -3 or less satisfies the inequality. We express the solution as $x > -3$. □

When finding the solution set of an inequality, we work with each member and do operations that are similar to those used in solving equations. That is,

*We may add the same number to each member, subtract the same number from each member, or multiply or divide each member by the same **positive** number and still maintain the same sense of the inequality.*

There is one operation which must be handled with special care:

▶   *If each member of the inequality is multiplied or divided by a negative number, the sense of the inequality is reversed.*

These operations are verified in the following examples.

EXAMPLE C   We know that $2 < 6$. If 3 is added to each member of the inequality, we obtain $2 + 3 < 6 + 3$, or $5 < 9$, which is still a correct inequality.

If 3 is subtracted from each member of the inequality $2 < 6$, we obtain $2 - 3 < 6 - 3$, or $-1 < 3$, which is a correct inequality.

If each member of the inequality $2 < 6$ is multiplied by 3, we obtain $2 \times 3 < 6 \times 3$, or $6 < 18$, which is a correct inequality.

If each member of the inequality $2 < 6$ is divided by 3, we obtain $\frac{2}{3} < \frac{6}{3}$, or $\frac{2}{3} < 2$, which is a correct inequality.

This example illustrates the principle that the sense of an inequality does not change if we add a number to each member, subtract a number from each member, or multiply or divide by a *positive* number.   □

EXAMPLE D   If we multiply each member of the inequality $2 < 6$ by -3, we obtain -6 for the left member and -18 for the right member. However, we know that $-6 > -18$. Thus we obtain $-6 > -18$; the sense of the inequality has been reversed.

If each member of the inequality $2 < 6$ is divided by -2, we obtain $-1 > -3$, and again the sense of the inequality is reversed.

This example illustrates the important principle that multiplication or division by a *negative* number *reverses* the sense of that inequality.   □

The goal in performing these operations is to isolate the unknown on one side of the inequality and the other numbers on the other side. In this way, we can solve the inequality and determine the values which satisfy it. This is the same as in solving an equation. The following examples illustrate solving inequalities.

EXAMPLE E

1. Solve the inequality $x + 6 < 9$.

    By subtracting 6 from each member, we obtain $x < 3$, which means that the values of x less than 3 satisfy the inequality. Thus the solution is $x < 3$.

2. Solve the inequality $x - 6 < 9$.

    By adding 6 to each member, we obtain $x < 15$, which is the solution.

    □

EXAMPLE F

1. Solve the inequality $\frac{x}{2} > 5$.

    By multiplying each member by 2, we obtain $x > 10$, which is the solution.

2. Solve the inequality $2x > 16$.

    By dividing each member by 2, we obtain $x > 8$, which is the solution.

    □

EXAMPLE G

Solve the inequality $6 - 2x < 20$.

▶ By first subtracting 6 from each member, we obtain $-2x < 14$. Then *if each member is divided by -2, we obtain $x > -7$, noting that the sense of the inequality has been reversed*. Thus the solution is $x > -7$. □

Example H illustrates the solution of an inequality that requires several basic operations. The operations performed are indicated.

EXAMPLE H

Solve the inequality $6(x - 5) \geq 10 + x$.

$6(x - 5) \geq 10 + x$	original inequality
$6x - 30 \geq 10 + x$	use distributive law
$6x \geq 40 + x$	add 30 to each member
$5x \geq 40$	subtract x from each member
$x \geq 8$	divide each member by 5

Thus the solution is $x \geq 8$. □

Example I illustrates the applied use of an inequality.

EXAMPLE I   A computer center is cooled by an air conditioner which is automatically activated if the temperature equals or exceeds 77°F. Since the Celsius and Fahrenheit scales are related by the equation

$$F = \frac{9}{5}C + 32$$

the air conditioner goes on when

$$\frac{9}{5}C + 32 \geq 77$$

If we wish to find the Celsius temperatures for which the air conditioner operates, we get

$\frac{9}{5}C + 32 \geq 77$	**original inequality**
$9C + 160 \geq 385$	**multiply each member by 5**
$9C \geq 225$	**subtract 160 from each member**
$C \geq 25$	**divide each member by 9**

The air conditioner goes on when the temperature equals or exceeds 25° on the Celsius scale.   ☐

EXAMPLE J   Another applied use of inequalities occurs with measurements. Measurements are not exact; they are only approximations, and the error associated with a measurement is often given. A surveyor, for example, might describe a particular distance (in meters) as 354.27 ± 0.05. This 0.05 error factor can be included in an inequality by describing the distance x (in meters) by

$$354.22 \leq x \leq 354.32$$

This inequality, which we note has *three* members, is read as "x is greater than or equal to 354.22 m *and* less than or equal to 354.32 m." This means that x is between or equal to these values, which has the same meaning as 354.27 m $\pm$ 0.05 m. Such inequalities are sometimes called *double inequalities* because they have two conditions that must be satisfied. We should stress that *both* conditions must be satisfied and that the inequality $354.22 \leq x \leq 354.32$ really means

$$354.22 \leq x \ and \ x \leq 354.32$$

Many computer programming languages do not allow a direct entry of a double inequality, but they do allow statements such as "$354.22 \leq x$ AND $x \leq 354.32$."   ☐

Exercises 5-3

In Exercises 1 through 12, perform the indicated operations on each member of the given inequalities.

1. (a) $3 > 1$, add 5   (b) $-2 < 4$, subtract 3
2. (a) $-2 < 8$, add 4   (b) $7 > 1$, subtract 4
3. (a) $1 > -2$, multiply by 3   (b) $5 < 10$, divide by 5
4. (a) $3 < 10$, multiply by 4   (b) $-8 > -16$, divide by 8
5. (a) $-1 < 0$, multiply by -2 (b) $6 < 9$, divide by -3
6. (a) $2 < 9$, multiply by -3 (b) $4 > -4$, divide by -1
7. (a) $-3 < -2$, multiply by -5 (b) $-12 < 8$, divide by -4
8. (a) $-6 < -4$, multiply by -3 (b) $-18 < -12$, divide by -6
9. $-2 < 8$, add 4 and then divide by -2
10. $11 > -3$, subtract 4 and then divide by -7
11. $8 \leq 12$, subtract 4 and then divide by -4
12. $6 \geq -8$, add 12 and then divide by -2

In Exercises 13 through 32, solve the given inequalities.

13. $x - 7 > 5$ 14. $x + 8 < 4$ 15. $x + 2 < 7$ 16. $x - 2 > 9$

17. $\dfrac{x}{3} > 6$ 18. $\dfrac{x}{4} < 5$ 19. $4x < -24$ 20. $3x > 6$

21. $-2x \geq -8$ 22. $-x \leq 5$ 23. $2x + 7 \leq 5$ 24. $5x - 3 \geq -13$

25. $2 - x > 6$ 26. $3 - x < 1$ 27. $4x - 1 < x - 7$ 28. $2x + 1 > 10 - x$

29. $3(x - 2) > -9$ 30. $6(x + 5) < x + 10$ 31. $2(2x + 1) \leq 5x - 4$ 32. $3(1 - 2x) \geq 1 - 4x$

In Exercises 33 through 40, set up inequalities from the given statements and then solve.

33. A company determines that it must spend less than $100,000 on new equipment. It is decided to spend $12,500 for a particular piece of new equipment. How much can be spent on other new pieces?

34. A highway sign requires maximum speeds of 55 mi/h. A driver knows that his speedometer always gives an indication which is 3 mi/h above the actual speed of the car. What speedometer indications correspond to true legal speeds?

35. What average speeds will allow a traveler to go more than 150 mi in 3 h?

36. A student has grades of 80, 92, 86, and 78 on the four tests in a certain course. The final examination in the course counts as two tests. What examination grades would cause her average (arithmetic mean) to fall below 80 for the course? (The arithmetic mean of several scores is found by adding the scores and dividing the total by the number of scores.)

37. A voltmeter is designed so that any reading is in error by no more than 1.3 V. If a measurement indicates 14.2 V, write an inequality which describes the true value of the voltage x.

38. A gauge is used to measure the pressure in a pipe used to transport natural gas. This gauge is designed so that any reading is in error by no more than 2.3 lb/in.^{2}. If the gauge indicates 64.9 lb/in.^{2}, write an inequality which describes the true value of the pressure P.

39. The rental cost for a machine is $100 for setup, plus $4 per hour of operation. How many hours of operation will result in total charges that exceed $240?

40. A sales representative earns a salary of $300 per week plus a commission of $6 for each unit sold. What is known about the number of units sold in one week if the total income is greater than $540?

In Exercises 41 through 44, use a scientific calculator to solve the given inequalities. (Assume all numbers are approximate.)

41. $3.14x - 6.28 \leq 1.29(3.66 - 1.22x)$

42. $19.17 - 13.14x > 12.07x - 32.11$

43. $17.76x - 16.23 \geq 12.28x + 64.29$

44. $805.25(0.65734x - 507.91) < 100.94 - 365.30x$

5-4 From Statement to Equation

Mathematics is particularly useful in technical areas because it can be used to solve many applied problems. Some of these problems are in formula form and can be solved directly. Many other problems are in verbal form and they must first be set up mathematically before direct methods of solution are possible. This section explains how to solve such problems. We will work primarily on the interpretation of verbal statements. It isn't possible to present specific rules for interpreting verbal statements for mathematical formulation. Certain conditions are *implied,* and these are the ones we must recognize. A very important step to the solution is a careful reading of the statement to make sure that all terms and phrases are understood.

In the following examples we develop a general procedure for solving such stated problems. After the first four examples have been presented, this procedure will be stated in outline form. Keep in mind that **the primary goal is to obtain an equation from the statement** through proper interpretation. Once this is accomplished, solving the equation is relatively simple.

EXAMPLE A   A rectangular plate is to be the base for a circuit board. Its length is 2 in. more than its width and it has a perimeter of 36 in. Find the dimensions.

We must first understand the meaning of all terms used. Here we must know what is meant by rectangle, perimeter, and dimensions. Once we are sure of the terms, we must recognize what is required. We are told to find the dimensions. This should mean "find the length and the width of the rectangle." Once this is established, we let x (or any appropriate symbol) represent one of these quantities. Thus we write:

Let $x =$ the width, in inches, of the rectangle.

Next we look to the statement for other key information. The order in which we find it useful may not be the order in which it is presented. For example, here we are told that the rectangle's length is 2 in. more than its width. Since we let x be the width, then

$x + 2 =$ the length, in inches, of the rectangle

We now have a representation of both required quantities. See Figure 5-1.

Now that we have identified the unknown quantities, we must look for information by which we can establish an equation. We look to the portion of the statement which is as yet unused. That is: "A rectangle . . . has a perimeter of 36 in." Recalling that the perimeter of a rectangle equals twice the length plus twice the width, we can now multiply the width x by 2, add

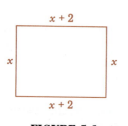

FIGURE 5-1

this to twice the length, $x + 2$, and then equate this sum to 36, the known value of the perimeter. This leads to the equation

perimeter
(algebraically) ──▶ $2x + 2(x + 2) = 36$ ◀── perimeter (numerically)

which we can now solve as follows:

$$2x + 2x + 4 = 36$$
$$4x = 32$$
$$x = 8$$

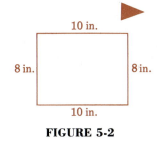

Therefore the width is 8 in. and the length is 10 in. See Figure 5-2. **Checking this result with the** statement of the problem, *not the derived equation,* we note that a rectangle with these dimensions has a perimeter of 36 in., which means that the solution checks.

Our analysis was complete once we established the equation. Probably a great deal of what was just stated seemed quite obvious. However, even more involved problems require a similar technique of reading and interpreting, which is the key to the solution. □

FIGURE 5-2

EXAMPLE B   A vending machine contains $1.40 in nickels and dimes. How many of each are there if there are 17 coins in all?

The phrase "how many of each" tells us that the number of nickels and the number of dimes are to be determined. Choosing one quantity as x, we write:

Let $x =$ the number of nickels.

Then the phrase "17 coins in all" means that

$17 - x =$ number of dimes

Since the total value of the coins is $1.40, by multiplying the number of nickels, x, by 5 and the number of dimes, $17 - x$, by 10, and equating this sum to 140 (the number of cents), we find the equation:

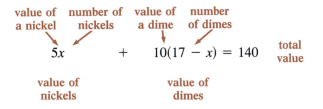

Solving this equation, we have

$$5x + 170 - 10x = 140$$
$$-5x = -30$$
$$x = 6$$

Therefore there are 6 nickels and $17 - 6 = 11$ dimes. We see that these coins have a total value of $1.40, which means that our solution checks. □

EXAMPLE C   Several 6-V and 12-V batteries are arranged so that their individual voltages combine to be a power supply of 84 V. How many of each type are present if the total number of batteries is 10?

From the phrase "how many of each type" tells us that the number of 6-V batteries and the number of 12-V batteries are to be determined. Choosing one quantity as x, we write:

Let x = the number of 6-V batteries

Then the phrase "the total number of batteries is 10" means that

$10 - x$ = the number of 12-V batteries

Since the total voltage is 84 V, by multiplying the number of 6-V batteries (x) by 6 and the number of 12-V batteries ($10 - x$) by 12 and equating this sum to 84 (the total voltage) we get

$$\underset{\substack{\text{voltage} \\ \text{of 6-V} \\ \text{batteries}}}{\underset{\substack{\text{voltage} \\ \text{per} \\ \text{battery}}}{6x} \quad \overset{\substack{\text{number} \\ \text{of 6-V} \\ \text{batteries}}}{} + \overset{\substack{\text{voltage} \\ \text{per} \\ \text{battery}}}{12} \underset{\substack{\text{voltage} \\ \text{of 12-V} \\ \text{batteries}}}{(10 - \overset{\substack{\text{number} \\ \text{of 12-V} \\ \text{batteries}}}{x})} = 84 \quad \substack{\text{total} \\ \text{voltage}}}$$

Solving this equation, we have

$$6x + 120 - 12x = 84 \qquad\qquad -6x = -36$$
$$120 - 6x = 84 \qquad\qquad\qquad x = 6$$

There are six 6-V batteries and four 12-V batteries. Checking this, we see that these 10 batteries produce a total voltage of 84 V.   □

EXAMPLE D   A machinist made 132 machine parts of two different types. He made 12 more type 1 than type 2 parts. How many of each did he make?

From the statement of the problem, we see that we are to determine the number of each type. Therefore we write:

Let x = number of type 1 parts and $132 - x$ = number of type 2 parts

From the statement "he made 12 more type 1 than type 2 parts," we write

$$x = (132 - x) + 12$$

which is the required equation. Solving this equation, we get

$$x = 132 - x + 12$$
$$2x = 144$$
$$x = 72$$

Therefore he made 72 type 1 parts and 60 type 2 parts. Note that this checks with the statement of the problem.   □

Having analyzed the statements of the first four examples, and having solved the resulting equations, we shall now state the basic steps that have been followed, and that should be followed in all such problems. This general procedure is as follows:

1. Read the statement of the problem. First read through quickly to get a general overview, then reread slowly and carefully.
2. List the information given and identify any questions being asked.
3. Clearly identify the unknown quantities, and then assign an appropriate letter to represent one of them, stating this choice clearly.
4. Specify each of the other unknown quantities in terms of the one specified in step 3.
5. If possible, construct a sketch using the known and unknown quantities.
6. Analyze the statement of the problem and construct the necessary equation. This is the most difficult step, but it does become easier with practice.
7. Solve the equation, clearly stating the solution.
8. Check the solution *with the original statement* of the problem.

We shall now present two more examples of stated problems. Note the way in which these steps are followed.

EXAMPLE E   A car travels 40 mi/h for 2 h along a certain route. Then a second car starts along the same route, traveling 60 mi/h. When will the second car overtake the first?

The word "when" means "at what time" or "for what value of t." We may let t represent the time either car has been traveling; so, choosing one quantity, we write:

Let t = the time, in hours, the first car has traveled.

The fact that the first car has been traveling for 2 h when the second car starts means that

$t - 2$ = time, in hours, the second car has traveled

The key to setting up the equation is the word "overtake," which implies, although it does not state explicitly, that *the cars will have gone the same distance.* Since distance equals rate times time, we establish the equation by equating the distance the first car travels, $40t$, to the distance the second car travels, $60(t - 2)$. Therefore the equation is

$$\underset{\substack{\text{distance}\\\text{traveled by}\\\text{first car}}}{\overset{\text{rate} \times \text{time}}{40t}} = \underset{\substack{\text{distance}\\\text{traveled by}\\\text{second car}}}{\overset{\text{rate} \times \text{time}}{60(t - 2)}}$$

(Continued on next page)

Solving this equation, we get

$$40t = 60t - 120$$
$$-20t = -120$$
$$t = 6$$

Therefore the first car travels 6 h and the second car 4 h. Note that a car traveling 40 mi/h for 6 h goes 240 mi, as does a car traveling 60 mi/h for 4 h. This confirms that the solution checks. ◻

EXAMPLE F Suppose 100 kg of a cement-sand mixture is 40% sand. How many kilograms of sand must be added so that the resulting mixture will be 60% sand? (Let n = number of kilograms of sand to be added.)

▶   We establish the equation by expressing *the number of kilograms of sand in the final mixture as a sum of the sand originally present,* **40 kg,** *and that which is added,* **n kg.** We then equate this to the amount in the final mixture, which is $0.60(100 + n)$. (The final mixture is 60% sand, and there is a total of $(100 + n)$ kg.) Thus

40%   number   60%   number
sand   of kg   sand   of kg

$$(0.40)(100) + n = 0.60(100 + n)$$

kg of sand   kg of   kg of sand
in original   sand   in final
mixture   added   mixture

$$40 + n = 0.60(100 + n)$$
$$40 + n = 60 + 0.60n$$
$$0.40n = 20$$
$$n = 50 \text{ kg}$$

Checking this with the original statement, we find that the final mixture will be 150 kg, of which 90 kg will be sand. Since

$$\frac{90}{150} = 0.60$$

the solution checks. ◻

Exercises 5-4

Solve each of the following. Follow the steps outlined after Example D.

1. When three resistors are connected in series, their resistances are added to produce a total resistance of 970 Ω. One of them has a resistance of 530 Ω, and the others have resistance levels equal to each other. Find the resistance levels of the other two resistors.

2. A power supply has two printed circuit boards which contain a combined total of 222 components. One board has 6 more than twice the number of components on the other board. How many components are in each board?

3. The cost of one floppy disk is 20¢ less than twice the cost of another. Given that 10 disks of each type cost a total of $34.00, find the cost of one of the cheaper disks.

4. An architect determines that if she reduces the dimensions of a square room by 2 ft on each side, the perimeter will be 56 ft. What is the length of the original room before the reduction?

5. The sum of two currents is 200 mA, and the larger current is 30 mA more than the smaller current. Find the value of the smaller current.

6. A 12-ft support stud is cut into two pieces so that one is three times as long as the other. How long is the longer piece?

7. An inlet pipe provides 4 gal/min more than a second pipe and 6 gal/min less than a third pipe. Together they supply 50 gal/min. How much does each provide?

8. A riverfront boat storage area is rectangular with fencing on all sides except the side along the river. If 550 m of fencing is used and the side along the river is 50 m shorter than the two longer sides, find the dimensions of the site.

9. One computer printer can type 2.5 times as fast as another. Together they can print 1120 lines per minute. What is the printing rate of each printer?

10. One computer has eight times the storage capacity of another and together they can store 18,432 bytes. What is the storage capacity of each computer?

11. Some resistors cost 5¢ and others cost 25¢. Twenty-one resistors cost $2.65. How many of each are there?

12. Eighteen resistors have a total value of $4.05. If some are worth 10¢ and the others are worth 25¢, how many are worth 10¢?

13. The cost of one type of smoke detector is $3 less than twice the cost of another type. If two units of each type are purchased at a total cost of $66, what is the cost of one of the more expensive smoke detectors?

14. A company plans to issue 24,500 shares of two different kinds of stock which will have a combined value of $800,000. One of the stocks is worth $100 per share while the other stock is worth $25 per share. How many shares of each stock will be issued?

15. Three different oil storage tanks have a combined capacity of 4400 gal. The largest tank holds three times as much as the smallest tank and twice as much as the other tank. What is the capacity of each tank?

16. One square cover plate has sides that are 5 mm longer than those of a second square cover plate. If the larger plate has a perimeter of 352 mm, find the perimeter of the smaller cover plate.

17. Two vans used for parts delivery are 420 mi apart when they begin traveling toward each other. One van goes 50 mi/h while the other goes 55 mi/h. How long do they drive before they meet?

18. A courier travels to and from a manufacturing plant in 7 h. His average speed to the plant is 80 km/h, and the return trip averages 20 km/h slower. How long did it take to reach the plant?

19. Two low-flying missiles are launched from the same location at sea and are sent in opposite directions. After 5 s they are 33,000 ft apart. If one missile travels 600 ft/s faster than the other, what is the speed of this faster missile?

20. An industrial cleaning solution is supposed to contain 70% water and 30% acid. How many quarts of pure acid must be added to 8 qt of a solution that is $\frac{1}{8}$ acid in order to make a solution that is 30% acid?

21. Two cars, originally 800 km apart, start at the same time and travel toward each other. One travels 10 km/h faster than the other, and they meet in 5 h. What is the speed of each?

22. How many milliliters of water must be added to 100 mL of a 50% solution of sulfuric acid to make a 10% solution?

23. Two gears have a total of 80 teeth, and one gear has 48 fewer teeth than the other. How many teeth does each gear have?

24. If you need 50 lb of an alloy that is 50% nickel, how many pounds of an alloy with 80% nickel must be mixed with an alloy that is 40% nickel?

25. According to Kirchhoff's current law, the sum of the currents into a node equals the current out of the node. The current out of a node is 650 mA and three currents go into it. The largest current is twice the smallest and 100 mA more than the other current. Find all three currents.

26. The base material for a certain roadbed is 80% crushed rock when measured by weight. How many tons of this material must be mixed with another material containing 30% rock to make 150 tons of material which is 40% crushed rock?

27. A sponsor pays a total of $4300 to run a commercial on two different television stations. One station charges $700 more than the other. What does each station charge to run the commercial?

28. A certain newspaper is published every day of the week. It prints 50,000 more copies on Sunday than on a weekday, and it prints a total of 1,100,000 copies in a week. What is the number of copies printed each day of the week?

29. Approximately 4.5 million wrecked cars are recycled in two consecutive years. There were 700,000 more recycled the second year than in the first year. How many are recycled each year?

30. A person pays $4800 in state and federal income taxes in a year. The federal income tax is five times as great as the state income tax. How much does the person pay on each of these income taxes?

31. A metallurgist melts and mixes 100 g of solder which is 50% tin with another solder that is 10% tin. How many grams of the second type must be used if the result is to be 25% tin?

32. A certain car's cooling system has an 8-qt capacity and is filled with a mixture that is 30% alcohol. How much of this mixture must be drained off and replaced with pure alcohol if the solution is to be 50% alcohol?

33. A company leases petroleum rights to 140 acres of land for $37,000. If part of the land leases for $200 per acre while the remainder leases for $300 per acre, find the amount of land leased at each price.

34. A certain type of engine uses a fuel mixture of 15 parts of gasoline to 1 part of oil. How much gasoline must be mixed with a gasoline-oil mixture, which is 75% gasoline, to make 8 L of the required mixture for the engine?

35. Driving at a certain speed, it takes a driver 4 h to travel from one city to another. If the speed could be increased by 16 mi/h, the trip could be done in 3 h. How fast did the driver travel?

36. An airplane takes 6 h to complete a trip in calm air. The return trip requires only 5 h because of a 40 mi/h tailwind. How long (in miles) is this trip?

37. A walkway 3 m wide is constructed along the outside edge of a square courtyard. If the perimeter of the courtyard is 320 m, what is the perimeter of the square formed by the outer edge of the walkway?

38. A company anticipates a budget surplus of $84,000. It is decided that this amount is to be distributed for additional advertising, with radio getting one share, newspapers getting four shares, and direct mail getting two shares. How much is spent for additional newspaper advertising?

39. If $8000 is invested at 7%, how much must be invested at 9% in order for the total interest amount to equal $1550?

40. Two stock investments totalled $15,000. One stock led to a 40% gain, but the other stock resulted in a 10% loss. If the net result is a profit of $2000, how much was invested in each stock?

5-5 Ratio and Proportion

This section introduces some of the terms so important in many applications of mathematics, including those in science and technology.

*The **ratio** of one number to another is the first number divided by the second number*. That is, the ratio of *a* to *b* is *a/b*. (Another notation is *a:b*.) By this definition we use division to compare two numbers.

EXAMPLE A

1. The ratio of 12 to 8 is $12 \div 8$, or $\frac{12}{8}$. Simplifying the fraction we obtain $\frac{3}{2}$, which means that the ratio of 12 to 8 is the same as the ratio of 3 to 2. It also means that 12 is $\frac{3}{2}$ as large as 8.

2. The ratio of 2 to 9 is $\frac{2}{9}$.

3. The ratio of 10 to 2 is $\frac{10}{2}$, or $\frac{5}{1}$. ☐

Every measurement is a ratio of the measured magnitude to an accepted unit of measurement. When we say that an object is 4 m long, we are saying that the length of the object is four times as long as the unit of length, the meter. Other examples of ratios are scales of measurements on maps, density (mass/volume), and pressure (force/area). As shown by these examples, ratios may compare quantities of the same kind or they may express divisions of magnitudes of different quantities.

Ratios are often used to compare like quantities. When they are used for this purpose we usually express each of the quantities in the same units. The following example illustrates this use of ratios.

EXAMPLE B

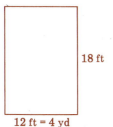

18 ft

12 ft = 4 yd

FIGURE 5-3

1. The length of a laboratory is 18 ft, and the width is 12 ft. See Figure 5-3. Therefore the ratio of the length of the laboratory to the width is 18 ft/12 ft, or $\frac{3}{2}$.

2. If the width of the room is expressed as 4 yd, we have the ratio as 18 ft/4 yd, or 9 ft/2 yd. However, this does not clearly show the ratio. It is preferable and more meaningful to change one of the measurements to the same units as the other. Changing the width from 4 yd to 12 ft, we express the ratio as 18 ft/12 ft or $\frac{3}{2}$, as above. From this ratio we easily see that the length of the room is $\frac{3}{2}$ as long as the width.

3. There are some special cases where ratios involve like quantities but different units. One drainage pipe specification requires a slope of $\frac{1}{4}$ in. for each foot of horizontal distance so that the ratio of vertical distance to horizontal distance is $\frac{1}{4}$ in./ft. ☐

As we have noted, ratios are often used to compare like quantities, but they may also be used to compare unlike quantities. When we do compare unlike quantities, we must attach the proper units to the ratio. This is illustrated in the following example.

EXAMPLE C   If a car travels 80 mi in 2 h and consumes 4 gal of gasoline, the ratio of distance to time is 80 mi/2 h, or 40 mi/h. Also, the ratio of distance to fuel consumed is 80 mi/4 gal, or 20 mi/gal. In each case we must note the units of the ratio. □

A statement of equality of two ratios is a **proportion.** By this definition,

$$\frac{a}{b} = \frac{c}{d}$$

is a proportion, equating the ratios a/b and c/d. (Another way of denoting a proportion is $a:b = c:d$, which is read as "a is to b as c is to d.") Thus we see that a proportion is an equation.

EXAMPLE D   1. The ratio $\frac{16}{6}$ equals the ratio $\frac{8}{3}$. We may state this equality by writing the proportion $\frac{16}{6} = \frac{8}{3}$.
2. $\frac{18}{12} = \frac{3}{2}$ and $\frac{80}{4} = \frac{20}{1}$ are proportions.
3. 80 km/2h = 40 km/1h = 40 km/h is a proportion. □

In a given proportion, if one ratio is known and one part of the other ratio is also known, then the unknown part can be found. This is done by letting an appropriate literal symbol represent the unknown part and then solving the resulting equation. This is illustrated in the following examples.

EXAMPLE E   The ratio of a given number to 3 is the same as the ratio of 16 to 6. Find the number.

First, let x = the number. We then set up the indicated proportion as

$$\frac{x}{3} = \frac{16}{6}$$

Solving for x, we multiply each side by 3:

$$\frac{3x}{3} = \frac{16(3)}{6}$$

which is simplified to

$$x = 8$$

Checking, we see that

$$\frac{8}{3} = \frac{16}{6}$$ □

EXAMPLE F   On a certain blueprint, a measurement of 25.0 ft is represented by 2.00 in. What is the actual distance between two points if they are 5.00 in. apart on the blueprint?

Let x = the required distance and then note that 25.0 is to 2.00 as x is to 5.00. Thus:

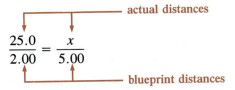

actual distances

$$\frac{25.0}{2.00} = \frac{x}{5.00}$$

blueprint distances

By multiplying each side by 5.00, we solve the equation. This leads to

$$\frac{5.00(25.0)}{2.00} = \frac{5.00x}{5.00}$$

or

$$x = 62.5 \text{ ft} \qquad \Box$$

EXAMPLE G   A certain alloy is 5 parts tin and 2 parts lead. How many kilograms of each are there in 35 kg of the alloy?

First, we let x = the number of kilograms of tin in the given amount of alloy. Next, we note that there are 7 total parts of alloy, of which 5 are tin. Thus 5 is to 7 as x is to 35. This gives us the equation

parts tin ⟶ $\dfrac{5}{7} = \dfrac{x}{35}$ ⟵ weight tin
total parts ⟶   ⟵ total weight

Multiplying each side by 35, we get

$$\frac{35(5)}{7} = \frac{35x}{35}$$

or

$$x = 25 \text{ kg}$$

There are 25 kg of tin and 10 kg of lead in the given amount of alloy. $\Box$

Exercises 5-5

In Exercises 1 through 8, set up the given ratios as fractions and simplify when possible.

1. 12 to 5;   3 to 23   **2.** 2 to 11;   13 to 7   **3.** 21 to 3;   2 to 12

4. 5 to 45;   60 to 15   **5.** 6 to 9;   12 to 18   **6.** 20 to 24;   5 to 30

7. 6 to 33;   8 to 28   **8.** 100 to 10;   72 to 156

In Exercises 9 through 20, find the indicated ratios.

9. 30 A to 8 A

10. 4 mi to 22 mi

11. 9 cm to 30 cm

12. 50 W to 16 W

13. 8 in. to 4 ft

14. 5 cm to 40 mm

15. 80 s to 3 min

16. 500 lb to 3 t

17. 12 m to 6 s

18. 2 in. to 100°C

19. 8 lb to 36 ft³

20. 25 L to 35 h

In Exercises 21 through 28, solve the given proportions for x.

21. $\dfrac{x}{2} = \dfrac{5}{8}$

22. $\dfrac{x}{3} = \dfrac{7}{12}$

23. $\dfrac{3}{14} = \dfrac{x}{4}$

24. $\dfrac{8}{15} = \dfrac{x}{9}$,

25. $\dfrac{3}{x} = \dfrac{9}{15}$

26. $\dfrac{12}{x} = \dfrac{1}{8}$

27. $\dfrac{4}{3} = \dfrac{12}{x}$

28. $\dfrac{25}{16} = \dfrac{30}{x}$

In Exercises 29 through 56, solve the given problems by first setting up the proper proportion.

29. The ratio of a number to 6 is the same as the ratio of 70 to 30. Find the number.

30. The ratio of a number to 15 is the same as the ratio of 17 to 60. Find the number.

31. The ratio of a number to 40 is the same as the ratio of 7 to 16. Find the number.

32. The ratio of a number to 44 is the same as the ratio of 8 to 33. Find the number.

33. The pitch of a roof is defined as the ratio of the rise to the span. If the rise is 6 ft and the span is 24 ft, find the pitch.

34. A stock is priced at $120 per share and annual earnings average to $8 per share. What is the price-to-earnings ratio?

35. If a substance of 908 g weighs 2.00 lb, how many grams of the substance would weigh 10.0 lb.?

36. 60 mi/h = 88 ft/s; what speed in miles per hour is 66 ft/s?

37. A rectangular picture 20 in. by 15 in. is to be enlarged so that the width of the enlargement is 25 in. What should be the length of the enlargement? See Figure 5-4.

38. If 40 lb of sodium hydroxide will neutralize 49 lb of sulfuric acid, how many pounds of sodium hydroxide are needed to neutralize 21 lb of sulfuric acid?

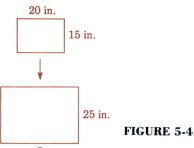

20 in.

15 in.

25 in.

FIGURE 5-4

?

39. A support beam 15 m long is cut into two pieces, the lengths of which are in the ratio of 2 to 3. What is the length of each piece?

40. The ratio of a person's federal income tax to state income tax is 10 to 3. If the person pays a total income tax of $2600, how much was paid to the federal government and how much to the state government?

41. A certain gasoline company sells lead-free gas and regular gas in the ratio of 5 to 2. If, in a month, the company sells a total of 14,000,000 gal, how many gallons of each type were sold?

42. A tablet of medication contains two substances in the ratio of 9 to 5. If a tablet contains 280 mg of medication, how much of each type is in the tablet?

43. A person 1.8 m tall is photographed with a 35 mm camera, and the film image is 20 mm. Under the same conditions, how tall is a person whose film image is 16 mm?

44. The relative density of a substance is the ratio of its density to the density of water. What is the relative density of aluminum if its density is 2.7 g/cm³? The density of water is 1.0 g/cm³.

45. The Mach number of a moving object is the ratio of its velocity to the velocity of sound (740 mi/h). Find the Mach number for a supersonic jet traveling 910 mi/h.

46. Find the Mach number of a jet with a velocity of 580 mi/h. (See Exercise 45.)

47. If the property tax on a $750,000 factory is $16,500, find the property tax on a $985,000 factory.

48. The density of a substance is the ratio of its mass to the volume it occupies. Find the density of gasoline if 317.45 cm^3 of gasoline has a mass of 216.21 g.

49. One shift of fire fighters constructs 900 m of fire line in 8 h. If another shift works at the same rate, how long will it take them to construct an additional 700 m of fire line?

50. Five clicks on an adjustment screw cause the inlet valve to change by 0.035 cm. How many clicks are required for a change of 0.021 cm?

51. Sixty gallons of oil flow through a pipe in 8.0 h. How long will it take 280 gal to flow through this same pipe?

52. A truck can go 245 mi on 35.0 gal of diesel fuel. How much diesel fuel would be required to go 1250 mi?

53. A shaft is designed to have a uniform taper. There is a taper of 3.5 mm in 14 cm of its length. How much taper is there in 37 cm of its length?

54. Under standard conditions in a 12 mi/h wind, a sugar pine seed will travel 77 m while a ponderosa pine seed will travel 130 m. In a wind of 19 mi/h the sugar pine seed travels 125 m. How far should the ponderosa pine seed travel in the same 19 mi/h wind?

55. For a hydraulic press, the mechanical advantage is the ratio of the large piston area to the small piston area. Find the mechanical advantage if the large and small pistons have areas of 27 cm^2 and 15 cm^2, respectively.

56. The shunt law, when applied to an electric circuit with parallel resistors, states that currents I_1, I_2, and resistances R_1, R_2 are related by

$$\frac{I_1}{I_2} = \frac{R_2}{R_1}$$

If $I_1 = 3$ A, $I_2 = 4$ A, and $R_1 = 12$ Ω, find R_2.

5-6 Variation

Where did we first get the formulas that are used in the various fields of technology? Part of the answer will be seen in the following chapter, when we begin the study of geometry. There, certain concepts are *defined* and a number of formulas for measuring quantities result. Many other formulas are derived through observation and experimental evidence. Since this method is very important in technology, this section includes illustrations showing how a number of formulas are established using this approach.

The ratio of one quantity to another remains the same in many applied situations. This relationship is the basis for determining many important formulas. Consider the illustrations in the following example.

EXAMPLE A

1. In a given length, the ratio of the length in inches to the length in feet is 12. This may be written as $\dfrac{i}{f} = 12$

2. Experimentation shows that the ratio of the electric resistance of a wire to the length of the wire is always the same. This may be written as $\dfrac{R}{l} = k$

3. It can be shown that for a given mass of gas, if the pressure remains constant, the ratio of the volume of the gas to the thermodynamic temperature remains constant. This may be written as $\dfrac{V}{T} = k$ ☐

In Section 4-1 we noted that *a* **variable** *is a quantity that may take on different values,* but *a* **constant** *takes on only one value during a given discussion.* We may express the fact that a ratio of two variables remains constant by the equation $\dfrac{y}{x} = k$

where x and y are the variables and k is the constant.

EXAMPLE B

1. In the first illustration of Example A, the number of inches i and the number of feet f are variables. Each may take on any value so long as their ratio remains 12. This means that when we consider many different distances, and therefore many different values of i and f, the ratio is always 12.

2. In the second illustration of Example A, the resistance R and the length l are the variables and k is the constant. The value of k may differ for different wires, but *for a given wire k* takes on a specific value, regardless of the length of the wire.

3. In the third illustration of Example A, the volume V and temperature T are the variables and k is the constant. Again k may differ from one body of gas to another, but it's a specific constant for any one given body of gas (although the volume and temperature can vary). ☐

By convention, letters near the end of the alphabet, such as x, y, and z, are generally used to denote variables. Letters near the beginning of the alphabet, such as a, b, and c, are generally used as constants. The meaning of other letters, such as k in this case, are specified in the problem.

If we solve the equation $y/x = k$ for y, we get

$$y = kx \qquad \textbf{direct variation}$$

(5-1)

This equation is read as "y is proportional to x" or "y varies directly as x." This type of relationship is known as **direct variation.**

Given a set of values for x and y, we can determine the value of the **constant of proportionality** k. Then we can substitute this value for k to obtain the relationship between x and y. Then we can find y for any other value of x. Conversely, we can find x for a given value of y. Consider the following example.

EXAMPLE C If y varies directly as x, and $x = 6$ when $y = 18$, find the value of y when $x = 5$.

First we express the relationship "y varies directly as x" as

$$y = kx$$

Next we substitute $x = 6$ and $y = 18$ into the equation and we get

$$18 = 6k$$

or $k = 3$. For this example *the constant of proportionality is 3, and this is substituted into* **y = kx,** *giving*

$$y = 3x$$

as the equation between **y and x.** Now for any given value of x we may find the value of y by substitution. For $x = 5$, we get

$$y = 3(5) = 15 \qquad \Box$$

In many problems, one variable will vary directly as another variable raised to some power. The following example illustrates this in the case of an applied problem.

EXAMPLE D The distance d that an object falls under the influence of gravity is proportional to the square of the time t of fall. If an object falls 64.0 ft in 2.00 s, how far does it fall in 6.00 s?

To express the fact that d varies directly as the square of t, we write

$$d = kt^2 \qquad \textit{directly} \text{ indicates multiplication of constant by variable}$$

Then, using the fact that $d = 64.0$ ft when $t = 2.00$ s, we get

$$64.0 = k(2.00)^2$$

which gives us $k = 16.0$ ft/s². In general,

$$d = 16.0t^2$$

We now substitute $t = 6.00$ s and we get

$$d = 16.0(6.00)^2 = 16.0(36.0) = 576 \text{ ft}$$

This means that an object falling under the influence of gravity will fall 576 ft in 6.00 s. $\qquad \Box$

Note that the constant of proportionality in Example D has a set of units associated with it. (In this case, ft/s^2.) This will be the case unless the quantities related by k have precisely the same units. *We can determine the units for k by solving the equation for k and noting the units on the other side of the equation.*

EXAMPLE E   In Example D, when we solve for k we can find its units as well as its value if we include the units in the calculation. In this case we have

$$d = kt^2$$
$$64.0 \text{ ft} = k(2.00 \text{ s})^2$$
$$k = \frac{64.0 \text{ ft}}{4.00 \text{ s}^2} = 16.0 \text{ ft/s}^2$$

☐

In the first illustration of Example A, the 12 has units of inches per foot. In the second illustration, if R is measured in ohms and l in feet, the units for k are ohms per foot. In the third illustration, if V is measured in liters and T in kelvins, the units for k are liters per kelvin. *The units actually used in a given problem will determine the units in which k will be measured for that problem.*

Another important type of variation is **inverse variation,** *which occurs when the product of two variables is constant,* as in $x \cdot y = k$. This can also be expressed by the equation

$$y = \frac{k}{x} \qquad \textbf{inverse variation} \tag{5-2}$$

Here y **varies inversely as** *x (or y is* **inversely proportional** *to x), and k is the constant of proportionality.* With direct variation in $y = kx$, the variable y increases as x increases and y decreases as x decreases. There is a direct relationship in the way that x and y vary. With the inverse variation of $y = k/x$, note that as x increases, y decreases. Also, if x decreases, y must increase. With inverse variation we have an opposite or inverse relationship between x and y.

EXAMPLE F   A business firm found that the volume V of sales of a certain item was inversely proportional to the price p.

This statement is expressed as

$$V = \frac{k}{p} \qquad \textit{inversely} \text{ indicates division of constant by variable}$$

If we know that $V = 1000$ sales/week for $p = \$5$, we get

$$1000 = \frac{k}{5}$$

or $k = 5000$ dollar-sales/week. This means that the equation relating V and p is

$$V = \frac{5000}{p}$$

For any value of p, we may find the corresponding value of V. For example, if $p = \$4$ we get

$$V = \frac{5000}{4} = 1250 \text{ sales/week} \qquad \square$$

Finally, it is possible to relate one variable to more than one other variable by means of **combined variation**. Let us consider the following example.

EXAMPLE G

1. The equation expressing the fact that the force F between two electrically charged particles, with charges q_1 and q_2, varies directly as the product q_1 and q_2 is

$$F = kq_1q_2 \qquad \text{\color{red}{\textit{F} varies directly as } \textit{q}_1 \times \textit{q}_2}$$

2. The equation expressing the fact that y varies directly as x and inversely as z is

$$y = \frac{kx}{z} \qquad y \begin{cases} \text{\color{red}{varies directly as } \textit{x}} \\ \text{\color{red}{varies inversely as } \textit{z}} \end{cases}$$

▶ Note that the word "and" appears in the statement, but the formula contains only products and quotients. The word *"and" is used only to note that y varies in more than one way; it does not imply addition.*

3. The equation expressing the fact that s varies directly as the square of t and inversely as the cube of v is

$$s = \frac{kt^2}{v^3}$$

Note again that only a product and quotient appear in the formula. $\qquad \square$

As before, in each case we can determine k by knowing one set of values of the variables. After replacing k by its known value, we can proceed to determine the value of one variable if we know the values of the others.

Exercises 5-6

In Exercises 1 through 12, express the given statements as equations.

1. y varies directly as t.

2. x varies directly as s.

3. y varies directly as the square of s.

4. s varies directly as the cube of t.

5. t varies inversely as y.

6. y varies inversely as the square of x.

7. y varies directly as the product st.

8. s varies directly as the product xyz.

9. y varies directly as s and inversely as t.

10. y varies directly as s and inversely as the square of t.

11. x varies directly as the product yz and inversely as the square of t.

12. v varies directly as the cube of s and inversely as t.

In Exercises 13 through 20, give the equation relating the variables after evaluating the constant of proportionality for the given set of values.

13. y varies directly as s, and $y = 25$ when $s = 5$.

14. y varies inversely as t, and $y = 2$ when $t = 7$.

15. s is proportional to the cube of t, and $s = 16$ when $t = 2$.

16. v is proportional to the product st, and $v = 18$ when $s = 2$ and $t = 3$.

17. u is inversely proportional to the square of d, and $u = 17$ when $d = 4$.

18. q is inversely proportional to the square root of p, and $q = 5$ when $p = 9$.

19. y is directly proportional to x and inversely proportional to t, and $y = 6$ when $x = 2$ and $t = 3$.

20. t is directly proportional to n and inversely proportional to p, and $t = 21$ when $n = 3$ and $p = 5$.

In Exercises 21 through 32, find the required value by setting up the general equation and then evaluating.

21. Find s when $t = 4$ if s varies directly as t and $s = 20$ when $t = 5$.

22. Find y when $x = 5$ if y varies directly as x and $y = 36$ when $x = 2$.

23. Find q when $p = 5$ if q varies inversely as p and $q = 8$ when $p = 4$.

24. Find y when $x = 6$ if y varies inversely as x and $y = 15$ when $x = 3$.

25. Find z when $x = 3$ if z is directly proportional to the square of x and $z = 64$ when $x = 4$.

26. Find d when $t = 3$ if d is directly proportional to the square of t and $d = 100$ when $t = 5$.

27. Find v when $t = 7$ if v varies inversely as the square of t and $v = 1$ when $t = 6$.

28. Find h when $r = 6$ if h varies inversely as the square of r and $h = 9$ when $r = 2$.

29. Find y when $x = 4$ and $z = 6$ if y varies directly as the product xz and $y = 9$ when $x = 3$ and $z = 5$.

30. Find s when $p = 75$ and $q = 5$ if s varies directly as p and inversely as the square of q and $s = 100$ when $p = 4$ and $q = 6$.

31. Find s when $v = 9$ and $t = 3$ if s varies directly as the square of v and inversely as t. We know that $s = 27$ when $v = 3$ and $t = 1$.

32. Find z when $x = 12$ and $y = 4$ if z varies inversely as the product xy. We know that $z = 4$ when $x = 2$ and $y = 3$.

In Exercises 33 through 44, solve the applied problems.

33. A fire science specialist studies the motion of an object projected upward from an explosion. The velocity v of the falling object is proportional to the time t of the fall. Find the equation relating v and t if $v = 96$ ft/s when $t = 3.0$ s.

34. An electric circuit has a fixed resistance so that the voltage E varies directly as the current I. If the voltage is 115 V when the current is 5.0 A, find the equation that relates E and I.

35. A certain automobile engine produces p cm^3 of carbon monoxide proportional to the time t that it idles. Find the equation relating p and t if $p = 60,000$ cm^3 for $t = 2$ min.

36. In chemistry, the general gas law states that the pressure P of a gas varies directly as the thermodynamic temperature T and inversely as the volume V. Express this statement as a formula. The constant of proportionality is called R.

37. According to Coulomb's law, the force F between two charges is directly proportional to the product of the charges (Q_1 and Q_2) and inversely proportional to the square of the distance s between them. Express this relationship as a formula.

38. The intensity of illumination I on a surface varies inversely as the square of the distance d from the source. If $I = 12.0$ footcandles when $d = 12.5$ ft, find I when $d = 10.0$ ft.

39. The horsepower necessary to propel a motorboat is proportional to the cube of the speed of the boat. If 10.5 hp is necessary to go 10.0 mi/h, what power is required for 15.0 mi/h?

40. The heat loss through rock wool insulation is inversely proportional to the thickness of the rock wool. If the loss through 6 in. of rock wool is 3200 Btu/h, find the loss through 2.5 in. of rock wool.

41. The electrical resistance R of a wire varies directly as the length l and inversely as the square of the diameter d. If a certain wire is 100.0 ft long with a diameter of 0.00200 ft, its resistance is 6.50 Ω. Find the resistance if the same material is used for a wire that is 25.0 ft long with a diameter of 0.00750 ft.

42. The kinetic energy E (energy due to motion) of an object varies directly as the square of its velocity v. Given that $E = 5000$ kg$\cdot$m^2/s^2 when $v = 20$ m/s, find E when $v = 50$ m/s.

43. An industrial robot is being designed, and there is a need for two gears that will mesh. The number N of teeth on the first gear varies directly as r_2 and inversely as r_1, where r_1 is the number of revolutions per minute of the first gear and r_2 is the number of revolutions per minute of the second gear. The first gear turns at 150 r/min while the second gear turns at 200 r/min. Find the number of revolutions per minute for the second gear if the first gear turns at the rate of 180 r/min.

44. A 300.0-m length of aluminum wire contracts in length by 0.230 m when cooled from 90.0°F to 30.0°F. Its decrease in length varies directly as the product of its original length L and the change in temperature $T_2 - T_1$. Find the constant of proportionality.

In Exercises 45 through 48, use a scientific calculator to solve the given problems.

45. The force F on the blade of a wind generator varies directly as the product of the blade's area A and the square of the wind velocity v. Find an equation relating force F, area A, and velocity v if $F = 19.16$ lb when $A = 3.720$ ft^2 and $v = 31.38$ ft/s.

46. The heat H (in joules) developed in an electric circuit varies directly as the product of time t (in seconds) and the square of the current I (in amperes). Find an equation relating heat H, time t, and current I if $H = 1630$ J when $t = 3.42$ s and $I = 9.76$ A.

47. The time t it takes a certain elevator to lift a load varies directly as the product of the weight w being lifted and the distance s. If it takes 21.23 s to lift 583.2 lb through 87.35 ft, how long will it take to lift 678.9 lb through 98.24 ft?

48. The force F applied to a body is directly proportional to the product of the mass of the body and the acceleration imparted to it. If a force of 37.936 N is applied to a body with a mass of 16.342 kg, the resulting acceleration is 2.3214 m/s^2. Find the acceleration if the force is 124.25 N and the mass is 87.329 kg.

Chapter 5 Formulas

$y = kx$	direct variation	(5-1)
$y = \dfrac{k}{x}$	inverse variation	(5-2)

5-7 Review Exercises for Chapter 5

In Exercises 1 through 12, find the solution for each equation. Check by substituting the value found in the original equation.

1. $x - 3 = 5$
2. $x + 14 = 11$
3. $3y = 27$
4. $26q = 13$
5. $4x + 21 = 5$
6. $2n - 1 = 17$
7. $3(x + 1) = x + 11$
8. $2(y - 1) = 19 - 5y$
9. $2(1 - 2t) = 11 - t$
10. $19 - 3x = 5(1 - 2x)$
11. $7(s - 1) + 2(s + 2) = 3s$
12. $8(t + 2) = 3(2t + 13)$

In Exercises 13 through 32, solve for the indicated letter in terms of the others. Fields of study in which each formula was developed are in parentheses.

13. $R = R_1 + R_2 + R_3$, for R_3 (electricity)

14. $F = \dfrac{wa}{g}$, for g (physics: force)

15. $r = \dfrac{ms_1}{s_2}$, for s_2 (mathematics: statistics)

16. $P = I^2R$, for R (electricity)

17. $d_m = (n - 1)A$, for n (physics: optics)

18. $T_2w = q(T_2 - T_1)$, for T_1 (chemistry: energy)

19. $M_1V_1 + M_2V_2 = PT$, for M_1 (economics)

20. $f(u + v_s) = f_su$, for v_s (physics: sound)

21. $R = \dfrac{wL}{H(w + L)}$, for H (interior design)

22. $A = \dfrac{-\mu R_0}{r + R_0}$, for r (electronics)

23. $W = T(S_1 - S_2) - H_1 + H_2$, for T (refrigeration)

24. $2p + dv^2 = 2d(C - W)$, for C (fluid flow)

25. $a(2 + 3x) = 3y + 2ax$, for y
26. $c(ax + c) = b(x + c)$, for a
27. $a(x + b) = b(x + c)$, for c
28. $2(y + a) - 3 = y(2 + a)$, for y
29. $3(2 - x) = a(a + b) - 2x$, for b
30. $r_1(r_1 - r_2) - 3r_3 = r_1r_2 + r_1^2$, for r_3

31. $3a(a + 2x) + a^2 = a(2 + a)$, for x

32. $\dfrac{x(a + 4)}{2} = 3(x + 5)$, for a

In Exercises 33 through 44, solve the given inequalities.

33. $x - 2 < 7$
34. $x + 4 < 2$
35. $2x \geq 10$
36. $6x \geq 18$
37. $3x + 10 < 1$
38. $5x - 1 < 19$
39. $6 - x > 10$
40. $5 - x > 4$
41. $5(x + 1) < x - 3$
42. $4(x + 2) < x - 4$
43. $12 - 2x \geq 3(x - 1)$
44. $9 - x \geq 3(x - 5)$

In Exercises 45 through 48, find the indicated ratios.

45. 2 ft to 36 in.
46. 80 mm to 15 cm
47. 4 min to 40 s
48. 75¢ to $3

In Exercises 49 through 52, solve the given proportions for x.

49. $\dfrac{x}{3} = \dfrac{8}{9}$
50. $\dfrac{x}{4} = \dfrac{17}{12}$
51. $\dfrac{3}{10} = \dfrac{x}{15}$
52. $\dfrac{6}{5} = \dfrac{x}{10}$

In Exercises 53 through 60, find the number.

53. One more than twice a number is nine.   **54.** Three less than a number is five.

55. Three times a number is eight more than the number.

56. Five more than a number is ten more than three times the number.

57. The ratio of a number to 5 is the same as the ratio of 7 to 15.

58. The ratio of a number to 7 is the same as the ratio of 25 to 28.

59. The ratio of a number to 12 is the same as the ratio of 7 to 8.

60. The ratio of a number to 25 is the same as the ratio of 13 to 15.

In Exercises 61 through 68, solve the given problems by using variation.

61. y varies directly as x; $y = 24$ when $x = 4$; find the resulting equation relating y and x.

62. s varies directly as the square of t; $s = 60$ when $t = 2$; find the resulting equation relating s and t.

63. m is inversely proportional to the square root of r; $m = 5$ when $r = 9$; find the resulting equation relating m and r.

64. v is inversely proportional to the cube of z; $v = 3$ when $z = 2$; find the resulting equation relating v and z.

65. f is directly proportional to m and inversely proportional to p; $f = 8$ when $m = 4$ and $p = 5$; find f when $m = 3$ and $p = 6$.

66. y is directly proportional to the cube of x and inversely proportional to the square of t; $y = 16$ when $x = 2$ and $t = 3$; find y when $x = 3$ and $t = 4$.

67. s varies directly as the product of t and u and inversely as v; $s = 27$ when $t = 3$, $u = 5$, and $v = 6$; find s for $t = 2$, $u = 3$, and $v = 4$.

68. v varies directly as the square root of n and inversely as the product of m and p; $v = 16$ when $n = 4$, $m = 3$, and $p = 6$; find v for $n = 9$, $m = 5$, and $p = 2$.

In Exercises 69 through 92, solve the given problems by first setting up the proper equation.

69. An architect designs a rectangular window such that the width is 18 in. less than the height. If the perimeter of the window is 180 in., what are its dimensions? See Figure 5-5.

70. A square tract of land is enclosed with fencing and then divided in half by additional fencing parallel to two of the sides. If 75 m of fencing is used, what is the length of one side of the tract? See Figure 5-6.

71. The cost of manufacturing one type of computer memory storage device is three times the cost of manufacturing another type. The total cost of producing two of the first type and three of the second type is $450. What is the cost of producing each type?

72. An air sample contains 4 ppm (parts per million) of two harmful pollutants. The concentration of one is four times the other. What is the concentration of each?

73. A rectangular security area is enclosed with 124 ft of fencing. If this rectangular area is arranged so that the ratio of the length to the width is $\frac{3}{2}$, find the dimensions.

74. If you get 68 on one test, 73 on a second test, and 84 on a third test, what score must you get on a fourth test in order to have an average (arithmetic mean) of 80?

75. An architect begins with a design of a square room, but she decides to make it rectangular by doubling the lengths of two opposite sides. If the resulting perimeter is 72 ft, what was the perimeter of the original room?

Height

Width

FIGURE 5-5

FIGURE 5-6

76. A company has three checking accounts for petty cash. One account has $3000 more than that of a second, which has $4500 more than the third account. The three accounts have a total of $21,600. How much is in each account?

77. An engineering firm constructing a bridge requires a certain number of 20-m steel girders to go the length of the span. If the girders were 16 m long, six more would be necessary. How long is the bridge?

78. The electric current in one transistor is three times that in another transistor. If the sum of the currents is 12 mA, what is the current in each?

79. An electric circuit requires 21 contacts. Some of the contacts cost 14¢ each while the others cost 22¢ each, and the total cost of the 21 contacts is $3.98. How many contacts of each type are there?

80. A vending machine accepts nickels, dimes, and quarters, but all nickels were given out as change. There are 140 coins remaining and they have a total value of $21.95. How many of each coin are there?

81. Two airplanes leave Chicago at the same time and fly in opposite directions. If one plane flies at 480 mi/h and the other flies at 220 mi/h, how long will it take them to be 1750 mi apart?

82. A jet took 6 h to fly against headwinds of 50 mi/h from city A to city B. It took 5 h on the return trip when the winds became tailwinds of 50 mi/h. How fast does the jet travel relative to the air?

83. Two jets, 12,000 km apart, start toward each other and meet 3 h later. One is going 300 km/h faster than the other. What is the speed of each?

84. An environmental scientist finds that a certain fish can swim 12 mi/h in still water. This fish swims upstream against a 3.0 mi/h current, then returns to its starting point by swimming downstream. The upstream trip took 0.20 h longer than the downstream trip. How far upstream did the fish swim?

85. How much water must be added to 15 L of alcohol to make a solution which is 60% alcohol?

86. A chemist has 30 mL of a 12% solution of sulfuric acid. How many milliliters of a 40% solution must be added to get a 15% solution?

87. A $12,500 account yields an interest rate of 8.00%. How much must be added to this account so that the total interest yield amounts to $1720?

88. If $6500 is invested at 6.00%, how much more must be invested at 8.00% in order for the total investment to yield an average of 7.50%?

89. The weight w on the end of a spring varies directly as the length x that the spring stretches. If a weight of 8.0 lb stretches the spring 2.3 in., what weight will stretch it 5.7 in.?

90. The cost C of operating an electric appliance varies as the product of its wattage w, the time t it is used, and the electric rate r. It costs $1.62 to run a 1200-W microwave oven for 15 h at a rate of 9¢/kWh. How much does it cost to operate a 60-W electric light bulb for 50 h at the same rate?

91. The period of a pendulum varies directly as the square root of its length. If a pendulum 1.00 ft long has a period of 1.11 s, find the period of a pendulum 10.0 ft long.

92. The amount of heat H which passes through a wall t feet thick is proportional to the product of the area A of the wall and $T_2 - T_1$, the difference in temperature of the surfaces of the wall, and is inversely proportional to t. Express this as an equation and solve for T_2.

In Exercises 93 through 96, use a scientific calculator to solve the given problems.

93. Solve for x: $\dfrac{1.23x}{8.17} = -3.17x + 8.15$

94. Solve for x: $3.127 - 6.218x < 5.200 + 4.018x$

95. If 60.0000 mi/h = 26.8224 m/s, what speed in miles per hour is 32.0000 m/s?

96. According to Hooke's law, the force on a spring varies directly as the distance x that the spring stretches. Find the equation relating force F and distance x if a force of 45.13 lb causes the spring to stretch 0.2690 ft.

CHAPTER SIX

Introduction to Geometry

Chapter 1 introduced a few of the basic geometric figures. We used the triangle, rectangle, and square to demonstrate some of the arithmetic operations. In this chapter we examine terms and formulas related to some of the elementary geometric figures. This chapter introduces some of the key and basic concepts of geometry so that they can be used in later chapters. Additional geometric concepts will be developed in Chapter 15.

The study of geometry allows us to solve many different practical and applied problems. This chapter includes many examples and exercises illustrating the use of geometry. For example, suppose that a pilot must fly around a circular restricted airport zone which has a radius of 5 mi. (See Figure 6-1.) If the pilot must fly around that zone as shown in the figure, how much longer is this route than the route which passes directly through the restricted zone? The solution to this problem will be given later in this chapter.

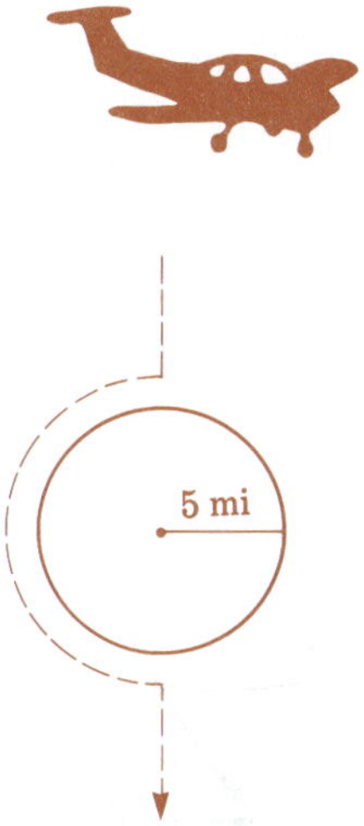

6-1 Basic Geometric Figures

Geometry deals with the properties and measurement of angles, lines, surfaces, and volumes and with the basic figures that are formed. In establishing the properties of the basic figures it is not possible to define every word and prove every statement. Certain words and concepts must

5 mi

FIGURE 6-1

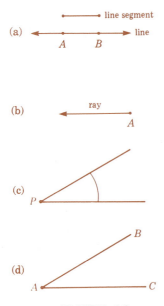

FIGURE 6-2

be used without definition. In general, *the words* **point, line,** *and* **plane** *are accepted in geometry without being defined.* This gives us a starting point for defining other terms.

In Figure 6-2a we show a line passing through points *A* and *B*. The line extends indefinitely far in both directions. *A* **line segment** *is a portion of a line and is bounded by two fixed points.* In Figure 6-2a, the line segment bounded by the points A and B is denoted by $\overline{AB}$. Although no line has a fixed length, every line segment does have a length that can be measured. *A* **ray** *consists of a fixed boundary point and the portion of a line to one of its sides* as in Figure 6-2b.

The amount of rotation of a ray about its endpoint is called an **angle.** (See Figure 6-2c.) Notation and terminology associated with angles are illustrated in the following example.

EXAMPLE A

In Figure 6-2d, the angle formed by *AB* and *AC* is denoted by $\angle BAC$. The **vertex** of the angle is the point *A*, and the **sides** of the angle are *AB* and *AC*.

One complete rotation (see Figure 6-3a) *of a ray about a point is defined to be an angle of 360* **degrees,** *written as 360°. A* **straight angle** *contains 180°* (see Figure 6-3b), *and a* **right angle** *(denoted, as in* Figure 6-3c, *by* ⌐) *contains 90°. If two lines meet so that the angle between them is a right angle, they are said to be* **perpendicular** (see Figure 6-4). □

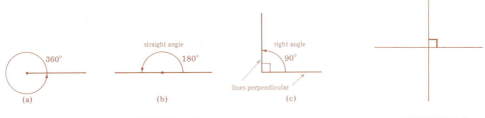

FIGURE 6-3   **FIGURE 6-4**

A common method of stating the magnitude of an angle is to give the **measure** of the angle. That is, the measure of a straight angle is 180°, and the measure of a right angle is 90°. However, in our discussions we shall refer to the angle rather than the measure of the angle. Also, we shall use the same symbol for the angle and the measure of the angle.

A device that can be used to measure angles approximately is a **protractor.** Figure 6-5 shows the protractor positioned to measure an angle of 50°.

When it is necessary to measure angles to an accuracy of less than 1°, **decimal parts of a degree** may be used. The use of decimal parts of a de-

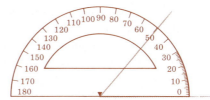

FIGURE 6-5

gree has become common with the extensive use of calculators. Previously the common way of expressing angles to such accuracy was to divide each degree into 60 equal parts, called **minutes,** and to divide each minute into 60 equal parts called **seconds.** Surveyors and astronomers frequently measure angles to the nearest second.

The notation for minutes is ′ and that for seconds is ″. Thus an angle of 32 degrees, 15 minutes, and 38 seconds is denoted as $32°15'38''$. Since $1° = 60'$, we find a decimal part of a degree in minutes by multiplying the decimal by $60'$. Also, since $1' = (\frac{1}{60})°$, we express minutes in terms of a decimal part of a degree by dividing by 60.

EXAMPLE B

1. $0.2° = (0.2)(60') = 12'$. Thus $17.2° = 17°12'$.

2. $36' = (\frac{36}{60})° = 0.6°$. Thus $58°36' = 58.6°$.

3. $6.35° = 6°21'$, since $(0.35)(60') = 21'$. ☐

EXAMPLE C For a person standing on the surface of the earth and looking at the moon, the angle made by the "top edge" of the moon, the viewer, and the center of the moon is about $(\frac{1}{4})°$. Express this angle in minutes.

Since $(\frac{1}{4})° = 0.25°$, we get $0.25° = (0.25)(60') = 15'$. ☐

EXAMPLE D A surveyor measures the angle between two land boundaries and gets $72°11'$. Express this angle in degrees and decimal parts of a degree.

Since $11' = (\frac{11}{60})° = 0.18°$ (to the nearest hundredth), we get $72°11' = 72.18°$. ☐

Many scientific calculators can be used to automatically change from degrees, minutes, and seconds to the corresponding degrees and decimal parts of a degree (or vice versa). The typical usage involves a key whose secondary purpose is to make these changes, and the key may be labeled as DMS-DD for degrees, minutes, seconds, and decimal degrees. Consult the manual of instructions for your particular calculator.

EXAMPLE E   Use a calculator to convert 48°30′9″ to the decimal format; then use a calculator to convert back to degrees, minutes, and seconds.

One very common use of the DMS-DD key is as follows. To convert 48°30′9″ to the decimal format, enter the following key sequence.

The display will show 48.5025, indicating that 48°30′9″ = 48.5025°. In the above key sequence, note the way that 48°30′9″ is entered in the calculator. Now let's reverse the process by using the calculator to convert 48.5025° to the degree-minute-decimal format. Enter

and the calculator display will be 48.3009. This is the calculator's way of displaying 48°30′9″, which is the correct result.   □

A **triangle** *is a closed plane figure which has three sides,* and these three sides form three interior angles. *In an* **equilateral triangle,** *the three sides are equal in length and each angle is a 60° angle* (see Figure 6-6a). *In an* **isosceles triangle,** *two of the sides are equal in length and the* **base angles** *(the angles opposite the equal sides) are also equal* (see Figure 6-6b). *In a* **scalene triangle,** *no two sides are equal in length and no two angles are equal* (see Figure 6-6c).

One of the most important triangles in scientific and technical applications is the **right triangle.** *In a right triangle, one of the angles is a right angle. The side opposite the right angle is called the* **hypotenuse,** *and the other two sides are called the* **legs** (see Figure 6-6d).

equilateral
triangle

(a)

isosceles
triangle

(b)

scalene
triangle

(c)

hypotenuse

right
triangle

(d)

FIGURE 6-6

EXAMPLE F

1. The triangle in Figure 6-7a, denoted as △ABC, is equilateral. Since AC = 5, we also know that AB = 5 and BC = 5. All of the angles are 60° angles, and we note this by writing ∠A = 60°, ∠B = 60°, and ∠C = 60°.

2. The triangle in Figure 6-7b, △DEF, is isosceles since DF = 7 and EF = 7. The base angles are ∠D and ∠E. Therefore, since ∠D = 70°, we know that ∠E = 70°.

3. The triangle in Figure 6-7c, △PQR, is scalene since none of the sides is equal in length to another side.

4. The triangle in Figure 6-7d, △LMN, is a right triangle since ∠M = 90°. The hypotenuse is the side LN, and the legs are MN and LM.   □

FIGURE 6-7

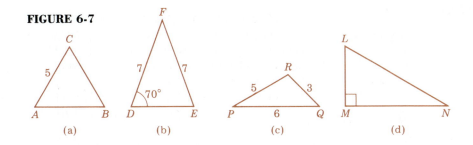

(a)   (b)   (c)   (d)

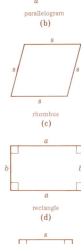

quadrilateral
(a)

parallelogram
(b)

rhombus
(c)

rectangle
(d)

square
(e)

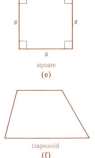

trapezoid
(f)

FIGURE 6-8

A **quadrilateral** *is a closed plane figure which has four sides,* and these four sides form four interior angles. In Figure 6-8a, a general quadrilateral is shown.

A **parallelogram** *is a quadrilateral in which the opposite sides are* **parallel.** *This means that extensions of these sides will not intersect.* In a parallelogram, opposite sides are equal and opposite angles are equal (see Figure 6-8b). A **rhombus** *is a parallelogram with four equal sides* (see Figure 6-8c).

A **rectangle** *is a parallelogram in which intersecting sides are perpendicular,* which means that all four interior angles are right angles. Also, opposite sides of a rectangle are equal in length and parallel (see Figure 6-8d). In a rectangle the longer side is usually called the **length** and the shorter side the **width.** A **square** *is a rectangle with four equal sides* (see Figure 6-8e).

A **trapezoid** *is a quadrilateral in which two sides are parallel. These parallel sides are called the* **bases** *of the trapezoid* (see Figure 6-8f).

EXAMPLE G

1. In parallelogram *ABCD*, shown in Figure 6-9a, opposite sides are parallel. This may be denoted as $AB \parallel DC$ and $AD \parallel BC$, where $\parallel$ means "parallel to." Also, opposite sides are equal, and opposite angles are equal. We write these as $AB = DC$, and $AD = BC$, and $\angle A = \angle C$ and $\angle B = \angle D$.

2. In rectangle *EFGH*, shown in Figure 6-9b, the opposite sides are equal so that $EF = HG$ and $EH = FG$. All interior angles are right angles. This means that $\angle E = \angle F = \angle G = \angle H = 90°$.

3. In quadrilateral *PQRS*, shown in Figure 6-9c, side *PQ* is parallel to side *SR*, which may be stated as $PQ \parallel SR$. It is a trapezoid with bases *PQ* and *SR*. The **base angles** of *PQ* are $\angle P$ and $\angle Q$, and the base angles of *SR* are $\angle S$ and $\angle R$.

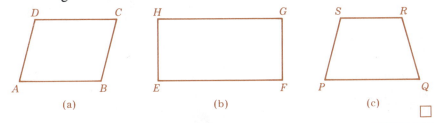

(a)   (b)   (c)

FIGURE 6-9

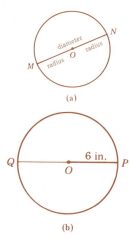

(a)

(b)

FIGURE 6-10

When we are identifying quadrilaterals, more than one designation may be technically correct. For example, a square can also be considered a rectangle (four right angles) or a rhombus (four equal sides). Only a square has both properties, however, and the word "square" should be used to identify the figure. It must also be kept in mind that a square has all the properties of a rectangle and a rhombus.

The last basic geometric figure we shall discuss in this section is the **circle.** *All the points on a circle are the same distance from a fixed point in the plane* (see Figure 6-10a). *This point O is the* **center** *of the circle. The distance ON (or OM) from the center to a point on the circle is the* **radius** *of the circle. The distance MN between two points on the circle and on a line passing through the center of the circle is the* **diameter** *of the circle. Thus the diameter is twice the radius, which we may write as* $d = 2r$.

EXAMPLE H

For the circle shown in Figure 6-10b, the radius OP is 6 in. long. This means that the diameter QP is 12 in. long. We may also show these as $r = 6$ in. and $d = 12$ in.   □

Exercises 6-1

In Exercises 1 through 4, first estimate the measure of the angles given in Figure 6-11, and then use a protractor to measure them to the nearest degree. (It may be necessary to extend the lines.)

1.

2.

3.

4.

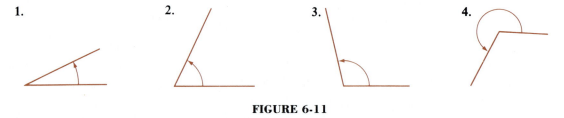

FIGURE 6-11

In Exercises 5 through 8, express each angle in degrees and minutes.

5. 56.4°   **6.** 18.9°   **7.** 136.45°   **8.** 79.05°

In Exercises 9 through 12, express each angle in degrees and decimal parts of a degree.

9. 156°15′   **10.** 33°48′   **11.** 67°6′   **12.** 16°57′

In Exercises 13 through 28, answer the given questions about the figures in Figure 6-12.

13. In Figure 6-12a, identify any right angles.

14. In Figure 6-12a, identify any straight angles.

15. In Figure 6-12b, identify the isosceles triangle.

16. In Figure 6-12b, identify the scalene triangle.

17. In Figure 6-12b, identify the right triangle.

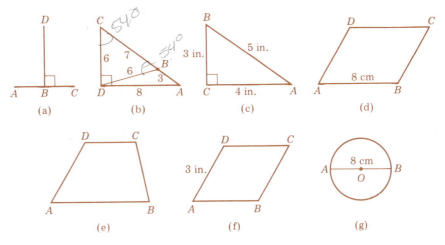

FIGURE 6-12

18. In Figure 6-12b, if $\angle DBC = 54°$ (the angle with sides DB and BC, with vertex at B), how many degrees are there in $\angle C$?

19. In Figure 6-12c, identify and determine the lengths of the legs of the right triangle.

20. In Figure 6-12c, identify and determine the length of the hypotenuse.

21. In the parallelogram in Figure 6-12d, identify the pairs of parallel sides.

22. In the parallelogram in Figure 6-12d, identify the pairs of opposite angles.

23. In the parallelogram in Figure 6-12d, if $\angle ABC = 135°$, what is $\angle CDA$?

24. In the parallelogram in Figure 6-12d, what is the length of side DC?

25. What are the bases of the trapezoid in Figure 6-12e?

26. What is the length of the side DC of the rhombus in Figure 6-12f?

27. For the circle in Figure 6-12g, identify a radius.

28. What is the radius of the circle in Figure 6-12g?

In Exercises 29 through 36, use a protractor and ruler to construct the figures required.

29. An equilateral triangle of side 2 in.

30. An isosceles triangle with equal sides of 3 cm and base angles of 75°.

31. A right triangle with hypotenuse of 4 cm and one of the other sides of 2 cm.

32. A parallelogram with sides of 4 in. and 2 in. and one interior angle of 60°.

33. A rectangle with sides of 4 in. and 2 in.

34. A trapezoid with bases of 4 cm and 3 cm.

35. A rhombus with sides of 3 cm and an interior angle of 30°.

36. A quadrilateral with interior angles of 35°, 122°, and 95°.

In Exercises 37 through 56, answer the questions.

37. (a) Is a square always a parallelogram? (b) Is a parallelogram always a square?

38. (a) Is an equilateral triangle always an isosceles triangle? (b) Is an isosceles triangle always an equilateral triangle?

39. Suppose that two opposite vertices of a rhombus are joined by a straight line. What are the figures into which it is divided?

40. Suppose that the opposite vertices of a figure are joined by a straight line and the figure is divided into two equilateral triangles. What is the figure?

41. If one leg of an isosceles right triangle is 10 cm in length, what is the length of the other leg?

42. The sum of the adjacent angles (vertices are at each end of a given side) of a parallelogram is equal to a straight angle. If one angle of a parallelogram is 65°, what is the adjacent angle?

43. The angle between a ray of light and a line perpendicular to a mirror is 36°. What is the angle between the ray of light and the plane of the mirror?

44. Data is stored on a circular floppy disk with a diameter of $5\frac{1}{4}$ in. What is the distance from the center of the disk to a point on the edge?

45. An airplane and a helicopter take off from the same spot. The path of the airplane makes an angle of 40° with the ground, while the helicopter goes straight up. What is the angle between their flight paths?

46. A parachutist glides to earth in a path that makes an angle of 80° with the ground. If she lands at the base of a telephone pole, what is the angle between her glide path and that of the telephone pole?

47. If a crater on Mercury is circular with a radius of 47 km, what is its diameter?

48. A circular access road is to be constructed around an office complex. If the distance between the center of the office complex and the road is 400 ft, what is the diameter of the circle formed by the road?

49. An A-frame house has a front described by Figure 6-13. Determine the angle located to the lower right side of the front.

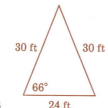

30 ft   30 ft

66°

FIGURE 6-13   24 ft   **FIGURE 6-14**

50. In analyzing the stresses on a roof truss, an engineer must begin with Figure 6-14 and complete the figure so that a parallelogram is formed. Complete Figure 6-14 so that it does form a parallelogram.

51. If a pilot is flying in a northerly direction and makes a 90° turn to the left, what is her new direction?

52. If a submarine captain is headed in an easterly direction and makes a 90° turn to the right, what is his new direction?

53. Use your calculator to express 20°60′ as degrees and decimal parts of a degree. Note that 20°60′ is equivalent to 21°.

54. Use your calculator to express 37°30′40″ in degrees and decimal parts of a degree.

55. Use your calculator to express 46.06° in degrees, minutes, and seconds.

56. Use your calculator to express 13.155° in degrees, minutes, and seconds.

6-2 Perimeter

Two of the most basic measures of plane geometric figures are perimeter and area. In this section we consider perimeter. Area will be presented in the following section.

In Chapter 1 we noted that *the* **perimeter** *of a plane figure is the distance around it*. In the following example, perimeters are found directly from this definition.

EXAMPLE A

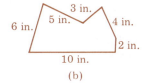

8 cm 7 cm

6 cm

(a)

3 in.

5 in. 4 in.

6 in.

2 in.

10 in.

(b)

FIGURE 6-15

1. To find the perimeter of the triangular cooling fin shown in Figure 6-15a, we add the lengths of the sides. We get the perimeter p as follows.

$$p = 6 \text{ cm} + 7 \text{ cm} + 8 \text{ cm} = 21 \text{ cm}$$

2. To find the perimeter of the figure shown in Figure 6-15b, we simply add the lengths of the individual sides, even though the figure may appear to be somewhat complicated. In this case we find the perimeter to be

$$p = 10 \text{ in.} + 6 \text{ in.} + 5 \text{ in.} + 3 \text{ in.} + 4 \text{ in.} + 2 \text{ in.} = 30 \text{ in.}$$

□

We can use the definition of perimeter to derive formulas for many specific plane figures. However, it isn't necessary to memorize most of these formulas because we can easily apply the basic and general definition of perimeter. Consider the formulas in the following example.

EXAMPLE B

For the figures shown in Figure 6-16, we have the following perimeter formulas:

See Appendix B for a computer program that calculates the perimeter of a triangle.

Figure	Perimeter	
(a) Triangle with sides a, b, c	$p = a + b + c$	**(6-1)**
(b) Equilateral triangle with side s	$p = 3s$	**(6-2)**
(c) Quadrilateral with sides a, b, c, d	$p = a + b + c + d$	**(6-3)**
(d) Rectangle of length l and width w	$p = 2l + 2w$	**(6-4)**
(e) Parallelogram with sides a and b	$p = 2a + 2b$	**(6-5)**
(f) Square of side s	$p = 4s$	**(6-6)**

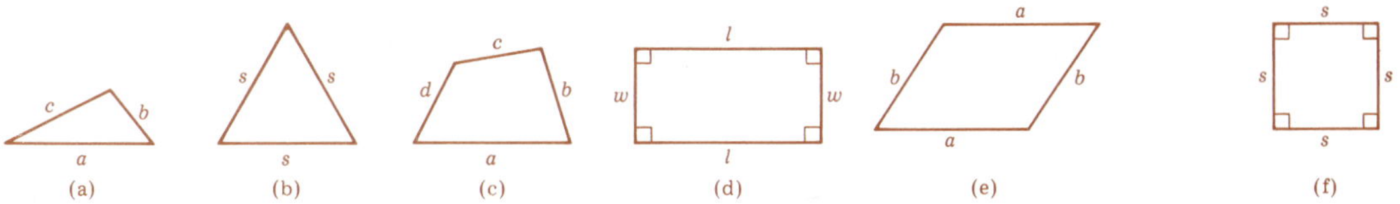

(a) (b) (c) (d) (e) (f)

FIGURE 6-16

□

EXAMPLE C

1. The perimeter of an equilateral triangle of side 18 ft is

$$p = 3(18 \text{ ft}) = 54 \text{ ft} \qquad \text{Using Eq. (6-2)}$$

2. The perimeter of a parallelogram with sides 21 cm and 15 cm is

$$p = 2(21 \text{ cm}) + 2(15 \text{ cm}) \qquad \text{Using Eq. (6-5)}$$
$$= 42 \text{ cm} + 30 \text{ cm} = 72 \text{ cm} \qquad \square$$

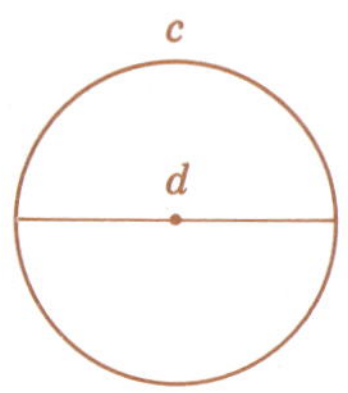

The perimeter of a circle is called the **circumference.** It cannot be found directly from the definition of perimeter. However, we shall use a basic geometric fact, without attempting to develop it, in order to arrive at the necessary formulas: *The ratio of the circumference to the diameter is the same for all circles.* This ratio is represented by π, a number equal to approximately 3.14. To eight significant digits, $\pi = 3.1415927$. We shall use the decimal form rather than the mixed-number form $3\frac{1}{7}$ (which is also only approximate) for π, since the decimal form is generally more convenient for calculations. (Many scientific calculators have a key whose secondary purpose is to provide the value of π expressed with as many significant digits as the calculator allows.)

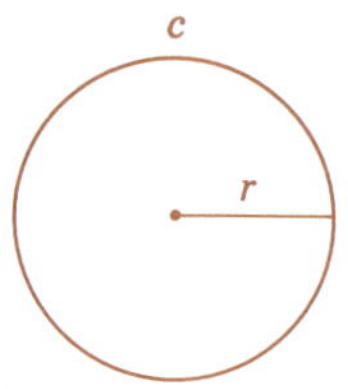

Since the ratio of the circumference c to the diameter d equals π, we have $c/d = \pi$. This leads to the formula

$$\boxed{c = \pi d} \qquad \text{circumference of circle} \qquad (6\text{-}7)$$

FIGURE 6-17

for the circumference. See Figure 6-17a. Also, since the diameter equals twice the radius, we have

$$\boxed{c = 2\pi r} \qquad \text{Note that } c = \pi d = \pi(2r) = 2\pi r \qquad (6\text{-}8)$$

See Appendix B for a computer program that calculates the circumference of a circle.

as a formula for the circumference in terms of the radius. See Figure 6-17b.

Although it isn't necessary to memorize most of the formulas for perimeter, it is necessary to memorize at least one of the above two formulas since they don't follow directly from the general perimeter definition.

EXAMPLE D

1. The circumference of a wheel with a diameter of 2.00 ft is

$$c = \pi(2.00 \text{ ft}) = 3.14(2.00 \text{ ft}) \qquad \text{Using Eq. (6-7)}$$
$$= 6.28 \text{ ft}$$

See Figure 6-18a.

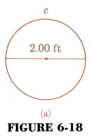

(a)

FIGURE 6-18

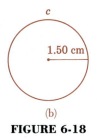

c

1.50 cm

(b)

FIGURE 6-18

2. The circumference of a washer with a radius of 1.50 cm is

$$c = 2\pi(1.50\text{ cm}) = 2(3.14)(1.50\text{ cm}) \quad \text{Using Eq. (6-6)}$$

$$= 9.42\text{ cm}$$

See Figure 6-18b. □

EXAMPLE E

Tubing

Oil

2.3 km

Spill

FIGURE 6-19

A circular oil spill has a diameter of 2.3 km. If this oil spill is to be enclosed within a length of specially designed flexible tubing, how long must this tubing be?

The length of the tubing must be as long as the circumference of the circle. Knowing the diameter, we can find the circumference of the circle as follows:

$$c = \pi(2.3\text{ km}) = 3.14(2.3\text{ km})$$

$$= 7.2\text{ km}$$

The length of the tubing required is therefore 7.2 km. See Figure 6-19. □

By using the definition of perimeter, we can find the perimeters of geometric figures that are combinations of basic figures. Consider the following examples.

EXAMPLE F

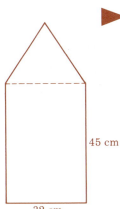

45 cm

32 cm

FIGURE 6-20

Figure 6-20 is a combination of a rectangle and an equilateral triangle. (The dashed line shows where the figures are joined, but *it should not be counted as part of the perimeter since it is not on the outside of the figure.*) Since the lower part of the figure is a rectangle, we see that the dashed line is 32 cm long. This tells us that each side of the triangle is 32 cm long, since it is equilateral. The perimeter can now be found by adding the 32 cm along the bottom to the two 32-cm lengths at the top to the two 45-cm lengths at the sides. We get

$$\begin{array}{ccc} \text{bottom} & \text{top} & \text{sides} \end{array}$$
$$p = 32\text{ cm} + 2(32\text{ cm}) + 2(45\text{ cm})$$
$$= 32\text{ cm} + 64\text{ cm} + 90\text{ cm}$$
$$= 186\text{ cm}$$

□

EXAMPLE G   A certain machine part is a square with a quarter circle removed (see Figure 6-21). The side of the square is 3.00 in. The perimeter of the part is to be coated with a special metal costing 25¢ per inch. What is the cost of coating this part?

Setting up a formula for the perimeter, we add the two sides of length s to *one-fourth of the circumference of a circle with radius* **s** to get

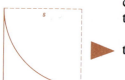

$s = 3.00$ in.

FIGURE 6-21

bottom and left $\searrow$ circular section

$$p = 2s + \frac{2\pi s}{4} = 2s + \frac{\pi s}{2}$$

where s is the side of the square as well as the radius of the circular part. Letting $s = 3.00$ in., we have

$$p = 2(3.00 \text{ in.}) + \frac{(3.14)(3.00 \text{ in.})}{2}$$

$$= 6.00 \text{ in.} + 4.71 \text{ in.} = 10.71 \text{ in.}$$

Since the coating costs 25¢ per inch, the cost is

$$c = 25p = (25¢/\text{in.})(10.71 \text{ in.}) = 268¢ = \$2.68$$

EXAMPLE H   A pilot must fly around a circular restricted airport zone which has a radius of 5 mi. (The 5 is exact.) See Figure 6-22. How much longer is this route than the route which passes directly through the restricted zone?

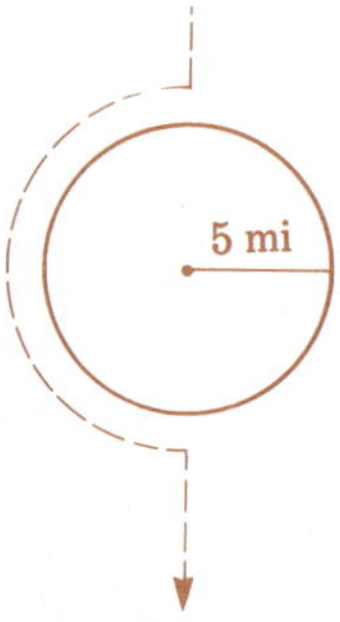

If the pilot could fly directly through the restricted zone, the distance traveled would be 10 mi (the diameter of the circle). Since the pilot must fly *around* the restricted zone, the distance traveled is one-half the circumference. Since

$$c = 2\pi r = 2\pi (5 \text{ mi})$$

$$= 10\pi \text{ mi} = 31.4 \text{ mi}$$

5 mi

we see that the pilot must travel $\frac{1}{2}(31.4 \text{ mi}) = 15.7 \text{ mi}$. This is 5.7 mi longer than the 10-mi route.

FIGURE 6-22

Exercises 6-2

In Exercises 1 through 24, find the perimeters (circumferences for circles) of the indicated figures for the given values.

1. The triangle shown in Figure 6-23a.
2. The triangle shown in Figure 6-23b.
3. The quadrilateral shown in Figure 6-23c.
4. The quadrilateral shown in Figure 6-23d.
5. Triangle: $a = 320$ m, $b = 278$ m, $c = 298$ m
6. Triangle: $a = 52$ yd, $b = 49$ yd, $c = 64$ yd

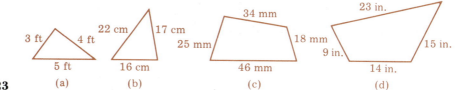

FIGURE 6-23   (a)   (b)   (c)   (d)

7. Quadrilateral: $a = 16.5$ in., $b = 17.3$ in., $c = 21.8$ in., $d = 29.2$ in.

8. Quadrilateral: $a = 6.92$ cm, $b = 8.26$ cm, $c = 9.93$ cm, $d = 8.07$ cm

9. Equilateral triangle: $s = 64.6$ mm

10. Equilateral triangle: $s = 128$ ft   *128×3 → p*

11. Isosceles triangle: equal sides of 15.3 in., third side of 26.5 in.

12. Isosceles triangle: equal sides of 36.2 cm, third side of 12.5 cm

13. Square: $s = 0.65$ m   14. Square: $s = 2.36$ yd

15. Rhombus: side of 0.15 mi   16. Rhombus: side of 178 mm

17. Parallelogram: $a = 47.2$ cm, $b = 36.8$ cm   18. Parallelogram: $a = 1.69$ in., $b = 1.46$ in.

19. Rectangle: $l = 68.7$ ft, $w = 46.6$ ft   20. Rectangle: $l = 4.57$ m, $w = 0.97$ m

21. Circle: $r = 15.1$ cm   22. Circle: $r = 10.6$ ft

23. Circle: $d = 6.74$ in.   24. Circle: $d = 42.0$ mm

In Exercises 25 through 32, find the perimeters of the indicated geometric figures shown in Figure 6-24. Dashed lines are used to identify figures but are not part of the figure of which the perimeter is to be found.

25. The figure in Figure 6-24a.

26. The figure in Figure 6-24b. All angles are right angles.

27. The figure in Figure 6-24c. All angles are right angles.

28. The figure in Figure 6-24d. All angles are right angles.

29. The figure in Figure 6-24e. A square is surmounted by a parallelogram.

30. The figure in Figure 6-24f. Half a square has been removed from a rhombus.   *3(2.83) + 2(2.00) = 2.490m*

31. The figure in Figure 6-24g. A square is surmounted by a semicircle.

32. The figure in Figure 6-24h. A quarter circle is attached to a rectangle.

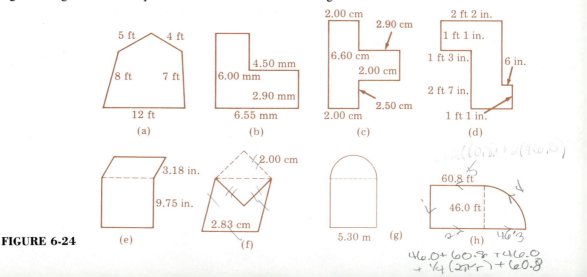

FIGURE 6-24   (e)   (f)

In Exercises 33 through 40, find a formula for the perimeter of the given figures.

33. An isosceles triangle with equal sides s and a third side a.

34. A rhombus with side s.

35. The semicircular figure shown in Figure 6-25a.

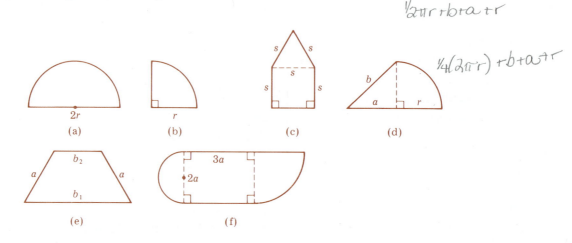

$\frac{1}{2}2\pi r + b + a + r$

$\frac{1}{4}(2\pi r) + b + a + r$

FIGURE 6-25

36. The quarter-circular figure shown in Figure 6-25b.

37. The figure shown in Figure 6-25c. A square is surmounted by an equilateral triangle.

38. The figure shown in Figure 6-25d. A quarter circle is attached to a triangle.

39. The isosceles trapezoid shown in Figure 6-25e.

40. The figure shown in Figure 6-25f. A semicircle and a quarter circle are attached to a rectangle.

In Exercises 41 through 64, solve the given problems.

41. If it costs 50¢/ft for rug binding, what is the cost of binding a rectangular rug that is 21 ft long and 12 ft wide?

42. A room's floor is rectangular with length 18 ft and width 9 ft. This room has three doors, each 3 ft wide. What is the total length of floor molding required for this room? See Figure 6-26.

FIGURE 6-26

43. The radius of the earth's equator is 3960 mi. What is the circumference?

44. A circular floppy disk with a diameter of $5\frac{1}{4}$ in. is to be enclosed in a square plastic container. What is the perimeter of the smallest square that can be used?

45. A park ranger wants to estimate the distance across a circular lake. If she measures the distance around the lake as 2800 ft, find the distance across the middle.

46. The frame of the top of a square card table is made from metal tubing which costs 20¢ per foot. What is the cost of the frames for two tabletops which are each 33 in. on a side?

47. What is the total cost (at $8 per meter) of fencing in a right-triangular plot of land? The legs are each 48 m and the hypotenuse is 68 m.

48. A certain machine part has the shape of a square with equilateral triangles attached to two sides (see Figure 6-27a). The side of the square is 2 cm. What is the perimeter of the machine part?

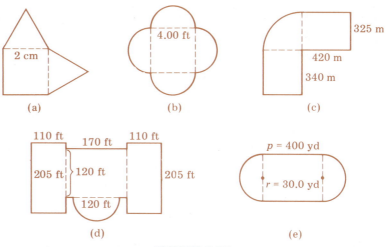

FIGURE 6-27

49. A certain flower bed is made up of semicircular areas attached to a square (see Figure 6-27b). The side of the square is 4.00 ft. What is the perimeter of the flower bed?

50. How much fencing is required to enclose the area shown in Figure 6-27c? The figure consists of two rectangles attached to a quarter circle.

51. To prevent a basement from being flooded, drainage pipe is to be put around the outside of a building whose outline is shown in Figure 6-27d, consisting of three rectangles and a semicircle. How much drainage pipe is required to go around the building? (Ignore the fact that the pipe is slightly away from the walls.)

52. The area within a racetrack is a rectangle with semicircles at each end (see Figure 6-27e). The radius of the circular parts is 30.0 yd, and the perimeter of the area within the track is 400 yd. How long is each straight section of the track?

53. As a wheel rolls along level ground, it makes one complete revolution as it travels a distance of 81.68 in. Find the diameter of the wheel.

54. If a certain car were to travel a distance of exactly one mile, one of its tires would make 830.0 revolutions. Find the radius of the tire in inches.

55. As a ball bearing rolls along a straight track, it makes 11.0 revolutions while traveling a distance of 109 mm. Find its radius.

56. A point on a rotating pulley travels a distance of 20,109 cm while the pulley makes 360 revolutions. Find the distance between that point and the center of the pulley.

57. If a wheel with a 13.00-in. radius and a wheel with a 14.00-in. radius both make one revolution on a road, how much farther does the 14.00-in. wheel travel?

58. The cross-sectional drawing of a machine part consists of a semicircle surmounted on an equilateral triangle with side 5.28 cm. See Figure 6-28. Find the perimeter.

59. Find the perimeter of the figure eight that is formed when a circle with diameter 0.558 cm is surmounted with a circle of diameter 0.902 cm.

FIGURE 6-28

FIGURE 6-29 |←22 cm→| |←22 cm→|

60. The centers of two pulleys are 88 cm apart and they both have diameters of 22 cm. How long is the belt that goes around the two pulleys? See Figure 6-29.

61. If your calculator has a key which can be used to display a value for π, first display that value and then use it to find the circumference of a circle with a radius 37.29 cm.

62. The value $3\frac{1}{7}$ (or $\frac{22}{7}$) is often used as an approximation for π. Use your calculator to enter $22 \div 7$ and display the result. Now subtract π by using the π key if it is available. (Otherwise, use 3.1415927.) What is the displayed difference between $\frac{22}{7}$ and π? When comparing the displays for $\frac{22}{7}$ and π, how many digits agree before the first discrepancy occurs?

63. Some calculators have a key for displaying the value of π with eight significant digits, but they actually store that value with greater precision. Use the π key to display the value of π, then subtract 3.14, then multiply by 1000. What does this display reveal about the stored value of π?

64. If your calculator has a key for displaying the value of π, how many digits does it display? In addition to the displayed value, how many additional digits are stored for additional precision? (See Exercise 63.)

6-3 Area

In addition to perimeter, another basic measure associated with a plane geometric figure is its **area,** which we also briefly introduced in Chapter 1. Although the concept of area is primarily intuitive, it is easily defined and calculated for the basic geometric figures. *Area gives us a measure of the surface of a geometric figure,* just as perimeter gives us the measure of the distance around it. There are many important real applications of area.

In finding the area of a geometric figure, we are finding the number of squares, one unit on a side, required to cover the surface of the figure. In the rectangle in Figure 6-30, we see that 12 squares, each 1 cm on a side, are needed to cover the surface. We therefore say that the area of the rectangle is 12 square centimeters, which we write as 12 cm^2.

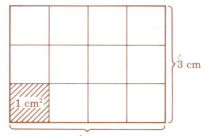

FIGURE 6-30   4 cm

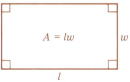

FIGURE 6-31

If we note the rectangle in Figure 6-30, we see that its area may be determined by multiplying its length by its width. Since the area of any rectangle may be found in this way, we define the area of a rectangle to be

$$A = lw \qquad \text{area of rectangle} \tag{6-9}$$

where **l** *and* **w** *must be measured in the same unit of length and* **A** *is in square units of length* (see Figure 6-31).

Since a square is a rectangle with all sides equal in length, its area is

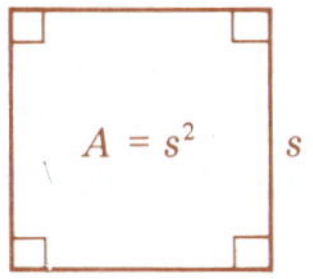

FIGURE 6-32

$$A = s^2 \qquad \text{area of square} \tag{6-10}$$

where **s** *is the length of one side* (see Figure 6-32).

EXAMPLE A

1. A given rectangle has a length of 9.0 ft and a width of 5.0 ft. With $l = 9.0$ ft and $w = 5.0$ ft the area of the rectangle is

$$A = (9.0 \text{ ft})(5.0 \text{ ft}) = 45 \text{ ft}^2 \qquad \text{using Eq. (6-9)}$$

2. A certain square has a side of 5.0 cm. With $s = 5.0$ cm the area of the square is

$$A = (5.0 \text{ cm})^2 = 25 \text{ cm}^2 \qquad \text{using Eq. (6-10)} \qquad \square$$

FIGURE 6-33

We may use the definition of the area of a rectangle to determine the area of a parallelogram. Examine the parallelogram in Figure 6-33 and note that a triangular area is formed to the left of the dashed line labeled h. This dashed line is perpendicular to the **base** b of the parallelogram. If we move the triangular area to the right of the parallelogram, a rectangle of length b and width h is formed. Since the area of the rectangle would be bh, the area of the parallelogram is

$$A = bh \qquad \text{area of parallelogram} \tag{6-11}$$

Here **h** *is called the* **height,** *or* **altitude,** *of the parallelogram. Remember: It is the perpendicular distance from one base* **b** *to the other. The value of* **h** *is* **not** *the length of the side of the parallelogram* (unless it is also a rectangle).

EXAMPLE B

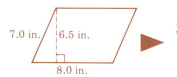

7.0 in. 6.5 in.

8.0 in.

FIGURE 6-34

1. The area of a parallelogram for which $h = 9.0$ m and $b = 11$ m is

$$A = (11 \text{ m})(9.0 \text{ m}) = 99 \text{ m}^2 \qquad \text{using Eq. (6-11)}$$

2. To find the area of the parallelogram shown in Figure 6-34, we need only the height (6.5 in.) and the base (8.0 in.). *The length of the other side is not used* (although it would be used in finding the perimeter). Thus

$$A = (8.0 \text{ in.})(6.5 \text{ in.}) = 52 \text{ in.}^2 \qquad \square$$

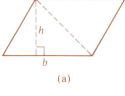

h

b

(a)

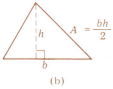

h $A = \dfrac{bh}{2}$

b

(b)

FIGURE 6-35

If we draw a **diagonal** (*line joining opposite vertices*) in a parallelogram, we divide it into two equal triangles, as shown in Figure 6-35a. Since the area of the triangle below the diagonal is one-half the area of the parallelogram, we have

$$A = \frac{1}{2}bh \qquad \text{area of triangle} \qquad (6\text{-}12)$$

as the area of the triangle (see Figure 6-35b). *Here* **h** *is the* **altitude,** *or* **height,** *of the triangle. It is the length of the perpendicular line from a vertex to the opposite base of the triangle.*

EXAMPLE C

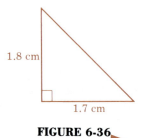

1.8 cm

1.7 cm

FIGURE 6-36

1. The area of a triangle for which $b = 16$ in. and $h = 12$ in. is

$$A = \frac{1}{2}(16 \text{ in.})(12 \text{ in.}) = 96 \text{ in.}^2 \qquad \text{using Eq. (6-12)}$$

It might be noted here that $\frac{1}{2}bh$ and $\dfrac{bh}{2}$ are the same.

2. The area of the right triangle shown in Figure 6-36 is

$$A = \frac{1}{2}(1.7 \text{ cm})(1.8 \text{ cm}) = 1.5 \text{ cm}^2$$

Since the legs of a right triangle are perpendicular, *either leg can be considered as the base and the other as the altitude.* $\square$

See Appendix B for a computer program that finds the area of a triangle using Eqs. (6-13) and (6-14).

Another useful formula for finding areas of triangles is Hero's formula given in Eq. (6-13). This formula is especially useful when we have *a triangle with three known sides and no right angle.*

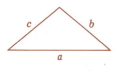

FIGURE 6-37

$$A = \sqrt{s(s-a)(s-b)(s-c)}$$ (6-13)

$$\text{where } s = \tfrac{1}{2}(a+b+c)$$ (6-14)

See Figure 6-37.

EXAMPLE D A triangular fin has sides $a = 12$ cm, $b = 15$ cm, and $c = 23$ cm. Its area is found by first determining the value of s in Eq. (6-14). See Figure 6-38.

$$s = \frac{1}{2}(12 \text{ cm} + 15 \text{ cm} + 23 \text{ cm}) = \frac{1}{2}(50 \text{ cm}) = 25 \text{ cm}$$

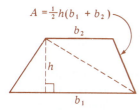

FIGURE 6-38

We can now proceed to use Eq. (6-13) as follows:

$$A = \sqrt{(25 \text{ cm})(25 \text{ cm} - 12 \text{ cm})(25 \text{ cm} - 15 \text{ cm})(25 \text{ cm} - 23 \text{ cm})}$$
$$= \sqrt{(25 \text{ cm})(13 \text{ cm})(10 \text{ cm})(2 \text{ cm})} = \sqrt{6500 \text{ cm}^4}$$
$$= 81 \text{ cm}^2$$

Note that in evaluating the square root, $\sqrt{\text{cm}^4} = \text{cm}^2$. □

When we join opposite vertices of a trapezoid with a diagonal, two triangles are formed, as in Figure 6-39. The area of the lower triangle is $\frac{1}{2}b_1 h$ and the area of the upper triangle is $\frac{1}{2}b_2 h$. The sum of the areas is the area of the trapezoid, and it may be expressed as

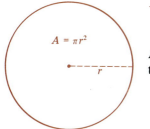

FIGURE 6-39

$$A = \frac{1}{2}h(b_1 + b_2)$$ area of trapezoid (6-15)

EXAMPLE E The floor area of a trapezoidal room for which $h = 6.0$ yd, $b_1 = 4.0$ yd, and $b_2 = 3.0$ yd is

$$A = \frac{1}{2}(6.0 \text{ yd})(4.0 \text{ yd} + 3.0 \text{ yd})$$ using Eq. (6-15)

$$= 21 \text{ yd}^2$$ □

The area of a circle cannot be found from the area of any quadrilateral. As in the case of the circumference, the area of a circle is expressed in terms of π. The area of a circle is given by

$$A = \pi r^2$$ area of circle (6-16)

FIGURE 6-40

See Figure 6-40.

EXAMPLE F

See Appendix B for a computer program that calculates the areas of triangles.

1. The area of a circle of radius 2.73 cm is

$$A = \pi(2.73 \text{ cm})^2 = (3.14)(7.4529 \text{ cm}^2) \qquad \text{using Eq. (6-16)}$$
$$= 23.4 \text{ cm}^2$$

2. To find the area of a circle of a given diameter, we *first divide the length of the diameter by 2 in order to obtain the length of the radius*. Thus if $d = 48.2$ in. then $r = 24.1$ in., and we get

$$A = \pi(24.1 \text{ in.})^2 = 3.14(581 \text{ in.}^2)$$
$$= 1820 \text{ in.}^2 \qquad \Box$$

So far we have included the units of measurement in the equations and algebraic work. Technically this should always be done, but it can be cumbersome (see Example D). Recognizing that all lengths are in a specific type of unit, we know that the result is in terms of that unit. Therefore, from this point on, when the units of measurement of the given parts and the result are specifically known, we shall not include them in the equation or algebraic work.

The determination of area is important in many applications involving geometric figures. Consider the following example.

EXAMPLE G

A painter charges 35¢ per square foot for painting house exteriors. One side of a house is a rectangle surmounted by a triangle. The base of the rectangle is 30.0 ft, the height of the rectangle is 9.00 ft, and the height of the triangle is 7.00 ft. What would his charge for this part of the job be? There are three windows, each 2.00 ft by 3.00 ft, in the side of the house (see Figure 6-41).

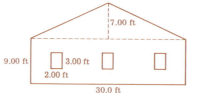

FIGURE 6-41

The area of the side of the house to be painted is the area of the rectangle plus that of the triangle minus the area of the windows. Thus

$$A = \underbrace{(30.0)(9.00)}_{\text{rectangle}} + \underbrace{\frac{1}{2}(30.0)(7.00)}_{\text{triangle}} - \underbrace{3(2.00)(3.00)}_{\text{windows}}$$

$$= 270 + 105 - 18.0 = 357 \text{ ft}^2$$

The charge is

$$c = 0.35(357) = \$124.95 \qquad \Box$$

Exercises 6-3

In Exercises 1 through 40, find the areas of the indicated geometric figures.

1. Rectangle: $l = 60.0$ cm, $w = 45.0$ cm
2. Rectangle: $l = 152$ ft, $w = 85.0$ ft
3. A rectangle with length 24 in. and width 7.0 in.
4. A rectangle with length and width of 2.34 m and 5.68 m, respectively
5. Square: $s = 7.60$ in.
6. Square: $s = 0.160$ km
7. A square with side 2.54 in.
8. A square with perimeter 8.0 cm
9. Parallelogram: $b = 72.0$ mm, $h = 34.0$ mm
10. Parallelogram: $b = 1.50$ yd, $h = 1.20$ yd
11. A parallelogram with base 5.30 cm and altitude 4.60 cm
12. A parallelogram with base and altitude of 12.4 ft and 20.3 ft, respectively
13. Rhombus: side $= 16.5$ in., altitude $= 6.40$ in.
14. Rhombus: side $= 240$ cm, altitude $= 150$ cm
15. A rhombus with a side of 8 ft and an altitude of 5 ft
16. A rhombus with a side and altitude of 12.4 m and 6.02 m, respectively
17. Triangle: $b = 0.750$ m, $h = 0.640$ m
18. Triangle: $b = 64.0$ in., $h = 14.5$ in.
19. A triangle with base 42 cm and altitude 16 cm
20. A triangle with base 14.0 ft and altitude 8.25 ft
21. Right triangle: legs $= 16.5$ ft and 28.8 ft
22. Right triangle: legs $= 396$ mm and 250 mm
23. A right triangle with legs 153 cm and 205 cm
24. A right triangle with legs 80.1 in. and 20.5 in.
25. Triangle: $a = 21.2$ cm, $b = 12.3$ cm, $c = 25.4$ cm
26. Triangle: $a = 2.45$ m, $b = 3.62$ m, $c = 3.97$ m
27. A triangle with sides 5.2 ft, 6.4 ft, and 4.7 ft
28. A triangle with sides 10.2 in., 12.8 in., and 14.9 in.
29. Trapezoid: $h = 0.0120$ km, $b_1 = 0.0250$ km, $b_2 = 0.0180$ km
30. Trapezoid: $h = 1.23$ yd, $b_1 = 3.74$ yd, $b_2 = 2.36$ yd
31. A trapezoid with altitude 6.33 m and bases of 4.55 m and 5.25 m
32. A trapezoid with altitude 1.05 ft and bases of 2.37 ft and 3.85 ft
33. Circle: $r = 0.478$ ft
34. Circle: $d = 5.38$ cm
35. Circle: $r = 23.21$ ft
36. Circle: $d = 93.7$ m
37. A circle with radius 2.625 in.
38. A circle with diameter 203 mm
39. A circle with circumference 4.00 ft
40. A circle with circumference 12.28 m

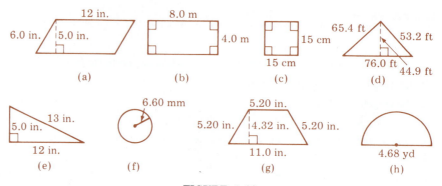

FIGURE 6-42

In Exercises 41 through 48, determine (a) the perimeter and (b) the area of the given figures.

41. Figure 6-42a   **42.** Figure 6-42b   **43.** Figure 6-42c   **44.** Figure 6-42d

45. Figure 6-42e   **46.** Figure 6-42f   **47.** Figure 6-42g   **48.** Figure 6-42h

In Exercises 49 through 52, find a formula for the areas of the given figures.

49. Figure 6-43a   **50.** Figure 6-43b   **51.** Figure 6-43c   **52.** Figure 6-43d

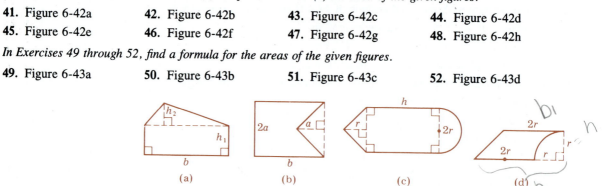

FIGURE 6-43

In Exercises 53 through 72, solve the given problems.

53. A rectangular swimming pool is 16.0 ft wide and 32.0 ft long. What is the cost of a solar cover which is sold for 35¢ per square foot?

54. A 32.0-ft long swimming pool is 3.00 ft deep at one end and 8.00 ft deep at the other end. If the bottom slopes uniformly from one end to the other, find the area of a wall that runs the length of the pool. See Figure 6-44.

FIGURE 6-44

55. A panel of sheetrock is rectangular with length 4.0 ft and width 8.0 ft. Two 1.0-ft diameter holes are cut out for heating ducts. What is the area of the remaining piece?

56. A gallon of paint will cover 300 ft². How much paint is required to paint the walls and ceiling of a room 12 ft by 16 ft by 8.0 ft, given that the room has three windows 2.0 ft by 3.0 ft and two doors 3.0 ft by 6.5 ft?

57. A window frame is rectangular with length 4 ft 4 in. and width 2 ft 2 in. This frame is to be covered with screen, and an additional 1 in. wide strip of screen must go around the outside of the frame. What is the area of the screen?

58. A metal plate is in the shape of a right triangle with sides of 3.56 cm, 4.08 cm, and 5.41 cm. Determine the area of the plate.

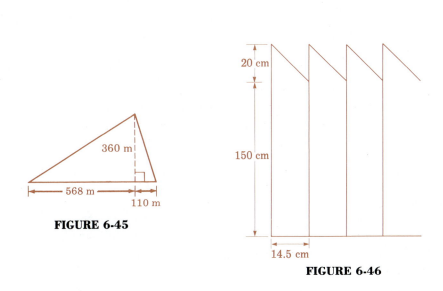

FIGURE 6-45

FIGURE 6-46

59. Find the area of the triangular tract of land shown in Figure 6-45.

60. A fence is to be made of trapezoidal slats as shown in Figure 6-46. What is the area of the fence if it is placed along a distance of 13.34 m?

61. A flat steel ring 38.0 cm in diameter has a hole 8.90 cm in diameter in the center. What is the area of one face of the ring?

62. How many pipes 1.00 in. in radius are required to carry as much water as a pipe 3.00 in. in radius?

63. What is the area of the largest circle which can be cut from a rectangular plate 21.2 cm long and 15.8 cm wide?

64. An oil spill forms a circle with diameter 2.34 km. If cleanup costs $3000/km², find the total cost of cleaning up this oil spill.

65. If the lengths of the sides of a rectangle are doubled, how does the area change?

66. If you triple the lengths of the sides of a rectangle, how does the area change?

67. Which figure encloses the larger area: a square with a perimeter of 12 cm, or a circle with a circumference of 12 cm?

68. Circular coins with radius 1 cm are to be punched out of a square sheet with area 100 cm². How many coins can be obtained from one of these sheets? What percentage of the sheet's area is used for the coins?

69. Use a calculator to find the area covered by a circular forest fire whose diameter is 7654 ft.

70. Use a calculator to find the area of a triangular parcel of land with sides of 35.50 m, 56.34 m, and 49.33 m.

71. Use a calculator to find the area of an equilateral triangle with side 1.045 m.

72. A wire is shaped to form a circle with an area of 4.658 m². If the same wire is reshaped to form a square, use a calculator to find the area of the square.

6-4 Volume

So far we have examined perimeter and area as measures of plane geometric figures. An important measure associated with a solid figure is its **volume**. Just as area gives us a measure of the surface of a plane geometric figure, *volume gives us a measure of the amount of space occupied by a solid figure*. In finding the volume of a solid geometric figure we are finding the number of cubes, one unit on an edge, required to fill the figure. In the **rectangular solid** in Figure 6-47, we see that 24 cubes, each 1 cm on an edge, are required to fill the solid. We say that the volume is 24 cubic centimeters, or 24 cm³.

If we note the rectangular solid in Figure 6-47, we see that the volume may be determined by finding the product of its length, width, and height. In general, the volume of a rectangular solid is

$$V = lwh$$ **volume of rectangular solid**   (6-17)

where **l, w,** *and* **h** *are in the same unit of length and* **V** *is in cubic units of length (see Figure 6-48).*

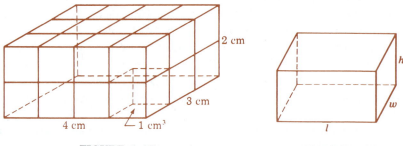

FIGURE 6-47   FIGURE 6-48

EXAMPLE A   A computer peripheral device is in the shape of a rectangular solid 18.0 cm long, 10.0 cm wide, and 9.00 cm high. Find its volume.

The volume is computed as

$V = lwh = (18.0)(10.0)(9.00) = 1620 \text{ cm}^3$ **using Eq. (6-17)**

 Note that the units of volume are always cubic units such as cm³, m³, in.³, ft³, and so on.   □

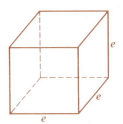

FIGURE 6-49

The **cube,** which we have already used in naming units of volume, *is a rectangular solid in which all edges are equal in length.* Since the length, width, and height equal the length of an edge e, the volume of a cube is

$$V = e^3$$

volume of cube   (6-18)

See Figure 6-49.

EXAMPLE B A farmer stores grain in a bin which is in the shape of a cube. If each edge of this cube is 8.0 ft, find the volume of the grain which can be stored.
 With $e = 8.0$ ft, we find the volume as follows:

$$V = e^3 = 8.0^3 = 510 \text{ ft}^3 \qquad \text{using Eq. (6-18)}$$

EXAMPLE C How many cubic feet are in 1 cubic yard?
 A cubic yard is a cube in which each edge has a length of 1 yd. Since 1 yd = 3 ft, that same volume can be represented by a cube in which each edge has a length of 3 ft, and the volume can now be computed as

$$V = e^3 = 3^3 = 27 \text{ ft}^3 \quad \text{Therefore, 1 yd}^3 = 27 \text{ ft}^3.$$

EXAMPLE D A rectangular lawn 150 ft long and 36 ft wide is to be covered with a 6-in. or 0.50-ft layer of topsoil. How much topsoil must be ordered? (Topsoil is usually ordered by the cubic yard.)
 When put into place, the topsoil will form a rectangular solid 150 ft long, 36 ft wide, and 6 in. (or 0.50 ft) high. Its volume is found by

$$V = (150)(36)(0.50) = 2700 \text{ ft}^3$$

We should now reduce the result to cubic yards. From Example C we see that 1 yd^3 = 27 ft^3 so that

$$2700 \text{ ft}^3 = \overset{100}{\cancel{2700 \text{ ft}^3}} \cdot \frac{1 \text{ yd}^3}{\cancel{27 \text{ ft}^3}} = 100 \text{ yd}^3$$

EXAMPLE E A chemical waste holding tank is in the shape of a rectangular solid which is 3.00 ft long, 3.00 ft wide, and 2.00 ft high. How many gallons of liquid waste can be stored in this tank? One cubic foot can contain 7.48 gal.
 The volume of the tank is

$$V = (3.00)(3.00)(2.00) = 18.0 \text{ ft}^3$$

Since each cubic foot holds 7.48 gal, we find the total capacity c as follows:

$$c = 18.0 \times 7.48 = 135 \text{ gal}$$

The tank can hold 135 gal of chemical waste.

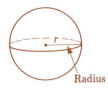

Radius

FIGURE 6-50

The **sphere** is the last solid figure we will consider in this chapter, but others will be included in Chapter 15. A sphere can be formed by rotating a circle about its diameter. The **radius** of the sphere is the length of the line segment connecting the center to a point on the surface. Like a circle, the **diameter** of a sphere is twice its radius (see Figure 6-50). The volume of the sphere is computed by using Eq. (6-19).

$$V = \frac{4}{3}\pi r^3 \qquad \text{volume of sphere} \qquad (6\text{-}19)$$

Here r is the radius of the sphere.

EXAMPLE F   The radius of a tennis ball is 1.25 in. Its volume is

$$V = \frac{4}{3}\pi(1.25)^3 = \frac{4}{3}(3.14)(1.95) \qquad \text{using Eq. (6-19)}$$

$$= 8.16 \text{ in.}^3 \qquad\qquad \Box$$

EXAMPLE G   The radius of the earth (which is approximately a sphere) is about 3960 mi. The volume of the earth is

$$V = \frac{4}{3}\pi(3960)^3$$

$$= 260{,}000{,}000{,}000 \text{ mi}^3$$

(In Chapter 10 we will see that such numbers can be concisely expressed in a form called *scientific notation*. This answer could be written as 2.60×10^{11} mi^3.) $\Box$

Exercises 6-4

In Exercises 1 through 8, find the volumes of the rectangular solids with the given dimensions.

1. $l = 73.0$ mm, $w = 17.2$ mm, $h = 16.0$ mm   **2.** $l = 3.21$ ft, $w = 5.12$ ft, $h = 6.34$ ft

3. $l = 0.87$ cm, $w = 0.61$ cm, $h = 0.15$ cm   **4.** $l = 1.2$ cm, $w = 3.9$ cm, $h = 9.7$ cm

5. $l = 16$ m, $w = 16$ m, $h = 18$ m   **6.** $l = 9.0$ in., $w = 12$ in., $h = 15$ in.

7. $l = 120$ cm, $w = 85$ cm, $h = 150$ cm   **8.** $l = 342$ ft, $w = 20.5$ ft, $h = 1.50$ ft

In Exercises 9 through 16, find the volume of the cubes with the given edges.

9. $e = 15$ cm   **10.** $e = 12$ ft   **11.** $e = 0.800$ in.   **12.** $e = 9.36$ m

13. $e = 0.22$ m   **14.** $e = 17.3$ in.   **15.** $e = 19.27$ cm   **16.** $e = 17.39$ cm

In Exercises 17 through 24, find the volume of the given sphere with radius r *or diameter* d.

17. $r = 27.3$ cm   **18.** $r = 78$ ft   **19.** $r = 15.2$ m   **20.** $r = 55.3$ mm

21. $d = 34$ in.   **22.** $d = 764$ cm   **23.** $d = 1.12$ mm   **24.** $d = 8.2$ km

In Exercises 25 through 48, answer the given questions.

25. How many cubic inches are in 1 ft³?

26. How many cubic inches are in 1 yd³?

27. How many cubic centimeters are in 1 m³?

28. How many cubic millimeters are in 1 m³?

29. In studying the effects of a poisonous gas, a fire science specialist must calculate the volume of a room 12 ft wide, 14 ft long, and 8.0 ft high. Find that volume.

30. An engine has a displacement equal to the volume of a cube with 7.4 cm as the length of an edge. Find the displacement.

31. A heating duct is 2.0 ft wide, 1.5 ft high, and 30 ft long. Find the volume of air contained in the duct. See Figure 6-51.

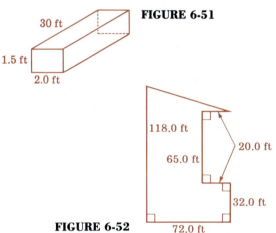

FIGURE 6-51

32. A beam in the shape of a rectangular solid is cut from a tree. The beam has a square end with each side 8.0 in. long, and the length of the beam is 30 ft. Find the volume of this beam.

33. In road construction, fill is measured in "yards" where a "yard" is understood to be one cubic yard. How many of these "yards" of fill can be carried by a truck whose trailer is 2.32 yd wide, 1.28 yd high, and 5.34 yd long?

34. A lawn has the shape shown in Figure 6-52. How many gallons of water fall on the lawn in a 0.75-in. rainfall? Assume that 1.00 ft³ contains 7.48 gal.

FIGURE 6-52

35. A book is 4.0 in. wide, 7.0 in. long, and 1.8 in. thick. Can 30,000 copies of this book be stored in a room which is 9.0 ft long, 12 ft wide, and 8.0 ft high?

36. In constructing the foundation for a house, a hole 6.0 ft deep, 52 ft long, and 25 ft wide must be dug. How many cubic *yards* of earth will be removed?

37. A weather balloon with diameter 2.08 m is filled with helium. Find its volume.

38. The diameter of a basketball is about 9.5 in. What is its volume?

39. What is the difference in volume between a baseball with diameter 2.75 in. and a soccer ball with diameter 8.50 in.?

40. A sample of an alloy is in the shape of a solid sphere with radius 6.40 cm. If this sphere is melted down and then cast into a cube, what is the length of an edge of the cube?

41. A sphere has a diameter of 4.26 in. If this sphere is made of brass, how much does it weigh? (Brass weighs 510 lb/ft³.)

42. Find the total volume of steel needed to make 150,000 ball bearings if each bearing has a diameter of 1.81 mm.

43. The radius of Jupiter is 69,200 km. What is its volume?

44. If the radius of a sphere is doubled, how does the volume change?

45. Use a calculator to find the number of gallons a water purification tank can hold. The tank is in the shape of a rectangular solid 10.21 m long, 3.97 m wide, and 1.98 m deep. Assume that 1.000 m³ contains 264.2 gal.

46. Use a calculator to find the number of cubic yards of "item 4" required for a section of a roadbed. Specifications require a 4.00 in. thick coating of this special "item 4" material. The roadbed is 8.00 ft wide and 1.20 mi long.

47. Use a calculator to find the volume of a spherical droplet of cyanoacrylate (superglue) with a diameter 1.344 mm.

48. If brass weighs 510 lb/ft³, what is the diameter of a brass sphere that weighs 100 lb?

Chapter 6 Formulas

Figure	*Perimeter*	
(a) Triangle with sides a, b, c	$p = a + b + c$	(6-1)
(b) Equilateral triangle with side s	$p = 3s$	(6-2)
(c) Quadrilateral with sides a, b, c, d	$p = a + b + c + d$	(6-3)
(d) Rectangle of length l and width w	$p = 2l + 2w$	(6-4)
(e) Parallelogram with sides a and b	$p = 2a + 2b$	(6-5)
(f) Square of side s	$p = 4s$	(6-6)

$c = \pi d$	circumference of circle	(6-7)
$c = 2\pi r$	circumference of circle	(6-8)
$A = lw$	area of rectangle	(6-9)
$A = s^2$	area of square	(6-10)
$A = bh$	area of parallelogram	(6-11)
$A = \dfrac{1}{2}bh$	area of triangle	(6-12)
$A = \sqrt{s(s - a)(s - b)(s - c)}$	area of triangle	(6-13)
where $s = \frac{1}{2}(a + b + c)$		(6-14)
$A = \dfrac{1}{2}h(b_1 + b_2)$	area of trapezoid	(6-15)
$A = \pi r^2$	area of circle	(6-16)
$V = lwh$	volume of rectangular solid	(6-17)
$V = e^3$	volume of cube	(6-18)
$V = \dfrac{4}{3}\pi r^3$	volume of sphere	(6-19)

6-5 Review Exercises for Chapter 6

In Exercises 1 through 8, change angles given in degrees and decimal parts of a degree to degrees and minutes. Change angles given in degrees and minutes to degrees and decimal parts of a degree.

1. 37.5°   **2.** 43.95°   **3.** 12.55°   **4.** 45.27°
5. 63°30′   **6.** 82°20′   **7.** 105°54′   **8.** 215°45′

In Exercises 9 through 24, find the perimeters of the indicated geometric figures.

9. Triangle: $a = 17.5$ in., $b = 13.8$ in., $c = 8.9$ in.

10. Quadrilateral: $a = 22.4$ cm, $b = 68.5$ cm, $c = 37.3$ cm, $d = 29.9$ cm

11. Equilateral triangle: $s = 8.5$ mm

12. Isosceles triangle: equal sides of 0.38 yd, third side of 0.53 yd

13. Square: $s = 6.8$ m   **14.** Rhombus: side of 15.2 in.

15. Parallelogram: $a = 692$ ft, $b = 207$ ft   **16.** Parallelogram: $a = 7.8$ m, $b = 6.2$ m

17. Rectangle: $l = 96$ cm, $w = 43$ cm   **18.** Rectangle: $l = 108$ in., $w = 92$ in.

19. Circle: $r = 4.25$ ft   **20.** Circle: $d = 38.0$ cm

21. The figure in Figure 6-53 (equilateral triangle attached to rectangle).

22. The figure in Figure 6-54 (square and rhombus attached).

23. The figure in Figure 6-55 (rectangle with a half circle removed).

24. The figure in Figure 6-56 (rectangle with a triangle and half-circle affixed).

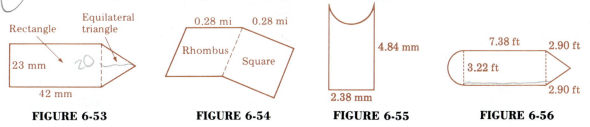

FIGURE 6-53   **FIGURE 6-54**   **FIGURE 6-55**   **FIGURE 6-56**

In Exercises 25 through 40, find the area of the given figures.

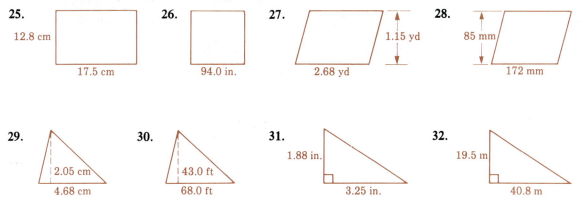

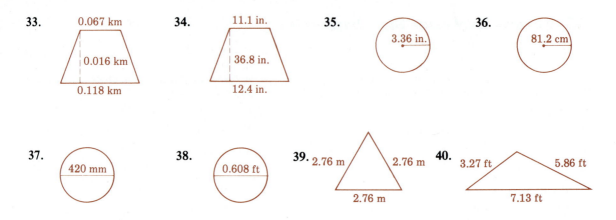

33. 0.067 km / 0.016 km / 0.118 km

34. 11.1 in. / 36.8 in. / 12.4 in.

35. 3.36 in.

36. 81.2 cm

37. 420 mm

38. 0.608 ft

39. 2.76 m / 2.76 m / 2.76 m

40. 3.27 ft / 5.86 ft / 7.13 ft

In Exercises 41 through 44, find the area of the indicated figures.

41. The figure in Figure 6-53 if h of the triangle is 20.0 mm.

42. The figure in Figure 6-54 if h of the rhombus is 0.22 mi.

43. The figure in Figure 6-55.   44. The figure in Figure 6-56.

In Exercises 45 through 52, find the volumes of the indicated figures.

45. Rectangular solid: $l = 2.00$ m, $w = 1.50$ m, $h = 1.20$ m

46. Rectangular solid: $l = 25.0$ in., $w = 10.0$ in., $h = 15.0$ in.

47. Cube: $e = 3.50$ yd   48. Cube: $e = 220$ mm

49. Sphere: $r = 5.45$ cm   50. Sphere: $d = 6.20$ ft

51. Sphere with diameter 3.74 yd   52. Sphere with radius 0.9874 m

In Exercises 53 through 60, set up the required formula.

53. The perimeter of the figure shown in Figure 6-57a (an equilateral triangle within an isosceles triangle).

54. The perimeter of the figure shown in Figure 6-57b (a trapezoid made up of a rhombus and an equilateral triangle).

55. The perimeter of the figure shown in Figure 6-57c (a trapezoid on a semicircle).

56. The perimeter of the figure shown in Figure 6-57d (two concentric—same center—semicircles).

57. The area of the figure shown in Figure 6-57a.   58. The area of the figure shown in Figure 6-57b.

59. The area of the figure shown in Figure 6-57c.   60. The area of the figure shown in Figure 6-57d.

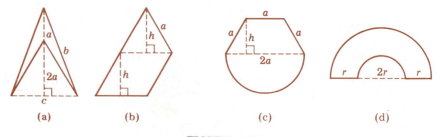

(a)   (b)   (c)   (d)

FIGURE 6-57

In Exercises 61 through 92, solve the given problems.

61. A tarpaulin is used to cover a rectangular field 360 ft long and 200 ft wide. What is the cost of the tarpaulin if it is priced at 27¢ per square foot?

62. A rectangular swimming pool is 48 ft long and 22 ft wide. It is bordered by a walkway that is 8.5 ft. wide. What is the perimeter around the pool? What is the perimeter around the outside edge of the walkway? See Figure 6-58.

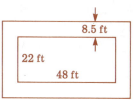

FIGURE 6-58

63. A retaining wall 1.50 m high is built around a rectangular piece of land that is 36.0 m long and 32.0 m wide. What is the area of the wall?

64. Specifications for a metal container indicate a shape that is a rectangular solid 16.0 cm long, 10.0 cm wide, and 8.75 cm high. The sheet metal for the top and bottom costs 8¢/cm² while the sides cost 5¢/cm². What is the total cost of 20 such containers?

65. Calculate the total weight of a brass sphere with radius 2.432 ft. Brass weighs 510 lb/ft³.

66. How many circular air ducts 8.00 cm in diameter have (at least) the same cross-sectional area as one rectangular duct of dimensions 24.0 cm by 36.0 cm?

67. An isosceles triangle has a perimeter of 75 cm. Each of the two equal sides is twice as long as the base. Find the length of the base.

68. A circle is drawn on a blueprint so that its circumference is 1088 mm longer than its diameter. Find the diameter of this circle.

69. If the diameter of the earth is 7920 mi and a satellite is in orbit at an altitude of 212 mi above the earth's surface, how far does the satellite travel in one rotation around the earth?

70. If water weighs 62.5 lb/ft³, find the total weight of water that fills a rectangular pool that is 32.0 ft long, 16.0 ft wide, and 4.00 ft deep.

71. A beam support in a building is in the shape of a parallelogram as shown in Figure 6-59. Find the area of the side of the beam shown.

72. Find the area of lots *A* and *B* shown in Figure 6-60. Lot *A* has a frontage on Main Street of 140 ft. Lot *B* has a frontage on Main Street of 84 ft. The boundary between the lots is 120 ft, and the right boundary line for lot *B* is 192 ft.

73. Find the area of the Norman window shown in Figure 6-61. It is a rectangle surmounted by a semicircle.

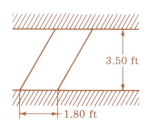

FIGURE 6-59

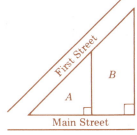

FIGURE 6-60

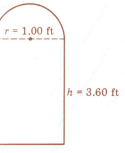

FIGURE 6-61

74. A rectangular box is to be used to store radioactive materials. The inside dimensions of the box are 12 in., 14 in., and 20 in., and it is to be lined with sheet lead 0.25 in. thick. Lead weighs 690 lb/ft³. What is the weight of the lead used?

75. A railroad boxcar is 11.2 m long, 2.50 m wide, and can be filled to a depth of 1.70 m. What volume of material can the boxcar hold?

76. A cubical box with an outside edge of 8.00 in. is made of wood 0.75 in. thick. What is the actual volume of the inside of the box?

77. Find the volume of an ice cube 2.10 cm on an edge.

78. The density of gold is 0.690 lb/in.^{3}. What is the weight of a cube of gold 3.00 in. on an edge?

79. A Styrofoam buoy is in the shape of a sphere with diameter 0.750 m. Find the total volume of eight such buoys.

80. A hollow metal sphere has an outside radius of 4.55 cm and an inside radius of 4.05 cm. What is the volume of the metal in the sphere?

81. A brick is placed in the water-holding tank of a toilet in order to conserve water. The brick is $2\frac{1}{4}$ in. high, $3\frac{3}{4}$ in. wide, and 8 in. long. How much water is saved in 500 flushes? Assume that 1.00 in.^{3} can hold 0.00433 gal.

82. How many acres of land are covered by a JFK Airport runway which is 14,572 ft long and 150 ft wide? (1 acre = 43,560 ft^2.)

83. How many cubic feet of cement are needed for a straight section of road which is exactly 1 mi long and 16 ft wide? The pavement is 9.0 in. thick.

84. A strip of steel is 50 ft long and 1.0 in. wide. If this strip is cut into sections which are connected to form a square, what is the length of a side of the square?

85. Standard bricks are $2\frac{1}{4}$ in. by $3\frac{3}{4}$ in. by 8 in. while English bricks are 3 in. by $4\frac{1}{2}$ in. by 9 in., and Roman bricks are $1\frac{1}{2}$ in. by 4 in. by 12 in. Find the difference in volume between the brick with the smallest volume and the type with the largest volume.

86. A cord of wood is a stack of logs which is 4 ft deep, 4 ft high, and 8 ft long. A cord of hickory will yield 12,300,000 Btu when burned in an airtight stove. Find the Btu yield for a stack of hickory logs which is 3 ft deep, $2\frac{1}{2}$ ft high, and 20 ft long.

87. A tree trunk has a circumference of 4.21 ft. Find its cross-sectional area.

88. A car travels 26.2 mi with tires having a radius of 15.0 in. How many revolutions are made by one of the tires?

89. What is the area of the rectangular label that is to fit tightly around a can 3.80 in. in diameter and 4.00 in. tall? Allow 0.410 in. overlap for pasting.

90. A running track is to be designed in the shape of a rectangle with semicircles affixed along the two widths, as in Figure 6-62. Find the width of the rectangle if its length is 320 ft and the perimeter of the track is supposed to be one-fourth of a mile.

320 ft

FIGURE 6-62

91. Find the volume of the smallest cube which will contain a sphere that has a volume of 14,800 cm^3.

92. A box of cereal is in the shape of a rectangular solid that is 2.50 in. wide, 7.50 in. long, and 11.75 in. high. By what amount would the total surface area be reduced if the same volume were packaged in a cube instead of the given rectangular solid?

In Exercises 93 through 96, use a scientific calculator to solve the given problems.

93. Find the volume of living space in a one-story rectangular house which is 24.00 ft wide and 53.52 ft long. All ceilings are 7 ft 11 in. high.

94. Express 37.23° in degrees, minutes, and seconds.

95. Express 51°23′46″ in degrees and decimal parts of a degree.

96. The second hand of a clock is 4.27 cm long. Find the area swept by this second hand in 1 day.

CHAPTER SEVEN

Basic Algebraic Operations

Section 4-3 introduced the basic operations in algebra. We have often used these operations in the last two chapters in working with equations and formulas from algebra and geometry. In this chapter we develop a more general use of these operations with algebraic expressions. We will learn how to add and subtract algebraic expressions in the first section. In Section 7-2 we will learn how to multiply and divide algebraic expressions that consist of a single term. We then proceed to study those same operations with more than one term.

The content of this chapter can be applied to real situations that require operations with algebraic expressions. For example, pilots know that it usually becomes colder at higher altitudes, and an analysis of the temperature changes might involve the algebraic expression

$$\left(\frac{9}{5}C_1 + 32\right) - \left(\frac{9}{5}C_2 + 32\right)$$

We will learn how to combine and simplify such expressions. This particular problem will be solved in Section 7-1.

7-1 Algebraic Addition and Subtraction

Most of the algebraic expressions included in this chapter are **polynomials.**
*A polynomial is an algebraic term or sum of terms, each of which is one of
the following:*

1. A constant.
2. A variable with an exponent that is a positive integer.
3. A product of constants and variables with exponents that are positive integers.

In a polynomial, the variable may not be in the denominator or under a radical. (In such cases, the exponent may be a negative number or a fraction, as we will see in Chapter 10.)

EXAMPLE A   The following expressions are polynomials:

$$2x^2 + x \qquad\qquad 4x^3 - x - 7 \qquad\qquad x^3y - 4xy^2 - 6$$

The following expressions are *not* polynomials:

$$2x^2 + \frac{1}{x} \qquad\qquad 4x^3 - \sqrt{x} - 7 \qquad\qquad \frac{1}{x^2 + 1}$$

$\square$

The **degree of a term** *in a polynomial is the sum of the powers of the variables.* (Remember, if a variable has no power indicated, that power is assumed to be 1.) For example, $-8x^2y^3$ is a term of degree 5, whereas $-32xyz^5$ is a term of degree 7. *The* **degree of a polynomial** *is found by combining like terms and then finding the term with the highest degree.* The degree of the polynomial $3x^2 + x + 5$ is 2 since the term with the highest degree is a term of degree 2. Similarly, $x^6y^2 + x^3y - 6y^4$ is a polynomial of degree 8.

From Section 4-3 we now know that in the addition of polynomials or other algebraic expressions *we may* **combine** *only like terms, but for unlike terms we can only* **indicate** *the sum.* Also, we learned how to use parentheses for grouping terms. The following example illustrates algebraic addition.

EXAMPLE B   Add $4xy + 3a - x$ and $2xy - 5a + 2x$.
This addition may be written as

$$
\begin{array}{r}
4xy + 3a - x \\
2xy - 5a + 2x \\
\hline
6xy - 2a + x
\end{array}
$$

where the sum is $6xy - 2a + x$. We are able to combine only the like terms. We note also that the sum of $3a$ and $-5a$ is $-2a$. A more common method of expressing addition in algebra is by means of parentheses. For the sum given above, we write

$$(4xy + 3a - x) + (2xy - 5a + 2x)$$

It is not possible to proceed with the addition until we write an equivalent expression without parentheses. Only then are we able to combine the like terms. Removing the parentheses from this expression, we have

$$4xy + 3a - x + 2xy - 5a + 2x$$

Combining like terms, we have $6xy - 2a + x$, which is the result given above. Thus we may state that

$$(4xy + 3a - x) + (2xy - 5a + 2x) = 4xy + 3a - x + 2xy - 5a + 2x$$
$$= 6xy - 2a + x \qquad \square$$

Generalizing the results of Example A, we see that:

When parentheses are preceded by a plus sign, and the parentheses are removed, the sign of each term within the parentheses is retained.

Recall from Section 3-2 that *when we subtract a signed number we may change the sign of a number and proceed as in addition.* For example, $8 - 3 = 8 + (-3) = 5$. This same principle is used to subtract one algebraic expression from another. The following example illustrates the use of this principle in subtraction of algebraic expressions.

EXAMPLE C   Subtract $2xy - 5a + 2x$ from $4xy + 3a - x$.
   Instead of using the horizontal arrangement of Example B, we will use a vertical arrangement to express the subtraction as

$$\begin{array}{r} 4xy + 3a - x \\ 2xy - 5a + 2x \\ \hline 2xy + 8a - 3x \end{array}$$

where the difference is $2xy + 8a - 3x$. In this subtraction, we may consider three separate subtractions. The first is $4xy - 2xy$, which results in $2xy$. The second is $(+3a) - (-5a)$, which, by the principle of subtracting signed numbers, becomes

$$(+3a) + (+5a) = 3a + 5a = 8a$$

The third subtraction is $(-x) - (+2x)$, which becomes

$$(-x) + (-2x) = -x - 2x = -3x$$

The combination of these three differences gives the result shown above.

(Continued on next page)

Using parentheses to indicate the subtraction, we have

$$(4xy + 3a - x) - (2xy - 5a + 2x)$$

which is again a more common algebraic expression. Since in each individual subtraction shown above it is necessary to change the sign of the number being subtracted, we remove the second set of parentheses here and also change the sign of *each term* within them. This leads to

$$4xy + 3a - x - 2xy + 5a - 2x = 2xy + 8a - 3x$$

which agrees with the result above. □

Generalizing the results of Example C, we see that:

▶ *When parentheses are preceded by a minus sign, and the parentheses are removed,* **the sign of each term within the parentheses is changed.**

This principle agrees with the distributive law in Section 4-3.

EXAMPLE D

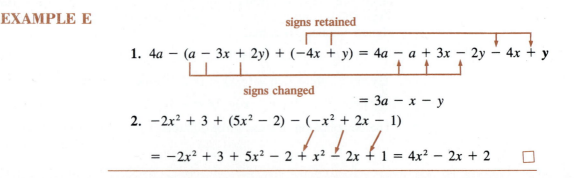

+ sign before parentheses

1. $3c + (2b - c) = 3c + 2b - c = 2b + 2c$

$2b = +2b$　　　　　signs retained

− sign before parentheses

2. $3c - (2b - c) = 3c - 2b + c = -2b + 4c$

$2b = +2b$　　　　　signs changed

3. $3c - (-2b + c) = 3c + 2b - c = 2b + 2c$

signs changed

Note in each case that the parentheses are removed and the sign before the parentheses is also removed. Also, in the first two illustrations we treat $2b$ as $+2b$. □

EXAMPLE E

signs retained

1. $4a - (a - 3x + 2y) + (-4x + y) = 4a - a + 3x - 2y - 4x + y$

signs changed

$$= 3a - x - y$$

2. $-2x^2 + 3 + (5x^2 - 2) - (-x^2 + 2x - 1)$

$$= -2x^2 + 3 + 5x^2 - 2 + x^2 - 2x + 1 = 4x^2 - 2x + 2 \qquad □$$

▶ Again we note carefully that *when a minus sign precedes the parentheses, the sign of* every **term** *is changed* in removing the parentheses.

Brackets [ ] *and* **braces** { } *are also used to group terms, particularly when a group of terms is contained within another group.* (Another symbol of grouping, the **bar** or **vinculum** ‾ , is used, especially for radicals and fractions. The expression $\sqrt{a + b}$ means the square root of the *quantity* $a + b$ and

$$\frac{c}{a + b}$$

means c divided by the quantity $a + b$.) In simplifying expressions containing more than one type of grouping symbols, we may remove the symbols one at a time. In general, it is better to *remove the innermost symbols first*. Consider the illustrations in the following examples.

EXAMPLE F Simplify the expression $5a - [2a - (3a + 6)]$.

The innermost grouping symbols are the parentheses, and we begin by removing them. After the parentheses are removed we simplify and then remove the brackets. This leads to

$$5a - [2a - (3a + 6)] = 5a - [2a - 3a - 6] \quad \text{parentheses removed}$$
$$= 5a - [-a - 6] \quad \text{simplify within brackets}$$
$$= 5a + a + 6 \quad \text{brackets removed}$$
$$= 6a + 6 \qquad \square$$

While a variety of grouping symbols can be used in algebraic expressions, the use of calculators and computers tends to result in parentheses nested within other parentheses as in the following example. Again, it is usually best to begin with the innermost set of parentheses and work outward.

EXAMPLE G In simplifying the expression $18 - ((-(a - 1) + 3a) - 4a)$, we begin with the innermost set of parentheses which are used to group $a - 1$. We get

inner parentheses removed

$$18 - ((-(a - 1) + 3a) - 4a) = 18 - ((-a + 1 + 3a) - 4a)$$
$$= 18 - ((2a + 1) - 4a) \quad \text{simplify}$$
$$= 18 - (2a + 1 - 4a) \quad \text{inner parentheses removed}$$
$$= 18 - (-2a + 1) \quad \text{simplify}$$
$$= 18 + 2a - 1 \quad \text{parentheses removed}$$
$$= 17 + 2a \qquad \text{simplify} \qquad \square$$

Algebraic addition and subtraction and the associated use of symbols of grouping are often encountered in equations. Consider the following example.

EXAMPLE H   Solve the equation $2 - (3a - 4x) = 2x + (7 - a)$ for x.

In this case, we first remove the parentheses and then collect terms with x on the left and the other terms on the right. This leads to the following solution:

$$2 - (3a - 4x) = 2x + (7 - a)$$
$$2 - 3a + 4x = 2x + 7 - a \qquad \text{parentheses removed}$$
$$4x - 2x = 7 - a - 2 + 3a$$
$$2x = 5 + 2a$$
$$x = \frac{5 + 2a}{2}$$

▶ We note here that *the 2's do not cancel* because the 2 of the numerator is a factor only of the term $2a$ and not of the entire numerator. ☐

EXAMPLE I   Pilots know that it usually becomes colder at higher altitudes. In the analysis of temperature changes at different altitudes, the expression

$$\left(\frac{9}{5}C_1 + 32\right) - \left(\frac{9}{5}C_2 + 32\right)$$

is found. We can simplify that expression by first removing parentheses to get

signs changed

$$\left(\frac{9}{5}C_1 + 32\right) - \left(\frac{9}{5}C_2 + 32\right) = \frac{9}{5}C_1 + 32 - \frac{9}{5}C_2 - 32$$
$$= \frac{9}{5}C_1 - \frac{9}{5}C_2$$

which is a simplified form of the original expression. ☐

Exercises 7-1

In Exercises 1 through 4, add the given expressions.

1. $2s - 3xy + 5a$
   $\underline{s + 2xy - 6a}$

2. $3t^2 - 4as - p$
   $+\ \underline{-9t^2 - 5as + 2h}$
   $-6t^2 - 9as + 2h - p$

3. $5y - 2x + 4a - 8xy$
   $\underline{4y - x - 8a - xy}$

4. $8u - rs + 2y - 3s$
   $\underline{-2u - 3rs - 7y - 6s}$

In Exercises 5 through 8, subtract the lower expression from the upper expression.

5. $5x - 6xy$
$\ 2x - 8xy$

6. $4as - 9py - s$
$\ 6as + 2py + 4s$

7. $4x^2 - 6xy - 4s$
$\ 7x^2 - 9xy - 2s$

8. $-7y + 2w - 5u + 7uy$
$\ -6y + 3w + 4u + 8uy$

In Exercises 9 through 24, remove the symbols of grouping and simplify the given expressions.

9. $4a + (3 - 2a)$

10. $6x + (-5x + 4)$

11. $3 - (4x + 7)$

12. $9y - (4y - 8)$

13. $(4s - 9) + (8 - 2s)$

14. $(5x - 2 + 3y) + (4 - 3x)$

15. $(t - 7 + 3y) - (2 - y + t)$

16. $-(6b^3 - 3as + 4x^2) - (2s - 3b^3 - 6x^2)$

17. $4 + [6x - (3 - 4x)]$

18. $3 + (5n - (2 - n))$

19. $3s - (2 - (4 - s))$

20. $7 - [2 - (4x^2 - 2)]$

21. $-(7 - x) - [(2x + 3) - (7x - 2)]$

22. $-(5 - x) - [(5x - 7) - (2 - 3x + b)]$

23. $(t - 5x) - \{[(6p^2 - x) - 9] - (6t + p^2 - x)\}$

24. $8 - \{6xy - [7 - (2xy - 5)] - [6 - (xy - 8)]\}$

In Exercises 25 through 28, perform the indicated operations.

25. Subtract $3a^2 - 7x + 2$ from the sum of $a^2 + x + 1$ and $4a^2 - 5x + 2$.

26. Subtract $5 - 3x$ from the sum of $6s + 5 - 2x$ and $12 + 7x$.

27. Subtract the sum of $4x + 2 - 3t$ and $x - 1 - t$ from the sum of $3x - t + 4t$ and $2x + 8 - 5t$.

28. Subtract the sum of $2a^2 - 7xy + 3x$ and $-a^2 + 10xy$ from the sum of $5a^2 - xy - 2x$ and $-3a^2 + xy + 5x$.

In Exercises 29 through 36, solve the given equations for x.

29. $5 - 7x = 2 - (5x - 1)$

30. $6 - (3 - x) = (5 + 4x) - (x - 1)$

31. $2x - (4 - x) = 6 + (x - 7)$

32. $-(x - 8) = 2x - (5 - 7x)$

33. $4a + (2x - a) = 5 - (a - x)$

34. $ax - (b - 2c) = c + (3b - 5ax)$

35. $(t - x) - (x + t) = (2x - 1) - (5 - x)$

36. $-(2a - x) - (4x - 5) = -[5x - (3 - x)]$

In Exercises 37 through 48, solve the given problems.

37. A company has two manufacturing plants with daily production levels of $6x + 20$ items and $4x - 8$ items, respectively. How many more items are produced daily by the first plant?

38. In a reforesting program, there are $2x + 1000$ trees planted in one area and $x - 2000$ trees planted in a second area. How many more trees are planted in the first area?

39. The resistance (in ohms) in one circuit is given by $5R - 20$ while a second circuit has a resistance of $3R - 50$. How much more resistance is in the first circuit?

40. One production method results in a cost (in dollars) of $12x - 400$ while a second method results in a cost of $10x - 300$. By how much does the cost of the first production method exceed that of the second production method?

41. In doing an analysis of the forces acting on a beam, we may encounter the following expression. Simplify it.

$$4M - (2M + 120) + 80$$

42. An industrial safety specialist analyzes the differences in velocities between two moving parts and obtains the expression

$$v_1 + at - (v_2 + at)$$

Simplify that expression.

43. In analyzing the difference in altitudes between two dropped objects, we find the expression

$$(16t_1^2 + 64) - (16t_2^2 + 32)$$

Simplify that expression.

44. In surveying, the difference between two particular angles is given by

$$[180 - (B_1 + C_1)] - [180 - (B_2 + C_2)]$$

Simplify that expression.

45. During the analysis of the physics of momentum, the following expression may be found. Solve for v_0.

$$I = -(mv - mv_0)$$

46. An acoustical engineer calculates a certain velocity of sound by using the equation

$$V = (0.6C + 1089.0) + (10.3 - 0.1C)$$

Solve for C.

47. A marketing manager calculates the net loss L on the sale of an item by using the formula

$$L = C - (P_2 - P_1)$$

where C is the overhead cost, P_2 is the selling price, and P_1 is the cost to the firm. Solve for P_2.

48. The width of a rectangular thermal blanket is 4 ft less than the length. Twice the length minus the width equals one-half of the perimeter. Find the dimensions of the thermal blanket.

In Exercises 49 through 52, use a scientific calculator to solve the given equations for x.

49. $1327 - 8315x = 6422x + 5512$ **50.** $-(-301.2x - 2.1) = -(206.9x + 27.3)$

51. $18.23x - 12.72 = 13.21 - (5.03x - 2.11)$ **52.** $-(2.013x + 8.035) = 0.034x - (6.000x - 3.187)$

7-2 Multiplication and Division of Monomials

In this section we consider the multiplication and division of algebraic expressions that are monomials (one term). Beginning with multiplication, recall that in Chapter 4 we saw that exponents relate directly to multiplication. For example, we have seen that $5 \times 5 \times 5 \times 5$ can be represented as 5^4. We shall now see a general expression for multiplication of powers of a number.

*The expression a^n is read as "**a** to the **n**th power," where **n** is the* **exponent** *and **a** is called the* **base.** The following example shows how to multiply factors with the same base.

EXAMPLE A $(a^3)(a^4) = (a \cdot a \cdot a)(a \cdot a \cdot a \cdot a) = a^7$

Also, $a^{3+4} = a^7$. We see that we can multiply a^3 by a^4 as follows:

$$(a^3)(a^4) = a^{3+4} = a^7$$

Generalizing the results of Example A, we see that if we multiply a^m by a^n, a^m has m factors of a, a^n has n factors of a, and therefore

$$\boxed{(a^m)(a^n) = a^{m+n}}$$ (7-1)

Sometimes it is necessary to take a number raised to a power, such as a^3, and raise it to a power, as in $(a^3)^4$. Consider the following example.

EXAMPLE B

$$(a^3)^4 = (a^3)(a^3)(a^3)(a^3)$$
$$= (a \cdot a \cdot a)(a \cdot a \cdot a)(a \cdot a \cdot a)(a \cdot a \cdot a) = a^{12}$$

Since $a^{3 \times 4} = a^{12}$, we see that we can raise a^3 to the fourth power by writing

$$(a^3)^4 = a^{3 \times 4} = a^{12}$$ □

Generalizing the results of Example B, we see that if we raise a^m to the nth power, we get

$$\boxed{(a^m)^n = a^{mn}}$$ (7-2)

The distinction between the meaning of Eqs. (7-1) and (7-2) should be carefully noted and understood. These equations will be extremely important in our future work, since they will be of direct use on many occasions.

EXAMPLE C

add exponents

1. $(b^4)(b^5) = b^{4+5} = b^9$

multiply exponents

2. $(b^4)^5 = b^{4 \times 5} = b^{20}$

add exponents

3. $bb^4 = b^{1+4} = b^5$ note that bb^4 is really $b^1 b^4$

└ has an exponent of 1   □

To multiply terms containing more than one base, the exponents of like bases are added. The following example demonstrates this.

EXAMPLE D

add exponents of a

$$(a^3b^2)(a^2b^4) = a^3b^2a^2b^4 = a^3a^2b^2b^4 = a^{3+2}b^{2+4} = a^5b^6$$

add exponents of b

▶ We can only indicate the multiplication of a^5 and b^6 as shown. **It is not possible to simplify further.** □

The power of a product leads to another important result in operating with exponents. The following example will allow us to develop a general expression for this case.

EXAMPLE E $(xy)^4 = (xy)(xy)(xy)(xy) = x^4y^4$
$(3c^2d^5)^3 = (3c^2d^5)(3c^2d^5)(3c^2d^5) = 27c^6d^{15}$

Also, $3^3(c^2)^3(d^5)^3 = 27c^6d^{15}$. □

We see from Example E that *if a product of factors is raised to a power* **n**, *the result is the product of each factor raised to the power* **n.** Generalizing this result, we have

$$(ab)^n = a^nb^n \qquad\qquad (7\text{-}3)$$

EXAMPLE F

1. $(3n)^4 = 3^4n^4 = 81n^4$

exponents multiplied

2. $(a^5b^6)^2 = (a^5)^2(b^6)^2 = a^{10}b^{12}$

exponents multiplied

3. $(3na^5)^4 = (3)^4(n)^4(a^5)^4 = 3^4n^4a^{20} = 81n^4a^{20}$

□

When we multiply monomials we use the notation of exponents and the rules for multiplying signed numbers. We first multiply the numerical coefficients and then the literal factors. The product of the new numerical coefficient and the literal factors is the required product.

EXAMPLE G
$$2ab^3(3a^2bc) = (2 \cdot 3)(a \cdot a^2 \cdot b^3 \cdot b \cdot c) = 6a^3b^4c$$
$$-2x^2y^3(4ax^2y) = -8ax^4y^4$$
$$(-p^2q)(3pq^2r)(-2qrs) = 6p^3q^4r^2s \qquad \square$$

The following example includes the use of Eq. (7-1), (7-2), and (7-3). Note the ways in which these equations are applied.

EXAMPLE H
$$(x^5)(x^3) + (x^5)^2 + (xy)^5 = x^8 + x^{10} + x^5y^5$$
$$(t^3)(t^5) + (t^2)^4 + (at)^8 = t^8 + t^8 + a^8t^8$$
$$= 2t^8 + a^8t^8 \qquad \square$$

Note the difference between expressions such as $(a^2 + b^2)^3$ and $(a^2b^2)^3$. In the first expression we have a *sum* raised to a power, while in the second expression we have a *product* raised to a power. Rule (7-3) works only when we are multiplying or dividing within the parentheses. We can express $(a^2b^2)^3$ as a^6b^6, but $(a^2 + b^2)^3$ is *not* $a^6 + b^6$.

In division, as in multiplication, the use of exponents plays an important role. The following example suggests the basic formula for the use of exponents in division.

EXAMPLE I   Divide a^5 by a^2.

We may express this division as a fraction, as stated in Section 4-2. This gives us

$$\frac{a^5}{a^2}$$

Now we use the principle stated in that section—namely, that if a fraction has a factor common to the numerator and denominator, this factor may be divided out. We note the common factor of a^2 in the numerator and the denominator, or

$$\frac{a^5}{a^2} = \frac{a^2a^3}{a^2}$$

Dividing out this factor of a^2, we get

$$\frac{a^5}{a^2} = a^3$$

Since $a^{5-2} = a^3$, we see that we can divide a^5 by a^2 as follows:

$$\frac{a^5}{a^2} = a^{5-2} = a^3$$

(Continued on next page)

Let us now divide a^2 by a^5. This is expressed as a fraction, and then the common factor of a^2 is divided out. This leads to

$$\frac{a^2}{a^5} = \frac{a^2}{a^2 a^3} = \frac{1}{a^3}$$

Here we see that

$$\frac{a^2}{a^5} = \frac{1}{a^{5-2}} = \frac{1}{a^3}$$

If we divide a^5 by a^5, we get

$$\frac{a^5}{a^5} = 1$$ $\square$

Generalizing the results of Example I, we see that if we divide a^m by a^n, we have

$$\frac{a^m}{a^n} = a^{m-n} \text{ if } m > n \text{ and } a \neq 0 \tag{7-4a}$$

$$\frac{a^m}{a^n} = \frac{1}{a^{n-m}} \text{ if } n > m \text{ and } a \neq 0 \tag{7-4b}$$

$$\frac{a^m}{a^n} = 1 \text{ if } m = n \text{ and } a \neq 0 \tag{7-4c}$$

EXAMPLE J

larger exponent

$$\frac{x^4}{x^3} = x^{4-3} = x \qquad\qquad \frac{x^3}{x^4} = \frac{1}{x^{4-3}} = \frac{1}{x}$$

subtract   larger exponent   subtract

$$\frac{3c^6}{c^2} = 3c^{6-2} = 3c^4 \qquad \frac{c^2}{3c^6} = \frac{1}{3c^{6-2}} = \frac{1}{3c^4}$$

equal exponents
$$\frac{x^3}{x^3} = 1 \qquad\qquad \frac{2s^2}{s^2} = 2(1) = 2$$ $\square$

EXAMPLE K When finding the dimensions of a container that will use the least amount of material for a given volume, we obtain the expression

$$\frac{6x^3}{x^2}$$

Simplify this expression.

Applying Rule (7-4a), we get

$$\frac{6x^3}{x^2} = 6x^{3-2} = 6x^1 = 6x$$ □

To divide one monomial by another, we use the basic principle introduced in Section 4-3 and divide out any factor common to numerator and denominator. In doing so we are using Eqs. (7-4). As in multiplication, we can combine only those exponents which have the same base.

EXAMPLE L

$$-16a^2b \div 4a = \frac{-16a^2b}{4a} = -4ab$$

$$36a^2b^3 \div (-12ab) = \frac{36a^2b^3}{-12ab} = -3ab^2$$

$$-18x^3y \div (-12xy^4) = \frac{-18x^3y}{-12xy^4} = \frac{3x^2}{2y^3}$$

$$-8ab^2x^5 \div (-14a^2b^2x) = \frac{-8ab^2x^5}{-14a^2b^2x} = \frac{4x^4}{7a}$$

Note that in the first two illustrations the exponents in the numerator are larger than those in the denominator. In the third illustration the exponent of x is larger in the numerator, and thus we use Eq. (7-4a) for x. The exponent of y is larger in the denominator, and hence we use Eq. (7-4b) for y. In the fourth illustration, b^2 appears as part of the common factor, and therefore b does not appear in the final result. □

While this section presented multiplication and division of monomials, the following two sections will include multiplication and division of multinomials.

Exercises 7-2

In Exercises 1 through 44, perform the indicated operations.

1. x^3x^7 **2.** n^2n^6 **3.** y^5y^2y **4.** $p^2p^3p^5$ **5.** $x^7 \div x^4$

6. $x^2 \div x^8$ **7.** $\dfrac{a^5}{a^4}$ **8.** $\dfrac{y^9}{y^2}$ **9.** $t^3t^5t^2$ **10.** $(x^3)^8$

11. $(n^2)^7$ **12.** $(t^3)^5$ **13.** $p^3 \div p^{13}$ **14.** $p \div p^9$ **15.** $\dfrac{8n^4}{2n}$

16. $\dfrac{9m}{6m^5}$ **17.** $(y^2y)^5$ **18.** $(rr^3)^8$ **19.** $(p^2p^3)^5$ **20.** $(x^6x^3)^4$

21. $(a^2x) \div (-a)$ **22.** $(-x^4y^3) \div (x^2y)$

23. $(-4ca^2t^4) \div (-2c^3a)$ **24.** $(9c^3yp^4) \div (-12cy^6p)$

25. $(a^3a^2)^8$ **26.** $(nn^3n^5)^2$ **27.** $(rr^6r^8)^{10}$ **28.** $(t^2t^3)(t^4t^5)$

29. $\dfrac{-5x^2yr}{20xyr^3}$ **30.** $\dfrac{-rst^4}{-rs^4t^2}$ **31.** $\dfrac{-7yt^3u}{-6yt}$ **32.** $\dfrac{9abc^4ds}{15bc^4d^4}$

33. $(-ax^2b)^2$ **34.** $(-a^3b)^3$ **35.** $(-at^2)^5$ **36.** $(-cx^2y)^6$

37. $\dfrac{-3x^2y^3}{9xy^5}$ **38.** $\dfrac{-12a^5b^3c^7}{-18a^5b^2c^8}$ **39.** $\dfrac{-28s^9t^{12}}{-84s^{12}t^9}$ **40.** $\dfrac{-r^6t^{10}}{2r^6s^{10}t^{10}}$

41. $(-4rs)(-3st^3)(-7rt)$ **42.** $(-2axy^3)(7ay^4)(-ax^4y)$

43. $(-2st^3x)^3$ **44.** $(-3axt^7)^4$

In Exercises 45 through 48, identify the given equations as true or false.

45. (a) $(2x)^3 = 2x^3$ **46.** (a) $t^5t^2 = t^{10}$ **47.** (a) $\dfrac{ax^2y^3}{axy} = xy^2$ **48.** (a) $x^6 \div x^3 = x^2$

(b) $x^8 \div x^2 = x^4$ (b) $s^4s^3 = s^7$ (b) $(at^3)^4 = a^4t^{12}$ (b) $(3x)^4 = 81x^4$

(c) $r \div r^3 = r^2$ (c) $\dfrac{6a^2x^3}{2ax} = 3ax^2$ (c) $a^2a^5 = a^7$ (c) $x^2x^4 = x^8$

(d) $(x^2y)^3 = x^5y^3$ (d) $\dfrac{8x^2y}{8x^2y} = 0$ (d) $\dfrac{ax^{10}y^8}{ax^{12}y^{10}} = \dfrac{x^2}{y^2}$ (d) $x^5 \div x = x^4$

(e) $(3x^2)^3 = 27x^6$ (e) $(-x^3)^4 = -x^{12}$ (e) $(x^2y^3) = (xy)^6$ (e) $x^2x^3 = x^5$

In Exercises 49 through 52, solve the given problems.

49. A rectangular solid container is made with a square top and bottom, each of side x. Its height is twice the length of a side on the square top. Find the expression for the volume of this container. See Figure 7-1.

50. In determining the total number of different positions for an arm on an industrial robot, it is necessary to multiply n^{12} and n^{16}. Find the expression representing that product.

51. In electric circuits, the power law can be expressed as

$$I = \dfrac{E^2}{R} \div E$$

Perform the indicated division and express the power law in simplified form.

FIGURE 7-1

52. A spherical ball bearing with radius r rests on the center of a flat circular disk with the same radius. In analyzing the average force per unit of area, we find the expression

$$\frac{4\pi cr^3}{3} \div \pi r^2$$

Perform the indicated division and express the result in simplified form.

7-3 Multiplication with Multinomials

In the preceding section, we developed the procedures and rules for multiplying monomials. In this section we use factors that are multinomials. We begin by considering the product of two factors, where one factor is a monomial and the other is a multinomial.

To multiply a monomial and a multinomial, we use the distributive law from Section 4-3. Recall that the distributive law is summarized in Eq. (4-3):

$$a(b + c) = ab + ac \tag{4-3}$$

This law indicates that we must *multiply each term of the multinomial by the monomial*.

EXAMPLE A

$$2ax(3ax^2 + 5x^3) = 2ax(3ax^2) + 2ax(5x^3)$$
$$= 6a^2x^3 + 10ax^4$$

$$3xy^3(x + 4x^2y + 2y) = 3xy^3(x) + 3xy^3(4x^2y) + 3xy^3(2y)$$
$$= 3x^2y^3 + 12x^3y^4 + 6xy^4 \qquad \square$$

Note that in the second part of Example A, the multinomial contained three terms so that the distributive law of Eq. (4-3) was extended. In general, the multinomial may contain any number of terms. The key point to remember is that every one of those terms must be multiplied by the monomial. One common algebraic error is to multiply the monomial with only the first term of the multinomial. Another common algebraic error is the failure to correctly assign signs to the terms appearing in the result. Example A included only positive signs so that an error in signs is not likely. However, we must be careful with expressions involving negative signs, such as those in Example B.

EXAMPLE B

$$-5x^2y(3x - 2xy + 4y) = -5x^2y(3x) - 5x^2y(-2xy) - 5x^2y(4y)$$
$$= -15x^3y + 10x^3y^2 - 20x^2y^2$$

$$-2ab(-3ab - a^3b + 4b^3) = -2ab(-3ab) - 2ab(-a^3b) - 2ab(4b^3)$$
$$= 6a^2b^2 + 2a^4b^2 - 8ab^4 \qquad \square$$

In the illustrations of Examples A and B, one factor was a monomial while the other was a multinomial. We now proceed to consider the multiplication of two multinomials.

To multiply one multinomial by another, we multiply each term of one by each term of the other, again using the rules outlined above. This is a result of the distributive law.

EXAMPLE C

$$(x - 2)(x + 7) = x(x) + x(7) + (-2)(x) + (-2)(7)$$
$$= x^2 + 7x - 2x - 14$$
$$= x^2 + 5x - 14$$

$$(x^2 - 2xy)(ab + a^2) = x^2(ab) + x^2(a^2) + (-2xy)(ab) + (-2xy)(a^2)$$
$$= abx^2 + a^2x^2 - 2abxy - 2a^2xy \qquad \square$$

EXAMPLE D The dimensions of a room (in feet) are given as $2x + 1$ and $3x - 2$. Find an expression for the area of that room.

Since area is the product of length and width (the given dimensions), we express the area (in square feet) as

$$(2x + 1)(3x - 2) = 2x(3x) + 2x(-2) + 1(3x) + 1(-2)$$
$$= 6x^2 - 4x + 3x - 2$$
$$= 6x^2 - x - 2 \qquad \square$$

EXAMPLE E

$$(2a + 3)(2a^2 + 3a - b) = (2a)(2a^2) + (2a)(3a) + (2a)(-b) + (3)(2a^2)$$
$$+ (3)(3a) + (3)(-b)$$
$$= 4a^3 + 6a^2 - 2ab + 6a^2 + 9a - 3b$$
$$= 4a^3 + 12a^2 - 2ab + 9a - 3b \qquad \square$$

EXAMPLE F

$$(x - y)(x^2 + xy + y^2) = x(x^2) + x(xy) + x(y^2) - y(x^2) - y(xy) - y(y^2)$$
$$= x^3 + x^2y + xy^2 - x^2y - xy^2 - y^3$$
$$= x^3 - y^3$$

Note that the product includes like terms that could be combined. $\qquad \square$

EXAMPLE G   The expression $(2a + b)^2$ is equivalent to $(2a + b)(2a + b)$. We therefore square $(2a + b)$ as follows:

$$(2a + b)^2 = (2a + b)(2a + b)$$
$$= 2a(2a) + 2a(b) + b(2a) + b(b)$$
$$= 4a^2 + 2ab + 2ab + b^2$$
$$= 4a^2 + 4ab + b^2$$

A common error in such cases is simply to square the individual terms. Here we see that $(2a + b)^2$ **is not** $4a^2 + b^2$. The result must include the $4ab$ term.

Exercises 7-3

In Exercises 1 through 36, perform the indicated multiplications.

1. $2a(a + 3x)$

2. $3x(2x - 5)$

3. $3a^2(2x - a^2)$

4. $4b^3(3 + 2b^2)$

5. $(-2st)(sx - t^2y)$

6. $(-8y)(8y + t^2)$

7. $(-3xy)(x^2y - 3axy^6)$

8. $(-5y^6)(-uy^7 - hpy)$

9. $(x - 3)(x - 1)$

10. $(t + 5)(t + 2)$

11. $(s - 2)(s + 3)$

12. $(x - 6)(x - 4)$

13. $(x + 1)(2x - 1)$

14. $(3a - 2)(5a + 4)$

15. $(5v + 1)(2v + 3)$

16. $(3v - 2)(4v - 3)$

17. $(a - x)(a - 2x)$

18. $(x + 2y)(x - y)$

19. $(2a - c)(3a + 2c)$

20. $(3s + 4t)(5s + 2t)$

21. $(2x - 5t)(x + 9)$

22. $(4x - 9uy)(2 + 3uy)$

23. $(2a - 9py)(2a + 9py)$

24. $(s - 3xu^2)(-xu^2 - 8s)$

25. $(2a + 1)(a^2 - 3a - 5)$

26. $(3x - 2)(2x^2 + x - 3)$

27. $(a - x)(a + 2xy - 3x)$

28. $(2 - x)(5 + x - x^2)$

29. $(x - 2)^2$

30. $(a + 5)^2$

31. $(x + 2y)^2$

32. $(2a - 3b)^2$

33. $(x - 2)(x + 3)(x - 4)$

34. $(2x - 3)(x + 1)(3x - 4)$

35. $(x + 1)^3$

36. $(2a - x)^3$

In Exercises 37 through 40, identify the equations as true or false.

37. (a) $(a + b)^2 = a^2 + b^2$

(b) $(x - 3)^2 = x^2 - 9$

(c) $(x - 1)^2 = x^2 - 2x + 1$

38. (a) $(2x + 3)^2 = 4x^2 + 9$

(b) $(x + y)^2 = x^2 + 2xy + y^2$

(c) $(x + y)^2 = x^2 + y^2$

39. (a) $(t + 1)^2 = t^2 + t + 1$

(b) $(t + 1)^2 = t^2 + 2t + 1$

(c) $(t + 1)(t - 1) = t^2 - 1$

40. (a) $x^2 - y^2 = (x - y)(x + y)$

(b) $x^2 - y^2 = (x - y)(x - y)$

(c) $x^2 + 9 = (x + 3)(x + 3)$

In Exercises 41 through 52, solve the given problems.

41. The length of a rectangular solar panel is $x + 3$ feet while its width is $x - 2$ feet. Find an expression for its area. See Figure 7-2.

$x - 2$ ft

FIGURE 7-2 $x + 3$ ft

42. In determining the optimal size of a container, a manufacturer encounters the equation

$$V = 2x(8 - x)(20 - 3x)$$

Multiply out the right member of this equation.

43. The total production cost of x pills used for lowering blood pressure is given by

$$C = 200(x + 3)(x + 5)$$

Multiply out the right member of this equation.

44. By multiplication, show that $(x + y)(x^2 - xy + y^2) = x^3 + y^3$. Then show by substitution that the same value is obtained for each side for the values $x = -2$ and $y = -3$.

45. Use multiplication to show that $(2x + y)(2x - y) = 4x^2 - y^2$. Then substitute 5 for x and 3 for y in both sides of the equation to show that the same value is obtained.

46. A parabolic antenna is filled with a liquid, and the force of the pressure leads to the expression

$$w(1 - x)(4 - x^2)$$

Multiply out this expression.

47. For the motion of an object, the following formula relates velocity v and time t for certain conditions. Multiply out the right side of the equation.

$$v = 16(t - 1)(t - 3)$$

48. In determining the focal length of a lens, we use the *lensmaker's equation*

$$\frac{1}{f} = (n - 1)(r_2 - r_1)$$

Multiply out the right side of this equation.

49. In analyzing the expansion of a steel rod, we find the equation

$$L = k(L_1 - L_2)(T_1 - T_2)$$

Multiply out the right side of this equation.

50. In determining the change in the cost of producing a number of buoys, we find the equation

$$C = n(V_0 - V_1)(p_1 - p_2)$$

Multiply out the right side of this equation.

51. The voltage across an electric resistor equals the product of the current and the resistance. The voltage across one resistor is twice that across a second resistor. The current is 3 A in the first resistor and 4 A in the second. Find the resistance (in ohms) if the second resistance is 5 Ω less than the first.

52. In determining the value of a savings account, we sometimes use the expression $P(1 + r)^2$. Multiply out this expression.

7-4 Division with Multinomials

In Section 7-2 we discussed the division of one monomial by another. In this section we begin by considering the division of a multinomial by a monomial.

In arithmetic, since

$$\frac{2+3}{7} = \frac{2}{7} + \frac{3}{7} \qquad \text{or} \qquad \frac{a+b}{c} = \frac{a}{c} + \frac{b}{c}$$

we see that we must divide each of the numbers of the numerator separately by the number of the denominator. Since algebraic expressions represent numbers, we have the following method of dividing a multinomial by a monomial: *Divide each term of the multinomial by the monomial and add the resulting terms algebraically to obtain the quotient.* Note that each and every term of the numerator must be divided by the monomial in the denominator.

EXAMPLE A

$$\frac{x^3 + x^2}{x} = \frac{x^3}{x} + \frac{x^2}{x} = x^2 + x$$

A very common algebraic error arises when a monomial is not divided into each term of the numerator. For example,

$$\frac{x^5 + 7}{x}$$

does **not** *equal* $x^4 + 7$ since each term of the numerator has not been divided by x. Actually, no further simplification is possible here.

EXAMPLE B

$$(2mn^2 - 3n) \div n = \frac{2mn^2 - 3n}{n}$$

$$= \frac{2mn^2}{n} - \frac{3n}{n} = 2mn - 3$$

each term of numerator divided by denominator

$$(4x^2y + 2xy^3) \div 2xy = \frac{4x^2y + 2xy^3}{2xy}$$

$$= \frac{4x^2y}{2xy} + \frac{2xy^3}{2xy} = 2x + y^2$$

$$(7a^3b^2 - 28a^3b^3 + 35a^2b^2) \div (-7ab^2) = \frac{7a^3b^2 - 28a^3b^3 + 35a^2b^2}{-7ab^2}$$

$$= \frac{7a^3b^2}{-7ab^2} + \frac{-28a^3b^3}{-7ab^2} + \frac{35a^2b^2}{-7ab^2}$$

$$= -a^2 + 4a^2b - 5a$$

In dividing by a multinomial, we shall deal exclusively with polynomials. (See Section 7-1.)

To divide one polynomial by another, we arrange both the dividend and the divisor in descending powers of the same literal factor. Then we divide the first term of the dividend by the first term of the divisor. The result of this division is the first term of the quotient. We then multiply the entire divisor by the first term of the quotient and subtract this product from the dividend. Now we repeat these operations by dividing the first term of the divisor into the first term of the difference just obtained. The result of this division is the second term of the quotient. We multiply the entire divisor by this second term and subtract the result from the first difference. We repeat this process until the difference is either zero or a quantity whose degree is less than that of the divisor.

EXAMPLE C   Divide $2x^2 + 5x - 3$ by $2x - 1$.

Since both the dividend and the divisor are already in descending powers of x, no rearrangement is necessary. Then we set up the division as follows:

$$2x - 1 \overline{)2x^2 + 5x - 3}$$

We now determine that $2x^2$ divided by $2x$ is x. Thus x becomes the first term of the quotient:

We now multiply $2x - 1$ by x and place the product below the dividend:

▶ The $2x^2 - x$ is now **subtracted** *from the dividend:*

We now determine that $2x$, the first term of the divisor, divided into $6x$, the first term of the remainder, is 3. Thus 3 becomes the second term of the quotient:

The divisor is now multiplied by 3, and the product is placed below the remainder and subtracted:

$$
\begin{array}{r}
x \ + \ 3 \quad\longleftarrow \text{ quotient} \\
2x - 1\overline{)2x^2 + 5x - 3} \\
\underline{2x^2 - x} \\
6x - 3 \\
\underline{6x - 3} \longleftarrow 3(2x - 1) \\
0 \longleftarrow [(6x - 3) - (6x - 3)] = 0
\end{array}
$$

Since the remainder is zero and all terms of the dividend have been used, the division is complete. We conclude that

$$(2x^2 + 5x - 3) \div (2x - 1) = x + 3$$

As a check, we can verify that $(x + 3)(2x - 1) = 2x^2 + 5x - 3$. □

EXAMPLE D   We can find the efficiency of a certain motor by dividing the output power by the input power. We find an expression for efficiency by dividing $3x^2 - 4x + x^3 - 12$ by $x + 2$.

The dividend is arranged in descending powers of x, and the division proceeds as shown:

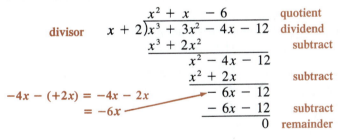

This division is exact.

The division may be checked by multiplication. In this case

$$(x + 2)(x^2 + x - 6) = x^3 + x^2 - 6x + 2x^2 + 2x - 12$$
$$= x^3 + 3x^2 - 4x - 12$$

Since the product equals the dividend, the result checks. □

EXAMPLE E   Divide $4y^3 + 6y^2 + 1$ by $2y - 1$.

Since no first-power term in y appears in the dividend, we insert one with a zero coefficient to simplify the division:

$$
\begin{array}{r}
2y^2 + 4y\ \ + 2 \\
2y - 1\overline{)4y^3 + 6y^2 + 0(y) + 1} \\
\underline{4y^3 - 2y^2} \quad\quad\quad\quad \\
8y^2 + 0(y) + 1 \\
\underline{8y^2 - 4y} \quad\quad \\
4y\ \ + 1 \\
\underline{4y\ \ - 2} \\
+ 3
\end{array}
$$

subtract

subtract

subtract

remainder

$0y - (-4y) = 4y$

The quotient in this case is

$$2y^2 + 4y + 2 + \frac{3}{2y - 1}$$

Note how the remainder is expressed as part of the quotient.   □

The solution of equations uses a combination of the basic operations. The following examples illustrate such solutions.

EXAMPLE F   Solve the equation $2(3 - x) = 8 - 4(x - a)$ for x.

The solution proceeds as follows:

$$2(3 - x) = 8 - 4(x - a)$$
$$6 - 2x = 8 - 4x + 4a$$
$$4x - 2x = 8 + 4a - 6$$
$$2x = 2 + 4a$$
$$x = \frac{2 + 4a}{2} = \frac{2}{2} + \frac{4a}{2} = 1 + 2a$$

□

EXAMPLE G   Solve the equation $x(b + 3) = b^3 + 5b^2 + 5b - 3$ for x.

Initially, we might want to begin by removing the parentheses by multiplying $b + 3$ by x. However, if we carefully examine the given equation, we see that the variable x can be isolated by simply dividing both sides of the equation by $b + 3$. We get

$$x = \frac{b^3 + 5b^2 + 5b - 3}{b + 3}$$

We can now proceed to perform the algebraic division indicated on the right-hand side of this last equation as follows:

$$\begin{array}{r} b^2 + 2b \ - 1 \\ b + 3\overline{)b^3 + 5b^2 + 5b \ - 3} \\ \underline{b^3 + 3b^2} \qquad\qquad \text{subtract} \\ 2b^2 + 5b \ - 3 \\ \underline{2b^2 + 6b} \qquad \text{subtract} \\ - b \ - 3 \\ \underline{- b \ - 3} \qquad \text{subtract} \\ 0 \end{array}$$

This division is exact and we can now express the solution as $x = b^2 + 2b - 1$. In this particular example, it would not have been wise to remove the grouping symbols. $\square$

Exercises 7-4

In Exercises 1 through 32, perform the indicated divisions.

1. $(6ab + 5a) \div a$

2. $(2x^2 - 3x) \div x$

3. $(9m^2 - 3m) \div 3m$

4. $(8s^3 + 2s^2) \div 2s$

5. $(a^3x^4 - a^2x^3) \div (-ax^2)$

6. $(-2xy^6 - 4x^2y^5) \div (2xy^3)$

7. $(3xy^4 - 6x^2y^5) \div (3xy^2)$

8. $(-14p^3q^5 + 49p^4q^2) \div (-7p^2q^2)$

9. $\dfrac{a^2b^2c^3 - a^3b^4c^6 - 2a^3bc^2}{a^2bc^2}$

10. $\dfrac{8rs^2t^5 - 18r^3st^4 - 16rst^3}{-2rst^2}$

11. $\dfrac{a^2b^3 - 2a^3b^4 - ab - ab^2}{-ab}$

12. $\dfrac{5m^2n^2y - 30mn^2 + 35m^2n^8}{-5mn^2}$

13. $(x^2 - 2x - 3) \div (x + 1)$

14. $(x^2 + 3x - 10) \div (x - 2)$

15. $(2x^2 - 5x - 3) \div (x - 3)$

16. $(3x^2 + 4x - 5) \div (x + 2)$

17. $(8x^2 + 6x - 7) \div (2x + 3)$

18. $(6x^2 + 7x - 3) \div (3x - 1)$

19. $\dfrac{8x^3 - 18x^2 - 7x + 12}{4x - 3}$

20. $\dfrac{4x^3 - 20x - 2x^2 - 12}{2x + 5}$

21. $(5x - 5x^2 + 2x^3 - 6) \div (x - 2)$

22. $(1 - x^2 + 6x^3) \div (1 + 2x)$

23. $\dfrac{x^4 - 1}{x - 1}$

24. $\dfrac{x^3 - 8}{x - 2}$

25. $\dfrac{5x^2 - 5}{x + 1}$

26. $\dfrac{x^6 - 3x^5 - x^3 - 3x^2 + x - 3}{x - 3}$

27. $\dfrac{6x^4 - 5x^3 + 7x^2 + x - 3}{3x - 1}$

28. $\dfrac{6x^4 + 15x^3 - 25x^2 - 22x + 21}{2x + 7}$

29. $\dfrac{a^2 + ab - 12b^2}{a + 4b}$

30. $\dfrac{2x^2 + xy - 6y^2}{x + 2y}$

31. $\dfrac{x^3 - 5x^2y + 2xy^2 + 2y^3}{x - y}$

32. $\dfrac{4s^3 - 8s^2t + 5st^2 - t^3}{2s - t}$

In Exercises 33 through 36, identify the given equations as true or false.

33. (a) $\dfrac{x^2 + y^2}{x} = x + y^2$

  (b) $\dfrac{x^2 - ax}{x} = x - a$

  (c) $\dfrac{x^2 - 4x + 4}{x - 2} = x - 2$

34. (a) $\dfrac{x^2 + 3xy}{x} = x^2 + 3y$

  (b) $\dfrac{x^2 + y^2}{x + y} = x + y$

  (c) $\dfrac{x^2 + 2xy + y^2}{x + y} = x + y$

35. (a) $\dfrac{10t^4 - 5t^2}{-5t} = -2t^3 + t$

  (b) $\dfrac{-12x^5 + 8x^2}{-4x} = 3x^4 - 2x$

  (c) $\dfrac{6x^2 - 8}{x^2} = -2$

36. (a) $\dfrac{8x^3 - 4ax^2}{2ax} = \dfrac{4x^2 - 2ax}{a}$

  (b) $\dfrac{x^2 + 4}{x + 2} = x + 2$

  (c) $\dfrac{x^2 - 6x + 9}{x - 3} = x - 3$

In Exercises 37 through 40, solve the given equations for x.

37. $(a - 1)x = a^2 + 4a - 5$

38. $(2c + 3)x = 2c^2 - 13c - 24$

39. $(a + 4)x = a^3 + 4a^2 - 5a - 20$

40. $(3b - 1)x = 3b^3 - b^2 + 3b - 1$

In Exercises 41 through 48, solve the given problems.

41. A total of $2x$ books weigh $4x^2 + 6x$ kg. How much does each book weigh? The books are all the same.

42. In considering the resistance in a parallel circuit with three resistors, we find the expression

$$\frac{R_2 R_3 + R_1 R_3 + R_1 R_2}{R_1 R_2 R_3}$$

Perform the indicated division and express the result as the sum of terms.

43. The volume of a gas is expanding at a rate (in cubic inches per minute) described by the expression

$$\frac{6r^2 + 8r + 10}{r + 2}$$

Perform the indicated division.

44. In attempting to determine the efficiency of a diesel engine, an automotive engineer encounters the algebraic expression

$$(x^2 + 5x + 6) \div (x + 3)$$

Perform the indicated division.

45. The shape of one section of a microwave-transmitting antenna is described by the expression

$$5x^2 + 4x$$

Divide this expression by $x + 2$.

46. If a satellite used for surveying travels $3x^2 - 8x - 28$ miles in $x + 2$ hours, find an expression for the speed in miles per hour.

47. Under certain conditions, when dealing with the electronics of coils, we may encounter the expression

$$\frac{60r^2}{6r + 1.2}$$ Perform the indicated division.

48. The surface area of a rectangular cover plate is represented by $x^2 + 8x + 15$. If its length is $x + 3$, what is its perimeter?

In Exercises 49 through 52, use a scientific calculator to perform the indicated divisions. Assume that all numbers are approximate.

49. $\dfrac{6.32x^5 - 2.17x^4 + 9.35x}{3.14x}$

50. $\dfrac{15.92x^7 - 12.63x^3 + 29.76x}{11.15x}$

51. $\dfrac{67.91x^5 - 8.305x^4 - 256.8x^3}{3.142x^3}$

52. $\dfrac{83.91x^8 + 12.67x^6 - 843.3x^5}{20.43x^5}$

Chapter 7 Formulas

$(a^m)(a^n) = a^{m+n}$ (7-1)

$(a^m)^n = a^{mn}$ (7-2)

$(ab)^n = a^n b^n$ (7-3)

$\dfrac{a^m}{a^n} = a^{m-n}$ if $m > n$ and $a \ne 0$ (7-4a)

$\dfrac{a^m}{a^n} = \dfrac{1}{a^{n-m}}$ if $n > m$ and $a \ne 0$ (7-4b)

$\dfrac{a^m}{a^n} = 1$ if $m = n$ and $a \ne 0$ (7-4c)

7-5 Review Exercises for Chapter 7

In Exercises 1 through 56, simplify the given expressions by performing the indicated operations.

1. $a - (x - 2a)$

2. $x - (3s - 2x)$

3. $2(x - 5y) - (3y - 7x)$

4. $-(y + 2s) - (5s - y)$

5. $-2 + (n + 4) - (6 - n)$

6. $t - (5 + 2t) + (t - 7)$

7. $2x - 3(x - 3y) - (y - x)$

8. $-8x + 4r - 2(-r - 3x)$

9. $(-3a^2b)(2ab^5)$

10. $(-2s^3t^2)(-4st^4)$

11. $(-2xy^2z)^3(-7xy^3z^5)$

12. $(-2x^2y)^2(-3xy^5)$

13. $(3ab^2)^3$

14. $(2a^4c)^4$

15. $(-2x^2yz^3)^4$

16. $(-x^3y^2z^4)^5$

17. $(-8a^2px^4) \div (2apx)$

18. $(-18rs^3t^5) \div (-24r^4st^6)$

19. $\dfrac{15x^2y^3z}{3xy^7z^4}$

20. $\dfrac{48ab^5c^2}{18a^7bc^6}$

21. $5 - [x - (3 - 4x)]$ $8-5y$

22. $3y + [(5y - 2) - 6y]$

23. $2x + [(2x - a) - (a - x)]$

24. $-(3xy - y) - 2[2x - (y - xy)]$

25. $2 - \{2 - [3x - (7 - x)]\}$

26. $4a + \{a - 3 + [2a - (3 - 2a)]\}$

27. $-\{2b - [b - (4 - 5b)] + (6 - b)\}$

28. $2x - y - 2\{3x - [y - (3y - 4x)]\}$

29. $2x^2(x^3 - 3x)$

30. $s^3(3s^4 - 2s)$

31. $-2a(a^2x - at)$

32. $3a^2j(-3j^4 + 4a - aj)$

33. $(x - 3)(2x + 7)$

34. $(3x - 2)(x + 5)$

35. $(2a - 5b)(3a + 2b)$

36. $(2x - y)(y - 5x)$

37. $(x + 1)(x^2 - x + 1)$

38. $(x - 3)(2x^2 + x - 2)$

39. $2(2 - x)(x^2 - x - 4)$ $-2x^3+6x^2+4x-16$

40. $-3(xy - q)(x - 2y + 3q)$

41. $(2x^3y^5 - 3x^6y^2) \div (-x^3y^2)$

42. $(-9a^3b^4 + 12ab^5) \div (-3ab^2)$

43. $\dfrac{h^2j^4 - 3hj^6 - 6h^4j^7}{hj^4}$

44. $\dfrac{-18f^3g^2k^2 + 24f^2gk^6 - 36fgk^4}{-6fgk^2}$

45. $(x^2 + x - 12) \div (x - 3)$

46. $(6x^2 - 7x - 5) \div (2x + 1)$

47. $\dfrac{2x^3 + x^2 - x - 8}{2x + 3}$

48. $\dfrac{6x^3 + x^2 - 12x + 7}{3x + 5}$

49. $\dfrac{x^3 + 1}{x + 1}$

50. $\dfrac{4x^3 + x - 1}{2x - 1}$

51. $\dfrac{2x^4 - x^3 - 2x - 8}{x^2 + 2}$

52. $\dfrac{6x^4 + x^3 + 5x^2 + 2}{2x^2 - x + 1}$

53. $\dfrac{3x^2 + 5xy - 2y^2}{3x - y}$

54. $\dfrac{-a^2b - 2b^2 + 3a^4}{a^2 - b}$

55. $\dfrac{x^3 - 1}{x - 1}$

56. $\dfrac{2x^3 - 7x^2 + 5x - 6}{2x^2 - x + 2}$

In Exercises 57 through 60, solve the given equations for x.

57. $ax - 3 - a(2 - x) = 3(4a - 1)$

58. $b(x - 3b) - (b^2 - x) = x - b(1 - b)$

59. $(a^2 - 1)x = a^4 - 1$

60. $(c + 2)x = 2c^2 + c - 6$

In Exercises 61 through 80, solve the given problems.

61. In analyzing the burning temperature of a plastic, a fire-science specialist encounters the expression

$$(1.8C + 32) - 0.05(1.8C - 32)$$

Simplify this expression.

62. In the analysis of the transfer of heat through two glass surfaces, the expression

$$8(T_1 - T_2) - 6(T_1 - T_2)$$

is encountered. Simplify this expression.

63. A certain absorption factor in radiation is determined by

$$C = 1 - 3(I_r + I_t)$$

Solve for I_r.

64. In electricity, when analyzing the voltage in a certain type of circuit, we may find the equation

$$V = i(R + r) - (E - E_1)$$

Solve for E_1.

65. When finding the center of mass of a particular area, we may encounter the expression

$$(2x - x^2) - (3x^2 - 6x)$$

Simplify this expression.

66. One factory produced $x + 12$ units in a given day, and a second factory produced $2x + 1$ units. How many more units were produced by the second factory?

67. The total revenue obtained by selling x units of a certain item is given by the expression

$$x(36 - 4x)$$

Perform the indicated multiplication.

68. Under certain conditions, the distance (in feet) that an object is above the ground is given by the expression

$$16t(5 - 2t)$$

Perform the indicated multiplication.

69. When dealing with the electron, physicists encounter the expression

$$m(v_2 - v_1)(v_2 + v_1)$$

Multiply out this expression.

70. Simplify the following expression, which is sometimes used in applications involving optics.

$$\frac{ID}{d}\left(n + \frac{1}{2}\right) - \frac{ID}{d}\left(n - \frac{1}{2}\right)$$

71. Perform the indicated multiplication in the following expression, which is sometimes used in applications involving the expansion of a heated surface.

$$lh(1 + at)^2$$

72. Simplify the following expression, which is sometimes used in applications involving the interference of light from a double source.

$$(2x + d)^2 - (2x - d)^2$$

73. Perform the indicated multiplication in the following expression, which is sometimes used in applications involving transistors.

$$[r_1 + (1 - a)r_2](1 - a)$$

74. A rectangular open-topped box is made from a square piece of aluminum 10 in. on each side by cutting out a square of side x from each corner and turning up the sides. Show that the volume of the box is

$$V = x(10 - 2x)^2$$

and then multiply out the right side of this equation. See Figure 7-3.

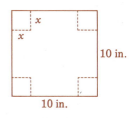

75. The revenue R obtained from selling x units at a price of p is related to x and p by the formula $R = px$. Under certain conditions, $R = p_0x - kx^2$, where p_0 is the highest price and k is a constant depending on the economic conditions. Substitute this expression for R and solve the resulting equation for p.

FIGURE 7-3

76. In the theory dealing with the efficiency of an aircraft engine, a technician finds the expression

$$\frac{Q_1 - Q_2}{Q_1}$$

By performing the indicated division, find an equivalent expression.

77. If the area of a rectangular panel is $2a^2 - a - 15$ and its length is $a - 3$, find its width.

78. In a study of geometric population growth, the equation

$$S(1 - r) = a(1 + r + r^2 + r^3) - a(r + r^2 + r^3 + r^4)$$

is used. Solve for S.

79. The centripetal force (in pounds) of a car on a curve is found from the equation $F = mv^2/r$. Find an expression for F if $m = 6x + 2$, $v = x + 5$, and $r = 3x + 1$. Also, find the force (in pounds) if $x = 10$.

80. Traveling at an average ground speed of 124 mi/h, a light aircraft travels a distance of $20x^2 - 202x - 84$ miles in $2x - 21$ hours. Find the actual distance traveled.

In Exercises 81 through 84, use a scientific calculator to perform the indicated operations. Assume all numbers are approximate in 81, 82, and 84, but the numbers in 83 are exact.

81. $(78.37x^2)(36.25x^6)$

82. $\dfrac{0.00342x^3 - 0.00164x^2 + 0.00203x}{0.000512x}$

83. $\dfrac{16384x^{14} - 4096x^{12} - 16x^4}{16x^4}$

84. Use a calculator to solve for x: $-(16.342x - 8.222) = -12.366x$

Factoring

If we multiply 3 and 7, the result of 21 is called the **product** while 3 and 7 are called **factors**. We sometimes begin with a product such as 21 and attempt to find the numbers which, when multiplied, will give us 21. **Factoring** is the process of breaking up a product into its component factors. Note that we are representing an expression as the *product* of factors, not as the sum or difference of terms. We used this concept in Chapter 1 when we considered a fraction and found the factors of the numerator and denominator so that we could simplify or reduce the fraction. There are also many times when we have an algebraic expression and we wish to find expressions which, when multiplied together, equal the given expression. As in arithmetic, algebraic expressions that multiply to yield a product are called factors of the product. The process of determining those algebraic factors is also referred to as **factoring**.

Factoring algebraic expressions has many uses of real and practical value. Complicated algebraic expressions can often be expressed in a simplified form through the process of factoring. In this chapter we consider the factoring of various types of algebraic expressions.

Among the applications included in this chapter is the following. An aircraft uses Venturi tubes for determining airspeed, which is displayed on the airspeed indicator. Analysis of the Venturi tubes leads to the expression

$$\frac{1}{2}krv_1^2 - \frac{1}{2}krv_2^2$$

This chapter presents general techniques for factoring such expressions, and we will show how that particular expression can be factored.

8-1 Common Monomial Factors

We begin this chapter with an example that illustrates the process of factoring.

EXAMPLE A   Since $3x + 6 = 3(x + 2)$, the quantities 3 and $x + 2$ are factors of $3x + 6$. In starting with the expression $3x + 6$ and determining that the factors are 3 and $x + 2$, we have factored $3x + 6$. Also, we refer to $3(x + 2)$ as the factored form of $3x + 6$.

Since $2ax - ax^2 = ax(2 - x)$, the quantities ax and $2 - x$ are factors of $2ax - ax^2$. In fact, ax itself has factors of a and x, which means that we actually have three factors: a, x, and $2 - x$.   □

In our work with factoring we shall consider only the factoring of polynomials (which include the integers), all terms of which will have integers as coefficients. When we extend the meaning of a prime number as stated in Section 1-4 and include the possibility of signed numbers, *we call a polynomial* **prime** *if it contains no factors other than +1 or −1, or plus or minus itself. We then say that an expression is* **factored completely** *if it is expressed as a product of its prime factors.* That is, an expression is factored completely when it cannot be factored any further.

Any set of factors may be checked by multiplication. That is, we can check that the factors have been correctly determined by multiplying them to verify that the product is the original expression.

EXAMPLE B   The expression $x + 2$ is prime since it cannot be factored. The expression $3x + 6$ is not prime since it can be factored as $3x + 6 = 3(x + 2)$.

The expression $2ax - ax^2$ is not prime since it can be factored. If we write $2ax - ax^2 = a(2x - x^2)$, we have factored the expression, but we have not factored it completely since $2x - x^2 = x(2 - x)$. Factoring $2ax - ax^2$ completely, we write $2ax - ax^2 = ax(2 - x)$. Here each factor is prime.   □

Methods of determining factors such as those in Example B will now be considered. Methods of factoring other types of expressions are considered in the sections which follow. However, we must be able to recognize prime factors when they occur.

Often an algebraic expression contains a factor or factors that are common to each term of the expression. *The first step in factoring any expression is to determine whether there is such a* **common monomial factor.** To do this we note the common factor by inspection and then use the reverse of the distributive law to show the factored form. For reference, the distributive law is

$$a(x + y) = ax + ay \qquad\qquad (8\text{-}1)$$

Consider the following example.

EXAMPLE C Factor $ax - 5a$.

We note that each term of the expression $ax - 5a$ contains a factor of a. Using the reverse of the distributive law, we have

$$ax - 5a = a(x - 5)$$

Here a is the common monomial factor and $a(x - 5)$ is the required factored form of $ax - 5a$. We **check this result by multiplying** and find that

$$a(x - 5) = ax - 5a$$

which is the original expression. ☐

EXAMPLE D Factor $6ax^2 - 18x^3$.

The numerical factor 6 and the literal factor x^2 are common to each term of the given polynomial so that 6 and x^2 are common monomial factors. Technically, 6 and x^2 can themselves be factored as $3 \cdot 2$ and $x \cdot x$, but the usual practice is not to reduce *monomial* factors to prime factors. We therefore say that $6x^2$ is the common monomial factor (instead of $3 \cdot 2 \cdot x \cdot x$). We now express the factoring of $6ax^2 - 18x^3$ as

$$6ax^2 - 18x^3 = 6x^2(a) + 6x^2(-3x) = 6x^2(a - 3x)$$

Multiplying these factors, we obtain the original expression to verify that the factoring is correct. ☐

When we factor by reversing the distributive law, it can be seen that the expression is written as the product of the common monomial factor and the quotient obtained by dividing the expression by the monomial factor. This is illustrated in the following example.

EXAMPLE E Factor $3ax^2 + 3x^3y - 6x^2z$.

The numerical factor 3 and the literal factor x^2 are common to each term of the polynomial so that $3x^2$ is the common monomial factor. Now dividing the polynomial by $3x^2$, we have

$$\frac{3ax^2 + 3x^3y - 6x^2z}{3x^2} = \frac{3ax^2}{3x^2} + \frac{3x^3y}{3x^2} - \frac{6x^2z}{3x^2} = a + xy - 2z$$

Therefore, the factoring of $3ax^2 + 3x^3y - 6x^2z$ is expressed as

$$3ax^2 + 3x^3y - 6x^2z = 3x^2(a + xy - 2z)$$ ☐

EXAMPLE F   Factor $4c^4x^2 - 8c^3x + 2c^3$.

The numerical factor 2 and the literal factor c^3 are common to each term, which means that $2c^3$ is the common monomial factor. Dividing the expression by $2c^3$, we have

$$\frac{4c^4x^2 - 8c^3x + 2c^3}{2c^3} = \frac{4c^4x^2}{2c^3} - \frac{8c^3x}{2c^3} + \boxed{\frac{2c^3}{2c^3}} = 2cx^2 - 4x + 1$$

Therefore,

$$4c^4x^2 - 8c^3x + 2c^3 = 2c^3(2cx^2 - 4x + 1)$$

Note the presence of the 1 in the second factor. As we see, it is the result of dividing the term $2c^3$ by the factor $2c^3$. Since the term and the common factor are the same, it is a common error to omit the 1 in the second factor. *It is incorrect to omit the* **1**, which can be seen through checking by multiplication. ☐

There is a systematic method of finding common monomial factors. *When each term of a polynomial is expressed as the product of its prime factors, the* **greatest common factor** *is the product of the factors common to all terms.* The following example illustrates this technique.

EXAMPLE G   Factor $4x^3 + 8x^2y - 24xy^2$.

Expressing each term as the product of prime factors, we get

$$4x^3 = 2 \cdot 2 \cdot x \cdot x \cdot x \qquad 8x^2y = 2 \cdot 2 \cdot 2 \cdot x \cdot x \cdot y$$
$$-24xy^2 = -1 \cdot 2 \cdot 2 \cdot 2 \cdot 3 \cdot x \cdot y \cdot y$$

We now note that 2 appears at least twice in each product and x appears at least once in each. Even though 2 appears three times in the second and third terms, it appears only twice in the first term; hence we may use only two factors of 2 in the greatest common factor. Similar arguments may be given for x and y. The greatest common factor is $2 \cdot 2 \cdot x = 4x$. The remaining factors of the first term give the product x^2; those of the second term give $2xy$; those of the third term give $-6y^2$. We get

$$4x^3 + 8x^2y - 24xy^2 = 4x(x^2 + 2xy - 6y^2)$$ ☐

As we have seen, we generally determine the common monomial factor of a polynomial by inspection. That is, we survey each term and identify any common numerical and literal factor. Example G illustrates a method for finding the *greatest* common factor, but we don't normally write out the factors of the individual terms. The following example gives several additional illustrations of factoring by inspection.

EXAMPLE H

$$2x - 2y = 2(x - y)$$
$$2x^2 + 4y = 2(x^2 + 2y)$$
$$4A^2B + A^2 = A^2(4B + 1)$$
$$3ax + 6a^2 = 3a(x + 2a)$$
$$6p^2q - 18pq^2 = 6pq(p - 3q)$$
$$4x^2 - 8xy - 20xy^2 = 4x(x - 2y - 5y^2)$$
$$6a^2b^3c - 9a^3bc^2 + 3a^2bc = 3a^2bc(2b^2 - 3ac + 1)$$

▶ Note the presence of the 1 in the last illustration. Again, since each factor of the third term of the original expression is common to all other terms, *we must include the factor of* 1. See Example F. □

EXAMPLE I When two resistors R_1 and R_2 are connected in parallel, they produce a total resistance of R_t. An equation describing the relationship among those resistances is

$$R_2 = \frac{R_1R_2}{R_t} - R_1$$

Factor the right side of this equation.

Examination of the right side of the equation shows that R_1 is the only common factor. We get

$$R_2 = R_1\left(\frac{R_2}{R_t} - 1\right)$$

As in Example H, we must again include the factor of 1. □

Exercises 8-1

In Exercises 1 through 36, factor the given expressions by determining any common monomial factors that may exist.

1. $5x + 5y$

2. $3x^2 - 3y$

3. $7a^2 - 14bc$

4. $3s - 12t$

5. $a^2 + 2a$

6. $3p + pq$

7. $2x^2 - 4x$

8. $5h + 10h^2$

9. $3ab - 3c$

10. $4x^2 - 4x$

11. $4p - 6pq$

12. $6s^3 - 15st$

13. $3y^2 - 9y^2z$

14. $12x - 48xy$

15. $abx - abx^2y$

16. $2xy - 8x^2y^2$

17. $6x - 18xy$

18. $12xy - 8axy$

19. $3a^2b + 9ab$

20. $6xy^3 - 9xy^2$

21. $a^2bc^2f - 4acf$

22. $2rs^2t - 8r^2st^2$

23. $ax^3y^2 + ax^2y^3$

24. $2a^2x^3y - 6a^2x^3$

25. $2x + 2y - 2z$

26. $3r - 3s + 3t$

27. $5x^2 + 15xy - 20y^3$

28. $4rs - 14s^2 - 16r^2$

29. $6x^2 + 4xy - 8x$

30. $5s^3 + 10s^2 - 20s$

31. $12pq^2 - 8pq - 28pq^3$

32. $18x^2y^2 - 24x^2y^3 + 54x^3y$

33. $35a^3b^4c^2 + 14a^2b^5c^3 - 21a^3b^2$

34. $15x^2yz^3 - 45x^3y^2z^2 + 16x^2y^2z$

35. $6a^2b - 3a + 9ab^2 - 12a^2b^2$

36. $4r^2s - 8r^3s^2 + 16r^4s - 4r^2s^3$

In Exercises 37 through 48, solve the given problems.

37. For a change in temperature for a gas we get the expression

$$\frac{nRT_2}{T_1} - nR$$

Factor this expression.

38. A transmitter used in a security system is to be contained in a cannister with a surface area given by

$$A = 2\pi r^2 + 2\pi rh$$

Factor the right side of this expression.

39. The total voltage output from three circuits is

$$IR_1 + IR_2 + IR_3$$

Factor this expression.

40. The heat energy produced by one hydroelectric generator is $I_1^2 R_1 t$ while the heat energy produced by a second generator is $I_2^2 R_2 t$. After determining the expression representing the total heat energy produced by both generators, factor that expression.

41. An electric car has four batteries whose total volume is expressed as

$$V = 6wh + 8wh + 4w^2h$$

Factor the right side of this formula.

42. A video game is programmed so that one path of a missile is described by the equation

$$y = 8x^4 - 6x^3 + 10x^2$$

Factor the right side of this equation.

43. The total surface area A of a microprocessor chip in the shape of a rectangular solid is given by the formula

$$A = 2lw + 2lh + 2hw$$

where l, w, and h are the length, width, and height of the chip. Factor the right side of this formula.

44. An air gun can shoot BBs upward so that the distance s (in feet) above the muzzle is given by the formula

$$s = 450t - 16t^2$$

where t is the time in seconds. Factor the right side of this formula.

45. A company found that its profit P for selling x items was given by

$$P = 2x^3 - 6x^2 + 10x$$

Factor the right side of this equation.

46. In discussing magnetic fields, we can sometimes use the equation

$$H = 2iMAC + 2iMBC - 10iMDC$$

Factor the right side of this equation.

47. The deflection of a certain beam of length L at a distance x from one end yields the expression

    $$wx^4 - 2wLx^3 + wL^3x$$

    Here w is the weight per unit length of the beam. Factor this expression.

48. In computing the value of a 6-month loan on which payments of R dollars per month are made, we use the equation

    $$P = Rv + Rv^2 + Rv^3 + Rv^4 + Rv^5 + Rv^6$$

    where v is a factor involving the interest which is paid. Factor the right side of this equation.

In Exercises 49 through 52, determine whether or not the expressions have been factored completely.

49. (a) $5x - 25x^2 = 5x(1 - 5x)$

    (b) $2x^2 - 8x = 2(x^2 - 4x)$

50. (a) $6ax^2 - 12a = 3a(2x^2 - 4)$

    (b) $4c^2y^2 - 2cy^2 = 2cy^2(2c - 1)$

51. (a) $3x^2y^3 + 6x^3y^2 = 3x^2y^2(y + 2x)$

    (b) $12axy^2 - 36ay = 12y(axy - 3a)$

52. (a) $6st - 24st^2 = 6st(1 - 4t)$

    (b) $16 + 36x^2 = 4(4 + 9x^2)$

In Exercises 53 through 56, determine whether or not the expressions are prime.

53. (a) $3x - 8$ (b) $4x - 8$ (c) $3x - 9$

54. (a) $x^2 + 2$ (b) $x^2 + 2x$ (c) $x^2 + x$

55. (a) $xy + y$ (b) $x^2y^2 + 9$ (c) $3xy + 9y$

56. (a) $3ax^2 + ay$ (b) $3ax^2 + y$ (c) $3ax^2y + y$

8-2 The Difference Between Two Squares

In Section 8-1 we discussed the factoring out of any common monomial factors. This should normally be the first step in any factoring process. In addition to factoring out common monomial factors, we must also do any factoring that involves more than one term. In this section we consider a very common and standard factoring procedure involving expressions that include a difference between two square quantities.

If we multiply $(x + y)$ by $(x - y)$, the product is $x^2 - y^2$. This product contains two perfect squares, x^2 and y^2. In general, *a **perfect square** is any quantity which is an exact square of a rational quantity.*

EXAMPLE A   The number 4 is a perfect square, since $4 = 2^2$. The quantity $9y^2$ is a perfect square, since $9y^2 = (3y)^2$. The quantity $25a^4$ is a perfect square, since $25a^4 = (5a^2)^2$. Other perfect squares are 16, $49x^2$, x^2y^2, b^4, p^2y^4, and $36x^8y^2$.

The number 7 is not a perfect square, since there is no rational number which, when squared, equals 7. The quantity x^3 is not a perfect square, since there is no integral power of x which makes the square of x^p equal to x^3. Also, $5x^2$ and $9x^3$ are not perfect squares. $\square$

Having noted that

$$(x + y)(x - y) = x^2 - y^2 \qquad\qquad\textbf{(8-2)}$$

we see that the multiplication results in the *difference* between the *squares* of the two terms which appear in the binomials. Therefore we can easily recognize the factors of the binomial $x^2 - y^2$, which are $(x + y)$ and $(x - y)$. It will be necessary to recognize the perfect squares represented by x^2 and y^2 when we factor expressions of this type. We can now apply the result of Eq. (8-2) to the problem of factoring the difference between any two squares.

EXAMPLE B

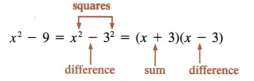

$$x^2 - 9 = x^2 - 3^2 = (x + 3)(x - 3)$$

Once we determine that an expression is the difference between two perfect squares, the factoring process becomes easy. First take the square root of each term:

$$\sqrt{x^2} = x \quad \text{and} \quad \sqrt{9} = 3$$

Now write the sum of these square roots $(x + 3)$ and their difference $(x - 3)$. We then indicate their product as $(x + 3)(x - 3)$. Additional illustrations of this procedure are given in Example C. □

EXAMPLE C

$$9 - 4y^2 = 3^2 - (2y)^2 = (3 + 2y)(3 - 2y)$$
$$36p^2 - 49q^2 = (6p)^2 - (7q)^2 = (6p + 7q)(6p - 7q)$$
$$4s^4 - 25t^4 = (2s^2)^2 - (5t^2)^2 = (2s^2 + 5t^2)(2s^2 - 5t^2)$$
$$4x^2 - 1 = (2x)^2 - 1^2 = (2x + 1)(2x - 1)$$
$$16y^4x^2 - p^6 = (4y^2x)^2 - (p^3)^2 = (4y^2x + p^3)(4y^2x - p^3)$$

In actually writing down the result of the factoring, we do not usually need to write the middle step shown in this example. The first illustration would be written simply as $x^2 - 9 = (x + 3)(x - 3)$. The middle steps are given here to indicate the perfect squares that are used. □

As indicated previously, if it is possible to factor out a common monomial factor, this should be done first. If the resulting factor is the difference between squares, the factoring is not complete until this difference is itself factored. This illustrates that **complete factoring** *often requires more than one step*. Also, when you are writing the result in complete factoring, be sure to include *all* factors.

The procedure applicable to the following examples can be summarized in these steps:

1. Factor out any common monomial factors.

2. To factor the difference between two squares, first find the square root of each of the two terms. Then one factor is the sum of the resulting square roots; the other factor is the difference of the resulting square roots.

3. Check the resulting factors to see if any additional factoring is possible.

EXAMPLE D

In factoring $3x^2 - 12$, we note that there is a common factor of 3 in each term. Therefore $3x^2 - 12 = 3(x^2 - 4)$. However, *the factoring is not yet complete* since $x^2 - 4$ is itself the difference between two perfect squares. Therefore $3x^2 - 12$ is completely factored as

common monomial factor   difference between squares

$$3x^2 - 12 = 3(x^2 - 4) = 3(x + 2)(x - 2)$$

EXAMPLE E

In factoring $x^4 - 16$, we note that we have the difference between two squares. Therefore $x^4 - 16 = (x^2 + 4)(x^2 - 4)$. However, the factor $x^2 - 4$ is again the difference between squares. Therefore

$$x^2 - 16 = (x^2 + 4)(x^2 - 4) = (x^2 + 4)(x - 2)(x + 2)$$

The factor $x^2 + 4$ is prime. It is not equal to $(x + 2)^2$ since

$$(x + 2)^2 = (x + 2)(x + 2) = x^2 + 2x + 2x + 4 = x^2 + 4x + 4$$

EXAMPLE F

In factoring $2a^4b - 2b$, we first factor out the common monomial of $2b$ to get $2b(a^4 - 1)$. However, the factor $a^4 - 1$ is the difference between two squares. Therefore

$$2a^4b - 2b = 2b(a^4 - 1) = 2b(a^2 + 1)(a^2 - 1)$$
$$= 2b(a^2 + 1)(a + 1)(a - 1)$$

Note that the factor $a^2 - 1$ is also the difference between two perfect squares. After factoring an expression, we should check the result to see if more factoring is necessary.

EXAMPLE G   An aircraft uses Venturi tubes for measuring airspeed, which is displayed on the airspeed indicator. Analysis of the Venturi tubes leads to the expression

$$\frac{1}{2}krv_1^2 - \frac{1}{2}krv_2^2$$

which is to be factored.

We first note that $\frac{1}{2}kr$ is a common factor of both terms. Factoring out $\frac{1}{2}kr$, we get

$$\frac{1}{2}kr(v_1^2 - v_2^2)$$

We now proceed to factor $v_1^2 - v_2^2$ as the difference between two squares. We get

$$\frac{1}{2}krv_1^2 - \frac{1}{2}krv_2^2 = \frac{1}{2}kr(v_1^2 - v_2^2) = \frac{1}{2}kr(v_1 + v_2)(v_1 - v_2)$$

The factoring is now complete.   □

EXAMPLE H   In the expression $(x + y)^2 - 25$, we have the difference between two squares. Treating $(x + y)$ as the first term, we get

$$(x + y)^2 - 25 = [(x + y) + 5][(x + y) - 5]$$
$$= (x + y + 5)(x + y - 5)$$

We expressed the final result by eliminating the unnecessary innermost sets of parentheses. Since only one type of grouping symbols was necessary, we use parentheses instead of brackets.   □

Equation (8-2) can be used to perform certain multiplications. The following example demonstrates how this is done.

EXAMPLE I   To multiply 49 by 51, we may express the product as

$$(49)(51) = (50 - 1)(50 + 1) = (50)^2 - 1^2 = 2500 - 1 = 2499$$

In Eq. (8-2) we are letting $x = 50$ and $y = 1$.

Other illustrations of this method of multiplication are:

$$(32)(28) = (30 + 2)(30 - 2) = (30)^2 - 2^2 = 900 - 4 = 896$$
$$(104)(96) = (100 + 4)(100 - 4) = (100)^2 - 4^2 = 10000 - 16 = 9984$$
$$(43)(57) = (50 - 7)(50 + 7) = (50)^2 - 7^2 = 2500 - 49 = 2451$$

The major value of performing multiplication in this way is to become better at working with the difference between squares. It is not often that we have the type of product which may be obtained in this manner, for it must be of a special form.   □

Factoring the difference between two squares, as discussed in this section, is a very common method of factoring and it is used frequently in algebra. It is therefore important to know well the product and its factors as given in Eq. (8-2).

Exercises 8-2

In Exercises 1 through 8, determine which of the given quantities are perfect squares. For those quantities which are perfect squares, determine the square root.

1. $9, 16, 3x^2$

2. $5, 4x, 4x^2$

3. $18, 121, 12x^2$

4. $x^4, 9x^4, 9ax^4$

5. $x^6, 16x^6, 16x^6y$

6. $x^8, 144x^8, 144a^2x^8$

7. $a^2b^4, x^2yz^4, 36ax^2$

8. $4r^6, s^2t^5x^2, 81rs^{12}$

In Exercises 9 through 16, determine the products by direct multiplication.

9. $(x + y)(x - y)$

10. $(a - 2)(a + 2)$

11. $(2a - b)(2a + b)$

12. $(3xy + 1)(3xy - 1)$

13. $(7ax^2 + p^3)(7ax^2 - p^3)$

14. $(5x^2 - 6y^5)(5x^2 + 6y^5)$

15. $(x + 2)(x + 2)$

16. $(a + b)(a + b)$

In Exercises 17 through 52, factor the given expression completely.

17. $a^2 - 1$

18. $x^2 - 4$

19. $t^2 - 9$

20. $E^2 - 16$

21. $16 - x^2$

22. $25 - y^2$

23. $4x^4 - y^2$

24. $25s^6 - 1$

25. $100 - a^4b^2$

26. $64 - x^2y^4$

27. $a^2b^2 - y^2$

28. $4q^2 - (rs)^2$

29. $81x^4 - 4y^6$

30. $49b^6 - 25c^4$

31. $5x^4 - 45$

32. $24x^2 - 54a^2$

33. $4x^2 - 100y^2$

34. $9x^2 - 81$

35. $x^4 - 1$

36. $a^4 - 81b^4$

37. $4x^2 + 36y^2$

38. $75a^2x^2 + 27b^4y^4$

39. $3s^2t^4 + 12s^2$

40. $9a^2b^6 + 81a^2b^2$

41. $16x^2y^2 + 24a^2b^2$

42. $12a^2x^3 - 4x^3$

43. $5ax^2 - 40a^2x$

44. $7x^2y^6 + 35y^6$

45. $(x + y)^2 - 1$

46. $(a - b)^2 - 9$

47. $25 - (x + y)^2$

48. $36 - (s + t)^2$

49. $(x + y)^2 - (x - y)^2$

50. $(x + 1)^2 - (x - 1)^2$

51. $a(x + y)^2 - a$

52. $ax^2(x + y)^2 - 9ax^2$

In Exercises 53 through 56, perform the indicated multiplications by means of Eq. (8-2). Check by direct multiplication. (See Example I.)

53. $(21)(19)$ **54.** $(78)(82)$ **55.** $(210)(190)$ **56.** $(144)(156)$

In Exercises 57 through 68, solve the given problems.

57. A surveyor uses a method of triangulation to find the distance between two points, and the square of the distance is expressed as

$$d^2 = c^2 - 64$$

Factor the right side of this equation.

58. The difference in volume between two types of brick is expressed as

$$a^2b - bc^2$$

Factor this expression.

59. A supersonic jet engine has two fuel intake pipes. The difference in the volume of fuel they supply is expressed as

$$c\pi r_1^2 - c\pi r_2^2$$

Factor this expression.

60. In the analysis of the deflection of a certain beam, the expression

$$L^2 - 4x^2$$

occurs. Factor this expression.

61. In economics, we sometimes use the following formula relating a quantity supplied, x, and the price p per unit quantity:

$$x = C^2 - k^2p^2$$

where C and k are constants depending on the economic conditions. Factor the right side of this formula.

62. In analyzing the energy of a moving object, we sometimes encounter the expression

$$mv_1^2 - mv_2^2$$

Factor this expression.

63. Under given conditions the force on a submerged plate is given by the equation

$$F = kwh_2^2 - kwh_1^2$$

Factor the right side of this equation.

64. The volume of steel used in a bushing sleeve is expressed as

$$\pi r_1^2 h - \pi r_2^2 h$$

Factor this expression.

65. The difference between the energy radiated by an electric light filament at temperature T_2 and that radiated by a filament at temperature T_1 is given by the formula

$$R = kT_2^4 - kT_1^4$$

Factor the right side of the formula.

66. Under certain conditions, the relativistic kinetic energy of an atom is described by the expression

$$T = a^2b^2c^2 - a^4b^4c^2$$

Factor the right side of this expression.

67. The difference between the intensities of two sounds is expressed as

$$2\pi dh_1^2 a_1^2 - 2\pi dh_2^2 a_2^2$$

Factor this expression.

68. One container is a cube with side x while a second container is a rectangular solid with sides x, y, and y. Find the expression representing the difference between the volumes of these two containers and then factor that expression.

8-3 Factoring Trinomials

In this section we discuss the factoring of trinomials of the form $ax^2 + bx + c$. As stated in Section 8-1, we shall consider only those cases in which a, b, and c are integers. Such a trinomial may or may not be factorable into two binomial factors with integer coefficients.

We first discuss the trinomial $x^2 + bx + c$; that is, the case in which $a = 1$. *The binomial factors of this trinomial (if they exist) will be of the form $(x + p)$ and $(x + q)$.* That is, we shall determine whether factors exist such that

$$x^2 + bx + c = (x + p)(x + q) \qquad \text{(8-3)}$$

Multiplying out the right-hand side of this equation, we obtain

$$(x + p)(x + q) = x^2 + px + qx + pq = x^2 + (p + q)x + pq$$

We now see that $b = p + q$ and $c = pq$. These relationships are shown below

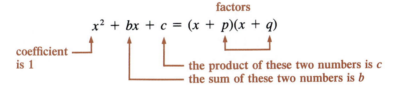

EXAMPLE A   In order to factor the trinomial $x^2 + 5x + 4$, we set up the factoring as

integers to be placed here

$$x^2 + 5x + 4 = (x \quad \blacksquare \quad)(x \quad \blacksquare \quad)$$

sum   product

and we are to find two integers such that their product is $+4$ and their sum is $+5$. The possible factors of $+4$ are

$$+2 \text{ and } +2, \quad -2 \text{ and } -2, \quad +4 \text{ and } +1, \quad -4 \text{ and } -1$$

Of these possibilities

▶   *only $+4$ and $+1$ have a sum of $+5$.*

Hence the required factors are $(x + 4)$ and $(x + 1)$. We conclude that

$$x^2 + 5x + 4 = (x + 4)(x + 1)$$

(Continued on next page)

When we factor $x^2 - 5x + 4$, an examination of the possible factors of $+4$ reveals that

▶ *the only factors whose sum is -5 are -4 and -1*

Therefore the factors are $(x - 4)$ and $(x - 1)$. Hence

$$x^2 - 5x + 4 = (x - 4)(x - 1)$$

When we try to factor $x^2 + 3x + 4$, an examination of the possible factors of $+4$, which are indicated in the first part of this example, reveals that none of the combinations has a sum of $+3$. Therefore

▶ **$x^2 + 3x + 4$ is prime and cannot be factored.**   □

EXAMPLE B In order to factor $x^2 + x - 6$ we must find two integers whose product is -6 and whose sum is $+1$. The possible factors of -6 are -1 and $+6$, $+1$ and -6, -2 and $+3$, and $+2$ and -3. Of these factors, only -2 and $+3$ have a sum of $+1$ so that

$$x^2 + x - 6 = (x - 2)(x + 3)$$

When we factor $x^2 - 5x - 6$, we see that the necessary factors of -6 are $+1$ and -6 since *their sum is -5*. Hence,

$$x^2 - 5x - 6 = (x + 1)(x - 6)$$

When we try to factor $x^2 + 4x - 6$, we see that this expression is prime, since *none of the pairs of factors of -6 has a sum of $+4$.*   □

We now consider factoring the trinomial $ax^2 + bx + c$, where a does not equal 1. The binomial factors of this trinomial will be of the form $(rx + p)$ and $(sx + q)$. That is, we shall determine whether factors exist such that

$$ax^2 + bx + c = (rx + p)(sx + q) \tag{8-4}$$

Multiplying out the right-hand side of this equation, we obtain

$$(rx + p)(sx + q) = rsx^2 + rqx + spx + pq$$
$$= rsx^2 + (rq + sp)x + pq$$

We now see that $a = rs$, $b = rq + sp$, and $c = pq$. It is therefore necessary to find four such integers, r, s, p, and q.

You probably realize that finding these integers is relatively easy so far as the ax^2 term and the c term are concerned. We simply need to find integers r and s whose product is a and integers p and q whose product is c. However, we must find the right integers so that the bx term is correct. This is usually the most troublesome part of factoring trinomials. We must remember that *the bx term will result from the sum of two terms* when the factors are multiplied together.

Summarizing this information on the factors of $ax^2 + bx + c$, where a does not equal 1, we have

$$a = rs \qquad c = pq$$
$$ax^2 + bx + c = (rx + p)(sx + q)$$
$$b = rq + ps$$

We see that *the "outer" product rqx and the "inner" product psx are the two terms which together form the bx term.* In finding the factors, we shall always take r and s to be positive if a is positive.

EXAMPLE C   When we factor $2x^2 - 11x + 5$, we take the factors of 2 to be $+2$ and $+1$ (we use only positive factors of a since it is $+2$). We now set up the factoring as

$$(2x \quad)(x \quad)$$

Since the product of the integers to be found is positive $(+5)$, only integers of the same sign need be considered. Also, since *the sum of the outer and inner products is negative (-11)*, the integers are negative. The factors of $+5$ are $+1$ and $+5$, and -1 and -5, which means that -1 and -5 is the only possible pair. Now trying the factors

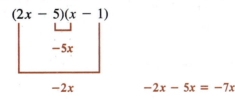

$$(2x - 5)(x - 1)$$
$$-5x$$
$$-2x \qquad\qquad -2x - 5x = -7x$$

we see that $-7x$ is not the correct sum for the middle term. Next trying

$$(2x - 1)(x - 5)$$
$$-x$$
$$-10x \qquad\qquad -10x - x = -11x$$

we have the correct sum of $-11x$. Therefore,

$$2x^2 - 11x + 5 = (2x - 1)(x - 5) \qquad\qquad \square$$

EXAMPLE D   Factor $4y^2 + 7y + 3$.

The positive factors of $+4$ are $+1$ and $+4$, and $+2$ and $+2$. Since $c = +3$, its factors must have the same sign, and since $b = +7$ *the sign is positive*. This means we need consider only positive factors of $+3$, and these factors are $+1$ and $+3$. Trying the factors

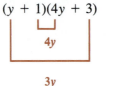

$$(y + 1)(4y + 3)$$

4y

3y   3y + 4y = 7y

we see that *we have the correct sum of $+7y$ for the outer and inner products*. This means

$$4y^2 + 7y + 3 = (y + 1)(4y + 3)$$

It is not necessary to try the other pair of factors $+2$ and $+2$.   □

EXAMPLE E   Factor $6x^2 + 5x - 4$.

First we find all factors of 6 (6 and 1, 2 and 3); then we find all factors of 4 (4 and 1, 2 and 2). We must choose a factor from each group so that the resulting pair can be used to form a 5. Since b is positive $(+5)$, we know that the outer or inner product of larger absolute value will be the product of positive numbers. This eliminates several possible combinations, such as

$$(6x + 1)(x - 4)$$

x
                                       $-24x + x = -23x$
$-24x$

We find that the factors $+6$ and $+1$ do not work with any of the factors of -4. Trying the factors $+3$ and $+2$, we find the combination

$$(3x + 4)(2x - 1)$$

8x
                                       $-3x + 8x = +5x$
$-3x$

has a sum of outer and inner products of $+5x$, the required value. Therefore

$$6x^2 + 5x - 4 = (3x + 4)(2x - 1)$$ □

▶   As mentioned in previous sections, *we must always* **be on the alert for a common monomial factor.** A check should be made for such a factor first. Consider the following example.

EXAMPLE F   The height (in feet) of a projectile is given by $16t^2 - 48t - 64$, where t is the time (in seconds) of travel. Factor this expression.

▶   At first it might appear that we should find factors of $+16$ and -64, but note that *there is a common factor of* **16** *in each term* so that we can write

$$16t^2 - 48t - 64 = 16(t^2 - 3t - 4)$$

It is now necessary to factor only $t^2 - 3t - 4$. The factors of this trinomial are $(t + 1)$ and $(t - 4)$ so that

$$16t^2 - 48t - 64 = 16(t + 1)(t - 4)$$ □

Many trinomials which are factorable by the methods of this section contain a literal factor in each term of the binomial factors. The following example illustrates this type of problem.

EXAMPLE G   Factor $4x^2 + 22xy - 12y^2$.

We first note that there is a common factor of 2 in each term. This means that $4x^2 + 22xy - 12y^2 = 2(2x^2 + 11xy - 6y^2)$. The problem now is to factor $2x^2 + 11xy - 6y^2$. Since the product of

$$(rx + py)(sx + qy) = rsx^2 + (rq + sp)xy + pqy^2$$

we see that this is essentially the same as the previous problems. The only difference is the presence of xy in the middle term and y^2 in the last term. However, the process of finding the values of r, s, p, and q is the same. We find that the arrangement

$$(2x - y)(x + 6y)$$

$$-xy$$

$$12xy \qquad 12xy - xy = 11xy$$

works for this combination. Therefore

$$4x^2 + 22xy - 12y^2 = 2(2x - y)(x + 6y)$$

In this example we began by factoring out the common monomial factor of 2. A common error is to forget such a factor when expressing the final result. Be sure to include any common monomial factors in the final result. Checking the result through multiplication will help eliminate many errors.   □

Much of the work done in factoring the trinomials in these examples can be done mentally. With experience, you will be able to write the factors of most such expressions by inspection.

Exercises 8-3

In Exercises 1 through 44, factor the given trinomials when possible.

1. $x^2 + 3x + 2$
2. $x^2 - x - 2$
3. $x^2 + x - 12$
4. $s^2 + 7s + 12$
5. $y^2 - 4y - 5$
6. $x^2 + 6x - 7$
7. $x^2 + 10x + 25$
8. $t^2 + 3t - 10$
9. $2q^2 + 11q + 5$
10. $2a^2 - a - 3$
11. $3x^2 - 8x - 3$
12. $2x^2 - 7x + 5$
13. $5c^2 + 34c - 7$
14. $3x^2 + 4x - 7$
15. $3x^2 - x - 5$
16. $7n^2 - 16n + 2$
17. $2s^2 - 13st + 15t^2$
18. $3x^2 - 14x + 8$
19. $5x^2 + 17x + 6$
20. $3x^2 - 17x - 6$
21. $4x^2 - 8x + 3$
22. $6x^2 + 19x - 7$
23. $12q^2 + 20q + 3$
24. $8y^2 + 5y - 3$
25. $6t^2 + 7tu - 10u^2$
26. $4x^2 + 33xy + 8y^2$
27. $8x^2 + 6x - 9$
28. $12x^2 - 7xy - 12y^2$
29. $4x^2 + 21x - 18$
30. $6s^2 - 7s - 20$
31. $8n^2 + 2n - 15$
32. $9y^2 + 28y - 32$
33. $2x^2 - 22x + 48$
34. $3x^2 - 12x - 15$
35. $4x^2 + 2xz - 12z^2$
36. $5x^2 + 15x + 25$
37. $2x^3 + 6x^2 + 4x$
38. $2x^4 + x^3 - 10x^2$
39. $10ax^2 + 23axy - 5ay^2$
40. $54a^2 - 45ab - 156b^2$
41. $3ax^2 + 6ax - 45a$
42. $2r^2x^2 - 7r^2x - 15r^2$
43. $14a^3x^2 - 7a^3x - 7a^3$
44. $5cx^2 + 5cx + 15c$

In Exercises 45 through 52, each of the given trinomials is a perfect square in the sense that they each have two identical factors. Factor each of these trinomials.

45. $x^2 + 2x + 1$
46. $x^2 - 6x + 9$
47. $x^2 - 8x + 16$
48. $x^2 + 12x + 36$
49. $4x^2 + 4x + 1$
50. $4x^2 + 12x + 9$
51. $9x^2 - 6x + 1$
52. $16x^2 - 24x + 9$

In Exercises 53 through 64, factor the indicated expression.

53. In analyzing projected population growth, a city planner uses the expression

$$N(r^2 + 2r + 1)$$

which is not completely factored. Factor this expression.

54. When determining the deflection of a certain beam of length L at a distance x from one end, we encounter the expression

$$x^2 - 3Lx + 2L^2$$

Factor this expression.

55. To find the total profit of an article selling for p dollars, it is sometimes necessary to factor the expression

$$2p^2 - 108p + 400$$

Factor this expression.

56. In calculating interest earned in an escrow account, the bank uses the expression

$$PR^2 + 2PR + P$$

Factor this expression.

57. In designing a parabolic mirror, an engineer uses the expression

$$y = 2x^2 - 24x + 64$$

Factor the right-hand side of this equation.

58. The resistance of a certain electric resistor varies with the temperature according to the equation

$$R = 10{,}000 + 600T + 5T^2$$

Factor the right-hand side of this equation.

59. An open box is made from a sheet of cardboard 12 in. square by cutting equal squares of side x from the corners and bending up the sides. Show that the volume of the box is given by the formula

$$V = 4x^3 - 48x^2 + 144x$$

Factor the right-hand side of this formula. See Figure 8-1.

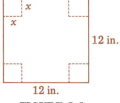

FIGURE 8-1

60. The total production cost for x core support plates used in nuclear reactors is expressed as

$$C = 6kx^2 + 11kx + 3k$$

Factor the right-hand side of this equation.

61. A manufacturer determines that the optimal size of a container yields the volume which is expressed as

$$V = 6x^3 - 15x^2 + 6x$$

Factor the right-hand side of this equation.

62. The change in surface area of a heated rectangular plate gives rise to the expression

$$Ak^2D^2 + 2AkD + A$$

Factor this expression.

63. In a study of pollution, an object is shot upward. The time this object is in the air can be found from the expression

$$16t^2 - 124t - 32$$

Factor this expression.

64. A rectangular panel is to be constructed so that wind pressures can be studied. The frame requires two wood sides of width x, but the bottom length will be along the ground so that no support is required there. With an enclosed area of 4000 ft² and wood sides totaling 180 ft, the dimensions can be found from the expression

$$2x^2 - 180x + 4000 = 0$$

Factor the left side of this equation. See Figure 8-2.

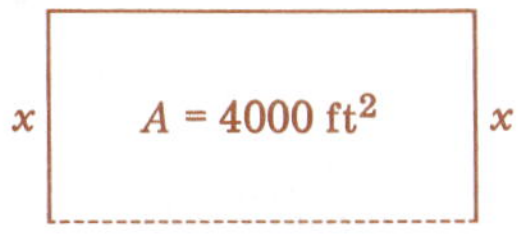

x | $A = 4000 \text{ ft}^2$ | x

FIGURE 8-2

8-4 The Sum and Difference of Cubes

In this chapter we have already discussed common monomial factors, the difference between two squares, and factoring trinomials. In this section we consider algebraic expressions that are either the sum of two perfect cubes or the difference between two perfect cubes. In general, a **perfect cube** is any quantity which is an exact cube of a rational quantity.

EXAMPLE A The number 8 is a perfect cube since $8 = 2^3$. The quantity $27x^3$ is a perfect cube since $27x^3 = (3x)^3$.

The number 4 is not a perfect cube since there is no rational number whose cube is 4. The quantity x^5 is not a perfect cube since there is no integral power p of x which makes the cube of x^p equal to x^5. □

We will base our factoring procedures on the equations

$$x^3 + y^3 = (x + y)(x^2 - xy + y^2) \tag{8-5}$$

$$x^3 - y^3 = (x - y)(x^2 + xy + y^2) \tag{8-6}$$

We can verify Eqs. (8-5) and (8-6) by multiplying out their right sides. After combining like terms and simplifying, both sides of each equation will agree. In Eqs. (8-5) and (8-6), neither $x^2 - xy + y^2$ nor $x^2 + xy + y^2$ is factorable; they are both prime.

EXAMPLE B

$$x^3 - 1 = x^3 - 1^3$$
$$= (x - 1)[(x)^2 + (1)(x) + (1)^2]$$
$$= (x - 1)(x^2 + x + 1)$$

$$27x^3 - 8 = (3x)^3 - 2^3$$
$$= (3x - 2)[(3x)^2 + (3x)(2) + (2)^2]$$
$$= (3x - 2)(9x^2 + 6x + 4)$$

$$x^3 + 8 = x^3 + 2^3$$
$$= (x + 2)[(x)^2 - 2x + (2)^2]$$
$$= (x + 2)(x^2 - 2x + 4)$$

$$8x^6 + 27 = (2x^2)^3 + 3^3$$
$$= (2x^2 + 3)[(2x^2)^2 - 6x^2 + (3)^2]$$
$$= (2x^2 + 3)(4x^4 - 6x^2 + 9)$$ □

As in the preceding sections of this chapter, we should again begin any factoring process by first checking for the presence of a monomial factor that is common to each term. If there is a common monomial factor, it should be factored out first. The procedure we are using can be summarized in these steps:

1. Factor out any common monomial factors.
2. To factor the sum or difference of two cubes, first find the cube root of each term. Refer to Eq. (8-5) or (8-6) to express the factored form in terms of those cube roots and their squares.
3. Check the resulting factors to see if any additional factoring is possible.

While none of the illustrations in Example B included a common monomial factor, the following examples do include such a factor.

EXAMPLE C In factoring $cx^3 + c$, we first note that c is a factor common to both terms and we factor it out to get $c(x^3 + 1)$. Now the factor $x^3 + 1$ is itself the sum of two perfect cubes and it is therefore factorable. We get

$$cx^3 + c = c(x^3 + 1) = c(x + 1)(x^2 - x + 1) \qquad \square$$

EXAMPLE D In factoring $ax^5 - ax^2$, we begin by factoring out the common monomial of ax^2. We get the factored form $ax^2(x^3 - 1)$. However, this expression has not been completely factored since $x^3 - 1$ is the difference between two cubes and is therefore factorable. Continuing, we refer to Eq. (8-6) and factor $x^3 - 1$ into $(x - 1)(x^2 + x + 1)$. We now know that

$$\begin{aligned} ax^5 - ax^2 &= ax^2(x^3 - 1) \\ &= ax^2(x - 1)(x^2 + x + 1) \qquad \square \end{aligned}$$

note

EXAMPLE E In factoring $a^3x^6 + a^3x^3y^3$, we first note that a^3x^3 is a monomial common to both terms. Factoring it out, we are left with $x^3 + y^3$, which can be factored according to Eq. (8-5). We get

$$\begin{aligned} a^3x^6 + a^3x^3y^3 &= a^3x^3(x^3 + y^3) \\ &= a^3x^3(x + y)(x^2 - xy + y^2) \qquad \square \end{aligned}$$

EXAMPLE F In factoring $x^6 - y^6$, we could actually consider this expression to be the difference between two squares or two cubes. While either approach will work, it is simpler to consider the given expression as the difference between two squares since that type of factoring is easier. We get

$$x^6 - y^6 = (x^3)^2 - (y^3)^2$$
$$= (x^3 + y^3)(x^3 - y^3)$$

Again, this expression has not been completely factored since $x^3 + y^3$ and $x^3 - y^3$ are both factorable. Applying Eqs. (8-5) and (8-6), we get

$$x^6 - y^6 = (x^3)^2 - (y^3)^2$$
$$= (x^3 + y^3)(x^3 - y^3)$$
$$= \overline{(x + y)(x^2 - xy + y^2)} \; \overline{(x - y)(x^2 + xy + y^2)}$$

EXAMPLE G The volume of material used to construct a steel bearing with a hollow core is given by the expression

$$\frac{4}{3}\pi x^3 - \frac{4}{3}\pi y^3$$

Factor this expression.

We should begin by factoring out the common monomial of $\frac{4}{3}\pi$. We get

$$\frac{4}{3}\pi x^3 - \frac{4}{3}\pi y^3 = \frac{4}{3}\pi(x^3 - y^3)$$

It is now necessary to factor only $x^3 - y^3$, which is the difference between two perfect cubes. Applying Eq. (8-6), we get

$$\frac{4}{3}\pi x^3 - \frac{4}{3}\pi y^3 = \frac{4}{3}\pi(x - y)(x^2 + xy + y^2)$$

In general, the procedures of this section closely parallel the procedures of Section 8-2, where we considered the difference between two squares. Whether we have the difference between two squares, the difference between two cubes, or the sum of two cubes, we must always begin by factoring out any common monomial factors. We must then recognize the remaining form as a special case that is factorable, and we must be able to properly execute the complete factoring.

Exercises 8-4

In Exercises 1 through 8, determine which of the given quantities are perfect cubes. For those quantities that are perfect cubes, determine the cube root.

1. $8; 25; 4x^3$

2. $27; 36; 8x^3$

3. $49; 64; 27x^3$

4. $100; 125; 64x^3$

5. $x^6; 8x^6; 8x^6y$

6. $x^8; 125x^9; 8a^3x^6$

7. $a^3x^9; x^3yz^6; 27ax^3$

8. $125x^8; 125x^6y^9; 27a^2x^3y^6$

In Exercises 9 through 16, multiply to find the indicated products.

9. $(x + y)(x^2 - xy + y^2)$

10. $(x - y)(x^2 + xy + y^2)$

11. $(x + 2)(x^2 - 2x + 4)$

12. $(x - 3)(x^2 + 3x + 9)$

13. $(2a - b)(4a^2 + 2ab + b^2)$

14. $(2a + 3b)(4a^2 - 6ab + 9b^2)$

15. $(x^2 - 2)(x^4 + 2x^2 + 4)$

16. $(x^2 + y^3)(x^4 - x^2y^3 + y^6)$

In Exercises 17 through 44, completely factor the given expressions.

17. $a^3 - 1$

18. $a^3 + 1$

19. $t^3 + 8$

20. $t^3 - 27$

21. $1 - x^3$

22. $8 + s^3$

23. $8x^3 + 27a^3$

24. $64s^3 - 27t^3$

25. $8x^6 - y^3$

26. $x^3 + 27y^9$

27. $a^3x^3 - y^6$

28. $a^3x^6 + b^6y^3$

29. $8x^4 + 8x$

30. $12xy - 12xy^4$

31. $ax^2 - ax^2y^3$

32. $4x^6 + 4$

33. $2kR_1^3 + 16kR_2^3$

34. $I_1^3R_1^3 + I_2^3R_2^3$

35. $54t_1^6 - 2t_2^3$

36. $3L_1^3 - 81L_2^{12}$

37. $125s^6 - 64t^9$

38. $250ax^4 - 128ax$

39. $a^3b^3 + c^{15}$

40. $a^9 + b^3c^{12}$

41. $1 - a^6x^6$

42. $x^{12} - y^6$

43. $1 - (x + y)^3$

44. $(x + y)^3 + 1$

In Exercises 45 through 56, solve the given problems.

45. In experimenting with different container sizes, a manufacturer reduces a cube from an edge of length x to an edge of length y. The total change in volume for N such containers is expressed as

$$D = Nx^3 - Ny^3$$

Factor the right-hand side of this equation.

46. At very low temperatures, the specific heat of solids is described by the Debye equation $c = kT^3$, where k is a constant for different materials and T is the Kelvin temperature. For a change in temperature, the corresponding change in the specific heat of a given solid is found from the expression

$$kT_1^3 - kT_2^3$$

Factor this expression.

47. Two different spheres are submerged in a fluid. The total volume displaced is expressed as

$$V = \frac{4}{3}\pi r_1^3 + \frac{4}{3}\pi r_2^3$$

Factor the right-hand side of this equation.

48. An alloy is made by combining two materials so that N cubes of edge s are mixed with N cubes of edge r. The total volume is

$$Ns^3 + Nr^3$$

Factor the right-hand side of this equation.

49. The volume of steel used in a number of hollow ball bearings is expressed as

$$V = kr_1^3 - kr_2^3$$

Factor the right-hand side of this equation.

50. In programming a video game, an image is made to follow a path described by the equation

$$y = 8x^3 + 27$$

Factor the right-hand side of this equation.

51. The cost of the materials used in rubber bushings is given by

$$C = k\pi r_1^3 - k\pi r_2^3$$

Factor the right-hand side of this equation.

52. The volume of gravel required at one site is equivalent to a cube of edge x while the volume required at a second site is equivalent to a cube of edge y. If the cost is C dollars per cubic yard, the total cost is given by

$$Cx^3 + Cy^3$$

Factor this expression.

53. A glass flask is filled with mercury and the temperature is then raised. The amount of mercury that will overflow is found from the expression

$$c_1^3 V(T_1 - T_2) - c_2^3 V(T_1 - T_2)$$

Factor this expression.

54. In a certain electric circuit, the current leaving a junction must be subtracted from the current entering that junction. Kirchhoff's rule states that the difference must be zero. If those currents change with time and the entering current (in amperes) is at^6 while the exiting current (in amperes) is at^3, factor the expression representing the difference.

55. A surveyor determines the difference between the corrections for sag in two different tapes by using the expression

$$\frac{nw^2 L_1^3}{24p^2} - \frac{nw^2 L_2^3}{24p^2}$$

Factor this expression.

56. A machine part requires that cubes of edge y are removed from each corner of a cube with edge x. Factor the expression that represents the remaining volume.

Chapter 8 Formulas

$$a(x + y) = ax + ay \tag{8-1}$$
$$(x + y)(x - y) = x^2 - y^2 \tag{8-2}$$
$$x^2 + bx + c = (x + p)(x + q) \tag{8-3}$$
$$ax^2 + bx + c = (rx + p)(sx + q) \tag{8-4}$$
$$x^3 + y^3 = (x + y)(x^2 - xy + y^2) \tag{8-5}$$
$$x^3 - y^3 = (x - y)(x^2 + xy + y^2) \tag{8-6}$$

8-5 Review Exercises for Chapter 8

In Exercises 1 through 64, factor the given expression if possible.

1. $5a - 5c$ **2.** $4r + 8s$ **3.** $3a^2 + 6a$ **4.** $6t^3 - 8t^2$

5. $12a^2b + 4ab$ **6.** $15t^3y - 10ty^2$ **7.** $8stu^2 - 24s^3tu^2$ **8.** $16x^2yz^4 - 4xyz^4$

9. $4x^2 - y^2$ **10.** $p^2 - 9u^2v^2$ **11.** $16y^4 - x^2$ **12.** $r^2s^2t^2 - 4x^2$

13. $x^2 + 2x + 1$ **14.** $x^2 - 2x - 3$ **15.** $x^2 - 7x + 6$ **16.** $x^2 + 2x - 63$

17. $ax^2 + 3a^2x - a^3$ **18.** $18r^2t - 9r^3t^2 - 6r^2t^3$

19. $2nm^3 - 4n^2m^2 + 6n^3m$ **20.** $8y^2 + 24y^3z - 32y^2z^4$

21. $4p^3t^2 - 12t^4 - 4t^2 + 4at^2$ **22.** $22r^2s^2t^2 - 121rst^2 - 22r^3st^2 + 33r^4s^2t^2$

23. $2xy^3 - 14x^2y^3 + 16xy^4 - 6x^3y^5$ **24.** $3st^2 - 6s^3t^3u - 12st^2u + 9st^3$

25. $(4rs)^2 - 9y^2$ **26.** $49r^4t^4 - y^6$ **27.** $36w^2x^2 + y^4$ **28.** $(a + b)^2 - c^2$

29. $2x^2 + 9x + 7$ **30.** $3y^2 + y - 10$ **31.** $5s^2 - 3s - 2$ **32.** $3a^2 + 20a - 7$

33. $14t^2 - 19t - 3$ **34.** $5x^2 + x - 4$ **35.** $9x^2 + 6x + 1$ **36.** $8r^2 + 2r - 15$

37. $x^2 + 3xy + 2y^2$ **38.** $6a^2 - 17ab + 12b^2$ **39.** $10c^2 + 23cd - 5d^2$ **40.** $4p^2 - 12pq + 9q^2$

41. $88x^2 - 19x - 84$ **42.** $16y^2 + 56y + 49$ **43.** $2x^2 - 18y^2$ **44.** $4r^2t^2 - 36p^2q^2$

45. $8y^4x^6 - 32y^2x^4$ **46.** $3m^5n - 27mn^3$ **47.** $3ax^2 + 3ax - 36a$ **48.** $36c^2x - 34cx - 30x$

49. $54r^3 - 45r^2s - 156rs^2$ **50.** $8c^2x^2 + 52c^2x + 72c^2$

51. $48y^3 - 64y^4 + 16y^5$ **52.** $18a^2u^2 + 23a^2u - 6a^2$

53. $5x^4 - 125$ **54.** $4a^8 - 64$ **55.** $16x^4 - 1$ **56.** $x^8 - 1$

57. $x^3 + 27$ **58.** $t^3 - 64$ **59.** $8x^3 + 1$ **60.** $8x^6 - 8y^3$

61. $ax^4y - axy^4$ **62.** $ab^4 + a^4b$ **63.** $125R_1^3 + 8a^3R_2^3$ **64.** $a^3x^{12} - a^6y^9$

In Exercises 65 through 80, factor the indicated expressions.

65. The total voltage from three circuits is given as

$$iR_1 + iR_2 + iR_3$$

Factor this expression.

66. An industrial safety specialist analyzes the upward motion of a projectile ejected from a machine. The distance s (in feet) above the machine and the time t (in seconds) are related by the formula

$$s = 64t - 16t^2$$

Factor the right-hand side of this equation.

67. In designing a gear for an industrial robot, an engineer determines the number of teeth by the expression

$$PN + 2P$$

Factor this expression.

68. In determining the flow velocity of a viscous fluid, we use the formula

$$v = k(R^2 - r^2)$$

Factor the right-hand side of this equation.

69. When determining the velocity of a fluid flowing through a pipe, we may encounter the expression

$$kD^2 - 4kr^2$$

Factor this expression.

70. When studying the pressure P and the volume V of a certain gas, we find the expression

$$CP(V_2 - V_1) + PR(V_2 - V_1)$$

where C and R are constants. Factor this expression.

71. When a jet of fluid strikes a surface, the work done by the jet is given by the expression

$$v_2^2 - 4v_1v_2 + 3v_1^2$$

Factor this expression.

72. In fluid dynamics, an analysis of the pressure difference between two tubes involves the expression

$$\frac{1}{2}rv_1^2 - \frac{1}{2}rv_2^2$$

Factor this expression.

73. In dealing with the theory of light reflection, we use the expression

$$\frac{u^2 - 2u + 1}{u^2 + 2u + 1}$$

Factor both the numerator and the denominator of this expression.

74. In the theory of magnetism, the expression

$$b(x^2 + y^2) - 2by^2$$

may be used. Factor this expression.

75. At an altitude of h ft above sea level, the boiling point of water is lower by a certain number of degrees than the boiling point at sea level, which is 212°F. The difference is given by the approximate equation

$$T^2 + 520T - h = 0$$

Assuming that $h = 5300$ ft, factor the left-hand side of this equation.

76. A newspaper advertisement is to be rectangular with the length equal to twice the width. The area A of the ad can be found from the equation $2x^2 - A = 0$. Assuming that $A = 50$ in.^{2}, factor the left side of this equation.

77. A glass container is filled with ethyl alcohol, which is then heated. The alcohol expands and spills out, with the lost volume given by

$$k_1^3 V (T_1 - T_2) - k_2^3 V (T_1 - T_2)$$

Factor this expression.

78. An electric circuit has an input and an output current that change with time. If their values (in amperes) are given by $c(t + 1)^3$ and ct^9, respectively, factor the expression that represents their difference. (Subtract the output current from the input current.)

79. One bearing requires 8 steel spheres of radius r_1 while a second bearing requires 27 steel spheres of radius r_2. The total volume of these spheres is described by the expression

$$8\left(\frac{4}{3}\right)\pi r_1^3 + 27\left(\frac{4}{3}\right)\pi r_2^3$$

Factor this expression.

80. A metal stock is in the shape of a solid cube of edge x. Three adjacent edges are ground down so that the stock is reduced to a cube with edge y. Find the expression for the amount by which the volume changes and then factor that expression.

CHAPTER NINE

Algebraic Fractions

We noted in Chapter 1 that a fraction indicates the division of one number by another. In Chapter 1 we developed the basic operations (addition, subtraction, multiplication, division) with fractions in which the numerator and denominator are whole numbers. In this chapter we extend those operations to fractions in which the numerator and denominator are algebraic expressions.

Algebra forms a basis which enables us to solve many applied problems. As an example, assume that a pilot knows that his plane will fly at 160 mi/h if there is no wind. He travels 350 mi with a headwind in the same time it takes to go 450 mi with a tailwind. We will show how to use the methods of this chapter for determining the speed of the wind.

9-1 Equivalent Fractions

Since algebraic expressions are representations of numbers, the basic operations on fractions from arithmetic form the basis of the algebraic operations. This means that the same basic rules of Chapter 1 will also apply to algebraic fractions.

There is one very important principle underlying the operations with fractions. It is the **fundamental principle of fractions:**

If the numerator and denominator of any fraction are multiplied or divided by the same quantity, not zero, the value of the fraction will remain unchanged.

This principle can be expressed in symbols as

$$\frac{a}{b} = \frac{ac}{bc}$$

where a, b, c can be numbers or algebraic expressions as long as c is not zero and b is not zero.

The use of the fundamental principle of fractions on any given fraction will yield an **equivalent fraction.** It is the same principle that we stated when we were working with equivalent fractions in arithmetic in Chapter 1. Note that this fundamental principle of fractions refers only to *multiplication* or *division* of *both* numerator and denominator.

EXAMPLE A From our work in Chapter 1, we know that if we multiply the numerator and denominator of the fraction $\frac{3}{4}$ by 2, we obtain the equivalent fraction $\frac{6}{8}$. If we divide the numerator and denominator of the fraction $\frac{18}{24}$ by 6, we obtain the equivalent fraction $\frac{3}{4}$. Therefore, $\frac{3}{4}$, $\frac{6}{8}$, and $\frac{18}{24}$ are equivalent. We may express this equivalence as follows:

$$\frac{3}{4} = \frac{6}{8} = \frac{18}{24}$$

$\square$

EXAMPLE B While we can multiply or divide both numerator and denominator by a quantity, *we cannot add or subtract a quantity*. If we subtract 1 from both numerator and denominator of the fraction $\frac{2}{3}$, we get $\frac{1}{2}$, but $\frac{2}{3}$ and $\frac{1}{2}$ are *not* equivalent fractions.

$\square$

Example A illustrates the use of the basic arithmetic operations on a fraction with whole numbers, as introduced in Chapter 1. Example C shows how those basic operations can be extended to algebraic fractions.

EXAMPLE C   If we multiply the numerator and denominator of the algebraic fraction

$$\frac{3x^2}{5x^3}$$

by x^2, we obtain the equivalent fraction

$$\frac{3x^4}{5x^5}$$

This can be seen by

$$\frac{(3x^2)(x^2)}{(5x^3)(x^2)} = \frac{3x^4}{5x^5}$$

If we divide the numerator and denominator of the fraction

$$\frac{3x^2}{5x^3}$$

by x^2, we obtain the equivalent fraction

$$\frac{3}{5x}$$

This can be shown as follows.

$$\frac{\dfrac{3x^2}{x^2}}{\dfrac{5x^3}{x^2}} = \frac{3}{5x}$$

We can now conclude that the algebraic fractions

$$\frac{3x^2}{5x^3}, \quad \frac{3x^4}{5x^5}, \quad \text{and} \quad \frac{3}{5x}$$

are equivalent, and we may write

$$\frac{3x^2}{5x^3} = \frac{3x^4}{5x^5} = \frac{3}{5x} \qquad \qquad \square$$

EXAMPLE D   If we take the square root of both numerator and denominator of the fraction

$$\frac{4x^2}{9y^2}$$

we get $2x/3y$, but these two fractions are *not* equivalent. Only multiplication or division of both numerator and denominator by a nonzero quantity will produce an equivalent fraction. $\qquad \square$

With an algebraic fraction, as with a fraction in arithmetic, *one of the most important operations is* **reducing** *it to its* **lowest terms,** *or* **simplest form.** As in the case of an arithmetic fraction, *we remove the factors which are common to both numerator and denominator by dividing both the numerator and the denominator by the common factors.* By finding the greatest common factor of the numerator and denominator, and dividing each by this factor, we obtain the simplest form, and we say that the fraction has been reduced to lowest terms.

EXAMPLE E   In simplifying the fraction $\frac{21}{28}$, we note that both numerator and denominator have a factor of 7. Therefore, dividing each by 7, we obtain $\frac{3}{4}$, the simplest form of the fraction $\frac{21}{28}$.

In order to simplify the fraction

$$\frac{15x^2y^3z}{20xy^4z}$$

we note that the greatest common factor of numerator and denominator is $5xy^3z$. Dividing both numerator and denominator by this greatest common factor, we get

$$\frac{(15x^2y^3z) \div (5xy^3z)}{(20xy^4z) \div (5xy^3z)} = \frac{3x}{4y} \qquad \square$$

EXAMPLE F   When we are simplifying the fraction

$$\frac{x^2 - x - 2}{x^2 + 3x + 2}$$

 we must first factor both numerator and denominator in order to determine whether there is a common factor. We obtain

$$\frac{x^2 - x - 2}{x^2 + 3x + 2} = \frac{(x - 2)(x + 1)}{(x + 2)(x + 1)} = \frac{x - 2}{x + 2}$$

We can now see that the greatest common factor is $(x + 1)$. We divide both numerator and denominator by this factor so that the indicated result is obtained. $\qquad \square$

Special note: *Remember that in simplifying fractions we must* **divide** *both numerator and denominator by the common* **factor.** *As in arithmetic, this process is sometimes called* **cancellation.** Many students are tempted to remove **terms** which appear in both numerator and denominator. This is an **incorrect** application of cancellation. The following examples illustrate common errors in the simplification of fractions.

EXAMPLE G   In attempting to simplify the expression

$$\frac{x^2 + 4}{x^2}$$ — a term, but not a factor, of numerator

a common error is to "cancel" the x^2 in both numerator and denominator to get a result of 4, but the given fraction is not equivalent to 4. *The original fraction cannot be further simplified* since the numerator and denominator have no common *factor* which can be canceled. **The x^2 in the numerator is not a common factor of the entire numerator.** ☐

EXAMPLE H   When simplifying the expression

$$\frac{x^2 - x - 2}{x^2 + 3x + 2}$$ terms, not factors, of numerator

terms, not factors, of denominator

many students would "cancel" the terms x^2 and 2, since they appear in both numerator and denominator. **This is incorrect.** Removing them in this way is equivalent to *subtracting* $x^2 + 2$ from the numerator and from the denominator, which violates the fundamental principle of fractions. The proper reduction of this fraction is shown in Example F.

When simplifying the expression

$$\frac{2x + 3}{2(x - 1)(x + 3)}$$ factor of $2x$, not of numerator
← term of numerator
term of $x + 3$

many students would "cancel" the 3's and 2's. Again, this is *incorrect*. Actually, *this expression cannot be further simplified,* since there is no common *factor* in the numerator and the denominator. **The 2 in the numerator is not a factor of the entire numerator, and the 3 is not a factor of either.** ☐

Example I gives additional illustrations of proper cancellation in the reduction of fractions to lowest terms.

EXAMPLE I   When simplifying the fraction

$$\frac{5a^2b^2c^4}{30ab^6c}$$

we note that the factor $5ab^2c$ is common to both numerator and denominator. Dividing the numerator and denominator by this greatest common factor, we get

$$\frac{5a^2b^2c^4}{30ab^6c} = \frac{ac^3}{6b^4}$$

When simplifying the fraction

$$\frac{2x^2 + 7x - 4}{4x^2 - 1}$$

we *first factor the numerator and denominator* to obtain

$$\frac{2x^2 + 7x - 4}{4x^2 - 1} = \frac{(2x - 1)(x + 4)}{(2x - 1)(2x + 1)}$$

Now we note the common factor of $2x - 1$, which is to be divided out (canceled). Proceeding, we get

$$\frac{2x^2 + 7x - 4}{4x^2 - 1} = \frac{\overset{1}{\cancel{(2x - 1)}}(x + 4)}{\underset{1}{\cancel{(2x - 1)}}(2x + 1)} = \frac{x + 4}{2x + 1}$$ □

There are cases, especially in the addition and subtraction of fractions, when a fraction in its simplest form must be changed to an *equivalent*, not the *simplest*, form. This is accomplished by the *multiplication* of both numerator and denominator by the same quantity.

When simplifying fractions, we often find that the numerator and denominator have factors which differ only in *sign*. We can show that

$$(a - b) = -(b - a) \tag{9-1}$$

by noting that $-(b - a) = -b + a = a - b$. *This equation shows us that we may change the signs of all the terms of a factor so long as we also introduce a factor of* -1. The following example illustrates the simplification of fractions where a change of signs is necessary.

EXAMPLE J We may simplify the fraction

$$\frac{x - 3}{3 - x}$$

 by expressing the denominator as $-(x - 3)$. After rewriting the denominator, we divide out the common factor of $x - 3$. This leads to

$$\frac{x - 3}{3 - x} = \frac{x - 3}{-(x - 3)} = \frac{1}{-1} = -1$$ □

EXAMPLE K Simplify the following fraction, which represents a mathematical model of the price-to-earnings ratio for a certain stock issue.

$$\frac{4 - x^2}{x^2 - 6x + 8}$$

The fraction

$$\frac{4 - x^2}{x^2 - 6x + 8} = \frac{(2 - x)(2 + x)}{(x - 2)(x - 4)}$$

has factors in the numerator and denominator which differ only in sign. Changing $(2 - x)$ to $-(x - 2)$, we get

$$\frac{(2 - x)(2 + x)}{(x - 2)(x - 4)} = \frac{-(x - 2)(2 + x)}{(x - 2)(x - 4)} = \frac{-(2 + x)}{x - 4}$$

This result can also be expressed as

$$\frac{2 + x}{4 - x}$$

by replacing $x - 4$ by $-(4 - x)$.

Acceptable forms of the simplest result are

$$\frac{-(2 + x)}{x - 4} = -\frac{2 + x}{x - 4} = \frac{2 + x}{-(x - 4)} = \frac{2 + x}{4 - x}$$

Changing signs of factors of fractions often causes students difficulty. This need not be the case if Eq. (9-1) is understood and used correctly. $\square$

Exercises 9-1

In Exercises 1 through 12, multiply the numerator and denominator of each given fraction by each of the two given factors and obtain the two equivalent fractions.

1. $\dfrac{2}{3}$;   2, 5

2. $\dfrac{7}{5}$;   3, 6

3. $-\dfrac{3}{4}$;   5, −5

4. $-\dfrac{5}{7}$;   −2, 2

5. $\dfrac{3ax}{b}$; $2a^2$, ab

6. $\dfrac{5b}{3x}$; $2x$, $7b^2x$

7. $\dfrac{5ax^2}{b}$; $-2x$, $2ax$

8. $\dfrac{3a^2x^3}{b^2}$; ax, $-ax^2$

9. $\dfrac{3}{x - 2}$; x, $x + 2$

10. $\dfrac{7}{a + 3}$; $3a$, $a + 1$

11. $\dfrac{x - y}{x + y}$; $x - y$, $x + y$

12. $\dfrac{2(x - 2y)}{2x - y}$; $x + 2y$, $2x + y$

In Exercises 13 through 24, divide the numerator and denominator of each given fraction by the given factor and obtain an equivalent fraction.

13. $\dfrac{16}{28};\quad 4$

14. $\dfrac{27}{39};\quad 3$

15. $\dfrac{6a^2x}{9a^3x};\quad 3ax$

16. $\dfrac{14r^2st}{28rst^2};\quad 7rt$

17. $\dfrac{2(x-1)}{(x-1)(x+1)};\quad x-1$

18. $\dfrac{(x+5)(x-3)}{3(x+5)};\quad x+5$

19. $\dfrac{2x^2+5x-3}{x^2+4x+3};\quad x+3$

20. $\dfrac{6x^2-11x+3}{2x^2-3x};\quad 2x-3$

21. $\dfrac{3x^2-12}{x^2+4x+4};\quad x+2$

22. $\dfrac{x^2-6x+9}{2x^2-2x-12};\quad x-3$

23. $\dfrac{4-x^2}{x^2-5x+6};\quad x-2$

24. $\dfrac{2x^2-12x+18}{9-x^2};\quad x-3$

In Exercises 25 through 56, determine the simplest form of each given fraction.

25. $\dfrac{8}{36}$

26. $\dfrac{42}{63}$

27. $\dfrac{2x}{6x}$

28. $\dfrac{5a}{30a}$

29. $\dfrac{18ax}{6a}$

30. $\dfrac{18at}{12t}$

31. $\dfrac{5a^2b}{20a}$

32. $\dfrac{12xyz^2}{21xyz^3}$

33. $\dfrac{16a^3}{18a^3}$

34. $\dfrac{12x^6}{15x^2}$

35. $\dfrac{18r^3st^2}{63r^5s^2t}$

36. $\dfrac{91a^3b^2c}{26ab^2c^5}$

37. $\dfrac{(2x-1)(x+4)}{(x+4)(x-2)}$

38. $\dfrac{(5x-3)(2x-1)}{(2x+1)(5x-3)}$

39. $\dfrac{(x-1)(x+1)(x+2)}{2(x-1)(x+3)}$

40. $\dfrac{3(x+5)(2x-1)(3x-2)}{9(5x+1)(3x-2)(2x-1)}$

41. $\dfrac{x^2-1}{x^2-2x+1}$

42. $\dfrac{x^2+5x+4}{x^2+3x-4}$

43. $\dfrac{3x^2-x}{3x^2+5x-2}$

44. $\dfrac{4x^2+9x-9}{4x^2-8x}$

45. $\dfrac{6x^2-19x+10}{8x^2-14x-15}$

46. $\dfrac{4x^2+10x+6}{12x^2+8x-4}$

47. $\dfrac{x^2-9y^2}{3xy-9y^2}$

48. $\dfrac{5a^2+39ab-8b^2}{5a^2+4ab-b^2}$

49. $\dfrac{(3x-1)(5-x)}{(x-5)(3x+1)}$

50. $\dfrac{(x-2)(x-3)(4-x)}{(3-x)(x-4)(2x-1)}$

51. $\dfrac{20-9x+x^2}{8+2x-x^2}$

52. $\dfrac{4ab-b^2}{b^2-16a^2}$

53. $\dfrac{3x-9x^2}{3x-1}$

54. $\dfrac{5a^2-10a}{4-a^2}$

55. $\dfrac{x^2-5x+6}{9-x^2}$

56. $\dfrac{x^2-6x+5}{20-4x}$

In Exercises 57 through 60, solve the given problems.

57. Evaluate the fraction.

$$\frac{4a-8}{4(a+2)}$$

for the value $a = 5$ before and after reducing it to simplest form.

58. Evaluate the fraction

$$\frac{2x^2 + 5xy - 3y^2}{2x^2 + 3xy - 2y^2}$$

for the values $x = 2$ and $y = 3$ before and after reducing it to simplest form.

59. Which of the following fractions are in simplest form?

$$\frac{x - 2}{x + 2} \qquad \frac{2x}{2x + 3} \qquad \frac{3x + 1}{3x + 5} \qquad \frac{x^2 - x + 2}{x^2 + 2} \qquad \frac{x^2 + x + 4}{x^2(x + 4)}$$

60. Evaluate the fraction

$$\frac{x - 2}{2 - x}$$

for the value $x = 7$ and compare it with the value obtained when it is reduced to lowest terms.

9-2 Multiplication and Division of Fractions

Recall from Chapter 1 that *the product of two fractions is a fraction whose numerator is the product of the numerators and whose denominator is the product of the denominators of the given fractions.* Since algebra is a generalization of arithmetic, the same definition holds in algebra. Symbolically, we write this as

$$\frac{a}{b} \cdot \frac{c}{d} = \frac{ac}{bd} \qquad (b \text{ and } d \text{ not zero}) \tag{9-2}$$

EXAMPLE A

$$\frac{5}{6} \cdot \frac{7a^2}{9x} = \frac{(5)(7a^2)}{(6)(9x)} = \frac{35a^2}{54x}$$

$$\frac{3a}{b} \cdot \frac{x^2}{5b} = \frac{(3a)(x^2)}{(b)(5b)} = \frac{3ax^2}{5b^2}$$

$$\frac{2x}{y} \cdot \frac{x^2}{y^3} = \frac{2x^3}{y^4} \qquad \square$$

Closely associated with the multiplication of fractions is the **power** of a fraction. To find $(a/b)^n$ we would have n factors of (a/b), which would result in a numerator of a^n and a denominator of b^n. Therefore

$$\left(\frac{a}{b}\right)^n = \frac{a^n}{b^n} \qquad (b \text{ not zero}) \tag{9-3}$$

Equation (9-3) does *not* state that a fraction is equivalent to another fraction obtained by raising both numerator and denominator to some power.

EXAMPLE B The fractions

$$\frac{3x}{4y} \quad \text{and} \quad \frac{9x^2}{16y^2}$$

are *not* equivalent, but the fractions

$$\left(\frac{3x}{4y}\right)^2 \quad \text{and} \quad \frac{9x^2}{16y^2}$$

are equivalent. □

EXAMPLE C
$$\left(\frac{2}{3}\right)^3 = \frac{2^3}{3^3} = \frac{8}{27}$$

$$\left(\frac{2a^2}{x}\right)^4 = \frac{(2a^2)^4}{x^4} = \frac{(2^4)(a^2)^4}{x^4} = \frac{16a^8}{x^4}$$ □

To divide one fraction by another, we recall the procedure developed for arithmetic fractions. *We invert the divisor (the fraction in the denominator) and multiply the dividend (the fraction in the numerator) by the inverted fraction.* We can explain this procedure on the basis of multiplying the numerator and denominator by the same quantity. If we want to divide a/b by c/d, we multiply the numerator and denominator by d/c. This may be written as

$$\frac{\dfrac{a}{b}}{\dfrac{c}{d}} = \frac{\dfrac{a}{b} \cdot \dfrac{d}{c}}{\dfrac{c}{d} \cdot \dfrac{d}{c}} = \frac{\dfrac{ad}{bc}}{\dfrac{cd}{dc}} = \frac{\dfrac{ad}{bc}}{1} = \frac{ad}{bc}$$

Showing this division of fractions symbolically, we get

$$\frac{\dfrac{a}{b}}{\dfrac{c}{d}} = \frac{ad}{bc} \quad \textbf{(b, c, and d not zero)} \qquad \textbf{(9-4)}$$

EXAMPLE D

$$\frac{2x}{5} \div \frac{3}{7c} = \frac{\dfrac{2x}{5}}{\dfrac{3}{7c}} = \frac{2x}{5} \cdot \frac{7c}{3} = \frac{(2x)(7c)}{(5)(3)} = \frac{14cx}{15}$$

multiply — invert divisor

$$\frac{\dfrac{4xy}{3b}}{\dfrac{7b}{xyz}} = \frac{4xy}{3b} \cdot \frac{xyz}{7b} = \frac{(4xy)(xyz)}{(3b)(7b)} = \frac{4x^2y^2z}{21b^2}$$ □

Another descriptive word which we met in connection with the division of fractions is **reciprocal.** Recalling that *the reciprocal of a number is 1 divided by that number*, we see that the reciprocal of a is $1/a$ (a not zero). Also, since

$$\frac{1}{\dfrac{a}{b}} = 1 \cdot \frac{b}{a} = \frac{b}{a}$$

we recall that the reciprocal of the fraction a/b is the inverted fraction b/a. This result is consistent with that found in Chapter 1.

EXAMPLE E

1. The reciprocal of $2a$ is $\dfrac{1}{2a}$.

2. The reciprocal of $\dfrac{1}{b}$ is $\dfrac{b}{1}$, or b.

3. The reciprocal of $-\dfrac{xy}{a}$ is $-\dfrac{a}{xy}$. When we are finding the reciprocal of a fraction, the sign of the fraction is unaltered.   □

One immediate use of the term "reciprocal" is in an alternative definition of the division of one fraction by another. We may state that *when we divide one fraction by another we multiply the dividend by the reciprocal of the divisor*.

As we stated in Section 9-1, results which are fractions are usually expressed in simplified form. Due to the process of multiplying fractions, it is usually easy to simplify the product or quotient of two fractions. Remember that in simplifying a fraction we must divide the numerator and denominator by factors which are common to both. Therefore, if *we factor the numerators and denominators of the fractions being multiplied or divided, and then* **indicate** *the appropriate product*, we can readily determine any common factors. If this is not done, it is very possible that the result obtained by multiplying out the fractions will be a fraction containing expressions which are extremely difficult to factor. This, in turn, will make the simplification difficult. Example F illustrates that the easier procedure is to *indicate* the product of factors; we should not multiply out the fractions and then attempt the reduction. In general, when multiplying algebraic fractions, we should:

1. *Indicate* the product of the numerators in factored form, and *indicate* the product of the denominators in factored form.
2. Divide out any common factors.
3. Leave the result in factored form.

The following examples illustrate the multiplication of algebraic fractions.

EXAMPLE F   Multiply $\frac{12}{32}$ by $\frac{16}{48}$.

If we *indicate* the multiplication as

$$\frac{12}{32} \cdot \frac{16}{48} = \frac{(12)(16)}{(32)(48)} \quad \longleftarrow \text{ indicated multiplication (not actually multiplied)}$$

we can easily see that both numerator and denominator are divisible by 12 and 16. Proceeding with reduction, we get

$$\frac{12}{32} \cdot \frac{16}{48} = \frac{\overset{1}{(\cancel{12})}\overset{1}{(\cancel{16})}}{\underset{2}{(\cancel{32})}\underset{4}{(\cancel{48})}} = \frac{(1)(1)}{(2)(4)} = \frac{1}{8}$$

If we perform the multiplication directly, by multiplying the numerator and denominator, we get

$$\frac{12}{32} \cdot \frac{16}{48} = \frac{192}{1536}$$

We can reduce $\frac{192}{1536}$ by determining that 192 and 1536 have 192 as the greatest common divisor. (Finding the greatest common divisor is not always easy.) Dividing the numerator and denominator of $\frac{192}{1536}$ by 192, we get $\frac{1}{8}$. This approach is much more difficult, and it shows that the better procedure is to first indicate the multiplication and then reduce. We don't want to multiply out to get cumbersome products which must be factored. Example G illustrates this same principle with an algebraic fraction. ☐

EXAMPLE G   The voltage (in volts) in an electric circuit is found by multiplying the current and the resistance. Under certain conditions, the voltage in a circuit is found by performing the multiplication shown below.

$$\frac{x^2 - y^2}{x + 2y} \cdot \frac{3x + 6y}{x - y}$$

If we perform the multiplication directly, by multiplying numerators together and denominators together, we obtain

$$\frac{3x^3 + 6x^2y - 3xy^2 - 6y^3}{x^2 + xy - 2y^2}$$

(Continued on next page)

Although the numerator is factorable, it is not of a simple form for factoring. The numerator is not factorable by methods we have developed. However, if we only *indicate the multiplication* (not actually perform it) as

$$\frac{x^2 - y^2}{x + 2y} \cdot \frac{3x + 6y}{x - y} = \frac{(x^2 - y^2)(3x + 6y)}{(x + 2y)(x - y)}$$

the numerator in this result is already partially factored and the denominator is completely factored. The additional work now required is considerably less than that required after direct multiplication. Completing the factoring of the numerator, we note that $x^2 - y^2 = (x + y)(x - y)$ and $3x + 6y = 3(x + 2y)$, and we get

$$\frac{x^2 - y^2}{x + 2y} \cdot \frac{3x + 6y}{x - y} = \frac{(x^2 - y^2)(3x + 6y)}{(x + 2y)(x - y)}$$

factors of $3x + 6y$

$$= \frac{3(x - y)(x + y)(x + 2y)}{(x + 2y)(x - y)}$$

Here we see that the greatest common factor of the numerator and denominator is $(x - y)(x + 2y)$. Dividing both numerator and denominator by this greatest common factor, we get the final result of

$$\frac{3(x + y)}{1} \quad \text{or} \quad 3(x + y) \qquad \Box$$

EXAMPLE H Multiply

$$\frac{8x^2y^3}{21abc} \quad \text{by} \quad \frac{15ax^2}{16b^2y^2}$$

$$\frac{8x^2y^3}{21abc} \cdot \frac{15ax^2}{16b^2y^2} = \frac{(8)(15)ax^4y^3}{(21)(16)ab^3cy^2}$$

Now we see that the greatest common factor of the numerator and denominator is $(8)(3)ay^2$. Dividing numerator and denominator by this factor, we have $5x^4y/14b^3c$. Therefore

$$\frac{8x^2y^3}{21abc} \cdot \frac{15ax^2}{16b^2y^2} = \frac{5x^4y}{14b^3c} \qquad \Box$$

EXAMPLE I Perform the division

$$\frac{b^2 - 2b + 1}{9b^2 - 1} \div \frac{5b - 5}{12b - 4}$$

We first indicate the product of the dividend and the reciprocal of the divisor. We then factor both the numerator and denominator. Next we divide out the greatest common factor. These steps lead to

$$\frac{b^2 - 2b + 1}{9b^2 - 1} \div \frac{5b - 5}{12b - 4} = \frac{(b^2 - 2b + 1)(12b - 4)}{(9b^2 - 1)(5b - 5)} \qquad \text{indicated multiplication}$$

$$= \frac{(b - 1)(b - 1)(4)(3b - 1)}{(3b + 1)(3b - 1)(5)(b - 1)} \qquad \begin{array}{l}\text{factor and show} \\ \text{cancellation}\end{array}$$

$$= \frac{4(b - 1)}{5(3b + 1)} \qquad \text{simplify}$$

The greatest common factor which was divided out was $(b - 1)(3b - 1)$. The result has been left in factored form. Although it would be correct to multiply out each of the numerators and denominators, the factored form is often more convenient and usually the preferred form of the result. ☐

Exercises 9-2

In Exercises 1 through 8, find the reciprocal of each expression.

1. $8n; \quad \dfrac{1}{13s}$

2. $a^2b; \quad \dfrac{a^2}{b}$

3. $\dfrac{a}{3b}; \quad \dfrac{3b}{a}$

4. $-\dfrac{2x^2}{y}; \quad -\dfrac{5cd}{3ax}$

5. $\dfrac{x + y}{x - y}; \quad \dfrac{x^2 + y^2}{x^2}$

6. $-\dfrac{s^2t^2}{s - t}; \quad \dfrac{x^2 - 9}{16 - y^2}$

7. $\dfrac{a}{a + b}; \quad \dfrac{-IR}{V}$

8. $\dfrac{4\pi r^2}{3}; \quad \dfrac{2L + 2H}{W}$

In Exercises 9 through 48, perform the indicated multiplications and divisions, expressing all answers in simplest form.

9. $\dfrac{12}{18} \cdot \dfrac{6}{9t}$

10. $\dfrac{15R}{20} \cdot \dfrac{40}{45R}$

11. $\dfrac{2}{9} \cdot \dfrac{3a}{5}$

12. $\dfrac{6x}{13} \cdot \dfrac{7}{a}$

13. $\dfrac{17rs}{12t} \cdot \dfrac{3t^2}{51s}$

14. $\dfrac{24xy^2}{5z} \cdot \dfrac{125z^2}{8y^3}$

15. $\left(\dfrac{3}{4}\right)^4$

16. $\left(\dfrac{7}{5}\right)^3$

17. $\left(\dfrac{a^2}{2x}\right)^5$

18. $\left(\dfrac{4xy^4}{8z^2}\right)^5$

19. $\left(\dfrac{ax^3}{b^2}\right)^3$

20. $\left(\dfrac{3x^2y^3}{6x^3}\right)^6$

21. $\dfrac{2}{5x} \div \dfrac{7c}{13}$

22. $\dfrac{6a}{17} \div \dfrac{5}{11c}$

23. $\dfrac{6x}{17} \div \dfrac{7}{68m}$

24. $\dfrac{15}{8z} \div \dfrac{25}{18y}$

25. $\dfrac{3x}{25y} \div \dfrac{27x^2}{5y^2}$

26. $\dfrac{3a^2x}{4ay^2} \div \dfrac{6ax}{5a^3x^2}$

27. $\dfrac{9a^2b^2}{10ab^3} \div \dfrac{72a^3b^4}{40}$

28. $\dfrac{4x}{3x^2y^2} \div \dfrac{20xy}{9y^3}$

29. $\dfrac{a - 5b}{a + b} \cdot \dfrac{a + 3b}{a - 5b}$

30. $\dfrac{5n + 10}{3n - 9} \cdot \dfrac{n - 3}{15}$

31. $\dfrac{x^2 - y^2}{14x} \cdot \dfrac{35x^2}{3x + 3y}$

32. $\dfrac{a + b}{4a} \cdot \dfrac{3a^2}{(a + b)^2}$

33. $\dfrac{x}{x - 1} \cdot \dfrac{x^2 - 1}{x^2 + 2x}$

34. $\dfrac{a + 2}{3} \cdot \dfrac{6a - 3}{a^2 + 2a}$

35. $\dfrac{x^2 + 2x - 3}{x^2 - 4} \cdot \dfrac{x^2 - x - 6}{x^2 - 5x + 4}$

36. $\dfrac{3x^2 + 10x - 8}{36x^2 - 16} \cdot \dfrac{9x^2 + 15x + 6}{x^2 + 3x - 4}$

37. $\dfrac{2b + 3}{5} \div \dfrac{4b + 6}{5b - 2}$

38. $\dfrac{6}{3x + 4} \div \dfrac{2x + 2}{9x^2 + 12x}$

39. $\dfrac{3a^2 - 3b^2}{a^2 - 4b^2} \div \dfrac{a^2 + 2ab + b^2}{a + 2b}$

40. $\dfrac{20x}{x^2 - 8x + 15} \div \dfrac{10x}{x^2 - 2x - 15}$

41. $\dfrac{s^2 - 5s - 14}{s^2 - 9s - 36} \div \dfrac{s^2 + 4s - 77}{s^2 + 10s + 21}$

42. $\dfrac{p^2 + 5pq + 6q^2}{2p^2 + pq - q^2} \div \dfrac{p^2 - 9q^2}{(p + q)^2}$

43. $\dfrac{4x^2 - 4}{3x^2 - 13x - 10} \div \dfrac{4x + 4}{x^2 - 6x + 5}$

44. $\dfrac{y^2 - y - 2}{2y^2 + y - 10} \div \dfrac{6y^2 - y - 7}{12y^2 + 16y - 35}$

45. $\dfrac{3a - 3b}{6a} \cdot \dfrac{(a + b)^2}{a^2 - b^2} \cdot \dfrac{4a^2}{2a^2 - ab - b^2}$

46. $\left(\dfrac{2x^2 + x - 15}{4x^2 - 5x - 21} \cdot \dfrac{x^2 - 6x + 9}{x^2 - 9}\right) \div \dfrac{4x^2 - 25}{4x + 7}$

47. $\left[\dfrac{1}{a^2(x - y)^3} \cdot a^3(3x - 3y)\right] \div \dfrac{1}{3^3(4x^2 - 9y^2)}$

48. $\left[(3x + 7)\left(\dfrac{1}{8x^2 - 10x - 25}\right)\right] \cdot [(14 - x - 3x^2) \div (x^2 - 4)]$

In Exercises 49 through 56, solve the given problems.

49. To find the focal length of a lens in terms of the object distance p and the image distance q, it is necessary to find the reciprocal of the expression

$$\dfrac{p + q}{pq}$$

Find this reciprocal.

50. In computing average energy per unit, the fraction

$$\dfrac{mv^2}{2}$$

must be divided by 3. Perform this operation.

51. The inductance of a resonant circuit is described by the fraction

$$\dfrac{1}{39.5 f^2 C}$$

If that inductance is halved, write an expression for the new inductance and express the result in simplest form.

52. In attempting to compute the minimum cost of a parts container, the product of the factors 3.2, x, and $x + 2$ must be divided by x^2. Perform the indicated operations and express the answer in simplest form.

53. Under certain conditions, the product of the price and the demand of a certain commodity is constant. If the price of each of x units is   and the demand for these is

$$\dfrac{2x + 6}{3} \qquad\qquad\qquad \dfrac{6000}{x^2 + 3x}$$

determine the expression for the product of the price and demand.

54. The optical reflection coefficient R in a semiconductor is given by

$$R = \left(\dfrac{1 - n}{1 + n}\right)^2$$

Perform the operation indicated on the right-hand side of this equation and express the result in simplest form.

55. In a certain electric circuit, the voltage and resistance vary with time according to the equations

$$V = \frac{t^2 - 1}{6t^2 + 3t}; \qquad R = \frac{t - 1}{2t^2}$$

The current is found by dividing the voltage by the resistance. Find a formula for the current I in terms of t.

56. In the development of the theory of relativity, we may find the expression

$$\left[\frac{\dfrac{c^2 - v^2}{c^2(c^2 - v^2)^2 - (c^2 - v^2)^2 v^2}}{(c^2 + v^2)^2} \right] \left(\frac{c^2 - v^2}{c^2 + v^2} \right)^2$$

Show that this expression equals 1.

9-3 The Lowest Common Denominator

Recall from Chapter 1 that *the sum of a set of fractions, all having the same denominator, is the sum of the numerators divided by the denominator.* We can express this in symbolic form as

$$\frac{a}{b} + \frac{c}{b} = \frac{a + c}{b} \qquad \text{where } b \neq 0$$

The reason for this is reviewed in the following example.

EXAMPLE A When we determine the sum $\frac{2}{7} + \frac{4}{7}$, we are finding the total number of one-sevenths of a unit. The first fraction represents two and the second fraction represents four of the one-sevenths of the unit. Thus the total is six one-sevenths. This we represent as

$$\frac{2}{7} + \frac{4}{7} = \frac{2 + 4}{7} = \frac{6}{7}$$

The same general method applies to algebraic fractions. □

EXAMPLE B $\dfrac{3a}{x^2} + \dfrac{4a}{x^2} = \dfrac{3a + 4a}{x^2} = \dfrac{7a}{x^2}$ □

We know that finding the sum of a set of fractions having the same denominator is a relatively easy task. However, if the fractions do not all have the same denominator, we also know that we must first change each to an equivalent fraction such that each resulting fraction has the same denominator. Since this step is of utmost importance in the addition and subtraction of fractions, and since it is also the step which causes the most difficulty with algebraic fractions, we shall devote this entire section to finding the common denominator of a given set of fractions. In Section 9-4, we discuss the complete method for the addition and subtraction of algebraic fractions.

From our previous work with fractions, we know that the most convenient and useful denominator for a set of fractions is the **lowest common denominator.** *We recall that this is the denominator that contains the smallest number of prime factors of the given denominators and is exactly divisible by each denominator. For algebraic fractions, it is the simplest expression into which all given denominators will divide evenly.*

EXAMPLE C   When we are finding the sum $\frac{3}{8} + \frac{5}{16}$, we know that the lowest common denominator of these fractions is 16. That is, 16 is the smallest number into which both 8 and 16 divide evenly.

When we are finding the sum

$$\frac{3}{8a} + \frac{5}{16a}$$

we note that both $8a$ and $16a$ will divide evenly into $16a$, which is therefore the lowest common denominator. If we use $16a^2$ as the common denominator, we do not have the simplest possible expression for the common denominator.   □

With algebraic fractions, just as with arithmetic fractions, to determine the lowest common denominator of a set of fractions *we first find all the prime factors of each denominator. We then form the product of these prime factors, giving each factor the largest exponent it has in any of the given denominators.* The process for finding the lowest common denominator can therefore be summarized by the steps:

1. For each denominator, find all of the prime factors.
2. Identify each of the different factors which appear in the denominators.
3. *Raise each factor to the* **largest** *exponent it has in any of the de-* **nominators.**
4. The lowest common denominator (l.c.d.) is the product of all of the results from step 3.

EXAMPLE D   1. When we are finding the lowest common denominator of $\frac{5}{24}$ and $\frac{7}{36}$, we first factor 24 and 36 into their prime factors. This gives

$$24 = 2 \cdot 2 \cdot 2 \cdot 3 = 2^3 \cdot 3 \quad \text{and} \quad 36 = 2 \cdot 2 \cdot 3 \cdot 3 = 2^2 3^2$$

The prime factors to be considered are 2 and 3. The largest exponent to which 2 appears is 3. The largest exponent to which 3 appears is 2. Therefore the lowest common denominator is $2^3 3^2 = 72$. No number smaller than 72 is divisible evenly by 24 and 36.

2. When we are finding the lowest common denominator of

$$\frac{1}{4x^2y} \quad \text{and} \quad \frac{7a}{6xy^4}$$

we factor the denominators into their prime factors. This gives

largest exponent
of 2   of *x*   of 3   of *y*

$$4x^2y = 2^2x^2y \quad \text{and} \quad 6xy^4 = 2\cdot3\cdot xy^4$$

The prime factors to be considered are 2, 3, x, and y. The largest exponent of 2 is 2; the largest exponent of 3 is 1; the largest exponent of x is 2; and the largest exponent of y is 4. Therefore the lowest common denominator is $2^2\cdot3\cdot x^2y^4 = 12x^2y^4$. This is the simplest expression into which $4x^2y$ and $6xy^4$ both divide evenly.   ☐

When we are finding the lowest common denominator of a set of algebraic fractions, and when some of the factors are binomials, *we* **indicate** *the multiplication of the necessary factors but do not multiply out the product*. This prevents us from losing the identity of the several factors and therefore makes the process of finding the appropriate equivalent fractions much easier. The reasoning for this will be shown in Section 9-4 when we actually add and subtract algebraic fractions. The following examples illustrate the procedure for finding the lowest common denominator for given sets of algebraic fractions.

EXAMPLE E   Find the lowest common denominator of the fractions

$$\frac{x - 9}{x^2 + x - 6} \quad \text{and} \quad \frac{x + 2}{x^2 - 5x + 6}$$

Factoring the denominators, we get

x − 2 appears in each, but is raised to first power only in each

$$x^2 + x - 6 = (x + 3)(x - 2) \qquad x^2 - 5x + 6 = (x - 2)(x - 3)$$

The necessary factors are $(x - 2)$, $(x - 3)$, and $(x + 3)$. Since the highest power to which each appears is 1, the lowest common denominator is

$$(x - 2)(x - 3)(x + 3)$$

 Here we note that $(x - 2)$ appears as a factor in both factorizations. Although it appears twice, its largest exponent is 1. Thus *we use only one factor of* $(x - 2)$ in the lowest common denominator.

(Continued on next page)

The product of these three factors is $x^3 - 2x^2 - 9x + 18$. It is obvious that we would lose the identity of the factors in the multiplication. In forming the equivalent fractions, we would have to divide this expression by each of the individual denominators. This would create much extra work, since in the factored form this division can be done by inspection. Therefore the factored form of $(x - 2)(x - 3)(x + 3)$ is preferable to the form $x^3 - 2x^2 - 9x + 18$. ☐

EXAMPLE F   Find the lowest common denominator of the fractions

$$\frac{2}{5x^2} \qquad \frac{3x}{10x^2 - 10} \qquad \frac{5}{x^2 - x}$$

Factoring the denominators, we get

$$5x^2 = 5x^2 \longleftarrow \text{largest exponent of } x$$
$$10x^2 - 10 = 10(x^2 - 1) = 2 \cdot 5(x - 1)(x + 1) \quad \boxed{\text{appears in each; largest exponent is 1}}$$
$$x^2 - x = x(x - 1) \qquad \bigsqcup 2^1 \qquad \bigsqcup (x + 1)^1$$

The prime factors of the lowest common denominator are 2, 5, x, $x - 1$, and $x + 1$. Each appears to the first power except x, which appears to the second power. *Although 5 and $(x - 1)$ appear in two factorizations, the largest exponent of each is 1.* Therefore the lowest common denominator is $2 \cdot 5x^2(x - 1)(x + 1) = 10x^2(x - 1)(x + 1)$. (Here we have multiplied the numerical factors in the coefficient. However, the others have been left in factored form.) ☐

EXAMPLE G   Three resistors are connected in series so that the total resistance can be found by adding the individual values. Under certain conditions, the resistances are described by the fractions given below. In order to add those fractions (as discussed in the following section), we must first find the lowest common denominator.

$$\frac{3x}{x^2 + 4x + 4} \qquad \frac{5}{x^2 - 4} \qquad \frac{x}{6x + 12}$$

Factoring the denominators, we get

$$x^2 + 4x + 4 = (x + 2)(x + 2) = (x + 2)^2$$
$$x^2 - 4 = (x - 2)(x + 2) \longleftarrow \quad \text{appears in each; largest exponent is 2}$$
$$6x + 12 = 6(x + 2) = 2 \cdot 3(x + 2)$$

The prime factors of the lowest common denominator are 2, 3, $x + 2$, and $x - 2$. All appear to the first power except $x + 2$, for which *the largest exponent is 2.* Therefore the lowest common denominator is

$$2 \cdot 3(x + 2)^2(x - 2) = 6(x + 2)^2(x - 2)$$

☐

The lowest common denominator of many simpler fractions, especially arithmetic fractions, can be determined by inspection. However, whenever the *lowest* common denominator is not obvious, the method outlined above should be followed.

Exercises 9-3

In the following exercises, find the lowest common denominator of each given set of fractions.

1. $\dfrac{a}{3}; \dfrac{a}{18}$

2. $\dfrac{x}{2}; \dfrac{3x}{7}$

3. $\dfrac{s}{12}; \dfrac{5s}{18}$

4. $\dfrac{r}{6}; \dfrac{7r}{40}$

5. $\dfrac{1}{4a}; \dfrac{1}{6a}$

6. $\dfrac{7}{15x}; \dfrac{8}{25x}$

7. $\dfrac{3a}{8t}; \dfrac{7b}{20t}$

8. $\dfrac{3V}{100R}; \dfrac{9V}{40R}$

9. $\dfrac{5}{18}; \dfrac{11}{45y}$

10. $\dfrac{7}{40}; \dfrac{7}{72n}$

11. $\dfrac{2}{3x}; \dfrac{5}{9x^2}$

12. $\dfrac{5}{56t^3}; \dfrac{17}{196t^2}$

13. $\dfrac{1}{4x^2}; \dfrac{3}{8x}$

14. $\dfrac{2}{15a^2}; \dfrac{3}{5a^3}$

15. $\dfrac{9}{60ax}; \dfrac{9}{28a}$

16. $\dfrac{3a}{8rst}; \dfrac{9b}{20st}$

17. $\dfrac{36}{125ax^2}; \dfrac{7}{15ax}$

18. $\dfrac{9}{16a^2b^2}; \dfrac{1}{12a^2b}$

19. $\dfrac{1}{25a^2}; \dfrac{7}{3a^3}$

20. $\dfrac{5}{4x^5}; \dfrac{9}{8x^2}$

21. $\dfrac{15}{32ab^3}; \dfrac{11}{12a^3b^2}$

22. $\dfrac{8}{27abc^5}; \dfrac{7x}{3a^2bc^4}$

23. $\dfrac{3}{5a^2}; \dfrac{2}{15a}; \dfrac{4}{3a^2}$

24. $\dfrac{5}{2x^3}; \dfrac{3}{8x}; \dfrac{1}{4x^2}$

25. $\dfrac{27}{4acx^3}; \dfrac{5}{12a^2cx}; \dfrac{13b}{20acx}$

26. $\dfrac{4}{25p^2q}; \dfrac{8}{15q^2r}; \dfrac{16}{27prs^2}$

27. $\dfrac{5}{4x-4}; \dfrac{3}{8x}$

28. $\dfrac{3}{2x+8}; \dfrac{5}{4x}$

29. $\dfrac{4}{3a+9}; \dfrac{5}{a^2+3a}$

30. $\dfrac{7}{6y^3-12y^2}; \dfrac{3y}{4y-8}$

31. $\dfrac{5}{3a^2x-9ax}; \dfrac{7x}{6a^2-18a}$

32. $\dfrac{9}{8a^2x^3+2a^3x^2}; \dfrac{26}{12x+3a}$

33. $\dfrac{3x}{2x-2y}; \dfrac{5}{x^2-xy}; \dfrac{7x}{6x^2-6y^2}$

34. $\dfrac{1}{x^4-x^3}; \dfrac{3}{4x^2-4}; \dfrac{a}{x^3+x^2}$

35. $\dfrac{a+3b}{a^2-ab-2b^2}; \dfrac{a+b}{a^2-4b^2}$

36. $\dfrac{x-5}{x^2-3x+2}; \dfrac{x}{2x^2-4x+2}$

37. $\dfrac{x-1}{4x^2-36}; \dfrac{x+7}{3x^2+18x+27}$

38. $\dfrac{7}{2t^2-5t-12}; \dfrac{5t}{2t^2+10t+6}$

39. $\dfrac{x-7y}{3x^2-7xy-6y^2}; \dfrac{x+y}{x^2-9y^2}; \dfrac{7}{2x-6y}$

40. $\dfrac{5}{6x^4-6y^4}; \dfrac{x+y}{4x^2+4y^2}; \dfrac{2x-5y}{3x^2-6xy+3y^2}$

9-4 Addition and Subtraction of Fractions

Having discussed the basic method of finding the lowest common denominator, we can perform the operations of addition and subtraction of algebraic fractions. As we have pointed out, if we want to add or subtract fractions whose denominators are the same, we place the sum of numerators over the denominator. *If the denominators differ, we must first change the fractions to equivalent ones with a common denominator equal to the lowest common denominator. We then combine the numerators of the equivalent fractions algebraically, placing this result over the common denominator.* The following examples illustrate the method.

EXAMPLE A   Combine:

$$\frac{7}{5ax} + \frac{3b}{5ax} - \frac{4b}{5ax}$$

Since the denominators are the same for each fraction, we need only to combine the numerators over the common denominator. We therefore have

$$\frac{7}{5ax} + \frac{3b}{5ax} - \frac{4b}{5ax} = \frac{7 + 3b - 4b}{5ax} = \frac{7 - b}{5ax}$$

Note that the signs of the terms of the numerators are the same as those of the fractions from which they were obtained.   □

EXAMPLE B   Combine:

$$\frac{3}{ax} + \frac{x}{2a^2} - \frac{7}{2x}$$

By examining the denominators, we see that the factors necessary in the lowest common denominator are 2, a, and x. Both 2 and x appear only to the first power and a appears squared. Thus the lowest common denominator is $2a^2x$. We now wish to write each fraction in an equivalent form with $2a^2x$ as a denominator. Since the denominator of the first fraction contains factors of a and x, it is necessary to introduce additional factors of 2 and a. In other words, we must multiply the numerator and denominator of the first fraction by $2a$. For similar reasons, we must multiply the numerator and denominator of the second fraction by x and the numerator and denominator of the third fraction by a^2. This leads to

$$\frac{3}{ax} + \frac{x}{2a^2} - \frac{7}{2x} = \frac{3(2a)}{(ax)(2a)} + \frac{x(x)}{(2a^2)(x)} - \frac{7(a^2)}{(2x)(a^2)}$$

change to equivalent
fractions each with l.c.d.

factors needed

$$= \frac{6a}{2a^2x} + \frac{x^2}{2a^2x} - \frac{7a^2}{2a^2x}$$

$$= \frac{6a + x^2 - 7a^2}{2a^2x} \qquad \text{combine numerators over l.c.d.}$$

□

EXAMPLE C   Two batteries are connected in series so that their combined voltage (in volts) can be found by adding the fractions given below. Perform the indicated addition to find an expression representing the total voltage.

$$\frac{4}{x + 3} + \frac{3}{x + 2}$$

Since there is only one factor in each denominator and they are different, the lowest common denominator of these fractions is the product $(x + 3)(x + 2)$. This in turn means that we must *multiply the numerator and denominator of the first fraction by $x + 2$ and those of the second fraction by $x + 3$* in order to make equivalent fractions with $(x + 3)(x + 2)$ as the denominator. Therefore, performing the addition, we get

$$\frac{4}{x + 3} + \frac{3}{x + 2} = \frac{4(x + 2)}{(x + 3)(x + 2)} + \frac{3(x + 3)}{(x + 2)(x + 3)}$$

change to equivalent fractions each with l.c.d.

factors needed

combine numerators over l.c.d.

$$= \frac{4(x + 2) + 3(x + 3)}{(x + 3)(x + 2)}$$

$$= \frac{4x + 8 + 3x + 9}{(x + 3)(x + 2)} \qquad \text{simplify}$$

$$= \frac{7x + 17}{(x + 3)(x + 2)}$$

□

EXAMPLE D   Combine:

$$\frac{x}{x + y} - \frac{2y^2}{x^2 - y^2}$$

Factoring the denominator of the second fraction, we have $x^2 - y^2 = (x + y)(x - y)$. Since the factor $x + y$ appears in the first fraction and there is no other (third) factor, the lowest common denominator is $(x + y)(x - y)$. Therefore we must *multiply the numerator and denominator of the first fraction by $x - y$, whereas the second fraction remains the same.* This leads to

(Continued on next page)

$$\frac{x}{x+y} - \frac{2y^2}{x^2-y^2} = \frac{x}{x+y} - \frac{2y^2}{(x+y)(x-y)}$$

change to equivalent
fractions each with l.c.d.

$$= \frac{x(x-y)}{(x+y)(x-y)} - \frac{2y^2}{(x+y)(x-y)}$$

factor needed

$$= \frac{x^2 - xy - 2y^2}{(x+y)(x-y)}$$

combine numerators
over l.c.d.

$$= \frac{(x+y)(x-2y)}{(x+y)(x-y)}$$

factor numerator

$$= \frac{x-2y}{x-y}$$

simplify

We note here that when the numerators are combined, the result is factorable. One of the factors also appears in the denominator, which means it is possible to reduce the resulting fraction. Remember: *We must express the result in simplest form.*   □

EXAMPLE E   Combine:

$$\frac{5}{a^2-a-6} + \frac{1}{a^2-5a+6} - \frac{2}{a^2-4a+4}$$

Factoring the denominators, we get

$$a^2 - a - 6 = (a-3)(a+2)$$ largest exponent of $a-3$ and $a+2$
$$a^2 - 5a + 6 = (a-3)(a-2)$$ is 1

$$a^2 - 4a + 4 = (a-2)(a-2) = (a-2)^2$$ largest exponent of $a-2$

The prime factors of the lowest common denominator are $a-3$, $a+2$, and $a-2$, where $a-2$ appears to the second power. The lowest common denominator is therefore $(a-3)(a+2)(a-2)^2$. To have equivalent fractions with the lowest common denominator, it is necessary to *multiply the numerator and the denominator of the first fraction by $(a-2)^2$, those of the second fraction by $(a-2)(a+2)$, and those of the third fraction by $(a+2)(a-3)$.* This leads to

$$\frac{5}{a^2-a-6} + \frac{1}{a^2-5a+6} - \frac{2}{a^2-4a+4}$$

$$= \frac{5}{(a-3)(a+2)} + \frac{1}{(a-3)(a-2)} - \frac{2}{(a-2)^2}$$ denominators factored

$$= \frac{5(a-2)^2}{(a-3)(a+2)(a-2)^2} + \frac{1(a-2)(a+2)}{(a-3)(a-2)(a-2)(a+2)} - \frac{2(a+2)(a-3)}{(a-2)^2(a+2)(a-3)}$$

change to equivalent
fractions with l.c.d.

$$= \frac{5(a-2)^2 + (a-2)(a+2) - 2(a+2)(a-3)}{(a-3)(a+2)(a-2)^2} \quad \text{combine numerators over l.c.d.}$$

$$= \frac{5(a^2 - 4a + 4) + a^2 - 4 - 2(a^2 - a - 6)}{(a-3)(a+2)(a-2)^2} \quad \text{simplify numerator}$$

$$= \frac{5a^2 - 20a + 20 + a^2 - 4 - 2a^2 + 2a + 12}{(a-3)(a+2)(a-2)^2}$$

$$= \frac{4a^2 - 18a + 28}{(a-3)(a+2)(a-2)^2}$$

$$= \frac{2(2a^2 - 9a + 14)}{(a-3)(a+2)(a-2)^2}$$

We should now check $2a^2 - 9a + 14$ to determine whether it is factorable. Since it is not factorable, we cannot further simplify the form of the result.

$\square$

Exercises 9-4

In Exercises 1 through 8, change the indicated sum of fractions to an indicated sum of equivalent fractions with the proper lowest common denominator. Do not combine.

1. $\dfrac{5}{9a} - \dfrac{7}{12a}$

2. $\dfrac{1}{40ax} + \dfrac{5}{84a}$

3. $\dfrac{5}{ax} + \dfrac{1}{bx} - \dfrac{4}{a}$

4. $\dfrac{6}{a^3b} - \dfrac{5}{4ab} - \dfrac{1}{6a^2}$

5. $\dfrac{4}{x^2 - x} - \dfrac{3}{2x^3 - 2x^2}$

6. $\dfrac{5b}{3ab - 6ac} - \dfrac{7c}{3b^2 - 12c^2}$

7. $\dfrac{x}{2x - 4} + \dfrac{5}{x^2 - 4} - \dfrac{3x}{x^2 + 4x + 4}$

8. $\dfrac{a-1}{3a-3} - \dfrac{8}{a^2 - 5a + 4} - \dfrac{1}{9}$

In Exercises 9 through 40, combine the given fractions, expressing all results in simplest form.

9. $\dfrac{2}{5x} + \dfrac{3}{2x}$

10. $\dfrac{3}{8s} - \dfrac{5}{12s}$

11. $\dfrac{2}{3a} - \dfrac{5}{3b}$

12. $\dfrac{7}{8s} + \dfrac{5}{20t}$

13. $\dfrac{6}{x} + \dfrac{3}{x^2}$

14. $\dfrac{4}{3a} - \dfrac{5}{2a^2}$

15. $\dfrac{1}{2} - \dfrac{5}{8b} + \dfrac{3}{20}$

16. $\dfrac{11}{12} + \dfrac{13}{40} - \dfrac{4}{15s}$

17. $\dfrac{8}{by} - \dfrac{1}{y^2}$

18. $\dfrac{1}{ax} + \dfrac{2}{a^2x}$

19. $\dfrac{3}{x^3y} + \dfrac{2}{3xy}$

20. $\dfrac{5}{p^2q} - \dfrac{1}{6pq}$

21. $\dfrac{2}{x} + \dfrac{5}{y} - \dfrac{3}{x^2y}$

22. $\dfrac{2}{a^2} - \dfrac{5}{6b} + \dfrac{3}{8ab}$

23. $\dfrac{7}{2x} - \dfrac{5}{4y} + \dfrac{1}{6z}$

24. $\dfrac{1}{6pq} + \dfrac{7}{3p} - \dfrac{9}{2p}$

25. $\dfrac{3}{a - 2} + \dfrac{5}{a + 2}$

26. $\dfrac{2}{2x - 1} + \dfrac{3}{2x + 3}$

27. $\dfrac{x}{2x - 6} - \dfrac{3}{4x + 12}$

28. $\dfrac{b}{a^2 - ab} + \dfrac{2}{a^3 + a^2b}$

29. $\dfrac{y - 1}{3y + 9} + \dfrac{2}{y - 3} - \dfrac{8}{3y - 9}$

30. $\dfrac{3}{4x - 16} - \dfrac{x}{4 - x} - \dfrac{1}{4x}$

31. $\dfrac{x - 1}{4x + 6} + \dfrac{2x}{6x + 9} - \dfrac{5}{6}$

32. $\dfrac{c + d}{5c - 2d} - \dfrac{c - d}{5c + 2d} - \dfrac{4c}{15c - 6d}$

33. $\dfrac{4}{4 - 9x^2} - \dfrac{x - 5}{2 + 3x}$

34. $\dfrac{q - 3}{q^2 - q - 12} + \dfrac{q + 1}{q^2 - 4q}$

35. $\dfrac{3x}{x^2 - 4} + \dfrac{2}{x - 2} - \dfrac{5x}{x + 2}$

36. $\dfrac{3}{2x - 2} + \dfrac{5}{x + 1} - \dfrac{x}{x^2 - 1}$

37. $\dfrac{2}{3} + \dfrac{3}{x + 5} - \dfrac{2}{x - 5}$

38. $\dfrac{4}{x} - \dfrac{1}{3x - 1} + \dfrac{x + 1}{3x + 1}$

39. $\dfrac{p + 2q}{2p^2 - 20pq + 50q^2} + \dfrac{7p}{4p - 20q} - \dfrac{7}{8}$

40. $\dfrac{2}{x^2 + 2x - 15} + \dfrac{1}{x^2 - 9} - \dfrac{7}{x^2 + 8x + 15}$

In Exercises 41 through 56, solve the given problems.

41. In a parallel circuit, the total resistance of two resistors is described by

$$\frac{1}{R} = \frac{1}{R_1} + \frac{1}{R_2}$$

Combine those two fractions on the right-hand side into one fraction.

42. One pump can empty a toxic waste holding tank in t_1 hours, a second pump can empty the same tank in t_2 hours, while a third tank requires t_3 hours. If all three pumps work together, their total pump rate is described by

$$\frac{1}{t_1} + \frac{1}{t_2} + \frac{1}{t_3}$$

of a tank per hour. Combine these three fractions.

43. A computer memory module is in the shape of a rectangular solid that is $\frac{3}{4}$ in. long, $\frac{5}{16}$ in. wide, and $\frac{7}{12}$ in. high. Find its surface area.

44. A survey determines that a parcel of land is in the shape of a trapezoid with a height of $\frac{3}{20}$ mi and bases of $\frac{5}{32}$ mi and $\frac{7}{48}$ mi. Find the area of this parcel.

45. Airplanes must be equipped with emergency locator transmitters (ELTs) which give a radio signal in the event of a crash. A value describing a characteristic of the transmitted signal is found from the expression

$$1 - \frac{x^2}{2} + \frac{x^4}{24} - \frac{x^6}{120}$$

Combine the fractions.

46. In attempting to determine the minimum cost of producing a component for a helium-neon laser, a scientist used the expression

$$26.500 + 5.080x + \frac{0.004}{x^2}$$

Combine and simplify.

47. In analyzing the heat transfer involved in a thermograph, we encounter the expression

$$\frac{R_2}{R_1} - 1$$

Combine that expression into a single fraction.

48. In the study of heat transfer, the thermal resistance of deposits is described by the formula

$$R_d = \frac{1}{U_d} - \frac{1}{U}$$

Combine the fractions on the right side of this equation.

49. In the study of absorption factors in radiation, the expression

$$I - \frac{I_r}{I_o} - \frac{I_t}{I_o}$$

is found. The variables I_o, I_r, I_t represent original intensity, reflected radiation, and transmitted radiation. Combine these three fractions into a single fraction.

50. In analyzing the behavior of two cars in a collision, investigators used the expression

$$r\left(\frac{m_1}{m_1 + m_2} - \frac{m_2}{m_1 + m_2}\right) - \frac{2rm_2}{m_1 + m_2}$$

Combine and simplify this expression.

51. When analyzing noise in an electric device, we use the expression

$$\frac{1}{8_m} + \frac{8}{8_m^2}$$

Combine and simplify.

52. Under certain conditions, the equation of the curve for the cable of a bridge is

$$y = \frac{wx^2}{2T_0} + \frac{kx^4}{12T_0}$$

Combine and simplify the terms on the right side of the equation.

53. In the theory dealing with the motion of the planets, we find the expression

$$\frac{p^2}{2mr^2} - \frac{gmM}{r}$$

Combine and simplify.

54. In the study of the stress and strain of deformable objects, the expression

$$a\left(l - \frac{l^2}{l + u} + 2u\right)$$

is found. First perform the multiplication and then combine and simplify.

55. In the theory dealing with a pendulum, we encounter the expression

$$\frac{P_1^2 + P_2^2}{2(h_1 + h_2)} + \frac{P_1^2 - P_2^2}{2(h_1 - h_2)}$$

Combine and simplify.

56. In determining the characteristics of a specific optical lens, we use the expression

$$\frac{2n^2 - n - 4}{2n^2 + 2n - 4} + \frac{1}{n - 1}$$

Combine and simplify.

9-5 Equations with Fractions

In science and technology, there are many equations that involve fractions. Solving these equations involves the use of the basic operations outlined in Chapter 5, but there is a basic procedure which can be used to eliminate the fractions and simplify the solution:

Multiply each term of the equation by the lowest common denominator.

This procedure effectively eliminates fractions so that they can be solved by methods previously discussed. This procedure is fundamentally different from the procedure used to combine fractions. When combining fractions we often multiply both numerator and denominator by some quantity, and the result is an equivalent fraction. With an equation involving fractions we "clear" the fractions by multiplying only the numerators of all terms. The following examples illustrate how to solve equations with fractions. We begin with an example which specifically illustrates the difference between combining fractions and solving an equation with fractions.

EXAMPLE A   Perform the indicated operation: $\dfrac{x}{2} + \dfrac{x}{3}$

In order to combine the given fractions, we must first convert them to equivalent fractions with a common denominator. We first note that 6 is the lowest common denominator. Now we change each fraction to an equivalent fraction with a denominator of 6.

$$\frac{x}{2} = \frac{(x)(3)}{(2)(3)} = \frac{3x}{6} \quad \text{and} \quad \frac{x}{3} = \frac{(x)(2)}{(3)(2)} = \frac{2x}{6}$$

so that

$$\frac{x}{2} + \frac{x}{3} = \frac{3x}{6} + \frac{2x}{6} = \frac{5x}{6}$$

However, in solving the equation

$$\frac{x}{2} + \frac{x}{3} = 10$$

we need not combine the fractions as shown above. Instead, we simply ▶ *multiply both sides of the equation by 6 to clear the fractions* and we get

$$\frac{x}{2}(6) + \frac{x}{3}(6) = 10(6)$$

$$3x + 2x = 60$$

$$5x = 60$$

$$x = 12$$

☐

EXAMPLE B   Solve for x:

$$\frac{x + 1}{6} + \frac{3}{2} = x$$

We first note that the lowest common denominator of the terms of the equation is 6. We *multiply each term by* 6 to get

$$\frac{6(x + 1)}{6} + \frac{6(3)}{2} = 6x$$

We now reduce each term to its lowest terms and solve the resulting equation:

$$(x + 1) + 9 = 6x \qquad \text{simplify}$$

$$x + 1 + 9 = 6x \qquad \text{remove parentheses}$$

$$10 = 5x \qquad \text{subtract } x \text{ from each side}$$

$$x = 2 \qquad \text{divide each side by 5 and reverse sides}$$

When we check this solution *in the original equation,* we obtain 2 on each side of the equation. Therefore the solution checks. ☐

It is very important to *check the solution* because even when correct procedures are followed, it sometimes happens that what appears to be a solution is actually discarded because it does not satisfy the original equation. (This will be illustrated in Example G.)

EXAMPLE C   Solve for x:

$$\frac{x}{a} - \frac{3}{4a} = \frac{1}{2}$$

We first determine that the lowest common denominator of the terms of the equation is $4a$. We then *multiply each term by* $4a$ and continue with the solution:

$$\frac{4a(x)}{a} - \frac{4a(3)}{4a} = \frac{4a(1)}{2} \qquad \text{multiply each term by } 4a$$

$$4x - 3 = 2a \qquad \text{simplify}$$

$$4x = 2a + 3 \qquad \text{add 3 to each side}$$

$$x = \frac{2a + 3}{4} \qquad \text{divide each side by 4}$$

Note that *we may not cancel a factor of* **2,** since 2 is not a factor of the numerator. Checking shows that upon substitution of $(2a + 3)/4$ for x, the left-hand side of the equation is $\frac{1}{2}$, which means that the solution checks. ☐

EXAMPLE D   Solve for x:

$$\frac{x}{a} - \frac{1}{2b} = \frac{3x}{ab}$$

The lowest common denominator of the terms of the equation is $2ab$. We therefore multiply each term of the equation by $2ab$ and proceed with the solution:

$$\frac{(2ab)x}{a} - \frac{2ab(1)}{2b} = \frac{2ab(3x)}{ab} \qquad \text{multiply each term by } 2ab$$

$$2bx - a = 6x \qquad \text{simplify}$$

$$2bx - 6x = a \qquad \text{add } a, \text{ subtract } 6x \text{ (each side)}$$

To isolate x, we must factor x from each term on the left. This leads to

$$x(2b - 6) = a \qquad \text{factor out } x \text{ on left}$$

$$x = \frac{a}{2b - 6} \qquad \text{divide each side by } 2b - 6$$

Checking will lead to the expression

$$\frac{3}{2b(b - 3)}$$

on each side of the original equation so that the solution is correct.   □

EXAMPLE E   Solve for t:

$$\frac{2}{t} = \frac{3}{t + 2}$$

After determining that the lowest common denominator of the terms of the equation is $t(t + 2)$ we *multiply each term by this expression* and proceed with the solution:

$$\frac{t(t + 2)(2)}{t} = \frac{t(t + 2)(3)}{t + 2} \qquad \text{multiply each term by } t(t + 2)$$

$$2(t + 2) = 3t \qquad \text{simplify}$$

$$2t + 4 = 3t \qquad \text{remove parentheses}$$

$$4 = t \quad \text{or} \quad t = 4 \qquad \text{subtract } 2t \text{ from each side}$$

Checking shows that with $t = 4$, each side of the original equation is $\frac{1}{2}$. □

EXAMPLE F For developing the equations which describe the motion of the planets, the equation

$$\frac{1}{2}v^2 - \frac{GM}{r} = -\frac{GM}{2a}$$

is used. Solve for M.

We first determine that the lowest common denominator of the terms of the equation is $2ar$. After *multiplying each of the three terms by 2ar* we get

$$arv^2 - 2aGM = -rGM$$

We can now proceed as usual. We first collect the terms containing M, and we then isolate M by factoring it out as follows.

$$rGM - 2aGM = -arv^2$$
$$M(rG - 2aG) = -arv^2 \qquad\qquad \text{factor out } M \text{ on left}$$
$$M = -\frac{arv^2}{rG - 2aG} \quad \text{or} \quad \frac{arv^2}{2aG - rG}$$

The second form of the result is obtained by using Eq. (9-1). Again note the use of factoring to arrive at the final result. ☐

EXAMPLE G Solve for x:

$$\frac{3}{x} + \frac{2}{x - 2} = \frac{4}{x^2 - 2x}$$

Factoring the denominator of the term on the right, $x^2 - 2x = x(x - 2)$, we find that the lowest common denominator of the terms of the equation is $x(x - 2)$. We get

$$\frac{3x(x - 2)}{x} + \frac{2x(x - 2)}{x - 2} = \frac{4x(x - 2)}{x(x - 2)} \qquad \text{multiply each term by } x(x - 2)$$
$$3(x - 2) + 2x = 4 \qquad\qquad \text{simplify}$$
$$3x - 6 + 2x = 4$$
$$5x = 10$$
$$x = 2$$

Substituting this value in the original equation, we obtain

$$\frac{3}{2} + \frac{2}{0} = \frac{4}{0}$$

Since division by zero is an undefined operation (see Section 3-4), the value $x = 2$ cannot be a solution. *We conclude that **there is no solution** to this equation.* ☐

Example G emphasizes the requirement that we must check solutions in the original equation. Another important conclusion from this example is that *whenever we multiply through by a lowest common denominator which contains the unknown, it is possible that a solution will be introduced into the resulting equation which is not a solution of the original equation. Such a solution is called an* **extraneous solution.** Only certain equations will lead to extraneous solutions, but we must be careful to identify them when they occur.

Many stated problems lead to equations that involve fractions. The following example illustrates such a problem.

EXAMPLE H   A plane can fly with a ground speed of 160 mi/h if there is no wind. It travels 350 mi with a headwind in the same time it takes to go 450 mi with a tailwind. Find the speed of the wind.

Let x = the speed of the wind. The ground speed of the plane as it flies into the headwind then becomes $160 - x$ mi/h. (The speed of 160 mi/h is reduced by the wind of x mi/h.) When flying with the tailwind, the speed becomes $160 + x$ mi/h.

This solution is based on the fundamental and important formula $d = rt$ (distance = rate × time). The key is to recognize that the upwind *time* is equal to the downwind *time*. With $d = rt$, we know that $t = d/r$, and we now express the equality of the upwind and downwind times as follows:

$$\text{upwind } \frac{d}{r} = \text{downwind } \frac{d}{r}$$

or

$$\frac{350}{160 - x} = \frac{450}{160 + x}$$

We now proceed to solve for x. We first clear the fractions by multiplying both sides of the equation by the lowest common denominator of $(160 - x)(160 + x)$ and we get this result.

$$350(160 + x) = 450(160 - x)$$

We can now eliminate the parentheses and proceed with the usual method of solution.

$$56{,}000 + 350x = 72{,}000 - 450x$$
$$800x = 16{,}000$$
$$x = 20 \text{ mi/h}$$

Checking the solution with the original statement, we see that the upwind and downwind speeds are 140 mi/h and 180 mi/h, respectively. Dividing 350 mi by 140 mi/h yields a time of 2.5 h. Dividing 450 mi by 180 mi/h also yields a time of 2.5 h so that the solution checks.   □

Example H involved distance, rate, and time so that it is necessary to know the formula $d = rt$. However, the solution also required the key observation that two values of time (or d/r) are equal.

The following example involves a situation that often occurs in real applications. The general key question is this: If one element can do a job in time t_1 while another element can do a job in time t_2, how long will the job take if both elements work together? This same general problem can be applied to two pumps emptying a tank, or two painters painting a house, and so on.

EXAMPLE I

One company makes a special boat which can remove the algae in a certain lake in 120 h. Another boat can remove the algae in 180 h. If both boats work together, how long would it take to remove the algae from this lake?

Relative to the lake under consideration, the first boat works at the *rate of $\frac{1}{120}$ lake per hour* while the rate of the second boat is $\frac{1}{180}$ *lake per hour*. When the boats work together, the algae is being removed at the combined rate of

$$\frac{1}{120} + \frac{1}{180}$$

lake per hour. The key to this problem is to identify the two *rates,* which are then combined through addition. Letting t represent the time required for both boats working together to clear this lake, we note that if we multiply t by the combined total *rate,* the result should be 1 (lake).

$$t\left(\frac{1}{120} + \frac{1}{180}\right) = 1 \qquad \text{hours}\left(\frac{\text{lake}}{\text{hour}}\right) = \text{lake}$$

We must now solve for t. We first multiply by t on the left side and then multiply both sides of the equation by the lowest common denominator of 360 to get

$$\frac{t}{120} + \frac{t}{180} = 1$$
$$3t + 2t = 360 \qquad \text{multiply each term by 360}$$
$$5t = 360 \qquad \text{simplify}$$
$$t = 72$$

Both boats working together can clear this lake in 72 h.

In general, if one element can do a job in time t_1 and a second element can do a job in time t_2, both elements working together can do the job in time t, where t is found by solving the equation

$$t\left(\frac{1}{t_1} + \frac{1}{t_2}\right) = 1 \quad \text{or} \quad t = \frac{1}{\dfrac{1}{t_1} + \dfrac{1}{t_2}}$$

This generalization can be easily extended to include cases involving more than two elements. ☐

Exercises 9-5

In Exercises 1 through 24, solve for s, t, v, x, or y.

1. $\dfrac{x}{8} = \dfrac{1}{2}$

2. $\dfrac{y}{6} = \dfrac{1}{3}$

3. $\dfrac{2s}{4} - 1 = s$

4. $\dfrac{x}{4} + 6 = x$

5. $\dfrac{2x}{5} + 3 = \dfrac{7x}{10}$

6. $\dfrac{3y}{8} - 2 = \dfrac{y}{2}$

7. $\dfrac{t+1}{2} + \dfrac{1}{4} = \dfrac{3}{8}$

8. $\dfrac{2x-1}{9} - \dfrac{1}{3} = 2x$

9. $\dfrac{x}{b} + 3 = \dfrac{1}{b}$

10. $\dfrac{y}{c} - 2 = \dfrac{3}{c}$

11. $\dfrac{y}{2a} - \dfrac{1}{a} = 2$

12. $\dfrac{x}{b} + \dfrac{1}{3b} = 1$

13. $\dfrac{1}{2a} - \dfrac{3s}{4} = \dfrac{1}{a^2}$

14. $\dfrac{2}{5c} + \dfrac{2t}{5} = \dfrac{1}{c^3}$

15. $\dfrac{2x}{b} + \dfrac{4}{b^2} = \dfrac{x}{4b}$

16. $\dfrac{3s}{4a} - \dfrac{s}{a^2} = \dfrac{1}{8}$

17. $\dfrac{1}{x+2} = \dfrac{1}{2x}$

18. $\dfrac{1}{2v-1} = \dfrac{1}{3}$

19. $\dfrac{2}{3s+1} = \dfrac{1}{4}$

20. $\dfrac{3}{3x+2} = \dfrac{2}{x}$

21. $\dfrac{5}{2x+4} + \dfrac{3}{x+2} = 2$

22. $\dfrac{2}{s-3} - \dfrac{3}{2s-6} = 4$

23. $\dfrac{1}{y^2-y} - \dfrac{1}{y} = \dfrac{1}{y-1}$

24. $\dfrac{2}{x^2-1} - \dfrac{2}{x+1} = \dfrac{1}{x-1}$

In Exercises 25 through 36, solve for the indicated letter.

25. $3 + \dfrac{1}{a} = \dfrac{1}{b}$, for a

26. $\dfrac{3}{a} - \dfrac{2}{c} = \dfrac{7}{ac}$, for c

27. $\dfrac{x}{n} + \dfrac{3}{4n} = \dfrac{x}{2}$, for x

28. $\dfrac{7}{p^2} = t - \dfrac{3}{2p} + \dfrac{t}{p}$, for t

29. An equation obtained in analyzing a certain electric circuit is

$$\dfrac{V-6}{5} + \dfrac{V-8}{15} + \dfrac{V}{10} = 0$$

Solve for V.

30. In biology, an equation used when analyzing muscle reactions is

$$P + \dfrac{a}{V^2} = \dfrac{k}{V}$$

Solve for a.

31. In optics, the thin lens equation is given as

$$\dfrac{1}{p} + \dfrac{1}{q} = \dfrac{1}{f}$$

where p is the distance of the object from the lens, q is the distance of the image from the lens, and f is the focal length of the lens. Solve for q.

32. In electricity, if resistors R_1 and R_2 are connected in parallel, their combined resistance R is given by the equation

$$\frac{1}{R} = \frac{1}{R_1} + \frac{1}{R_2}$$

Solve for R.

33. A formula relating the number of teeth N of a gear, the outside diameter D_0 of the gear, and the pitch diameter D_p is

$$D_p = \frac{D_0 N}{N + 2}$$

Solve for N.

34. An equation found in the thermodynamics of refrigeration is

$$\frac{W}{Q_1} = \frac{T_2}{T_1} - 1$$

where W is the work input, Q_1 is the heat absorbed, T_1 is the temperature within the refrigerator, and T_2 is the external temperature. Solve for T_2.

35. In hydrodynamics, the equation

$$\frac{p}{d} = \frac{P}{d} - \frac{m^2}{2\pi^2 r^2}$$

is used. Solve for d.

36. An equation which arises in the study of the motion of a projectile is

$$m^2 - \frac{2v^2 m}{gx} + \frac{2v^2 y}{gx^2} + 1 = 0$$

Solve for g.

In Exercises 37 through 44, solve the given problems.

37. An investor has $6000 in one account and $8000 in another account. Each account is increased by the same amount so that they have a ratio of $\frac{4}{5}$. Find that amount.

38. An advertiser wants a rectangular ad with the length and width in the ratio of $\frac{3}{2}$. If the ad has a border with a total length of 25 in., find the dimensions.

39. Water is pumped through a section of pipe so that its speed is 3.0 mi/h. If the speed is increased to 4.0 mi/h, it would take the same time for the water to travel an additional 0.60 mi. How long is the original section of pipe?

40. A jet travels 900 mi between two cities. It then proceeds to travel at the same speed for 1125 mi in a time that is $\frac{1}{2}$ h longer than the first trip. How much time did the first trip require?

41. A fire fighter has two pumps available for draining a flooded basement. One pump can do the job in 4.0 h, while the other pump would require only 3.0 h. If both pumps are used, how long does it take to drain the basement?

42. If a certain car's lights are left on, the battery will be dead in 4.0 h. If only the radio is left on, the battery will be dead in 18 h. If a driver leaves the lights and radio on, how long will the battery last?

43. One painting team can paint a bridge in 600 h while a different team can do the same job in 400 h. If both teams work together, how long will it take them to paint the bridge?

44. One particular employee can clean a suite of offices in 6.00 h while another employee can do the job in 8.00 h. If both employees work together, how long will it take them to clean these offices?

Chapter 9 Formulas

$$(a - b) = -(b - a) \tag{9-1}$$

$$\frac{a}{b} \cdot \frac{c}{d} = \frac{ac}{bd} \quad \text{(b and d not zero)} \tag{9-2}$$

$$\left(\frac{a}{b}\right)^n = \frac{a^n}{b^n} \quad \text{(b not zero)} \tag{9-3}$$

$$\frac{\dfrac{a}{b}}{\dfrac{c}{d}} = \frac{ad}{bc} \quad \text{(b, c, and d not zero)} \tag{9-4}$$

9-6 Review Exercises for Chapter 9

In Exercises 1 through 12, reduce the given fractions to simplest form.

1. $\dfrac{9rst^6}{3s^4t^2}$

2. $\dfrac{-14y^2z^3}{84y^5z^2}$

3. $\dfrac{2a^2bc}{6ab^2c^3}$

4. $\dfrac{76x^3yz^3}{19xz^5}$

5. $\dfrac{4x + 8y}{x^2 - 4y^2}$

6. $\dfrac{a^2x - a^2y}{ax^2 - ay^2}$

7. $\dfrac{p^2 + pq}{3p + 2p^3}$

8. $\dfrac{4a - 12ab}{5b - 15b^2}$

9. $\dfrac{2a^2 + 2ab - 2ac}{4ab + 4b^2 - 4bc}$

10. $\dfrac{p^2 - 3p - 4}{p^2 - p - 12}$

11. $\dfrac{6x^2 - 7xy - 3y^2}{4x^2 - 8xy + 3y^2}$

12. $\dfrac{2y^2 - 14y + 20}{7y - 2y^2 - 6}$

In Exercises 13 through 20, find the lowest common denominator of each of the given sets of fractions.

13. $\dfrac{x}{6}; \ \dfrac{x}{9}$

14. $\dfrac{a}{2x^2}; \ \dfrac{b}{5x^2}$

15. $\dfrac{4}{3t^2}; \ \dfrac{3r}{4t}$

16. $\dfrac{c}{bt^3}; \ \dfrac{c}{at^2}$

17. $\dfrac{7}{12bt}; \ \dfrac{9}{16b^2t}$

18. $\dfrac{cx}{ay^2z^3}; \ \dfrac{cy}{byz^4}$

19. $\dfrac{2}{4x^3 - 8x^2}; \ \dfrac{3x}{5x - 10}$

20. $\dfrac{x - 1}{x^2 + 2x - 15}; \ \dfrac{x + 4}{x^2 + 8x + 15}$

In Exercises 21 through 56, perform the indicated operations.

21. $\dfrac{2x}{3a} \cdot \dfrac{5a^2}{x^3}$

22. $\dfrac{5b^3}{6c} \cdot \dfrac{3c^5}{10b}$

23. $\dfrac{3x^2}{4y^3} \cdot \dfrac{5y^4}{x^3}$

24. $\dfrac{7p}{12q^3} \cdot \dfrac{20q^4}{35p^3}$

25. $\dfrac{3a}{4} \div \dfrac{a^2}{8}$

26. $\dfrac{ab^3}{bc} \div a^2c$

27. $\dfrac{au^2}{4bv^2} \div \dfrac{a^2u}{8b^2v}$

28. $\dfrac{20m^3n^2}{12mn^4} \div \dfrac{27mn}{8m^2n^3}$

29. $\dfrac{2}{a^2} - \dfrac{3}{5ab}$

30. $\dfrac{3x}{4y} - \dfrac{5y}{6x}$

31. $\dfrac{5}{c} - \dfrac{3}{c^2d} + \dfrac{1}{2cd}$

32. $\dfrac{3}{2x^2} + \dfrac{1}{4x} - \dfrac{a}{6x^3}$

33. $\left(\dfrac{2}{3}\right)^3$

34. $\left(\dfrac{I}{Rt}\right)^3$

35. $\left(\dfrac{ax}{3y^2}\right)^4$

36. $\left(\dfrac{2x^2y}{xyz^3}\right)^5$

37. $\dfrac{2x}{x^2-1} \cdot \dfrac{x-1}{x^2}$

38. $\dfrac{5a}{a^2-a-6} \cdot \dfrac{a+2}{10a^2}$

39. $\dfrac{x^2-2x-15}{x^2-9} \cdot \dfrac{x^2-6x+9}{4x-12}$

40. $\dfrac{2x^2-14x+24}{x^2-4x+3} \cdot \dfrac{3x^2-27}{4x-20}$

41. $\dfrac{a^2-1}{4a} \div \dfrac{2a+2}{8a^2}$

42. $\dfrac{6x}{x^2+x-2} \div \dfrac{3}{2x+4}$

43. $(9x^2-4y^2) \div \dfrac{3x-2y}{y-2x}$

44. $\dfrac{6r^2-rs-s^2}{4r^2-16s^2} \div \dfrac{2r^2+rs-s^2}{r^2+3rs+2s^2}$

45. $\dfrac{3}{x^2} + \dfrac{2}{x^2+3x}$

46. $\dfrac{1}{a^2-2a} - \dfrac{3}{2a}$

47. $\dfrac{2x}{x-2} - \dfrac{x^2-3}{x^2-4x+4}$

48. $\dfrac{3}{3x+y} - \dfrac{7}{3x^2-5xy-2y^2}$

49. $\dfrac{2x-3}{2x^2-x-15} - \dfrac{3x}{x^2-9}$

50. $\dfrac{2x}{x^2+2x-8} + \dfrac{3x}{2x+8}$

51. $\dfrac{3}{x^2+5x} - \dfrac{x}{x^2-25} + \dfrac{2}{2x^2+9x-5}$

52. $\dfrac{3}{8x+16} - \dfrac{1-x}{4x^2-16} + \dfrac{5}{2x-4}$

53. $\left(\dfrac{2}{a-b} \cdot \dfrac{a+b}{5}\right) \div \left(\dfrac{a^2+2ab+b^2}{a^2-b^2}\right)$

54. $\left(\dfrac{x^4-1}{(x-1)^2} \div (x^2+1)\right) \cdot \left(\dfrac{x-1}{x+1}\right)$

55. $\left(\dfrac{1}{x-2} \div \dfrac{3+x}{x+1}\right) \div \left(\dfrac{2+x}{x^2-x-6}\right)$

56. $\left(\dfrac{2s}{s-1} + \dfrac{s^2}{s^2-1}\right) \div \left(\dfrac{s^3}{s-1}\right)$

In Exercises 57 through 64, solve the given equations.

57. $\dfrac{1}{2} - \dfrac{x}{6} + 3 = \dfrac{2x}{9}$

58. $\dfrac{6}{7} - \dfrac{x}{4} = 2x - \dfrac{5}{14}$

59. $\dfrac{bx}{a} - \dfrac{1}{4} = x + \dfrac{3}{2a}$, for x

60. $\dfrac{4(x-y)}{5b} - \dfrac{x}{b^2} = \dfrac{y}{5b}$, for x

61. $\dfrac{5}{x+1} - \dfrac{3}{x} = \dfrac{-5}{x(x+1)}$

62. $\dfrac{20}{y^2-25} - \dfrac{1}{y+5} = \dfrac{2}{y-5}$

63. $\dfrac{5x}{x^2+2x} + \dfrac{1}{x} = \dfrac{3}{4x+8}$

64. $\dfrac{2}{2t+1} - \dfrac{t-1}{4t^2-4t-3} = \dfrac{1}{2t-3}$

In Exercises 65 through 84, solve the given problems.

65. In the study of the conduction of electrons in metals, we find the expression

$$\dfrac{1}{2}\left(\dfrac{eE}{m}\right)(ne)\left(\dfrac{l}{v}\right)$$

Perform the indicated multiplication.

66. The following expression is found in the study of the resistance that a pipe gives to a liquid flowing through it.

$$(p_1 - p_2) \div \left(\frac{(p_1 - p_2)\pi a^4}{8lu} \right)$$

Perform the indicated division.

67. A sphere of radius r floats in a liquid. When determining the height h which the sphere protrudes above the surface, the expression

$$\frac{1}{4r^2} - \frac{h}{12r^3}$$

is found. Combine and simplify.

68. In biology, when discussing the theory of x-rays, the expression

$$\frac{ds}{L} - \frac{ds^3}{24L^3}$$

arises. Combine the terms of this expression.

69. In the theory related to photographic developing, the expression

$$1 - \frac{2g_1}{g_2} + \frac{g_1^2}{g_2^2}$$

arises. Combine the terms of this expression.

70. When considering the impedance of an electronic amplifier, the expression

$$1 - \frac{2Z_1}{Z_1 + Z_2}$$

is encountered. Combine and simplify.

71. In the theory of electricity, the expression

$$L^2\omega^2 - \frac{2L}{C} + \frac{1}{C^2\omega^2}$$

arises (ω is the Greek letter omega). Combine these fractions.

72. In the study of the velocity of light, the expression

$$\frac{2d}{c}\left(1 + \frac{v^2}{2c^2} + \frac{3v^4}{4c^4}\right)$$

arises. Combine and simplify.

73. When considering the amplification of an electronic device, the expression

$$\left[\left(\frac{\mu}{\mu + 1} \right) R \right] \div \left[\frac{r}{\mu + 1} + R \right]$$

(μ is the Greek letter mu) is found. Simplify this expression.

74. If a certain stock always pays a dividend of $1, and this is always a 5% dividend, the present value P of the stock is found from the formula

$$P = \frac{\dfrac{1}{1+i}}{1 - \dfrac{1}{1+i}}$$

where i is the percentage expressed as a decimal. Simplify the right side of this formula.

75. In economics, when determining discounts the equation

$$r = \frac{A}{Pt} - \frac{1}{t}$$

arises. Solve for A.

76. In the study of the convection of heat, the equation

$$\frac{1}{U} = \frac{x}{k} + \frac{1}{h}$$

is used. Solve for x.

77. In photography, when analyzing the focusing of cameras we may use the equation

$$\frac{q_2 - q_1}{d} = \frac{f + q_1}{D}$$

Solve for q_1.

78. An electric circuit has capacitors in series and parallel so that the combined capacitance is given by

$$\frac{1}{C} = \frac{1}{C_2} + \frac{1}{C_1 + C_3}$$

Solve for C_1.

79. In the study of the motion of the planets, the equation

$$\frac{mv^2}{2} - \frac{kmM}{r} = -\frac{kmM}{2a}$$

is found. Solve for r.

80. In chemistry, when dealing with ideal gas vapors the equation

$$\frac{1}{P} = \frac{1}{P_A} + \frac{x}{P_B} - \frac{x}{P_A}$$

arises. Solve for P_A.

81. One computer output device can display a certain number of lines in 6.0 min. A second device requires 8.0 min to display the same number of lines. How long will it take the two devices together to display this number of lines?

82. One pump can empty an oil tanker in 5.0 h. It takes a second pump 8.0 h. How long will it take the two pumps working together?

83. A company allocates $60,000 more for research than for manufacturing. If there is a $\frac{5}{3}$ ratio of research to manufacturing, find the amount spent on research.

84. One computer printer can produce a document in 80 min while another printer requires only 48 min. If both printers work together, how much time is required for the document?

CHAPTER TEN

Exponents, Roots, and Radicals

In earlier chapters we have already introduced several basic rules involving exponents and roots of numbers. We begin this chapter with a brief review of those rules, and we then extend them to cases involving zero and negative exponents. We then use those extended rules in developing scientific notation, which is commonly used in science and technology. We proceed to consider roots, radicals, and fractional exponents.

There are many applications in science and technology that involve exponents of different types. As one example, aircraft wings must produce lift of about 1000 N/m². In order to determine the velocity V_1 of air over the upper wing surface, we use the equation

$$V_1 = (2L + RV_2^2)^{1/2} R^{-1/2}$$

where L is the lift, R is the air density, and V_2 is the velocity on the lower wing surface. In this chapter we evaluate such an expression, and that evaluation will involve the rules for exponents such as those given here.

10-1 Zero and Negative Exponents

In Example A we illustrate some of the rules of exponents that were introduced in Chapter 7.

EXAMPLE A

$$(a^m)(a^n) = a^{m+n} \qquad\qquad (2^3)(2^4) = 2^7 \qquad \text{(7-1)}$$

$$(a^m)^n = a^{mn} \qquad\qquad (2^3)^5 = 2^{15} \qquad \text{(7-2)}$$

$$(ab)^n = a^n b^n \qquad\qquad (4x)^2 = 16x^2 \qquad \text{(7-3)}$$

$$\frac{a^m}{a^n} = a^{m-n} \quad \text{if} \quad m > n \quad \text{and} \quad a \neq 0 \qquad \frac{x^8}{x^3} = x^5 \qquad \text{(7-4a)}$$

$$\frac{a^m}{a^n} = \frac{1}{a^{n-m}} \quad \text{if} \quad m < n \quad \text{and} \quad a \neq 0 \qquad \frac{x^3}{x^8} = \frac{1}{x^5} \qquad \text{(7-4b)}$$

$$\frac{a^m}{a^n} = 1 \quad \text{if} \quad m = n \quad \text{and} \quad a \neq 0 \qquad \frac{a^6}{a^6} = 1 \qquad \text{(7-4c)}$$

$\square$

We now need to further develop those concepts so that we will be prepared to consider some of the topics which will arise in later sections.

While the above rules involved only positive integer exponents, this section shows how negative integers and zero may also be used as exponents. Such considerations are necessary for Section 10-2 in which scientific notation is discussed. Chapter 12 also involves the use of negative integer and zero exponents, and many calculator and computer computations require an understanding of their use.

We begin by applying Eq. (7-4a) to the case in which the exponent in the denominator is larger so that $m - n$ is a negative integer. This is illustrated in the following example.

EXAMPLE B

Applying Eq. (7-4a) to the fraction a^3/a^5, we get

$$\frac{a^3}{a^5} = a^{3-5} = a^{-2}$$

Using Eq. (7-4b) leads to the result

$$\frac{a^3}{a^5} = \frac{1}{a^{5-3}} = \frac{1}{a^2}$$

If these results are to be consistent, it must follow that

$$\frac{1}{a^2} = a^{-2} \qquad\qquad\qquad \square$$

Generalizing the results in Example B, we find that if we define

$$a^{-n} = \frac{1}{a^n} \qquad \text{if } a \neq 0 \qquad\qquad\qquad \text{(10-1)}$$

then all results obtained by the use of Eqs. (7-4a), (7-4b), and (10-1) are consistent. Therefore Eq. (10-1) shows us that we may use negative integers as exponents. Also, we see that *we may move a* **factor** *from the numerator to the denominator or from the denominator to the numerator by changing the sign of its exponent*. Note that only the sign of the exponent changes with such a move, but the sign of the fraction remains the same. Consider the illustrations of the following example.

EXAMPLE C   In the division $b^4 \div b^7$, we may write

$$\frac{b^4}{b^7} = b^{4-7} = b^{-3} = \frac{1}{b^3}$$

As other illustrations of the use of Eq. (10-1), we have

$$x^{-2} = \frac{1}{x^2} \qquad \frac{1}{x^{-2}} = \frac{1}{\dfrac{1}{x^2}} = 1 \cdot \frac{x^2}{1} = x^2$$

$$a^{-1} = \frac{1}{a} \qquad a = a^{-(-1)} = \frac{1}{a^{-1}} \qquad\qquad \Box$$

As with positive exponents, we must be careful to note just what number or expression has the negative exponent. Consider the illustrations in the following example.

EXAMPLE D ▶   1. $4x^{-2} = \dfrac{4}{x^2}$ since *only x has the exponent of* -2.

     2. $(4x)^{-2} = \dfrac{1}{(4x)^2} = \dfrac{1}{16x^2}$ since the quantity $4x$ has the exponent of

     -2, as indicated by the parentheses.

     3. $\dfrac{1}{3a^{-1}} = \dfrac{1}{3\left(\dfrac{1}{a}\right)} = \dfrac{1}{\dfrac{3}{a}} = 1 \cdot \dfrac{a}{3} = \dfrac{a}{3}$ $\Box$

In general, with certain specific exceptions, *negative exponents are not used in the expression of a final result*. They are, however, used in intermediate steps in many operations.

EXAMPLE E   In reducing the fraction

$$\frac{x^2yz^3}{xy^4z^2}$$

to its simplest form, we may place all factors in the numerator by the use of negative exponents and combine exponents of the same base to determine the result. This leads to

$$\frac{x^2yz^3}{xy^4z^2} = x^2x^{-1}yy^{-4}z^3z^{-2} = x^{2-1}y^{1-4}z^{3-2} = xy^{-3}z = \frac{xz}{y^3}$$

In dividing a^m by a^n, where $m = n$, we note that

$$\frac{a^m}{a^m} = 1 \quad \text{a quantity divided by itself equals 1.} \qquad \frac{a^m}{a^m} = a^{m-m} = a^0 \quad \text{from Eq. (7-4a)}$$

This leads to Eq. (10-2)

$$\boxed{a^0 = 1 \quad \text{if} \quad a \neq 0} \tag{10-2}$$

This equation states that *any algebraic expression which is not zero and which is raised to the zero power is* 1.

EXAMPLE F

$$5^0 = 1 \qquad\qquad x^0 = 1$$
$$(rs)^0 = 1 \qquad\qquad (ax - p)^0 = 1$$
$$5x^0 = 5(1) = 5 \qquad\qquad (5x)^0 = 1$$

▶ Note that in the expression $5x^0$ *only x has the exponent zero.*

Zero and negative exponents may be used in the other laws of exponents just as any positive exponent. For reference, we now list the laws of exponents we have encountered:

$$(a^m)(a^n) = a^{m+n} \tag{7-1}$$

$$(a^m)^n = a^{mn} \tag{7-2}$$

$$(ab)^n = a^nb^n \tag{7-3}$$

$$\frac{a^m}{a^n} = a^{m-n} \quad (a \neq 0) \tag{7-4}$$

$$\left(\frac{a}{b}\right)^n = \frac{a^n}{b^n} \quad (b \neq 0) \tag{9-3}$$

$$a^{-n} = \frac{1}{a^n} \quad (a \neq 0) \tag{10-1}$$

$$a^0 = 1 \quad (a \neq 0) \tag{10-2}$$

EXAMPLE G

1. Using Eq. (7-1): $(a^5)(a^0) = a^{5+0} = a^5$

2. Using Eqs. (7-1) and (10-1): $a^{-2}b^{-1}a^5 = a^{5-2}b^{-1} = \dfrac{a^3}{b}$

3. Using Eqs. (7-2) and (10-1): $(a^5)^{-1} = a^{-5} = \dfrac{1}{a^5}$

4. Using Eqs. (7-3) and (10-1):

$$(x^{-2}y^3)^{-2} = x^{(-2)(-2)}y^{(3)(-2)} = x^4y^{-6} = \dfrac{x^4}{y^6}$$

5. Using Eq. (7-4): $\dfrac{x^{-2}}{x^{-7}} = x^{-2-(-7)} = x^{-2+7} = x^5$

6. Using Eqs. (7-4) and (10-1):

$$\dfrac{t^3}{t^7} = t^{3-7} = t^{-4} = \dfrac{1}{t^4}$$

7. Using Eqs. (9-3) and (10-1):

$$\left(\dfrac{s^{-2}}{t^3}\right)^4 = \dfrac{s^{(-2)(4)}}{t^{(3)(4)}} = \dfrac{s^{-8}}{t^{12}} = \dfrac{1}{s^8t^{12}}$$

8. Using Eqs. (10-1), (10-2), and (7-1):

$$\dfrac{a^0b^{-5}b^7c^2}{c^6} = \dfrac{(1)(b^{-5+7})}{c^{-2}c^6} = \dfrac{b^2}{c^{-2+6}} = \dfrac{b^2}{c^4}$$

$\square$

EXAMPLE H   In developing a computer program to search for data, we find that we must solve the equation

$$\dfrac{1}{2^n} = 8$$

for n. We proceed as follows

equal $\begin{bmatrix} \dfrac{1}{2^n} = 2^{-n} \\ 8 = 2^3 \end{bmatrix}$ equal

We conclude that $-n = 3$, or $n = -3$. $\square$

Exercises 10-1

In Exercises 1 through 44, express each of the given expressions in the simplest form. Write all answers with only positive exponents.

1. t^{-5} 2. R^{-3} 3. x^{-4} 4. s^{-8}

5. $\dfrac{1}{x^{-3}}$ **6.** $\dfrac{1}{a^{-7}}$ **7.** $\dfrac{1}{R_1^{-3}}$ **8.** $\dfrac{1}{t^{-8}}$

9. $3c^{-2}$ **10.** $(3c)^{-1}$ **11.** $\dfrac{1}{3c^{-1}}$ **12.** $\dfrac{1}{(3c)^{-1}}$

13. 5^0 **14.** $\dfrac{1}{5^0}$ **15.** $(c-5)^0$ **16.** $\dfrac{1}{(c-5)^0}$

17. $5c^0$ **18.** $3x^0$ **19.** $\dfrac{9x^0}{y^{-2}}$ **20.** $\dfrac{2x^{-1}}{3y^0}$

21. $\dfrac{3}{3^{-5}}$ **22.** $\dfrac{7^{-2}}{7^6}$ **23.** $\dfrac{6^{-4}}{6^{-5}}$ **24.** $\dfrac{9^{-5}}{9^{-4}}$

25. $a^2x^{-1}a^{-3}$ **26.** $b^{-2}c^{-3}b^5$ **27.** $2(c^{-2})^4$ **28.** $6(x^{-1})^{-1}$

29. $\dfrac{x^{-5}y^{-1}}{xy^{-3}}$ **30.** $\dfrac{3^{-1}a}{3a^{-1}}$ **31.** $\dfrac{5^{-2}x}{5x^{-1}}$ **32.** $\dfrac{2a^{-3}}{2^{-1}a}$

33. $\dfrac{st^{-2}}{(2t)^0}$ **34.** $\dfrac{(m^{-2}n)^0}{m^3n^{-4}}$ **35.** $\dfrac{2^{-2}x^{-2}y^{-4}}{2x^{-6}y^0}$ **36.** $\dfrac{6^{-1}a^{-2}b}{6a^{-3}b^{-2}}$

37. $\dfrac{(3a)^{-1}b^2}{3a^0b^{-5}}$ **38.** $\dfrac{(3p^0)^{-2}}{3p^2q^{-8}}$ **39.** $\left(\dfrac{2a^{-1}}{5b}\right)^2$ **40.** $\left(\dfrac{x^2}{y^{-1}z}\right)^{-2}$

41. $\dfrac{(xy^{-1})^{-2}}{x^2y^{-3}}$ **42.** $\dfrac{r^0st^{-2}}{(r^{-1}s^2t)^{-3}}$ **43.** $\dfrac{(3a^{-2}bc^{-1})^{-1}}{6a^{-3}(bc)^{-1}}$ **44.** $\dfrac{(axy^{-1})^{-2}}{(a^{-2}x^{-1}y)^{-3}}$

In Exercises 45 through 56, solve the given problems.

45. The resistance in a certain parallel electric circuit is described as
$$R_1^{-1} + R_2^{-1}$$
Rewrite this expression so that it contains only positive exponents.

46. The fundamental lens equation is given as
$$F^{-1} = u^{-1} + v^{-1}$$
Rewrite this equation so that it contains only positive exponents.

47. In the study of radioactivity the equation
$$N = N_0 e^{-kt}$$
is found. Express this equation without the use of negative exponents.

48. In certain economic circumstances, the price p for each of x commodities is given by the formula
$$p = 10(3^{-x})$$
Express this formula without the use of negative exponents.

49. The impedance Z of a particular electronic circuit may be expressed as
$$Z = \left(h + \dfrac{1}{R}\right)^{-1}$$
Express the right side as a simple fraction without negative exponents.

50. In the theory dealing with electricity, the expression

$$e^{-i(\omega t - ax)}$$

(ω is the Greek omega) is encountered. Rewrite this expression so that it contains no negative signs in the exponent.

51. Negative exponents are often used to express the units of a quantity. For example, velocity can be expressed in units of ft/s or $\text{ft} \cdot \text{s}^{-1}$. In this same manner, express the units of density in g/cm^3 and the units of acceleration m/s^2 by the use of negative exponents.

52. Hall's constant for certain semiconductors is given as

$$2 \times 10^{-3} m^3 c^{-1}$$

Express this constant using only positive exponents.

53. In determining heating requirements, a solar engineer uses the expression

$$362 \; \text{Btu} \cdot \text{h}^{-1} \cdot \text{ft}^{-2}$$

Write this expression using only positive exponents.

54. When using scientific notation (see Section 10-2) and logarithms (see Chapter 12), you will find that negative and zero powers of 10 are important. Since $100 = 10^2$,

$$\frac{1}{100} = \frac{1}{10^2} = 10^{-2}$$

Similarly, express $\frac{1}{10}$ and $\frac{1}{1000}$ as negative powers of 10.

55. In discussing electronic amplifiers, the expression

$$\left(\frac{1}{r} + \frac{1}{R} \right)^{-1}$$

is found. Simplify this expression.

56. In determining the number of seedlings that must be planted now in order to produce a certain number of trees later, a forester encounters the equation

$$\left(\frac{1}{2} \right)^n = 16$$

Find a value for n which makes this equation true.

10-2 Scientific Notation

In technical and scientific work we often use numbers that are extremely large or extremely small in magnitude. For example, a light-year is equivalent to a distance of 9,460,530,000,000,000 m. In addition to simply representing very large and very small numbers, we must often perform calculations with them. For example, the frequency (in hertz) in a certain electric circuit is found by evaluating $4,260,000 \div 0.0009425$. In this section we begin with the representation of such numbers, and we then proceed to consider calculations involving them.

EXAMPLE A   Television signals travel at about 30,000,000,000 cm/s. The mass of the earth is about 6,600,000,000,000,000,000,000 t. A typical protective coating used on aluminum is about 0.0005 in. thick. The wavelength of some x-rays is about 0.000000095 cm. In chemistry, the number of molecules in a gas in standard conditions is 602,300,000,000,000,000,000,000. ☐

Expressing such numbers with conventional notation is often cumbersome. Also, limitations of computers and calculators require that we develop a more efficient convention if we are to work with such numbers. The notation referred to as **scientific notation** is normally used to represent these numbers of extreme magnitudes.

A number written in scientific notation is expressed as the product of a number between 1 *and* 10 *and a power of* 10. Symbolically this can be written as

$$P \times 10^k$$

where P is between 1 and 10 (or possibly equal to 1) and k is an integer. The procedure for converting a number from ordinary notation to scientific notation is summarized by the following steps:

1. Move the decimal point to the right of the first nonzero digit.
2. Multiply this new number by a power of 10 whose exponent k is determined as follows:
   (a) If the original number is greater than 1, k is the number of places the decimal point was moved to the left.
   (b) If the original number is between 0 and 1, k is negative and corresponds to the number of places the decimal point was moved to the right.

The following example illustrates how numbers are written in scientific notation.

EXAMPLE B   To change 85,000 to scientific notation, first move the decimal point to the right of the first nonzero digit, which in this case is 8. We get 8.5. Now multiply 8.5 by 10^k where $k = 4$ since 85,000 is greater than 1 and we moved the decimal point 4 places to the left. Our final result is 8.5×10^4.

(Note that $10^4 = 10,000$ and $8.5 \times 10,000 = 85,000$ so that the value of the original number does not change.) ☐

EXAMPLE C   To change 0.0000824 to scientific notation, first move the decimal point to the right of the first nonzero digit to get 8.24. Now multiply this result by 10^k where $k = -5$ since the original number is between 0 and 1 and the decimal point was moved 5 places to the right. We get 8.24×10^{-5}. ☐

In Example D we see a clear pattern evolving. Shifting the decimal point to the left gives us a positive exponent while a shift to the right yields a negative exponent.

EXAMPLE D

$$56000 = 5.6 \times 10^4$$

4 places to left

$$0.143 = 1.43 \times 10^{-1}$$

1 place to right

$$0.000804 = 8.04 \times 10^{-4}$$

4 places to right

$$2.97 = 2.97 \times 10^0$$

0 places

To change a number from scientific notation to ordinary notation, the procedure described above is reversed. The following example illustrates the method of changing from scientific notation to ordinary notation.

EXAMPLE E   To change 2.9×10^3 to ordinary notation, we must move the decimal point 3 places to the right. Therefore additional zeros must be included for the proper location of the decimal point. We get

$$2.9 \times 10^3 = 2900$$

This is reasonable when we consider that $10^3 = 1000$ so that 2.9×10^3 is really 2.9×1000, or 2900.

To change 7.36×10^{-2} to ordinary notation, we must move the decimal point 2 places to the left. Again we must include additional zeros so that the decimal point can be properly located. We get

$$7.36 \times 10^{-2} = 0.0736$$

The importance of scientific notation is demonstrated by the use of *prefixes* on units to denote certain powers of 10 in the metric system. These prefixes were introduced in Section 2-1, where only positive powers of 10 were used in describing the prefixes. However, some prefixes can be properly described by the use of negative exponents. A complete table of metric prefixes is found in Table 6 in Appendix D at the end of the book.

EXAMPLE F   In Section 2-1 we noted the use of the metric prefixes for million, thousand, hundredth, thousandth, millionth, billionth, and trillionth. Since "centi" means hundredth, we see that

$$5 \text{ cm} = 0.05 \text{ m} = 5 \times 10^{-2} \text{ m}$$

In the same way, we have

$$6 \text{ mg} = 0.006 \text{ g} = 6 \times 10^{-3} \text{ g} \qquad 3 \text{ } \mu\text{g} = 0.000003 \text{ g} = 3 \times 10^{-6} \text{ g}$$

$$6.5 \text{ ns} = 0.0000000065 \text{ s} = 6.5 \times 10^{-9} \text{ s} \qquad 8.1 \text{ km} = 8100 \text{ m} = 8.1 \times 10^3 \text{ m}$$

$$4.2 \text{ Gg} = 4,200,000,000 \text{ g} = 4.2 \times 10^9 \text{ g}$$

As we have seen from these examples, scientific notation provides an important application of the use of exponents, positive and negative. Also, we have seen its use in the metric system. Another use of scientific notation is in handling calculations involving numbers of very large or very small magnitudes. This can be seen by the fact that a feature on many calculators is that of scientific notation. Also, typical programming languages incorporate a form of scientific notation. In BASIC, for example, 9.62×10^{12} is entered 9.62E12 and 5.38×10^{-7} can be entered as 5.38E−7.

Calculations with very large or very small numbers can be made by first expressing all numbers in scientific notation. Then the actual calculation can be performed on numbers between 1 and 10, using the laws of exponents to find the proper power of 10 for the result. It is acceptable to leave the result in scientific notation.

EXAMPLE G   To evaluate $9,620,000,000,000 \div 0.0000000000328$, we may set up the calculation as

$$\frac{9.62 \times 10^{12}}{3.28 \times 10^{-11}} = \left(\frac{9.62}{3.28}\right) \times 10^{23} = 2.93 \times 10^{23}$$

with the exponent arising from $12 - (-11)$.

The power of 10 in this result is large enough so that we would normally leave it in this form. Even on most calculators with the scientific notation feature, it would be necessary to express the numerator and denominator in scientific notation before performing the calculation.

Here is how this example might be done on a typical calculator. First, 9.62×10^{12} is entered using a key commonly labeled EE for "enter exponent." The key sequence of

$$9.62 \boxed{\text{EE}}\ 12$$

causes 9.62×10^{12} to be entered. Then the operation of division is indicated by the key labeled $\div$ and the denominator is entered as follows:

$$3.28 \boxed{\text{EE}}\ 11 \boxed{+/-}$$

Note that the exponent of -11 is entered through use of the change sign key which follows the 11. The sequence

$$9.62 \boxed{\text{EE}}\ 12 \boxed{\div} 3.28 \boxed{\text{EE}}\ 11 \boxed{+/-} \boxed{=} \quad \boxed{\mathit{2.9329268\ 23}}$$

gives a display which is interpreted as 2.9329268×10^{23}. Note that the calculator display of 2.9329268 23 includes the space as a means of separating the exponent.

In this example, the dividend and divisor both have three significant digits. Recall that when we multiply or divide approximate numbers, *the result should be rounded off* so that it has as many significant digits as are contained in the original number with the fewest significant digits. This rule indicates that the result of $9,620,000,000,000 \div 0.0000000000328$ should be rounded off to three significant digits as in 2.93×10^{23}. Again, the calculator result of 2.9329268×10^{23} is potentially misleading since it implies a degree of accuracy which is not justified. *Calculator results should be rounded off to the appropriate degree of accuracy or precision.*   □

EXAMPLE H   In evaluating the product $(7.50 \times 10^9)(6.44 \times 10^{-3})$, we obtain

$$(7.50 \times 10^9)(6.44 \times 10^{-3}) = 48.3 \times 10^6$$

However, since a number in scientific notation is expressed as the product of a number between 1 and 10 and a power of 10, we should rewrite this result as

$$48.3 \times 10^6 = (4.83 \times 10)(10^6)$$
$$= 4.83 \times 10^7$$

Exercises 10-2

In Exercises 1 through 4, find the value of k which makes the equation true.

1. $100,000 = 10^k$ **2.** $\dfrac{1}{100} = 10^k$ **3.** $\dfrac{1}{10,000} = 10^k$ **4.** $1,000,000 = 10^k$

In Exercises 5 through 12, change the given numbers from scientific notation to ordinary notation.

5. 4×10^6 **6.** 3.8×10^9 **7.** 8×10^{-2} **8.** 7.03×10^{-11}

9. 2.17×10^0 **10.** 7.93×10^{-1} **11.** 3.65×10^{-3} **12.** 8.04×10^3

In Exercises 13 through 24, change the given numbers from ordinary notation to scientific notation.

13. 3000   **14.** 420,000   **15.** 0.076   **16.** 0.0029

17. 0.704   **18.** 0.0108   **19.** 9.21   **20.** 10.3

21. 0.000053   **22.** 1,006,000   **23.** 2,010,000,000   **24.** 0.0004923

In Exercises 25 through 36, perform the indicated calculations by first expressing all numbers in scientific notation. (See Example G.)

25. $(6700)(23,200)$ **26.** $(4510)(9700)$ **27.** $(0.0153)(0.608)$

28. $(79,500)(0.00854)$ **29.** $\dfrac{3740}{80,500,000}$ **30.** $\dfrac{0.0186}{0.0000665}$

31. $0.00005820 \div 8635$ **32.** $0.0000385 \div 0.000000903$ **33.** $\dfrac{(6.80)(8,040,000)}{4,200,000}$

34. $\dfrac{(0.0753)(73,900)}{0.0000811}$ **35.** $\dfrac{(54,300)(19,356)}{(82,500)(0.00101)}$ **36.** $\dfrac{(0.000439)(865,000)}{(307,000)(0.00000883)}$

In Exercises 37 through 44, express each of the given metric units in terms of scientific notation without the use of the prefix. (See Example F and Table 6 in Appendix D.)

37. 6.5 Mg   **38.** 3.8 Tm   **39.** 45 mm   **40.** 650 μL

41. 3.92 cL   **42.** 1.15 ps   **43.** 80.6 μs   **44.** 29.1 fg

In Exercises 45 through 64, change numbers in ordinary notation to scientific notation and change numbers in scientific notation to ordinary notation. Units and symbols may be found in Table 5 in Appendix D.

45. In UHF television broadcasting, one station transmits with a frequency of 2,000,000,000 Hz.

46. The speed of a communications satellite is about 17,500 mi/h.

47. The mass of an electron is 0.00000000000000000000000000091 g.

48. The wavelength of red light is about 0.00000065 m.

49. The sun weighs about 4×10^{30} lb.

50. It takes about 5×10^4 lb of water to grow one bushel of corn.

51. Special boring machines can produce finishes to about 10^{-6} in.

52. Some computers perform an addition in 1.5×10^{-9} s.

53. The area of the oceans of the earth is about 360,000,000 km^2.

54. A typical capacitor has a capacitance of 0.00005 F.

55. The power of a radio signal received from a laser beam probe is 1.6×10^{-12} W.

56. To attain an energy density equal to that in some laser beams, an object would have to be heated to about 10^{30} °C.

57. The planet Mercury is an average of 36,000,000 mi away from the sun.

58. The half-life of uranium 235 is 710,000,000 years.

59. The faintest sound which can be heard has an intensity of about 10^{-12} W/m^2.

60. Under certain conditions, the pressure of a gas is 7.5×10^{-2} Pa.

61. One electron volt is approximately 0.0000000000000000006 J.

62. One nanometer is 0.000000001 m.

63. The speed of light in a vacuum is 3×10^{10} cm/s.

64. A certain computer has 1.05×10^6 bytes of memory.

In Exercises 65 through 68, perform the indicated calculations by first expressing all numbers in scientific notation. Units and symbols may be found in Table 5 in Appendix D.

65. There are about 161,000 cm in 1 mi. What is the area in square centimeters of 1 mi^2?

66. A particular virus is a sphere 0.000000175 m in diameter. What volume does the virus occupy?

67. The transmitting frequency of a television signal is given by the formula $f = \dfrac{v}{\lambda}$, where v is the velocity of the signal and λ (the lowercase Greek lambda) is the wavelength. Determine f (in hertz) if $v = 3.00 \times 10^{10}$ cm/s and $\lambda = 4.95 \times 10^2$ cm.

68. The final length L of a pipe which has increased in length from L_0 due to an increase of temperature T is found by use of the formula $L = L_0(1 + \alpha T)$, where α (the Greek alpha) depends on the material of the pipe. Find the final length of an iron steam pipe if $L_0 = 195.0$ m, $T = 85$°C, and $\alpha = 9.8 \times 10^{-6}$/°C.

In Exercises 69 through 72, perform the indicated operations by using a scientific calculator. Express answers in scientific notation.

69. $(1.65 \times 10^{12})(2.97 \times 10^7)$

70. $(2.35 \times 10^{-8}) \div (1.35 \times 10^6)$

71. $(3.2 \times 10^{20})(5.6 \times 10^{-4}) \div (2.6 \times 10^6)$

72. $(6.358 \times 10^{10})(3.977 \times 10^4) \div (9.864 \times 10^{22})$

10-3 Roots of a Number

In Chapter 1, we introduced the concept of the root of a number and we examined some ways of determining the square root of a number. In this chapter we further develop these topics so that we will be able to do the manipulations required in the chapters that follow.

We defined *a square root of a given number to be that number which, when squared, equals the given number*. Considering now our work in Chapter 3 with signed numbers, we see that there are actually two numbers which could be considered to be the square root of a given positive number. Also, there are no numbers among those we have discussed so far which could be the square root of a negative number.

EXAMPLE A   Since $(+3)^2 = 9$ and $(-3)^2 = 9$, we see that, by the definition we have used up to this point, the square root of 9 can be either $+3$ or -3. Also, since the square of $+3$ and the square of -3 are both 9, we have no apparent result for the square root of -9.   □

Since there are two numbers, one positive and the other negative, whose squares equal a given number, we define *the* **principal square root** *of a positive number to be positive*. That is:

$$\boxed{\sqrt{N} = x}$$ **(10-3)**

▶   where $\sqrt{N}$ is the principal square root of N and **equals (positive)** x. Also, of course, $x^2 = N$. This avoids the ambiguity of two possible answers for a square root.

EXAMPLE B $\sqrt{16} = 4$ $\sqrt{25} = 5$ $\sqrt{81} = 9$ $\sqrt{0.09} = 0.3$
We do *not* accept answers such as $\sqrt{144} = -12$.   □

By defining the principal square root of a number in this way, we do not mean that problems involving square roots cannot have negative results. By $-\sqrt{N}$ we indicate the negative of the principal value, which means the negative of a positive signed number, and this is a negative number.

EXAMPLE C $-\sqrt{16} = -(+4) = -4$ $-\sqrt{25} = -5$
$-\sqrt{81} = -9$ $-\sqrt{0.09} = -0.3$
We do *not* accept answers such as $-\sqrt{144} = -(-12) = 12$.   □

Returning to the problem of the square root of a negative number, we find it necessary to define a new kind of number to provide the required result. We define

$$\boxed{j^2 = -1 \quad \text{or} \quad j = \sqrt{-1}}$$ **(10-4)**

The number j is called the **imaginary unit.** (It is sometimes referred to as the *j-operator*.) Some fields of science, technology, and mathematics normally represent the imaginary unit by *i* instead of *j*, but we will use *j* since electric current is also represented by *i*. Using the imaginary unit *j*, we now define

$$(bj)^2 = -b^2 \qquad\qquad\qquad (10\text{-}5)$$

$$(bj)^2 = b^2j^2 = b^2(-1) = -b^2$$

where *bj is called an* **imaginary number.** (The word "imaginary" is simply the name of the number; these numbers are not imaginary in the usual sense of the word.) Imaginary numbers are discussed at greater length in Chapter 20.

Since j^2 is defined to be -1, it follows that

$$\sqrt{-a} = \sqrt{a(-1)} = \sqrt{aj^2} = \sqrt{a}\,\sqrt{j^2} = \sqrt{a}\,j$$

where *a* is a positive number. That is,

$$\sqrt{-a} = \sqrt{a}\,j \qquad \textbf{(where } a \, > \, \textbf{0)} \qquad\qquad (10\text{-}6)$$

EXAMPLE D

$$\sqrt{-16} = 4j \qquad\qquad\qquad \sqrt{-25} = 5j$$

$$-\sqrt{-9} = -3j \qquad\qquad\qquad -\sqrt{-100} = -10j$$

Also,

$$(6j)^2 = 6^2j^2 = 36(-1) = -36$$

$$(2j)^2 = 2^2j^2 = 4(-1) = -4$$

$$(-3j)^2 = (-3)^2j^2 = 9(-1) = -9$$

$\square$

Most of our work with roots will deal with square roots, but we shall have some occasion to consider other roots of a number. In general: *The* **principal *n*th root** *of a number N is designated as*

$$\sqrt[n]{N} = x \qquad\qquad (10\text{-}7)$$

▶ *where $x^n = N$. The number n is called the* **index** *while N is called the* **radicand.** *(If $n = 2$, it is usually not written, as we have already seen.)* ***If N is positive, x is positive. If N is negative and n is odd, x is negative.*** That is, odd roots of negative numbers are negative. We shall not consider the case of *N* being negative while *n* is even and larger than 2, as in expressions such as $\sqrt[4]{-16}$ or $\sqrt[6]{-64}$.

EXAMPLE E

$$\sqrt[3]{64} = 4 \qquad \text{since} \quad 4^3 = 64$$
$$\sqrt[3]{-64} = -4 \qquad \text{since} \quad (-4)^3 = -64$$
$$\sqrt[4]{81} = 3 \qquad \text{since} \quad 3^4 = 81$$
$$\sqrt[5]{32} = 2 \qquad \text{since} \quad 2^5 = 32$$
$$\sqrt[5]{-32} = -2 \qquad \text{since} \quad (-2)^5 = -32$$

□

If a negative sign precedes the radical sign, we express the result as the negative of the principal value. That is, we first determine the value of the *n*th root and we then consider the sign preceding the radical. This is illustrated in the following example.

EXAMPLE F

$$-\sqrt[3]{64} = -4 \qquad\qquad -\sqrt[3]{-64} = -(-4) = 4$$

$$-\sqrt[4]{16} = -2 \qquad\qquad -\sqrt[5]{-32} = -(-2) = 2$$

$$-\sqrt[7]{-128} = -(-2) = 2$$

□

EXAMPLE G   Find the surface area of a cube with volume 8.00 ft³.
We must use the formula for the volume of a cube and solve for *e*. We get

$$e^3 = 8.00 \quad \text{or} \quad e = \sqrt[3]{8.00} = 2.00 \text{ ft}$$

Since each edge is 2.00 ft, each of the six square sides must have an area of 4.00 ft² so that the total surface area is 24.0 ft².

□

There are ways to use calculators and computers to evaluate expressions such as $-\sqrt[5]{17.38}$. Some calculators have a special key labeled $\boxed{\sqrt[x]{y}}$ which can be used for that purpose. In Section 10-6 we will discuss a general procedure for using calculators with such expressions.

We conclude this section with a brief discussion of the type of number which the root of a number represents. In Section 3-4 we noted that a *rational number is a number which can be expressed as the division of an integer by another integer*. If the root of a number can be expressed as the ratio of one integer to another, then it is **rational.** Otherwise, it is **irrational.** All of the rational and irrational numbers combined constitute the **real numbers.** The **imaginary numbers** introduced in this section constitute another type of number.

EXAMPLE H

$\sqrt{4} = 2$, and since 2 is an integer, it is rational.

$\sqrt[3]{0.008} = 0.2 = \frac{1}{5}$, and since $\frac{1}{5}$ is the ratio of one integer to another, it is rational.

$\sqrt{2}$ is approximately 1.414, but it is not possible to find two integers whose ratio is exactly $\sqrt{2}$. Therefore $\sqrt{2}$ is irrational.

$\sqrt[3]{7}$ and $\sqrt[5]{-39}$ are irrational since it is impossible to express either number as the ratio of one integer to another.

$\sqrt{-4} = 2j$, which we have already found to be an imaginary number.

$\square$

In discussing roots of numbers we are considering numbers of which many—in fact most—are irrational.

Exercises 10-3

In Exercises 1 through 32, find the indicated principal roots.

1. $\sqrt{49}$
2. $\sqrt{36}$
3. $-\sqrt{144}$
4. $-\sqrt{81}$
5. $\sqrt{0.16}$
6. $\sqrt{0.01}$
7. $-\sqrt{0.04}$
8. $-\sqrt{0.36}$
9. $\sqrt{400}$
10. $\sqrt{900}$
11. $-\sqrt{1600}$
12. $-\sqrt{3600}$
13. $\sqrt[3]{8}$
14. $\sqrt[3]{27}$
15. $\sqrt[3]{-8}$
16. $\sqrt[3]{-27}$
17. $-\sqrt[3]{125}$
18. $-\sqrt[3]{-125}$
19. $\sqrt[3]{0.125}$
20. $\sqrt[3]{-0.001}$
21. $\sqrt[4]{16}$
22. $-\sqrt[4]{625}$
23. $\sqrt[5]{243}$
24. $-\sqrt[5]{-243}$
25. $\sqrt[6]{64}$
26. $-\sqrt[8]{256}$
27. $\sqrt{-4}$
28. $\sqrt{-49}$
29. $-\sqrt{-400}$
30. $-\sqrt{-900}$
31. $\sqrt{-0.49}$
32. $-\sqrt{-0.0001}$

In Exercises 33 through 40, perform the indicated operations.

33. $(5j)^2$
34. $5j^2$
35. $(5j)(4j)$
36. $(-7j)^2$
37. $(-3j)(-4j)$
38. $(-8j)^2$
39. $-(8j)^2$
40. $-8j^2$

In Exercises 41 through 44, classify each given number as rational, irrational, or imaginary.

41. $\sqrt{4}$, $\sqrt{3}$, $\sqrt[3]{8}$, $\sqrt[3]{-8}$, $\sqrt{-8}$
42. $\sqrt[3]{27}$, $\sqrt{15}$, $\sqrt{50}$, $\sqrt[5]{-32}$, $\sqrt{-9}$
43. $\sqrt[3]{100}$, $\sqrt[4]{0.0001}$, $\sqrt{-1}$, $\sqrt[5]{64}$, $\sqrt{0.16}$
44. $\sqrt{144}$, $\sqrt[4]{250}$, $\sqrt[3]{-0.027}$, $\sqrt{-81}$, $\sqrt{0.64}$

In Exercises 45 through 52, solve the given problems.

45. A water tank is in the shape of a cube with a volume of 216 ft³. What is the length of an edge of this tank? See Figure 10-1.

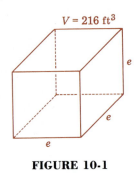

$V = 216 \text{ ft}^3$

46. A cubical storage container has a volume of 1728 ft³. Find the area of the bottom side.

47. The period T, in seconds, of a pendulum of length l, in feet, is given by the equation

$$T = \frac{\pi}{2}\sqrt{\frac{l}{2}}$$

What is the period of a pendulum with a length of 18.0 ft?

FIGURE 10-1

48. The ratio of the rates of diffusion of two gases is given by

$$\frac{r_1}{r_2} = \frac{\sqrt{m_2}}{\sqrt{m_1}}$$

where m_1 and m_2 are the masses of the molecules of the gases. Find the ratio r_1/r_2 if $m_1 = 25$ units and $m_2 = 81$ units.

49. The ratio between successive speeds of a six-speed gearbox is $\sqrt[5]{1024/243}$, given that the maximum speed is 1024 r/min and the minimum speed is 243 r/min. Determine this ratio. (The symbol r denotes revolution.)

50. In a study of the population growth of a bacteria culture, the number of bacteria currently present is found to be 1000 times the fourth root of the number present in the original culture. If the original culture had 810,000 bacteria, find the number currently present.

51. The annual yield (in decimal form) of an investment is found by evaluating

$$\sqrt[3]{\frac{V}{I}} - 1$$

where V is the value of the investment after three years and I is the original amount invested. Evaluate this expression for $V = \$532.40$ and $I = \$400$.

52. The diameter (in millimeters) of a water droplet can be found by using the formula

$$d = \sqrt[3]{0.95493V}$$

where V is the volume. Find d if $V = 28.275$ mm³.

10-4 Simplifying Radicals

We can often simplify radicals by using a basic property which states that *the root of a product equals the product of the roots of its factors*. That is

$$\boxed{\sqrt{ab} = \sqrt{a}\sqrt{b} \qquad \textbf{(where } a \textbf{ and } b \textbf{ are positive)}} \qquad \text{(10-8)}$$

To avoid difficulties with imaginary numbers, we assume that all letters represent positive real numbers. The development of Eq. (10-8) is based on observations similar to the one given in Example A, which follows.

EXAMPLE A   We know that $\sqrt{36} = 6$. If we now consider

$$\sqrt{36} = \sqrt{4 \cdot 9} = \sqrt{4}\sqrt{9} = 2 \cdot 3$$

we see that the same result of 6 is obtained. □

In Eq. (10-8), *if the root of either a or b can be found exactly, we may simplify the radical*. Consider the following example.

EXAMPLE B   By writing $\sqrt{48} = \sqrt{16 \cdot 3}$, we get the following result:

$$\sqrt{48} = \sqrt{16 \cdot 3} = \sqrt{16}\sqrt{3} = 4\sqrt{3}$$

Thus $\sqrt{48}$ can be expressed in terms of $\sqrt{3}$. Also, in the same way we have

$$\sqrt{175} = \sqrt{25 \cdot 7} = \sqrt{25}\sqrt{7} = 5\sqrt{7}$$

□

► In using Eq. (10-8) to simplify the radical, it is essential that we are able to factor any given number into two factors, **one of which is a perfect square.** The radical is not simplified completely until the number under the radical contains no perfect square factors.

EXAMPLE C   In Example B we could have written $\sqrt{48} = \sqrt{6 \cdot 8}$. However, with these factors no simplification can be done because neither 6 nor 8 is a perfect square. If we noted that $\sqrt{48} = \sqrt{4 \cdot 12} = \sqrt{4}\sqrt{12} = 2\sqrt{12}$, the radical has been simplified, *but not completely.* We must then note that $\sqrt{12} = \sqrt{4 \cdot 3} = \sqrt{4}\sqrt{3} = 2\sqrt{3}$, which means that $\sqrt{48} = 2\sqrt{12} = 2(2\sqrt{3}) = 4\sqrt{3}$. Radicals should always be expressed in simplest form.

Other illustrations of the use of Eq. (10-8) in simplifying radicals follow:

perfect square

$$\sqrt{72} = \sqrt{36 \cdot 2} = \sqrt{36}\sqrt{2} = 6\sqrt{2}$$

$$\sqrt{54} = \sqrt{9 \cdot 6} = \sqrt{9}\sqrt{6} = 3\sqrt{6}$$

$$\sqrt{126} = \sqrt{9 \cdot 14} = \sqrt{9}\sqrt{14} = 3\sqrt{14}$$

$$\sqrt{-25} = \sqrt{25 \cdot (-1)} = \sqrt{25}\sqrt{-1} = 5j$$

With a number such as 126, the perfect square factor of 9 might not be immediately obvious, and we may have to experiment with different combinations of factors until the right combination is identified. This could involve finding all prime factors by methods used in Section 1-4. □

Square roots of products involving literal symbols can also be simplified by use of Eq. (10-8). The simplification is based on the fact that *the square root of the square of a positive real number is that number*. That is,

$$\sqrt{a^2} = a \qquad (a > 0)$$

(10-9)

By use of Eq. (10-8) an algebraic product can be expressed as the product of square roots. Then, for any of these products for which Eq. (10-9) is applicable, the expression can be simplified. The following example gives three illustrations.

EXAMPLE D

$$\sqrt{a^2b^2c} = \sqrt{a^2}\sqrt{b^2}\sqrt{c} = ab\sqrt{c}$$

$$\sqrt{25a^4} = \sqrt{25}\sqrt{a^4} = 5\sqrt{(a^2)^2} = 5a^2$$

$$\sqrt{6a^2b^6x} = \sqrt{6}\sqrt{a^2}\sqrt{b^6}\sqrt{x} = (\sqrt{6})(a)(\sqrt{(b^3)^2})(\sqrt{x})$$

$$= (\sqrt{6})(a)(b^3)(\sqrt{x})$$

not perfect squares

$$= ab^3\sqrt{6}\sqrt{x} = ab^3\sqrt{6x}$$

□

Another operation frequently performed on fractions with radicals is to write the fraction so that no radicals appear in the denominator. Before the extensive use of calculators, one reason for this was ease of calculation, and therefore this reason is no longer important. However, the procedure of writing a fraction with radicals in a form in which no radicals appear in the denominator, called **rationalizing the denominator,** is also useful for other purposes. Therefore, the following examples illustrate the rationalization of denominators.

EXAMPLE E   The following expressions involve denominators which are already rationalized:

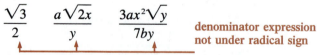

$$\frac{\sqrt{3}}{2} \qquad \frac{a\sqrt{2x}}{y} \qquad \frac{3ax^2\sqrt{y}}{7by}$$

denominator expression not under radical sign

The following expressions involve denominators which are not rationalized:

$$\sqrt{\frac{3}{2}} \qquad \frac{\sqrt{3}}{\sqrt{2}} \qquad 3a\sqrt{\frac{x}{4}}$$

denominator expression under radical sign

□

In some cases we can rationalize the denominator of a fraction by using Eq. (10-8). If the denominator has the form $\sqrt{a}$, then we can multiply the numerator and denominator by $\sqrt{a}$ so that the denominator simplifies to $(\sqrt{a})(\sqrt{a}) = \sqrt{a^2} = a$.

EXAMPLE F

$$\frac{3}{\sqrt{2}} = \frac{3}{\sqrt{2}} \frac{\sqrt{2}}{\sqrt{2}} = \frac{3\sqrt{2}}{2}$$

choose a multiplying factor that eliminates the radical in the denominator

$$\frac{a}{\sqrt{5}} = \frac{a}{\sqrt{5}} \frac{\sqrt{5}}{\sqrt{5}} = \frac{a\sqrt{5}}{5}$$

choose the smallest factor that yields a perfect square

$$\frac{a}{\sqrt{b}} = \frac{a}{\sqrt{b}} \frac{\sqrt{b}}{\sqrt{b}} = \frac{a\sqrt{b}}{b}$$

Other denominators may be rationalized by using the following principle:

$$\sqrt{\frac{a}{b}} = \sqrt{\frac{a \cdot b}{b \cdot b}} = \frac{\sqrt{ab}}{\sqrt{b^2}} = \frac{\sqrt{ab}}{b} \qquad (10\text{-}10)$$

The use of Eq. (10-10) is demonstrated in the following examples.

EXAMPLE G

$$\sqrt{\frac{1}{3}} = \sqrt{\frac{1 \cdot 3}{3 \cdot 3}} = \frac{\sqrt{3}}{\sqrt{3^2}} = \frac{\sqrt{3}}{3}$$

$$\sqrt{\frac{12}{5}} = \sqrt{\frac{4 \cdot 3}{5}} = \sqrt{\frac{4 \cdot 3 \cdot 5}{5 \cdot 5}} = \frac{2\sqrt{3 \cdot 5}}{\sqrt{5^2}} = \frac{2\sqrt{15}}{5}$$

EXAMPLE H

$$\sqrt{\frac{a^2}{b}} = a\sqrt{\frac{1}{b}} = a\sqrt{\frac{b}{b^2}} = \frac{a\sqrt{b}}{b}$$

$$\sqrt{\frac{a^4 b}{c^3}} = \sqrt{\frac{a^4 b}{c^2 c}} = \frac{a^2}{c}\sqrt{\frac{b}{c}} = \frac{a^2}{c}\sqrt{\frac{bc}{c^2}} = \frac{a^2\sqrt{bc}}{c^2}$$

$$\sqrt{\frac{4a^3}{7b}} = 2a\sqrt{\frac{a}{7b}} = 2a\sqrt{\frac{a \cdot 7b}{(7b)^2}} = \frac{2a\sqrt{7ab}}{7b}$$

EXAMPLE I   The time (in seconds) it takes an object to fall 1000 ft can be expressed as

$$\sqrt{\frac{1000}{16}}$$

We can simplify this expression as follows.

$$\sqrt{\frac{1000}{16}} = \sqrt{\frac{100 \cdot 10}{16}} = \sqrt{\frac{100}{16}} \cdot \sqrt{10} = \frac{10\sqrt{10}}{4} = \frac{5\sqrt{10}}{2}$$ □

So far, we have considered the simplification of expressions involving square roots, since they are important in sections that follow. However, we can carry out similar operations with expressions involving other roots. We can generalize Eq. (10-8) and get $\sqrt[n]{ab} = \sqrt[n]{a}\sqrt[n]{b}$. The following example illustrates this.

EXAMPLE J $$\sqrt[3]{16} = \sqrt[3]{8}\sqrt[3]{2} = 2\sqrt[3]{2}$$
                     perfect cube

$$\sqrt[4]{3a^7} = \sqrt[4]{3}\sqrt[4]{a^4}\sqrt[4]{a^3} = (\sqrt[4]{3})(a)(\sqrt[4]{a^3}) = a\sqrt[4]{3a^3}$$
                           perfect fourth power

$$\sqrt[5]{32x^8} = (\sqrt[5]{32})(\sqrt[5]{x^5})(\sqrt[5]{x^3}) = (2)(x)(\sqrt[5]{x^3}) = 2x\sqrt[5]{x^3}$$
                           perfect fifth powers

In these three illustrations, we have taken out the factors which are a perfect cube, a perfect fourth power, and a perfect fifth power, respectively. Note that the powers must be taken into account. We cannot simplify $\sqrt[3]{25}$ since no perfect cube factor is present. It is incorrect to conclude that $\sqrt[3]{25}$ simplifies to 5.   □

Exercises 10-4

In Exercises 1 through 12, rationalize the denominators and simplify.

1. $\dfrac{1}{\sqrt{2}}$ **2.** $\dfrac{1}{\sqrt{3}}$ **3.** $\dfrac{2}{\sqrt{5}}$ **4.** $\dfrac{3}{\sqrt{6}}$

5. $\dfrac{1}{\sqrt{a}}$ **6.** $\dfrac{a}{\sqrt{b}}$ **7.** $\dfrac{\sqrt{a}}{\sqrt{b}}$ **8.** $\dfrac{\sqrt{r}}{\sqrt{2s}}$

9. $\sqrt{\dfrac{3}{5}}$ **10.** $\sqrt{\dfrac{2}{7}}$ **11.** $\sqrt{\dfrac{a^3}{3}}$ **12.** $\sqrt{\dfrac{a^4}{bc}}$

In Exercises 13 through 36, express the given radicals in simplest form. If a radical appears in the denominator, rationalize the denominator.

13. $\sqrt{12}$ **14.** $\sqrt{27}$ **15.** $\sqrt{28}$ **16.** $\sqrt{44}$

$2\sqrt{11}$

17. $\sqrt{45}$ 18. $\sqrt{99}$ 19. $\sqrt{150}$ 20. $\sqrt{98}$

21. $\sqrt{147}$ 22. $\sqrt{162}$ 23. $\sqrt{243}$ 24. $\sqrt{640}$ $= 8\sqrt{10}$

25. $\sqrt{ac^2}$ 26. $\sqrt{3a^4}$ 27. $\sqrt{a^3b^2}$ 28. $\sqrt{12a^5}$

29. $\sqrt{4a^2bc^3}$ 30. $\sqrt{9a^3b}$ 31. $\sqrt{80x^4yz^5}$ 32. $\sqrt{240xy^7z^6}$

33. $\sqrt{\dfrac{ab^2}{12}}$ 34. $\sqrt{\dfrac{c^3e^5}{44}}$ 35. $\sqrt{\dfrac{2x^2y}{5a^8}}$ 36. $\sqrt{\dfrac{13xy^5}{40a^3}}$

In Exercises 37 through 48, simplify the given radicals.

37. $\sqrt[3]{54}$ 38. $\sqrt[3]{24}$ 39. $\sqrt[3]{8a^4}$ 40. $\sqrt[3]{6a^{10}}$

41. $\sqrt[4]{16a^9}$ 42. $\sqrt[4]{81a^4b^5}$ 43. $\sqrt[4]{243a^{11}}$ 44. $\sqrt[5]{64x^7}$

45. $\sqrt[4]{162a^{10}x^{12}}$ 46. $\sqrt[5]{243r^6s^{10}t^{12}}$ 47. $\sqrt[7]{256r^7s^{14}t^{16}}$ 48. $\sqrt[8]{a^9b^{16}c^5}$

In Exercises 49 through 60, solve the given problems.

49. In discussing forces in fluid dynamics, the expression

$$\frac{V}{\sqrt{gl}}$$

is found. Express this in rationalized form.

50. The acoustical velocity of a gas is described by

$$a = \sqrt{\frac{Lrt}{M}}$$

Express this equation in rationalized form.

51. A radar antenna is mounted on a circular plate with area A. The diameter of the plate is given by

$$d = 2\sqrt{\frac{A}{3.14}}$$

Express this equation in rationalized form.

52. The expression for the resonant frequency of a computer's electric circuit is given by

$$f = \frac{1}{2\pi}\sqrt{\frac{1}{LC}}$$

Express this equation in rationalized form.

53. A parking lot is planned as a rectangle with a length twice as long as its width. If its area is 12,800 ft^2, find its dimensions.

54. A beam is in the shape of a rectangular solid with a square top and bottom. The sides are rectangles with the length equal to 36 times the width. If the surface area of a side is 576 in.^{2}, find the volume of the beam.

55. In statistics, the standard deviation of a set of numbers represents a special value. When the standard deviation is added to the mean and subtracted from the mean, the two results contain about two-thirds of all numbers in the original set. In a study of voltage levels, 40 numbers produce a mean of 110.0 and a standard deviation of

$$\sqrt{\frac{250}{40}}$$

Find the levels between which two-thirds of the numbers lie.

56. An accident investigator knows that the stopping distance x of a car is proportional to the square of the speed s of the car before the brakes are applied. One particular car can stop in 400 ft if it is traveling at a speed of 100 ft/s. First determine the equation relating x and s. Then find the speed of the car if it stops in 800 ft. (Assume three-significant-digit accuracy in all values.)

57. The open-sea speed of an oil tanker is determined by

$$V = k\sqrt[3]{\frac{P}{W}}$$

Express this equation in rationalized form.

58. The radius of a spherical droplet of jet fuel can be found by using the formula

$$r = \sqrt[3]{\frac{3V}{4\pi}}$$

Express this equation in rationalized form.

59. The diameter of a crankshaft for a diesel engine is determined by

$$d = k\sqrt[3]{\frac{16J}{C}}$$

Express this equation in rationalized form.

60. In analyzing the rate of fluid flow through a pipe, the expression

$$r = \sqrt[4]{\frac{8nLR}{\pi D}}$$

is encountered. Express this equation in rationalized form.

10-5 Basic Operations with Radicals

Now that we have seen how radicals may be simplified, we can show how the basic operations of addition, subtraction, multiplication, and division are performed on them. These operations follow the basic algebraic operations, the only difference being that radicals are involved. Also, these radicals must be expressed in simplest form so that the final result is in simplest form.

In adding and subtracting radicals, we add and subtract like radicals, just as we add and subtract like terms in the algebraic expressions we have already encountered. Here, like radicals are those that have the same simplest radical form. Consider Example A.

EXAMPLE A   similar radicals

$$\sqrt{3} + 4\sqrt{5} - 3\sqrt{3} + \sqrt{5} = (\sqrt{3} - 3\sqrt{3}) + (4\sqrt{5} + \sqrt{5})$$

similar radicals $= -2\sqrt{3} + 5\sqrt{5}$

However, when working with radicals, we must be certain that all radicals are in simplest form. This is necessary because many radicals that do not appear to be similar in form actually *are* similar, as shown in Example B.

EXAMPLE B $\sqrt{3} + \sqrt{80} - \sqrt{27} + \sqrt{5}$ appears to have four different types of terms, none of them similar. However, noting that

$$\sqrt{80} = \sqrt{16 \cdot 5} = 4\sqrt{5}$$

and

$$\sqrt{27} = \sqrt{9 \cdot 3} = 3\sqrt{3}$$

we have the equivalent expression $\sqrt{3} + 4\sqrt{5} - 3\sqrt{3} + \sqrt{5}$, which is the same expression found in Example A. Now we can see that the third term and the first term are similar, as are the second and fourth terms. We get

$$\sqrt{3} + \sqrt{80} - \sqrt{27} + \sqrt{5} = \sqrt{3} + 4\sqrt{5} - 3\sqrt{3} + \sqrt{5}$$
$$= -2\sqrt{3} + 5\sqrt{5} \quad \square$$

The following examples further illustrate the method of addition and subtraction of radicals.

EXAMPLE C

$$\sqrt{8} + \sqrt{18} - 6\sqrt{2} + 3\sqrt{32} = \sqrt{4 \cdot 2} + \sqrt{9 \cdot 2} - 6\sqrt{2} + 3\sqrt{16 \cdot 2}$$
$$= 2\sqrt{2} + 3\sqrt{2} - 6\sqrt{2} + 12\sqrt{2} \longleftarrow \text{ all similar radicals}$$
$$= 11\sqrt{2}$$

$$\sqrt{20} + 3\sqrt{5} - 7\sqrt{12} + 2\sqrt{45} = \sqrt{4 \cdot 5} + 3\sqrt{5} - 7\sqrt{4 \cdot 3} + 2\sqrt{9 \cdot 5}$$
$$= 2\sqrt{5} + 3\sqrt{5} - 14\sqrt{3} + 6\sqrt{5}$$
$$= 11\sqrt{5} - 14\sqrt{3}$$
$$\underset{\uparrow}{} \text{ not similar to other radicals} \quad \square$$

EXAMPLE D

$$\sqrt{a^2 b} + \sqrt{9b} - 2\sqrt{16c^2 b} = a\sqrt{b} + 3\sqrt{b} - 2(4c)\sqrt{b}$$
$$= (a + 3 - 8c)\sqrt{b}$$

$$\sqrt{3a^3 c^2} - \sqrt{12ab^2} + \sqrt{24ac^3} = \sqrt{(3a)(a^2 c^2)} - \sqrt{(3a)(4b^2)} + \sqrt{(6ac)(4c^2)}$$
$$= ac\sqrt{3a} - 2b\sqrt{3a} + 2c\sqrt{6ac} \quad \square$$

not similar to others
due to factor c

When multiplying expressions containing radicals, we proceed just as in any algebraic multiplication. When simplifying the result, we use the relation

$$\sqrt{a}\sqrt{b} = \sqrt{ab}$$

which is actually Eq. (10-8). The following examples illustrate this principle.

EXAMPLE E

$$\sqrt{3}\sqrt{7} = \sqrt{21}$$
$$\sqrt{2}\sqrt{t} = \sqrt{2t}$$

$$\sqrt{2}(\sqrt{6} + 3\sqrt{5}) = \sqrt{2}\sqrt{6} + 3\sqrt{2}\sqrt{5}$$
$$= \sqrt{12} + 3\sqrt{10}$$
$$= 2\sqrt{3} + 3\sqrt{10}$$

In Example F we illustrate the procedure for multiplying binomial expressions that contain radical terms. We must be careful to use the distributive law correctly so that we include all terms of the result.

EXAMPLE F

$5\sqrt{3}\sqrt{2} - 6\sqrt{3}\sqrt{2}$

$$(\sqrt{3} - 2\sqrt{2})(3\sqrt{3} + 5\sqrt{2}) = 3\sqrt{3}\sqrt{3} - \sqrt{3}\sqrt{2} - 10\sqrt{2}\sqrt{2}$$
$$= 3\sqrt{9} - \sqrt{6} - 10\sqrt{4}$$
$$= 3(3) - \sqrt{6} - 10(2)$$
$$= 9 - \sqrt{6} - 20$$
$$= -11 - \sqrt{6}$$

$\sqrt{a}\sqrt{b} - 6\sqrt{a}\sqrt{b}$

$$(\sqrt{a} - 3\sqrt{b})(2\sqrt{a} + \sqrt{b}) = 2\sqrt{a}\sqrt{a} - 5\sqrt{a}\sqrt{b} - 3\sqrt{b}\sqrt{b}$$
$$= 2\sqrt{a^2} - 5\sqrt{ab} - 3\sqrt{b^2}$$
$$= 2a - 5\sqrt{ab} - 3b$$
$$(\sqrt{2} - 3\sqrt{x})(\sqrt{2} + \sqrt{3}) = \sqrt{2}\sqrt{2} + \sqrt{2}\sqrt{3} - 3\sqrt{x}\sqrt{2} - 3\sqrt{x}\sqrt{3}$$
$$= 2 + \sqrt{6} - 3\sqrt{2x} - 3\sqrt{3x}$$

not similar

In Section 10-4 we learned how to rationalize a fraction in which a radical appears in the denominator. However, we restricted our attention to the simpler cases of rationalization. Example G illustrates rationalizing the denominator of a fraction in which the numerator is a sum of terms.

EXAMPLE G

$$\frac{\sqrt{3} + 5}{\sqrt{2}} = \frac{(\sqrt{3} + 5)}{\sqrt{2}} \cdot \frac{\sqrt{2}}{\sqrt{2}} = \frac{\sqrt{3}\sqrt{2} + 5\sqrt{2}}{\sqrt{4}}$$

$$= \frac{\sqrt{6} + 5\sqrt{2}}{2}$$

If the denominator is the sum of two terms, one or both of which are radicals, the fraction can be rationalized by multiplying both numerator and denominator by the difference of the same two terms.

This is so because

$$(\sqrt{a} + \sqrt{b})(\sqrt{a} - \sqrt{b}) = a - b \qquad \qquad (10\text{-}11)$$

which yields an expression without radicals. Equation (10-11) is actually a result of Eq. (8-2), which states that $(x + y)(x - y) = x^2 - y^2$. *The* **conjugate** *of $a + b$ is $a - b$, and the conjugate of $a - b$ is $a + b$.* In rationalizing a fraction with a denominator having a sum or difference that involves radicals, we can multiply the numerator and denominator by the conjugate of the denominator. We illustrate this in Example H.

EXAMPLE H

1. Rationalize $\dfrac{\sqrt{3}}{\sqrt{2} - \sqrt{5}}$.

   In rationalizing this expression we multiply the numerator and the denominator by $\sqrt{2} + \sqrt{5}$, which is found from the denominator by changing the sign between the terms.

$$\frac{\sqrt{3}}{\sqrt{2} - \sqrt{5}} = \frac{\sqrt{3}}{(\sqrt{2} - \sqrt{5})} \cdot \frac{(\sqrt{2} + \sqrt{5})}{(\sqrt{2} + \sqrt{5})} = \frac{\sqrt{6} + \sqrt{15}}{2 - 5} = -\frac{\sqrt{6} + \sqrt{15}}{3}$$

$$\text{conjugate}$$

2. $$\frac{\sqrt{a}}{2\sqrt{a} + 3} = \frac{\sqrt{a}}{(2\sqrt{a} + 3)} \cdot \frac{(2\sqrt{a} - 3)}{(2\sqrt{a} - 3)} = \frac{2\sqrt{a}\sqrt{a} - 3\sqrt{a}}{2 \cdot 2 \cdot \sqrt{a} \cdot \sqrt{a} - 3 \cdot 3}$$

$$\text{conjugate}$$

$$= \frac{2a - 3\sqrt{a}}{4a - 9}$$

EXAMPLE I   The resistance (in ohms) in an electric circuit can be expressed as

$$\frac{\sqrt{60}}{\sqrt{6} + \sqrt{10}}$$

We can rationalize this expression by using the conjugate as shown.

$$\frac{\sqrt{60}}{(\sqrt{6} + \sqrt{10})} \cdot \frac{(\sqrt{6} - \sqrt{10})}{(\sqrt{6} - \sqrt{10})} = \frac{\sqrt{360} - \sqrt{600}}{6 - 10} = \frac{6\sqrt{10} - 10\sqrt{6}}{-4}$$

$$\underset{\text{conjugate}}{\underbrace{\qquad\qquad\qquad}}$$

$$= \frac{3\sqrt{10} - 5\sqrt{6}}{-2} \qquad \square$$

Exercises 10-5

In Exercises 1 through 48, perform the indicated operations, expressing each answer in simplest form. In Exercises 39 through 48, rationalize the denominators.

1. $\sqrt{7} + 3\sqrt{7} - 2\sqrt{7}$
2. $\sqrt{3} - 2\sqrt{3} + 3\sqrt{3}$
3. $\sqrt{7} - \sqrt{5} + 3\sqrt{7} + 2\sqrt{5}$
4. $\sqrt{11} + 8\sqrt{11} - \sqrt{17} + 3\sqrt{11}$
5. $2\sqrt{3} + \sqrt{27}$
6. $3\sqrt{5} + \sqrt{75}$
7. $2\sqrt{40} + 5\sqrt{10}$
8. $3\sqrt{44} - 2\sqrt{11}$
9. $\sqrt{2}\sqrt{50}$
10. $\sqrt{3}\sqrt{9}$
11. $\sqrt{5}\sqrt{15}$
12. $\sqrt{2}\sqrt{24}$
13. $\sqrt{8} + \sqrt{18} + \sqrt{32}$
14. $\sqrt{12} + \sqrt{27} + \sqrt{48}$
15. $\sqrt{28} - 2\sqrt{63} + 5\sqrt{7}$
16. $2\sqrt{24} + \sqrt{6} - 2\sqrt{54}$
17. $2\sqrt{8} - 2\sqrt{12} - \sqrt{50}$
18. $2\sqrt{20} - \sqrt{44} + 3\sqrt{11}$
19. $\sqrt{a} + \sqrt{9a}$
20. $\sqrt{5x} + \sqrt{20x}$
21. $\sqrt{2a} + \sqrt{8a} + \sqrt{32a^3}$
22. $\sqrt{ac} + \sqrt{4ac} + \sqrt{16a^3c}$
23. $a\sqrt{2} + \sqrt{72a^2} - \sqrt{12a}$
24. $\sqrt{x^2yz} - \sqrt{y^3z} + 2x\sqrt{y^2z}$
25. $\sqrt{3}(\sqrt{7} - 3\sqrt{6})$
26. $\sqrt{5}(\sqrt{20} - 6\sqrt{3})$
27. $\sqrt{2}(\sqrt{8} - \sqrt{32} + 5\sqrt{18})$
28. $\sqrt{7}(\sqrt{14} + 2\sqrt{6} - \sqrt{56})$
29. $\sqrt{a}(\sqrt{ab} + 3\sqrt{ac})$
30. $\sqrt{2a}(\sqrt{8} + \sqrt{6a^3})$
31. $(\sqrt{2} + \sqrt{3})(2\sqrt{2} - \sqrt{3})$
32. $(\sqrt{7} - 2\sqrt{5})(3\sqrt{7} + \sqrt{5})$
33. $(\sqrt{5} - 3\sqrt{3})(2\sqrt{5} + \sqrt{27})$
34. $(2\sqrt{11} - \sqrt{8})(3\sqrt{2} + \sqrt{22})$
35. $(\sqrt{a} - 3\sqrt{c})(2\sqrt{a} + 5\sqrt{c})$
36. $(\sqrt{2b} + 3)(\sqrt{2} - \sqrt{b})$

37. $(\sqrt{3} + 2)^2$
38. $(2\sqrt{3} - 1)^2$
39. $\dfrac{\sqrt{3} + \sqrt{2}}{\sqrt{2}}$
40. $\dfrac{\sqrt{3} + \sqrt{5}}{\sqrt{3}}$

41. $\dfrac{\sqrt{8} + \sqrt{50}}{\sqrt{2}}$
42. $\dfrac{\sqrt{45} - \sqrt{12}}{\sqrt{3}}$
43. $\dfrac{\sqrt{7}}{\sqrt{3} + \sqrt{7}}$
44. $\dfrac{2\sqrt{5}}{\sqrt{5} - \sqrt{11}}$

45. $\dfrac{\sqrt{3} + \sqrt{6}}{2\sqrt{3} - \sqrt{6}}$ **46.** $\dfrac{\sqrt{2} - 3\sqrt{7}}{2\sqrt{2} + \sqrt{7}}$ **47.** $\dfrac{\sqrt{a}}{\sqrt{a} + 2\sqrt{b}}$ **48.** $\dfrac{\sqrt{x}}{2\sqrt{x} - 3\sqrt{y}}$

In Exercises 49 through 56, solve the given problems.

49. The resistance in an electric circuit is described by

$$\frac{\sqrt{R_1 R_2}}{\sqrt{R_1} + \sqrt{R_2}}$$

Rationalize this expression.

50. In determining the pressure of a load placed on two bridge supports, the expression

$$\frac{\sqrt{W}}{\sqrt{L_1} + \sqrt{L_2}}$$

occurs. Rationalize this expression.

51. A surveyor determines the area of a certain parcel of land by evaluating

$$\frac{1}{2}\sqrt{a}(\sqrt{a} + 3\sqrt{a})$$

Find the simplest form of this expression.

52. In discussions of fluid flow through pipes, the expression

$$\frac{h_1 - h_2}{(\sqrt{h_1} + \sqrt{h_2})\sqrt{2g}}$$

arises. Rationalize this expression.

53. In the theory of waves in wires, we may encounter the expression

$$\frac{\sqrt{d_1} - \sqrt{d_2}}{\sqrt{d_1} + \sqrt{d_2}}$$

Evaluate this expression for $d_1 = 10$ and $d_2 = 3$.

54. In the theory dealing with vibratory motion, the equation

$$a^2 - 2al + k^2 = 0$$

is found. Show that $a = l + \sqrt{l^2 - k^2}$ and $a = l - \sqrt{l^2 - k^2}$ both satisfy the equation.

55. Three square pieces of land are along a straight road, as shown in Figure 10-2.
Find the distance x along the road in simplest radical form.

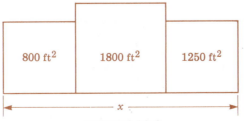

FIGURE 10-2

56. In analyzing the critical speed for a cantilevered shaft, the expression

$$W = \sqrt{\frac{k}{M + 0.23m}}$$

is encountered. Rationalize this equation.

In Exercises 57 through 60, use a calculator to evaluate the given expression. Express all results with three significant digits.

57. $\sqrt{3}(\sqrt{5} + \sqrt{7})$

58. $(\sqrt{3} + \sqrt{10})(\sqrt{3} - \sqrt{2})$

59. $\dfrac{\sqrt{5}}{\sqrt{11} + \sqrt{2}}$

60. $\dfrac{\sqrt{5} - \sqrt{2}}{\sqrt{21} + \sqrt{3}}$

10-6 Fractional Exponents

So far, the only numbers used for exponents have been positive integers, negative integers, and zero. In this section we will show how fractions may also be used as exponents. We will see that these fractional exponents are very convenient in certain topics of mathematics. We discuss this subject at this time because, as we shall see, an expression raised to a fractional exponent can be interpreted as a radical. This allows us to represent many radical expressions as expressions raised to fractional exponents. This principle enables us to use calculators and computers for evaluating many expressions which, if left in radical form, would be extremely difficult to manipulate or evaluate.

For fractional exponents to be meaningful, they must satisfy the basic laws of exponents already established. The significance of fractional exponents is established in the following example.

EXAMPLE A   If $a^{1/3}$ is to have a meaning consistent with the previous laws of exponents, then

$$a^{1/3}a^{1/3}a^{1/3} = a^{1/3 + 1/3 + 1/3} = a$$

and

$$(a^{1/3})^3 = a$$

However, we have already established that

$$(\sqrt[3]{a})^3 = a$$

This means that $a^{1/3}$ must be interpreted as $\sqrt[3]{a}$ for the basic laws of exponents to hold.   □

Generalizing the conclusion of Example A, *we define $a^{1/n}$ to mean the nth principal root of a, or*

$$a^{1/n} = \sqrt[n]{a} \qquad\qquad\qquad (10\text{-}12)$$

The following example gives several illustrations of the meaning of Eq. (10-12).

EXAMPLE B

$$4^{1/2} = \sqrt{4} = 2 \qquad\qquad (27)^{1/3} = \sqrt[3]{27} = 3$$

$$(-32)^{1/5} = \sqrt[5]{-32} = -2 \qquad (x^4)^{1/2} = \sqrt{x^4} = \sqrt{(x^2)^2} = x^2$$

Normally it is not necessary, and in fact it is often cumbersome, to include the radical interpretation when evaluating expressions with fractional exponents. Thus the preceding results would usually be written as

$$4^{1/2} = 2 \qquad (27)^{1/3} = 3 \qquad (-32)^{1/5} = -2 \qquad (x^4)^{1/2} = x^2 \qquad \square$$

EXAMPLE C Scientific calculators usually have a special square root key, but many lack keys for cube root, fourth root, fifth root, and so on. To evaluate $\sqrt[4]{6561}$ using a calculator, first express that value using a fractional exponent as in $6561^{1/4}$. We now see that finding the fourth root of 6561 is equivalent to raising 6561 to the exponent of $\frac{1}{4}$, or 0.25. Entering the sequence of keys

6561 $\boxed{y^x}$ 0.25 $\boxed{=}$ $\boxed{\qquad\qquad 9.}$

shows that $\sqrt[4]{6561} = 9$.

We can evaluate $\sqrt[3]{3375}$ by entering

3375 $\boxed{y^x}$ 3 $\boxed{1/x}$ $\boxed{=}$ $\boxed{\qquad\qquad 15.}$

In this example the exponent of $\frac{1}{3}$ was entered by pressing the key for 3 and then pressing the reciprocal key (labeled $1/x$) so that the exponent becomes the equivalent of $\frac{1}{3}$. We conclude that $\sqrt[3]{3375} = 15$. $\square$

When an expression of the form $a^{1/n}$ is raised to some power, say m, the basic laws of exponents should still be valid. Thus

$$(a^{1/n})^m = a^{(1/n)(m)} = a^{m/n}$$

The meaning of $a^{m/n}$ is shown in the following example.

EXAMPLE D $(a^{1/3})^2 = a^{2/3}$, which can be written as $(a^2)^{1/3}$. Thus $(a^{1/3})^2 = (a^2)^{1/3}$, or $(\sqrt[3]{a})^2 = \sqrt[3]{a^2}$. ☐

Generalizing the result of Example D, we see that

$$a^{m/n} = (\sqrt[n]{a})^m = \sqrt[n]{a^m}$$ (10-13)

The meaning of $a^{m/n}$ is further illustrated in the following examples.

EXAMPLE E

$8^{2/3} = (8^{1/3})^2 = 2^2 = 4$ $\qquad$ $4^{7/2} = (4^{1/2})^7 = 2^7 = 128$

$(27)^{4/3} = [(27)^{1/3}]^4 = 3^4 = 81$ $\qquad$ $(x^4)^{3/2} = [(x^4)^{1/2}]^3 = (x^2)^3 = x^6$ ☐

In Example E, although $(8^{1/3})^2 = (8^2)^{1/3}$, it is usually easier to take the root first and then raise the result to the power.

EXAMPLE F $8^{2/3}$ written in radical form is $\sqrt[3]{8^2}$ or $(\sqrt[3]{8})^2$. Since $\sqrt[3]{8^2} = \sqrt[3]{64} = 4$ and $(\sqrt[3]{8})^2 = 2^2 = 4$, we see that $8^{2/3} = \sqrt[3]{8^2} = (\sqrt[3]{8})^2$.
Also, $4^{7/2} = \sqrt{4^7} = (\sqrt{4})^7$. ☐

EXAMPLE G Scientific calculators can be used to evaluate expressions of the form $a^{m/n}$. To evaluate $25^{3/2}$, enter the following sequence of keys:

25 $\boxed{y^x}$ $\boxed{(}$ 3 $\boxed{\div}$ 2 $\boxed{)}$ $\boxed{=}$ $\boxed{\quad\quad\quad 125.\quad}$

The result will be 125. In this sequence of keys, we used parentheses to group the $3 \div 2$ which represents the exponent of $\frac{3}{2}$.

To use a calculator for evaluating $\sqrt[5]{32^3}$, use Eq. (10-13) to express this value as $32^{3/5}$ and then enter the key sequence

32 $\boxed{y^x}$ $\boxed{(}$ 3 $\boxed{\div}$ 5 $\boxed{)}$ $\boxed{=}$ $\boxed{\quad\quad\quad 8.\quad}$

to get a result of 8. Therefore $\sqrt[5]{32^3} = 8$.

To use a calculator for evaluating $\sqrt[9]{37.8^4}$, we first express that number as $37.8^{4/9}$ and then enter the key sequence

37.8 $\boxed{y^x}$ $\boxed{(}$ 4 $\boxed{\div}$ 9 $\boxed{)}$ $\boxed{=}$ $\boxed{\quad 5.0246691\quad}$

to get a result of 5.0246691. Since the original number has three significant digits, our result should also have three significant digits and we therefore interpret the calculator display to be 5.02. ☐

The following examples illustrate the use of the basic laws of exponents with expressions which have fractional exponents.

EXAMPLE H

$$8^{-2/3} = \frac{1}{8^{2/3}} = \frac{1}{(8^{1/3})^2} = \frac{1}{2^2} = \frac{1}{4}$$

$$(8^{4/5})^0 = 1$$

$$2^{1/2}2^{1/3} = 2^{1/2 + 1/3} = 2^{5/6}$$

$$\frac{2^{1/2}}{2^{1/3}} = 2^{1/2 - 1/3} = 2^{1/6}$$

$$\frac{4^{-1/2}}{2^{-5}} = \frac{2^5}{4^{1/2}} = \frac{32}{2} = 16$$

☐

EXAMPLE I

$$x^{1/2}x^{1/4} = x^{1/2 + 1/4} = x^{3/4}$$

$$\frac{a}{a^{1/2}b^{-2}} = a^{1 - 1/2}b^2 = a^{1/2}b^2$$

$$\frac{x^{-1/3}y^{2/3}}{x^{2/3}y^{-2/3}} = \frac{y^{2/3 + 2/3}}{x^{2/3 + 1/3}} = \frac{y^{4/3}}{x}$$

☐

EXAMPLE J

In order for an aircraft wing to produce the required lift of 1000 N/m², the velocity of air over the upper wing surface is found by applying Bernoulli's equation and evaluating

$$V_1 = (2L + RV_2^2)^{1/2} R^{-1/2}$$

Find V_1 (in meters per second) if the velocity on the lower wing surface is $V_2 = 108$ m/s, the air density is $R = 1.29$ kg/m³, and the lift is $L = 1000$ N/m².

First, evaluating $2L + RV_2^2$, we get $2(1000) + (1.29)(108)^2 = 17050$. We now evaluate

$$V_1 = (2L + RV_2^2)^{1/2} R^{-1/2}$$

$$= (17050)^{1/2}(1.29)^{-1/2} = \frac{17050^{1/2}}{1.29^{1/2}}$$

$$= \frac{\sqrt{17050}}{\sqrt{1.29}} = 115 \text{ m/s}$$

The velocity of air over the upper wing surface must be 115 m/s.

☐

Exercises 10-6

In Exercises 1 through 8, change the given expressions to radical form.

1. $5^{1/2}$ **2.** $7^{1/3}$ **3.** $a^{1/4}$ **4.** $b^{1/5}$

5. $x^{3/5}$ **6.** $x^{3/4}$ **7.** $R^{7/3}$ **8.** $s^{9/7}$

In Exercises 9 through 16, use fractional exponents to express the following:

9. $\sqrt[3]{a}$ **10.** $\sqrt[7]{b}$ **11.** $\sqrt{x}$ **12.** $\sqrt[5]{ax}$

13. $\sqrt[3]{x^2}$ **14.** $\sqrt[4]{a^3}$ **15.** $\sqrt[5]{b^8}$ **16.** $\sqrt[10]{s^5}$

In Exercises 17 through 48, evaluate the given expressions.

17. $9^{1/2}$ **18.** $36^{1/2}$ **19.** $64^{1/3}$ **20.** $125^{1/3}$

21. $16^{1/4}$ **22.** $243^{1/5}$ **23.** $(-8)^{1/3}$ **24.** $(-64)^{1/3}$

25. $4^{3/2}$ **26.** $9^{3/2}$ **27.** $8^{4/3}$ **28.** $27^{2/3}$

29. $81^{3/4}$ **30.** $16^{3/4}$ **31.** $(-8)^{2/3}$ **32.** $(-27)^{5/3}$

33. $36^{-1/2}$ **34.** $100^{-1/2}$ **35.** $16^{-1/4}$ **36.** $32^{-2/5}$

37. $-(-32)^{-1/5}$ **38.** $-64^{-2/3}$ **39.** $(4^{1/2})(27^{2/3})$ **40.** $(25^{1/2})(100^{3/2})$

41. $\dfrac{64^{1/2}}{64^{1/3}}$ **42.** $\dfrac{81^{1/4}}{81^{1/2}}$ **43.** $2^{-1/2} \cdot 2^{3/2}$ **44.** $\dfrac{5^{1/2}}{5^{-1/2}}$

45. $(4^{-2})(27^{-1/3})$ **46.** $\dfrac{25^{-3/2}}{81^{3/4}}$ **47.** $49^{-1/2} - 3(14^{-1})$ **48.** $(125)^{-2/3} + 4(25)^{-1/2}$

In Exercises 49 through 56, use the laws of exponents to simplify the given expressions. Express all answers with positive exponents.

49. $a \cdot a^{1/2}$ **50.** $\dfrac{b}{b^{1/3}}$ **51.** $\dfrac{ab}{a^{1/4}}$ **52.** $\dfrac{xyz^2}{z^{-3/2}}$

53. $\dfrac{x(x^{1/3})^4}{x^{2/5}}$ **54.** $\dfrac{xy^{1/2}}{(y^{2/3})^{-2}}$ **55.** $\dfrac{x^{-1/2}x^2}{x^{2/3}}$ **56.** $\dfrac{xy^{-1/2}z}{z^{-2/3}x^{-1/2}}$

In Exercises 57 through 68, solve the given problems.

57. A restaurant finds that it can make better use of its floor space by using tables with shapes described by the equation

$$\sqrt{x^5} + \sqrt{y^5} = k$$

Express this equation with fractional exponents instead of radicals.

58. In determining the magnetic field within a coil of wire, the expression

$$kNia^2(a^2 + b^2)^{-3/2}$$

is used. Rewrite this expression in radical form.

59. Use fractional exponents to express the side x of a square in terms of area A of the square.

60. Use fractional exponents to express the edge x of a cube in terms of the volume V of the cube.

61. An expression which arises in Einstein's theory of relativity is

$$\sqrt{\frac{c^2 - v^2}{c^2}}$$

Express this with the use of fractional exponents and then simplify.

62. In analyzing curvature in the deflection of a beam, the equation

$$R = \frac{(1 + D_1^2)^{3/2}}{D_2}$$

is found. Express this equation in radical form.

63. An expression encountered with a certain electronic device is

$$r_p = \frac{2K^{-2/3}I^{-1/3}}{3D}$$

Write this expression without negative exponents.

64. Considering only the volume of gasoline vapor and air mixture in the cylinder, the efficiency of an internal combustion engine is approximately

$$E = 100\left(1 - \frac{1}{R^{2/5}}\right)$$

where E is in percent and R is the compression ratio of the engine. Rewrite this equation in radical form.

65. In determining the number of electrons involved in a certain calculation with semiconductors, the expression

$$9.60 \times 10^{18}\, T^{3/2}$$

is used. Evaluate this expression for $T = 289$ K.

66. Three lead cubes are used for ballast. Their volumes are V_1 cm³, V_2 cm³, V_3 cm³, and they are stacked on top of each other. Express the height of the pile using fractional exponents.

67. In studying the magnetism of a circular loop, the equation

$$B = \frac{2\pi kIR^2}{(x^2 + R^2)^{3/2}}$$

is found. Simplify this expression for the center of the loop, where $x = 0$.

68. Kepler's third law of planetary motion may be stated as follows: *The mean radius (about the sun) of any planet is proportional to the $\frac{2}{3}$ power of the period of that planet.* Using the facts that the earth has a mean radius of about 9.3×10^7 mi and its period is 1 year, calculate the mean radius of Saturn, assuming that its period is 27 years. (Actually it is about 29.5 years, but the answer obtained is only about 6% in error.)

In Exercises 69 through 76, use a scientific calculator to evaluate the given expressions. Express all results to the nearest tenth.

69. $\sqrt[4]{19.6}$ **70.** $\sqrt[3]{257.1}$ **71.** $\sqrt[5]{23.6^3}$ **72.** $\sqrt[7]{34.56^4}$

73. $121.4^{2/3}$ **74.** $67.9^{7/4}$ **75.** $(32)^{2/5}(17)^{1/3}$ **76.** $\dfrac{123.5^{4/9}}{203.4^{5/7}}$

Chapter 10 Formulas

$$(a^m)(a^n) = a^{m+n} \qquad\qquad\qquad\qquad\qquad\qquad (7\text{-}1)$$

$$(a^m)^n = a^{mn} \qquad\qquad\qquad\qquad\qquad\qquad (7\text{-}2)$$

$$(ab)^n = a^n b^n \qquad\qquad\qquad\qquad\qquad\qquad (7\text{-}3)$$

$$\frac{a^m}{a^n} = a^{m-n} \qquad (a \neq 0) \qquad\qquad\qquad\qquad (7\text{-}4)$$

$$\left(\frac{a}{b}\right)^n = \frac{a^n}{b^n} \qquad (b \neq 0) \qquad\qquad\qquad\qquad (9\text{-}3)$$

$$a^{-n} = \frac{1}{a^n} \qquad (a \neq 0) \qquad\qquad\qquad\qquad (10\text{-}1)$$

$$a^0 = 1 \qquad (a \neq 0) \qquad\qquad\qquad\qquad (10\text{-}2)$$

$$\sqrt{N} = x \qquad\qquad\qquad\qquad\qquad\qquad\qquad (10\text{-}3)$$

$$j^2 = -1 \text{ or } j = \sqrt{-1} \qquad\qquad\qquad\qquad\qquad (10\text{-}4)$$

$$(bj)^2 = -b^2 \qquad\qquad\qquad\qquad\qquad\qquad\qquad (10\text{-}5)$$

$$\sqrt{-a} = \sqrt{a}\,j \qquad\qquad\qquad\qquad\qquad\qquad (10\text{-}6)$$

$$\sqrt[n]{N} = x \qquad\qquad (\text{where } x^n = N) \qquad\qquad (10\text{-}7)$$

$$\sqrt{ab} = \sqrt{a}\sqrt{b} \qquad (a \text{ and } b \text{ are positive}) \qquad (10\text{-}8)$$

$$\sqrt{a^2} = a \qquad (a > 0) \qquad\qquad\qquad\qquad\qquad (10\text{-}9)$$

$$\sqrt{\frac{a}{b}} = \sqrt{\frac{a \cdot b}{b \cdot b}} = \frac{\sqrt{ab}}{\sqrt{b^2}} = \frac{\sqrt{ab}}{b} \qquad\qquad (10\text{-}10)$$

$$(\sqrt{a} + \sqrt{b})(\sqrt{a} - \sqrt{b}) = a - b \qquad\qquad\qquad (10\text{-}11)$$

$$a^{1/n} = \sqrt[n]{a} \qquad\qquad\qquad\qquad\qquad\qquad\qquad (10\text{-}12)$$

$$a^{m/n} = (\sqrt[n]{a})^m = \sqrt[n]{a^m} \qquad\qquad\qquad\qquad (10\text{-}13)$$

10-7 Review Exercises for Chapter 10

In Exercises 1 through 32, evaluate the given expressions.

1. 10^{-1}

2. 2^{-4}

3. $\dfrac{1}{3^{-2}}$

4. $\dfrac{1}{4^{-1}}$

5. $3^0 6^{-1}$

6. $\dfrac{9^0}{4^{-3}}$

7. $\sqrt{169}$

8. $-\sqrt{900}$

9. $\sqrt[3]{125}$ 10. $-\sqrt[3]{-125}$ 11. $-\sqrt{256}$ 12. $\sqrt{0.0196}$

13. $\sqrt{\dfrac{1}{16}}$ 14. $\sqrt{\dfrac{4}{25}}$ 15. $-\sqrt{\dfrac{9}{121}}$ 16. $-\sqrt{\dfrac{144}{169}}$

17. $100^{1/2}$ 18. $625^{1/2}$ 19. $1000^{1/3}$ 20. $8000^{1/3}$

21. $49^{3/2}$ 22. $243^{3/5}$ 23. $121^{3/2}$ 24. $8^{7/3}$

25. $\dfrac{16^{3/4}}{27^{2/3}}$ 26. $\dfrac{25^{3/2}}{5^{-1}}$ 27. $\dfrac{(-216)^{2/3}}{3^{-2}}$ 28. $\dfrac{81^{-3/4}}{32^{2/5}}$

29. $\sqrt{-81}$ 30. $\sqrt{-144}$ 31. $-\sqrt{-0.64}$ 32. $-\sqrt{-0.01}$

In Exercises 33 through 40, write the given numbers in scientific notation.

33. 570 34. 60800 35. 3.25 36. 0.103

37. 0.000769 38. 7750000 39. 86.95 40. 1024

In Exercises 41 through 44, change the given numbers from scientific notation to ordinary notation.

41. 3×10^3 42. 6×10^{-4} 43. 3.14×10^5 44. 6.75×10^{-5}

In Exercises 45 through 60, write the given expressions in simplest form, expressing all results with positive exponents.

45. $3a^{-2}b$ 46. $2xy^{-1}$ 47. $\dfrac{mn^{-2}}{m^{-3}}$ 48. $\dfrac{2rs^{-1}}{t^{-5}}$

49. $\dfrac{2x^3y^{-1}}{3x^{-2}y^2}$ 50. $\dfrac{(2a)^0(b^{-1}c)}{4a^2bc^{-3}}$ 51. $\dfrac{(-a)^{-2}xy^0}{(-x)^{-1}y^{-3}}$ 52. $\dfrac{3^{-1}(ab^{-1})^{-2}}{(ab^0)^{-1}}$

53. $a^{1/4}a^{1/3}$ 54. $x^{1/5}x^{2/3}$ 55. $\dfrac{a^{2/3}}{a^{-1/2}}$ 56. $\dfrac{b^{-1}}{b^{-1/2}}$

57. $(xy^{-2})^{1/2}$ 58. $(8x^{-3}y^{3/2})^{1/3}$ 59. $\dfrac{(st^{1/2})^{2/3}}{t^{-2}}$ 60. $\dfrac{(16c^2)^{3/4}}{ac^{-1/5}}$

In Exercises 61 through 84, express the given radicals in simplest form. Rationalize the denominator if a radical appears in the denominator.

61. $\sqrt{44}$ 62. $\sqrt{27}$ 63. $\sqrt{72}$ 64. $\sqrt{54}$

65. $\sqrt{128}$ 66. $\sqrt{124}$ 67. $\sqrt[3]{40}$ 68. $\sqrt[3]{108}$

69. $\sqrt{\dfrac{1}{11}}$ 70. $\sqrt{\dfrac{6}{11}}$ 71. $\sqrt{\dfrac{4}{7}}$ 72. $\sqrt{\dfrac{36}{37}}$

73. $\sqrt{4a^2}$ 74. $\sqrt{28a}$ 75. $\sqrt{125b^2c}$ 76. $\sqrt{90b^3}$

77. $\sqrt{\dfrac{6}{a}}$ 78. $\sqrt{\dfrac{3}{a^2}}$ 79. $\sqrt{\dfrac{28}{3a}}$ 80. $\sqrt{\dfrac{400,000}{ab}}$

81. $\sqrt[3]{16a^3}$ 82. $\sqrt[3]{81x^2y^4}$ 83. $\sqrt[5]{64a^8}$ 84. $\sqrt[7]{64a^8}$

In Exercises 85 through 108, perform the indicated operations and simplify. Rationalize the denominator if a radical appears in the denominator.

85. $\sqrt{63} - 2\sqrt{28}$ 86. $2\sqrt{45} - 3\sqrt{80}$ 87. $3\sqrt{7} - 2\sqrt{6} + \sqrt{28}$

88. $5\sqrt{3} - \sqrt{27} - \sqrt{15}$ 89. $2\sqrt{40} - 3\sqrt{90} + \sqrt{70}$ 90. $\sqrt{12} - 3\sqrt{27} + 2\sqrt{80}$

91. $\sqrt{2a^2} + 3\sqrt{8} - \sqrt{32a^2}$ 92. $\sqrt{20a} + 4\sqrt{5a^3} - 3\sqrt{45a}$ 93. $\sqrt{2}(\sqrt{6} - 2\sqrt{24})$

94. $\sqrt{3}(3\sqrt{6} + \sqrt{54})$ 95. $2\sqrt{5}(\sqrt{3} - 3\sqrt{5})$ 96. $3\sqrt{2}(5\sqrt{2} + 2\sqrt{6})$

97. $\sqrt{a}(\sqrt{ab} - 3\sqrt{5b})$

98. $\sqrt{ab}(\sqrt{b} - 3\sqrt{a})$

99. $(\sqrt{6} - \sqrt{5})(2\sqrt{6} - 3\sqrt{5})$

100. $(3\sqrt{7} + 4\sqrt{2})(\sqrt{7} - 3\sqrt{2})$

101. $(2\sqrt{6} - 3\sqrt{7})(5\sqrt{6} - \sqrt{7})$

102. $(3\sqrt{11} - \sqrt{3})(\sqrt{11} + 2\sqrt{3})$

103. $(\sqrt{a} - 3\sqrt{b})(2\sqrt{a} + \sqrt{b})$

104. $(\sqrt{ab} - \sqrt{c})(2\sqrt{ab} + \sqrt{c})$

105. $\dfrac{\sqrt{2}}{\sqrt{5} - \sqrt{2}}$

106. $\dfrac{3}{\sqrt{7} - 2\sqrt{3}}$

107 $.\dfrac{\sqrt{2} - 1}{2\sqrt{2} + 3}$

108. $\dfrac{\sqrt{11} - \sqrt{5}}{2\sqrt{11} + \sqrt{5}}$

In Exercises 109 through 132, solve the given problems.

109. The communications link to a certain space capsule operates at a frequency of 27,500,000,000 Hz. Write this number in scientific notation.

110. A computer can retrieve 15,600,000 units of data in one second. Write this number in scientific notation.

111. A film of oil on water is approximately 0.0000002 in. thick. Write this number in scientific notation.

112. One calendar year consists of exactly 31,536,000 s. Round this number to three significant digits and express the result in scientific notation.

113. The charge on an electron is about 1.6×10^{-19} C. Write this number in ordinary notation.

114. Light travels 3.1×10^{16} ft in one year. Write this number in ordinary notation.

115. Some gamma rays have wavelengths of 5×10^{-13} m. Write this number in ordinary notation.

116. The density of steam is about 6×10^{-7} kg/m³. Write this in ordinary notation.

117. A 10-kg ball dropped from a height of 100 m reaches the ground with a kinetic energy of 9.8×10^3 J. Write this in ordinary notation.

118. The center of gravity of a half-ring is found by using the expression

$$(r^2 + 4R^2)(2\pi R)^{-1}$$

where r is the radius of the cross section and R is the radius of the ring. Write this expression without negative exponents.

119. An expression for the focal length of a certain lens is

$$[(\mu - 1)(r_1^{-1} - r_2^{-1})]^{-1}$$

Rewrite this expression without the use of negative exponents.

120. The speed of sound through a medium is given by

$$v = \sqrt{\frac{E}{d}}$$ where E and d are constants depending on the medium. Rationalize this expression.

121. In analyzing electrode properties in a cathode ray tube, the expression

$$v = \sqrt{\frac{2eV}{m}}$$ is found. Express this equation so that the right-hand side is in rationalized form.

122. In studying the effects of charges propelled through a magnetic field, the following expression is found.

$$y = Bx^2\left(\frac{q}{8mV}\right)^{1/2}$$

Rewrite the right-hand side of this equation using radicals and then express it in rationalized form.

123. The radius (in inches) of a floppy disk is expressed in terms of its top surface area as follows.

$$r = \frac{\sqrt{441\pi}}{\sqrt{64\pi}}$$

Rationalize and simplify the right-hand side of this equation.

124. An expression encountered in electronics is

$$\frac{1}{\sqrt{1 + \left(\dfrac{\omega_0}{\omega}\right)^2}}$$ Express this in simplest rationalized form.

125. The voltage across a resistor is found by using the formula $V = IR$, where V is the voltage, I is the current in amperes, and R is the resistance in ohms. Find V given that $I = 7.4$ mA and $R = 8.6 \times 10^2$ Ω.

126. To compute the volume V of a helium-filled balloon which can support a mass m, we use the formula

$$V = \frac{m}{d_a - d_h}$$

Here d_a is the density of air and d_h is the density of helium. Find V, given that $m = 650$ kg, $d_a = 1.29 \times 10^{-3}$ kg/m³, and $d_h = 0.18 \times 10^{-3}$ kg/m³.

127. In studying the properties of biological fluids, the expression $0.036M^{3/4}$ is used. Evaluate this expression for $M = 1.6 \times 10^5$.

128. An expression encountered in the study of the flow of fluids in pipes is

$$\frac{0.38}{N^{1/5}}$$ Evaluate this expression, given that $N = 3.2 \times 10^6$.

129. In analyzing the plastic deformation within single crystals, the expression

$$\left(\frac{a}{\sqrt{2}}\right) \div \left(\frac{a\sqrt{6}}{2}\right)$$ is derived. Simplify and express in rationalized form.

130. The velocity of sound in air, in meters per second, is proportional to the square root of $(273 + T)/273$, where T is the temperature in degrees Celsius. Given that $v = 331$ m/s when $T = 0°$C, find the equation relating the velocity of sound and temperature in rationalized form.

131. The density of an object equals its weight divided by its volume. A certain metal has a density of 1331 lb/ft³. What is the edge of a cube of this metal which weighs 8.00 lb?

132. The thermodynamic temperature of the filament of a light bulb equals approximately 1000 times the fourth root of the wattage. Find the temperature in degrees Celsius ($273°$ less than the thermodynamic temperature) of the filament of a 25-W bulb.

In Exercises 133 through 144, use a scientific calculator to perform the indicated operations.

133. $(23.26 \times 10^4)(32.15 \times 10^{12})$ (Express answer in scientific notation.)
134. $(0.0132 \times 10^{-3})(0.586 \times 10^{-4})$ (Express answer in scientific notation.)
135. $(132.45 \times 10^{-7}) \div (131.86 \times 10^{-9})$ (Express answer in scientific notation.)
136. $(3.294 \times 10^5)(2.685 \times 10^7) \div (9.773 \times 10^9)$ (Express answer in scientific notation.)
137. $\sqrt[5]{136.73}$ (Express answer to the nearest hundredth.)
138. $\sqrt[3]{1008.6}$ (Express answer to the nearest tenth.)
139. $\sqrt[4]{19.76^3}$ (Express answer to the nearest hundredth.)
140. $\sqrt[8]{37.235^2}$ (Express answer to the nearest hundredth.)
141. $154.3^{2/3}$ (Express answer with four significant digits.)
142. $263^{-3/4}$ (Express answer with three significant digits.)
143. $365^{-1.2}$ (Express answer with three significant digits.)
144. $\dfrac{25^{2/3}}{17^{4/11}}$ (Express answer with two significant digits.)

Quadratic Equations

In Chapter 5 we introduced the basic methods of solving simple equations. Since then, additional algebraic operations have been discussed so that we are now ready to solve the important **quadratic equation**. In Section 11-1 we will identify quadratic equations and discuss what is meant by their solution. In Sections 11-2 and 11-3 we will discuss basic methods of solving such equations.

Quadratic equations are used in many applications. For example, consider an aviation accident investigator who is analyzing an aircraft accident caused by an exploding Thermos of hot tea. It is known that at an altitude of h ft above sea level, the boiling point of water is lower by an amount (in degrees Fahrenheit) that can be found by solving the quadratic equation

$$T^2 + 520T - h = 0$$

If a light aircraft climbs to 5300 ft from sea level, what is the boiling point of water? We will solve this and other such applied problems in this chapter.

11-1 The Quadratic Equation

Given that a, b, and c are constants, the equation

$$ax^2 + bx + c = 0 \quad \text{where } a \neq 0 \qquad (11\text{-}1)$$

is called **the general quadratic equation in x**.

Since it is the term in x^2 which distinguishes the quadratic equation from other types of equations, *if $a = 0$ the equation is not considered to be quadratic*. However, either b, or c, or both may be zero and the equation is quadratic. Equation (11-1) therefore has this stipulation: $a \neq 0$, but b and c may be any numbers.

EXAMPLE A   Since the equation $2x^2 + 3x + 7 = 0$ is in the form of Eq. (11-1), it is a quadratic equation, where $a = 2$, $b = 3$, and $c = 7$.

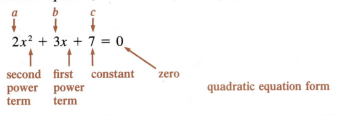

The equation $4x^2 - 7x + 9 = 0$ is also in the form of Eq. (11-1), despite the presence of the minus sign, since we may identify $a = 4$, $b = -7$, and $c = 9$. That is,

$$4x^2 - 7x + 9 = 4x^2 + (-7)x + 9.$$

The equation $x^3 - 9x + 8 = 0$ is *not* of the form of Eq. (11-1) because of the term x^3.   □

EXAMPLE B   The equation $4x - 9 = 0$ is not a quadratic equation. It would fit the form of Eq. (11-1) only if $a = 0$, and such an equation is not quadratic.

The equation $3x^2 - 19 = 0$ is a quadratic equation, where $a = 3$, $b = 0$, and $c = -19$.

The equation $x^2 - 8x = 0$ is a quadratic equation, where $a = 1$, $b = -8$, and $c = 0$.

The equation $x^2 = 0$ is a quadratic equation, where $a = 1$, $b = 0$, and $c = 0$.   □

When a quadratic equation is already in the form of Eq. (11-1) with descending powers of x, it is said to be in **standard form**.

If basic operations can be performed on an equation so that it may be written in a form equivalent to Eq. (11-1), it is then a quadratic equation. Once this standard form is found, the values of a, b, and c may be determined. Consider the illustrations of the following example.

EXAMPLE C   The equation $3x^2 - 6 = 7x$ is not of the form of Eq. (11-1), but it may easily be put in this standard form by subtracting $7x$ from each side of the equation. Performing this operation, we get

$$3x^2 - 7x - 6 = 0$$
$$\uparrow \quad \uparrow \quad \uparrow$$
$$a \quad \ b \quad \ \ c$$

which means $a = 3$, $b = -7$, and $c = -6$.

The equation $2x^2 = (x - 8)^2$ is not in the form of Eq. (11-1). To determine whether or not it may be put in this form, we must square the right-hand side as indicated. By then collecting terms on the left, we can establish the form of the equation. This leads to

$$2x^2 = x^2 - 16x + 64$$

or

$$x^2 + 16x - 64 = 0 \qquad \text{standard form}$$
$$\uparrow \qquad \ \uparrow \qquad \ \ \uparrow$$
$$a = 1 \qquad b \qquad \ \ c$$

This last equation is in the standard form of Eq. (11-1), where $a = 1$, $b = 16$, and $c = -64$.

The equation $3x^2 = (3x - 1)(x + 2)$ becomes

$$3x^2 = 3x^2 + 5x - 2 \qquad \text{or} \qquad -5x + 2 = 0$$

We see that this last form is not the same as Eq. (11-1) so that the equation is not quadratic. □

We recall from Chapter 5 that *a solution of an equation is a value of the variable which, when substituted into the equation, makes the two sides of the equation equal.* If a value of x is substituted into Eq. (11-1), the left-hand side must result in zero if that value of x is a solution. As we shall see in Section 11-2, *the solution of a quadratic equation is generally a pair of numbers, although occasionally only one number satisfies the equation.* In any case, there cannot be more than two numbers which satisfy a quadratic equation. The following example illustrates checking possible values of x as solutions of a quadratic equation. Note that we are not yet finding solutions to quadratic equations. This next example simply illustrates how to *check* potential solutions after they have been identified.

EXAMPLE D   The shape of a parabolic antenna is related to the equation $2x^2 + 5x - 3 = 0$. Determine which, if any, of the given values $x = 1$, $x = -3$, $x = \frac{1}{2}$, $x = 2$ are solutions of that equation.

Testing $x = 1$, we get

$$\overbrace{}^{x = 1}$$
$$2(1)^2 + 5(1) - 3 = 2 + 5 - 3 = 4$$

Since the resulting value is not zero, substitution of 1 for x does *not* make the equation true and $x = 1$ is not a solution.

Testing $x = -3$, we get

$$2(-3)^2 + 5(-3) - 3 = 2(9) - 15 - 3 = 18 - 15 - 3 = 0$$

Since the value is zero, substitution of -3 for x does make the equation true and $x = -3$ is a solution.

Testing $x = \frac{1}{2}$, we get

$$2\left(\frac{1}{2}\right)^2 + 5\left(\frac{1}{2}\right) - 3 = 2\left(\frac{1}{4}\right) + \frac{5}{2} - 3 = \frac{1}{2} + \frac{5}{2} - 3 = 3 - 3 = 0$$

Since the value is zero, substitution of $\frac{1}{2}$ for x does make the equation true and $x = \frac{1}{2}$ is also a solution. Since we have now found two such values, we know that the solutions to this equation are $x = -3$ and $x = \frac{1}{2}$. Any other value of x cannot be a solution since there can be at most two different solutions. The value of $x = 2$, for example, cannot be a solution since we already have two solutions. $\qquad\square$

Our primary concern is with quadratic equations which have real roots. However, it is possible that the solution of a quadratic equation can contain numbers with $j = \sqrt{-1}$. (Such numbers are discussed at greater length in Chapter 20.) It is also possible that the real solutions of a quadratic equation are equal, so that only one value satisfies the equation.

EXAMPLE E The equation $x^2 + 4 = 0$ has the solutions of $2j$ and $-2j$. This can be verified by substitution:

$$(2j)^2 + 4 = 4j^2 + 4 = 4(-1) + 4 = -4 + 4 = 0$$
$$(-2j)^2 + 4 = (-2)^2 j^2 + 4 = 4(-1) + 4 = -4 + 4 = 0$$

The equation $x^2 + 4x + 4 = 0$ has the solution $x = -2$ only. The two solutions here are equal. The reason for this will be seen in Section 11-2. $\qquad\square$

Exercises 11-1

In Exercises 1 through 8, write a quadratic equation in the form of Eq. (11-1) by using the given values for the coefficients a, b, c.

1. $a = 4, b = -3, c = 2$ **2.** $a = 1, b = 0, c = 9$ **3.** $a = -1, b = 8, c = 0$

4. $a = 9, b = 6, c = 12$ **5.** $a = 1, b = 0, c = 0$ **6.** $a = 2, b = 0, c = 0$

7. $a = 1, b = -1, c = -1$ **8.** $a = -3, b = 8, c = -7$

In Exercises 9 through 20, determine whether or not the given equations are quadratic by performing algebraic operations that could put each in the form of Eq. (11-1). If the standard quadratic form of Eq. (11-1) is obtained, identify a, b, and c.

9. $x^2 - 7x = 4$

10. $3x^2 = 5 - 9x$

11. $x^2 = (x - 1)^2$

12. $2x^2 - x = 2x(x + 8)$

13. $(x + 2)^2 = 0$

14. $x(x^2 - 1) = x^3 + x^2$

15. $x^2 = x(1 - 6x)$

16. $x(1 + 2x) = 3x - 2x^2$

17. $x^2(1 - x) = 4$

18. $x(x^2 + 2x - 1) = 0$

19. $3x(x^2 - 2x + 1) = 1 + 3x^3$

20. $x(x^3 + 6) - 1 = x^2(x^2 + 1)$

In Exercises 21 through 36, if the equation is not of the form of Eq. (11-1), put it in that form. Then test the given values to determine which, if any, are solutions of the equation.

21. $x^2 - 5x + 6 = 0$; $x = 1, x = 2, x = 3$

22. $x^2 + x - 6 = 0$; $x = 0, x = 2, x = 3$

23. $x^2 - 4x + 4 = 0$; $x = 2, x = -2, x = 4$

24. $x^2 + 1 = 0$; $x = 0, x = -1, x = 1$

25. $x^2 - x - 2 = 0$; $x = 1, x = -1, x = 2$

26. $x^2 + 4x + 3 = 0$; $x = -1, x = 1, x = -3$

27. $2x^2 - 3x + 1 = 0$; $x = -1, x = 0, x = \dfrac{1}{2}, x = 1$

28. $3x^2 - x - 2 = 0$; $x = 1, x = 2, x = -\dfrac{2}{3}, x = -1$

29. $V^2 + V = 12$; $V = -1, V = -3, V = 3, V = -4$

30. $y^2 = 3y + 10$; $y = 2, y = -1, y = 5$

31. $(3x - 4)(x - 2) = 0$; $x = -2, x = 1, x = 3$

32. $3t^2 + t + 6 = 4(1 - t)$; $t = -\dfrac{2}{3}, t = -1, t = 1$

33. $s^2 = 4(s - 1)$; $s = -2, s = 1, s = 2$

34. $x(x + 4) = 2x$; $x = -3, x = -2, x = 0, x = 1$

35. $n^2 + 6n - 3 = 6(n + 1)$; $n = -3, n = -1, n = 3$

36. $x^2 + 16 = 0$; $x = -4, x = 2, x = 4$

In Exercises 37 through 44, solve the given problems.

37. A computer simulation program involves a particle whose path is described by the equation

$$y = -x^2 + 6x - 8$$

Given that $y = -3$, find the resulting quadratic equation. Show that $x = 5$ is a solution of the quadratic equation.

38. A manufacturer approximates the cost of producing x items by the equation

$$c = -5x^2 + 8x + 600$$

Given that $c = 468$, find the resulting quadratic equation. Show that $x = 6$ is a solution of the quadratic equation.

39. An object is shot upward so that its distance above the ground is given by

$$s = 96t - 16t^2$$

where t is the time in seconds. Given that $s = 128$ ft, find the resulting quadratic equation. Show that $t = 2.00$ s is a solution of the equation.

40. The value of an account after two years is described by

$$V = P(1 + r)^2$$

where P is the initial amount invested and r is the interest rate in decimal form. If $P = 500$ and $V = 605$, show that the resulting equation is quadratic and show that $r = 0.1$ is a solution.

41. Conforming to a zoning ordinance, a store owner wants to make a rectangular sign with an area of 48.0 ft². He wants the length l to exceed the width w by 2.0 ft. Show that the resulting equation involving the width is quadratic. See Figure 11-1.

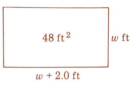

42. To find the current in a certain alternating-current circuit, it is necessary to solve the equation

$$m^2 + 10m + 2500 = 0$$

Show that

$$-5 + 15j\sqrt{11} \quad \text{and} \quad -5 - 15j\sqrt{11}$$

are the solutions of this equation.

FIGURE 11-1

43. A hole of radius r is drilled in the center of a circular disk with radius $2r$ and the remaining area is 4 cm². Show that this statement leads to a quadratic equation. See Figure 11-2.

44. If the sides of a square of side s are doubled, the area is increased by 12 m². Show that this statement leads to a quadratic equation. See Figure 11-3.

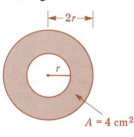

FIGURE 11-2

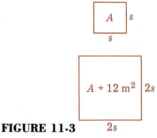

FIGURE 11-3

11-2 Solution by Factoring

In solving a quadratic equation by the method of factoring, we should begin by putting the quadratic equation in the standard form of Eq. (11-1) ($ax^2 + bx + c = 0$). If the left-hand side is factorable, the solution is easily obtained. Once the factors are found, each is individually set equal to zero and the solution reduces to solving two simple equations of the type discussed in Chapter 5. The reason for this is that *the product of the factors is zero if at least one of the factors is zero.* That is, if $pq = 0$ then $p = 0$ or $q = 0$. The procedure is summarized in these steps:

1. Arrange the quadratic equation so that it is in the standard form of Eq. (11-1) ($ax^2 + bx + c = 0$).
2. Factor the left-hand side.
3. Individually *set each of the two factors equal to zero and solve.*
4. Check both solutions by substituting them in the original equation.

EXAMPLE A   Solve the equation $x^2 - x - 2 = 0$.

We first write the left side in factored form as follows.

$$(x - 2)(x + 1) = 0$$

The product of these two factors is zero if at least one of them is zero. Setting each factor equal to zero and solving the resulting equations, we get

$$x - 2 = 0 \qquad x + 1 = 0$$
$$x = 2 \qquad\qquad x = -1$$

The solutions are $x = 2$ and $x = -1$. Checking each of these values, we get

$$2^2 - 2 - 2 = 4 - 2 - 2 = 0$$

and

$$(-1)^2 - (-1) - 2 = 1 + 1 - 2 = 0$$

Checking the solution is strongly recommended.   □

In Example A, if $x - 2 = 0$, the left-hand side may be written $0 \cdot (x + 1)$, and this product is zero regardless of the value of $x + 1$. Therefore we see that setting this factor equal to zero should give us a solution to the original equation. The same is true if $x + 1 = 0$. We must keep in mind, however, that *the equation must be written in the standard form of Eq. (11-1) before we factor the left-hand side. This is necessary since we must have zero on the right*. If any number other than zero appears on the right, this method doesn't work.

EXAMPLE B   Solve the equation $x^2 - 10x = -21$.

The correct procedure is to first get a form that is equivalent to Eq. (11-1).

$$x^2 - 10x + 21 = 0$$

which factors into $(x - 7)(x - 3) = 0$. In this way, we get the correct solutions $x = 7$ and $x = 3$.

Many students would be tempted to factor the left-hand side into $x(x - 10)$ and then set $x = -21$ and $x - 10 = -21$. The "solutions" obtained, $x = -21$ and $x = -11$, are not correct, as can be verified by substitution.   □

EXAMPLE C Solve the equation $6x^2 = 5 - 7x$.

Following the procedure described above, we get

$$6x^2 + 7x - 5 = 0 \longleftarrow \begin{array}{l}\text{to have 0 on right}\\ \text{add } 7x - 5 \text{ to each side}\end{array}$$

$$(2x - 1)(3x + 5) = 0 \qquad \text{factor}$$

$$2x - 1 = 0 \qquad 3x + 5 = 0 \qquad \text{set each factor} = 0$$

$$2x = 1 \qquad\qquad 3x = -5 \qquad \text{solve}$$

$$x = \frac{1}{2} \qquad\qquad x = -\frac{5}{3}$$

To be certain that no improper algebraic steps have been taken, it is best to check each solution in the original equation, even though it is not in the form of Eq. (11-1). Checking, we obtain the following:

$$6\left(\frac{1}{2}\right)^2 \overset{?}{=} 5 - 7\left(\frac{1}{2}\right) \qquad 6\left(-\frac{5}{3}\right)^2 \overset{?}{=} 5 - 7\left(-\frac{5}{3}\right)$$

$$6\left(\frac{1}{4}\right) \overset{?}{=} 5 - \frac{7}{2} \qquad 6\left(\frac{25}{9}\right) \overset{?}{=} 5 + \frac{35}{3}$$

$$\frac{3}{2} = \frac{3}{2} \qquad\qquad \frac{50}{3} = \frac{50}{3}$$

Since the values on each side are equal, the solutions check. That is, substitution of the solutions will make the original equation true so that the solutions are verified. ☐

In Section 11-1 we mentioned that *the two solutions, or* **roots** *of the equation, can be equal. This is true when the two factors are the same.* The following example illustrates this type of quadratic equation.

EXAMPLE D Solve the equation $4x^2 - 12x + 9 = 0$.

Factoring this equation, we obtain

$$(2x - 3)(2x - 3) = 0$$

Setting each factor equal to zero gives

$$2x - 3 = 0$$

$$x = \frac{3}{2}$$

Since both factors are the same, the equation has a **double root** of $\frac{3}{2}$. Checking verifies the solution. ☐

Many students improperly solve a quadratic equation in which $c = 0$. The following example illustrates this type and the error often made in its solution.

EXAMPLE E   Solve the equation $x^2 - 6x = 0$.

Noting that the two terms of this equation contain x, *a common error is to divide through by x*. This results in the equation $x - 6 = 0$ and the solution $x = 6$. However, *the solution $x = 0$ was lost through the division by x*. Instead of dividing through by x, we should factor the equation into

$$x(x - 6) = 0$$

By setting each factor equal to zero, we obtain

$$x = 0 \qquad x - 6 = 0$$
$$x = 6$$

There are *two solutions*: $x = 0$ and $x = 6$. Checking verifies these roots. □

A number of different verbally stated problems may lead to quadratic equations. The following example illustrates setting up and solving such problems.

EXAMPLE F   Above sea level, the boiling point of water is a lower temperature than at sea level where it is 212°F. The difference (in Fahrenheit degrees) is given by the approximate equation

$$T^2 + 520T - h = 0$$

where h is the altitude in feet. Find the approximate boiling point of water in a light airplane flying at 5300 ft.

With $h = 5300$ ft, the quadratic equation becomes

$$T^2 + 520T - 5300 = 0$$

Factoring is not easy with numbers of this type, but the above expression does factor as follows

$$(T - 10)(T + 530) = 0 \qquad \text{factor}$$

Proceeding, we set each factor equal to zero and solve.

$$T - 10 = 0 \qquad T + 530 = 0 \qquad \text{set each factor} = 0$$
$$T = 10 \qquad\quad T = -530 \quad \text{solve}$$

We are finding the amount by which the temperature is *lowered* so that only the positive solution of $T = 10$ is reasonable. Since the 212°F boiling point is lowered by 10°F, we have a boiling point of 202°F at 5300 ft. (Some pilots have been injured by exploding Thermos containers when their hot contents began to boil at higher altitudes.) □

Exercises 11-2

In Exercises 1 through 32, solve the given quadratic equations by factoring.

1. $x^2 - 9 = 0$
2. $4x^2 - 25 = 0$
3. $x^2 + x - 2 = 0$
4. $x^2 - 5x + 6 = 0$
5. $t^2 - 3t = 10$
6. $s^2 = 6s + 7$
7. $3x^2 + 5x - 2 = 0$
8. $4x^2 - 7x + 3 = 0$
9. $2x^2 + 3x = 2$
10. $3y^2 = 8 - 10y$
11. $6x^2 + 13x - 5 = 0$
12. $6x^2 - 7x - 20 = 0$
13. $6n^2 - n = 2$
14. $2v^2 + 30 = 17v$
15. $5x^2 + 4 = 21x$
16. $9p^2 + 20 = -27p$
17. $2x^2 = 9 + 7x$
18. $6x^2 + 11x = 10$
19. $x^2 + 4x + 4 = 0$
20. $x^2 + 9 = 6x$
21. $R^2 - 8R = 0$
22. $3t^2 + 5t = 0$
23. $8m^2 + 24m = 14$
24. $18x^2 - 3x = 6$
25. $4x^2 + 49 = 28x$
26. $9x^2 + 30x + 25 = 0$
27. $R^2 = 7R$
28. $4r^2 = 48r$
29. $8m + 3 = 3m^2$
30. $3 - x = 4x^2$
31. $x^2 - 4a^2 = 0$ (*a* is constant)
32. $2x^2 - 8ax + 8a^2 = 0$ (*a* is constant)

In Exercises 33 through 44, solve any resulting quadratic equations by factoring.

33. The number x of items produced by a certain company and the corresponding profit are related by the equation

$$P = -x^2 + 19x - 34$$

Find the "break-even" point. That is, find the number of items for which the profit P is zero.

34. The kinetic energy of a moving car can be found from

$$k = \frac{1}{2}mv^2$$

where k is the kinetic energy, m is the mass of the car, and v is its velocity. If $k = 18{,}519$ kg·m²/s² and $m = 1200$ kg, find the velocity (in m/s) of the car.

35. Under certain conditions, the motion of an object suspended by a helical spring requires the solution of the equation

$$D^2 + 8D + 12 = 0$$

Solve for D.

36. The perimeter of a rectangular garden is 70 m and the area is 250 m². Find the dimensions of the garden. See Figure 11-4.

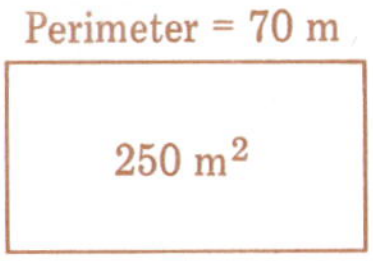

37. At an altitude of h ft above sea level, the boiling point of water is lower by a certain number of degrees than the boiling point at sea level, which is 212°F. The *difference* is given by the approximate equation

$$T^2 + 520T - h = 0$$

Compute the approximate boiling point of water on the top of Mt. Baker in Washington (altitude about 10,800 ft). See Example F.

Perimeter = 70 m

250 m²

FIGURE 11-4

38. The distance an object falls due to gravity is given by $s = 16t^2$, where s is the distance in feet and t is the time in seconds. How long does it take an object to fall 100 ft?

39. Find the dimensions of a cube that has a total surface area of 150 cm².

40. Under specified conditions, the deflection of a beam requires the solution of the equation

$$4Lx - x^2 - 4L^2 = 0$$

where x is the distance from one end and $2L$ is the length of the beam. Solve for x in terms of L.

41. By increasing its speed by 200 mi/h, a jet can cover 6000 mi in 1 h less. What is the speed of the jet?

42. A nuclear power plant supplies a fixed power level at a constant voltage, and the current is determined from the equation

$$100I^2 - 1700I + 1600 = 0$$

Solve for I.

43. A box for shipping sheet metal screws must be designed so that it contains a volume of 84 ft³. Various considerations require that the box be 3 ft wide, and the height must be 3 ft less than the length. Find the dimensions of the box. See Figure 11-5.

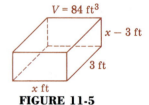

44. A circular floppy disk has a radius of 2.625 in. What radius would cause the surface area to be quadrupled?

FIGURE 11-5

11-3 The Quadratic Formula

See Appendix B for a computer program that solves quadratic equations.

Many quadratic equations cannot be solved by factoring, but we can use a special formula to solve any quadratic equation. We begin by presenting this formula and illustrating its use, and then we shall justify it and show how it is developed.

Equation (11-2) is called the **quadratic formula**. If an equation is

$$x = \frac{-b \pm \sqrt{b^2 - 4ac}}{2a} \tag{11-2}$$

in the form of Eq. (11-1) ($ax^2 + bx + c = 0$), then direct substitution of the appropriate values a, b, and c into Eq. (11-2) gives the solutions. Examination of Eq. (11-2) shows that the symbols $\pm$ precede the radical. This indicates that *there are two solutions, one for the + sign and one for the − sign* of the $\pm$ which precedes the radical. The following examples illustrate the use of Eq. (11-2).

EXAMPLE A Solve the equation $x^2 - 5x + 6 = 0$ by the quadratic formula.

$$a = 1 \quad b = -5 \quad c = 6$$

Since the equation is in the proper form, we recognize that $a = 1$, $b = -5$, and $c = 6$. Therefore

$$x = \frac{-(-5) \pm \sqrt{(-5)^2 - 4(1)(6)}}{2(1)} = \frac{5 \pm \sqrt{25 - 24}}{2} = \frac{5 \pm 1}{2}$$

Therefore

$$x = \frac{5 + 1}{2} = 3 \quad \text{and} \quad x = \frac{5 - 1}{2} = 2$$

These roots could have been found by factoring, but we see that the same results are obtained by using the formula. The roots are easily checked through substitution in the original equation. □

A very common error in applying Eq. (11-2) is to place the denominator of $2a$ under the radical, but not under the $-b$ term in the numerator. Such an error in Example A would have resulted in the *incorrect* solutions of

$$x = 5 + \frac{1}{2} = 5.5 \quad \text{and} \quad x = 5 - \frac{1}{2} = 4.5$$

▶ *It is important to always divide the denominator of $2a$ into* **both** *terms of the numerator, not just the radical term.*

EXAMPLE B Solve $2x^2 + 7x = 3$ by the quadratic formula.

Before the formula can be used, the equation must be put in the form of Eq. (11-1). This gives

$$2x^2 + 7x - 3 = 0$$

so that $a = 2$, $b = 7$, and $c = -3$. The solutions are

$$x = \frac{-7 \pm \sqrt{49 - 4(2)(-3)}}{2(2)} = \frac{-7 \pm \sqrt{49 + 24}}{4} = \frac{-7 \pm \sqrt{73}}{4}$$

This form of the result is generally acceptable. If decimal approximations are required, then, using $\sqrt{73} = 8.544$, we get

$$x = \frac{-7 + 8.544}{4} = 0.386 \quad \text{and} \quad x = \frac{-7 - 8.544}{4} = -3.886 \quad □$$

The calculations shown at the end of Example B can be easily performed on a scientific calculator. When using a calculator for solving quadratic equations with the quadratic formula, it is best to begin by evaluating $b^2 - 4ac$. Then take the square root of that result before proceeding. The solution described by

$$x = \frac{-b + \sqrt{b^2 - 4ac}}{2a}$$

is depicted below.

$$\underbrace{\sqrt{b^2 - 4ac}} \qquad \underbrace{+ \, (-b)} \qquad \div \qquad \underbrace{2a}$$

$$b\,\boxed{x^2}\,\boxed{-}\,4\,\boxed{\times}\,a\,\boxed{\times}\,c\,\boxed{=}\,\boxed{\sqrt{x}}\,\boxed{+}\,b\,\boxed{+/-}\,\boxed{=}\,\boxed{\div}\,2\,\boxed{\div}\,a\,\boxed{=}$$

The other solution is obtained by making two changes in the given sequence of key strokes: First, replace the $\boxed{+}$ key by the $\boxed{-}$ key. Second, press the change sign key $\boxed{+/-}$ at the end. Also, memory storage keys can be used to store the value of $b^2 - 4ac$ so that it can be used for the second solution.

EXAMPLE C
▶

Solve the equation $3x^2 = 2x - 5$ by the quadratic formula.

Putting the equation in the proper form as $3x^2 - 2x + 5 = 0$, we have $a = 3$, $b = -2$, and $c = 5$. Therefore

$$x = \frac{-(-2) \pm \sqrt{(-2)^2 - 4(3)(5)}}{2(3)} = \frac{2 \pm \sqrt{4 - 60}}{6}$$

$$= \frac{2 \pm \sqrt{-56}}{6} = \frac{2 \pm 2\sqrt{-14}}{6} = \frac{1 \pm \sqrt{-14}}{3}$$

Since $\sqrt{-14} = j\sqrt{14}$, we see that the result contains imaginary numbers. *There are no real roots.* The two solutions can be expressed as

$$\frac{1 + j\sqrt{14}}{3} \quad \text{and} \quad \frac{1 - j\sqrt{14}}{3}$$

or as

$$\frac{1}{3} + j\frac{\sqrt{14}}{3} \quad \text{and} \quad \frac{1}{3} - j\frac{\sqrt{14}}{3} \qquad \square$$

When the solution to a quadratic equation involves imaginary numbers, the two solutions will always be identical except for the sign change which precedes the imaginary part. Such complex numbers that differ only in the sign which precedes the imaginary part are called **conjugates**. The complex numbers $a + bj$ and $a - bj$ are conjugates.

There are many real problems that require quadratic equations. In Example D we illustrate one such problem.

EXAMPLE D A nuclear waste holding facility is situated on a rectangular parcel of land which is 100 m long and 80 m wide. A safety zone in the form of a uniform strip constitutes the outer limits of this parcel as shown in Figure 11-6. The facility itself has the interior area of 6000 m². Find the width of the safety zone. (Assume an accuracy of two significant digits.)

In Figure 11-6 we let x = the width of the safety zone. The interior area of the facility itself is rectangular with length = $100 - 2x$, width = $80 - 2x$, and area = 6000 m². Using the area formula for a rectangle we get

$$\text{length} \quad \times \quad \text{width} \quad = \text{area}$$
$$(100 - 2x)(80 - 2x) = 6000$$

Simplifying, we get

$$8000 - 200x - 160x + 4x^2 = 6000$$
$$4x^2 - 360x + 2000 = 0$$
$$x^2 - 90x + 500 = 0$$

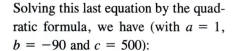

FIGURE 11-6

Solving this last equation by the quadratic formula, we have (with $a = 1$, $b = -90$ and $c = 500$):

$$x = \frac{-(-90) \pm \sqrt{(-90)^2 - 4(1)(500)}}{2(1)}$$
$$= \frac{90 \pm \sqrt{8100 - 2000}}{2}$$
$$= \frac{90 \pm \sqrt{6100}}{2} = \frac{90 \pm 10\sqrt{61}}{2} = 45 \pm 5\sqrt{61}$$

Since $\sqrt{61} = 7.810$, it follows that $5\sqrt{61} = 39.05$. Thus $x = 6.0$ and $x = 84$. However, x cannot be 84 since that would make the safety zone wider than the whole parcel, which means that this answer cannot be true. If the safety zone is 6.0 m wide, the interior area is 88 m by 68 m, and this area is 6000 m². This means that the safety zone is approximately 6.0 m wide. Although the mathematical solution yields two answers, only one of them is valid in this situation. □

So far we have presented and illustrated the use of Eq. (11-2), the quadratic formula. In the next example we will solve a quadratic equation by another method. Using this particular method, we will be able to derive the very important quadratic formula.

EXAMPLE E   Solve the equation $x^2 + 4x - 21 = 0$.

This equation can easily be solved by factoring. However, there is an alternative *method* in which we are interested.

The terms $x^2 + 4x$ are two of the terms we may obtain by expanding $(x + 2)^2$. In fact, if 4 is added to $x^2 + 4x$, we have $(x + 2)^2$ exactly. Therefore let us write the given equation as

$$x^2 + 4x \qquad = 21$$

Adding 4 to each side gives us

$$x^2 + 4x + 4 = 21 + 4$$

This means that another form of the original equation is

$$(x + 2)^2 = 25$$

If we now take the square roots of each side, we get

$$x + 2 = +5 \quad \text{or} \quad x + 2 = -5$$

here we have $+5$ and -5 since the principal square root and its negative both satisfy the *original equation*. Solving each of these equations gives us the solutions $x = 3$ and $x = -7$, which can be verified by checking.

The key to this solution is that the left-hand side is a perfect square, which allows us to solve simple equations after we find the square root of both sides.   □

The method used to find the solution in Example E is called **completing the square**. The basic idea of this method is to create the perfect square of a simple algebraic expression on the left, allowing the right to have any numerical value. Since $(x + a)^2 = x^2 + 2ax + a^2$, we see that the square of one-half the coefficient of x added to $x^2 + 2ax$ gives such a perfect square. We shall now use this method to derive the quadratic formula.

First we start with the general quadratic equation

$$ax^2 + bx + c = 0 \qquad\qquad \textbf{(11-1)}$$

Next we divide through by a, obtaining

$$x^2 + \frac{b}{a}x + \frac{c}{a} = 0$$

Let us subtract c/a from each side, which gives

$$x^2 + \frac{b}{a}x = -\frac{c}{a}$$

▶ Now *to complete the square we take one-half of b/a, which is b/2a, square it, obtaining b²/4a², and add this to each side of the equation to* get

$$x^2 + \frac{b}{a}x + \frac{b^2}{4a^2} = \frac{b^2}{4a^2} - \frac{c}{a}$$

The left-hand side is now the perfect square of $(x + b/2a)$. Indicating this and combining fractions on the right, we have

$$\left(x + \frac{b}{2a}\right)^2 = \frac{b^2 - 4ac}{4a^2}$$

Taking the square root of each side, we get

$$x + \frac{b}{2a} = \frac{\sqrt{b^2 - 4ac}}{2a} \quad \text{and} \quad x + \frac{b}{2a} = \frac{-\sqrt{b^2 - 4ac}}{2a}$$

Subtracting $b/2a$ from each side and combining fractions we get

$$x = \frac{-b + \sqrt{b^2 - 4ac}}{2a} \quad \text{and} \quad x = \frac{-b - \sqrt{b^2 - 4ac}}{2a}$$

These two solutions are combined into one expression which is Eq. (11-2), the quadratic formula. When we apply the quadratic formula, we eliminate the intermediate steps which are included in completing the square, but the results will be the same.

In this chapter we have presented three methods for solving quadratic equations:

1. Factoring.
2. Quadratic formula (Eq. 11-2).
3. Completing the square.

As a practical consideration, the method of completing the square is rarely used since the quadratic equation is more efficient and yields the same results. In choosing between factoring and the quadratic formula, it is generally a good strategy to use factoring if the quadratic expression is factorable. If the quadratic equation cannot be solved by factoring, then the quadratic formula should be used. Here is a quick test which reveals if the quadratic equation can be solved by factoring: First put the equation in the form of Eq. (11-1) and calculate the value of $b^2 - 4ac$. *If that value is zero or a positive number which is a perfect square, then factoring can be used; otherwise the quadratic formula should be used.* Figure 11-7 on the next page summarizes the strategy for solving quadratic equations. This strategy helps to prevent wasted time and effort with attempts to factor expressions which cannot be factored.

EXAMPLE F Following the strategy of Figure 11-7, we can solve the quadratic equation $3x^2 - 5x + 2 = 0$ by factoring since $b^2 - 4ac = 25 - 4(3)(2) = 1$ is a perfect square. However, the quadratic equation $5x^2 - 3x - 4 = 0$ can be solved with the quadratic formula since $b^2 - 4ac = 9 - 4(5)(-4) = 9 + 80 = 89$ is not a perfect square so that *factoring will not be possible.*

☐

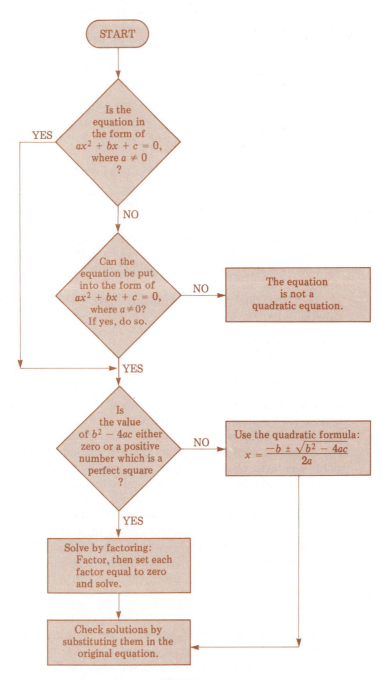

FIGURE 11-7

Exercises 11-3

In Exercises 1 through 28, solve the quadratic equations by using the quadratic formula.

1. $x^2 - 2x - 3 = 0$ **2.** $x^2 + 3x - 10 = 0$ **3.** $2x^2 + 7x + 3 = 0$ **4.** $3x^2 - 5x - 2 = 0$

5. $x^2 + 5x + 3 = 0$ **6.** $x^2 - 3x - 1 = 0$ **7.** $s^2 = 4s + 2$ **8.** $n^2 - 2n = 6$

9. $4t^2 = 8t - 3$ **10.** $3x^2 = x + 10$ **11.** $9x^2 - 16 = 0$ **12.** $3R^2 = 75$

13. $4x^2 - 12x = 7$ **14.** $2x^2 + 7x + 2 = 0$ **15.** $R^2 + 2R = -5$ **16.** $t^2 - t = -2$

17. $t^2 - t = 8$ **18.** $x^2 - 8 = 5x$ **19.** $4x^2 = 8 - 2x$ **20.** $3u^2 = 18 - 6u$

21. $I^2 - 7I = 0$ **22.** $5s^2 = 7s$ **23.** $2t^2 - 3t = -8$ **24.** $4x^2 = 9x - 6$

25. $10t + 8 = 3t^2$ **26.** $7 - 15x = -2x^2$

27. $6a^2x^2 + 11ax + 3 = 0$ (*a* is constant)   **28.** $2x^2 + (a + 2)x + a = 0$ (*a* is constant)

In Exercises 29 through 32, solve the quadratic equations by the method of completing the square. In Exercise 32, use p, q, and r as constants.

29. $x^2 + 2x - 3 = 0$ **30.** $x^2 + 6x - 7 = 0$ **31.** $2x^2 + 3x + 1 = 0$ **32.** $px^2 + qx + r = 0$

In Exercises 33 through 44, solve by means of the quadratic formula any quadratic equations which may arise.

33. Find two consecutive positive integers whose product is 272.

34. When a number is added to its square, the result is 240. Find the number.

35. The perimeter of a rectangular sign is 58 cm and its area is 204 cm². Find its dimensions.

36. Two different machines are mounted on square plates. The side of one square exceeds the side of the other square by 2.00 ft. The total area of the two squares is 50.0 ft². Find the side of each square plate.

37. Under certain conditions, the distance *s* that an explosion propels an object above the ground is given by

$$s = 143t - 16t^2$$

where *t* is the time in seconds. After what amount of time is the object 60.0 ft above the ground?

38. Under specified conditions, the power developed in an element of an electric circuit in a smoke alarm is

$$P = EI - RI^2$$

where *P* is the power, *E* is a specified voltage, and *R* is a specified resistance. Assuming that *P*, *E*, and *R* are constants, solve for *I*.

39. In studying the effects of gasoline vapors, a fire fighter finds that under certain conditions, the partial pressure *P* of a certain gas (in atmospheres) is found by solving the equation

$$P^2 - 3P + 1 = 0$$

Solve and then choose the value of *P* which is less than one atmosphere.

40. To find the current in an alternating-current circuit containing an inductance *L*, a resistance *R*, and a capacitance *C*, it is necessary to solve the equation

$$LCm^2 + RCm + 1 = 0$$

for *m*. Solve this equation.

41. The mechanical power developed in an electric motor is given by

$$P = T - I^2R$$

Find the current *I* (in amperes) if the power *P* is 448 W, the total power *T* is 480 W, and the resistance *R* is 2.00 Ω.

42. If the radius of a circular solar cell is doubled, the area increases by 85 cm². Find the original radius.

43. A farmer has a rectangular field 400 m by 500 m. On this field, a uniform strip on the outside will be left unplanted so that half of the total area will be planted. How wide is the strip? See Figure 11-8.

44. A rectangular piece of sheet metal has a length that is twice the width. In each corner, a square of side 3 in. is cut out, and the outer strips are then bent up to form an open box with volume 168 in.³. Find the dimensions of the original sheet. See Figure 11-9.

FIGURE 11-8   500 m   2x   **FIGURE 11-9**

In Exercises 45 through 48, use a scientific calculator and the quadratic formula to solve the given quadratic equations. Round all answers to the nearest hundredth.

45. $3.22x^2 + 7.34x + 2.11 = 0$ **46.** $17.16x^2 + 29.77x - 8.36 = 0$

47. $3.89x^2 - 16.15x - 0.83 = 0$ **48.** $127.36x^2 - 85.11x = 101.55$

Chapter 11 Formulas

$$ax^2 + bx + c = 0 \quad \text{where } a \neq 0 \qquad \text{general quadratic equation} \qquad \text{(11-1)}$$

$$x = \frac{-b \pm \sqrt{b^2 - 4ac}}{2a} \qquad \text{quadratic formula} \qquad \text{(11-2)}$$

11-4 Review Exercises for Chapter 11

In Exercises 1 through 12, solve the given quadratic equations by factoring.

1. $x^2 + 7x + 12 = 0$ **2.** $x^2 - 7x - 8 = 0$ **3.** $2x^2 - 11x - 6 = 0$

4. $4x^2 = 4x + 3$ **5.** $6n^2 - 35n - 6 = 0$ **6.** $8s^2 - 9s = 0$

7. $16x^2 - 24x + 9 = 0$ **8.** $4p^2 + 49 = 28p$ **9.** $15R^2 + 21R = 0$

10. $9r^2 - 55r + 6 = 0$ **11.** $t^2 + t = 110$ **12.** $2x^2 + 39x = 20$

In Exercises 13 through 20, solve the given quadratic equations by (a) factoring and (b) the quadratic formula.

13. $x^2 + 5x - 24 = 0$ **14.** $x^2 - 4x - 21 = 0$ **15.** $x^2 + 9 = -6x$

16. $2x^2 + 3 = 5x$ **17.** $6m^2 + 5 = 11m$ **18.** $7x^2 + 24x = 16$

19. $3x^2 - 6x + 3 = 0$ **20.** $5x^2 + 40x = 0$

In Exercises 21 through 32, solve the given quadratic equations by using the quadratic formula.

21. $x^2 + 4x + 2 = 0$ **22.** $x^2 + x - 1 = 0$ **23.** $2x^2 + 5x - 1 = 0$

24. $3r^2 + 4r - 2 = 0$ **25.** $y^2 - 6y = 6$ **26.** $5t^2 = 5t + 3$

27. $2x^2 + x + 5 = 0$ **28.** $3u^2 - u = -9$ **29.** $x^2 + 4 = 0$

30. $4y^2 = -25$ **31.** $4x^2 = 36$ **32.** $5p^2 = 35p$

In Exercises 33 through 40, solve by factoring if possible. Otherwise, use the quadratic formula. Follow the strategy outlined in Figure 11-7.

33. $x^2 + 8x - 9 = 0$ **34.** $x^2 + 3x = 5$ **35.** $2x^2 - 5 = 2x$ **36.** $4x^2 + 16 = 16x$

37. $x^2 = 7 + 2x$ **38.** $2x^2 = 3x + 5$ **39.** $2x + 8 = 5x^2$ **40.** $3x^2 + 4x = 4$

Many equations containing algebraic fractions may be solved as quadratic equations. By first clearing fractions (see Section 9-5), the quadratic form may be derived. Exercises 41 through 44 illustrate this type of equation. In each, solve for x ($\neq$ means "does not equal").

41. $\dfrac{1}{x} + x = 4$ $(x \neq 0)$ **42.** $\dfrac{1}{3} + \dfrac{1}{x} = x$ $(x \neq 0)$

43. $\dfrac{1}{x} - \dfrac{2}{x - 1} = 2$ $(x \neq 0, x \neq 1)$ **44.** $\dfrac{1}{2}(x + 2) - \dfrac{3}{x} = \dfrac{3}{2}$ $(x \neq 0)$

In Exercises 45 through 64, solve the given or resulting quadratic equations by any appropriate method unless otherwise specified.

45. Solve $x^2 + 2x + 2 = 0$ by completing the square.   **46.** Solve $x^2 - 4x = 8$ by completing the square.

47. Two positive numbers differ by 14 and their product is 351. Find the larger number.

48. Two consecutive positive integers have a product of 420. Find these integers.

49. The length of a rectangle is 1 in. more than the width. See Figure 11-10. If 3 in. is added to the length and 1 in. to the width, the new area is twice the original area. Find the dimensions of the original rectangle.

FIGURE 11-10

50. If the radius of a circle is doubled, its area increases by 147π m². Find the radius of the circle.

51. Under certain frictional conditions, the motion of an object on a spring requires the solution of the equation

$$D^2 + 2rD + k^2 = 0$$

where r and k are constants of the system. Solve for D.

52. The relation of cost C (in dollars) and units x (in hundreds) produced by a certain company is

$$C = 3x^2 - 12x + 13$$

For what value of $x (x < 2)$ is $C = \$3$?

53. A projectile is fired vertically into the air. The distance (in feet) above the ground, in terms of the time in seconds, is given by the formula

$$s = 96t - 16t^2$$

How long will it take to hit the ground?

54. In a certain electric circuit there is a resistance R of 5.0 Ω and a voltage E of 110 V. The equation relating current i (in amperes), E, and R is

$$i^2R + iE = 17395$$

What current ($i > 0$) flows in the circuit?

55. A general formula for the distance s traveled by an object, given an initial velocity v and acceleration a in time t, is

$$s = vt + \tfrac{1}{2}at^2$$

Solve for t.

56. In the theory concerning the motion of biological cells and viruses, the equation

$$n = 2.5p - 12.6p^2$$

is used. Solve for p in terms of n.

57. A company finds that its profit for producing x units weekly is

$$p = 25x - x^2 - 100$$

For which values of x is the profit zero?

58. In analyzing the forces on a certain beam, the equation

$$x^2 - 3Lx + 2L^2 = 0$$

is used. Solve for x in terms of L, where L represents the length of the beam.

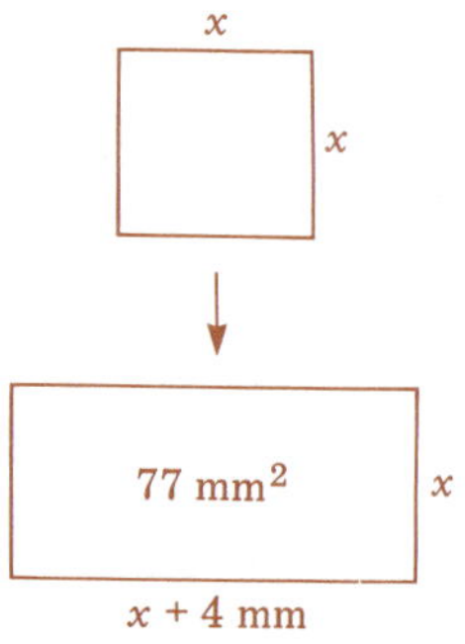

FIGURE 11-11

59. A machine designer finds that if she changes a square part to a rectangular part by increasing one dimension of the square by 4 mm, the area becomes 77 mm². Determine the side of the square part. See Figure 11-11.

60. A metal cube expands when heated. If the volume changes by 9.509 mm³ and each edge is 0.10 mm longer after the cube is heated, what was the original length of the edge of the cube?

61. An electric utility company is placing utility poles along a road. It is determined that one less pole per mile would be necessary if the distance between poles was increased by 5 ft. How many poles are being placed each mile?

62. An item costs P dollars. If it is sold for $30, the profit is $2P$ percent. Find the original cost.

63. After flying 1200 mi, a pilot determines that he can make the return trip in 1 h less if he increases his average speed by 100 mi/h. What was his original average speed?

64. A rectangular duct in a ventilating system is made of sheet metal 7.0 ft wide and has a cross-sectional area of 3.0 ft². What are the cross-sectional dimensions of the duct? See Figure 11-12.

In Exercises 65 through 72, use a scientific calculator to solve the given problems. Round all answers to the nearest hundredth.

65. $0.0052x^2 - 0.0034x - 0.0018 = 0$

66. $1008.7x^2 + 1008.2x - 1008.5 = 0$

67. $376.5x^2 - 255.8x - 108.6 = 0$

68. $3.14x^2 - 19.37x + 1.03 = 0$

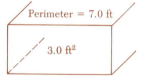

FIGURE 11-12

69. When the sides of a square are doubled, the area increases by 112.36 cm². Find the length of a side of the original smaller square.

70. A hole is drilled with diameter 0.47625 cm. It is later enlarged so that the area increases by 0.13855 cm². Find the diameter of the enlarged hole.

71. A square has perimeter p ft and area $p + 6$ ft². Find the length of a side.

72. It has been found that a visually appealing rectangle has length and width in the ratio of 1.618 to 1. If such a rectangle is to have an area of 6.00 m², find its dimensions. See Figure 11-13.

FIGURE 11-13

CHAPTER TWELVE

Logarithms

Logarithms were used extensively before calculators and computers became common. While the general use of logarithms is now less than it once was, they continue to play an important role in mathematics. Technicians and scientists should know at least the fundamental concepts of logarithms, and this chapter is designed to provide those concepts. Applications include the decibel sound scale, computer applications, radioactive decay, and the Richter scale used for measuring the intensity of earthquakes. In chemistry, the distinction between a base and an acid is defined in terms of logarithmic units. In electronics and mechanical systems, mathematical functions related to logarithms are often used. Apart from these particular applications, there are many numerical calculations that require the use of logarithms. Try to calculate 1024^{64} on your calculator and you are likely to find an error display because the result is too large, but logarithms can be used to determine an approximate result. (See Section 12-2 for a solution.)

In Section 12-1 we learn the basic definition of a logarithm. In the following sections we work directly with logarithms and calculations involving logarithms. We also examine the general method of interpolating values (finding values from tables when the desired values are between those that are actually listed). Although methods of interpolation with logarithms can be avoided by using calculators, Section 12-5 provides a general technique which can be applied to many other mathematical and physical tables.

This chapter includes several different applications. One example refers to an electric circuit used in an airplane's navigational system. The time (in seconds) it takes the current in the circuit to reach 2.0 A is given by

$$t = \frac{25\left(\ln\dfrac{2.0}{15}\right)}{16}$$

We show how to evaluate such expressions. That particular expression will be evaluated later in the chapter. The aviation industry makes extensive use of electronics and the industry term "avionics" (for aviation electronics) reflects that widespread use.

12-1 The Logarithm of a Number

The concept of logarithm is closely linked to the concept of exponents. In fact, *a logarithm is actually an exponent*. Recall that in an expression such as b^x, *the number b is called the* **base**, while *the number x is called the* **exponent**. We now define the logarithm of a number as follows.

If $b^x = N$, then $\log_b N = x$. (where $b > 0$ and $b \neq 1$)

From the above definition we see that $\log_b N$ is the exponent to which b must be raised to get N. There are two different ways to express the same relationship, as illustrated in the following examples.

EXAMPLE A The equation $2^5 = 32$ can be expressed in

logarithmic form as $\log_2 32 = 5$. $2^5 = 32$

The equation $10^3 = 1000$ can be expressed in

logarithmic form as $\log_{10} 1000 = 3$.

The equation $10^{-1} = 0.1$ can be expressed in

logarithmic form as $\log_{10} 0.1 = -1$.

The equation $\log_5 25 = 2$ can be expressed in

exponential form as $5^2 = 25$.

The equation $\log_{10} 100 = 2$ can be expressed in

exponential form as $10^2 = 100$.

The equation $\log_{10} 0.001 = -3$ can be expressed in

exponential form as $10^{-3} = 0.001$.

In this example we can see that the logarithm of a number yields a result that is actually an exponent. It would be helpful to stop reading at this point and practice converting back and forth between the two forms given in the above definition. It is essential that we understand that an expression such as $\log_2 32 = 5$ is equivalent to $2^5 = 32$. We must be able to easily convert from one form to the other.

While *the base b can be any* **positive** *number except 1,* there are special values which are often used. The base 2 has a special importance for computer applications. Another special base will be discussed in Section 12-3. We now consider the base 10, which is so common that it is referred to as the **common logarithm**. Because the base 10 is used so often, we have a *special notation that consists of omitting the subscript.* See the following example.

EXAMPLE B $\log 10000 = 4$ since $10^4 = 10000.$ (log 10000 is the same as $\log_{10} 10000$)

$\log 10 = 1$ since $10^1 = 10.$

$\log 0.01 = -2$ since $10^{-2} = 0.01.$ □

In the remainder of this section we consider only common logarithms. So far, all of the logarithms involved bases and results that could be determined by inspection. We can evaluate log 100 because we know that $10^2 = 100$; we get log 100 = 2. However, we can't use inspection to evaluate a number such as log 23 since we don't know a number x that will satisfy the equation $10^x = 23$. In evaluating such numbers, we can use tables, calculators, or computers.

The key point to recognize is that we can raise 10 to powers that are fractions or decimals.

EXAMPLE C   From Table 1 in Appendix D we find that $\sqrt{10} = 3.162$. Since $\sqrt{10} = 10^{1/2}$, we get $10^{1/2} = 3.162$ or $\log 3.162 = 1/2$. Using a calculator, we can enter the key sequence

$10 \boxed{y^x} \boxed{(} 1 \boxed{\div} 3 \boxed{)} \boxed{=} \boxed{2.1544347}$

to determine that $\sqrt[3]{10} = 2.154$ (rounded off) so that

$$10^{1/3} = 2.154 \quad \text{or} \quad \log 2.154 = \frac{1}{3}$$

Using tables, calculators, and laws of exponents, we can establish the following.

$10^{0.0000} = 1.000$ so that $\log 1.000 = 0.0000$
$10^{0.2500} = 1.778$ so that $\log 1.778 = 0.2500$
$10^{0.7500} = 5.623$ so that $\log 5.623 = 0.7500$ □

Through advanced mathematics, it can be proved in general that *given a positive number N, it is possible to find an exponent x such that* $10^x = N$ *(or log N = x)*.

Table 2 in Appendix D gives four-place logarithms (the logarithms are written with four significant digits) for numbers with three significant digits from 1.00 to 9.99. We know that these **logarithms are powers to which 10 must be raised in order to equal the given number** *N*. The method of finding the logarithm of a number from 1.00 to 9.99 from this table is illustrated in the following example.

EXAMPLE D   To find the logarithm of 8.36 using Table 2 in Appendix D, we first look under *N* to find 83 (the first two significant digits). To the right of 83 and under 6 (the third significant digit) we find the digits 9222. Therefore

decimal point not shown in table

$$\log 8.36 = 0.9222$$

logarithm is the exponent

This means that $10^{0.9222} = 8.36$.

To find the logarithm of 8.36 using a *scientific calculator*, the usual procedure is to first enter 8.36 and then press the key labeled log *x* or log. (On some calculators this is a key whose secondary purpose is to find logarithms. Consult the manual for your particular calculator.) The sequence

8.36 [log] **0.92220628**

shows that log 8.36 = 0.922.   ☐

Using a scientific calculator we are able to find the logarithm of any positive number, but *Table 2 in Appendix D directly gives us only logarithms of numbers between 1.00 and 9.99.* Examples E and F show how the table can be used to find logarithms of numbers that are not between 1.00 and 9.99.

EXAMPLE E   To find the logarithm of 5340, we first write

$$5340 = 5.34 \times 10^3$$

From Table 2 in Appendix D, we find the logarithm of 5.34 to be 0.7275, which means that $5.34 = 10^{0.7275}$. Thus

add

$$5340 = 5.34 \times 10^3 = (10^{0.7275})(10^{3.0000}) = 10^{3.7275}$$

Since 3.7275 is the required power to which 10 must be raised to equal 5340, we have

$$\log 5340 = 3.7275$$ ☐

EXAMPLE F $0.0272 = 2.72 \times 10^{-2}$. Therefore

$$0.0272 = (2.72 \times 10^{-2}) = 10^{0.4346} \times 10^{-2} \quad \text{from Table 2}$$
$$\log 0.0272 = 0.4346 - 2 = -1.5654 \qquad \square$$

In Examples E and F we first wrote the number in scientific notation. Then we were able to find the logarithm of the number by combining the logarithm found from the table with the power of 10. By definition, *the* **characteristic** *of the logarithm is the integral power of* 10 *of the number in scientific notation*, and *the* **mantissa** *of the logarithm is the logarithm of the number between* 1 *and* 10 *of this form.*

If tables are used, the proper determination of the characteristic and mantissa of a logarithm is very important. However, it is expected that most logarithms will be determined by use of a calculator. This is further illustrated in the following examples.

EXAMPLE G   To find log 5340 by means of a calculator, we use the sequence 5340 $\boxed{\log}$. The resulting display is 3.7275413.

Using a calculator to find log 0.0272, the sequence 0.0272 $\boxed{\log}$ gives the display -1.5654311.

Compare these results with those of Examples E and F. $\square$

EXAMPLE H   Sound from normal conversation is found to be 1,000,000 times a standard reference level. The intensity level (in decibels) is found by evaluating

$$10(\log 1000000)$$

Using a calculator, we enter the key sequence

10 $\boxed{\times}$ 1000000 $\boxed{\log}$ $\boxed{=}$ $\boxed{ \mathit{60.}}$

to get a display of 60. Normal conversation produces sound with an intensity level of 60 dB. $\square$

Exercises 12-1

In Exercises 1 through 12, write the logarithmic form of the given exponential equations.

1. $10^2 = 100$

2. $10^3 = 1000$

3. $10^{-2} = 0.01$

4. $10^{-4} = 0.0001$

5. $10^{3.4600} = 2884$

6. $10^{-1.57} = 0.02692$

7. $10^{-3.4444} = 0.0003594$

8. $10^{0.6667} = 4.6419$

9. $2^{10} = 1024$

10. $3^4 = 81$

11. $6^3 = 216$

12. $5^4 = 625$

In Exercises 13 through 24, write the exponential form of the given logarithmic equations.

13. $\log 10 = 1$

14. $\log 100 = 2$

15. $\log 1000 = 3$

16. $\log 1,000,000 = 6$

17. $\log 0.01 = -2$

18. $\log 0.0001 = -4$

19. $\log 567 = 2.7536$

20. $\log 0.00344 = -2.4634$

21. $\log_2 8 = 3$

22. $\log_2 64 = 6$

23. $\log_3 243 = 5$

24. $\log_5 125 = 3$

In Exercises 25 through 36, find the value of the given logarithms without using a calculator.

25. $\log 10,000$

26. $\log 100,000$

27. $\log 0.001$

28. $\log 0.00001$

29. $\log 10^6$

30. $\log 10^{12}$

31. $\log 10^{-3}$

32. $\log 10^{-8}$

33. $\log_2 16$

34. $\log_3 9$

35. $\log_4 64$

36. $\log_6 1296$

In Exercises 37 through 48, determine the logarithm of each given number using any method.

37. 7.26

38. 8.15

39. 93.6

40. 2600

41. $83,000$

42. 208

43. $50,000$

44. $256,000,000$

45. 0.0533

46. 0.00358

47. 0.0305

48. 0.0000637

In Exercises 49 through 60, find the value of x.

49. $\log_x 36 = 2$

50. $\log_x 27 = 3$

51. $\log_{10} 100 = x$

52. $\log_2 32 = x$

53. $\log_5 x = 4$

54. $\log_2 x = 8$

55. $\log_x 0.01 = -2$

56. $\log_x 0.5 = -1$

57. $\log_9 81 = x$

58. $\log_{10} 0.001 = x$

59. $\log x = 5$

60. $\log x = -5$

In Exercises 61 through 68, determine the logarithm of the numbers given in each problem.

61. A certain radio station broadcasts at a frequency of 1,340,000 Hz.

62. The Grand Coulee Dam contains about 10,600,000 yd^3 of concrete.

63. A research chemist notes that the magnification of her microscope is 20,000 to 1.

64. The mass of the Milky Way is estimated to be about 6×10^{41} lb.

65. The intensity of sound of a power lawn mower is 9.7×10^9 times that of rustling leaves.

66. The current in a certain electric circuit is increasing at the rate of 0.125 A/s.

67. Under certain conditions the partial pressure of a vapor is 1.6×10^{-8} Pa.

68. The coefficient of thermal expansion for aluminum is 2.2×10^{-5} per degree Celsius.

12-2 Properties of Logarithms

Logarithms were once extremely important for scientific and technical calculations, but the widespread availability of calculators and computers has reduced this use considerably. However, there remain many calculations that require logarithms. If we attempt to evaluate 1024^{64} with a calculator, we will probably get a display indicating an error because the result is too large for most calculators. Yet we can evaluate 1024^{64} using logarithms. Also, by doing basic calculations with logarithms we can establish certain very important properties of logarithms. In this section we illustrate the use of logarithms for finding products, quotients, powers, and roots of numbers, and thereby determine these important properties of logarithms.

EXAMPLE A We know from Eq. (7-1) that $2^3 \cdot 2^4 = 2^7$. We also know that $2^3 = 8$, $2^4 = 16$, and $2^7 = 128$. In the display below we show those values and their logarithmic forms to illustrate a basic property of logarithms that is generalized in Eq. (12-1).

$$8 \cdot 16 = 128$$
$$2^3 \cdot 2^4 = 2^{3+4} \qquad 3 = \log_2 8 \qquad 4 = \log_2 16 \qquad 3 + 4 = \log_2 128$$
$$\log_2 128 = \log_2 8 + \log_2 16$$

We see that $128 = 8 \cdot 16$ and that $\log_2 128 = \log_2 8 + \log_2 16$. ☐

Following the method of Example A, we shall now find a general expression for the logarithm of the product of two numbers. Considering the numbers P and Q, which may be expressed as 10^x and 10^y, respectively, we have

$$PQ = (10^x)(10^y) = 10^{x+y}$$

where

$$x = \log P \qquad y = \log Q \qquad x + y = \log PQ$$

or

$$\boxed{\log PQ = \log P + \log Q} \qquad\qquad (12\text{-}1)$$

From Eq. (12-1) we see that

The logarithm of the product of two numbers is equal to the sum of the logarithms of the numbers.

In finding the product we must determine the number whose logarithm has been found by addition. This number, found from a known logarithm, is called an **antilogarithm**. We now have the following procedure for multiplication by use of logarithms:

1. Determine the logarithms of the numbers to be multiplied.

2. Find the sum of these logarithms.

3. The required product is the antilogarithm of this sum.

We have already noted that we can use a calculator to find the logarithm of a number by entering the number and pressing the ⌊log⌋ key. To find the antilogarithm of a number, enter the number and then press ⌊INV⌋ ⌊log⌋. (On some calculators we could use the ⌊10^x⌋ key instead.) Examples B and C illustrate the method of using logarithms for finding products of numbers.

EXAMPLE B   Find the product of 1.30 and 120 by use of logarithms.

$$\log(1.30 \times 120) = \log 1.30 + \log 120$$

$$\log 1.30 = 0.1139$$

$$\log 120 \ = \underline{2.0792} \quad \text{add}$$

$$ 2.1931 \quad = \textbf{log of the product}$$

$$ \textbf{characteristic}$$

Using a calculator, we enter the key sequence

2.1931 [INV] [log] (*155.99116*)   **take antilogarithm**

We should round the result to 156.

If we use Table 2 in Appendix D we find that a mantissa of 0.1931 is the logarithm of 1.56. Since the characteristic of the logarithm is 2, the antilogarithm of 2.1931 is 1.56×10^2, so we have $1.30 \times 120 = 1.56 \times 10^2 = 156$.   □

EXAMPLE C   Find the product (0.00472)(60.8) by logarithms.

$$\log 0.00472 = 0.6739 - 3 \qquad \textbf{4.72} \times \textbf{10}^{-3}$$

$$\log 60.8 \quad\ = \underline{1.7839} \quad 2 - 3$$

$$ 2.4578 - 3 = 0.4578 - 1$$

From Table 2 we find that the mantissa nearest to 0.4578 is 0.4579. Using this value, we find the antilogarithm of the mantissa to be 2.87. We see that *to get the proper characteristic, we let* $2 - 3 = -1$. Thus

$$(0.00472)(60.8) = 2.87 \times 10^{-1} = 0.287$$

We could have done all operations on a calculator as in Example B.   □

Following the same line of reasoning used to develop the method for multiplication, we can use the following property to find the logarithm of the quotient P/Q:

$$\log\left(\frac{P}{Q}\right) = \log P - \log Q \tag{12-2}$$

Equation (12-2) means that:

> *The logarithm of the quotient of two numbers is equal to the logarithm of the numerator minus the logarithm of the denominator.*

In order to divide one number by another using logarithms, we subtract the logarithm of the denominator from the logarithm of the numerator and then find the antilogarithm of the result. The following example illustrates this procedure.

EXAMPLE D Find the quotient $\dfrac{629}{27.0}$ by using logarithms.

┌─ **indicates quotient**

$$\log \left(\frac{629}{27.0} \right) = \log 629 - \log 27.0 \qquad \begin{array}{l} \log 629 = 2.7987 \\ \log 27.0 = \underline{1.4314} \qquad \text{subtract} \\ \qquad\qquad 1.3673 \end{array}$$

└──────── **characteristic**

Using a calculator, we enter 1.3673 and press the $\boxed{\text{INV}}\,\boxed{\text{log}}$ keys to get 23.297, which is rounded to 23.3. Using Table 2 in Appendix D we find that the mantissa nearest to 0.3673 is 0.3674, which is the logarithm of 2.33. The characteristic is 1 so that $\dfrac{629}{27.0} = 2.33 \times 10^1 = 23.3$ □

To find a power of a number P, which may be expressed as 10^x, we have

$$P = 10^x \qquad P^n = (10^x)^n = 10^{nx}$$

where $x = \log P$ and $nx = \log P^n$. This means that

$$\boxed{\log P^n = n \log P} \tag{12-3}$$

Therefore:

> *The logarithm of the nth power of a number is equal to n times the logarithm of the number.*

▶ The exponent may be integral or fractional. ***If we wish to find the root of a number, we interpret n as a fractional exponent.*** The following examples illustrate the use of Eq. (12-3) for finding powers and roots of numbers.

EXAMPLE E Evaluate 2^4 using logarithms.

Since $\log P^n = n \log P$ we get $\log 2^4 = 4 \log 2$. From Table 2 in Appendix D we find that $\log 2 = 0.3010$ so that $4 \log 2 = 4(0.3010) = 1.2040$. Again using Table 2, we see that the mantissa nearest to 0.2040 is 0.2041, which is the logarithm of 1.6. The characteristic is 1. We now have $2^4 = 1.6 \times 10^1 = 16$. □

$\log 2^4 = 4 \log 2$

In Example F the value of 1024^{64} cannot be directly determined with most calculators and computers because the result is a number beyond their ordinary limits.

EXAMPLE F   In computer design, one configuration has 64 different sequences of 10 binary digits each so that the total number of different possible states is $(2^{10})^{64} = 1024^{64}$. Evaluate 1024^{64} using logarithms. Since $\log P^n = n \log P$, we know that $\log 1024^{64} = 64 \log 1024$. While most calculators won't directly evaluate 1024^{64}, we can use them to determine that $\log 1024 = 3.0103$. We now multiply $(64)(3.0103)$ to get the product 192.6592. Using Table 2 or a calculator, we can determine that the antilogarithm of 0.6592 is 4.56. The characteristic is 192, so

$$1024^{64} = 4.56 \times 10^{192}$$

The above steps can be summarized as follows.

Let $N = 1024^{64}$
$\log N = \log(1024^{64})$
$\log N = 64 \log 1024$ from Eq. (12-3)
$\log N = 64(3.0103)$
                                     characteristic
$\log N = 192.\underline{6592}$
                                     mantissa
$N = 4.56 \times 10^{192}$ antilogarithm of 0.6592 is 4.56
$1024^{64} = 4.56 \times 10^{192}$

This is the most convenient form of the answer. If we write it out in ordinary notation, we would write 456 followed by 190 zeros. That's a big number! ☐

In Example F we see that while we cannot use a calculator to directly evaluate 1024^{64}, we could use our knowledge of logarithms to find that value on the calculator. However, *we must know the meaning and use of logarithms* in order to use the calculator for this computation.

EXAMPLE G   Evaluate $\sqrt[3]{0.916}$ using logarithms.
Since $\sqrt[3]{0.916} = (0.916)^{1/3}$ we have $\log \sqrt[3]{0.916} = \log (0.916)^{1/3}$.

                                                    mantissa   characteristic
                                                       ↓   ↓
$$\log \sqrt[3]{0.916} = \frac{1}{3} \log 0.916 \qquad \log 0.916 = 0.9619 - 1$$

▶ In using the table, in order to have an integer subtracted from the mantissa after the division by 3, *we write the characteristic as 2 − 3, placing the 2 before the mantissa*. Therefore, $\log 0.916 = 2.9619 - 3$ or

$$\frac{1}{3}(\log 0.916) = \frac{1}{3}(2.9619 - 3) = 0.9873 - 1$$

We must now find the antilogarithm. In Table 2 in Appendix D the mantissa nearest to 0.9873 is 0.9872, which is the logarithm of 9.71. The characteristic is $9 - 10 = -1$. Therefore

$$\sqrt[3]{0.916} = 9.71 \times 10^{-1} = 0.971$$

☐

In Example G we could use a calculator to directly evaluate $\sqrt[3]{0.916}$. The key sequence

$$0.916 \; \boxed{y^x} \; \boxed{(} \; \boxed{1} \; \boxed{\div} \; \boxed{3} \; \boxed{)} \; \boxed{=} \; \boxed{\quad 0.97117723 \quad}$$

which shows that $\sqrt[3]{0.916} = 0.97117723$. However, there are numbers too large or too small for calculator entry so that the procedures of Example G must be used. We could evaluate $\sqrt[3]{0.916}$ on a calculator by using logarithms as follows.

$$0.916 \; \boxed{\text{log}} \; \boxed{\div} \; \boxed{3} \; \boxed{=} \; \boxed{\text{INV}} \; \boxed{\text{log}} \; \boxed{\quad 0.97117723 \quad}$$

The above key sequence involves taking the logarithm of 0.916, dividing by 3, and then finding the antilogarithm of the result. The calculator display of 0.97117723 should be rounded to three significant digits as 0.971.

EXAMPLE H   Evaluate

$$\frac{(65.1)(\sqrt{804})}{6.82}$$

by logarithms.

The numerator is a product of two factors, so we determine its logarithm by adding the logarithms of its factors. Since $\sqrt{804} = 804^{1/2}$ and $\log P^n = n \log P$, we find the logarithm of $\sqrt{804}$ by taking one-half the logarithm of 804. The solution is as follows:
Let

$$N = \frac{(6.51)(\sqrt{804})}{6.82}$$

then

$$\log N = \log 65.1 + \frac{1}{2} \log 804 - \log 6.82$$

$$\log N = 1.8136 + \frac{1}{2}(2.9053) - 0.8338$$

$$\log N = 2.4325$$

In Table 2 in Appendix D the mantissa nearest to 0.4325 is 0.4330, which is the logarithm of 2.71. The characteristic is 2. Therefore the result is $2.71 \times 10^2 = 271$.

☐

Most of the calculations shown in this section can be quickly and easily performed with a calculator. However, if a calculator is not available or if the results are beyond the range of values which can be directly handled, logarithms provide a method by which such calculations can be made. In this section we have developed some very important properties of logarithms. Also, it can be seen that powers and roots of numbers can be easily determined by use of logarithms. We should also note that calculators and computers use logarithms to perform many of these types of calculations.

Exercises 12-2

In Exercises 1 through 32, perform the indicated calculations by means of logarithms.

1. (6.29)(7.36)

2. (25.2)(3.08)

3. (0.613)(89.2)

4. (0.0726)(846)

5. (4000)(3.14)

6. (0.00105)(0.429)

7. (78000)(1.47)

8. (8080)(0.0715)

9. $\dfrac{9.33}{4.27}$

10. $\dfrac{84600}{45.2}$

11. $\dfrac{0.841}{0.617}$

12. $\dfrac{0.00913}{6.66}$

13. $\dfrac{2.38}{80.4}$

14. $\dfrac{27.2}{8030}$

15. $\dfrac{0.0984}{0.0076}$

16. $\dfrac{156000}{79400}$

17. $\dfrac{(45.1)(6.12)}{73.8}$

18. $\dfrac{3.14}{(0.184)(61.3)}$

19. $\dfrac{(5.03)(89.1)}{(46.1)(2.36)}$

20. $\dfrac{(0.0187)(816)}{(51.5)(0.0845)}$

21. 4.56^2

22. 2.35^{10}

23. 0.493^2

24. 7.21^8

25. $\sqrt{26.8}$

26. $\sqrt[3]{56.9}$

27. $\sqrt[4]{31.9}$

28. $\sqrt{0.592}$

29. $5.36^{0.3}$

30. $8.45^{2.1}$

31. $(\sqrt{56.9})(\sqrt[3]{18.8})$

32. $\dfrac{(73.8)(\sqrt{0.854})}{0.308}$

In Exercises 33 through 48, perform the indicated calculations by means of logarithms.

33. To find the area of a car tire that is in contact with the ground, use the formula

$$A = \frac{W}{4p}$$

Find the area A for a car with weight $W = 3270$ lb and tire air pressure $p = 28.0$ lb/in.^{2}.

34. If the yield on an investment is 7.75% annually, what is the amount earned on $12,500?

35. Using the Richter scale for measuring the magnitude of earthquake intensity I, we have

$$R = \log\left(\frac{I}{I_0}\right)$$

where I_0 is the minimum base level used for comparison. Find the Richter scale reading of the San Francisco earthquake for which

$$I = (2.5 \times 10^8)I_0.$$

36. The loudness of a sound, measured in decibels (dB), is given by

$$10 \log\left(\frac{I}{I_0}\right)$$

where I is the intensity of the sound and I_0 is the intensity of the faintest sound that can be heard. A busy city street has a noise with an intensity level $I = 10^7 I_0$. Evaluate to find the decibel level.

37. Balsa wood has a density of 8.25 lb/ft³. Lead has a density of 705 lb/ft³. How many cubic feet of balsa wood have the same weight as 1 ft³ of lead?

38. If seawater is 3.25% salt by weight, how many pounds of seawater must be evaporated to obtain 852 lb of salt?

39. The voltage V across an electric resistor R in which a current I is flowing is given by $V = IR$. Find V, given that $I = 3.75 \times 10^{-2}$ A and $R = 56.8$ Ω.

40. The cutting speed of a milling machine is the product of the circumference of the cutter and the number of revolutions per minute it makes. Find the cutting speed of a milling cutter which is 5.25 in. in diameter, given that it makes 35.0 r/min.

41. In a particular year, a farmer has 535 acres of land in wheat. If 14,300 bushels are harvested, what is the average production per acre?

42. The general gas law of chemistry and physics, $PV = RT$, gives the relation among the pressure, volume, and temperature of a body of gas (R is a constant equal to about 8.31 J/mol·K). If the pressure of a certain gas is 3.67×10^5 Pa and its temperature is 373 K, what is its volume (in cubic meters)?

43. The velocity of sound in air is given by

$$v = \sqrt{1.410 p/d}$$

where p is the pressure and d is the density. Given that $p = 1.01 \times 10^5$ Pa and $d = 1.29$ kg/m³, find v (in meters per second).

44. The radius of a weather balloon is 12.2 ft. Calculate the volume of the gas it contains.

45. The maximum range R of a rocket fired with a velocity v is given by

$$R = \frac{v^2}{g}$$

where g is the acceleration due to gravity. Find the maximum range of a rocket if $v = 9850$ ft/s and $g = 32.2$ ft/s².

46. Under certain conditions the efficiency (in percent) of an internal combustion engine is found by evaluating

$$100\left(1 - \frac{1}{R^{0.4}}\right)$$

where R is the compression ratio. Compute the efficiency of an engine with a compression ratio of 6.55.

47. The ratio between successive speeds of a six-speed gearbox is $\sqrt[5]{625/178}$ if the maximum speed is 625 r/min and the minimum speed is 178 r/min. Determine this ratio.

48. When a fluid flows through a pipe, frictional effects cause a loss of pressure. A numerical factor used in the determination of pressure losses, under certain conditions, equals $0.316/\sqrt[4]{5000}$. Calculate this factor.

In Exercises 49 through 56, use a scientific calculator and logarithms to perform the indicated calculations.

49. 16^{160}

50. 64^{1024}

51. 2^{500}

52. 8^{128}

53. $\sqrt{16^{256}}$

54. $\sqrt{3^{213}}$

55. $(16^{256})(455^{40})$

56. $\sqrt[5]{1753^{300}}$

12-3 Natural Logarithms

We have seen that in the basic definition of the logarithm of a number, the base b of the logarithm may be any positive number except 1. Most of our work has been with common logs (base 10), but there is another number that is very useful as a base of logarithms. This number is e, an irrational number equal to approximately 2.718. *Logarithms to the base e are called* **natural logarithms**. We denote the natural logarithm of x by writing $\ln x$.

If $e^x = N$,   then $\ln N = x$.

Whenever we see the "ln" notation, we must remember that the base is the number e (about 2.718). Many formulas in science and technology incorporate these natural logarithms.

EXAMPLE A

Since $e^0 = 1$, $\ln 1 = 0$. Since $e^1 = e$, $\ln e = 1$.
In the same way, $\ln e^{1/2} = \frac{1}{2}$, or $\ln e^{0.5} = 0.5$, and $\ln e^2 = 2$. Using the approximate value of $e = 2.718$, we have

$$\ln e = \ln 2.718 = 1$$
$$\ln e^{0.5} = \ln 2.718^{0.5} = \ln 1.649 = 0.5$$
$$\ln e^2 = \ln 2.718^2 = \ln 7.389 = 2$$

☐

In Example A we evaluated natural logarithms by knowing the correspondence between the logarithmic and exponential forms of a relationship. We can also obtain values of natural logarithms by using a scientific calculator or a table. With a calculator we use the key usually labeled ln x or ln.

EXAMPLE B Use a scientific calculator to find the value of $\ln 52.6$.
By pressing the sequence of keys

52.6 [ln] (3.9627161)

we get the display shown.
We can also determine the value of $\ln 0.03482$ by entering the sequence

0.03482 [ln] (−3.3575633)

☐

It is sometimes necessary to find antilogarithms of natural logarithms. That is, we seek a value of N where $\ln N$ is equal to some given value. The next example shows how a calculator can be used for problems of this type.

EXAMPLE C   Use a scientific calculator to find N if $\ln N = -0.3562$.

Depending on your particular calculator, one of the following two key sequences should normally produce the desired result of 0.70033254.

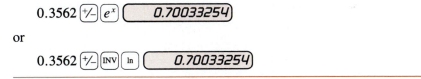

or

0.3562 $\boxed{+/-}$ $\boxed{\text{INV}}$ $\boxed{\text{ln}}$ $\boxed{\quad\quad 0.70033254 \quad}$ $\square$

To determine values of natural logarithms using a table, we can refer to Table 4 in Appendix D. Since we cannot express numbers in terms of powers of e in the same way we can express them in terms of powers of 10, this is not a table of mantissas. The natural logarithms listed may be used directly only with the given numbers.

EXAMPLE D   Using Table 4 in Appendix D to find $\ln 7.5$, we look opposite 7.5 in the N column and find 2.0149 in the $\ln N$ column. This shows that

$\ln 7.5 = 2.0149$ which means $e^{2.0149} = 7.5$

Also,

$\ln 60 = 4.0943$ which means $e^{4.0943} = 60$

and

$\ln 0.7 = 9.6433 - 10 = -0.3567$ which means $e^{-0.3567} = 0.7$ $\square$

It is possible to use interpolation (see Section 12-4) in Table 4 in Appendix D to obtain reasonable approximations of natural logarithms which are not listed. Such values are correct to about three or four significant digits.

In developing common logarithms for calculation, we obtained certain important properties of logarithms. *These properties are not dependent on the base which is used and are therefore valid for any base* so that the following properties apply to natural logarithms:

$$\ln PQ = \ln P + \ln Q \tag{12-4}$$

$$\ln \frac{P}{Q} = \ln P - \ln Q \tag{12-5}$$

$$\ln P^n = n \ln P \tag{12-6}$$

Note that Eq. (12-4) expresses for natural logarithms the same property Eq. (12-1) expressed for common logarithms. That similarity also exists between Eqs. (12-2) and (12-5), and between Eqs. (12-3) and (12-6). We illustrate these properties for natural logarithms in the following examples which refer to Table 4 in Appendix D for the necessary values.

EXAMPLE E Using the fact that $600 = 60 \times 10$, find $\ln 600$.
From Eq. (12-4), we have

$$\ln 600 = \ln (60 \times 10) = \ln 60 + \ln 10$$

multiplication

Thus

$$\ln 600 = 4.0943 + 2.3026$$
$$= 6.3969$$

EXAMPLE F Using Eq. (12-5), we find that

indicates division

$$\ln \frac{7}{4} = \ln 7 - \ln 4$$
$$= 1.9459 - 1.3863$$
$$= 0.5596$$

Thus

$$\ln \frac{7}{4} = \ln 1.75 = 0.5596.$$

EXAMPLE G Using Eq. (12-6), we find that

$$\ln (1.5)^2 = 2 \ln 1.5$$
$$= 2(0.4055)$$
$$= 0.8110$$

Thus

$$\ln (1.5)^2 = \ln 2.25 = 0.8110$$

These examples show that we can find the natural logarithms of many numbers which are not listed.

The following examples illustrate the use of natural logarithms in applied problems.

EXAMPLE H   If Q_0 mg of radium decays through radioactivity, an equation relating the amount Q which remains after t years is

$$t = \frac{\ln Q_0 - \ln Q}{4.27 \times 10^{-4}}$$

Find the "half-life" of radium by finding the number of years for 500 mg of radium to decay such that 250 mg remains.

From the given information $Q_0 = 500$ mg and $Q = 250$ mg. Thus

$$t = \frac{\ln 500 - \ln 250}{4.27 \times 10^{-4}}$$

We can evaluate ln 500 and ln 250 or we can use Eq. (12-5). Using Eq. (12-5) we get $\ln 500 - \ln 250 = \ln \frac{500}{250}$ so that

$$t = \frac{\ln \dfrac{500}{250}}{4.27 \times 10^{-4}} = \frac{\ln 2.00}{4.27 \times 10^{-4}}$$

$$= \frac{0.6931}{4.27 \times 10^{-4}}$$

$$= 1.62 \times 10^3 \text{ years}$$

It therefore takes 1620 years for 500 mg of radium to decay such that 250 mg remains. That is, the "half-life" of radium is 1620 years. ☐

EXAMPLE I   A certain electric circuit is used in an airplane's navigational system. The time (in seconds) it takes the current in the circuit to reach 2.0 A is given by

$$t = -\frac{25\left(\ln \dfrac{2.0}{15}\right)}{16}$$

Find that time.

We note that $\ln \frac{2.0}{15} = \ln 2.0 - \ln 15$ by Eq. (12-5). Since $\ln 2.0 = 0.6931$ and $\ln 15 = 2.7081$, we get

$$t = -\frac{25(0.6931 - 2.7081)}{16}$$

$$= -\frac{25(-2.0150)}{16} = \frac{50.3750}{16}$$

$$= 3.1 \text{ s}$$

The given value of t could have been computed on a calculator by using the following sequence of key strokes.

$$2 \boxed{\div} \; 15 \boxed{=} \boxed{\text{ln}} \boxed{\times} \; 25 \boxed{\div} \; 16 \boxed{=} \boxed{+/-} \; \boxed{3.1482860}$$ ☐

Exercises 12-3

In Exercises 1 through 4 find the indicated natural logarithms from Table 4 in Appendix D.

1. ln 8.4   **2.** ln 18   **3.** ln 0.9   **4.** ln 0.2

In Exercises 5 through 12, use any method to find the value of each of the given natural logarithms.

5. ln 200   **6.** ln 640   **7.** $\ln \frac{3}{8}$ **8.** ln 1000

9. ln 0.001   **10.** ln 0.012   **11.** $\ln (3.7 \times 10^3)$ **12.** $\ln (3.47 \times 10^{-4})$

In Exercises 13 through 20, use natural logarithms to perform the indicated calculations.

13. (6.92)(1.43)   **14.** (0.082)(7.4)   **15.** $\dfrac{81350}{267.4}$ **16.** $\dfrac{0.0208}{0.00345}$

17. $\sqrt{787}$ **18.** $\sqrt[3]{28.36}$ **19.** 2.37^{50} **20.** 0.901^{80}

In Exercises 21 through 28, use ln 5 = 1.6094, ln 3 = 1.0986, and Eqs. (12-4), (12-5), and (12-6) to determine the value of each of the following: (Do not use the tables or a calculator for finding logarithms.)

21. Given that $15 = 3 \times 5$, find ln 15.   **22.** Given that $0.6 = \frac{3}{5}$, find ln 0.6.

23. Given that $5^3 = 125$, find ln 125.   **24.** Given that $3^5 = 243$, find ln 243.

25. Given that $3^4 = 81$, find ln 81.   **26.** Given that $5^2 = 25$, find ln 25.

27. Given that $2025 = 5^2 \times 3^4$, find ln 2025.   **28.** Given that $3.24 = 3^4 \div 5^2$, find ln 3.24.

In Exercises 29 through 40, solve the applied problems.

29. A formula used by technicians for analyzing convection heat transfer is

$$C = \frac{4.92}{(\ln R)^{2.584}}$$

A first step in determining the value of C is to determine the value of ln R for a given value of R. Find ln R if $R = 22$. (Do not solve for C.)

30. In studying the population of birds in a certain forest, the equation

$$t = \frac{-\ln \dfrac{P}{P_0}}{0.05}$$

is used. Solve for t if $P = 400$ and $P_0 = 500$.

31. The age in years of a certain fossil is estimated by a method of carbon dating based on the half-life of a carbon isotope. Determine the estimated age by evaluating

$$t = \frac{\ln \dfrac{3}{2}}{2.70 \times 10^{-5}}$$

32. The work (in joules) done by a gas as it expands from volume V_1 to volume V_2 is described by

$$W = nRT \ln \frac{V_2}{V_1}$$

Find the work W if $n = 1$ mole (exactly), $R = 8.31$ J/mole·K, $T = 295$ K, $V_1 = 5.00$ L, and $V_2 = 12.0$ L.

33. Assuming daily compounding, the approximate time (in years) it takes for a deposit of P dollars to grow to a value of V dollars is given by

$$t = \frac{\ln V - \ln P}{r}$$

where r is the interest rate in decimal form. How long does it take for a $2500 deposit to grow to $3000 if the interest rate is 6.00%?

34. In treating a coaxial cable between two cities as a capacitor, the capacitance is found from the formula

$$C = \frac{2\pi kL}{\ln \dfrac{b}{a}}$$

Find C (in microfarads) if $a = 0.90$ cm, $b = 1.2$ cm, $L = 5.0 \times 10^7$, and $k = 5.0 \times 10^{-7}$.

35. In how many years will 1000 mg of radium decay such that 750 mg remains? (See Example H.)

36. If interest is compounded continuously (daily compounded interest closely approximates this), a bank account can double in t years according to the equation

$$i = \frac{\ln 2}{t}$$

where i is the interest rate expressed in decimal form. What interest rate is required for an account to double in 7 years?

37. Under certain conditions, the electric current i in a circuit containing a resistance R and an inductance L is given by

$$\ln \frac{i}{I} = -\frac{Rt}{L}$$

where I is the current at $t = 0$. Calculate how long, in seconds, it takes i to reach 0.430 A if $I = 0.750$ A, $R = 7.50$ Ω, and $L = 1.25$ H.

38. For the electric circuit in Exercise 37, find how long it takes the current i to reach 0.1 of its original value.

39. A savings account is set up so that compounding is done continuously. If P dollars are deposited for time t (in years), the interest rate r (in decimal form) is found from

$$r = \frac{\ln \dfrac{V}{P}}{t}$$

where V is the latest value of the account. Find r if a $500 deposit results in a value of $645.23 after exactly 3 years.

40. Insulation resistance R is described by

$$R = \frac{10^6 t}{C(\ln V_0 - \ln V)}$$ Find the resistance if $t = 120$, $C = 0.080$, $V_0 = 140$, and $V = 110$.

In Exercises 41 through 44, use a scientific calculator to determine each value.

41. $\ln 564.7$ 42. $\ln 0.003825$ 43. $\ln (55)^{62}$ 44. $\ln (6.022 \times 10^{23})$

In Exercises 45 through 48, the natural logarithms of numbers are given. Use a scientific calculator to determine the original numbers.

45. 5.76   46. 0.584   47. −0.3992   48. −2.13645

12-4 Interpolation Technique

In this section we discuss a general method of *interpolation* which becomes important when many different mathematical and physical tables are used. *Interpolation allows us to find values from a table even when the values are between those values that are actually listed.* We begin with examples involving logarithms so that this technique can be illustrated, but the method is general and can be used with a wide variety of different tables. In fact, interpolation with logarithms can be avoided by the use of calculators or computers, but there are many other tables that do require a knowledge of interpolation. We will also consider examples of interpolation involving tables other than tables of logarithms. The importance of this section extends far beyond its applicability to logarithms alone.

When using tables, we can often select a closest value if an exact value cannot be found. However, more accurate results can be obtained if we interpolate.

EXAMPLE A   Use Table 2 in Appendix D to determine the value of log 2.465.

From the table we see that log 2.46 = 0.3909 and log 2.47 = 0.3927. Since 2.465 is midway between 2.460 and 2.470, we conclude that log 2.465 is midway between 0.3909 and 0.3927. Thus log 2.465 = 0.3918. (We found the value midway between 0.3909 and 0.3927 by adding those numbers and then dividing the sum by 2.)   □

This approach, referred to as **linear interpolation**, is not strictly correct, but it does yield excellent approximations. In the next example we use a more formal and general technique for this same method.

EXAMPLE B   Find the logarithm of 3.267.

The logarithm of 3.267 lies between log 3.260 and log 3.270. Since 3.267 lies $\frac{7}{10}$ of the way from 3.260 to 3.270, we assume that log 3.267 lies $\frac{7}{10}$ of the way from log 3.260 to log 3.270. The value of log 3.267 can be found by use of the following diagram:

$$
10 \left[\begin{array}{c} 7 \left[\begin{array}{c} \log 3.260 = 0.5132 \\ \\ \log 3.267 = \ ? \end{array} \right. \\ \log 3.270 = 0.5145 \end{array} \right] x \right] 13
$$

First we note that 13 is the **tabular difference** (the actual position of the decimal point is not important here) between the values of the logarithms. Next, from the diagram we *set up the proportion*

$$\frac{7}{10} = \frac{x}{13}$$

which gives the value of $x = 9.1$. Since the mantissa of the required logarithm may have only four significant digits, we round off the value of x to $x = 9$. This is then added to the logarithm of 3.260 to yield

log 3.267 = 0.5141
          5132 + 9

EXAMPLE C   Find the number whose logarithm is 0.4554.

The nearest mantissas in the table are 0.4548 and 0.4564. This leads to the following diagram:

$$10 \begin{bmatrix} x \begin{bmatrix} \log 2.850 = 0.4548 \\ \\ \log\ ? = 0.4554 \end{bmatrix} 6 \\ \log 2.860 = 0.4564 \end{bmatrix} 16$$

We now have the proportion

$$\frac{x}{10} = \frac{6}{16}$$

which leads to $x = 3.75$. Rounding this off to the nearest integer, we have $x = 4$. Adding 4 to 2850, we get 2854 so that

log 2.854 = 0.4554

and the required result is 2.854.

As mentioned earlier, we may use interpolation with a wide variety of different tables. In the following examples, interpolation is used with a table which lists the velocity of sound in water at certain temperatures. These examples do not involve logarithms.

EXAMPLE D

v (m/s)	T (°C)
1430	0
1455	10
1487	20
1528	30
1580	40
1645	50
1722	60

In an experiment designed to measure the velocity of sound in water, the values in the given table were found. By use of interpolation, determine the velocity of sound in water, in meters per second, at a temperature of 23°C.

Since 23°C is $\frac{3}{10}$ of the way from 20°C to 30°C, we assume that the velocity of sound at 23°C is $\frac{3}{10}$ of the way between 1487 m/s and 1528 m/s. The difference between these velocities is 41 m/s, which means we have the proportion

$$\frac{x}{41} = \frac{3}{10} \quad \text{or} \quad 10x = 123$$

or

$$x = 12.3$$

Thus we add 12 m/s (rounded off) to 1487 m/s to find that the velocity is 1499 m/s at 23°C, according to the results of this experiment. □

EXAMPLE E

Using the same table of values given in Example D, find the temperature which corresponds to a velocity of 1550 m/s.

Since 1550 is $\frac{22}{52}$ of the way from 1528 to 1580, we assume that the temperature is $\frac{22}{52}$ of the way between 30 and 40. This suggests the proportion

$$\frac{22}{52} = \frac{x}{10} \quad \text{or} \quad 52x = 220 \quad \text{or} \quad x = 4.23$$

$$52\begin{bmatrix} 22\begin{bmatrix} 1528 & 30 \\ 1550 & ---- \\ 1580 & 40 \end{bmatrix}x \end{bmatrix}10$$

We therefore add 4°C (rounded off) to 30°C to find that the temperature corresponding to 1550 m/s is 34°C. □

Exercises 12-4

In Exercises 1 through 8, use linear interpolation and Table 2 in Appendix D to find the common (base 10) logarithm of each given number.

1. 7.125   **2.** 4385   **3.** 0.1532   **4.** 0.006027

5. 25.41   **6.** 310.6   **7.** 3.141×10^4 **8.** 6.028×10^{-3}

In Exercises 9 through 16, use linear interpolation and Table 2 in Appendix D to find the antilogarithm of each given logarithm.

9. 0.4807   **10.** 0.7814   **11.** 1.8472   **12.** 2.2666

13. 3.5260   **14.** $9.8706 - 10$ **15.** $5.4567 - 10$ **16.** $19.1049 - 20$

In Exercises 17 through 24, refer to the table given in Example D. Use interpolation to find the indicated values of v and T.

17. Find v for $T = 5°C$.   **18.** Find v for $T = 35°C$.

19. Find v for $T = 18°C$.   **20.** Find v for $T = 41°C$.

21. Find T for $v = 1440$ m/s.   **22.** Find T for $v = 1700$ m/s.

23. Find T for $v = 1620$ m/s.   **24.** Find T for $v = 1547$ m/s.

In Exercises 25 through 32, refer to the accompanying table which describes the radioactive decay of polonium, where N is the number of milligrams of undecayed radioactive nuclei and t is the decay time in days. Use interpolation to find the indicated values of N and t.

N	t
500	0
390	50
305	100
238	150
186	200

25. Find N for $t = 25$.

26. Find N for $t = 30$.

27. Find N for $t = 160$.

28. Find N for $t = 187$.

29. Find t for $N = 400$.

30. Find t for $N = 375$.

31. Find t for $N = 263$.

32. Find t for $N = 250$. (This value of t is the *half-life* of polonium since it is the time required for half the number of nuclei to disintegrate.)

In Exercises 33 through 36, perform the indicated calculations by means of logarithms. Use linear interpolation and Table 2 in Appendix D for finding any logarithms or antilogarithms.

33. $(86.05)(0.5425)$

34. $\sqrt{27.34}$

35. 76.82^4

36. $\dfrac{\sqrt[3]{0.8137}}{66.08}$

In Exercises 37 through 40, perform the indicated calculations by means of logarithms. Use linear interpolation and Table 2 in Appendix D for finding any logarithms or antilogarithms.

37. A certain alloy is 62.73% iron. How many pounds of iron are there in 169.3 lb of this alloy?

38. In one year, 2,731,000 farms had a total of 1.063×10^9 acres of land. Find the average number of acres per farm.

39. The surface area of a sphere is given by $A = 4\pi r^2$, where r is the radius of the sphere. Calculate the surface area of the earth, assuming that $r = 3959$ mi and $\pi = 3.142$. See Figure 12-1.

40. The effective value of the current I in an alternating-current circuit is given by

$$I = \frac{I_m}{\sqrt{2}}$$

where I_m is the maximum value of the current. Determine I, given that $I_m = 3.627$ A.

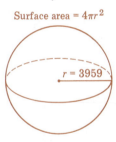

Surface area = $4\pi r^2$

$r = 3959$

FIGURE 12-1

Chapter 12 Formulas

$$\log PQ = \log P + \log Q \tag{12-1}$$

$$\log \left(\frac{P}{Q}\right) = \log P - \log Q \tag{12-2}$$

$$\log P^n = n \log P \tag{12-3}$$

$$\ln PQ = \ln P + \ln Q \tag{12-4}$$

$$\ln \frac{P}{Q} = \ln P - \ln Q \tag{12-5}$$

$$\ln P^n = n \ln P \tag{12-6}$$

12-5 Review Exercises for Chapter 12

In Exercises 1 through 8, write the logarithmic form of the given exponential equation.

1. $10^4 = 10000$ **2.** $10^{-3} = 0.001$ **3.** $6^4 = 1296$ **4.** $5^{-3} = 0.008$

5. $10^1 = 10$ **6.** $10^{1.23} = 17.0$ **7.** $10^{-2.54} = 0.00288$ **8.** $10^{0.4} = 2.5$

In Exercises 9 through 16, write the exponential form of the given logarithmic equations.

9. $\log_2 128 = 7$ **10.** $\log_3 81 = 4$ **11.** $\log 100000 = 5$ **12.** $\log 0.1 = -1$

13. $\ln 20 = 3.0$ **14.** $\ln 0.25 = -1.4$ **15.** $\log_4 1024 = 5$ **16.** $\log_5 0.04 = -2$

In Exercises 17 through 48, perform all calculations by means of common logarithms (base is 10). When referring to Table 2 in Appendix D, use interpolation for finding the logarithm of a number with more than three significant digits.

17. $(2.07)(3.45)$ **18.** $(651)(2.93)$ **19.** $(8.75)(0.205)$ **20.** $(0.0521)(12300)$

21. $(0.00482)(0.00367)$ **22.** $(18100)(645)$ **23.** $(27.34)(5.628)$ **24.** $(0.02455)(9.768)$

25. $\dfrac{8.13}{2.37}$ **26.** $\dfrac{64520}{830}$ **27.** $\dfrac{85.4}{2.76}$ **28.** $\dfrac{0.249}{5.63}$

29. $\dfrac{7.14}{9310}$ **30.** $\dfrac{0.0862}{0.174}$ **31.** $\dfrac{89.15}{9.176}$ **32.** $\dfrac{15010}{0.9007}$

33. $(10.8)^3$ **34.** $(1.08)^{10}$ **35.** $(76.1)^4$ **36.** $(0.405)^5$

37. $(6.184)^4$ **38.** $(1.034)^{0.3}$ **39.** $\sqrt{23.1}$ **40.** $\sqrt{84.9}$

41. $\sqrt[3]{1.17}$ **42.** $\sqrt[3]{789}$ **43.** $\sqrt{1308}$ **44.** $\sqrt[3]{0.9006}$

45. $(61.2)(\sqrt{128})$ **46.** $\dfrac{\sqrt{86000}}{45.8}$ **47.** $\dfrac{(0.721)(98.4)}{\sqrt[3]{8.17}}$ **48.** $\dfrac{(67.11)(9.004)}{(8.114)^{0.3}}$

In Exercises 49 through 52, find the indicated natural logarithms from Table 4 in Appendix D.

49. $\ln 4.1$ **50.** $\ln 12$ **51.** $\ln 0.6$ **52.** $\ln 0.1$

In Exercises 53 through 56, use $\ln 2 = 0.6931$, $\ln 3 = 1.0986$, and the properties of natural logarithms (Eqs. 12-4, 12-5, 12-6) to find the given values. (Do not use the tables or a calculator for finding these logarithms.)

53. Given that $6 = 2 \times 3$, find $\ln 6$.   **54.** Given that $8 = 2^3$, find $\ln 8$.

55. Given that $1.5 = 3 \div 2$, find $\ln 1.5$.   **56.** Given that $32 = 2^5$, find $\ln 32$.

In Exercises 57 through 60, use Table 4 in Appendix D and the method of linear interpolation to find the following values:

57. $\ln 42.5$ **58.** $\ln 72$ **59.** $\ln 99$ **60.** $\ln 9.08$

In Exercises 61 through 80, solve the given problems. Perform all calculations by means of logarithms.

61. The magnitudes of tsunamis (popularly known as tidal waves) are measured by

$$m = 3.32 \log_{10} h$$

where h is the maximum wave height (in meters) measured at the coast. Find the magnitude of a tsunami that has a 6.02-m height. (Such a tsunami hit Hawaii in 1946.)

62. An earthquake is measured at 5.2 on the Richter scale so that

$$5.2 = \log \frac{I}{I_0}$$

where I is the earthquake's intensity and I_0 is the reference level used for comparison. Find the value of the ratio I/I_0.

63. The loudness of a sound, measured in decibels (dB), is given by

$$10 \log \frac{I}{I_0}$$

where I is the intensity of the sound and I_0 is the intensity of the faintest sound that can be heard. At a distance of 100 ft, the noise of a jet has an intensity level of $I = 1 \times 10^{14} I_0$. Find the decibel level.

64. The number of bacteria in a certain culture after t hours is given by

$$N = (1000)10^{0.0451t}$$

How many are present after 3.00 h?

65. In the study of light, a basic law gives the amount transmitted through a medium. This amount is expressed as a power of 10. The percent of light transmitted in a particular experiment is $100(10^{-0.3010})$. What percent is this?

66. In chemistry, the pH value of a solution is a measure of its acidity. The pH value is defined by the relation

$$\text{pH} = -\log (\text{H}^+)$$

where H^+ is the hydrogen-ion concentration. If the pH of a certain wine is 3.4065, find the hydrogen-ion concentration. (If the pH value is less than 7, the solution is acid. If the pH value is above 7, the solution is basic.)

67. Find the value of the hydrogen-ion concentration in pure water. Pure water has a pH of 7. (See Exercise 66.)

68. Under certain conditions the temperature T (in degrees Celsius) of a cooling object is given by the equation

$$T = 50.0(10^{-0.1t})$$

where t is the time in minutes. Find the temperature T after 5.00 min.

69. The coefficient of performance of a refrigerator is defined as the heat removed divided by the work done. Find the coefficient of performance for a refrigerator if 575 Btu (British thermal units) of heat is removed while 125 Btu of work is being done.

70. Milk is about 87.3% water. How many kilograms of water are there in 5.37 kg of milk?

71. A car averages 15.8 mi/gal of gasoline. At a particular stop it takes 9.60 gal of gasoline to fill the tank. Assuming that this is the amount used since it was last filled, how far has the car traveled between gasoline stops?

72. Light travels at the rate of 186,000 mi/s. The earth is about 92,900,000 mi from the sun. How long does it take the light from the sun to reach the earth?

73. With a gross income of $42,540, a person pays 32.6% in state and federal taxes. What amount was paid in taxes?

74. The maximum bending moment of a beam supported at both ends is given by the formula

$$M = \frac{wl^2}{8}$$

where w is the weight per unit length and l is the distance between supports. Find M for a beam that is 48.35 ft long, is supported at each end, and weighs 144.5 lb.

75. If an object is falling due to gravity, its velocity v in terms of the distance h fallen is given by

$$v = \sqrt{64.4h}$$

What is the velocity, in feet per second, of an object which has fallen 410 ft?

76. The velocity v of a rocket is given by the formula

$$v = 2.30u \log \frac{m_0}{m}$$

where u is the exhaust velocity, m_0 is the initial mass of the rocket, and m is the final mass of the rocket. Calculate v, given that $u = 2.05$ km/s, $m_0 = 1250$ Mg, and $m = 6.35$ Mg.

77. A formula for the area of a triangle with sides a, b, and c is

$$A = \sqrt{s(s - a)(s - b)(s - c)}$$

where s is half the perimeter. Calculate the area of a triangular land parcel with sides 45.78 m, 56.81 m, and 32.17 m.

78. Under specific conditions, an equation relating the pressure P and volume V of a gas is

$$\ln P = C - 1.50 \ln V$$

Find P (in atmospheres) if $C = 3.00$ and $V = 2.20$ ft³.

79. If an electric capacitor C is discharging through a resistor R, the time t of discharge is given by

$$t = RC \ln \frac{I_0}{I}$$

where I is the current and I_0 is the current for $t = 0$. Find the time in seconds if $I_0 = 0.0528$ A, $I = 0.00714$ A, $R = 100$ Ω, and $C = 100$ μF.

80. In computer programming, the job size of the quicksort algorithm is given by

$$N \log_2 N$$

where N is the size of an array. Find the job size for an array of size 256.

In Exercises 81 through 92, use a scientific calculator to find the indicated value.

81. $\log 37.587$ **82.** $\log 0.03294$

83. $\ln 16.183$ **84.** $\ln 0.00088726$

85. Find N if $\log N = 2.34567$ **86.** Find N if $\log N = -2.461426$

87. Find N if $\ln N = 1.34444$ **88.** Find N if $\ln N = -1.0055$

89. 204^{52} **90.** 3^{250}

91. $\sqrt{5^{455}}$ **92.** $(908^{40}) \div (526^{60})$

CHAPTER THIRTEEN

Graphs

In this chapter we develop the basic and standard procedures used in constructing graphs. A **graph** is a drawing that depicts some type of relationship between variable quantities.

We begin in Section 13-1 with the concept of a *function*, and we continue by graphing functions and other types of relationships.

This chapter will include several applied examples and exercises that illustrate the use of graphs. One of the examples involves the loading of a light aircraft. It is essential that the pilot, passengers, baggage, and fuel fall within prescribed limits of weight and balance. In Figure 13-1 we show the graph included in the operating manual for one particular light

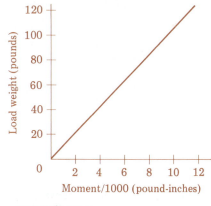

FIGURE 13-1

aircraft. If the baggage weighs 104 lb, what is the corresponding moment/ 1000 (pound-inches)? We will show how to solve such problems. This particular solution will be included later in this chapter.

13-1   Functions

In the earlier chapters we established the basic operations of algebra and discussed the solution of certain types of equations. We also introduced various basic formulas from geometry and other fields. In these formulas and equations one quantity can be expressed in terms of one or more other quantities by the use of the basic operations. It can then be seen that the various quantities of a formula relate in a specific way as denoted by the formula.

If a scientific experiment were to be performed to determine whether or not a relationship exists between the distance an object falls and the time of its fall, observation of the results should indicate (approximately at least) that $s = 16t^2$, where s is the distance (in feet) and t is the time (in seconds) of fall. The experiment would show that the distance and the time are related.

In attempting to find a new oil reserve, it is found that the cost of drilling a well can be determined from the depth of that well. The time it takes a particular missile to reach its target can be computed from the distance to the target.

The percentage of chromium in iron-base alloys affects the rate of corrosion of the alloy. In general, as the percentage of chromium in the alloy increases, the rate of corrosion decreases. It is possible to set up an approximate equation or chart to show the relation of the chromium percentage and corrosion rate.

Considerations such as these lead us to the important mathematical concept of a **function**. In general:

If two variables are related so that for each value of the first variable a single value of the second variable can be determined, the second variable is a function of the first variable.

With $s = 16t^2$ expressing the relationship between distance s and time t, we have the variables s and t related so that for each value of t there is a single value for s. Therefore this equation describes a function. With $s = 16t^2$ it is easier to substitute values for t and calculate the corresponding values for s. If we assign values to t and calculate the corresponding values of s, then t is called the **independent variable** while s is called the **dependent variable**. In general, we assign a value to the independent variable, and the value of the dependent variable is then determined. Functions require that for each value of the independent variable, the dependent variable will have only one corresponding value.

EXAMPLE A The formula that expresses the area of a circle in terms of its radius is $A = \pi r^2$. Here A is a function of r. The variable r is the independent variable, and A is the dependent variable. We can see that the words "dependent" and "independent" are appropriately chosen. As the formula is written, we choose a value of r, and once this is done, the value of A is determined by that choice. For this example, we say that the area of a circle is a function of the radius of the circle. □

EXAMPLE B The stretching force of a steel spring is directly proportional to the elongation of the spring. This can be stated in an equation as $F = kx$. Here x is the independent variable and F is the dependent variable. (k is not a variable; it is a constant.) □

We might wonder why the word "function" is introduced, since the formulas we have been dealing with seem sufficient. *In mathematics the word "function" specifically refers to the operation which is performed on the independent variable to find the dependent variable.* Consider the following example.

EXAMPLE C The distance s (in feet) which an object falls in t seconds is given by $s = 16t^2$. The electric power developed in a 16-Ω resistor by a current of i amperes is given by $P = 16i^2$. Here s is a function of t and P is a function of i. However, the function is the same. That is, to evaluate the dependent variable we square the independent variable and multiply this result by 16. *Even though the letters are different, the* operation *is the same.* □

For convenience of notation and to emphasize the importance of the operational meaning of a function, the phrase "function of x" is written $f(x)$. (*This is a special notation, and does not mean f times x.*) Therefore, to indicate that "y is a function of x", we write $y = f(x)$.

EXAMPLE D For the equation $y = x^2 - 5$, we say that $y = f(x)$, where $f(x) = x^2 - 5$. We now have $y = x^2 - 5$ and $f(x) = x^2 - 5$ as different ways of describing the same function. □

One good use of the functional notation $f(x)$ is to express the value of a function for a specified value of the independent variable. *Thus "the value of the function f(x) when x = a" is expressed as f(a).*

EXAMPLE E   If $f(x) = 5 - 2x$, $f(0)$ is the value of the function for $x = 0$ so that

$$f(0) = 5 - 2(0) = 5 \qquad \text{substitute 0 for } x$$

To find $f(-1)$, we have

$$f(-1) = 5 - 2(-1) = 5 + 2 = 7 \qquad \text{substitute } -1 \text{ for } x$$

We now know that $f(0) = 5$ and $f(-1) = 7$. Note that to find the value of the function for a number specified on the left, we substitute this number into the function on the right. Also, if we state that $y = f(x)$, then $y = 5$ for $x = 0$ and $y = 7$ for $x = -1$.   □

EXAMPLE F   If $f(x) = 4x - 2x^2$, find $f(-3)$.
To find $f(-3)$, we substitute -3 for x in the function. This gives us

$$
\begin{aligned}
f(-3) &= 4(-3) - 2(-3)^2 \qquad \text{substitute } -3 \text{ for } x \\
&= -12 - 2(9) \\
&= -12 - 18 \\
&= -30
\end{aligned}
$$
□

EXAMPLE G   If $f(x) = x^2 + 1$, find $f(a)$, $f(-a)$, and $f(a + 1)$.
To find $f(a)$, we substitute a for x in the given equation to get

$$f(a) = a^2 + 1 \qquad \text{substitute } a \text{ for } x$$

We can find $f(-a)$ by substituting $-a$ for x as follows:

$$f(-a) = (-a)^2 + 1 = a^2 + 1 \qquad \text{substitute } -a \text{ for } x$$

Finally, we evaluate $f(a + 1)$ by the same procedure.

$$\text{substitute } a + 1 \text{ for } x$$
$$f(a + 1) = (a + 1)^2 + 1 = a^2 + 2a + 1 + 1 = a^2 + 2a + 2$$
□

It is possible to identify functions from verbal statements. Such functions may be based upon known formulas or a proper interpretation of a given statement.

EXAMPLE H   The cost of fencing in a square prison farm is $20 per foot of perimeter. Express the total cost as a function of the length of a side.

The perimeter of a square is four times the length of a side and the total length of that perimeter is to be multiplied by $20. This is expressed as

cost per foot of perimeter
perimeter

$$C = 20(4s)$$
$$= 80s$$

The total cost, as a function of the length of a side, is $C = 80s$.   □

In the definition of a function, it was stipulated that any value for the independent variable must yield only a single value of the dependent variable. This requirement is stressed in more advanced and theoretical courses. *If a value of the independent variable yields more than one value of the dependent variable, the relationship is called a* **relation** instead of a function. A relation involves two variables related so that values of the second variable can be determined from values of the first variable. A function is a relation in which each value of the first variable yields only one value of the second. A function is therefore a special type of relation. However, there are relations that are not functions.

EXAMPLE I   For $y^2 = x^2$, if $x = 2$, then y can be either $+2$ or -2. Since a value of x yields more than one single value for y, we have a relation, not a function.   □

One last comment regarding functions is in order at this time. Not all functions are valid for all values of the independent variable, nor are all values of the dependent variable always possible. The following example illustrates this point.

EXAMPLE J   In Example A where the area of a circle is given by the function $A = \pi r^2$, there is no real meaning for negative values of the radius or the area. Thus we would restrict values of r and A to zero and greater. That is, $r \geq 0$ and $A \geq 0$.

If we consider only real numbers, $f(x) = \sqrt{x - 1}$ is valid only for values of x greater than or equal to 1. That is, $x \geq 1$. Also, since *the positive square root is indicated*, the values of the function are zero or greater.

If $f(x) = 1/x$, all values of x are possible except $x = 0$, because that value would necessitate division by zero, which is undefined.

The **domain** *of a function is the set of all numbers that can be substituted for the independent variable*, while *the* **range** *is the set of all possible values that the dependent variable can assume.* For $f(x) = \sqrt{x - 1}$, the domain is described by $x \geq 1$ while the range is the collection of all values zero or greater ($y \geq 0$).   □

Exercises 13-1

In Exercises 1 through 4, identify the dependent and the independent variable of each function.

1. $y = 3x^4$

2. $s = -16t^2$

3. $p = \dfrac{c}{V}$ (*c* is a constant)

4. $v = a(1.05)^t$ (*a* is a constant)

In Exercises 5 through 8, state the basic operation of the function which is to be performed on the independent variable. For example, with $f(x) = x^2$ we "square the value of the independent variable."

5. $f(x) = 3x$

6. $f(y) = y + 3$

7. $f(r) = 2 - r^2$

8. $f(s) = s + s^2$

In Exercises 9 through 12, indicate the function expressed by each of the given equations by replacing the dependent variable by the $f(x)$ type of notation. (See Example D.)

9. $y = 5 - x$

10. $F = 6q^2$

11. $v = t^2 - 3t$

12. $s = 10^{-5r}$

In Exercises 13 through 28, find the indicated values of the given functions whenever possible.

13. $f(x) = x$; $f(0), f(3)$

14. $f(x) = x + 2$; $f(0), f(-3)$

15. $f(t) = 2t - 1$; $f(4), f(-2)$

16. $f(r) = 2 - r$; $f(3), f(-3)$

17. $f(x) = 3x^2 - 2$; $f(0), f\left(\dfrac{1}{2}\right)$

18. $f(z) = z^2 - z$; $f(2), f(-2)$

19. $f(x) = 3 - x^2$; $f(2), f(-0.3)$

20. $f(x) = 2x - 3x^2$; $f(-1), f(-3)$

21. $f(s) = s^3$; $f(-1), f(2)$

22. $f(p) = p^3 + 2p - 1$; $f(a)$

23. $f(t) = 3t - t^3$; $f(1), f(-2)$

24. $f(x) = x^3 - x^4$; $f(3), f(-3)$

25. $f(q) = \dfrac{q}{q - 3}$; $f(-3), f(3)$

26. $f(v) = v + \dfrac{1}{v}$; $f\left(\dfrac{1}{5}\right), f(0)$

27. $f(x) = x - 2x^2$; $f(a^2), f\left(\dfrac{1}{a}\right)$

28. $f(x) = 4^x$; $f(-2), f\left(\dfrac{1}{2}\right)$

In Exercises 29 through 32, determine which of the given relations are also functions.

29. $y = x^3$

30. $y = x^4$

31. $y = \pm\sqrt{x}$

32. $|y| = |x|$

In Exercises 33 through 44, find the indicated functions.

33. Express the circumference c of a circle as a function of its radius r.

34. Express the volume V of a cube as a function of an edge e.

35. Express the area A of a square as a function of its perimeter p.

36. Express the area A of a circle as a function of its diameter d.

37. The number of tons t of air pollutants emitted by burning a certain fuel is 0.02 times the number of tons n of the fuel which are burned. Write the function suggested by this statement.

38. If the sales tax on a drill press is 5%, express the total cost C as a function of the list price P.

39. Express the distance s in kilometers traveled in 3 h as a function of the velocity v in kilometers per hour.

40. A company installs underground cable at a cost of $500 for the first 50 ft and $5 per foot thereafter. Express the cost C as a function of the length l of underground cable.

41. One computer printer can produce 1500 more lines per minute than a second printer. Express the rate R_1 of the first printer as a function of the rate R_2 of the second printer.

42. The corn production of one field is 5000 bushels less than twice the production of a second field. Express the production p_1 of the first field as a function of the production p_2 of the second field.

43. The voltage V across an electric resistor varies directly as the current i in the resistor. Given that the voltage is 10 V when the current is 2 A, express the current as a function of the voltage.

44. The stopping distance d of a car varies directly as the square of its speed s. Express d as a function of s, given that $d = 80$ m when $s = 100$ km/h.

13-2 The Rectangular Coordinate System

In this section we discuss the basic system for representing the graph of a function. In Chapter 3 we saw how numbers can be represented on a line. Since a function represents two related sets of numbers, it is necessary to use such a number line for each of these sets. This is most conveniently done if the lines intersect at right angles.

One line is placed horizontally and labeled the x **axis**. *It is labeled just as the number line in Chapter 3, with a point chosen as zero and labeled 0, the* **origin**. *The positive values are to the right and the negative values to the left of the origin.* It is customary to use the numbers on the x axis for the values of the independent variable.

Through the origin, and perpendicular to the x axis, another line is placed and labeled the **y axis**. *The origin is also zero for values on the y axis, with positive values above and negative values below.* The y axis is customarily used for values of the dependent variable.

The x axis and y axis as described above constitute what is known as the **rectangular coordinate system** *or* **Cartesian coordinate system**. *The four regions into which the plane is divided by the axes are known as* **quadrants** and are numbered I, II, III, IV as shown in Figure 13-2.

Any point P in the plane of the rectangular coordinate system is designated by the pair of numbers (x, y) always written in this order. Here x is the value of the independent variable and y is the corresponding value of the dependent variable. *It can be seen that the x value (called the* **abscissa**) *is the perpendicular distance of P from the y axis, and the y value (called the* **ordinate**) *is the perpendicular distance of P from the x axis. The values of x and y are called the* **coordinates** *of the point P.*

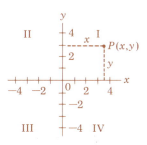

FIGURE 13-2

EXAMPLE A Locate the points $A(2, 3)$ and $B(-4, -3)$ on the rectangular coordinate system. (The point $A(2, 3)$ is the point A which has $(2, 3)$ as its coordinates.)

The coordinates $(2, 3)$ for point A means that the point is 2 units to the *right* of the y axis and 3 units *above* the x axis, as shown in Figure 13-3. The coordinates $(-4, -3)$ for B means that the point is 4 units to the *left* of the y axis and 3 units *below* the x axis, as shown.

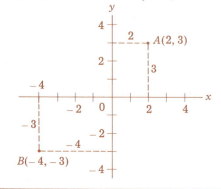

FIGURE 13-3

EXAMPLE B In attempting to analyze the relationship between an index of hardness and an index of color durability for different paint samples, it becomes necessary to plot the points $P(4, 1)$, $Q(-2, 5)$, $R(-3, -2)$, $S(4, -3)$, $T(-5, 0)$, and $V(0, 2)$. Those points are plotted in Figure 13-4. Note that this representation of points allows *only one point for any pair of values* (x, y). Conversely, any pair of values (x, y) will correspond to only one point.

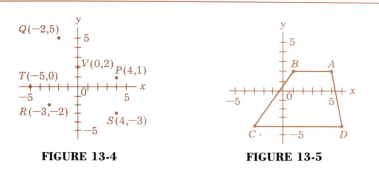

FIGURE 13-4   **FIGURE 13-5**

EXAMPLE C See Figure 13-5. If we let the points $A(5, 2)$, $B(1, 2)$, $C(-3, -4)$, and $D(6, -4)$ be the vertices of a quadrilateral and connect them in order, we see that the resulting geometric figure is a trapezoid. The upper base is seen to be parallel to the x axis, since the y coordinates are the same. This is also true of the lower base. Since they are both parallel to the x axis, they are parallel to each other.

EXAMPLE D What is the sign of the ratio of the abscissa to the ordinate of a point in the second quadrant?

Since any point in the second quadrant is above the x axis, the y value (ordinate) is positive. Also, since any point in the second quadrant is to the left of the y axis, the x value (abscissa) is negative. The ratio of a negative number to a positive number is negative, which is the required answer. See Figure 13-6.

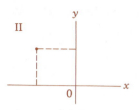

FIGURE 13-6

EXAMPLE E In what quadrant does a point (a, b) lie if $a < 0$ and $b < 0$?

A point for which the abscissa is negative (which is the meaning of $a < 0$) must be in either the second or third quadrant. A point for which the ordinate is negative is in either the third or fourth quadrant. See Figure 13-7. The x coordinate must be in the second or third quadrant, and the y coordinate must be in the third or fourth quadrant. This implies that the point must be in the third quadrant.

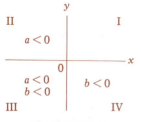

FIGURE 13-7

Exercises 13-2

In Exercises 1 through 8, determine (at least approximately) the coordinates of the points shown in Figure 13-8.

1. A, B
2. C, D (-2,-2)
3. E, F
4. G, H
5. I, J
6. K, L
7. M, N
8. P, Q

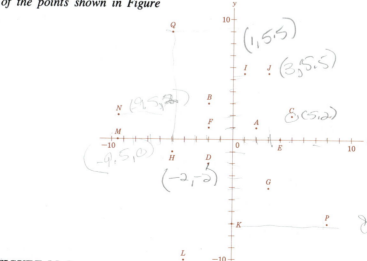

FIGURE 13-8

In Exercises 9 through 16, plot (at least approximately) the given piants.

9. $A(0, 3)$; $B(3, 0)$ 10. $A(-3, -6)$; $B(2, -4)$ 11. $A(-1, 4)$; $B(2, 5)$

12. $A(0, -6)$; $B(-3, 4)$ 13. $A(\frac{1}{2}, 5)$; $B(-\frac{1}{2}, -5)$ 14. $A(0, \frac{1}{2})$; $B(\frac{1}{4}, 2)$

15. $A(2.5, -3.5)$; $B(0, -3.4)$ 16. $A(-5.5, 8.1)$; $B(\frac{3}{2}, -3)$

In Exercises 17 through 20, identify the quadrant in which each point is located.

17. $(2, 12)$; $(-3, 4)$ 18. $(5, -2)$; $(-2, -8)$ 19. $(-3, -5)$; $(-2, 1)$ 20. $(7, -10)$; $(-9, 2)$

In Exercises 21 through 24, draw the geometric figures, the coordinates of whose vertices are given (do not cross lines).

21. Triangle: $A(0, 3)$; $B(5, -1)$; $C(2, 3)$

22. Quadrilateral: $A(-1, -2)$; $B(0, -3)$; $C(1, 3)$; $D(-2, 4)$

23. Rectangle: $A(-2, -2)$; $B(5, -2)$; $C(5, 4)$; $D(-2, 4)$

24. Rhombus: $A(-1, -2)$; $B(2, 2)$; $C(7, 2)$; $D(4, -2)$

In Exercises 25 through 36, answer the questions.

25. What is the value of x for each point on the y axis?

26. What is the value of y for each point on the x axis?

27. In which quadrants is the ratio of the abscissa to the ordinate positive?

28. In which quadrants is the ratio of the abscissa to the ordinate negative?

29. In what quadrant does a point (a, b) lie if $a > 0$ and $b < 0$?

30. In what quadrant does a point (a, b) lie if $a < 0$ and $b > 0$?

31. What are the coordinates of the vertices of a square with side 2.5 in., whose center is at the origin and whose sides are parallel to the axes?

32. Describe the type of triangle whose vertices are at $(0, 3)$, $(0, 0)$, and $(-2, 0)$.

33. Describe the type of triangle whose vertices are at $(-2, 0)$, $(2, 0)$, and $(0, 3)$.

34. Three vertices of a rectangle are $(1, -3)$, $(7, -3)$, and $(7, 2)$. What are the coordinates of the fourth vertex?

35. The vertices of the base of an isosceles triangle are $(-2, -3)$ and $(8, -3)$. What is the abscissa of the third vertex?

36. Describe the location of all points whose abscissas equal the ordinates.

13-3 The Graph of a Function

Having introduced the rectangular coordinate system in the last section, we are now ready to sketch the graph of a function. In this way we can get a "picture" of the function, and this picture allows us to see the behavior and properties of the function. In this section we restrict our attention to functions whose graphs are either **straight lines** or **parabolas** (to be explained soon).

The graph of a function consists of all points whose coordinates (x, y) satisfy the functional relationship y = f(x). By choosing a specific value for x, we can then find the corresponding value for y by evaluating $f(x)$. In this way we obtain as many points as necessary to plot the graph of the function.

Usually we need to find only enough points to get a good approximation to the graph by joining these points with a smooth curve.

We shall first consider the graph of a **linear function**, which is a function of the form

$$f(x) = ax + b \qquad\qquad (13\text{-}1)$$

Here a and b are constants. *It is called* **linear**, *since the graph of such a function is always a straight line*. The following example illustrates the basic technique.

EXAMPLE A Graph the function described by $f(x) = 3x + 2$.

Since $f(x) = 3x + 2$, by letting $y = f(x)$ we may write $y = 3x + 2$. Now by substituting numbers for x, we can find the corresponding values for y. If $x = 0$, then

$$y = f(0) = 3(0) + 2 = 2$$

so that the point $(0, 2)$ is on the graph of $f(x) = 3x + 2$. If $x = 1$, then

$$y = f(1) = 3(1) + 2 = 5$$

This means that the point $(1, 5)$ is on the graph of $f(x) = 3x + 2$. This information will be most helpful when it appears in tabular form. In preparing the table, *list the values of x in order* so that the points indicated can be joined in order. After the table has been set up, the indicated points are plotted on a rectangular coordinate system and then joined. The table and graph for this function are shown in Figure 13-9. Note that the line in the graph is extended beyond the points found in the table. This indicates that we know it continues in each direction.

The table of x and y values can be constructed either vertically or horizontally. An advantage of the vertical format is that it becomes easier to read the coordinates of the different points.

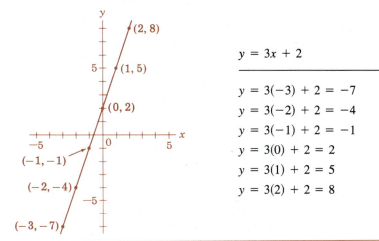

$y = 3x + 2$	x	y
$y = 3(-3) + 2 = -7$	-3	-7
$y = 3(-2) + 2 = -4$	-2	-4
$y = 3(-1) + 2 = -1$	-1	-1
$y = 3(0) + 2 = 2$	0	2
$y = 3(1) + 2 = 5$	1	5
$y = 3(2) + 2 = 8$	2	8

FIGURE 13-9

The knowledge that a function of the form $y = ax + b$ is a straight line can be used to definite advantage. By finding two points, we can draw the line. Two special points easily found are those where the curve crosses each of the axes. *These points are known as the* **intercepts** *of the line.* The reason these are easily found is that one of the coordinates is zero. We find the **y intercept** by setting $x = 0$ and determining the corresponding value of y where the line crosses the y axis. We find the **x intercept** by setting $y = 0$ and determining the corresponding value of x where the line crosses the x axis. A third point should be found as a check.

EXAMPLE B   Graph the function $f(x) = -2x - 5$.

By setting $y = -2x - 5$ and then letting $x = 0$, we obtain $y = -5$. This means that the graph passes through the point $(0, -5)$. Now, setting $y = 0$ and solving for x, we have $x = -\frac{5}{2}$. This means the graph passes through the point $(-\frac{5}{2}, 0)$. As a check, we let $x = 1$ and find $y = -7$, meaning that the line passes through $(1, -7)$. The table follows and the graph is shown in Figure 13-10.

x	y
$-\frac{5}{2}$	0
0	-5
1	-7

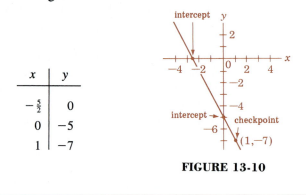

FIGURE 13-10

If a line passes through the origin, it has only this one intercept $(0, 0)$. In this case it is necessary to find at least one point other than the intercept in order to plot the graph. However, the origin and two other points (one as a check) are easily found. Exercises 1 through 4 at the end of this section illustrate this type of straight line.

We now consider the graph of the **quadratic function.**

$$f(x) = ax^2 + bx + c \quad \text{where} \quad a \neq 0 \qquad (13\text{-}2)$$

Here a, b, and c are constants. *The* **graph** *of this function is a* **parabola.** The examples that follow illustrate this function and its graph. Note that the right-hand side of Eq. (13-2) is the same expression as that appearing in the quadratic equations of Chapter 11.

EXAMPLE C   A parabolic antenna is to be constructed so that its cross section is described by the equation $y = \frac{1}{2}x^2$. Graph the function $y = \frac{1}{2}x^2$.

Choosing values of x and obtaining corresponding values of y, we can determine the coordinates of a set of representative points on the graph. In finding the y values we must *be careful in handling negative values of x*: Remember that the square of a negative number is a positive number. For example, if $x = -3$,

$$y = \frac{1}{2}(-3)^2 = \frac{1}{2}(9) = \frac{9}{2}$$

With these ideas in mind we obtain the following table for $y = \frac{1}{2}x^2$: Note that the points are connected in Figure 13-11 with a smooth curve, not straight-line segments.

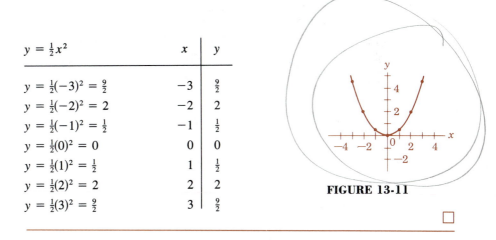

$y = \frac{1}{2}x^2$	x	y
$y = \frac{1}{2}(-3)^2 = \frac{9}{2}$	-3	$\frac{9}{2}$
$y = \frac{1}{2}(-2)^2 = 2$	-2	2
$y = \frac{1}{2}(-1)^2 = \frac{1}{2}$	-1	$\frac{1}{2}$
$y = \frac{1}{2}(0)^2 = 0$	0	0
$y = \frac{1}{2}(1)^2 = \frac{1}{2}$	1	$\frac{1}{2}$
$y = \frac{1}{2}(2)^2 = 2$	2	2
$y = \frac{1}{2}(3)^2 = \frac{9}{2}$	3	$\frac{9}{2}$

FIGURE 13-11

Figure 13-11 shows the basic shape of a parabola, which is the graph of a quadratic function. The parabola may be shifted right or left, and up or down, or it may open down. When graphing a parabola, it is important to find where the curve stops falling (or rising) and begins to rise (or fall). *The point at which this change in vertical direction occurs is called the* **vertex.** Instead of randomly plotting points on the parabola, we can use Eq. (13-3) to find

$$x = \frac{-b}{2a} \quad \text{where } y = ax^2 + bx + c \qquad \text{\small x coordinate of vertex} \qquad (13\text{-}3)$$

the x coordinate of the vertex. The y coordinate of the vertex can be found by substituting the x coordinate into the original quadratic function. The points we plot should include the vertex and points to its right and left. This is illustrated in the following example.

EXAMPLE D   Graph the function $y = 2x^2 - 4x - 7$.

Instead of selecting points at random, we begin by using Eq. (13-3) to find the x coordinate of the vertex. From the given quadratic function we have $a = 2$, $b = -4$, and $c = -7$ so that

$$x = \frac{-b}{2a} = \frac{-(-4)}{2(2)} = 1$$

The x coordinates we select should include $x = 1$ and some values greater than 1 and less than 1. We get the following table which is used to develop the graph shown in Figure 13-12.

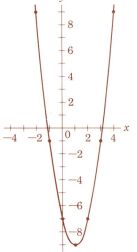

FIGURE 13-12

$y = 2x^2 - 4x - 7$	x	y
$y = 2(-2)^2 - 4(-2) - 7 = 9$	-2	9
$y = 2(-1)^2 - 4(-1) - 7 = -1$	-1	-1
$y = 2(0)^2 - 4(0) - 7 = -7$	0	-7
$y = 2(1)^2 - 4(1) - 7 = -9$	1	-9
$y = 2(2)^2 - 4(2) - 7 = -7$	2	-7
$y = 2(3)^2 - 4(3) - 7 = -1$	3	-1
$y = 2(4)^2 - 4(4) - 7 = 9$	4	9

In Example D, if we include more values of x, we may have to adjust the *scale* of the vertical axis. That is, the intervals along the y axis may be changed so that the graph will fit.

EXAMPLE E   Graph the function $y = 4 - 2x - x^2$.

The given function is quadratic with $a = -1$, $b = -2$, and $c = 4$ (note that a is the coefficient of x^2 even though in this case it is the third term) so that we can find the x coordinate of the vertex as follows:

$$x = \frac{-b}{2a} = \frac{-(-2)}{2(-1)} = -1$$

The table should incorporate $x = -1$ and values above and below -1. The table and corresponding graph are shown in Figure 13-13.

In finding values for the table, *special care must be used in determining y values for negative values of x*. For example, for $x = -2$, we have

$$y = 4 - 2(-2) - (-2)^2 = 4 + 4 - (4) = 4$$

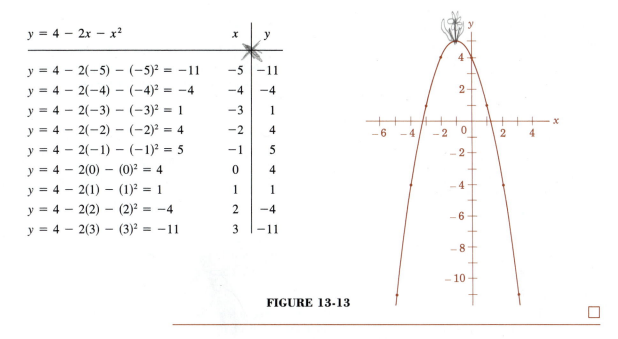

$y = 4 - 2x - x^2$

	x	y
$y = 4 - 2(-5) - (-5)^2 = -11$	-5	-11
$y = 4 - 2(-4) - (-4)^2 = -4$	-4	-4
$y = 4 - 2(-3) - (-3)^2 = 1$	-3	1
$y = 4 - 2(-2) - (-2)^2 = 4$	-2	4
$y = 4 - 2(-1) - (-1)^2 = 5$	-1	5
$y = 4 - 2(0) - (0)^2 = 4$	0	4
$y = 4 - 2(1) - (1)^2 = 1$	1	1
$y = 4 - 2(2) - (2)^2 = -4$	2	-4
$y = 4 - 2(3) - (3)^2 = -11$	3	-11

FIGURE 13-13

Examples C, D, and E confirm that when graphing $y = ax^2 + bx + c$, if $a > 0$ the parabola opens upward and the vertex is a minimum point. If $a < 0$, the parabola opens downward and the vertex is a maximum point.

In addition to computers, there are now calculators that can be used to display graphs of functions. Such calculators have rectangular display windows capable of producing graphs as well as other data. We could use such a calculator to produce the graph of $y = 4 - 2x - x^2$, and the result is essentially everything shown in Figure 13-13 with the actual numbers and axis labels excluded.

Exercises 13-3

In Exercises 1 through 28, graph the given functions.

1. $y = x$ **2.** $y = -x$ **3.** $y = -2x$ **4.** $y = 4x$

5. $y = 3x - 5$ **6.** $y = 2x + 14$ **7.** $y = 9 - 4x$ **8.** $y = 8 - 2x$

9. $y = 6x - 1$ **10.** $y = -5x + 3$ **11.** $y = \frac{1}{2}x + 1$ **12.** $y = \frac{1}{3}x - 2$

13. $y = -x + \frac{1}{3}$ **14.** $y = 2x - \frac{1}{2}$ **15.** $y = \frac{2 - x}{3}$ **16.** $y = \frac{4 + 3x}{2}$

17. $y = x^2$ **18.** $y = -x^2$ **19.** $y = x^2 - 4$ **20.** $y = 4 - x^2$

21. $y = 2x - x^2$ **22.** $y = x^2 - 2x$ **23.** $y = \frac{1}{3}x^2$ **24.** $y = \frac{1}{4}x^2$

25. $y = 1 - \frac{1}{2}x^2$ **26.** $y = 2 + \frac{1}{3}x^2$ **27.** $y = 2x^2 - 3x + 2$ **28.** $y = 3 - x - x^2$

In Exercises 29 through 32, find the coordinates of the intercepts of each line.

29. $y = x + 3$ **30.** $y = 2x + 8$ **31.** $y = -3x + 6$ **32.** $y = -5x - 8$

In Exercises 33 through 44, graph the functions by plotting the dependent variable along the y axis and the independent variable along the x axis. In the applied problems, be certain to determine whether or not negative values of the variables and the scales on each axis are meaningful.

33. A spring is stretched x in. by a force F. The equation relating x and F is

$$F = kx$$

where k is a constant. A force of 10 lb stretches a given spring 2 in. Plot the graph for F and x.

34. The *mechanical advantage* of an inclined plane is the ratio of the length of the plane to its height. This can be expressed in a formula as

$$M = \frac{L}{h}$$

Suppose that the height of a given plane is 2 m. Plot the graph of the mechanical advantage and length.

35. The electric resistance of wire resistors varies with the temperature according to the relation

$$R_2 = R_1 + R_1\alpha(T_2 - T_1)$$

where R_2 is the resistance at temperature T_2, R_1 is the resistance at temperature T_1, and α (the Greek alpha) is a constant depending on the type of wire. Plot the graph of R_2 and T_2 for a copper wire resistor ($\alpha = 0.004/°C$), given that $R_1 = 20\ \Omega$ and $T_1 = 10°C$.

36. Under certain conditions, the price p for each of x units of a commodity is given by

$$p = -\frac{3}{4}x + 6$$

Plot the graph.

37. A firm can produce up to 500 units per day of an item which sells for $2. Fixed costs are $200 daily and each item costs 50¢ to produce. Find the equation relating the profit p and the number x of units produced daily. Plot the graph.

38. The formula relating the Celsius and Fahrenheit temperature scales is

$$F = \frac{9}{5}C + 32$$

Plot the graph of this function.

39. The cross-sectional shape of a radar antenna is parabolic with a shape described by the equation

$$y = 0.025x^2$$

Plot the graph of this function.

40. The shape of the surface of a fluid in a rotating container is parabolic. The equation of the parabola is

$$y = \frac{\omega^2}{2g}x^2$$

where ω (the Greek omega) is the angular velocity of the container and g is the acceleration due to gravity. Plot the shape of the surface of a liquid [the origin is at the lowest (middle) point of the surface], given that $\omega = 10$ r/s and that $g = 32$ ft/s². Assume that the container is 4 ft in diameter.

41. Under certain conditions, the relation between the mass W per unit area of an oxide forming on metal and the time t of oxidation is

$$t = kW^2$$

where k is a constant depending on the metal. Plot the graph for W and t for a metal where $k = 0.25 \text{ h} \cdot \text{m}^4/\text{g}^2$, t is measured in hours, and W is in grams per square meter.

42. Express the total surface area of a cube as a function of an edge and plot the graph.

43. The distance h (in feet) above the surface of the earth of an object as a function of the time (in seconds) is given by

$$h = 60t - 16t^2$$

if the object is given an initial upward velocity of 60 ft/s. Plot the graph.

44. The mass per unit length m of a bridge at a distance of x meters from the center of the bridge is

$$m = 150 + 0.8x^2$$

Plot the graph of m and x, given that the bridge is 100 m long and m is measured in kilograms per meter.

In Exercises 45 through 48, use a calculator or computer capable of displaying a graph to obtain graphs of the given functions.

45. $y = 5x - 3$ 46. $y = -2.7x + 1.4$ 47. $y = x^2$ 48. $y = -2.3x^2 - 1.1x - 2.0$

13-4 Graphs of Other Functions

In the preceding section we discussed the method of plotting the graph of a function. However, we limited our discussion to only linear and quadratic functions. This section introduces the graphs of several other functions.

We use the basic procedure presented in the last section. That is, we select specific values of x and find the corresponding values of y in order to find particular points which lie on the graph of the function. Following are four examples of plotting the graphs of functions which are not linear or quadratic.

EXAMPLE A

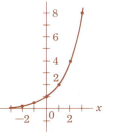

FIGURE 13-14

A population growth model is described by the equation $y = 2^x$. Graph the function $y = 2^x$.

$y = 2^x$

	x	y
$y = 2^{-2} = \frac{1}{2^2} = \frac{1}{4}$	-2	$\frac{1}{4}$
$y = 2^{-1} = \frac{1}{2}$	-1	$\frac{1}{2}$
$y = 2^0 = 1$	0	1
$y = 2^1 = 2$	1	2
$y = 2^2 = 4$	2	4
$y = 2^3 = 8$	3	8

In order to plot the graph of this function, *we must deal with powers of 2, including both positive and negative powers*. Note the evaluations shown with the table at the left, from which the points are plotted in Figure 13-14.

This is an example of the graph of an **exponential function**.   □

In general, an exponential function will be of the form $y = b^x$ where b is a positive constant. The shape of any other exponential function will be similar to the graph of Figure 13-14, although it may be downward instead of upward.

EXAMPLE B

Graph the function $y = 2 + \dfrac{1}{x}$.

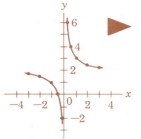

FIGURE 13-15

In finding the points for this graph we must be careful not to set $x = 0$, because division by zero is undefined. To get an accurate graph, however, we must choose values of x near zero. Consequently, *we include fractional values of x* in the following table. For example, if $x = \frac{1}{2}$, then

$$y = 2 + \frac{1}{\frac{1}{2}} = 2 + 2 = 4$$

This graph is known as a **hyperbola** (see Figure 13-15).

x	-3	-2	-1	$-\frac{1}{2}$	$-\frac{1}{4}$	$\frac{1}{4}$	$\frac{1}{2}$	1	2	3
y	$\frac{5}{3}$	$\frac{3}{2}$	1	0	-2	6	4	3	$\frac{5}{2}$	$\frac{7}{3}$

☐

EXAMPLE C

Graph the function $y = x^3 - 3x$.

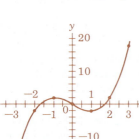

FIGURE 13-16

Proper use of signed numbers and their powers is essential in finding the values of y for values of x, *particularly negative values*. For example, if $x = -2$, then

$$y = (-2)^3 - 3(-2) = -8 + 6 = -2$$

The following table gives the points used to plot the graph in Figure 13-16.

x	-3	-2	-1	0	1	2	3
y	-18	-2	2	0	-2	2	18

This curve is known as a **cubic curve**. Note that *the scale of the y axis is different from the scale of the x axis*. This is done when the range of values used differs considerably.

☐

EXAMPLE D

Graph the function $y = \sqrt{x + 1}$.

In finding the y values of this graph, we must be careful to use only values of x which will give us real values for y. This means that the value under the radical sign must be zero or greater. For example, if $x = -5$, we would have $y = \sqrt{-4}$, which is an imaginary number. Therefore x cannot be -5. In fact, *we cannot have any value of x which is less than -1*. Also, we must note that values of y will be greater than zero, except for $x = -1$.

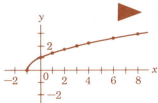

FIGURE 13-17

This is due to the fact that *y is equal to the principal (positive) square root of x + 1*. Thus we find the following points for the graph:

x	-1	0	1	2	3	4	6	8
y	0	1	1.4	1.7	2	2.2	2.6	3

The graph is plotted in Figure 13-17. This graph is half of a parabola. All quadratic functions ($y = ax^2 + bx + c$) graph as parabolas that open up or down, but this function graphs as half of a parabola that opens to the right. □

Examples A, B, C, and D illustrated an exponential function, a hyperbola, a cubic curve, and a parabola, respectively. There are many other functions that do not belong to one of these categories. Some of the following exercises involve functions that we have not illustrated, but the same basic procedure applies. Substitute values for x, find the corresponding values of y, and plot the points that result. Then connect the points with a smooth curve.

Exercises 13-4

In Exercises 1 through 24, graph the given functions.

1. $y = 3^x$

2. $y = \left(\frac{1}{2}\right)^x$

3. $y = 3^{2x}$

4. $y = 2^{-x}$

5. $y = \frac{1}{x}$

6. $y = 1 + \frac{2}{x}$

7. $y = 2 - \frac{1}{x}$

8. $y = \frac{3}{x} - 4$

9. $y = \sqrt{x}$

10. $y = \sqrt{x + 4}$

11. $y = \sqrt{1 - x}$

12. $y = \sqrt{4 - x}$

13. $y = x^3$

14. $y = -\frac{1}{2}x^3$

15. $y = 2x^3 - 10$

16. $y = 6x - x^3$

17. $y = \frac{1}{4}x^4$

18. $y = x^4 - 2x^2$

19. $y = \frac{1}{x^2}$

20. $y = \frac{1}{x^2 + 1}$

21. $y = \sqrt{x^2 + 1}$

22. $y = \sqrt{25 - x^2}$

23. $y = \frac{1}{\sqrt{x}}$

24. $y = \frac{1}{\sqrt{1 - x}}$

In Exercises 25 through 36, graph the functions by plotting the dependent variable along the y axis and the independent variable along the x axis. In the applied problems, be certain to determine whether or not negative values of the variables and the scales on each axis are meaningful.

25. The number of bacteria in a certain culture increases by 50% each hour. The number N of bacteria present after t hours is

$$N = 1000\left(\frac{3}{2}\right)^t$$

given that 1000 were originally present. Graph the function.

26. Under the condition of constant temperature, it is found that the pressure p (in kilopascals) and the volume V (in cubic centimeters) in an experiment on air are related by

$$p = \frac{1000}{V}$$

Plot the graph.

27. The electric current I in a circuit with a voltage of 50 V, a constant resistor of 10 Ω, and a variable resistor R is given by the equation

$$I = \frac{50}{10 + R}$$

Plot the graph of I and R.

28. An object is p in. from a lens of a focal length of 5 in. The distance from the lens to the image is given by

$$q = \frac{5p}{p - 5}$$

Plot the graph.

29. The total profit P a manufacturer makes in producing x units of a commodity is given by

$$P = x^3 - 3x^2 - 5x - 150$$

Plot the graph, using values of 0 through 10 for x.

30. Under certain conditions the deflection d of a beam at a distance of x ft from one end is

$$d = 0.05(30x^2 - x^3)$$

where d is measured in inches. Plot the graph of d and x, given that the beam is 10 ft long.

31. After t years, the population of trees planted as a forestry experiment is given by

$$P = 100(1.02)^t$$

Plot the graph.

32. The amount N of a radioactive material present after t seconds is found by using the formula

$$N = N_0 e^{-kt}$$

where N_0 is the amount originally present, $e = 2.7$ (see Section 12-3), and k is a constant depending on the material. Given that 200 g of a certain isotope of uranium, for which $k = 0.01$ per second, was originally present, establish the function relating N and t and plot the graph. Use logarithms for the calculations and use 20-s intervals for t.

33. The gravitational force of attraction between two objects is inversely proportional to the square of the distance between their centers. Assuming the constant of proportionality to be 8 units, plot the graph.

34. In analyzing the magnitude of earthquake intensities, it becomes necessary to use the function

$$R = \log Q$$

Plot the graph of that function.

35. In a direct-current circuit, capacitance is determined by the equation

$$C = \frac{Q}{V}$$

where C is the capacitance in farads, Q is the charge in coulombs, and V is the voltage in volts. Plot the graph if Q is a constant 120 μC.

36. If an amount of P dollars is deposited in a bank, the value V of the account after time t (in years) is given by

$$V = P(1 + r)^t$$

where r is the interest rate in decimal form. Let $P = 100$, let $r = 0.05$, and plot the graph.

13-5 Graphical Interpretations

In the previous sections of this chapter we have seen how the graph of a function is constructed. Another important aspect of working with graphs is being able to read information from a graph. This section demonstrates how this is done and also how equations can be solved by the use of graphs.

The procedure for reading values from a graph is essentially the reverse of plotting the coordinates of points. The following examples illustrate the method.

EXAMPLE A   When operating a light aircraft, it is essential that the pilot, passengers, baggage, and fuel do not exceed the maximum allowable safe load. It is also essential that those items are distributed so that they fall within prescribed center-of-gravity limitations. One particular aircraft operating manual provides the loading graph for baggage shown in Figure 13-18. If the baggage weighs 104 lb, find the corresponding value of the moment/1000 (pound-inches).

Knowing that the baggage weighs 104 lb, we locate that weight on the vertical scale and we then move horizontally to meet the graph of the line at the point shown. At that point, the horizontal scale has a coordinate of about 9.9 pound-inches. This procedure is repeated for the graphs representing the pilot, passengers, and fuel, and the combined results are then used to determine whether the aircraft is correctly balanced. Every pilot must be able to use this procedure.

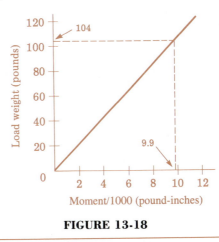

FIGURE 13-18

EXAMPLE B   From the graph shown in Figure 13-19, determine the value of y for $x = 3.5$.

  We first locate 3.5 on the x axis and then construct a line (dashed line in the figure) perpendicular to the x axis *until it crosses the curve.* From this point on the curve, we draw another line perpendicular to the y axis *until it crosses the y axis.* The value at which this line crosses is the required answer. Therefore $y = 1.4$ (approximated from the graph) for $x = 3.5$. In general, when given the x coordinate of a point of a graph, we move vertically (up or down) until we meet the graph and we can then approximate the y coordinate. Also, if we know the y coordinate, we can move horizontally (right or left) until we meet the curve. The corresponding x coordinate can then be approximated.   □

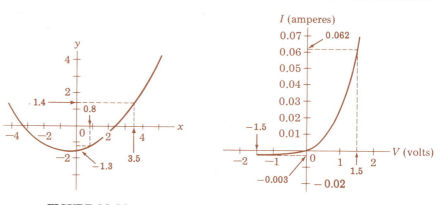

FIGURE 13-19   FIGURE 13-20

EXAMPLE C   The current as a function of the voltage for a typical type of transistor is shown in Figure 13-20. We can see that for a voltage of -1.5 V (voltage and current can be considered as having direction) the current is -0.003 A. For a voltage of 1.5 V the current is about 0.062 A.   □

EXAMPLE D   A chemical explosion propels an object upward and its distance (in feet) above ground level is related to time t (in seconds) after the explosion by the function $d = -16t^2 + 128t$. From the graph of this function determine how long it takes for the projectile to return to ground level. Also find the maximum height achieved by this projectile.

  Since *negative values of t are meaningless* in this situation, our table of values begins with $t = 0$ as follows:

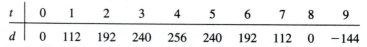

t	0	1	2	3	4	5	6	7	8	9
d	0	112	192	240	256	240	192	112	0	-144

We also ignore the last pair of values (9 and -144) since d cannot be negative. The resulting graph is shown in Figure 13-21. From the graph we see

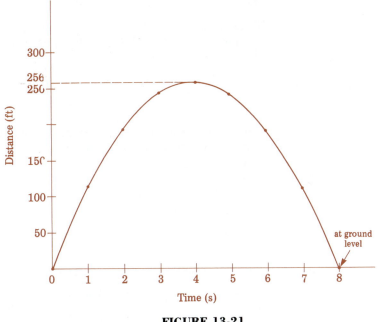

FIGURE 13-21

that the projectile returns to ground level ($d = 0$) after 8 s. Also, the graph shows that the maximum value of d is approximately 256 ft.

Noting that the function is quadratic, we can determine the t coordinate of the vertex at $-b/2a$. This gives us $t = -128/(-32) = 4$s, which confirms our conclusion that $d = 256$ ft is the maximum height. (When $t = 4$ s, $d = 256$ ft.) $\square$

EXAMPLE E   Solve the equation $x^2 - 4x + 2 = 0$ graphically.

In solving this equation, we wish to *find those values of x which make the left side zero*. By setting $y = x^2 - 4x + 2$ and finding those values of x for which y is zero, we have found the solutions to the equation. Therefore we graph

$$y = x^2 - 4x + 2$$

for which the table is

x	-1	0	1	2	3	4	5
y	7	2	-1	-2	-1	2	7

From Figure 13-22 we see that the graph crosses the x axis at approximately $x = 0.6$ and $x = 3.4$. Use of the quadratic formula verifies these values, which are the required solutions. $\square$

FIGURE 13-22

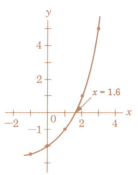

FIGURE 13-23

EXAMPLE F

Solve the equation $2^x - 3 = 0$ graphically.

First we set $y = 2^x - 3$ and graph this function by constructing the following table:

x	-1	0	1	2	3
y	-2.5	-2	-1	1	5

From the graph in Figure 13-23, we see that the required solution is $x = 1.6$, which is the value for which $y = 0$, or the curve crosses the x axis. ☐

Exercises 13-5

In Exercises 1 through 4, find the approximate values for y for the indicated values of x from the given figures.

1. Find y for $x = -1$ and $x = 2$ from Figure 13-24(a).
2. Find y for $x = 1.5$ and $x = 4.7$ from Figure 13-24(b).
3. Find y for $x = -2.3$ and $x = 1.8$ from Figure 13-24(c).
4. Find y for $x = -15$ and $x = 37$ from Figure 13-24(d).

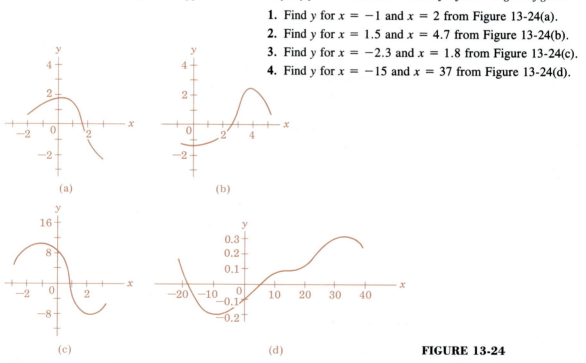

FIGURE 13-24

In Exercises 5 through 12, first graph the given functions and then use the graph to determine the approximate values of y for the indicated values of x.

5. $y = 3x + 2$; $x = 1.5, x = 3.2$
6. $y = -x + 3$; $x = -1.5, x = 1.2$
7. $y = 8 - 3x$; $x = -0.4, x = 2.1$
8. $y = 2 - x^2$; $x = -1.8, x = 1.8$

9. $y = 2x^2 - 5x + 1$; $x = -1.1, x = 2.7$ 10. $y = 6x - x^3$; $x = -0.7, x = 2.3$

11. $y = \dfrac{3}{x - 3}$; $x = 1.6, x = 3.6$ 12. $y = \sqrt{2x + 4}$; $x = 0.8, x = 3.1$

In Exercises 13 through 24, solve the given equations graphically.

13. $7x - 5 = 0$ 14. $3x + 13 = 0$ 15. $2x^2 - x - 4 = 0$ 16. $x^2 + 3x - 5 = 0$

17. $x^3 - 5 = 0$ 18. $2x^3 - 5x + 4 = 0$ 19. $\sqrt{2x + 9} - x = 0$ 20. $2x - \sqrt{x + 6} = 0$

21. $\sqrt[3]{2x + 1} - 2 = 0$ 22. $2\sqrt{x} + x - 2 = 0$ 23. $3^x - 5 = 0$ 24. $3x - 2^x = 0$

In Exercises 25 through 32, determine the required values from the appropriate graph.

25. Figure 13-25 shows the graph of the charge on a capacitor and the time. Determine the charge on the capacitor at $t = 0.005$ s and $t = 0.050$ s.

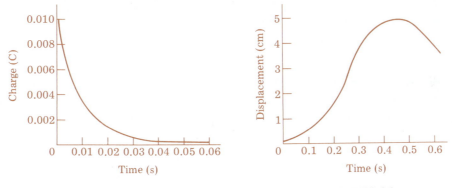

FIGURE 13-25   FIGURE 13-26

26. Figure 13-26 shows the graph of the displacement of a valve and the time. Determine the valve displacement at $t = 0.25$ s and $t = 0.60$ s.

27. Figure 13-27 shows the graph of the number of grams of a certain compound which will dissolve in 100 g of water and the temperature of the water. Determine the number of grams which dissolve at 33°C and 58°C.

28. Figure 13-28 shows the graph of milligrams of new mass on a certain plant and the time in days. How much new mass is on the plant after 12 days? After 22 days?

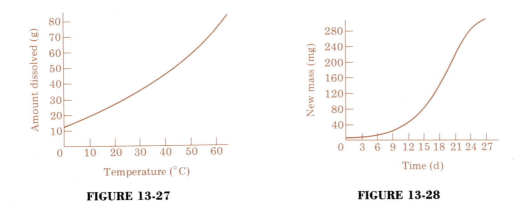

FIGURE 13-27   FIGURE 13-28

29. The velocity of sound, in meters per second, for a given temperature, in degrees Celsius, is given by the formula

$$v = 331 + 0.61\,T$$

Determine the velocity of sound for $T = 15°C$, $25°C$, $37°C$, and $48°C$.

30. The displacement of a particular beam as a function of the distance from a fixed end is

$$d = 0.2(x^2 - 10x)$$

where d is measured in inches and x in feet. Find the displacement for $x = 2.8$ ft, 4.7 ft, 5.3 ft, and 8.7 ft.

31. Given that the power in an electric circuit is constant at 0.02 W, the resistance in the circuit as a function of the current is given by

$$R = \frac{0.02}{I^2}$$

Find the resistance (in ohms) for $I = 0.005$ A, 0.016 A, 0.037 A, and 0.042 A.

32. The velocity of the flow of a liquid from a container is given by the equation

$$v = \sqrt{64h}$$

where h is the distance of the opening in the container below the surface of the liquid. Determine the velocity of flow (in feet per second), for an opening that is 3.2 ft, 5.8 ft, 6.3 ft, and 7.9 ft below the surface.

In Exercises 33 through 36, solve the given equations graphically.

33. Under certain conditions the force F (in newtons) on an object is found from the equation

$$0.8F - 22 = 0$$

Solve for F.

34. The distance d (in feet) that an object is above the surface of the earth as a function of time is given by

$$d = 85 + 60t - 16t^2$$

When will it hit the ground? [*Hint:* What does d equal when the object is on the ground?]

35. To find the radius (in inches) of a 1-qt container which requires the least amount of material to make, we must solve the equation

$$2\pi r - \frac{57.8}{r^2} = 0$$

Solve for r and thereby determine the required radius.

36. The current i (in amperes) in a particular circuit as a function of time t (in seconds) is given by

$$i = 0.002(1 - e^{-80000t})$$

For what value of t is $i = 0.190$ mA? (Use values of t in microseconds.)

13-6 Graphing Inequalities

In Chapter 5 we discussed solutions of algebraic inequalities and in this section we consider graphs of inequalities. We introduce the fundamental approach in the following example.

EXAMPLE A   Graph the inequality $x > 3$ on the rectangular coordinate system.

We begin by graphing $x = 3$ which is a vertical line through $x = 3$ on the x axis. Every point along that line satisfies the equation $x = 3$, regardless of the value of the y coordinate. All points to the right of that line have an x coordinate greater than 3 and therefore satisfy the inequality $x > 3$. In Figure 13-29 we depict our graph by shading the region containing those points. It is customary to **represent the boundary with** **dashes** **whenever** **the boundary itself is** **not** **included as part of the solution**. (Solid boundaries are included as part of the solution.) The graph of $x > 3$ is therefore represented by the shaded region of Figure 13-29. ☐

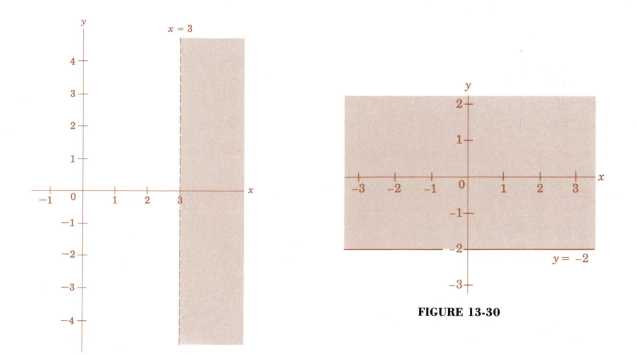

FIGURE 13-30

FIGURE 13-29

EXAMPLE B   Graph the inequality $y \geq -2$ on the rectangular coordinate system.

Since $y \geq -2$ means y is greater than or equal to -2, we **begin by** **showing the line** $y = -2$ **as a** solid **line since the points on the line itself** **do satisfy the given inequality**. All points on or above this line also satisfy the original inequality. We therefore shade the region shown in Figure 13-30. ☐

EXAMPLE C   Graph $y < 3 - x$.

We first draw the line $y = 3 - x$, which has intercepts $(0, 3)$ and $(3, 0)$. Since points on that line will not satisfy the given inequality, the boundary will not be included, so a dashed line is used (see Figure 13-31). We now choose any point on one side of the line as a test point; the origin is usually a good choice. Choosing the origin, we substitute zero for x and zero for y in the inequality to get $0 < 3$, *which is true*. Since the coordinates of the origin satisfy the given inequality, *the graph of $y < 3 - x$ is the region on the side of the line which contains the origin*. We therefore get the shaded region shown in Figure 13-31.

If the test point had not satisfied the given inequality, then the region representing the graph of $y < 3 - x$ would have been on the other side of the line. □

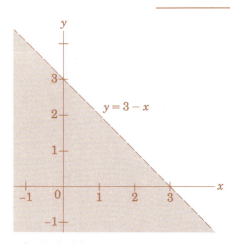

FIGURE 13-31

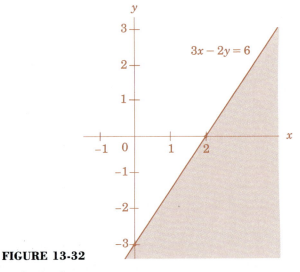

FIGURE 13-32

EXAMPLE D   Graph the inequality $3x - 2y \geq 6$.

We begin by graphing the line $3x - 2y = 6$, which has intercepts $(0, -3)$ and $(2, 0)$, as shown in Figure 13-32. We use a solid line because the points on the line itself do satisfy the given inequality. Using the origin $(0, 0)$ as a test point, we substitute zero for x and zero for y in the given inequality to get $0 \geq 6$, *which is false*. We therefore *shade the side of the line which does not contain the origin* and we get the graph shown in Figure 13-32. □

EXAMPLE E   Graph the inequality $y \geq x^2$.

We first graph the parabola $y = x^2$ by using the following table of values:

x	-3	-2	-1	0	1	2	3
y	9	4	1	0	1	4	9

We *draw the parabola using a solid curve* (instead of dashes) because the points on the parabola itself do satisfy the given inequality. In this case, we cannot use the origin as a test point because it lies on the parabola; we must *choose a point not on the boundary* itself. Choosing $(0, -1)$, we substitute zero for x and -1 for y in the given inequality to get $-1 \geq 0$, *which is false*. We therefore shade the region not containing the test point of $(0, -1)$ and we get the graph shown in Figure 13-33. □

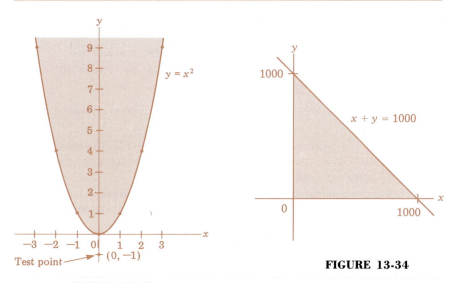

FIGURE 13-33

FIGURE 13-34

EXAMPLE F   In any given day, a refinery can produce x gallons of gasoline and y gallons of diesel fuel, in any combination. However, the available equipment allows a maximum total output of 1000 gal. Make a graph of the different possible production combinations.

Since the maximum total output is 1000 gal, we get $x + y \leq 1000$. We graph the line $x + y = 1000$ and use the origin as a test point (see Figure 13-34). Substituting zero for both x and y in the inequality, we get $0 \leq 1000$, which is true. We therefore shade the region below the line. However, we include only the first quadrant points since $x \geq 0$ and $y \geq 0$. (Negative values for x or y would be meaningless in the context of this problem.) The shaded region therefore represents the graph of all the different possible production combinations. □

Example F illustrates an application of graphing inequalities. By graphing additional constraints, we can proceed to determine how to best distribute resources. This procedure is used in a branch of applied mathematics called *linear programming*.

Exercises 13-6

In Exercises 1 through 28, graph the given inequalities on the rectangular coordinate system.

1. $x > 0$
2. $x \leq 2$
3. $y \geq 3$
4. $y \leq 0$
5. $x \leq -3$
6. $y \leq -4$
7. $y > -5$
8. $x > -8$
9. $y > 1 - x$
10. $y < x + 1$
11. $y < 2x - 1$
12. $y > 3x + 4$
13. $y \leq 2x + 3$
14. $y \geq 3x - 5$
15. $\frac{1}{2}y < 4$
16. $\frac{1}{3}y < 6$
17. $3y < 8 - 6x$
18. $2y \geq 3 + 2x$
19. $y - 3 > x$
20. $y - 1 \leq 2x$
21. $3x + 4y < 3$
22. $4x + 3y \geq 5$
23. $2x - 5y \geq 10$
24. $6x - y < 6$
25. $y < x^2$
26. $y > x^2 + 1$
27. $y < 3 - x^2$
28. $y < x^3$

In Exercises 29 through 36, use the rectangular coordinate system to graph the inequality suggested by the stated conditions.

29. A company can produce x resistors and y capacitors in 1 h. The number of resistors must be less than or equal to 50. With no constraints on the number of capacitors, graph the possible values of x and y. Exclude from the graph any negative values of x or y.

30. A computer can allocate x minutes of CPU time to administrative uses and y minutes to research uses. The research time must be less than 480 min. Assume that there are no limitations on x, but exclude any negative values of x or y and graph the possible values of x and y.

31. A company can invest up to x hours in research and up to y hours in development, but a maximum of 400 h are available. Graph the possible values of x and y.

32. A supplier must provide x gallons of hydrochloric acid and y gallons of sulfuric acid each week. The total amount of acid must be at least 40 gal. Given only the preceding constraints and the fact that neither x nor y can be negative, graph the possible values of x and y.

33. A chemist must mix a solution that contains x liters of water and y liters of ethyl alcohol. The total mixture must be at least 24 L. Graph the possible values of x and y.

34. An investor plans to put an amount of money into stock purchases and a savings account. The amount x invested in stocks must be less than or equal to the amount y allocated to the savings account. Given only this constraint and the fact that neither x nor y can be negative, graph the possible values of x and y.

35. A wood products company processes lumber and plywood. Market demand requires that the number of lumber units x must be at least twice the number of plywood units y. Given only this constraint and the fact that neither x nor y can be negative, graph the possible values of x and y.

36. A small manufacturer finds that it earns a net profit of $2 on each box of floppy disks and $5 on each box of computer paper. If x boxes of disks and y boxes of paper are produced, the profit is less than $5000. Graph the possible values of x and y.

13-7 Other Types of Graphs

In the preceding sections of this chapter, we have considered graphs which use the rectangular coordinate system for representing relationships between sets of numbers. Sets of data are often depicted with other types of graphs which represent statistical sets of data.

Of the numerous graphical methods used for representing data, we shall discuss four: the **circle graph** (often called a **"pie chart"**), the **broken-line graph**, the **bar graph**, and the **histogram**.

The first of these, the circle graph, is particularly useful for showing the relationship of a whole category to the various parts of the category. This is done by determining the percentage that each part is of the whole. Since there are 360° in a circle, we multiply each percentage by 360° to get the number of degrees of the circle to be used for each part. Consider the following example.

EXAMPLE A   In one year the liquid petroleum products produced in the United States were as follows: 1,700,000,000 barrels of motor fuel, 1,400,000,000 barrels of fuel oil, 300,000,000 barrels of jet fuel, 200,000,000 barrels of liquid petroleum gases, and 400,000,000 barrels of other products. Now 1,700,000,000 is 42.5% of the total production of 4,000,000,000 barrels, and 42.5% of 360° is 153°. This means that the part of the circle used to represent motor-fuel production will have an angle of 153°. In this way we may set up the following table which in turn is used for constructing the graph in Figure 13-35.

Product	*Percent of production*	*Angle*
Motor fuel	42.5	153°
Fuel oil	35	126°
Jet fuel	7.5	27°
Liquid gases	5	18°
Others	10	36°
check by adding	100%	360°

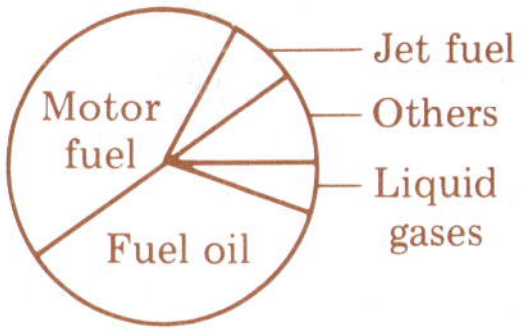

FIGURE 13-35

Another type of graph is the broken-line graph. We make a graph of this kind by connecting a series of points by straight-line segments. *It is used primarily to show how one variable changes as another variable (the equivalent of the independent variable of a function) changes continuously.* Appropriate scales should be chosen for each variable, and those scales should be properly labeled. Usually, the scale of the independent variable is placed horizontally and that of the other variable (the dependent variable) is placed vertically. The points are then located and joined.

EXAMPLE B   The temperature recorded for a certain city from noon to midnight is shown in the following table. The broken-line graph of these data is shown in Figure 13-36.

Temperature (°C)	15	18	20	26	25	23	20	17	12	13	10	8	6
Time (P.M.)	12	1	2	3	4	5	6	7	8	9	10	11	12

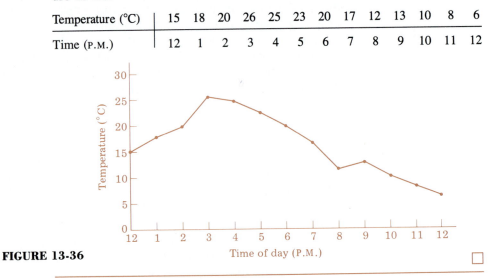

FIGURE 13-36

A type of graph valuable for showing the relative magnitude of data is the bar graph. In constructing a bar graph, we choose a scale suitable for the data and then determine the length of each bar. The bars are then constructed either horizontally or vertically and clearly labeled. The following example describes a bar graph.

EXAMPLE C   A supplier of computer components produces special logic gates and a typical production run consists of 50 "AND" gates, 40 "OR" gates, 30 "NAND" gates, 25 "NOR" gates, and 40 "NOT" gates. We represent this data with the bar graph shown in Figure 13-37.

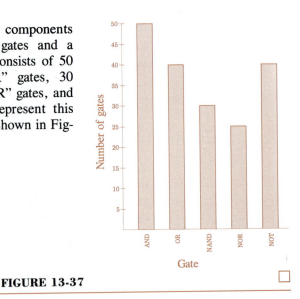

FIGURE 13-37

The last type of graph we shall consider in this section is the histogram. While the bar graph is used to compare items with different qualities, *the histogram is useful for displaying the distribution of items with the same quality.*

EXAMPLE D In Figure 13-38 we show the histogram which corresponds to the table below. This table describes test results for the breaking points (in amperes) of 20 fuses. Figure 13-38 shows the distribution of breaking-point values.

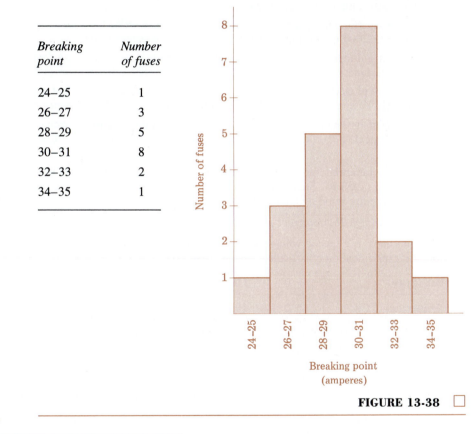

Breaking point	Number of fuses
24–25	1
26–27	3
28–29	5
30–31	8
32–33	2
34–35	1

FIGURE 13-38 □

Exercises 13-7

In Exercises 1 through 6, make a circle graph to show the given data.

1. For one year, the income of the federal government (in billions of dollars) was 112 from individual income taxes, 41 from corporate income taxes, 67 from social security contributions, 18 from excise taxes, and 12 from miscellaneous items.

2. Of the surface area of the earth, 71% is covered by water, 9% by the continent of Asia, 6% by the continent of Africa, 5% by the continent of North America, 3% by the continent of South America, and 6% by the remaining land areas, including the continents of Europe, Antarctica, and Australia.

3. A certain type of brass is made with 85% copper, 5% zinc, 5% tin, and 5% lead.

4. The electric motors shipped to a company during a given year have the following horsepower ratings: 2 hp (120), 2.5 hp (75), 3 hp (155), 3.25 hp (25), 3.5 hp (180), 4 hp, (110).

5. Of accidental deaths in the United States in a typical year, about 46000 are due to motor vehicle accidents, 20000 to home accidents, 11000 to work accidents, and 14000 to other types of accidents.

6. In a certain company, 22 employees are involved in research, 32 in development, 18 in manufacturing, and 8 are administrators.

In Exercises 7 through 12, make a bar graph to represent the given data.

7. The number of calories per one-cup serving of various beverages are: apple juice (125), orange juice (110), ginger ale (80), milk (165), cocoa (230).

8. The steel production for a particular plant for one year is described as follows: first quarter (25,000 t), second quarter (32,000 t), third quarter (45,000 t), fourth quarter (28,000 t).

9. The sales (in millions of dollars) of a corporation during a given year are found to be: heavy machinery (58), light machinery (30), chemicals (14), special tools (10), other sources (15).

10. In testing a set of electric resistors it is found that (to the nearest ohm) 5 had a resistance of 5 Ω, 8 a resistance of 6 Ω, 17 a resistance of 7 Ω, 4 a resistance of 8 Ω, and 2 a resistance of 9 Ω.

11. The enrollment in the departments of a certain technical institute for a given year is as follows: electrical (360), mechanical (170), drafting (220), construction (190), chemical (40), business (620), data processing (110).

12. Heights of the tallest buildings are as follows: Sears Tower (1454 ft), World Trade Center (1350 ft), Empire State Building (1250 ft), Standard Oil (1136 ft), John Hancock Center (1127 ft), Texas Commerce Tower (1002 ft), Chrysler Building (1046 ft).

In Exercises 13 through 18, construct a broken-line graph to represent the data.

13. A thermocouple is a special voltage source for an electric circuit. The voltage is a function of the temperature. An experiment on a particular thermocouple gives the following results:

Voltage (v)	0	2.9	5.9	9.0	12.3	15.8
Temperature (°C)	0	10	20	30	40	50

14. An oil burner propels air which is heated to 90°C, and that temperature drops as the distance away from the burner increases. The following table lists measured values:

Temperature (°C)	90	85	78	72	60
Distance (m)	0	10	20	30	40

15. When $500 is invested at a rate of 6% compounded annually, the value is described by the following table.

Value ($)	500	530	562	596	631	669	709	752	797
t (years)	0	1	2	3	4	5	6	7	8

16. In an experiment in which pressure and volume of air are measured, the following results are obtained:

Volume (cm³)	200	180	160	140	120	100
Pressure (kPa)	200	220	260	300	360	450

17. The amount x that a cam lifts a valve varies with the angle θ (the Greek theta) through which the cam is rotated. For a particular cam the values of x and θ are as follows:

$\theta(°)$	0	30	60	90	120	150	180	210	240
x (in.)	0.00	0.08	0.26	0.38	0.32	0.20	0.12	0.09	0.00

18. The number of isotopes of some of the heavier elements (the atomic number of each element is the first number given) are: 82 (lead) 16; 84 (polonium) 19; 86 (radon) 12; 88 (radium) 11; 90 (thorium) 12; 92 (uranium) 14; 94 (plutonium) 11.

In Exercises 19 through 24, construct a histogram to represent the data.

19. In a study in which 42 similar circuits are tested for the voltage level at a certain point, the following results (rounded to the nearest tenth) are obtained.

Voltage (V)	11.7	11.8	11.9	12.0	12.1	12.2
Number of circuits	1	6	10	15	8	2

20. Samples of cement mixtures are tested for the ratio of water to cement (rounded to hundredths) with the following results:

Ratio	0.60–0.64	0.65–0.69	0.70–0.74	0.75–0.79	0.80–0.84
Number of samples	20	22	18	16	2

21. A forest ranger measures the diameters of spruce trees (rounded to tenths) which were planted together. The results are summarized in the following table:

Diameter (cm)	8.0–8.9	9.0–9.9	10.0–10.9	11.0–11.9	12.0–12.9	13.0–13.9
Number of trees	6	14	21	19	12	5

22. Scientific thermometers are tested for their true readings at 10°C and the following results (rounded to tenths) are obtained:

Temperature (°C)	9.0–9.4	9.5–9.9	10.0–10.4	10.5–10.9	11.0–11.4
Number of thermometers	4	3	8	0	1

23. The diameters of ball bearings used in car generators are measured (to the nearest hundredth). The results are summarized in the following table:

Diameter (mm)	3.00–3.01	3.02–3.03	3.04–3.05	3.06–3.07	3.08–3.09
Number of ball bearings	30	22	40	31	20

24. A chemical supply company provides hydrochloric acid in containers labeled 4 L, but several such containers are measured and the following results (rounded to hundredths) are obtained:

Liters	3.80–3.89	3.90–3.99	4.00–4.09	4.10–4.19
Number of containers	12	30	6	4

Chapter 13 Formulas

$$f(x) = ax + b \qquad \text{linear function (straight line)} \qquad\qquad (13\text{-}1)$$

$$f(x) = ax^2 + bx + c \quad \text{where} \quad a \neq 0 \qquad \text{quadratic function (parabola)} \quad (13\text{-}2)$$

$$x = \frac{-b}{2a} \quad \text{where } y = ax^2 + bx + c \qquad x \text{ coordinate of vertex of a parabola} \quad (13\text{-}3)$$

13-8 Review Exercises for Chapter 13

In Exercises 1 through 12, find the indicated values of the given functions.

1. $f(x) = x + 3;\quad f(0), f(-1)$

2. $f(s) = 2s - 1;\quad f(-1), f(2)$

3. $f(x) = 2 - x;\quad f(-2), f(\frac{1}{3})$

4. $f(y) = 4y - 2;\quad f(-5), f(-2)$

5. $f(x) = 2x^2 - 1;\quad f(\sqrt{2}), f(-\frac{1}{2})$

6. $f(t) = -t^2;\quad f(3), f(-4)$

7. $f(z) = z^2 - 2z - 3;\quad f(-3), f(0.2)$

8. $f(x) = 7 - x - 4x^2;\quad f(-1), f(-v)$

9. $f(r) = -r^3;\quad f(0), f(-2)$

10. $f(n) = 12 - n^3;\quad f(-2), f(3)$

11. $f(x) = \sqrt{4x + 1};\quad f(0), f(6)$

12. $f(s) = \dfrac{2}{5 - \sqrt{s}};\quad f(4), f(16)$

In Exercises 13 through 16, plot the given points, connect them to form a geometric figure, and identify the figure.

13. $(0, 4);\quad (-1, 2);\quad (3, -2)$

14. $(-1, 0);\quad (0, 1);\quad (1, 2)$

15. $(2, 4);\quad (3, 6);\quad (-1, 2);\quad (0, 0)$

16. $(4, 1);\quad (0, -1);\quad (2, 2);\quad (1, -1)$

In Exercises 17 through 36, graph the given functions.

17. $y = 5x - 1$

18. $y = 3 - 4x$

19. $y = 3 - \dfrac{1}{2}x$

20. $y = 4x - \dfrac{2}{3}$

21. $y = \dfrac{5x - 1}{4}$

22. $y = \dfrac{1}{5}x + \dfrac{1}{2}$

23. $y = 3x^2 - 4$

24. $y = 3 - 2x^2$

25. $y = \dfrac{x^2 - 2}{4}$

26. $y = 1 - 3x - x^2$

27. $y = 8 - x^3$

28. $y = 2x^2 - x^3$

29. $y = \left(\dfrac{1}{4}\right)^{2x}$

30. $y = 2^{x-1}$

31. $y = 4 - \dfrac{1}{x}$

32. $y = \dfrac{1}{x - 1}$ **33.** $y = 2\sqrt{x} - 3$ **34.** $y = \dfrac{1}{\sqrt{x}}$

35. $y = \sqrt{x^2 + 9}$ **36.** $y = \dfrac{1}{\sqrt{8 - x}}$

In Exercises 37 through 46, graph the given inequalities.

37. $x \geq 5$ **38.** $y \leq -1$ **39.** $y > 2 + 3x$ **40.** $y < 2x - 4$

41. $8y \leq 4 - x$ **42.** $4y \geq -2x + 1$ **43.** $y \geq -x^2 + 2$ **44.** $y \leq 4 + 3x^2$

45. $y < 2x^2 - 3$ **46.** $y > x^3 - 1$

In Exercises 47 through 52, solve the given equations graphically.

47. $7x - 9 = 0$ **48.** $6x + 11 = 0$ **49.** $3x^2 - x - 2 = 0$

50. $5 - 7x - x^2 = 0$ **51.** $x^3 - x^2 + 2x - 1 = 0$ **52.** $x^4 - 2x^2 - x + 2 = 0$

In Exercises 53 through 60, construct the required graphs.

53. The atmosphere is a mixture of about 78% nitrogen, 21% oxygen, and 1% other gases. Make a circle graph to represent these data.

54. A certain weather station records the rainfall by season with the following results: spring (18 in.), summer (8 in.), fall (12 in.), winter (14 in.). Make a circle graph to represent these data.

55. Of the mountains in the world which are higher than 15,000 ft, there are 42 in Asia, 7 in Europe, 14 in North America, 28 in South America, and 7 elsewhere. Use a bar graph to represent these data.

56. A study of a certain musical sound yields the following data for the frequency (a measure of the pitch) and the amplitude (a measure of loudness):

Frequency (Hz)	220	440	660	880	1100
Amplitude (mm)	10	2	3	1	0.5

Use a bar graph to represent these data.

57. The increase in length of a certain metal rod is measured as a function of the temperature with the following results:

Increase in length (mm)	0.5	2.2	4.7	8.2	12
Increase in temperature (°C)	50	100	150	200	250

Use a broken-line graph to represent these data.

58. In an experiment it is found that the time required for milk to curdle depends on the temperature at which it is kept.

Temperature milk kept (°F)	40	50	60	70	80	90
Time to reach curdling point (h)	82	72	45	37	32	29

Make a broken-line graph to represent these data.

59. A hydrometer is used to measure the freezing level of coolant solutions in a fleet of cars with the following results:

Freezing level (°F)	−50 to −41	−40 to −31	−30 to −21	−20 to −11	−10 to −1	0 to +9
Number of cars	3	18	12	5	2	1

Construct a histogram for this collection of data.

60. The temperature of water near a discharge point of a nuclear power plant is measured at noon on different days and the following results are obtained:

Temperature (°C)	23.0–23.4	23.5–23.9	24.0–24.4	24.5–24.9	25.0–25.4	25.5–25.9
Number of days	1	3	10	7	2	2

Construct a histogram for this collection of data.

In Exercises 61 through 68, set up the required functions.

61. A right triangle has one leg of 6 in. and the other of x in. Express the area as a function of x. See Figure 13-39.

62. A rectangle has a width of 4 cm and a length of x cm. Express its perimeter p as a function of x. See Figure 13-40.

FIGURE 13-39 x in. x cm   **FIGURE 13-40**

63. The sales tax in a certain state is 6%. Express the tax T on a certain item as a function of the cost C of the item.

64. A salesperson earns $200 plus 3% commission on monthly sales. Express the salesperson's monthly income I as a function of the monthly dollar sales S.

65. The rate H at which heat is developed in a filament of an electric light bulb is proportional to the square of the electric current I. A current of 0.5 A produces heat at the rate of 60 W. Express H as a function of I.

66. A kitchen exhaust fan should remove each minute a volume of air that is proportional to the floor area. A properly operating fan removes 20 m³/min for a kitchen of 10 m² of floor area. Express the volume of air that is removed per minute for a properly operating fan as a function of the floor area A.

67. The vertical side of the rectangle of a Norman window (a semicircle surmounted on a rectangle) is 2 ft. Express the area of the window as a function of the radius of the circular part. See Figure 13-41.

68. A number of white mice are subjected to various doses of radioactivity. The time T each lives afterward is inversely proportional to the square of the intensity I of the dose. Assuming that, on the average, a mouse lives 10 h after having received a dose of 4 units, express T as a function of I.

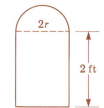

In Exercises 69 through 72, find the required values graphically.

FIGURE 13-41

69. We can determine the cost of using a computer owned by a certain company by the graph in Figure 13-42. How much does it cost to use the computer for 5 h in a given month? For 16 h? For 28 h?

70. The velocity of a satellite is a function of its distance from the surface of the earth. The graph of the velocity and distance for a particular satellite is shown in Figure 13-43. The distance from the earth to the satellite varies from 200 to 1600 mi. What is the velocity of the satellite when it is 400 mi from the earth? When it is 750 mi? When it is 1400 mi?

71. The length of a cable hanging between two equal supports 100 ft apart is given by

$$L = 100(1 + 0.0003y^2)$$

where y is the sag (vertical distance from top of support to bottom of cable) in the cable. Determine the length of a cable for which the sag is 5 ft; 10 ft; 20 ft.

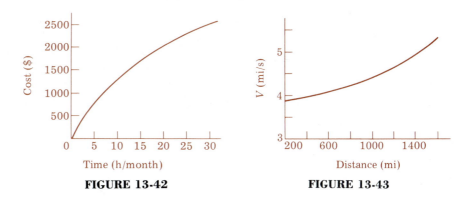

FIGURE 13-42   FIGURE 13-43

72. Under certain conditions, the distance from a source of light where the illumination is least is found by solving the equation

$$x^3 + 8(x - 100)^3 = 0$$

Find the required distance (in meters).

In Exercises 73 through 76, find the required functions and graphs.

73. A temperature in degrees Fahrenheit (°F) equals 32 more than $\frac{9}{5}$ the number of degrees Celsius (°C). Express F as a function of C and plot the graph. From the graph determine the temperature at which $F = C$.

74. The path of a projectile is approximately parabolic. Given that y is the vertical distance from the ground to the projectile and x is the horizontal distance traveled, the equation

$$y = x - 0.0004x^2$$

is the equation of the path of a given projectile, where distances are measured in feet. Plot the graph, using units of 50 ft for x, and determine how far the projectile travels. (Assume level ground.)

75. A rectangular tract of land is to have a perimeter of 800 m. Express the area as a function of its width and plot the graph. See Figure 13-44.

Perimeter = 800 m

FIGURE 13-44

76. Under certain circumstances, the resonant frequency in an electric circuit is a function of the capacitance in the circuit. This can be expressed by the equation

$$f = \frac{1}{2\pi\sqrt{C}}$$

Plot the graph of f and C, given that C varies from 10^{-5} to 10^{-4} F and f is measured in hertz.

CHAPTER FOURTEEN

Simultaneous Linear Equations

In preceding chapters we have presented methods for solving one equation with one variable. In this chapter we introduce ways of solving systems of linear equations. There are many applied problems in science, technology, and other fields that involve separate equations with the same variables. The equations that often result are referred to as **simultaneous equations** or a **system of equations.** When forces on a structure are analyzed, simultaneous equations often result. When determining electric currents, we often use simultaneous equations. Many types of stated problems from a wide variety of different fields also lead to simultaneous equations.

Armed with an ability to solve systems of equations, we will be able to solve many applied problems. For example, suppose that an airplane begins a flight with 36.0 gal of fuel stored in two separate wing tanks. During a flight, one-fourth of the fuel in one tank is consumed while three-eighths of the fuel in the other tank is consumed, and the total amount of fuel consumed is 11.25 gal. How much fuel is left in each tank? We will solve this problem later in the chapter.

14-1 Graphical Solution of Two Simultaneous Equations

We begin this section by considering a situation which leads to the construction of a system of two simultaneous equations with two unknowns.

EXAMPLE A   Two batteries are connected so that the sum of their voltages is 7.5 V. If one of the batteries is reversed, the difference between their voltages is 1.5 V. To determine the voltage of each battery, we could use the equations

$$x + y = 7.5$$
$$x - y = 1.5$$

Here x and y are the voltage levels of the two batteries.

Two planes leave the same airport at the same time and fly in opposite directions. One plane travels 200 km/h faster than the other, and at the end of 4 h they are 4400 km apart. The speeds of the planes can be found by solving the simultaneous equations

$$v_1 - v_2 = 200$$
$$4v_1 + 4v_2 = 4400$$

Here v_1 and v_2 are the speeds of the planes.   □

We now consider the problem of solving for the variables when we have two simultaneous equations with two variables. This chapter includes three methods for solving such equations. In this section we shall see how the solution may be found graphically. The following sections discuss other methods of solution.

In Chapter 13 we graphed several different kinds of functions, including the linear function. We saw that *the graph of $y = ax + b$ is always a straight line*. The number of points on a line is infinite, but if a second line were to cross a given line, there would be only one point in common. The coordinates of this point satisfy the equations of both lines simultaneously.

EXAMPLE B   Using intercepts or other points, we graph the line $y = 3x - 3$ and $2x + 3y = 6$ as shown in Figure 14-1. The exact coordinates of the point of intersection are $(\frac{15}{11}, \frac{12}{11})$. These exact coordinates were obtained by an algebraic method we will discuss in Section 14-2. This result can be shown to be correct, since these coordinates satisfy the equation of each line. For the first line, we have

$$y = 3x - 3$$

substituting

$$\frac{12}{11} = 3\left(\frac{15}{11}\right) - 3 = \frac{45}{11} - \frac{33}{11} = \frac{12}{11}$$

and for the second line, we have

$$2x + 3y = 6$$

substituting

$$2\left(\frac{15}{11}\right) + 3\left(\frac{12}{11}\right) = \frac{30}{11} + \frac{36}{11} = \frac{66}{11} = 6$$

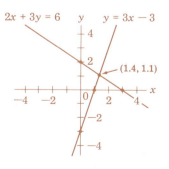

FIGURE 14-1

(Continued on next page)

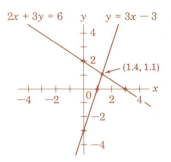

FIGURE 14-1

Even though we cannot obtain this accuracy for the coordinates of the point of intersection from the graph, *we can see from the graph that the point of intersection is about (1.4, 1.1).* This is the type of solution we shall attempt to obtain in this section. *The answers we get using this graphic approach might not be exact,* but they should be reasonably close to the exact solutions. If the values we obtain are not exact, they will not satisfy the equations exactly. The estimated values are acceptable if, upon substitution, both sides of each equation become *approximately* equal.   □

To solve a pair of simultaneous linear equations, we graph the two equations on the same set of axes and determine the point of intersection. The x coordinate of this point is the desired value of x; the y coordinate of this point is the desired value of y. Together they are the solution to the system of two equations in two variables. Since the coordinates of no other point satisfy both equations, the solution is unique.

EXAMPLE C   Graphically solve the system of equations

$$x - y = 6$$
$$2x + y = 3$$

We determine the intercepts and one checkpoint for each equation. Then we graph each equation as shown in Figure 14-2.

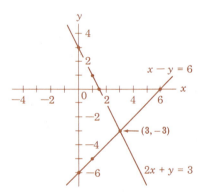

$x - y = 6$	
x	y
0	−6
6	0
1	−5

$2x + y = 3$	
x	y
0	3
$\frac{3}{2}$	0
1	1

FIGURE 14-2

We see from the graph that the lines cross at about (3, −3) so that the solution to this set of equations is x = 3, y = −3. (Actually this solution is exact, although this cannot be shown without substitution of the values into the given equations.)   □

EXAMPLE D   A navigational technique involves two synchronized radio signals from two different sources. A position is determined from the sum and difference of the two signals by solving the equations

$$x + y = 9$$
$$x - y = 2$$

Graphically solve this system of equations.

　　We begin by graphing the two lines from the intercepts and checkpoints indicated in the following tables:

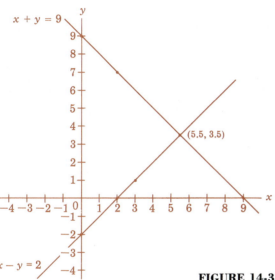

$x + y = 9$	
x	y
0	9
9	0
2	7

$x - y = 2$	
x	y
0	-2
2	0
3	1

FIGURE 14-3

　　The resulting graph is shown in Figure 14-3. From the graph we estimate the coordinates of the point of intersection as (5.5, 3.5) so that $x = 5.5$ and $y = 3.5$. □

　　In Section 13-3 we noted that there are calculators capable of displaying graphs of functions. Some calculators can simultaneously display the graphs of two or more functions. This allows us to use calculators to get the graphical solutions of a system of equations.

　　There are two special circumstances in which two lines do not intersect at just one point. This is the case (a) if the lines are parallel and do not intersect at all or (b) if the lines are coincident, all the points of one being the same as all the points of the other. *When the lines are parallel, the system is called* **inconsistent,** *and when they are coincident the system is called* **dependent.** The following examples illustrate these cases. Example E illustrates an inconsistent system of equations while Example F illustrates a dependent system.

EXAMPLE E   Graphically solve the system of equations

$$x + y = 3$$
$$2x + 2y = 9$$

We set up the following tables to indicate the intercepts and checkpoint for each line. The lines are then graphed as in Figure 14-4.

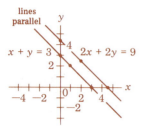

$x + y = 3$	
x	y
0	3
3	0
1	2

$2x + 2y = 9$	
x	y
0	$\frac{9}{2}$
$\frac{9}{2}$	0
2	$\frac{5}{2}$

FIGURE 14-4

We observe that within the limits of accuracy of the graphing, the lines appear to be parallel. (They are, in fact, parallel. This is always the case when one equation of the system can be multiplied through by a constant so that the coefficients of the variables are the same, respectively, as those in the other equation. In this case, if the first equation is multiplied by 2, we have $2x + 2y = 6$, which is the same as the second equation except for the constant.) *Since the lines are parallel, the system is inconsistent and there is no solution.* ☐

EXAMPLE F   Graphically solve the system of equations

$$x + 2y = 6$$
$$3x + 6y = 18$$

We set up the following tables to indicate the intercepts and checkpoint for each line. The lines are then graphed as in Figure 14-5.

$x + 2y = 6$	
x	y
0	3
6	0
2	2

$3x + 6y = 18$	
x	y
0	3
6	0
2	2

FIGURE 14-5

We note that *both intercepts are the same.* The two lines are coincident. *Hence every point on the two lines is a solution,* since the coordinates satisfy both equations. This, in turn, means that there is an infinite number of solutions. *Since no unique solution may be determined, we call the system dependent.* ☐

Exercises 14-1

In Exercises 1 through 24, solve the given systems of equations graphically. Where possible, estimate the coordinates of the point of intersection to the nearest 0.1 of a unit. If variables other than x and y are used, plot the first along the x axis and the second along the y axis.

1. $x + y = 3$
   $x - y = 3$

2. $x + y = 5$
   $2x + y = 7$

3. $x + 2y = 4$
   $x - 2y = 0$

4. $3x - y = 1$
   $2x + y = 4$

5. $r - x = 7$
   $2r + x = 5$

6. $m + n = 6$
   $3m + n = 2$

7. $x + 3y = 8$
   $x - y = 0$

8. $x + y = 1$
   $2x - 8y = 1$

9. $3y + 2x = 4$
   $4y + 5x = 3$

10. $2R - T = 7$
    $4R + 3T = 9$

11. $x + 4y = 4$
    $-2x + 2y = 7$

12. $3x - 2y = 4$
    $-9x - 4y = 13$

13. $a + 2b = -3$
    $3a - 5b = -3$

14. $2u - 3v = 8$
    $u + 4v = 6$

15. $s + t = 7$
    $3s + 3t = 5$

16. $2x - 5y = 10$
    $-6x + 15y = -30$

17. $3p - q = 6$
    $2p + 2q = 7$

18. $2x - 4y = 7$
    $3x + 2y = 3$

19. $3x - y = 8$
    $2x - 5y = 15$

20. $7r - 2s = 14$
    $2r + 3s = 9$

21. $0.5x - 1.6y = 3.2$
    $1.2x + 3.3y = 6.6$

22. $0.3x - 0.2y = 1.2$
    $1.3x + 2.5y = 5.0$

23. $\frac{1}{2}x + y = 4$
    $x + \frac{1}{3}y = 2$

24. $\frac{1}{3}x - 2y = 8$
    $3x - \frac{1}{5}y = 2$

In Exercises 25 through 32, each system of equations is either inconsistent or dependent. Identify those systems which are inconsistent and identify those which are dependent.

25. $x + y = 4$
    $x + y = 5$

26. $x + 2y = 6$
    $-3x - 6y = -18$

27. $2x - 3y = 6$
    $12x - 18y = 36$

28. $-3x + 7y = 8$
    $6x - 14y = 16$

29. $0.2x - 0.3y = 2.4$
    $x - 1.5y = 12$

30. $1.1x + 2.0y = 3.0$
    $-2.2x - 4.0y = 6.0$

31. $ax + by = 4$
    $ax + by = 2$

32. $ax + by = c$
    $3ax + 3by = 3c$

In Exercises 33 through 40, solve the given problems graphically.

33. Two hangar rods are connected to a box beam which supports a concrete deck. Forces acting on the rods sometimes push in the same direction and sometimes they push in opposite directions. The forces x and y can be found by solving the equations

$$x + y = 120 \qquad x - y = 70$$

34. An airplane flies into a head-wind to a destination and it then returns with the tail-wind. The speed of the plane P and the speed of the wind W can be found by solving the equations

$$P - W = 80 \qquad P + W = 140$$

35. One alloy is 70% lead and 30% zinc while another alloy is 40% lead and 60% zinc. We can find the amount (in pounds) of each alloy needed to make 100 lb of another alloy 50% lead and 50% zinc by solving the equations

$$x + y = 100 \qquad 0.7x + 0.4y = 50$$

36. A printer requires 17 min to produce 8 pages of letters and 5 pages of tables. It takes 14 min to produce 5 pages of each. The time required to print 1 page of tables can be found by solving the following equations for y.

$$8x + 5y = 17 \qquad 5x + 5y = 14$$

37. A person collecting water samples rows downstream 10 mi in 2 h and upstream 8 mi in 4 h. The rate at which this person rows in still water (r_1) and the rate of the current of the stream (r_2) can be found by solving the equations

$$2r_1 + 2r_2 = 10 \qquad 4r_1 - 4r_2 = 8$$

38. An electrician rents a generator and a heavy-duty drill for 5 h at a total cost of $50. On another job, there is a cost of $56 for renting the generator for 4 h and the drill for 8 h. The hourly rates g and d can be found by solving the equations

$$5g + 5d = 50 \qquad 4g + 8d = 56$$

39. A current of 2 A passes through a resistor R_1 and a current of 3 A passes through a resistor R_2; the total voltage across the resistors is 8 V. Then the current in the first resistor is changed to 4 A and that in the second resistor is changed to 1 A; the total voltage is 11 V. The resistances (in ohms) can be found by solving the equations

$$2R_1 + 3R_2 = 8 \qquad 4R_1 + R_2 = 11$$

40. If two ropes with tensions T_1 and T_2 support a 20-lb sign, the tensions can be found by solving the equations

$$0.7T_1 - 0.6T_2 = 0$$
$$0.7T_1 + 0.8T_2 = 20$$

14-2 Algebraic Substitution in Two Equations

We have just seen how a system of two linear equations can be solved graphically. This technique is good for obtaining a "picture" of the solution of two equations. One of the difficulties of graphical solutions is that the solutions are not exact. If exact solutions are required, we must use algebraic methods. Also, the graphical method tends to be slower than other methods. This section presents one basic algebraic method of solution, and we shall discuss another one in Section 14-3.

The method of this section is called **algebraic substitution**. *We first solve one of the equations for one of the two variables and then substitute the result into the other equation for that variable.* The result is a simple equation in one variable. We solve that equation to determine the value of one of the variables. We can then substitute that value into either equation so that the value of the other variable can be found. The following examples illustrate the method of algebraic substitution.

EXAMPLE A   Use substitution to solve the system of equations

$$2x + y = 4$$
$$3x - y = 1$$

The first step is to solve one of the equations for one of the variables. We might quickly examine both equations to determine which equation and which variable would involve the easiest steps. In this system it is slightly easier to solve for y than for x because fractions can be avoided. Inspection shows that one algebraic step is all that is required to solve the first equation for y. Consequently, the first equation would be a good choice. It should be emphasized that *either equation can be solved for either variable*, and the final result will be the same. Solving the first equation for y, we have $y = 4 - 2x$. ***This is now substituted into the second equation***, giving

$$3x - (4 - 2x) = 1$$

— in second equation, y is replaced by $4 - 2x$

We now have an equation with only one variable, and we proceed to solve this equation for x:

$$3x - 4 + 2x = 1$$
$$5x = 5$$
$$x = 1$$

Now ***the value of y which corresponds to $x = 1$ is found by substituting $x = 1$ into either equation.*** Since the first has already been solved for y, we have

$$y = 4 - 2(1) = 2$$

We now know that the solution is $x = 1$ and $y = 2$. We *check this by substituting both values into the other equation*. We get

$$3(1) - 2 = 1$$

so that the solution checks. □

EXAMPLE B   Use substitution to solve the system of equations

$$2x - 4y = 23$$
$$3x + 5y = -4$$

With these two equations it doesn't make any difference which variable or equation we select. Simply choosing to solve the first equation for x, we have $2x = 23 + 4y$, or

$$x = \frac{23 + 4y}{2}$$

Substituting this expression into the second equation and then solving for y, we get

$$3\left(\frac{23 + 4y}{2}\right) + 5y = -4$$

— in second equation, x is replaced by $\dfrac{23 + 4y}{2}$

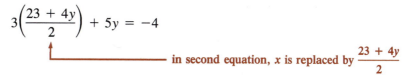

(Continued on next page)

We now multiply both sides by the lowest common denominator of 2 to get

$$3(23 + 4y) + 10y = -8$$
$$69 + 12y + 10y = -8$$
$$22y = -77$$
$$y = -\frac{7}{2}$$

Substituting this value into the solution for x in the first equation, we get

$$x = \frac{23 + 4\left(-\dfrac{7}{2}\right)}{2} = \frac{23 - 14}{2} = \frac{9}{2}$$

The solution is $x = \frac{9}{2}$ and $y = -\frac{7}{2}$. Checking this solution *in the second equation,* we have

$$3\left(\frac{9}{2}\right) + 5\left(-\frac{7}{2}\right) = \frac{27}{2} - \frac{35}{2} = -\frac{8}{2} = -4$$

This means that the solution checks.   □

EXAMPLE C Two electric currents I_1 and I_2 (in amperes) can be found by solving the equations

$$-5I_1 + I_2 = 6 \qquad 10I_1 + 3I_2 = -2$$

Solve by substitution.

Solving for I_2 in the first equation we get

$$I_2 = 6 + 5I_1$$

We now substitute $6 + 5I_1$ for I_2 in the second equation to get

$$10I_1 + 3(6 + 5I_1) = -2$$
$$10I_1 + 18 + 15I_1 = -2$$
$$25I_1 = -20 \qquad I_1 = -\frac{20}{25} = -\frac{4}{5}$$

Substituting this value into the solution for I_2 we get

$$I_2 = 6 + 5\left(-\frac{4}{5}\right) = 6 - 4 = 2$$

We obtain the solution $I_1 = -\frac{4}{5}$ A and $I_2 = 2$ A. We check this solution in the second equation. Since

$$10\left(-\frac{4}{5}\right) + 3(2) = -8 + 6 = -2$$

we conclude that the solution is correct.   □

When we are trying to solve a system which turns out to be dependent, we can reduce the equation to $0 = 0$ after substitution. This seems reasonable, since $0 = 0$ regardless of the value of x, and the coordinates of all points which satisfy one equation of a dependent system also satisfy the other equation, regardless of the value of x. *If the system is inconsistent, the equation after substitution may be reduced to $0 = a$, where a is not zero.* This is also reasonable, since zero does not equal a nonzero number and there are no points whose coordinates satisfy both equations.

EXAMPLE D Use substitution to solve the system of equations

$$2x - y = 3$$
$$4x - 2y = 6$$

Solving the first equation for y and substituting this expression into the second equation, we get

$$y = 2x - 3$$
$$4x - 2(2x - 3) = 6$$
$$4x - 4x + 6 = 6$$
$$6 = 6$$

By subtracting 6 from each side, we have $0 = 0$. **The system is dependent,** and has an infinite number of solutions (see Example F of Section 14-1). $\square$

EXAMPLE E Use substitution to solve the system of equations

$$3x - 2y = 4$$
$$9x - 6y = 2$$

Solving the first equation for x, we get

$$x = \frac{2y + 4}{3}$$

Substituting this expression into the second equation, we obtain

$$9\left(\frac{2y + 4}{3}\right) - 6y = 2$$
$$3(2y + 4) - 6y = 2$$
$$6y + 12 - 6y = 2$$
$$12 = 2$$
$$10 = 0$$

Since this cannot be true, the system is inconsistent and has no solution (see Example E of Section 14-1). $\square$

Exercises 14-2

In Exercises 1 through 24, solve the given systems of equations by substitution.

1. $x = y - 3$
   $x + y = 13$

2. $y = x + 2$
   $x + 2y = 7$

3. $x - y = 2$
   $2x - y = 8$

4. $-x + y = -6$
   $-x - 2y = 6$

5. $x + y = 1$
   $5x + 10y = 8$

6. $y + x = 1$
   $2y - 8x = 1$

7. $2x + y = 0$
   $3x + 2y = 1$

8. $2x = 3y$
   $x - 4y = 0$

9. $x - 2y = 0$
   $2x - 3y = 1$

10. $a - 2b = 4$
    $2a + b = 6$

11. $3x = 2y$
    $-x + 4y = 0$

12. $2x + 3y = 6$
    $x - y = 4$

13. $x - 3y = 6$
    $3x = 6y + 18$

14. $x = y + 1$
    $x - y = 5$

15. $k - 8u = 4$
    $2k + u = 2$

16. $z + 2u = 3$
    $3z - u = 1$

17. $\frac{1}{3}x + y = 9$
    $x - \frac{1}{3}y = 6$

18. $\frac{1}{2}x + y = 8$
    $x + 4y = 7$

19. $2x - 5y = 1$
    $3x - 2y = -4$

20. $2u + 3w = 6$
    $4u + 2w = 8$

21. $2r + 3s = 2$
    $3r - 2s = 16$

22. $3m + 2n = 7$
    $5m + 3n = 10$

23. $3x - 4k = 12$
    $4x + 5k = -7$

24. $3y - 2z = 6$
    $2y + 9z = 8$

In Exercises 25 through 32, solve the given systems of equations by substitution.

25. A space shuttle is used to launch a communications satellite. The satellite is launched either in the same direction or the opposite direction of the shuttle, and the speed, in kilometers per hour, of both objects can be found by solving the equations

$$x + y = 1200 \qquad x - y = 800$$

26. A manager allocates x hours to manufacturing and y hours to machine maintenance. The values of x and y can be found by solving the equations

$$x + y = 480 \qquad 3x - 5y = 0$$

27. Two furnaces consume fuel oil at the rates of r_1 gal/h and r_2 gal/h, respectively. The total consumption is 2.00 gal/h, and one furnace burns oil at a rate that is 0.50 gal/h more than the other. The two rates can be found by solving the equations

$$r_1 + r_2 = 2.00 \qquad r_1 = r_2 + 0.50$$

28. To determine how many milliliters of a 30% solution of hydrochloric acid should be drawn off from 100 mL and replaced by a 10% solution to give an 18% solution, we use the equations

$$x + y = 100 \qquad 0.30x + 0.10y = 18$$

Here we must determine the value of the variable y.

29. The voltage between two points in an electric circuit is 60 V. The contact point of a voltage divider is placed between these two points so that the voltage is divided into two parts; one of these parts is twice the other. These two voltages V_1 and V_2 can be found by solving the equations

$$V_1 = 2V_2 \qquad V_1 + V_2 = 60$$

30. The purchasing agent for a company acquires hardware and software for a computer system with a total selling price of $5000. The hardware has a 5% sales tax while the software has a 4% sales tax, and the total tax is $230. The prices x and y of the hardware and software, respectively, can be found by solving the equations

$$x + y = 5000$$
$$0.05x + 0.04y = 230$$

31. A roof truss is in the shape of an isosceles triangle. The perimeter of the truss is 50 ft, and the base doubled is 9 ft more than three times the length of a rafter (neglecting overhang). See Figure 14-6. The length of the base b and a rafter r can be found by solving the equations

$$b + 2r = 50$$
$$2b = 3r + 9$$

FIGURE 14-6

32. Under certain conditions, the tensions T_1 and T_2 (in newtons) supporting a derrick are found by solving the equations

$$0.68T_1 + 0.57T_2 = 750$$
$$0.73T_1 - 0.82T_2 = 0$$

In Exercises 33 through 36, use a scientific calculator to solve the systems of equations by substitution.

33. $x + 3.77y = 4.13$
      $2.67x + 4.11y = 2.93$

34. $3.1416x - y = 2.3333$
    $3.1416x + y = 1.3427$

35. $3.725x + 4.113y = 6.857$
    $5.235x - 1.030y = -3.116$

36. $8.37x - 9.14y = 3.29$
    $17.33x + 19.10y = 12.26$

14-3 Addition-Subtraction Method in Two Equations

So far we have discussed the graphical method and substitution method for solving a system of two simultaneous linear equations. In this section we introduce a second algebraic method known as the **addition-subtraction** method. *The basis of this method is that if one of the variables appears on the same side of each equation, and if the coefficients of this variable are numerically the same, it is possible to add (or subtract) the left-hand sides (along with the right-hand sides) in order to obtain an equation with only one of the variables remaining.* The resulting equation with only one variable can then be solved by the methods introduced in Chapter 5. The following two examples illustrate this addition-subtraction method.

EXAMPLE A Use the addition-subtraction method to solve the system of equations

$$2x - y = 1$$
$$3x + y = 9$$

We note that if the left-hand sides of the two equations are added, the terms $+y$ and $-y$ will be combined to produce a zero so that the result will not contain a term which includes y. The two sides of each equation are equal. If the left-hand sides are added, the sum will equal the sum of the right-hand sides. Since y is not present, the resulting equation will contain only x and can then be solved for x. We may then find y by substituting the value of x into either of the original equations and then solving for y. By adding the left-hand sides and equating this sum to the sum of the right-hand sides, we get

$$
\begin{aligned}
2x - y &= 1 \\
3x + y &= 9 \qquad \text{add} \\
2x + 3x \longrightarrow 5x \quad\;\; &= 10 \longleftarrow 1 + 9 \\
-y + y \longrightarrow \quad x &= 2
\end{aligned}
$$

Substituting $x = 2$ into the second equation, we have

$$3(2) + y = 9$$
$$6 + y = 9$$
$$y = 3$$

The solution is $x = 2$ and $y = 3$. Since we used the second equation to solve for y, we check this solution in the first equation and get

$$2(2) - 3 = 1$$

which means that the solution is correct. □

EXAMPLE B Use the addition-subtraction method to solve the system of equations

$$2x + 3y = 7$$
$$2x - y = -5$$

If we subtract the left-hand side of the second equation from the left-hand side of the first equation, x will not appear in the result. Subtracting the respective sides of the two equations we get

$$
\begin{aligned}
2x + 3y &= 7 \\
2x - y &= -5 \qquad \text{subtract} \\
2x - 2x = 0 \longrightarrow \quad 4y &= 12 \longleftarrow 7 - (-5) \\
3y - (-y) \longrightarrow \quad y &= 3
\end{aligned}
$$

Substituting $y = 3$ into the first equation, we obtain

$$2x + 3(3) = 7$$
$$2x + 9 = 7$$
$$2x = -2$$
$$x = -1$$

The solution is $x = -1$, $y = 3$. Checking this solution in the second equation, we have $2(-1) - (3) = -2 - 3 = -5$. ☐

In many systems it is necessary to multiply one (or both) of the equations by constants so that the resulting coefficients of one unknown are numerically the same. The following two examples illustrate this procedure.

EXAMPLE C Use the addition-subtraction method to solve the system of equations

$$2x + 3y = -5$$
$$4x - y = 4$$

If each term of the second equation is multiplied by 3, it will be possible to eliminate y by addition and continue to the solution. Multiplying each term, ***including the constant on the right,*** of the second equation by 3, we have

$$2x + 3y = -5 \qquad \text{first equation}$$
$$12x - 3y = 12 \qquad \text{each term of second equation multiplied by 3}$$

Adding the equations, we get

$$14x = 7$$

$$3y + (-3y) = 3y - 3y = 0$$

$$x = \frac{1}{2}$$

Substituting $x = \frac{1}{2}$ into the first equation, we get

$$2\left(\frac{1}{2}\right) + 3y = -5$$
$$1 + 3y = -5$$
$$3y = -6$$
$$y = -2$$

The solution is $x = \frac{1}{2}$, $y = -2$. Checking this solution by substituting in the second equation, we have

$$4\left(\frac{1}{2}\right) - (-2) = 2 + 2 = 4$$

so that the solution is verified. ☐

EXAMPLE D A power supply for a space shuttle requires x inductors and y capacitors, where x and y are found by solving the equations

$$2x + 5y = 46$$
$$3x - 2y = 12$$

It is not possible to multiply just one of these equations by an integer to make the coefficients of one of the variables numerically the same. We must multiply both equations by appropriate integers. We can multiply the first equation by 3 and the second equation by 2 in order to make the coefficients of x equal. Or we may multiply the first equation by 2 and the second one by 5 in order to eliminate y. That is, we must find the least common multiple of the coefficients of one variable. Choosing the least common multiple of 5 and -2, we obtain a new set of equations by making both coefficients of y equal to 10. We get

$$4x + 10y = 92 \qquad \text{multiply first equation by 2}$$
$$\underline{15x - 10y = 60} \qquad \text{multiply second equation by 5}$$
$$19x \qquad\quad = 152 \qquad \text{add}$$

$$10y + (-10y) = 10y - 10y = 0$$

$$x = \frac{152}{19} = 8$$

$$2(8) + 5y = 46 \qquad \text{substitute } x = 8 \text{ in first equation}$$
$$5y = 46 - 16$$
$$5y = 30$$
$$y = 6$$

Checking, we get

$$3(8) - 2(6) = 24 - 12 = 12$$

The solution of $x = 8$ and $y = 6$ has therefore been verified. That is, the power supply requires 8 inductors and 6 capacitors. □

If the system is dependent or inconsistent, we have the same type of result as that mentioned in Section 14-2. That is, if the system is dependent, the addition-subtraction method will result in $0 = 0$ after the equations are combined, or the result will be $0 = a$ (a not zero) if the system is inconsistent.

EXAMPLE E   Use the addition-subtraction method to solve the system of equations

$$3x - 6y = 8 \qquad -x + 2y = 3$$

Multiplying the second equation by 3 and adding equations, we get

$$
\begin{array}{l}
3x - 6y = 8 \\
\underline{-3x + 6y = 9} \\
 0 = 17 \quad \textbf{add}
\end{array}
$$

Since this result is not possible, the system is *inconsistent*.   ☐

Exercises 14-3

In Exercises 1 through 24, solve the given systems of equations by the addition-subtraction method.

1. $x + y = 7$
$x - y = 3$

2. $x + y = 5$
$x - 2y = 1$

3. $2x + y = 5$
$x - y = 1$

4. $x + y = 11$
$2x - y = 1$

5. $m + n = 12$
$2m - n = 0$

6. $2x + y = 9$
$2x - y = -1$

7. $d + t = 3$
$2d + 3t = 10$

8. $2r + s = 1$
$r + 2s = -1$

9. $x + 2n = 11$
$3x - 5n = -22$

10. $2x + y = 2$
$5x + y = 8$

11. $x - 2y = 5$
$2x + y = 20$

12. $4t - x = 0$
$3t - x = -2$

13. $7x - y = 5$
$7x - y = 4$

14. $3x - y = 4$
$-6x + 2y = -8$

15. $a + 7b = 15$
$3a + 2b = 7$

16. $-8x + 7y = 2$
$3x - 5y = 9$

17. $13p - 18q = 21$
$4p + 12q = 5$

18. $11k + 15t = 1$
$2k + 3t = 1$

19. $\frac{2}{7}x - \frac{1}{3}y = -3$
$\frac{5}{14}x + \frac{2}{3}y = -10$

20. $\frac{7}{8}x + \frac{5}{12}y = -4$
$x + y = 8$

21. $2m = 10 - 3n$
$3m = 12 - 4n$

22. $62x = 4y + 43$
$15x + 12y = 17$

23. $\dfrac{1}{x} + \dfrac{2}{y} = 3$

$\dfrac{1}{x} - \dfrac{2}{y} = -1$

24. $\dfrac{4}{x} - \dfrac{1}{y} = 1$

$\dfrac{6}{x} + \dfrac{2}{y} = 5$

In Exercises 25 through 36, use the addition-subtraction method to solve the stated problems.

25. Two batteries produce a total voltage of 7.5 V. The difference in their voltages is 4.5 V. (See Figure 14-7.) We can determine the two voltage levels by solving the equations

$$V_1 + V_2 = 7.5 \qquad V_1 - V_2 = 4.5$$

26. The perimeter of a rectangular sign is 26 ft and the length is 3 ft longer than the width. See Figure 14-8. The dimensions can be found by solving the equations

$$2x + 2y = 26 \qquad x - y = 3$$

7.5 V

4.5 V   **FIGURE 14-7**

Perimeter = 26 ft

y ft

x ft   **FIGURE 14-8**

27. In determining the optimal oil-fuel mixture for a chain saw, the amount of oil x and gasoline y can be found by solving the equations

$$16x + y = 2 \qquad 32x - y = 1$$

28. One line printer can produce x lines per minute; a second line printer can produce y lines per minute. They print 7500 lines if the first prints for 2 min and the second prints for 1 min. They can print 9000 lines if the second prints for 2 min and the first for 1 min. We can determine the printing rates x and y by solving the equations

$$2x + y = 7500 \qquad x + 2y = 9000$$

29. While a pulley belt is making one complete revolution, one of the pulley wheels makes 6 revolutions and the other makes 15 revolutions. The circumference of one wheel is 2 ft more than twice the circumference of the other wheel. The circumferences c_1 and c_2 can be found by solving the equations

$$6c_1 = 15c_2 \qquad c_1 - 2c_2 = 2$$

30. A rocket is launched so that it averages 2000 km/h. An hour later, an interceptor rocket is launched along the same path at an average speed of 2500 km/h. To find the time of flights of the rockets t_1 and t_2 before the interceptor rocket overtakes the first rocket, we must solve the following equations:

$$2000t_1 = 2500t_2 \qquad t_1 = t_2 + 1$$

31. A prospective employee must take a test of verbal skills and a test of mathematical skills. The total score is 1150, and the math score exceeds the verbal score by 150. The two scores can be found by solving the equations

$$x + y = 1150 \qquad x = y + 150$$

32. If x dollars are invested at an annual rate of 5% and y dollars are invested at 7%, their total annual interest amounts to $405. The amount y exceeds x by $1500. The two amounts can be found by solving the equations

$$0.05x + 0.07y = 405 \qquad y = x + 1500$$

33. A person wants to determine how many liters m of milk containing 3% butterfat and how many liters c of cream containing 15% butterfat should be mixed to give 20 L of milk containing 6% butterfat. The following equations can be used:

$$m + c = 20 \qquad 0.03m + 0.15c = 1.2$$

34. Under certain conditions two electric currents (in amperes) I_1 and I_2 can be found by solving the equations

$$3I_1 + 4I_2 = 3 \qquad 3I_1 - 5I_2 = -6$$

35. Under certain conditions, when balancing weights x and y on a board weighing 20 lb, we can find x and y by solving the equations

$$5x - 7y = 20 \qquad 4x - 3.2y = 40$$

36. A concrete deck is to be supported with x box beams and y "I" beams. The values of x and y can be found by solving the equations

$$2x + 3y = 40 \qquad 5x - 4y = 31$$

In Exercises 37 through 40, use a scientific calculator to solve the given systems of equations by the addition-subtraction method.

37. $4.7x - 3.9y = 8.2$
    $6.2x - 4.9y = 5.3$

38. $6.23x + 9.33y = 8.12$
    $0.37x - 0.49y = 6.15$

39. $2.363x - 1.487y = 9.363$
    $1.005x + 1.005y = 3.212$

40. $-58.34x - 12.11y = 10.02$
    $16.19x - 18.00y = -10.34$

14-4 Determinants in Two Equations

In many systems of linear equations, the methods we have discussed are difficult because of the numbers involved or because the systems are simply too large. In this section we describe a method that is more systematic and easier to implement in a computer program. We first describe the procedure for two equations with two unknowns, and then we explain why it works.

The method we use is based on **determinants.** For systems of two linear equations with two unknowns we use **second-order determinants** defined as follows.

$$\begin{vmatrix} a_1 & b_1 \\ a_2 & b_2 \end{vmatrix} = a_1 b_2 - a_2 b_1 \tag{14-1}$$

A determinant is a number associated with a square array of numbers. The order of a determinant refers to the number of rows (or columns) in the array. When finding the value of a second-order determinant, we can easily remember the definition of Eq. (14-1) if we think of it as representing the product of the two numbers along the downward diagonal, minus the product of the two numbers along the upward diagonal. See Figure 14-9.

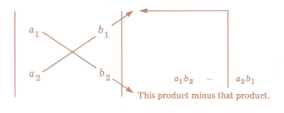

FIGURE 14-9

EXAMPLE A   Find the values of the given determinants.

$$\begin{vmatrix} 1 & 2 \\ 3 & 4 \end{vmatrix} = 1(4) - 3(2) = 4 - 6 = -2$$

$$\begin{vmatrix} 2 & 5 \\ 3 & -9 \end{vmatrix} = 2(-9) - 3(5) = -18 - 15 = -33$$

$$\begin{vmatrix} 2 & 1 \\ -3 & 7 \end{vmatrix} = 2(7) - (-3)(1) = 14 + 3 = 17$$

Let's now consider a general system of two equations with two unknowns.

$$a_1 x + b_1 y = c_1$$
$$a_2 x + b_2 y = c_2$$

(14-2)

The solutions for x and y can be expressed using determinants as follows.

$$x = \frac{\begin{vmatrix} c_1 & b_1 \\ c_2 & b_2 \end{vmatrix}}{\begin{vmatrix} a_1 & b_1 \\ a_2 & b_2 \end{vmatrix}} \quad \text{and} \quad y = \frac{\begin{vmatrix} a_1 & c_1 \\ a_2 & c_2 \end{vmatrix}}{\begin{vmatrix} a_1 & b_1 \\ a_2 & b_2 \end{vmatrix}}$$

(14-3)

See Appendix B for a computer program that uses determinants to solve systems of equations.

We should examine the patterns in these two equations. Note that the denominators are the same. Both denominators use the coefficients of x and y in the original system of equations. In the numerators, the first variable x has the first column made up of the column of c's; the second variable y has the second column consisting of c's. If an unknown is missing from either equation, its coefficient is zero, and we should enter zero in the determinant.

This result of using determinants to solve systems of equations is referred to as **Cramer's rule.** The following example illustrates this procedure.

EXAMPLE B   Use determinants to solve the given system of equations.

$$2x + \ y = 0$$
$$3x - 2y = 7$$

We should make a correspondence between the above system of equations and Eqs. (14-3). Noting that the solutions for both x and y involve the same denominator, it would be wise to first evaluate that particular determinant which uses the coefficients of the variables.

$$\begin{vmatrix} 2 & 1 \\ 3 & -2 \end{vmatrix} = 2(-2) - 3(1) = -4 - 3 = -7$$

To find the value of the determinant in the numerator for x, we start with the same coefficients but we replace the *first* column with the constants on the right-hand side of the equations. We get

$$\begin{vmatrix} 0 & 1 \\ 7 & -2 \end{vmatrix} = 0(-2) - 7(1) = 0 - 7 = -7$$

To find the value of the determinant in the numerator for y, we start with the coefficients and replace the *second* column with the constants 0 and 7 to get

$$\begin{vmatrix} 2 & 0 \\ 3 & 7 \end{vmatrix} = 2(7) - 3(0) = 14 - 0 = 14$$

Finally, we express the solutions as shown below.

$$x = \frac{\begin{vmatrix} 0 & 1 \\ 7 & -2 \end{vmatrix}}{\begin{vmatrix} 2 & 1 \\ 3 & -2 \end{vmatrix}} = \frac{-7}{-7} = 1$$

$$y = \frac{\begin{vmatrix} 2 & 0 \\ 3 & 7 \end{vmatrix}}{\begin{vmatrix} 2 & 1 \\ 3 & -2 \end{vmatrix}} = \frac{14}{-7} = -2$$

We now know that $x = 1$ and $y = -2$. We should again check solutions by substituting them in the original equations. $\square$

EXAMPLE C Use determinants to solve the given system of equations.

$$3x + 4y = -6$$
$$5x - 2y = 16$$

Applying Eqs. (14-3) we get

$$x = \frac{\begin{vmatrix} -6 & 4 \\ 16 & -2 \end{vmatrix}}{\begin{vmatrix} 3 & 4 \\ 5 & -2 \end{vmatrix}} = \frac{-6(-2) - 16(4)}{3(-2) - 5(4)} = \frac{12 - 64}{-6 - 20} = \frac{-52}{-26} = 2$$

$$y = \frac{\begin{vmatrix} 3 & -6 \\ 5 & 16 \end{vmatrix}}{\begin{vmatrix} 3 & 4 \\ 5 & -2 \end{vmatrix}} \quad \text{same} \quad = \frac{3(16) - 5(-6)}{-26} = \frac{48 + 30}{-26} = \frac{78}{-26} = -3$$

We conclude that $x = 2$ and $y = -3$. Checking, we substitute those values in the original equations and simplify to get $-6 = -6$ and $16 = 16$, so the solution is verified. $\square$

This method of using determinants becomes especially useful when the coefficients make the solution more complicated. In the next example, a calculator can be used to simplify the evaluation of the determinants.

EXAMPLE D   The currents (in amperes) of two circuits can be found from the equations given below. Use determinants to solve the given system of equations.

$$1.34x - 2.73y = 9.44$$
$$-8.35x + 7.22y = 5.36$$

We apply Eqs. (14-3) to get

$$x = \frac{\begin{vmatrix} 9.44 & -2.73 \\ 5.36 & 7.22 \end{vmatrix}}{\begin{vmatrix} 1.34 & -2.73 \\ -8.35 & 7.22 \end{vmatrix}} = \frac{9.44(7.22) - 5.36(-2.73)}{1.34(7.22) - (-8.35)(-2.73)} = \frac{82.7896}{-13.1207} = -6.31$$

$$y = \frac{\begin{vmatrix} 1.34 & 9.44 \\ -8.35 & 5.36 \end{vmatrix}}{\begin{vmatrix} 1.34 & -2.73 \\ -8.35 & 7.22 \end{vmatrix}} = \frac{1.34(5.36) - (-8.35)(9.44)}{-13.1027} = \frac{86.0064}{-13.1207} = -6.56$$

The solutions $x = -6.31$ A and $y = -6.56$ A were rounded to three significant digits. When checking these solutions by substituting them in the original equations, we cannot expect to get identities, but both sides of each equation should be approximately the same.   □

In some systems of equations, the determinant used in both denominators of Eqs. (14-3) may be zero. Since division by zero is undefined, no solution will be found. This indicates that the system is either inconsistent or dependent. With a zero denominator determinant, the system is inconsistent if the numerators are nonzero; it is dependent if the numerators are also zero.

We have described the procedure for using determinants to solve systems of two linear equations with two unknowns. We now explain why this procedure works. Again consider Eqs. (14-2)

$$a_1 x + b_1 y = c_1$$
$$a_2 x + b_2 y = c_2$$ (14-2)

If we multiply the first equation by b_2 and the second by b_1, we get

$$a_1 b_2 x + b_1 b_2 y = c_1 b_2$$
$$a_2 b_1 x + b_2 b_1 y = c_2 b_1$$

Subtracting, we obtain

$$a_1 b_2 x - a_2 b_1 x = c_1 b_2 - c_2 b_1$$

Factoring out x, we get

$$(a_1 b_2 - a_2 b_1)x = c_1 b_2 - c_2 b_1$$

Solving for x we find that

$$x = \frac{c_1 b_2 - c_2 b_1}{a_1 b_2 - a_2 b_1}$$

which corresponds exactly with

$$x = \frac{\begin{vmatrix} c_1 & b_1 \\ c_2 & b_2 \end{vmatrix}}{\begin{vmatrix} a_1 & b_1 \\ a_2 & b_2 \end{vmatrix}} = \frac{c_1 b_2 - c_2 b_1}{a_1 b_2 - a_2 b_1}$$

By similar reasoning, we can show that

$$y = \frac{a_1 c_2 - a_2 c_1}{a_1 b_2 - a_2 b_1}$$

which corresponds exactly with the expression for y that results from Eqs. (14-3).

Exercises 14-4

In Exercises 1 through 12, evaluate the given determinants.

1. $\begin{vmatrix} 4 & 3 \\ 2 & 1 \end{vmatrix}$

2. $\begin{vmatrix} 4 & -3 \\ 2 & 1 \end{vmatrix}$

3. $\begin{vmatrix} 2 & 3 \\ -5 & 8 \end{vmatrix}$

4. $\begin{vmatrix} 2 & -4 \\ -5 & 1 \end{vmatrix}$

5. $\begin{vmatrix} 8 & 12 \\ 4 & 6 \end{vmatrix}$

6. $\begin{vmatrix} -9 & 5 \\ 8 & -6 \end{vmatrix}$

7. $\begin{vmatrix} 2 & -5 \\ -4 & 9 \end{vmatrix}$

8. $\begin{vmatrix} -4 & -2 \\ -9 & 8 \end{vmatrix}$

9. $\begin{vmatrix} -5 & -8 \\ -3 & -6 \end{vmatrix}$

10. $\begin{vmatrix} -4 & -6 \\ -2 & -9 \end{vmatrix}$

11. $\begin{vmatrix} 10 & -4 \\ -2 & 6 \end{vmatrix}$

12. $\begin{vmatrix} -5 & -9 \\ 12 & 14 \end{vmatrix}$

In Exercises 13 through 32, use determinants to solve the given systems of equations.

13. $x + y = 6$
    $x - y = -4$

14. $x + 2y = 7$
    $x - y = -2$

15. $2x - y = 1$
    $x + y = 5$

16. $2x + 2y = -2$
    $2x - y = 4$

17. $2s = 8$
    $s - t = 1$

18. $2R_2 = 16$
    $2R_1 - R_2 = -2$

19. $2v_1 - v_2 = 7$
    $3v_1 - 2v_2 = 9$

20. $m + n = 0$
    $3m - 2n = -20$

21. $x + 2y = 8$
    $4x - 5y = -59$

22. $2x - y = -17$
    $x + 5y = 41$

23. $x + 2y = 30$
    $2x + y = 33$

24. $4x - 3y = 14$
    $3x + 2y = 36$

25. $8x - 9y = -6$
    $-3x + 4y = -4$

26. $7x + 12y = -9$
    $4x + 8y = 4$

27. $3x + 4y = 11$
    $5x - 2y = 14$

28. $5x + y = 3$
    $10x - 3y = -4$

29. $5x + 10y = 4$
    $x + y = 0.5$

30. $3x - 2y = 2.1$
    $4x + y = 1.7$

31. $\frac{1}{2}x + \frac{1}{2}y = \frac{3}{8}$
    $x - y = \frac{1}{4}$

32. $x + y = \frac{5}{6}$
    $x - y = -\frac{1}{6}$

In Exercises 33 through 40, use determinants to solve the stated problems.

33. Two different thermistors have resistances R_1 and R_2 (in ohms) that can be found from the following equations.

$$R_1 + R_2 = 32000$$
$$R_1 - R_2 = 16000$$

34. One machine produces x parts per hour while a second machine produces y parts per hour. Those values can be found from the following equations.

$$x + y = 410$$
$$8x + 4y = 2440$$

35. One solution of coolant consists of 30% water and 70% ethyl alcohol. A second solution is 60% water and 40% ethyl alcohol. A more desirable mixture is obtained by mixing x liters of the first solution with y liters of the second solution. Those amounts can be found by solving the following equations.

$$0.3x + 0.6y = 8$$
$$0.7x + 0.4y = 8$$

36. A company pays 4% sales tax on a microcomputer costing x dollars, and it pays 5% sales tax on a printer costing y dollars. Those costs can be found by solving the following equations.

$$0.04x + 0.05y = 340$$
$$0.04x - 0.05y = 140$$

37. A receipt shows that it cost a total of $29.25 to purchase 6 files and 3 blades. Another receipt shows that 2 files and 4 blades cost $12.00. The cost of 1 file and the cost of 1 blade can be found from the following equations.

$$6f + 3b = 29.25$$
$$2f + 4b = 12.00$$

38. Four pine seedlings and 3 fir seedlings cost $29.00. Two pine seedlings and 7 fir seedlings cost $42.00. Their individual costs can be found by solving the following equations.

$$4p + 3f = 29.00$$
$$2p + 7f = 42.00$$

39. The sum of two resistances is 55.7 Ω while their difference is 24.7 Ω. Their individual values (in ohms) can be found by solving the following equations.

$$R_1 + R_2 = 55.7$$
$$R_1 - R_2 = 24.7$$

40. The perimeter of a triangular roof truss is 55.2 ft, and the base is 11.4 ft longer than a rafter. The lengths of the base and rafter can be found from the following equations.

$$b + 2r = 55.2$$
$$b - r = 11.4$$

In Exercises 41 through 44, use determinants and a calculator to solve the given systems of equations.

41. $2.56x - 3.47y = 5.92$
    $3.76x + 1.93y = 4.11$

42. $345x + 237y = -412$
    $-207x + 805y = 623$

43. $3725x - 4290y = 16.25$
    $4193x + 2558y = 29.36$

44. $23.07x - 19.13y = 5736$
    $18.12x + 12.14y = 2053$

14-5 Algebraic Methods in Three Equations

The previous sections of this chapter involved systems of simultaneous equations consisting of two linear equations with two unknowns. In this section we consider systems with three unknowns in three linear equations. With three unknowns, the graphical method becomes impractical since we would have to sketch graphs in a coordinate system with x, y, and z axes. That would be extremely difficult. However, the two algebraic methods presented in Sections 14-2 and 14-3 can be easily extended to three equations with three unknowns.

We begin by illustrating the substitution method as it applies to these larger systems of equations.

EXAMPLE A   Use substitution to solve the system of equations

$$x - y - z = 2$$
$$x + 2y - 2z = 3$$
$$3x - 2y - 4z = 5$$

As in Section 14-2, the first step is to solve one of the equations for one of the variables. The first equation can be solved for x with the result $x = y + z + 2$. In the two other equations we now replace x by the expression $y + z + 2$ and obtain

$$(y + z + 2) + 2y - 2z = 3$$
$$3(y + z + 2) - 2y - 4z = 5$$

Note that both of these equations involve only the variables y and z. We used parentheses to emphasize that x was replaced by the expression enclosed within those parentheses. We can now remove the parentheses, combine like terms, and carry any constants to the right-hand side of each equation. We get

$$3y - z = 1$$
$$y - z = -1$$

Since we now have two equations with two unknowns, we could use any of the previous methods to solve for y and z. We will continue with the substitution method. In the *second* equation above, we can solve for y to get $y = z - 1$. In the *first* equation above, we now replace y by $z - 1$ to get

$$3(z - 1) - z = 1$$
or $$3z - 3 - z = 1$$
or $$2z = 4$$
or $$z = 2$$

(Continued on next page)

We know that $z = 2$, but we also know that $y = z - 1$. As a result, we solve for y by calculating $y = 2 - 1 = 1$. We now seek the value of x. We have already established that $x = y + z + 2$, so we can determine the value of x by substituting 1 and 2 for y and z, respectively. We get

$$x = y + z + 2 = 1 + 2 + 2 = 5$$

We conclude that $x = 5$, $y = 1$, and $z = 2$. We can check this solution by substituting those values in the three original equations. We would get $2 = 2$, $3 = 3$, and $5 = 5$, so the solution is verified.   □

In general, when we use the substitution method to solve a system of three equations with three unknowns, we should

1. Select one of the equations and use it alone to solve for one of the unknowns in terms of the others.

2. In each of the other two equations, replace the unknown selected in step 1 by the equivalent expression obtained in that first step.

3. The result should be two equations with two unknowns. Combine like terms, simplify, and solve the smaller system by using the methods already presented in this chapter.

4. The value of the last unknown can be found by substituting the two known values into any one of the three original equations.

5. Check the solution by substituting the values into the original equations.

In addition to the method of algebraic substitution, we can also use the addition-subtraction method to solve a system of three equations with three unknowns. We begin by selecting two of the equations and using the addition-subtraction method to eliminate one of the three unknowns. Then we proceed to select a *different* pair of equations from the original system, and we use the addition-subtraction method to eliminate the same unknown. These steps should result in two new equations with only two unknowns. We can solve this smaller system by using the methods in the preceding sections. The value of the third remaining unknown can be solved by substituting the two known values into any of the three original equations. The following example illustrates this method.

EXAMPLE B   Use the addition-subtraction method to solve the system of equations

$$x + 2y + 2z = 6 \qquad \text{Eq. (A)}$$
$$-x + y + 3z = 6 \qquad \text{Eq. (B)}$$
$$4x - 3y - 2z = 5 \qquad \text{Eq. (C)}$$

We begin the addition-subtraction method by choosing Eqs. (A) and (B). Adding the corresponding sides, we eliminate the unknown x as shown below.

$$x + 2y + 2z = 6 \qquad \text{Eq. (A)}$$
$$\underline{-x + y + 3z = 6} \qquad \text{Eq. (B)}$$
$$3y + 5z = 12 \qquad \text{add}$$

We must now choose a different pair of equations. Selecting Eqs. (B) and (C), we proceed to again eliminate the same unknown x. This can be accomplished by multiplying Eq. (B) by 4 so that addition will eliminate x as shown below.

$$-4x + 4y + 12z = 24 \qquad \text{Eq. (B) multiplied by 4}$$
$$\underline{4x - 3y - 2z = 5} \qquad \text{Eq. (C)}$$
$$y + 10z = 29 \qquad \text{add}$$

We have produced two new equations that do not include x. We can now concentrate on solving this system of two equations with two unknowns.

$$3y + 5z = 12 \qquad \text{from Eqs. (A) and (B)}$$
$$y + 10z = 29 \qquad \text{from Eqs. (B) and (C)}$$

Continuing with the addition-subtraction approach, we will eliminate z by doubling the first equation and then subtracting the equations as follows.

$$6y + 10z = 24$$
$$\underline{y + 10z = 29}$$
$$5y = -5 \qquad \text{subtract}$$
$$\text{or} \qquad y = -1$$

Substituting -1 for y in the equation $y + 10z = 29$, we get $-1 + 10z = 29$ or $10z = 30$, so $z = 3$. Knowing that $y = -1$ and $z = 3$, we can now solve for x by substituting those known values in any one of the original equations. Selecting Eq. (A), we substitute to get

$$x + 2(-1) + 2(3) = 6$$
$$\text{or} \quad x - 2 + 6 = 6$$
$$\text{or} \qquad\qquad\qquad x = 2$$

We now know that $x = 2$, $y = -1$, and $z = 3$. We should check this solution to verify that it simultaneously satisfies each of the three original equations. After substituting the solutions for x, y, z and simplifying, Eq. (A) reduces to $6 = 6$, Eq. (B) reduces to $6 = 6$, and Eq. (C) reduces to $5 = 5$, so the solution is correct. $\square$

In Example A we used only the substitution method, and in Example B we used only the addition-subtraction method. There is no requirement that only one method must be used throughout a given solution. We can mix methods in order to follow the easiest or most convenient path. In Example C we begin with the substitution method when we replace x by $y + 7$, but we continue with the addition-subtraction method when we solve for y.

EXAMPLE C   The values of three forces (in pounds) acting on a beam can be found by solving the equations given below. Use any method to solve the system of equations

$$x - y = 7 \rightarrow x = y + 7$$
$$2x + 3y - 8z = 3$$
$$-3x + 2y + 8z = -6$$

↓ Substitute $y + 7$ for x in the 2nd and 3rd equations.

$$2(y + 7) + 3y - 8z = 3$$
$$-3(y + 7) + 2y + 8z = -6$$

↓ Remove parentheses, combine like terms, and simplify.

$$
\begin{array}{r}
5y - 8z = -11 \\
-y + 8z = 15 \\
\hline
4y = 4
\end{array}
$$ add

or $y = 1$

Since $x = y + 7$, we now have $x = 1 + 7 = 8$. Substituting 8 for x and 1 for y in the second equation of the original system, we get

$$2(8) + 3(1) - 8z = 3$$
or $16 + 3 - 8z = 3$
or $-8z = -16$
or $z = 2$

We conclude that the forces are $x = 8$ lb, $y = 1$ lb, and $z = 2$ lb. Again, we should check this solution by substituting these values in the original equations. Making those substitutions and simplifying the resulting equations, we get $7 = 7$, $3 = 3$, and $-6 = -6$, so the solution is verified. □

In this section we extended the methods of Sections 14-2 and 14-3 to include systems of three equations with three unknowns. It would not be too difficult to extend the use of determinants to systems of three equations with three unknowns. These same methods can also be extended to four equations with four unknowns and to larger systems as well. However, when the systems become much larger, these methods become much more difficult, and we should consider using special computer programs or other methods.

Exercises 14-5

In Exercises 1 through 24, solve the given systems of equations by any method.

1. $x + y = 3$
    $x + y + z = 6$
    $-x + y + 2z = 7$

2. $x - y = 4$
    $2x + y - 3z = 5$
    $-x + 2y + 4z = 0$

3. $x + y + z = 3$
    $-x + 3z = 7$
    $2x - y - z = -6$

4. $x - 2y + z = 10$
$\quad\quad y - z = -6$
$\quad 3x - y + z = 15$

5. $x + y + z = 2$
$\quad x - y - z = 0$
$\quad x + 2y + 3z = 5$

6. $x - y + z = 4$
$\quad x + y + 5z = 8$
$\quad -x + 3y + z = -2$

7. $2x - 2y + z = 1$
$\quad 3x + 3y + z = 26$
$\quad 4x + 2y - 3z = 31$

8. $3x - y + 4z = 7$
$\quad 2x + y - 2z = -8$
$\quad x + 3y - 3z = -13$

9. $-3x - 2y + 3z = -1$
$\quad 2x + 5y - 3z = -6$
$\quad 4x + 3y + 3z = 22$

10. $2x - 3y + 4z = 6$
$\quad -2x + 4y - 3z = -2$
$\quad 2x + 5y - 2z = 10$

11. $2x + 5y - 4z = 3$
$\quad 8x + 3y + 4z = 15$
$\quad 3x + 4y - 3z = 4$

12. $x + 2y + 7z = 0$
$\quad -3x + 5y + 6z = 0$
$\quad 4x - 5y + 2z = 0$

13. $x - 2y + 5z = -12$
$\quad 2x - 3y - 4z = 9$
$\quad -2x + 2y + 3z = -12$

14. $2x + 4y + 7z = 7$
$\quad 3x + 4y - 8z = 2$
$\quad -x - 3y - 3z = 2$

15. $3x + 9y - 5z = 22$
$\quad -x + 3y - 8z = 6$
$\quad 4x + 9y + 9z = 1$

16. $3x - 8y + 2z = 6$
$\quad 2x - 8y + 3z = 1$
$\quad 5x - 9y - 5z = -2$

17. $5x - 3y + 3z = -10$
$\quad 4x + 4y + 4z = 56$
$\quad 3x - 2y + 5z = -34$

18. $7x - 6y + 10z = 50$
$\quad 5x - 5y + 5z = 20$
$\quad 4x - 8y + 3z = -6$

19. $10x + 7y + 3z = 0$
$\quad 9x + 8y + 4z = -7$
$\quad 8x + 9y + 2z = 10$

20. $5x + 6y - 4z = 29$
$\quad 7x + 3y - 5z = 19$
$\quad 4x + 2y - 3z = 11$

21. $2x - y + \frac{1}{2}z = 6$
$\quad \frac{1}{2}x + \frac{1}{2}y + z = 18$
$\quad 2x + \frac{1}{2}y - z = 0$

22. $\frac{1}{3}x + \frac{1}{3}y + z = 3$
$\quad \frac{1}{2}x + y + \frac{1}{2}z = 11$
$\quad x + y + z = 13$

23. $\frac{1}{2}x + \frac{1}{3}y + \frac{1}{3}z = 3$
$\quad x + y + \frac{1}{3}z = 9$
$\quad \frac{1}{2}x + 2y + 2z = 8$

24. $\frac{1}{2}x + \frac{1}{3}y - \frac{1}{4}z = -2$
$\quad \frac{1}{2}x - \frac{1}{2}y + \frac{1}{2}z = 12$
$\quad \frac{1}{3}x + \frac{1}{2}y - \frac{3}{4}z = -10$

In Exercises 25 through 32, use any method to solve the stated problems.

25. Under certain conditions, three electric currents (in amperes) I_1, I_2, and I_3 can be found by solving the equations

$$I_1 + 3I_2 \quad\quad = 16$$
$$I_1 + 3I_2 + I_3 = 22$$
$$4I_1 - 3I_2 + 2I_3 = 1$$

26. Three forces (in pounds) act on a beam, and their values can be found from the following system of equations.

$$F_x + F_y + F_z = 61$$
$$0.5F_y + 0.5F_z = 18$$
$$F_x + F_y - F_z = 21$$

27. The volumes of copper x, nickel y, and zinc z (all in cm³) needed to produce a certain alloy can be found from the equations

$$0.5x + 0.2y + 0.1z = 17$$
$$0.2x + 0.4y + 0.1z = 24$$
$$0.3x + 0.4y + 0.1z = 25$$

28. Three different water pumps are used to drain a pond. They remove water at the rates of r_1, r_2, and r_3, respectively. (The rates are in gal/min.) Those rates can be found by solving the system of equations

$$r_1 + r_2 + r_3 = 400$$
$$r_1 + r_2 \quad\quad = 200$$
$$r_1 + 2r_2 + 3r_3 = 880$$

29. Under standard conditions, three engines consume gasoline at the rates of r_1, r_2, and r_3 (measured in mi/gal). These rates can be found by solving the following system of equations.

$$r_1 + \ \ r_2 + r_3 = 52$$
$$r_1 + 3r_2 - r_3 = 36$$
$$2r_1 + 2r_2 + r_3 = 80$$

30. Three different manufacturing teams produce floppy disks. The number of disks produced by each team in one hour can be found by solving these equations.

$$x + \ \ y + \ \ z = \ \ 410$$
$$8x + 8y \ \ \ \ \ \ \ = 1840$$
$$8y + 8z = 2480$$

31. If \$500 is invested at an annual rate of x, \$800 is invested at rate y, and \$1000 is invested at rate z, those rates can be found (in decimal form) by solving this system of equations.

$$500x + 800y + 1000z = 149$$
$$500x \ \ \ \ \ \ \ \ \ \ \ + 1000z = \ \ 85$$
$$800y + 1000z = 124$$

32. A contractor rents a backhoe, pump, and generator at a daily cost of B, P, and G, respectively. The dollar values of B, P, and G can be found by solving this system of equations.

$$5B + 5P + 5G = 1250$$
$$B + 5P + 5G = \ \ 450$$
$$2B + \ \ P + 3G = \ \ 480$$

14-6 Stated Problems

In the first section of this chapter we mentioned that many types of stated problems lead to systems of simultaneous equations. This has also been illustrated with examples and exercises in each of the previous sections. Now that we have seen how such systems are solved, we shall set up equations from stated problems and then solve these equations. The following examples illustrate the procedure.

EXAMPLE A   In a certain house, the living room (rectangular) has a perimeter of 78 ft. The length is 3 ft less than twice the width. Find the dimensions of the room. See Figure 14-10.

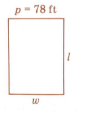

First we let l = length of the room and w = width of the room. Since *the perimeter is 78 ft,* we know that

$$2l + 2w = 78 \qquad p = 2l + 2w$$

Since *the length is 3 ft less than twice the width,* we have

$$l = 2w - 3$$

FIGURE 14-10

Dividing through the first equation by 2 and rearranging the second equation, we have

$$l +\ w = 39$$
$$l - 2w = -3$$

Subtracting the sides of the second equation from the corresponding sides of the first equation, we get

$$3w = 42$$
$$w = 14 \text{ ft}$$

Having found the value of w, we now proceed to determine the value of l as follows.

$$l = 2(14) - 3 = 25 \text{ ft}$$

Thus the length is 25 ft and the width is 14 ft. This checks with the *statement of the problem*. □

EXAMPLE B   A manager has 10 employees working 35 h each in one week. The weekly employee time must be allocated so that the machine maintenance time is $\frac{1}{5}$ of the manufacturing time. Find the number of hours spent on maintenance and on manufacturing.

Let x = manufacturing time and let y = maintenance time. Since *there are* $10 \times 35 = 350$ *h to be allocated,* we get

$$x + y = 350$$

Also, since *the maintenance time is $\frac{1}{5}$ of the manufacturing time,* we get

$$y = \frac{1}{5}x$$

Choosing the method of substitution, we get

$$x + \left(\frac{1}{5}x\right) = 350 \qquad \text{in first equation, } y \text{ is replaced by } \frac{1}{5}x$$
$$5x + x = 1750 \qquad \text{multiply all terms by 5}$$
$$6x = 1750$$
$$x = 291.7$$

Since $x + y = 350$ and $x = 291.7$, we get the following equation for y.

$$291.7 + y = 350$$
$$y = 58.3$$

The sum of x and y is 350 h and 58.3 is $\frac{1}{5}$ of 291.7 so that the solution checks. The manager should allocate 291.7 h to manufacturing and 58.3 h to maintenance. □

EXAMPLE C Two types of electromechanical carburetors are being assembled. Each type *A* carburetor requires 15 min of assembly time, whereas type *B* carburetors require 12 min each. Type *A* carburetors require 2 min of testing whereas type *B* carburetors require 3 min each. There are 222 min of assembly time and 45 min of testing time available. How many of each type should be assembled and tested if all of the available time is to be utilized?

▶ We let x = number of type *A* carburetors and let y = number of type *B* carburetors. ***All of the type A carburetors require a total of 15x min for assembly while all of the type B carburetors require a total of 12y min for assembly.*** Since *the assembly time available is 222 min*, we get $15x + 12y = 222$. By similar reasoning we express the testing time as $2x + 3y = 45$. We now rewrite the first equation and multiply the second equation by 4 to get

$$15x + 12y = 222 \qquad \text{assembly time}$$
$$8x + 12y = 180 \qquad \text{testing time}$$

Subtracting the second equation from the first we get

$$7x = 42$$
$$x = 6$$

Also,

$$15(6) + 12y = 222 \qquad \text{first equation with } x = 6$$
$$90 + 12y = 222$$
$$12y = 132$$
$$y = 11$$

With the time available, 6 type *A* carburetors and 11 type *B* carburetors should be processed. Checking, we see that these results agree with the *statement of the problem.* □

EXAMPLE D Two jet planes start at cities *A* and *B*, 6400 km apart, traveling the same route toward each other. They pass each other 2 h later. The jet which starts from city *A* travels 200 km/h faster than the other. How far are they from city *A* when they pass each other? See Figure 14-11.

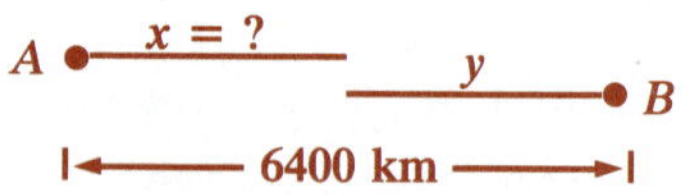

FIGURE 14-11

Since we wish to determine the distance from city A to the place at which they pass each other, we let x = distance from city A to the point at which they pass and y = distance from city B to the point at which they pass.

Since *the cities are 6400 km apart,* we have

$$x + y = 6400$$

The other information we have is that one jet travels 200 km/h faster than the other. ***It is necessary to set up the other equation in terms of speed.*** The speed of each jet is the distance traveled divided by the time. Since the jet from city A travels x km in 2 h, its speed is $x/2$. In the same way, the speed of the other jet is $y/2$. Since *the jet from city A travels 200 km/h faster,* we have

$$\frac{x}{2} = \frac{y}{2} + 200$$

We now have the necessary two equations.

Proceeding with the solution, we repeat the first equation and multiply through the second equation by 2. This gives

$$x + y = 6400$$
$$x = y + 400$$

We may easily use either the substitution method or the addition-subtraction method of solution. Using the addition-subtraction method, we have

$$
\begin{array}{ll}
x + y = 6400 & \text{first equation} \\
\underline{x - y = 400} & \text{from second equation} \\
2x \phantom{{}+y} = 6800 & \text{add} \\
x = 3400 \text{ km}
\end{array}
$$

This means that $y = 3000$ km. The faster jet traveled at 1700 km/h, and the slower one at 1500 km/h, so the solution checks with the statement of the problem.

An alternative method would be to let u = speed of the jet from city A and v = speed of the jet from city B. Even though the required distance from city A is not one of the unknowns, if we find u we may find the distance by multiplying u by 2, the time that elapses before the planes meet. Using these unknowns, we have the first equation $u = v + 200$ from the fact that the jet from city A travels 200 km/h faster. Also, since distance equals speed times time, the distance from city A to the meeting place is $2u$ and that from city B is $2v$. The total of these distances is 6400 km. Therefore $2u + 2v = 6400$ is the other equation. $\square$

EXAMPLE E   How many pounds of sand must be added to a mixture which is 50% cement and 50% sand in order to get 200 lb of a mixture which is 80% sand and 20% cement?

Since the unknown quantities are the amounts of sand and original mixture, we let x = number of pounds of sand required and y = number of pounds of original mixture required.

Now we know that the final mixture will have a total weight of 200 lb. Therefore

$$x + y = 200 \qquad \text{total weight}$$

Next, the amount of sand, x, and the amount of sand present in the original mixture, $0.5y$, equals the amount of sand in the final mixture, $0.8(200)$. This gives us

$$x + 0.5y = 0.8(200) \qquad \text{weight of sand}$$

Rewriting the first equation and multiplying the second equation by 2, we get

$$x + y = 200$$
$$2x + y = 320$$

Subtracting, we obtain

$$x = 120 \text{ lb}$$

which also gives us

$$y = 80 \text{ lb}$$

Since half the original mixture is sand, the final mixture has 160 lb of sand, and this is 80% of the 200 lb. The solution checks, and the required answer is 120 lb of sand.   □

EXAMPLE F   When three batteries are connected in series, their voltages are added and the total is 9.0 V. The first two batteries have the same voltage levels, and their combined total is one-half the voltage of the third battery. Find the voltage level of each battery.

We can represent the three voltage levels by x, y, and z. Since their total is 9.0 V, we know that

$$x + y + z = 9.0$$

Since the first two batteries have the same voltage levels, we know that $x = y$. Also, the first two batteries have a combined total that is one-half of the third battery so that

$$x + y = \tfrac{1}{2}z$$

With $x = y$ we can use substitution to rewrite the other two equations as

$$x + x + z = 9.0$$
$$x + x \quad\;\; = \tfrac{1}{2}z$$

The two equations are easily changed into the following forms.

$$2x + \;\; z = 9.0$$
$$2x - \tfrac{1}{2}z = 0$$

Subtracting the last equation from the one above it gives us

$$\tfrac{3}{2}z = 9.0$$
or $\quad z = 6.0$

With $z = 6.0$, the equation $2x + z = 9.0$ becomes

$$2x + 6.0 = 9.0$$
or $\qquad\quad 2x = 3.0$
or $\qquad\quad\; x = 1.5$

Also, since $x = y$, we get $y = 1.5$. Our solution is $x = 1.5$ V, $y = 1.5$ V, and $z = 6.0$ V. Checking this with the original statement of the problem will show that the solution is correct. □

EXAMPLE G   An airplane begins a flight with a total of 36.0 gal of fuel stored in two separate wing tanks. During a flight, $\tfrac{1}{4}$ of the fuel in one tank is consumed while $\tfrac{3}{8}$ of the fuel in the other tank is consumed; the total amount of fuel consumed is 11.25 gal. How much fuel is left in each tank?

We let $x =$ original amount of fuel in one tank and $y =$ original amount of fuel in the other tank. Since the plane began with 36.0 gal, we know that $x + y = 36.0$. We get

$$x + y = 36.0 \qquad \text{\color{brown}The two tanks began with a total of 36.0 gal.}$$
$$\tfrac{1}{4}x + \tfrac{3}{8}y = 11.25 \qquad \text{\color{brown}fuel consumed}$$

Using determinants we get

$$x = \frac{\begin{vmatrix} 36.0 & 1 \\ 11.25 & \tfrac{3}{8} \end{vmatrix}}{\begin{vmatrix} 1 & 1 \\ \tfrac{1}{4} & \tfrac{3}{8} \end{vmatrix}} = \frac{36.0(\tfrac{3}{8}) - 11.25(1)}{1(\tfrac{3}{8}) - \tfrac{1}{4}(1)} = \frac{2.25}{0.125} = 18.0$$

Although we could now solve for y using determinants, we might also note that since $x + y = 36.0$ and $x = 18.0$, it is obvious that $y = 18.0$. Each tank began with 18.0 gal. One tank has $\tfrac{3}{4}$ of the 18.0 gal remaining, so it contains 13.5 gal. The other tank has $\tfrac{5}{8}$ of the 18.0 gal remaining, so it contains 11.25 gal. □

Exercises 14-6

In Exercises 1 through 32, solve the problems by designating the unknown quantities by appropriate letters and then solving the system of equations which is found.

1. At one location in an electric circuit, the sum of two currents is 5 A while at another location their difference is 11 A. Find the two currents.

2. A quality control test involves a total area of 252 ft² covered with two different types of carpet. If the area covered by the type A carpet is 34 ft² more than the area covered by the type B carpet, find the area covered by each type.

3. A cable is 23 ft. long. Where must it be cut so that the longer piece is 2 ft longer than the shorter piece?

4. The sum of two voltages is 100 V. If the higher voltage is doubled and the other halved, the sum becomes 155 V. What are the voltages?

5. In a certain process, five drills of one type will last as long as two drills of another type. The sum of the average number of hours of operation for the two drills is 105 h. What is the average number of hours each can be used?

6. Two wind turbines are used to supply a total of 50 kW of electricity. If the larger of the two wind turbines supplies twice as much electrical power as the smaller one, find the amount supplied by each.

7. Two spacecraft (on separate missions) spend a total of 22 days circling the earth. One spends 2 days less than twice the number of days the other spends in space. How long does each spend in space?

8. The side of one square metal plate is 5 mm longer than twice the side of another square plate. The perimeter of the larger plate is 52 mm more than that of the smaller. Find the sides of the two plates. See Figure 14-12.

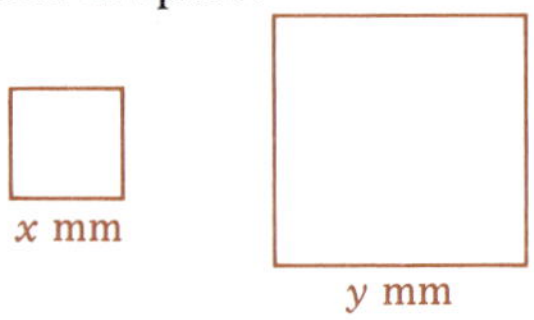

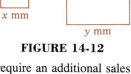

9. A company maintains two separate accounts for the investment of $45,000. Part of this amount is invested at 6% and the remainder at 8%. If the total annual income is $3400, how much is invested at each rate?

FIGURE 14-12

10. A purchase order includes items that cost a total of $8200. Some of those items require an additional sales tax of 4% while the remaining items require a 5% sales tax. If the total sales tax is $330, how much is taxed at each rate?

11. A rectangular field which is 100 ft longer than it is wide is divided into three smaller fields by placing two dividing fences parallel to those along the width. A total of 2600 ft of fencing is used. What are the dimensions of the field? See Figure 14-13.

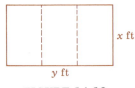

12. A supplier has 5000 gal of oil stored in two tanks. If 3% of the amount in the first tank is removed, and 4% of the amount in the second tank is removed, the total amount removed is 170 gal. How much oil is left in each tank?

FIGURE 14-13

13. A fixed amount of a certain fuel contains 150,000 Btu of potential heat. Part is burned at 80% efficiency and part is burned at 70% efficiency so that the total amount of heat actually delivered is 114,000 Btu. Find the number of Btu burned at 80% and at 70%.

14. A firm pays $500 each year for service contracts on a computer and copying machine. The service contract on the copying machine is $20 more than three times the cost of the computer service contract. How much does each service contract cost?

15. In tests on batteries used in digital watches, three batteries of one type last as long as two batteries of another type, and all five batteries provide 6 years of service. How long is the service time provided by one of each type of battery?

16. It takes a machine 2 h to process six items of one type and two items of a second type. It takes the same machine 4 h to process five items of each type. Find the time required to process each of the two types of items.

17. Based upon estimates of air pollution, fuel consumption for transportation contributes a percentage of pollution which is 16% less than all other sources combined. What percentage of air pollution is a result of this one source?

18. The relation between the Celsius and Fahrenheit temperature scales is that the number of Fahrenheit degrees equals 32 more than $\frac{9}{5}$ the number of Celsius degrees. At what reading are they equal?

19. The voltage across an electric resistor equals the current times the resistance. The sum of two resistances is 14 Ω. When a current of 3 A passes through the smaller resistor and 5 A through the larger resistor, the sum of the voltages is 60 V. What is the value of each resistor?

20. The speed of one train is 5 mi/h more than that of another train. They start at the same time at stations 425 mi apart and arrive at the same station at the same time 5 h later. What is the speed of each?

21. Two trains are traveling at 80 km/h and 72 km/h, respectively. The faster train travels for a certain length of time and the slower train travels 2 h longer. The total distance traveled by the two trains is 1360 km. Find the time each travels.

22. A vending machine has $3.50 in dimes and quarters, and there are as many dimes as quarters. How many of each are there?

23. Two gears together have 64 teeth. The larger gear has three times as many teeth as the smaller one. Find the number of teeth in each gear.

24. Suppose that 120 children and 140 adults attend a show for which the total receipts are $530. If a parent and two young children pay $5.50, how much are the adult tickets and how much are the children's tickets?

25. By mass, one alloy consists of 60% copper and 40% zinc. Another is 30% copper and 70% zinc. How many grams of each alloy are needed to make 120 g of an alloy which consists of 50% of each metal?

26. How many liters of a mixture containing 70% alcohol should be added to a mixture containing 20% alcohol to give 16L of a mixture containing 50% alcohol?

27. The relative density of an object may be defined as its weight in air divided by the difference between its weight in air and its weight when submerged in water. The sum of the weights in water and in air of an object of relative density 10 is 30 lb. What is its weight in air?

28. After a laboratory experiment, a student reports that the difference between two resistances is 8 Ω and that twice one of the resistances is 3 Ω less than twice the other. Is this possible?

29. The sum of three currents is 120 A. The first current is three times that of the second. The first two currents total 40 A. Find the value (in amperes) of each current.

30. A large corporation assigns 175 employees to a special project. That total is the sum of 2% of the research department, 5% of the sales department, and 10% of the manufacturing department. If 5% were selected from each department, the total would be the same. If 5% of the research and manufacturing departments are selected, the total becomes 160. How many employees were originally in each of the three departments?

31. When three different pumps work together, they can remove 90 gal/min from a waste holding tank. If the third pump is turned off, the rate becomes 80 gal/min. If the second and third pumps work alone, they can remove 70 gal/min. What is the individual removal rate for each pump?

32. When measured by volume, one alloy is 50% copper, 20% nickel, and 30% zinc. A second alloy is half copper and half nickel. A third alloy is 60% copper, 30% nickel, and 10% zinc. Find the volume of each alloy that must be used to produce a mixture containing 31.5 cm³ of copper, 18.0 cm³ of nickel, and 5.5 cm³ of zinc.

Chapter 14 Formulas

$$\begin{vmatrix} a_1 & b_1 \\ a_2 & b_2 \end{vmatrix} = a_1b_2 - a_2b_1 \qquad \text{determinant}$$

(14-1)

$$a_1x + b_1y = c_1$$
$$a_2x + b_2y = c_2$$

(14-2)

$$x = \frac{\begin{vmatrix} c_1 & b_1 \\ c_2 & b_2 \end{vmatrix}}{\begin{vmatrix} a_1 & b_1 \\ a_2 & b_2 \end{vmatrix}} \quad \text{and} \quad y = \frac{\begin{vmatrix} a_1 & c_1 \\ a_2 & c_2 \end{vmatrix}}{\begin{vmatrix} a_1 & b_1 \\ a_2 & b_2 \end{vmatrix}}$$

(14-3)

14-7 Review Exercises for Chapter 14

In Exercises 1 through 32, solve the given systems algebraically.

1. $2x + y = 6$
   $3x - y = 6$

2. $3x + y = 12$
   $3x - y = 24$

3. $x + 2y = 12$
   $x + 3y = 16$

4. $x + 2y = 12$
   $2x - y = 4$

5. $2x - y = 4$
   $x + 3y = 16$

6. $y + 2x = -3$
   $3y - 5x = -3$

7. $2p + 7q = 10$
   $3p - 2q = 10$

8. $m = n + 3$
   $m = 5 - 3n$

9. $2u - 3v = 5$
   $-u + 4v = -5$

10. $4x + 3y = -1$
    $5x + 2y = 4$

11. $3a + 7b = 15$
    $2a - 5b = 39$

12. $2p = 4x + 1$
    $2p = 2x - 3$

13. $3x = -16 - y$
    $7x = -8 - 5y$

14. $3y = 11 - 5h$
    $15y - 15h = 7$

15. $-6y + 4z = -8$
    $5y + 6z = 2$

16. $7x + 9y = 3$
    $5x + 4y = 1$

17. $3n + d = 20$
    $6n + 2d = 40$

18. $x - 2y = 6$
    $-2x + 4y = 5$

19. $3x + y = -5$
    $6x + 8y = 2$

20. $4x - y = 7$
    $8x + 3y = -26$

21. $2x - 3y = 8$
    $5x + 2y = 9$

22. $8x - 9y = 3$
    $3x + 4y = 2$

23. $8s + 5t = 6$
    $7s - 9t = 8$

24. $17x - 9y = 10$
    $8x + 7y = -13$

25. $0.03x + 2y = 1$
    $0.02x - 3y = 5$

26. $0.10x - 0.05y = 3$
    $0.03x + 0.20y = 2$

27. $2x - 13y = 5$
    $15x + 3y = 8$

28. $8r - 9t = 9$
    $6r + 12t = 13$

29. $\dfrac{2}{3}x + \dfrac{3}{4}y = 26$

$\dfrac{1}{3}x - \dfrac{3}{4}y = -14$

30. $\dfrac{4x + 5y}{20} = 3$

$\dfrac{1}{2}x - \dfrac{1}{3}y = 2$

31. $\dfrac{9}{r} - \dfrac{5}{s} = 2$

$\dfrac{7}{r} - \dfrac{3}{s} = 6$

32. $\dfrac{2}{x} - \dfrac{2}{y} = 4$

$\dfrac{1}{x} - \dfrac{3}{y} = -2$

In Exercises 33 through 40, solve the given systems of equations graphically.

33. $x + 3y = 2$
$6x - 3y = 5$

34. $2x + 3y = 0$
$x + y = 2$

35. $3x - y = 6$
$x - 2y = 4$

36. $4x - y = 9$
$4x + 3y = 0$

37. $4u + v = 2$
$3u - 4v = 30$

38. $r + 4s = 6$
$r + 6s = 7$

39. $3x - y = 5$
$-6x + 2y = 3$

40. $m + 3n = 7$
$3m - 2n = 5$

In Exercises 41 through 44, evaluate the given determinants.

41. $\begin{vmatrix} 2 & 3 \\ 4 & 5 \end{vmatrix}$

42. $\begin{vmatrix} -4 & 2 \\ 6 & 4 \end{vmatrix}$

43. $\begin{vmatrix} 3 & -4 \\ 2 & -5 \end{vmatrix}$

44. $\begin{vmatrix} -12 & 9 \\ 7 & 8 \end{vmatrix}$

In Exercises 45 through 48, use determinants to solve the given systems of equations.

45. $x + 3y = -1$
$2x + 8y = 2$

46. $5x + y = -3$
$8x + 2y = -8$

47. $2x + 3y = 13$
$6x - 2y = -5$

48. $0.2x + 0.4y = -1.6$
$0.7x - 0.3y = 4.6$

In Exercises 49 through 52, solve the given systems of equations by any method.

49. $x + y = 0$
$x + y + 3z = 9$
$2x - 3y + 2z = 16$

50. $x + y + z = 4$
$2x + y = 7$
$4x - y + 3z = -7$

51. $2x + y + 4z = 4$
$-3x + 2y + 5z = -25$
$4x - 2y + 4z = 4$

52. $3r + 2s + t = 11$
$4r + 3s - t = -6$
$2r + 2s + 2t = 14$

In Exercises 53 through 60, solve the applied problems by any appropriate method.

53. When analyzing a certain electric circuit, the equations

$$4i_1 - 10i_2 = 3$$
$$-10i_2 - 5(i_1 + i_2) = 6$$

are obtained. Solve for the indicated electric currents i_1 and i_2 (in amperes).

54. The equation $s = v_0 t + \frac{1}{2}at^2$ is used to relate the distance s traveled by an object in time t, given that the initial velocity is v_0 and the acceleration is a. For an object moving down an inclined plane, it is noted that $s = 32$ ft when $t = 2$ s and $s = 63$ ft when $t = 3$ s. Find v_0 (in feet per second) and a (in feet per second squared).

55. Nitric acid is produced from air and nitrogen compounds. To determine the size of the equipment required, chemists often use a relationship between the air flow rate m (in moles per hour) and exhaust nitrogen rate n. For a certain operation, the following equations

$$1.58m + 41.5 = 38.0 + 2.00n$$
$$0.424m + 36.4 = 189 + 0.0728n$$

are obtained. Solve for m and n.

56. Two ropes support a weight. The equations used to find the tensions in the ropes are

$$0.866T_2 - 0.500T_3 = 0$$
$$0.500T_2 + 0.866T_3 = 50$$

Find T_2 and T_3 (in pounds).

57. When applying Kirchhoff's laws to a certain electric circuit, the following equations result. The currents (in amperes) can be found by solving this system of equations.

$$I_1 + I_2 + I_3 = 0$$
$$2I_1 - 3I_2 \qquad = 4$$
$$2I_1 \qquad + I_3 = 3$$

58. A manager allocates time t_1 to manufacturing, t_2 to preventive maintenance, and t_3 to repairs (all in hours). For a particular job, the optimal allocation of time can be found by solving the following equations.

$$t_1 + t_2 + t_3 = 48$$
$$t_1 + 6t_2 - t_3 = 52$$
$$t_1 + 3t_3 = 60$$

59. Three different alloys contain different percentages of gold, silver, and copper. The amounts (in grams) of the alloys are represented by x, y, and z, respectively. Those values can be found from the following equations.

$$0.020x + 0.040y + 0.10z = 36$$
$$0.80x + 0.80y + 0.60z = 848$$
$$0.18x + \qquad 0.30z = 132$$

60. A department's $8000 weekly budget consists of an amount x for research, an amount y for development, and an amount z for maintenance. Those values (in dollars) can be found by solving these equations.

$$x + y + z = 8000$$
$$2x - y + 2z = 2500$$
$$x + 3y - 4z = 12000$$

In Exercises 61 through 76, set up the appropriate systems of equations and solve.

61. A welder works for 3 h on a particular job and his apprentice works on it for 4 h. Together they weld 255 spots. If the hours worked are reversed, they would have welded 270 spots together. How many spots can each one weld in an hour?

62. Two men together make 120 castings in one day. If one of them turns out half again as many as the other, how many does each make in a day?

63. The circumference of a circular lawn area is 5.60 ft more than the perimeter of a nearby square area. The diameter of the circle is 10.0 ft more than the side of the square. Find the diameter of the circle and the side of the square. (Use $\pi = 3.14$ and three significant digits in working this problem.) See Figure 14-14.

64. The perimeter of a rectangular area is 24 km and the length is 6 km longer than the width. Find the dimensions of the rectangle. See Figure 14-15.

65. The billing department of a large manufacturer reviews 884 invoices and finds that 700 require no changes. If the number of invoices requiring additional payments is 32 less than twice the number of invoices requiring refunds, how many invoices require additional payment and how many require refunds?

FIGURE 14-14   FIGURE 14-15

66. Two thermometer readings differ by 48°C. Four times the smaller reading exceeds twice the larger reading by 2°C. What are the readings?

67. The cost of booklets at a printing firm consists of a fixed charge and a charge for each booklet. The total cost of 1000 booklets is $550; the total cost of 2000 booklets is $800. What is the fixed charge and the cost of each booklet?

68. While a 40-ft pulley belt is making one complete revolution, one of the pulley wheels makes one more revolution than the other. The smaller wheel is replaced by another wheel which has half the radius; the replacement wheel makes six revolutions more than the larger wheel for each revolution of the belt. What are the circumferences of the two original wheels?

69. A car travels 355 mi by going at one speed for 3 h and at another speed for 4 h. If it travels 3 h at the second speed and 4 h at the first speed, it would have gone 345 mi. Find the two speeds.

70. An airplane travels 1200 km in 6 h with the wind. The trip takes 8 h against the wind. Determine the speed of the plane relative to the air and the speed of the wind.

71. For proper dosage a certain drug must be a 10% solution. How many milliliters each of a 5% solution and a 25% solution should be mixed to obtain 1000 mL of the required solution?

72. How many milliliters each of a 10% solution and a 25% solution of nitric acid must be used to make 95 mL of a 20% solution?

73. The sum of three resistors is 75 Ω. The first resistance is one-fourth the sum of the other two. The second resistance is one-half of the third. Find the value of each resistance.

74. Three different machines are used to produce shear pins. Together they produce 40 pins per hour. The first machine produces as much as the other two combined. The first two machines produce a total that is four times that of the third machine. Find the output of each machine for 1 h of operation.

75. A part requires $2.28 worth of nuts, bolts, and washers. The nuts cost 5¢ each, the bolts cost 10¢ each, and the washers cost 2¢ each. There are as many nuts as bolts, and there are twice as many washers as bolts. How many of each are there?

76. A computer simulation of population growth begins with three villages having a combined population of 9000 people. The first village has an annual growth rate of 5%, the second village has a 6% growth rate, while the third grows at a 10% rate. In the first year the population increases by 770 people. If the third village begins with a population that is twice the combined population of the other two, find the original population of each village.

In Exercises 77 through 80, use a scientific calculator to solve the given systems of equations.

77. $0.0358x - 0.0429y = 0.136$
    $-0.0587x + 0.0308y = 0.902$

78. $12.34x - 17.98y = 16.37$
    $25.00x + 32.34y = -19.11$

79. $8.929x - 9.300y = 5.374$
    $9.632x + 5.314y = 6.282$

80. $-256.3x + 872.5y = 643.8$
    $516.0x - 327.9y = -212.3$

Additional Topics from Geometry

In Chapter 6 we introduced some basic concepts of geometry so that they could be used in Chapters 7 through 14. In this chapter we expand upon those topics in order to acquire a more comprehensive understanding of geometry. Section 15-1 is devoted to introducing additional terminology related to various types of angles. The remaining sections introduce topics related to basic geometric figures.

The content of this chapter will allow us to solve many more applied problems. The examples and exercises include a wide variety of problems such as this one: While flying directly to a destination, a pilot is instructed to make a 90° left turn to avoid other air traffic. After several minutes, air traffic control instructs the pilot to make a left turn of 150°. What is the angle of a third left turn that will allow the plane to return to its original direction? We will solve this problem in this chapter.

15-1 Angles

As stated in Chapter 6, geometry deals with the properties and measurement of angles, lines, surfaces, and volumes, as well as the basic figures which are formed. We noted that certain concepts, such as *a* **point**, *a* **line**, *and a* **plane**, *are accepted as being understood intuitively*. We shall

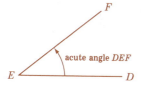

FIGURE 15-1

obtuse angle *GHJ*

FIGURE 15-2

make no attempt to define these terms, although we shall use them in defining and describing other terms. This in itself points out an important aspect in developing a topic: Not everything can be defined or proved; some concepts must be accepted and used as a basis for studying geometry.

In Chapter 6 we defined an **angle** and how it is measured in terms of degrees, minutes, seconds, and decimal parts of a degree. *We also defined a* **straight angle** *as an angle of* 180° *and a* **right angle** *as an angle of* 90°.

Two other basic types of angles are identified by whether or not they are less or greater than 90°. *An angle between* 0° *and* 90° *is an* **acute angle** (see Figure 15-1). *An angle greater than* 90° *but less than* 180° *is an* **obtuse angle** (see Figure 15-2).

EXAMPLE A

An angle of 73° is an acute angle since 73° is between 0° and 90°.

An angle of 154° is an obtuse angle since 154° is between 90° and 180°.

In geometry, it is often necessary to refer to specific pairs of angles. *Two angles which have a common vertex and a side common between them are known as* **adjacent angles.** *If two lines cross to form equal angles on opposite sides of the point of intersection, which is the common vertex, these equal angles are called* **vertical angles.**

EXAMPLE B In Figure 15-3, ∠*BAC* and ∠*CAD* have a common vertex at *A* and the common side *AC* between them so that ∠*BAC* and ∠*CAD* are adjacent angles.

In Figure 15-4, lines *AB* and *CD* intersect at *O*. Here ∠*AOC* and ∠*BOD* are vertical angles, and they are equal. Also, ∠*BOC* and ∠*AOD* are vertical angles and are equal.

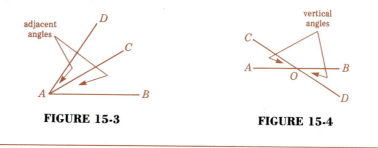

FIGURE 15-3   **FIGURE 15-4**

If the sum of two angles is 180°, *the angles are called* **supplementary angles.** *If the sum of two angles is* 90°, *the angles are called* **complementary angles.** These types of angles are illustrated in the following example.

EXAMPLE C   In Figure 15-5(a), $\angle BAC = 55°$, and in Figure 15-5(b) $\angle DEF = 125°$. Since $\angle BAC + \angle DEF = 55° + 125° = 180°$, $\angle BAC$ and $\angle DEF$ are supplementary angles.

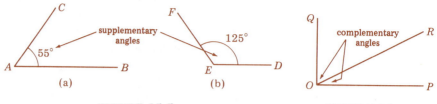

FIGURE 15-5   FIGURE 15-6

In Figure 15-6, $\angle POQ$ is a right angle, which means that $\angle POQ = 90°$. Since $\angle POR + \angle ROQ = \angle POQ = 90°$, $\angle POR$ and $\angle ROQ$ are complementary angles.   □

Since the sum of supplementary angles is 180°, if one angle is known, the other angle (its **supplement**) can be found by subtracting the known angle from 180°. Also, since the sum of complementary angles is 90°, one angle can be found by subtracting the other angle (its **complement**) from 90°.

EXAMPLE D   If one angle is 36°, then its supplement is 144°, since

$$180° - 36° = 144°$$

The complement of an angle of 36° is 54°, since

$$90° - 36° = 54°$$ □

In Chapter 6 we defined parallel lines as lines whose extensions will not meet. In a plane, *if a line crosses two parallel or nonparallel lines, it is called a* **transversal**. In Figure 15-7, $AB \parallel CD$, which means that AB is parallel to CD, and the transversal of these parallel lines is the line EF.

When a transversal crosses a pair of parallel lines, certain pairs of equal angles result. In Figure 15-7, the **corresponding angles** are equal. (That is, $\angle 1 = \angle 5$, $\angle 2 = \angle 6$, $\angle 3 = \angle 7$, and $\angle 4 = \angle 8$.) Also, the **alternate interior** angles are equal ($\angle 3 = \angle 6$ and $\angle 4 = \angle 5$). The **alternate exterior** angles are also equal ($\angle 1 = \angle 8$ and $\angle 2 = \angle 7$).

FIGURE 15-7

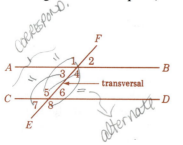

EXAMPLE E

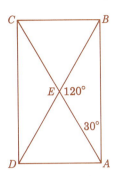

FIGURE 15-8

A surveyor has established that the tract of land shown in Figure 15-8 has the angles shown. Also, it is known that the lot is rectangular. Find $\angle DEC$, $\angle CEB$, $\angle ECD$, and $\angle DAE$.

$\angle DEC = 120°$ since $\angle DEC$ and $\angle AEB$ are vertical angles, which are known to be equal. Since $\angle AEB = 120°$ (as shown), it follows that $\angle DEC = 120°$.

Since $\angle AEC$ is a straight angle of $180°$, it follows that $\angle AEB$ and $\angle CEB$ are supplementary and their sum is therefore $180°$. Since $\angle AEB$ is known to be $120°$, it follows that $\angle CEB = 60°$.

Since $ABCD$ is a rectangle, we know that $AB \parallel DC$. Therefore, $\angle ECD = 30°$ since $\angle ECD$ and $\angle EAB$ are alternate interior angles, which are equal.

$\angle DAE = 60°$ since $\angle DAE$ and $\angle EAB$ are complementary angles, and we already know that $\angle EAB = 30°$. We know that those angles are complementary because $\angle DAB$ is one of the right angles in a rectangle. □

Exercises 15-1

In Exercises 1 through 4, determine the values of the described angles.

1. The complement of $37°$
2. The complement of $83°$
3. The supplement of $159°$
4. The supplement of $27°$

In Exercises 5 through 8, use Figure 15-9. Identify the indicated angles.

5. An acute angle with one side CE
6. The obtuse angle
7. Two pairs of adjacent angles
8. One pair of complementary angles

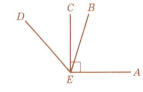

FIGURE 15-9

In Exercises 9 through 12, use Figure 15-10. Identify the indicated angles.

9. One pair of supplementary right angles
10. Two pairs of adjacent angles
11. The complement of $\angle DBE$
12. The supplement of $\angle CBD$

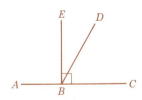

FIGURE 15-10

In Exercises 13 through 16, use Figure 15-11. Determine the indicated angles.

13. $\angle DBE$ 14. $\angle EBF$ 15. $\angle DBA$ 16. $\angle FBA$

$AC \perp BE$
$FB \perp DB$
$\angle CBD = 65°$

FIGURE 15-11

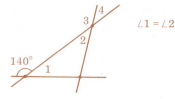

FIGURE 15-12

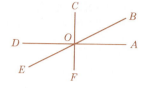

FIGURE 15-13

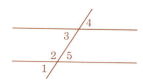

FIGURE 15-14

In Exercises 17 and 18, use Figure 15-12. Determine the indicated angles.

17. $\angle 4$ **18.** $\angle 3$

In Exercises 19 and 20, use Figure 15-13. Determine the indicated angles ($DA \perp CF$, $\angle AOB = 28°$).

19. $\angle EOF$ **20.** $\angle EOC$

In Exercises 21 through 24, use Figure 15-14. Identify the indicated pairs of angles from those which are numbered.

21. Two pairs of vertical angles   **22.** One pair of alternate interior angles

23. One pair of corresponding angles   **24.** One pair of supplementary angles

In Exercises 25 through 28, use Figure 15-15. Lines which appear to be parallel are parallel. Determine the indicated angles.

25. $\angle 1$ **26.** $\angle 2$ **27.** $\angle 4$ **28.** $\angle 3$

In Exercises 29 through 32, use Figure 15-16. Determine the indicated angles.

29. $\angle FCE$ **30.** $\angle ECD$ **31.** $\angle BCE$ **32.** $\angle BFC$

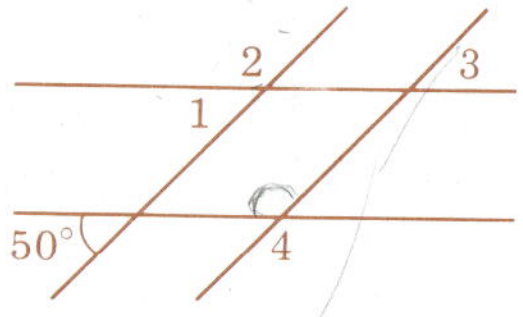

FIGURE 15-15

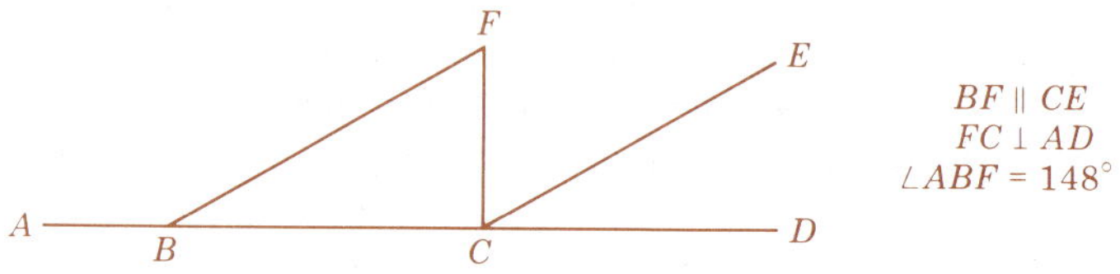

FIGURE 15-16

$BF \parallel CE$
$FC \perp AD$
$\angle ABF = 148°$

In Exercises 33 through 40, use Figure 15-17 and determine the indicated angles. In Figure 15-17 the line through points A, B, C, and D is the deck of a bridge. Lines BF and CE represent pole supports and they form right angles with the bridge deck. All other lines are support cables. Lines which appear to be parallel are parallel.

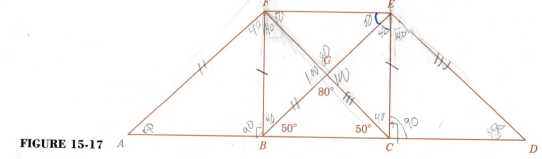

FIGURE 15-17

33. ∠*FBG*   **34.** ∠*FEG*   **35.** ∠*CDE*   **36.** ∠*CED*

37. ∠*BEC*   **38.** ∠*FGE*   **39.** ∠*BGF*   **40.** ∠*BED*

In Exercises 41 through 44, use a calculator to determine the values of the described angles.

41. The complement of 52°37′42″

42. The complement of 58′56″

43. The supplement of 172°59′47″

44. The supplement of 4°12′3″

15-2 Properties of Triangles, Quadrilaterals, and the Circle

Convex polygon

FIGURE 15-18

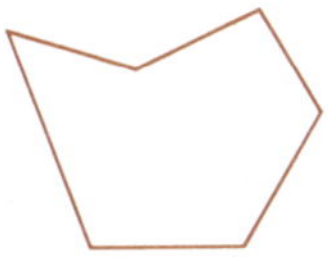

Concave polygon

FIGURE 15-19

Regular octagon

FIGURE 15-20

When part of a plane is bounded and closed by straight-line segments, it is called a **polygon.** *In general, polygons are named according to the number of sides they have. A* **triangle** *has three sides, a* **quadrilateral** *has four sides, a* **pentagon** *has five sides, a* **hexagon** *has six sides, and so on.* The polygons of greatest importance are the triangle and the quadrilateral. This section is devoted to properties of triangles and quadrilaterals, as well as the properties of the circle.

A polygon in which none of the interior angles equals or is greater than 180° is called a **convex polygon** (see Figure 15-18). If a polygon is not convex, it is **concave** (see Figure 15-19). *In a* **regular polygon,** *all sides are equal in length and all interior angles are equal.* In Figure 15-20, a regular octagon is shown.

In a polygon, *a line segment that joins any two nonadjacent vertices is called a* **diagonal.** From this definition, we can see that a triangle cannot have diagonals whereas polygons of four or more sides do have diagonals. In Figure 15-21 the diagonals of a hexagon are shown as dashed lines.

In Chapter 6 we discussed certain types of triangles. These included the equilateral triangle, isosceles triangle, scalene triangle, and right triangle. We shall now develop certain additional important properties of triangles.

One extremely important property of a triangle is that *the sum of its interior angles is 180°.* We can prove this as follows. In Figure 15-22, $EC \parallel AB$. Since ∠1, ∠2, and ∠3 constitute a straight angle,

$$\angle 1 + \angle 2 + \angle 3 = 180°$$

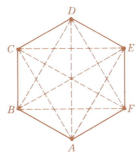

FIGURE 15-21

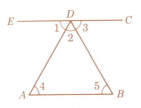

FIGURE 15-22

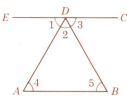

FIGURE 15-22

Also, $\angle 1 = \angle 4$ since these angles are alternate interior angles. Similarly, $\angle 3 = \angle 5$. Thus by substitution

$$\angle 4 + \angle 2 + \angle 5 = 180°$$

Therefore if two angles of a triangle are known, the third may be determined by subtracting the sum of the first two from 180°.

EXAMPLE A　In the triangle shown in Figure 15-23, we may find $\angle A$ as follows:

$$\angle B + \angle C = 55° + 80° = 135°$$
$$\angle A = 180° - 135°$$
$$= 45° \qquad \square$$

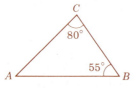

FIGURE 15-23

EXAMPLE B　In the triangle shown in Figure 15-24, side AB is parallel to DC. We may find $\angle ACB$ as follows:

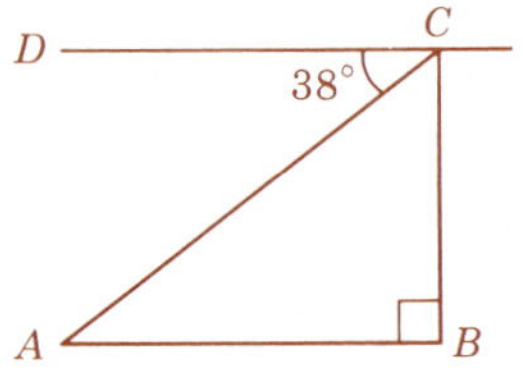

$$\angle A = 38° \qquad \text{$\angle A$ and $\angle DCA$ are alternate interior angles}$$
$$\angle B = 90° \qquad \text{it is a right angle}$$
$$\angle A + \angle B = 38° + 90° = 128°$$
$$\angle ACB = 180° - 128° = 52°$$

FIGURE 15-24　We now know the three angles in triangle ABC. $\qquad \square$

EXAMPLE C　A forest ranger uses orienteering to find the location of two known bird sanctuaries. The path followed is triangular as shown in Figure 15-25. Find the angle at point B.

Since all triangles have the property that the sum of the three angles is 180°, we get

$$\angle B + 37° + 93° = 180° \qquad \text{sum of angles of triangle}$$
$$\angle B + 130° = 180°$$
$$\angle B = 50°$$

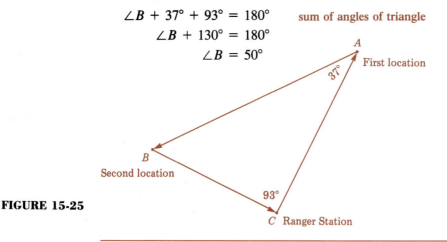

FIGURE 15-25

EXAMPLE D   An airplane is flying in a certain direction and then makes a 90° left turn. After several minutes, a left turn of 150° is made. What is the angle of a third left turn which will cause the plane to return to the original direction? (See Figure 15-26.)

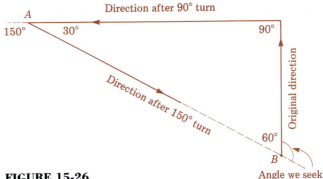

FIGURE 15-26

From Figure 15-26 we see that the interior angle of the triangle at *A* is the supplement of 150°, or 30°. Since the sum of the interior angles of the triangle is 180°, the interior angle at *B* is

$$\angle B = 180° - (90° + 30°) = 60°$$

This means that the angle we seek is the supplement of 60°, which is 120°. □

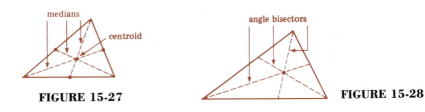

FIGURE 15-27   **FIGURE 15-28**

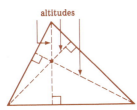

FIGURE 15-29

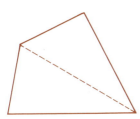

FIGURE 15-30

Another property of triangles is that *the three* **medians**—*line segments drawn from a vertex to the midpoint of the opposite side—meet at a single point* (see Figure 15-27). *This point is called the* **centroid** *of the triangle. Also, the three* **angle bisectors** *meet at a common point* (see Figure 15-28). *The three* **altitudes (heights)**, *which are drawn from a vertex to the opposite side (or its extension), meet at a common point* (see Figure 15-29). These three points are generally not the same point for a given triangle.

We will now consider quadrilaterals. In Chapter 6 we discussed the rectangle, square, parallelogram, rhombus, and trapezoid. Here we consider an important property of all quadrilaterals.

Consider a quadrilateral to be divided into two triangles as shown in Figure 15-30. The sum of the angles of each triangle is 180°. Since these angles give the total of the angles of the quadrilateral, we conclude that: *The sum of the angles of a quadrilateral is* 360°.

EXAMPLE E

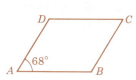

FIGURE 15-31

In the parallelogram in Figure 15-31, we know that $\angle A = \angle C$ and $\angle D = \angle B$ since they are opposite angles. Since $\angle A = 68°$, we get $\angle C = 68°$. The four angles total 360° and

$$\angle A + \angle C = 68° + 68° = 136°$$

so that $\angle D + \angle B = 360° - 136° = 224°$. Since $\angle D = \angle B$, we know that $\angle D = 112°$ and $\angle B = 112°$. ☐

We conclude this section with an extension of the discussion of the circle. In Chapter 6 we defined the **radius** and **diameter** of the circle and gave formulas for finding the area and circumference. We now note that *two circles with the same center are* **concentric,** as shown in Figure 15-32.

A line segment having its endpoints on a circle is a **chord.** *A* **tangent** *is a line that touches a circle at one point (**does not pass through**). A* **secant** *is a line that passes through two points of a circle.* These are illustrated in the following example.

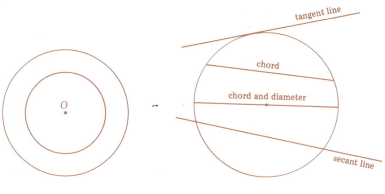

FIGURE 15-32   **FIGURE 15-33**

EXAMPLE F

For the circle shown in Figure 15-33, the top line is a tangent line, the two middle line segments are both chords (the chord through the center is also a diameter), and the bottom line is a secant line. ☐

A **central angle** *of a circle is an angle with its vertex at the center of the circle. An arc of a circle consists of that part of the circle between and containing two specific points on the circle. There are two such arcs on a circle: the* **minor arc** *and the* **major arc.** *An arc is measured by its central angle.* If an arc has a central angle of 50°, then the measure of the arc is also 50°. These arcs are illustrated in the following example.

EXAMPLE G

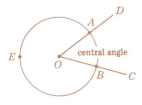

FIGURE 15-34

For the circle in Figure 15-34, $\angle DOC$ is a central angle. (We could also have designated it as $\angle AOB$.) Also, that part of the circle between and including A and B is the arc AB. There are two arcs: the minor arc AB (designated as $\overset{\frown}{AB}$) and the major arc AEB (designated as $\overset{\frown}{AEB}$). We shall use two-letter arc designations only for minor arcs. If $\angle DOC = 50°$, then $\overset{\frown}{AB} = 50°$ and $\overset{\frown}{AEB} = 310°$. In general, a minor arc is less than $180°$ and a major arc is between $180°$ and $360°$. (If both arcs are exactly $180°$, we call them *semicircles*.) ☐

An angle is **inscribed** in an arc (see Figure 15-35) if the sides of the angle contain the endpoints of the arc and the vertex of the angle is a point (not an endpoint) of the arc. An important property associated with inscribed angles is this: *An inscribed angle is one-half of its intercepted arc.* This is illustrated in the following example.

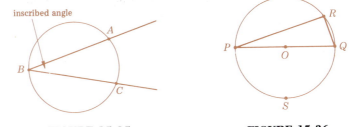

FIGURE 15-35   **FIGURE 15-36**

EXAMPLE H

In the circle shown in Figure 15-35, $\angle ABC$ is inscribed in $\overset{\frown}{ABC}$, and it intercepts $\overset{\frown}{AC}$. If $\overset{\frown}{AC} = 60°$, $\angle ABC = 30°$.

In the circle shown in Figure 15-36, PQ is a diameter and $\angle PRQ$ is inscribed in semicircular $\overset{\frown}{PRQ}$. Since $\overset{\frown}{PSQ} = 180°$, $\angle PRQ = 90°$. We conclude that: *An angle inscribed in a semicircle is a right angle.* ☐

An important property of a tangent line to a circle is: *A tangent to a circle is perpendicular to the radius drawn to the point of contact.* This is illustrated in the following example.

EXAMPLE I

In the figure shown in Figure 15-37, O is the center of the circle and PQ is tangent at Q. If $\angle OPQ = 25°$, find $\angle POQ$.

Since the center of the circle is point O, OQ is a radius of the circle. A tangent is perpendicular to a radius at the point of tangency, which means that $\angle OQP = 90°$ so that

$$\angle OPQ + \angle OQP = 25° + 90° = 115°$$

Since the sum of the angles of a triangle is $180°$, we have

$$\angle POQ = 180° - 115° = 65°$$ ☐

FIGURE 15-37

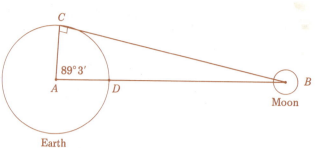

FIGURE 15-38   Earth

EXAMPLE J In Figure 15-38, an observer at point C sees the moon on the horizon while someone at point D sees the moon directly overhead. By knowing the distance along the surface of the earth between points C and D, it is found that the angle at A is 89°3′. Find the angle at B.

The angle at C is 90° since a line of sight on the horizon means the line of sight is tangent and the tangent meets the radius at a right angle. The sum of the angles at A, C, and B must be 180° so we get

$$\angle A + \angle B + \angle C = 180°$$
$$89°3′ + \angle B + 90° = 180°$$
$$\angle B + 179°3′ = 180°$$
$$\angle B = 0°57′$$

Exercises 15-2

In Exercises 1 through 12, find $\angle A$ in the indicated figures.

1. Figure 15-39(a)   **2.** Figure 15-39(b)   **3.** Figure 15-39(c)   **4.** Figure 15-39(d)

5. Figure 15-40(a)   **6.** Figure 15-40(b)   **7.** Figure 15-40(c)   **8.** Figure 15-40(d)

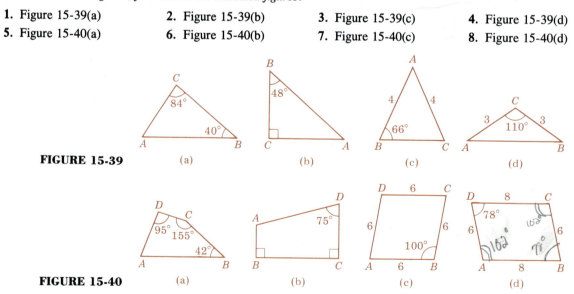

FIGURE 15-39   (a)   (b)   (c)   (d)

FIGURE 15-40   (a)   (b)   (c)   (d)

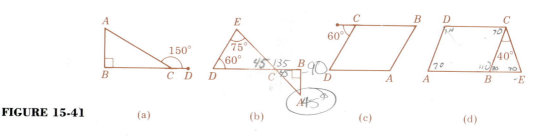

FIGURE 15-41   (a)   (b)   (c)   (d)

9. Figure 15-41(a)   **10.** Figure 15-41(b)   **11.** Figure 15-41(c); *ABCD* is a parallelogram.

12. Figure 15-41(d); *ABCD* is a parallelogram. *CE* = *CB*.

In Exercises 13 through 16, find the measure of the indicated angle or arc.

13. Find $\angle AOB$ in Figure 15-42(a); $\widehat{AB}$ = 32°.   **14.** Find $\widehat{AB}$ in Figure 15-42(b); $\angle AOB$ = 118°.

15. Find $\widehat{AC}$ in Figure 15-42(c); $\angle ABC$ = 38°.   **16.** Find $\angle ABC$ in Figure 15-42(d); $\widehat{AC}$ = 180°.

In Exercises 17 through 24, use Figure 15-43. In the figure, O is the center of the circle. Identify the following:

17. Two secant lines   **18.** A tangent line   **19.** Two chords

20. An inscribed angle   **21.** An acute central angle   **22.** An obtuse central angle

23. Two minor arcs   **24.** Two major arcs

In Exercises 25 through 28, use Figure 15-43 ($\angle BOG$ = 60°). Determine each of the following:

25. $\widehat{BG}$ **26.** $\widehat{CG}$ **27.** $\angle BCG$ **28.** $\angle GAO$

In Exercises 29 through 32, use Figure 15-44. In the figure, O is the center of the circle, line BT is tangent to the circle at B, and $\angle ABC$ = 55°. Determine the indicated arcs and angles.

29. $\widehat{AC}$ **30.** $\widehat{CAB}$ **31.** $\angle CBT$ **32.** $\angle BTC$

In Exercises 33 through 36, answer the given questions.

33. What is the maximum number of acute angles a triangle may contain?

34. What is the maximum number of acute angles a quadrilateral may contain?

35. How many diagonals does a pentagon have?

36. How many diagonals can be drawn from a single vertex of an octagon?

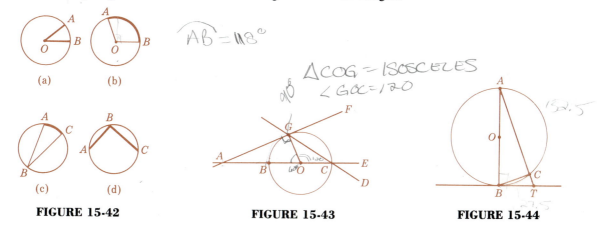

(a)   (b)

(c)   (d)

FIGURE 15-42   **FIGURE 15-43**   **FIGURE 15-44**

In Exercises 37 through 48, solve the given problems.

37. A tooth on a saw is in the shape of an isosceles triangle. If the angle at the point is 36°, find the two base angles.

38. A transmitting tower is supported by a wire which makes an angle of 50° with the ground. What is the angle between the tower and the wire?

39. Three ships are positioned at sea so that the distances between them are all equal. Find the angles of the triangle formed by these ships.

40. The surface of a road makes an angle of 2.4° with level ground (see Figure 15-45). If a survey stake is located at the high side, find the angle A between the stake and the road surface.

41. If a stop sign is made in the shape of a regular octagon, find the interior angles. See Figure 15-46.

42. If a building is in the shape of a regular pentagon, find its interior angles. See Figure 15-47.

FIGURE 15-45   FIGURE 15-46   FIGURE 15-47

43. The side view of an airplane wheel chock is in the shape of a quadrilateral with two right angles and an acute angle of 42°. Find the measure of the obtuse angle.

44. A cathedral ceiling rises in height from the front of a room to the rear. If the ceiling makes an angle of 110° with the front wall, find the angle between this ceiling and the rear wall. See Figure 15-48.

45. The streets in a certain city meet at the angles shown in Figure 15-49. Find the angle x between the indicated streets.

FIGURE 15-48   FIGURE 15-49   FIGURE 15-50

46. Metal braces support a beam as shown in Figure 15-50. Find the indicated angle.

47. The velocity of an object moving in a circular path is always directed tangent to the circle in which it is moving. A child whirls a stone on the end of a string in a vertical circle. The string was initially in a vertical position, and the stone makes $5\frac{1}{2}$ revolutions before the string breaks. In what direction does the stone travel at that instant?

48. The gear in Figure 15-51 has 24 teeth. Find the indicated angle.

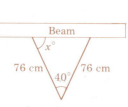

FIGURE 15-51

15-3 The Pythagorean Theorem

See Appendix B for a computer program that uses the Pythagorean theorem to find the third side of a triangle.

One property of a right triangle is so important that we will devote this section to developing it and showing some of its applications. This property is stated in the **Pythagorean theorem:**

In a right triangle, the square of the length of the hypotenuse equals the sum of the squares of the lengths of the other two sides.

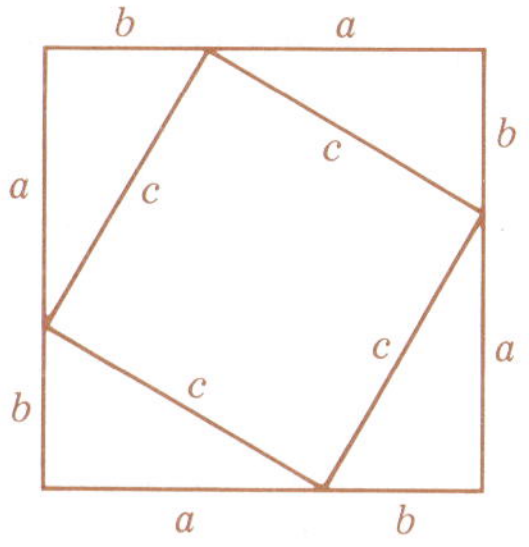

Recall that in a right triangle, the hypotenuse is the side opposite the right angle. First we shall show a derivation of this theorem which makes use of both geometry and algebra. In Figure 15-52 a square of side c is inscribed in a square of side $a + b$ as shown. The area of the outer square $(a + b)^2$ minus the area of the four triangles of sides a, b, and c equals the area of the inner square. This is expressed as

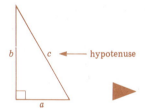

FIGURE 15-52

$$(a + b)^2 - 4\left(\frac{1}{2}ab\right) = c^2$$

$$a^2 + 2ab + b^2 - 2ab = c^2$$

$$\boxed{a^2 + b^2 = c^2} \qquad \text{Pythagorean theorem} \qquad (15\text{-}1)$$

We see that this result is the Pythagorean relation for each of the four triangles of sides a, b and c. For any right triangle with sides a, b, and c, *where c is the hypotenuse,* Eq. (15-1) is valid. See Figure 15-53.

We now present five examples illustrating the use of the Pythagorean theorem.

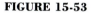

FIGURE 15-53

EXAMPLE A

For a right triangle ABC with the right angle at C, if $AC = 5$, and $BC = 12$, find AB (see Figure 15-54).

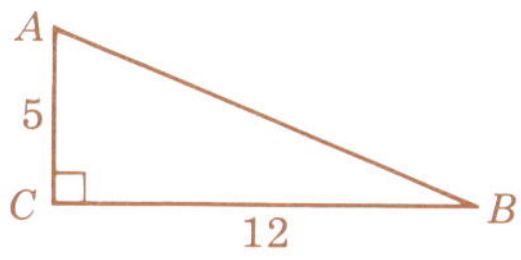

Letting AC, BC, and AB denote the sides, we find that the Pythagorean theorem, applied to this triangle, is

$$(AC)^2 + (BC)^2 = (AB)^2$$

legs —— hypotenuse

FIGURE 15-54

Substituting the values for AC and BC, we get

$$5^2 + (12)^2 = (AB)^2$$
$$25 + 144 = (AB)^2$$
$$169 = (AB)^2$$
$$13 = AB$$

We find the final value $AB = 13$ by taking the square root of both sides. We use the positive square root since lengths are considered positive. ☐

EXAMPLE B

A pole is on level ground and perpendicular to the ground. Guy wires, which brace the pole on either side, are attached at the top of the pole. Each guy wire is 25.0 ft long and the pole is 20.0 ft high. How far are the grounded ends of the guy wires from each other? (See Figure 15-55.)

From the figure we see that we are to find AD. In order to find AD we shall find AC, which in turn equals CD, and therefore $AD = 2AC$. From the Pythagorean theorem, we have $(AC)^2 + (BC)^2 = (AB)^2$. Substituting the known values of AB and BC, we get

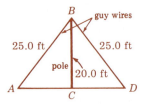

FIGURE 15-55

$$\overset{\text{hypotenuse}}{}$$

$$(AC)^2 + (20.0)^2 = (25.0)^2$$
$$(AC)^2 + 400 = 625$$
$$(AC)^2 = 225$$
$$AC = 15.0 \text{ ft}$$

Therefore $AD = 30.0$ ft.   □

EXAMPLE C

A jet travels 7.50 km while gaining altitude at a constant rate. If it travels between horizontal points 5.80 km apart, what is its gain in altitude?

From the statement of the problem, we set up the diagram in Figure 15-56. We are to find the vertical distance x:

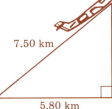

FIGURE 15-56

$$(5.80)^2 + x^2 = (7.50)^2$$
$$33.64 + x^2 = 56.25$$
$$x^2 = 56.25 - 33.64$$
$$= 22.61$$
$$x = 4.75 \text{ km}$$

The jet gains 4.75 km in altitude.   □

EXAMPLE D

A surveyor wants to determine the distance between two points, but a large building is an obstruction. However, the distance can be found by selecting a convenient third point which can be used to form a right triangle. Find the distance between points A and B (see Figure 15-57).

Since the right angle is at C, the hypotenuse is the required distance AB. Applying the Pythagorean theorem we get

$$(103)^2 + (192)^2 = (AB)^2$$
$$10609 + 36864 = (AB)^2$$
$$47473 = (AB)^2$$
$$AB = 218 \text{ m}$$

The distance between points A and B is 218 m.

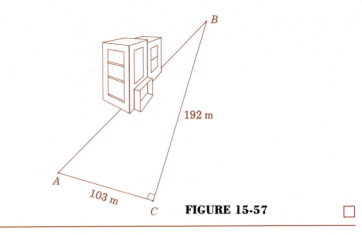

FIGURE 15-57

EXAMPLE E — A cable is attached to the top of a 22.0-ft pole. A man holding the cable moves a certain distance from the pole and notes that there are 30.0 ft of cable from the ground to the top of the pole. He then moves another 10.0 ft from the pole. How long must the cable be to reach the ground at his feet? (See Figure 15-58.)

From the figure, we see that if we first find his original distance x from the pole, we can then proceed to calculate the required distance y. Applying the Pythagorean theorem to the triangle with sides x, 22.0, and 30.0, we get the following solution.

$$x^2 + (22.0)^2 = (30.0)^2$$
$$x^2 + 484 = 900$$
$$x^2 = 416$$
$$x = 20.4 \text{ ft}$$

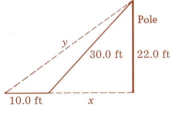

FIGURE 15-58

Now we add 10.0 to 20.4 and apply the Pythagorean theorem to the triangle with sides 30.4, 22.0, and y. We get

$$(30.4)^2 + (22.0)^2 = y^2$$
$$924 + 484 = y^2$$
$$y^2 = 1408 \qquad y = 37.5 \text{ ft}$$

When applying the Pythagorean theorem, we must be sure that we are working with a *right* triangle. Also, we must take care to ensure that Eq. (15-1) is arranged correctly with the hypotenuse and legs in their proper locations. In a right triangle, the hypotenuse is always opposite the right angle and it is always the longest side.

Exercises 15-3

In Exercises 1 through 20, find the indicated sides of the right triangle shown in Figure 15-59. Where necessary, round off results to three significant digits.

FIGURE 15-59

	a	b	c
1.	3	4	?
2.	9	12	?
3.	8	15	?
4.	24	10	?
5.	6	?	10
6.	2	?	4
7.	5	?	7
9.	?	12	16
11.	?	15	32
13.	56	?	82
15.	5.62	40.5	?
17.	0.709	?	2.76
19.	?	16.5	42.4

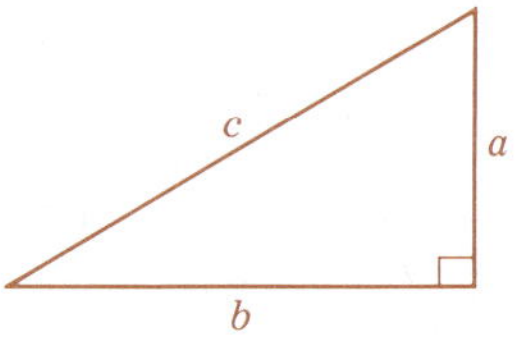

	a	b	c
8.	3	?	9
10.	?	10	18
12.	?	5	36
14.	?	125	230
16.	23.5	4.33	?
18.	?	0.0863	0.145
20.	73.7	?	86.1

In Exercises 21 through 52, set up the given problems and solve by use of the Pythagorean theorem.

21. What is the length of a diagonal of a square of side 4.00 cm?

22. What is the length of the diagonal of a rectangle 8.50 in. long and 4.60 in. wide?

23. The shortest side of a triangle whose angles are 30°, 60°, and 90° is 8.00 in. For this special right triangle, the shortest side is one-half the hypotenuse. Find the perimeter and the area.

24. A square is inscribed in a circle. See Figure 15-60. (All four vertices of the square touch the circle.) Find the length of the side of the square if the radius of the circle is 8.50 m.

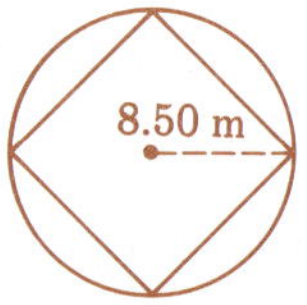

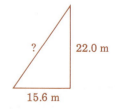

FIGURE 15-60

25. An airplane is 3000 ft directly above one end of a 10,000-ft runway. What is the distance between the airplane and a glide-slope indicator located on the ground at the other end of the runway?

26. A ramp for the disabled must be designed so that it rises a total of 1.2 m over a flat distance of 7.8 m. How long is the third side of the ramp? See Figure 15-61.

27. One steel support beam is to be placed along the diagonal of a rectangular walk-way which has dimensions 6.0 ft by 11 ft. Find the length of the beam.

FIGURE 15-61

28. A parachutist jumps from a helicopter at a point directly above a target. If the helicopter was 3000 ft above the target and the target is missed by 600 ft, how far did the parachutist travel after leaving the helicopter? (Assume that the path of the parachutist is a straight line.)

29. In calculating the effect of wind on an airplane's flight path, a pilot must find the length of the hypotenuse in a right triangle with legs 50 and 200. Find that length.

30. A 22.0-m-high tree casts a shadow 15.6 m long. How far is it from the top of the tree to the tip of the shadow? See Figure 15-62.

FIGURE 15-62

31. A searchlight is 520 ft from a wall, and its beam reaches a point 38.0 ft up the wall. What is the length of the beam? See Figure 15-63.

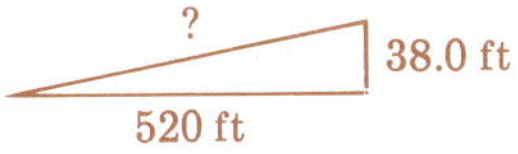

38.0 ft
520 ft

FIGURE 15-63

32. A motorist travels 25.0 mi due east of a town. How far north must the motorist now travel to be 28.0 mi on a direct line from the town?

33. A railing is 5.50 m long and extends along a stairway between floors which are 2.80 m apart. What is the horizontal distance between the ends of the railing?

34. A man rows across a river that is 600 m wide. The current carries him downstream 55.0 m from the point directly across from his starting point. How far does he actually travel?

35. The guy wires bracing a telephone pole on a level ground area and the line along the ground between the grounded ends of the wires form an equilateral triangle whose side is 20.0 ft. Find the height of the point at which the wires are attached to the pole.

21.0 ft
6.50 ft

FIGURE 15-64

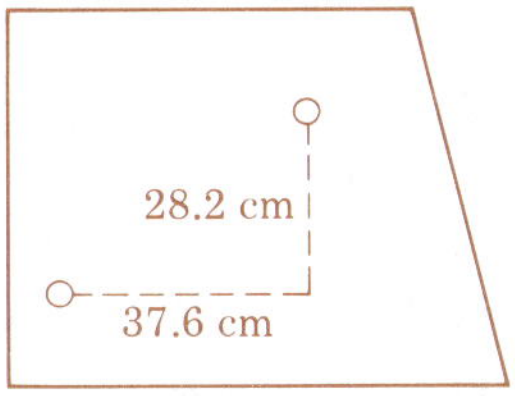

28.2 cm
37.6 cm

FIGURE 15-65

36. Figure 15-64 shows a roof truss. The rafters are 21.0 ft long, including a 1.5-ft overhang, and the height of the truss is 6.50 ft. Determine the length of the base of the truss.

37. Figure 15-65 shows a metal plate with two small holes bored in it. What is the center-to-center distance between the holes? The given distances are measured center to center.

38. In an alternating-current circuit containing a resistor and a capacitor, the capacitor contributes an effective resistance to the current, called the *reactance*. The total effective resistance in the circuit, called the *impedance Z*, is related to the resistance R and reactance X in exactly the same way that the hypotenuse of a right triangle is related to the sides. Find Z for a circuit in which $R = 17.0 \ \Omega$ and $X = 8.25 \ \Omega$.

39. The electric intensity at point P due to an electric charge Q is in the direction from the charge shown in Figure 15-66. The electric intensity E at P is equivalent to the two intensities, E_h and E_v, which are horizontal and vertical, respectively. Given that $E = 35.0$ kV/m and $E_h = 17.3$ kV/m, find E_v.

40. Two forces, F_1 and F_2, acting on an object are at right angles to each other. Their net resultant force F on the object is related to F_1 and F_2 in the same way that the hypotenuse of a right triangle is related to the sides. Given that $F_1 = 865$ lb and $F_2 = 225$ lb, find F.

41. A source of light L, a mirror M, and a screen S are situated as shown in Figure 15-67. Find the distance a light ray travels in going from the source to the mirror and then to the screen.

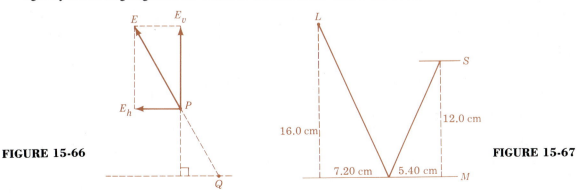

FIGURE 15-66

E E_v
E_h P
Q

L
S
16.0 cm
12.0 cm
7.20 cm 5.40 cm
M

FIGURE 15-67

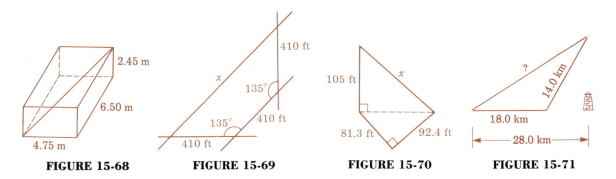

FIGURE 15-68   **FIGURE 15-69**   **FIGURE 15-70**   **FIGURE 15-71**

42. A rectangular dining room is 6.50 m long, 4.75 m wide, and 2.45 m high. What is the distance from a corner on the floor to the opposite corner at the ceiling? See Figure 15-68.

43. Figure 15-69 shows four streets of a city. What is the indicated distance?

44. Figure 15-70 shows a quadrilateral tract of land. What is the length of the indicated side?

45. A ship 28.0 km due west of a lighthouse travels 18.0 km toward the lighthouse and then turns. After traveling another 14.0 km in a straight line, it is due north of the lighthouse. How far, on a direct line, is the ship from its starting point? See Figure 15-71.

46. The hypotenuse of a right triangle is 24.0 cm, and one leg of the triangle is twice the other. Find the perimeter of the triangle.

47. One leg of a right triangle is 15 ft long and the area of the triangle is 180 ft². What is the length of the hypotenuse?

48. How long is the side of an isosceles right triangle whose hypotenuse is s?

49. Find a formula for the altitude to the side of an equilateral triangle of side s.

50. Find a formula for the length of a diagonal of a square with side s.

51. Two airplanes depart from the same airport at 3:00. One plane flies north at a speed of 188 mi/h while the other plane flies east at a speed of 324 mi/h. How far apart are the planes at 4:30?

52. A steel rod is used as a stiffener in a box which is in the shape of a cube. If the length of each edge is 1.0 m, find the length of the longest steel rod which can fit in this cube.

15-4 Similar Triangles

In our discussion of geometry in Chapter 6 and in the preceding sections, we studied several important basic properties of geometric figures, such as the concepts of perimeter and area, which deal with the actual size of a figure. Also, we saw that the length of the sides and the size of the angles, respectively, are important in defining the properties of figures. In this section, we consider the properties of triangles that have the same basic shape although not necessarily the same size.

Two triangles are **similar** *if they have the same shape (but not necessarily the same size).* There are two very important properties of similar triangles:

1. The corresponding angles of similar triangles are equal.
2. The corresponding sides of similar triangles are proportional.

Before considering some implications of these two properties, we will first illustrate what is meant by corresponding angles and corresponding sides.

EXAMPLE A   In Figure 15-72 the triangles are similar and are lettered so that corresponding angles have the same letters. That is, angles A and A' are corresponding angles, B and B' are corresponding angles, and C and C' are corresponding angles. Also, the sides between these vertices are corresponding sides. That is, AB corresponds to side $A'B'$, and so forth.   □

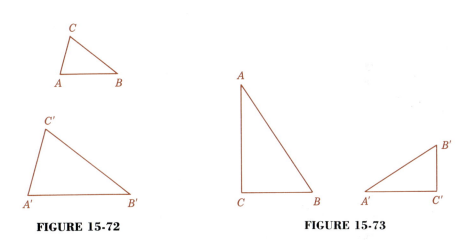

FIGURE 15-72   FIGURE 15-73

EXAMPLE B   In Figure 15-73 we show another pair of triangles which are also similar, even though these triangles are not drawn so that corresponding parts are in the same position relative to the page. The triangles are lettered so that the corresponding parts can be identified. Angles A and A' are corresponding angles, B and B' are corresponding angles, and C and C' are corresponding angles. Sides AB and $A'B'$ are corresponding sides, and so on.   □

If we wish to show that two triangles are similar, we must show that one of the two properties listed above is valid. It is a characteristic of triangles that *if one of these two conditions is valid, then the other condition must also be valid*.

That is, if we can show that the corresponding angles of the two triangles are equal, then we can conclude that the corresponding sides are proportional. If the corresponding sides are proportional, the corresponding angles are equal.

EXAMPLE C The two triangles in Figure 15-72 have been drawn so that corresponding angles are equal. This means that the *corresponding sides are also proportional which is shown as*

$$\frac{AB}{A'B'} = \frac{BC}{B'C'} = \frac{CA}{C'A'} \longleftarrow \text{sides of } \triangle ABC$$
$$\longleftarrow \text{sides of } \triangle A'B'C'$$

If we know that two triangles are similar, we can use the two listed properties to determine unknown parts of one triangle from the known parts of the other. The following example illustrates how this can be done.

EXAMPLE D In Figure 15-74, given triangle ABC where DE is parallel to BC, show that triangle ADE is similar to triangle ABC.

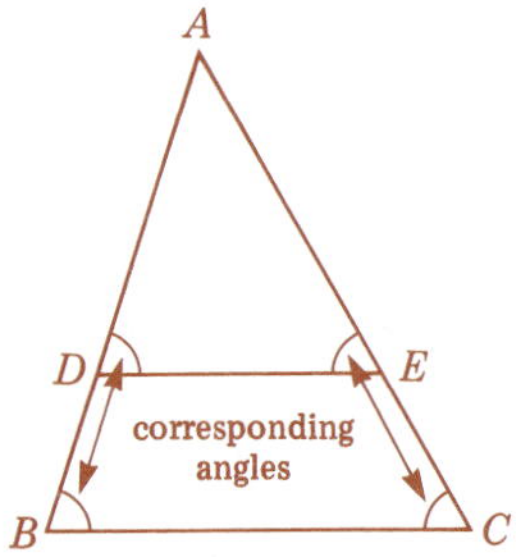

We shall review certain symbols that are commonly used in geometry. The symbol $\triangle$ denotes "triangle." $\sim$ denotes "similar," and as we noted earlier, $\parallel$ denotes "parallel." The statement above can be stated as: Given $\triangle ABC$ where $DE \parallel BC$, show that $\triangle ADE \sim \triangle ABC$.

To show that the triangles are similar, we will show that the corresponding angles are equal. Recall that corresponding angles of parallel lines cut by a transversal are equal. This implies that

$$\angle ADE = \angle ABC \quad \text{and} \quad \angle AED = \angle ACB$$

Since $\angle DAE$ is common to both triangles, we have one angle in each triangle equal to one angle in the other triangle. *Since the corresponding angles are equal,* $\triangle ADE \sim \triangle ABC$.

FIGURE 15-74

corresponding angles

EXAMPLE E In Figure 15-75, in right $\triangle ABC$ with right angle at C, $CD \perp AB$. (As we noted earlier, $\perp$ means "perpendicular.") Show that $\triangle ADC \sim \triangle CDB$.

First, both triangles contain a right angle. That is, $\angle CDA = \angle CDB = 90°$. Since the sum of the angles in a triangle is $180°$, *the sum of the other two angles in* $\triangle ADC$ *is* $90°$, or $\angle CAD + \angle ACD = 90°$. Also, $\angle ACB$ is a right angle, which means that $\angle ACD + \angle DCB = 90°$. These two equations can be written as

$$\angle CAD = 90° - \angle ACD \qquad \angle DCB = 90° - \angle ACD$$

FIGURE 15-75

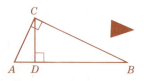

Since the right-hand sides of these equations are equal, we conclude that $\angle CAD = \angle DCB$. We have now shown that *two angles are respectively equal in the two triangles.* Since, in any triangle, the sum of the angles is $180°$, *the remaining angles must also be equal.* Therefore $\angle ACD = \angle CBD$. The triangles are therefore similar. We do note, however, that the corresponding sides appear in different positions with respect to the page. This is indicated by writing the ratio of corresponding sides as

$$\frac{AD}{CD} = \frac{DC}{DB} = \frac{AC}{BC} \longleftarrow \text{sides of } \triangle ADC$$
$$\longleftarrow \text{sides of } \triangle CDB$$

where side CD (or DC) is part of both triangles.

EXAMPLE F In Figure 15-76, $\triangle ABC$ and $\triangle DEF$ were designed to be similar, and the known lengths of certain sides are as shown. Find the lengths of sides CB and AB.

Since the triangles are similar, the corresponding sides are proportional so that

$$\frac{AC}{DF} = \frac{CB}{FE} = \frac{BA}{ED}$$

Substituting the known values, we get

$$\frac{6}{4} = \frac{CB}{3} = \frac{BA}{2}$$

▶ Since *the middle and right ratios are both equal to* $\frac{6}{4}$, we can solve for the unknown in each case. We get

$$\frac{CB}{3} = \frac{6}{4} \quad \text{or} \quad CB = \frac{6(3)}{4} = \frac{9}{2} \text{ cm}$$

$$\frac{BA}{2} = \frac{6}{4} \quad \text{or} \quad BA = \frac{6(2)}{4} = 3 \text{ cm}$$

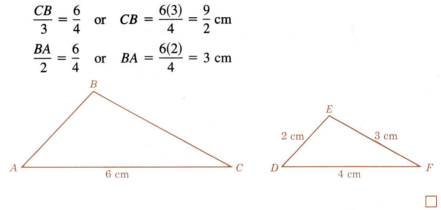

FIGURE 15-76

EXAMPLE G On level ground a silo casts a shadow 24 ft long. At the same time, a pole 4.0 ft high casts a shadow 3.0 ft long. How tall is the silo? (See Figure 15-77.)

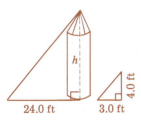

FIGURE 15-77

The rays of the sun are essentially parallel. The two triangles indicated in Figure 15-77 are similar, since *each has a right angle and the angles at the tops are equal.* The lengths of the hypotenuses are of no importance in this problem, so we use only the other sides in stating the ratios of corresponding sides. Denoting the height of the silo by h, we get

$$\frac{h}{4.0} = \frac{24}{3.0} \qquad h = 32 \text{ ft}$$

We conclude that the silo is 32 ft tall.

One of the most practical uses of similar figures is that of **scale drawings**. Maps, charts, blueprints, and most drawings which appear in books are familiar examples of scale drawings. Actually, there have been many scale drawings used in this book in the previous sections.

In any scale drawing, all distances are drawn a certain ratio of the distances they represent and all angles equal the angles they represent. Consider the following example.

EXAMPLE H   In drawing a map of the area shown in Figure 15-78, a scale of 1 cm = 200 km is used. In measuring the distance between Chicago and Toronto on the map, we find it to be 3.5 cm. The actual distance x between Chicago and Toronto is found from the proportion

$$\text{actual distance} \longrightarrow \frac{x}{3.5 \text{ cm}} = \frac{200 \text{ km}}{1 \text{ cm}} \quad \text{or} \quad x = 700 \text{ km}$$

where the top arrow points to "scale" (200 km / 1 cm) and the left labels are "actual distance" and "distance on map".

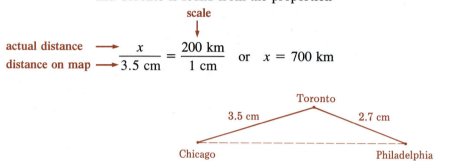

FIGURE 15-78

If we did not have the scale but knew that the distance between Chicago and Toronto is 700 km, then by measuring distances on the map between Chicago and Toronto (3.5 cm) and between Toronto and Philadelphia (2.7 cm), we could find the distance between Toronto and Philadelphia. It is found from the following proportion, determined by use of similar triangles:

$$\frac{700 \text{ km}}{3.5 \text{ cm}} = \frac{y}{2.7 \text{ cm}}$$

$$y = \frac{2.7(700)}{3.5} = 540 \text{ km}$$

EXAMPLE I   A satellite photograph reveals three missile silos an equal distance apart. The images of the silos on the photograph are 2.80 mm apart. Analysis of the photograph and other satellite data implies that the scale is 1.00 mm = 0.429 km. Find the actual distance between the silos.

The triangle formed by the silos in the photograph is similar to the triangle formed by the actual silos. Representing the actual distance between the silos as x, we get

$$\frac{\text{actual distance}}{\text{distance in photo}} = \frac{x}{2.80 \text{ mm}} = \frac{0.429 \text{ km}}{1.00 \text{ mm}} \quad \text{or} \quad x = 1.20 \text{ km}$$

where "scale" labels the fraction 0.429 km / 1.00 mm.

Similarity requires *equal* angles and *proportional* sides. *If the corresponding angles and the corresponding sides of two triangles are equal, then the two triangles are said to be* **congruent**. As a result of this definition, the areas and perimeters of congruent triangles are also equal. Informally, we can say that similar triangles have the same shape whereas congruent triangles have the same shape and same size.

EXAMPLE J A right triangle with legs of 2 in. and 4 in. is congruent to any other right triangles with legs of 2 in. and 4 in. However, it is similar to any right triangle with legs of 5 in. and 10 in., since the corresponding sides are proportional. See Figure 15-79(a).

One equilateral triangle of side 6 cm is congruent to any other equilateral triangle of side 6 cm; and it is similar to any other equilateral triangle, regardless of the length of the side. We know that corresponding angles are equal and that the ratios of corresponding sides must be equal. See Figure 15-79(b).

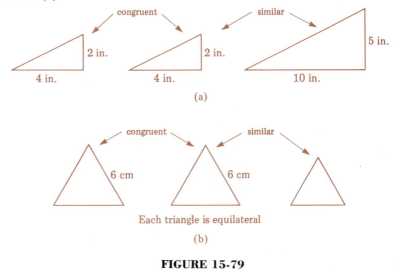

congruent   similar

2 in.   2 in.   5 in.

4 in.   4 in.   10 in.

(a)

congruent   similar

6 cm   6 cm

Each triangle is equilateral

(b)

FIGURE 15-79

Exercises 15-4

In Exercises 1 through 8, assume that $\triangle ABC \sim \triangle A'B'C'$ *with angles A and A' corresponding, angles B and B' corresponding, and angles C and C' corresponding. Find the indicated missing parts.*

1. If $\angle A = 20°$ and $\angle B = 100°$, find angles $\angle A'$, $\angle B'$, and $\angle C'$.
2. If $AB = 5$, $BC = 6$, and $A'B' = 15$, find $B'C'$.
3. If $AB = 10.0$, $BC = 11.0$, $AC = 13.0$, and the shortest side of $\triangle A'B'C'$ is 15.0, find the other two sides of $\triangle A'B'C'$.

4. If $\angle B = 30°$ and $\angle A' = 80°$, find $\angle C$.

5. If $\angle B = 65°$ and $\angle C = 75°$, find angles $\angle A'$, $\angle B'$, $\angle C'$.

6. If $AB = 3.0$, $BC = 4.0$, $AC = 5.0$, and $A'B' = AC$, find $B'C'$ and $A'C'$.

7. If $AC = BC = A'C' = 12.7$ and $\angle C = 40°$, find $B'C'$ and angles $\angle A'$, $\angle B'$.

8. If $A'B' = 20.2$, $B'C' = 15.3$, and $A'C' = AB = 10.9$, find AC and BC.

In Exercises 9 through 12, determine whether or not the triangles in the given figures are (a) similar and (b) congruent. Angles that are equal are marked in the same way, and so are equal sides.

9. Figure 15-80   **10.** Figure 15-81   **11.** Figure 15-82   **12.** Figure 15-83

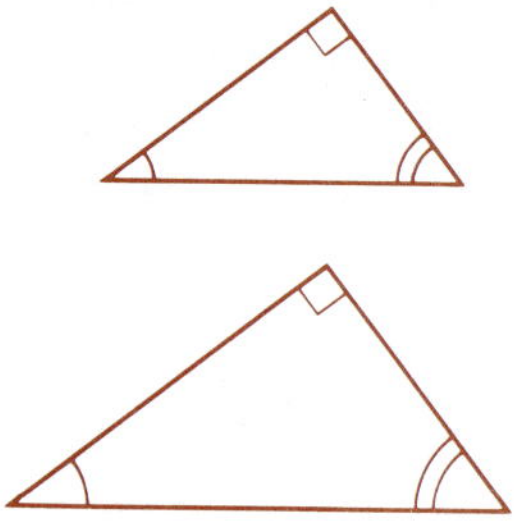

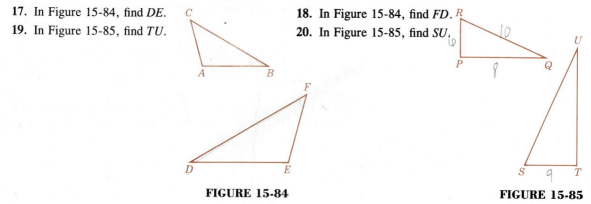

FIGURE 15-80 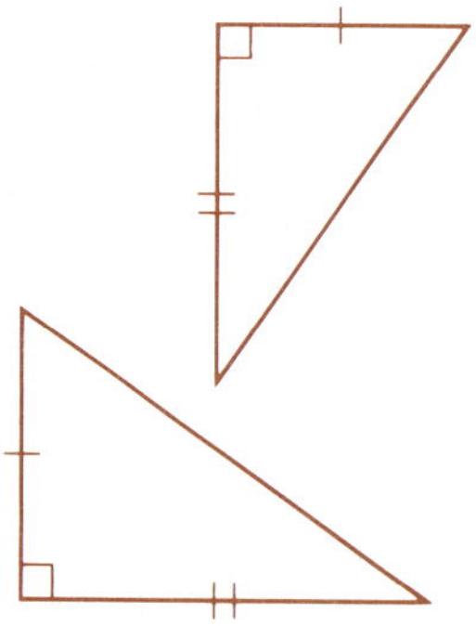   **FIGURE 15-81** 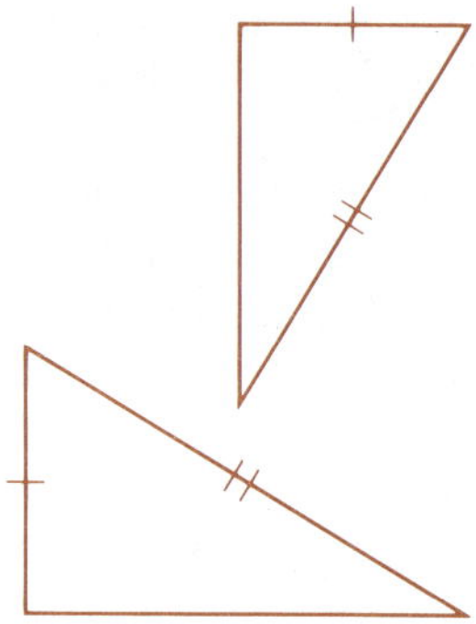   **FIGURE 15-82** 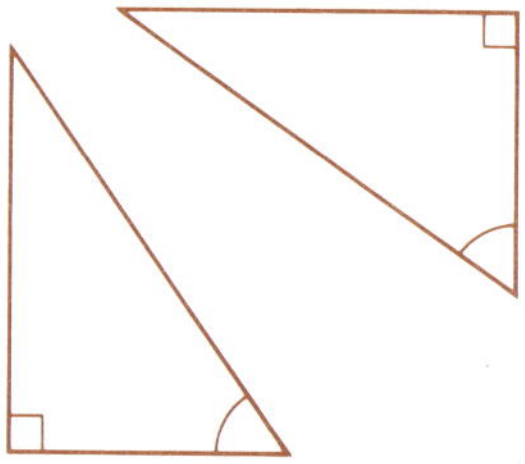   **FIGURE 15-83**

In Exercises 13 through 16, for the given pair of triangles, identify the corresponding angles and corresponding sides to the given angles and sides. The triangles in each of the figures are similar.

13. In Figure 15-84, $\angle A$ corresponds to ——— and side AC corresponds to ———.

14. In Figure 15-84, $\angle F$ corresponds to —C—— and side DF corresponds to —BC—.

15. In Figure 15-85, $\angle Q$ corresponds to ——— and side RP corresponds to ———.

16. In Figure 15-85, $\angle S$ corresponds to ——— and side UT corresponds to ———.

In Exercises 17 through 20, for the given sides of the triangles of Figures 15-84 and 15-85, find the indicated sides. In Figure 15-84: $AB = 5$, $BC = 7$, $AC = 4$, and $FE = 8$. In Figure 15-85: $RP = 6$, $PQ = 8$, $RQ = 10$, and $ST = 9$.

17. In Figure 15-84, find DE.

19. In Figure 15-85, find TU.

18. In Figure 15-84, find FD.

20. In Figure 15-85, find SU.

FIGURE 15-84   **FIGURE 15-85**

In Exercises 21 through 24, use Figure 15-86. In this figure BD ∥ AE, AE = 18, and DB = 6.

21. If $DC = 7$, find CE.

22. If $CA = 27$, find CB.

23. If $CB = 10$, find BA.

24. If $CE = 24$, find DE.

In Exercises 25 and 26, use △ABC in Figure 15-87. In the figure AC ⊥ BC and CD ⊥ AB.

25. Given that $\triangle ABC \sim \triangle ACD$, find AB if $AD = 9$ and $AC = 12$.

26. Given that $\triangle ABC \sim \triangle CBD$, find BD if $BC = 6$ and $AB = 9$.

In Exercises 27 and 28, solve for the unknown side by use of an appropriate proportion. Parallel lines are marked with arrows.

27. In Figure 15-88, $KM = 6$, $MN = 9$, and $MO = 12$. Find LM.

28. In Figure 15-89, $BD = 5$, $BE = 8$, and $BA = 10$. Find BC.

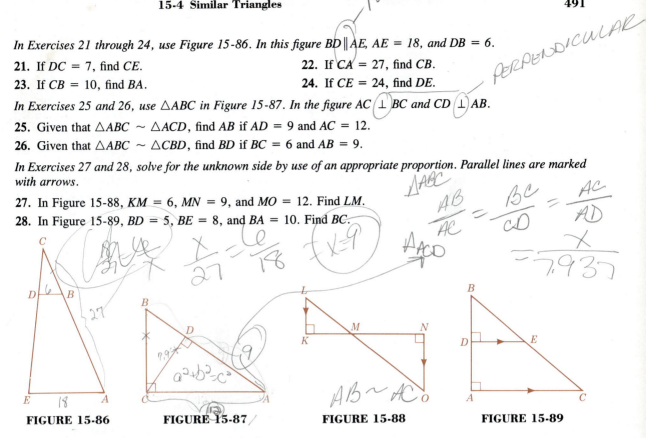

FIGURE 15-86   **FIGURE 15-87**   **FIGURE 15-88**   **FIGURE 15-89**

In Exercises 29 and 30, draw the appropriate figures.

29. Draw $\triangle ABC$ such that $AB \perp BC$ and $\angle ACB = \angle CAB$.

30. Draw $\triangle ABC$ with D on AB and E on AC such that $DE \parallel BC$.

In Exercises 31 and 32, find the required values.

31. Two triangles are similar. The sides of the larger triangle are 3.0 cm, 5.0 cm, and 6.0 cm, and the shortest side of the other triangle is 2.0 cm. Find the remaining sides of the smaller triangle.

32. Two triangles are similar. The angles of the smaller triangle are 50°, 100°, and 30°, and the sides of the smaller triangle are 7.00 in., 9.00 in., and 4.57 in. The longest side of the larger triangle is 15.0 in. Find the other two sides and the three angles of the larger triangle.

In Exercises 33 and 34, show that the required triangles are similar.

33. In Figure 15-90, show that $\triangle XYK \sim \triangle NFK$ ($XY \parallel FN$)

34. In Figure 15-87, show that $\triangle ACB \sim \triangle ADC$.

FIGURE 15-90

In Exercises 35 and 36, find the required values.

35. In Figure 15-91, $\triangle ABC \cong \triangle EDC$ ($\cong$ means "congruent"). If $AD = 16$ in., how long is AC?

FIGURE 15-91

36. In Figure 15-92, $\triangle ABC \cong \triangle ADC$. If $\angle CAD = 40°$, how many degrees are there in $\angle CAB + \angle ABC$?

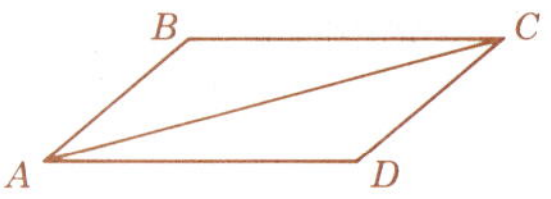

In Exercises 37 through 56, solve the given problems.

FIGURE 15-92

37. One stake used as a snow marker casts a shadow 5.0 ft long. A second stake casts a shadow 3.0 ft long. If the second stake is 4.0 ft tall, how tall is the first stake? (Both stakes are on level ground.) See Figure 15-93.

38. On level ground an oak tree casts a shadow 36 m long. At the same time, a pole 9 m high casts a shadow 12 m long. How high is the tree?

39. A 1-m stick is placed vertically in the shadow of a vertical pole such that the ends of their shadows are at the same point. If the shadow of the meter stick is 80 cm long and that of the pole is 280 cm long, how high is the pole? See Figure 15-94.

40. In constructing a metal support in the form of $\triangle ABC$ as shown in Figure 15-95, it is deemed necessary to strengthen the support with an added brace DE which is parallel to BC. How long must the brace be if $AB = 20$ in., $AD = 14$ in., and $BC = 25$ in.?

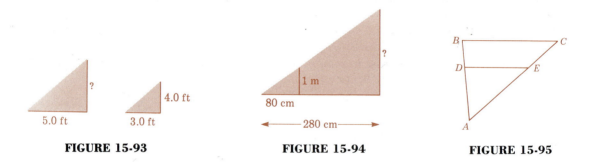

FIGURE 15-93   **FIGURE 15-94**   **FIGURE 15-95**

41. A certain house blueprint has a scale of $1\frac{1}{4}$ in. = 10 ft. The living room is 18 ft long. What distance on the blueprint represents this length?

42. On a map, 12 cm = 100 mi. What is the distance between two cities if the distance between them on the map is 7.5 cm?

43. From Figure 15-78, find the distance from Chicago to Philadelphia.

44. On a map the scale is 1 cm = 50 mi. If the actual distance between Albany and New York is 125 mi, find the distance between these cities on the map.

45. A satellite photograph includes the images of three different military bases. In the photograph, the distances between the bases are 10.4 cm, 12.3 cm, and 5.2 cm. If the photograph has a scale of 1.000 cm = 146.7 km, find the actual distances between the military bases.

46. In a book on aerial photography, it is stated that the typical photograph scale is 1/18450. In an 8.0 in. by 10.0 in. photograph with this scale, what is the longest distance between two locations included in the photograph? Express the answer in miles and round to the nearest tenth.

47. A radio transmitter antenna casts a shadow 37.3 m long at the same time that a meter stick casts a shadow that is 1.80 m long. How high is the antenna?

48. A model of the Lockheed YF-12A Interceptor is $\frac{1}{48}$ scale. A side view of an inlet spike on this model is an isosceles triangle with equal sides of 2.05 in. and base 0.875 in. Find the corresponding sides on the full-scale jet.

49. A 4.0 ft wall stands 2.0 ft from a building. The ends of a straight pole touch the building and the ground 6.0 ft from the wall. A point on the pole touches the wall's top. How high on the building does the pole touch? See Figure 15-96.

50. To find the width ED of a river, a surveyor places markers at A, B, C, and D (see Figure 15-97). He places them so that $AB \parallel ED$, $BC = 50.0$ m, $DC = 300$ m, and $AB = 80.0$ m. How wide is the river?

FIGURE 15-96   **FIGURE 15-97**

51. A 30.0-m pole on level ground is supported by two 60.0-m guy wires attached at its top. The guy wires are on opposite sides of the pole. How far is it between the grounded ends of the wires?

52. Town B is due east of town A, and town E is due east of town D. The direct routes from A to E and B to D cross at town C. The route DC is 15 mi, DB is 45 mi, and DE is 24 mi. How far is it from town A to town B? See Figure 15-98.

53. Figure 15-99 shows a roof truss. Assume that all parts which appear equal *are* equal and that $AC = 6.72$ ft and $BE = 6.70$ ft. Find the base of the truss AD.

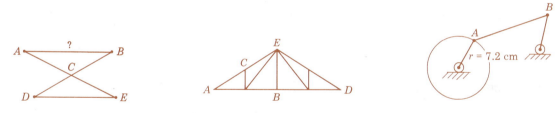

FIGURE 15-98   **FIGURE 15-99**   **FIGURE 15-100**

54. On the blueprint of a bridge, a support plate is rectangular with length 1.32 cm and width 0.20 cm. The scale used is 1.00 cm = 5.00 ft. What is the actual area of the plate?

55. Figure 15-100 shows a crank-lever mechanism. From the figure determine the scale and then determine the length AB.

56. A photograph 6.00 in. wide and 9.00 in. long is enlarged so that the length of the enlargement is 15.0 in. Find the ratio of the area of the enlargement to that of the original photograph.

15-5 Solid Geometric Figures: Prisms and Cylinders

Except for a brief mention of rectangular solids and spheres in Chapter 6, we have restricted our attention to plane geometry. Plane geometry deals with figures which lie in a plane or flat surface. In addition to plane figures, there are many important geometric solid figures. In this and the following sections we discuss the determination of surface area and volume of the most important of these solid figures.

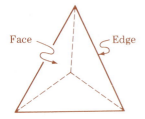

FIGURE 15-101

FIGURE 15-102

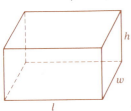

FIGURE 15-103

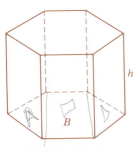

FIGURE 15-104

A **polyhedron** *is a geometric solid figure which is bounded by planes. The plane surfaces of the polyhedron are called* **faces,** *and the intersections of the faces are called* **edges.** In Figure 15-101 a polyhedron of four faces and six edges is illustrated. The cube is another example of a polyhedron.

A **prism** *is a polyhedron satisfying these conditions:*

1. The bases (top and bottom) are made of equal and parallel polygons.

2. The sides are parallelograms.

A prism with triangular bases is shown in Figure 15-102. The cube is an example of a prism with square bases. We shall consider only prisms in which the side edges are perpendicular to the bases. (Such prisms are called **right prisms.**)

A prism of particular importance is a **rectangular solid,** which we introduced in Chapter 6 and illustrate in Figure 15-103. In Chapter 6 we defined the **volume** of a rectangular solid of length *l*, width *w*, and height *h* as

$$V = lwh \qquad \text{volume of rectangular solid} \qquad (15\text{-}2)$$

We now note that the bottom face of the rectangular solid, which is of length *l* and width *w*, is the base of the solid. If *B* is the area of the base, which means that $B = lw$ for the rectangular solid, we have

$$V = Bh \qquad \text{volume of prism} \qquad (15\text{-}3)$$

as the formula for the volume. Equation (15-3) can be used to find the volume of a rectangular solid, as well as the volume of any prism with the base area *B*.

EXAMPLE A

The prism shown in Figure 15-104 has a regular hexagon as a base. If we are able to determine the area of the hexagon, we can use Eq. (15-3) to find the volume. If, for example, $B = 12$ in.^{2} and $h = 5.0$ in., the volume is

$$V = (12)(5.0) = 60 \text{ in.}^3 \qquad \text{using Eq. (15-3)} \qquad \square$$

EXAMPLE B

The base of the prism shown in Figure 15-105 is a right triangle with legs of 14.6 cm and 10.9 cm. The height of the prism is 12.3 cm. To find the volume we first find the area of the base to be one-half the product of the

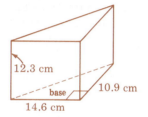

FIGURE 15-105

lengths of the legs of the triangle. Thus

$$B = \frac{1}{2}(14.6)(10.9) \qquad \text{using Eq. (6-12)}$$

$$= 79.6 \text{ cm}^2$$

Having found the area of the base B, we can now find the volume as follows.

$$V = Bh = (79.6)(12.3) \qquad \text{using Eq. (15-3)}$$

$$= 979 \text{ cm}^3 \qquad\qquad\qquad \square$$

The **lateral area** L *of a prism is the area of the side faces (not including the bases).* The **total area** A *of a prism is the lateral area plus the area of the bases.* Consider the following example.

EXAMPLE C ▶ For the prism in Figure 15-105, the lateral area is the sum of the areas of the front face, right face, and back face. To find each of these areas, we multiply length by width. For the back face we *use the Pythagorean theorem to find the length.*

In the following expression, the three terms represent the areas of the three sides.

$$L = \underset{\text{front face}}{(14.6)(12.3)} + \underset{\text{right face}}{(10.9)(12.3)} + \underset{\text{back face}}{(\sqrt{14.6^2 + 10.9^2})(12.3)}$$

$$= 180 + 134 + \underset{\text{length (dashed line)}}{(18.2)(12.3)} = 314 + 224$$

$$= 538 \text{ cm}^2$$

The lateral area is 538 cm² (rounded off to three significant digits). Noting that the length of each face is multiplied by the height of the prism, we see that: *The lateral area may be found by multiplying the perimeter of the base by the height.*

The total area is found by adding the area of the bases to the lateral area. We now determine the total surface area as follows.

$$A = 538 + \overset{\text{two bases (top and bottom)}}{2\left(\frac{1}{2}\right)(14.6)(10.9)}$$

$$= 538 + 159$$

$$= 697 \text{ cm}^2$$

The total area is 697 cm². $\qquad\qquad\qquad \square$

In the case of a rectangular solid (see Figure 15-103), there are three pairs of equal rectangular faces. A formula for the total area of a rectangular solid is

$$A = 2lw + 2wh + 2lh$$ area of rectangular solid   (15-4)

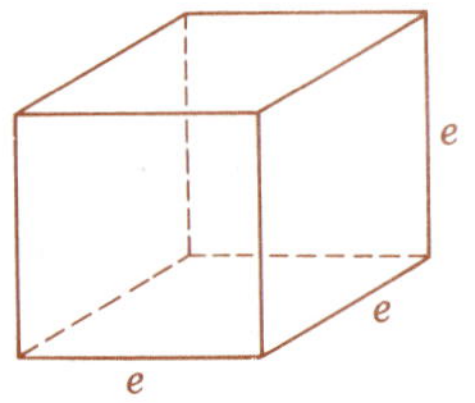

For a cube, each of the six faces is a square so that the total area of a cube is

$$A = 6e^2$$ area of cube   (15-5)

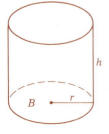

FIGURE 15-106

Here e is the length of the edge of the cube. See Figure 15-106.

EXAMPLE D   For a rectangular solid for which $l = 260$ mm, $w = 130$ mm, and $h = 150$ mm, as shown in Figure 15-107, the total area is

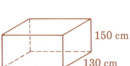

FIGURE 15-107

    bottom and top   right and left   front and back
$$A = 2(260)(130) + 2(130)(150) + 2(260)(150)$$ using Eq. (15-4)
$$= 67,600 + 39,000 + 78,000$$
$$= 184,600 \text{ mm}^2$$

The total area of a cube 6.82 in. on an edge is

                 6 equal faces
$$A = 6(6.82)^2 = 6(46.5)$$ using Eq. (15-5)
$$= 279 \text{ in.}^2$$ □

A **right circular cylinder** *is generated by revolving a rectangle about one of its sides. In a right circular cylinder each base is a circle, and the cylindrical surface is perpendicular to the bases* (see Figure 15-108).

As with the prism, *the volume of a right circular cylinder is the product of the area of the base and the height, or altitude, of the cylinder.* Since the base is a circle, the area of the base is πr^2, and we have the formula

FIGURE 15-108

$$V = \pi r^2 h$$ volume of cylinder   (15-6)

for the volume, where r is the radius of the base and h is the height. Recall that the height is also referred to as the altitude.

EXAMPLE E   Find the volume of a container in the shape of a right circular cylinder for which $d = 16.5$ in. and $h = 6.75$ in. See Figure 15-109.

In this case we are given the diameter of the base, which means that the radius r is 8.25 in. The volume is found as follows.

FIGURE 15-109

$$V = \pi(8.25)^2(6.75) \qquad \text{using Eq. (15-6)}$$
$$= (3.14)(68.1)(6.75)$$
$$= 1440 \text{ in.}^3$$

The lateral area of a prism is the product of the perimeter of a base and the height. In a similar manner, *the lateral area of a cylinder is the circumference of a base times the height*. We can think of this as the label on a can being removed and flattened out as demonstrated in Figure 15-110. Since the circumference of a base is $2\pi r$, the lateral area of a cylinder is

$$L = 2\pi rh \qquad \text{lateral area of cylinder} \qquad (15\text{-}7)$$

The total area of a right circular cylinder is the sum of the areas of the bases and the lateral area. Since each base is a circle, we have

$$A = 2\pi r^2 + 2\pi rh = 2\pi r(r + h) \qquad \text{area of cylinder} \qquad (15\text{-}8)$$

as the formula for the total area.

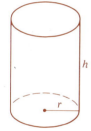

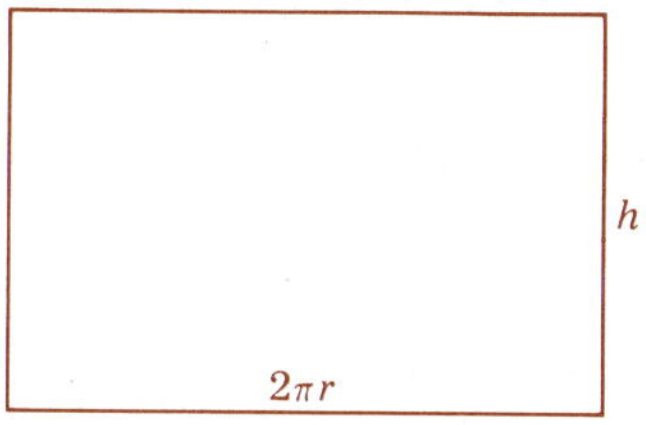

FIGURE 15-110

EXAMPLE F An oil can in the shape of a right circular cylinder has a radius of 12.5 cm and is 30.0 cm high. What is (a) the lateral area and (b) the total area of the can? See Figure 15-111.

Using Eq. (15-7), with $r = 12.5$ and $h = 30.0$, we get

$$L = 2\pi(12.5)(30.0) = 2(3.14)(12.5)(30.0)$$
$$= 2355 \text{ cm}^2$$

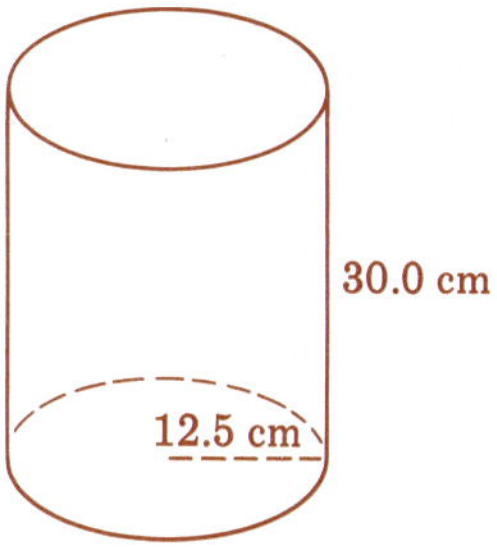

Using Eq. (15-8), we have

$$\overset{\text{top and bottom}}{A = 2\pi(12.5)^2} + \overset{\text{lateral area}}{2\pi(12.5)(30.0)}$$
$$= 2(3.14)(156.3) + 2(3.14)(12.5)(30.0)$$
$$= 981 + 2355$$
$$= 3336 \text{ cm}^2$$

To three significant digits, $L = 2360 \text{ cm}^2$ and $A = 3340 \text{ cm}^2$.

FIGURE 15-111

EXAMPLE G   A storage tank for a kerosene heater is in the shape of a right circular cylinder with diameter 1.50 ft and height 4.16 ft. How many gallons can be stored in this tank? (1.00 ft^3 = 7.48 gal)

Since the diameter is 1.50 ft, we know that $r = \frac{1}{2}(1.50)$, or 0.750 ft. Using Eq. (15-6), we get

$$V = \pi r^2 h$$
$$= (3.14)(0.750)^2(4.16)$$
$$= 7.35 \text{ ft}^3$$

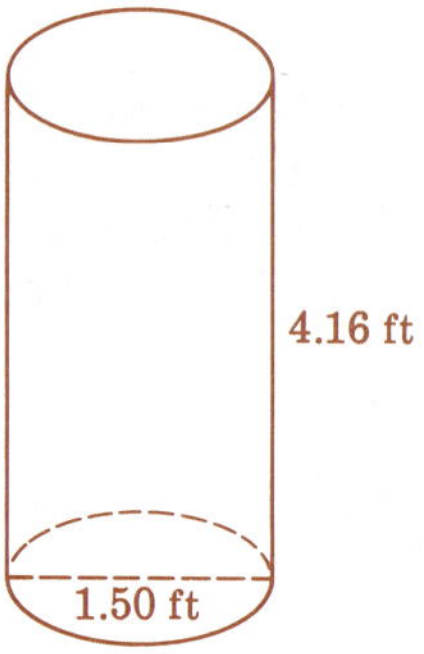

4.16 ft

Since each cubic foot can hold 7.48 gal, we get the total number of gallons, the capacity C of the tank, by

$$C = 7.48V = 7.48(7.35)$$
$$= 55.0 \text{ gal}$$

See Figure 15-112.

1.50 ft

FIGURE 15-112

Exercises 15-5

In Exercises 1 through 8, determine the volumes of the prisms with the given dimensions.

 1. Triangular base: $B = 460$ ft^2, $h = 15.0$ ft
 2. Triangular base: $B = 37.8$ m^2, $h = 5.27$ m
 3. Quadrilateral base: $B = 2750$ cm^2, $h = 36.0$ cm.
 4. Polygon base: $B = 8630$ in.^{2}, $h = 40.2$ in.
 5. Base is a right triangle with legs 2.37 ft and 4.22 ft; $h = 1.77$ ft.
 6. Base is a right triangle with legs 36.3 mm and 58.2 mm; $h = 12.8$ mm.
 7. Base is a parallelogram for which base is 53.7 cm and height is 17.7 cm; $h = 23.5$ cm.
 8. Base is a parallelogram for which base is 1.37 yd and height is 1.13 yd; $h = 1.27$ yd.

In Exercises 9 through 16, determine (a) the lateral areas and (b) the total areas of the prisms with the given dimensions.

 9. Rectangular solid: $l = 16.3$ ft, $w = 11.5$ ft, $h = 14.3$ ft
10. Rectangular solid: $l = 72.0$ mm, $w = 32.3$ mm, $h = 10.3$ mm
11. Cube: $e = 0.820$ m
12. Cube: $e = 2.73$ ft
13. Quadrilateral base for which $B = 3270$ in.^{2}, $p = 238$ in.; $h = 51.8$ in.
14. Polygon base for which $B = 7360$ cm^2, $p = 421$ cm; $h = 42.9$ cm.
15. Base is a right triangle with legs 4.25 cm and 6.50 cm; $h = 3.10$ cm.
16. Base is a right triangle with legs 17.0 in. and 10.5 in.; $h = 3.25$ in.

In Exercises 17 through 24, determine the volumes of the cylinders with the given values of r, d (diameter of base), and h.

17. $r = 20.0$ cm; $h = 15.0$ cm

18. $r = 7.00$ in.; $h = 4.00$ in.

19. $r = 15.0$ ft; $h = 3.60$ ft

20. $r = 1.58$ m; $h = 8.48$ m

21. $d = 366$ mm; $h = 140$ mm

22. $d = 0.634$ yd; $h = 0.156$ yd

23. $d = 24.2$ in.; $h = 32.3$ in.

24. $d = 22.0$ cm; $h = 12.2$ cm

In Exercises 25 through 32, determine (a) the lateral areas and (b) the total areas of the cylinders with the given values of r, d, and h.

25. $r = 300$ mm; $h = 120$ mm

26. $r = 60.0$ ft; $h = 80.0$ ft

27. $r = 8.20$ in.; $h = 2.40$ in.

28. $r = 2.30$ m; $h = 1.10$ m

29. $d = 24.0$ cm; $h = 8.50$ cm

30. $d = 3.80$ in.; $h = 7.50$ in.

31. $d = 86.4$ ft; $h = 12.4$ ft

32. $d = 84.2$ mm; $h = 123$ mm

In Exercises 33 through 44, solve the given problems. All involve prisms or cylinders.

33. How many cubic meters of concrete are needed for a driveway 20.0 m long, 2.75 m wide, and 0.100 m thick?

34. A swimming pool is 50.0 ft wide, 80.0 ft long, 3.00 ft deep at one end, and 8.00 ft deep at the other end. How many cubic feet of water will it hold? (Assume that the slope on the bottom is constant.) See Figure 15-113.

35. A rectangular box is to be used to store radioactive materials. The inside of the box is 12.0 in. long, 9.00 in. wide, and 9.00 in. deep. What is the area of sheet lead which must be used to line the inside of the box?

36. A glass prism used in the study of optics has a right triangular base. The legs of the right triangle are 3.00 cm and 4.00 cm. The prism is 8.50 cm high. What is the total surface area of the prism? See Figure 15-114.

FIGURE 15-113   **FIGURE 15-114**

37. A glass prism is used for conducting experiments with lasers. Each base is an equilateral triangle with area 3.90 cm² and the prism is 14.5 cm tall. Find its volume.

38. What is the weight of 1.00 km of copper wire which is 0.500 cm in diameter? The density of copper is 8.90 g/cm³.

39. A cylindrical grain storage container 82.0 ft high has a radius of 24.3 ft. One bushel of grain occupies about 1.24 ft³. How many bushels can be stored in the container?

40. At 75¢ per square yard, how much would it cost to paint the outside of the storage container of Exercise 39? (The bottom is not to be painted.)

41. A lawn roller is a cylinder 96.0 cm long and 30.0 cm in radius. What area is rolled in one complete revolution of the roller?

42. The cylinder of the lawn roller in Exercise 41 is filled with water for weight. How many liters of water does the cylinder hold? ($1000 \text{ cm}^3 = 1$ L)

43. A wing of a building is supported by four concrete columns in the shape of prisms with regular hexagons as bases. If these hexagons each have an area of 12.5 ft^2 and the columns are each 12 ft tall, find the number of cubic feet of concrete needed for all four columns.

44. A section of a pipeline is 2.00 mi long and the pipe has a diameter of 3.00 ft. How many gallons are in this section of pipe when it is filled? ($1.00 \text{ ft}^3 = 7.48$ gal)

15-6 Solid Geometric Figures: Pyramids, Cones, and Spheres

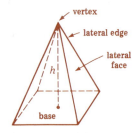

FIGURE 15-115

The geometric solid figures discussed in the previous section have parallel and equal bases. In this section we discuss other types of solids. Applications of these figures are found in architecture, engineering, and astronomy, as well as a variety of other technical fields.

The base of a **pyramid** *is a polygon. The other faces, the* **lateral faces,** *are triangles which meet at a common point, the* **vertex.** *The edges where the lateral faces meet are the* **lateral edges.** *The* **height,** *or* **altitude,** *of the pyramid is the perpendicular distance from the vertex to the base* (see Figure 15-115).

If we compare the volumes of a prism and a pyramid of the same base and height, we can see that the volume of the pyramid is considerably less that that of the prism (see Figure 15-116). Although we shall not prove it here, the volume of the pyramid is exactly one-third of the volume of the prism. This is expressed in Eq. (15-9), where

$$V = \frac{1}{3}Bh \qquad \text{volume of pyramid} \qquad (15\text{-}9)$$

is the formula for the volume of a pyramid with base area B and height h.

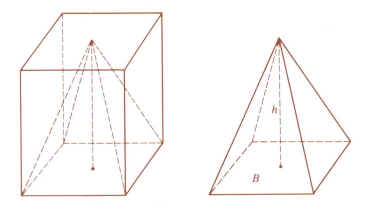

FIGURE 15-116

EXAMPLE A

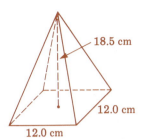

FIGURE 15-117

The base of a pyramid is a square 12.0 cm on a side as shown in Figure 15-117. The height of the pyramid is 18.5 cm. To find the volume we must note that the base area B is $(12.0 \text{ cm})^2$ so that

square base

$$V = \frac{1}{3}(12.0)^2(18.5) \qquad \text{using Eq. (15-9)}$$

$$= 888 \text{ cm}^3 \qquad \qquad \square$$

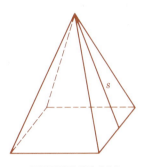

FIGURE 15-118

A **regular pyramid** *has identical congruent triangles for its lateral or side faces.* In Figure 15-118 we show one example of a regular pyramid. That particular regular pyramid has a square base, but other regular pyramids might have bases that are equilateral triangles, regular pentagons, regular hexagons, and so on. *For each of the triangles that make up the lateral faces, we denote the altitude or height by s and refer to it as the* **slant height** *of the pyramid.* The area of each triangular lateral face is one-half the product of the slant height s and the base of the triangle. Instead of expressing the lateral area as the sum of the areas of individual triangles with the same base and slant height, we use the fact that the sum of the bases of the triangular faces is equal to the perimeter p of the base of the pyramid. *The* **lateral area** *is the sum of the areas of the lateral faces of the pyramid and it is found by*

$$\boxed{L = \frac{1}{2}ps} \qquad \text{lateral area of pyramid} \qquad \textbf{(15-10)}$$

Here p is the perimeter of the base and s is the slant height.

EXAMPLE B

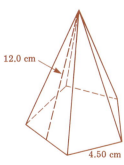

FIGURE 15-119

The base of the pyramid shown in Figure 15-119 is a regular pentagon in which each of the five sides is 4.50 cm long. The slant height of the pyramid is 12.0 cm. To find the lateral area, we note that the perimeter of the base is $5(4.50 \text{ cm}) = 22.5 \text{ cm}$; thus

$$L = \frac{1}{2}(22.5)(12.0) = 135 \text{ cm}^2 \qquad \text{using Eq. (15-10)} \qquad \square$$

A **conical surface** *is generated by rotating a straight line about one of its points. A* **cone** *is a solid geometric figure bounded by the conical surface and a plane which cuts the surface.* Although many types of cones may be generated, one type is of particular importance. It is *the* **right circular cone,** *which is generated by rotating a right triangle about one of its legs.*

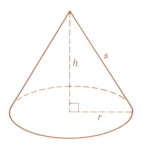

FIGURE 15-120

Thus the **base** of the cone is a circle and the **slant height** is the hypotenuse of the right triangle (see Figure 15-120). The **height,** or **altitude,** of the cone is one leg of the right triangle. The radius of the base is the length of the other leg. We shall restrict our attention to right circular cones, and whenever we use "cone" we mean right circular cone.

As with the pyramid and prism, if we compare the volumes of a cone and cylinder with the same base and height, we see that the volume of the cone is considerably less than that of the cylinder. In fact, the volume of the cone is exactly one-third of the volume of the cylinder, just as with the pyramid and prism. The volume of a right circular cone is

$$V = \frac{1}{3}\pi r^2 h \qquad \text{volume of cone} \qquad \textbf{(15-11)}$$

▶ Here r is the radius of the base and h is the height of the cone. *Note the difference between the height* h *and slant height* s.

EXAMPLE C   The volume of a cone shown in Figure 15-121 for which $r = 5.50$ in. and $h = 3.75$ in. is

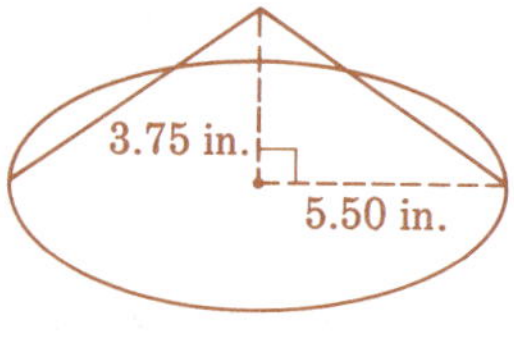

$$V = \frac{1}{3}\pi(5.50)^2(3.75) \qquad \text{using Eq. (15-11)}$$

$$= \frac{1}{3}(3.14)(30.25)(3.75)$$

$$= 119 \text{ in.}^3 \qquad\qquad \square$$

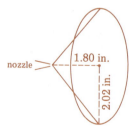

3.75 in.

5.50 in.

FIGURE 15-121

EXAMPLE D   A high-pressure atomizing oil burner propels a spray of oil in the shape of a cone with radius 2.02 in. and height 1.80 in. What is the volume of this cone? See Figure 15-122.

Using Eq. (15-11), we get

$$V = \frac{1}{3}\pi(2.02)^2(1.80)$$

$$= \frac{1}{3}(3.14)(4.080)(1.80)$$

$$= 7.69 \text{ in.}^3 \qquad\qquad \square$$

nozzle

1.80 in.

2.02 in.

FIGURE 15-122

It is reasonable that the lateral area of a cone follows the same basic formula as that of a pyramid. Recalling from Figure 15-120 that the slant height of a cone is the hypotenuse of the right triangle with legs r and h, and that the perimeter of a circle is its circumference, we can use Eq. (15-10) to get

$$L = \frac{1}{2}ps = \frac{1}{2}(2\pi r)s$$

Simplifying, we can express the lateral area of a cone as

$$\boxed{L = \pi r s} \qquad \text{lateral area of cone} \qquad (15\text{-}12)$$

Here r is the radius of the base and s is the slant height.

EXAMPLE E

A tent is in the shape of a cone with a radius of 3.20 m and height of 3.60 m. What is the surface area of the tent?

We must first find the slant height s. *This is done by use of the Pythagorean theorem.* As noted above, the radius and height are the legs of a right triangle with the slant height as the hypotenuse (see Figure 15-123) so that

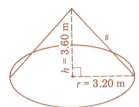

FIGURE 15-123

$$s^2 = r^2 + h^2 \qquad \text{Pythagorean theorem}$$
$$= (3.20)^2 + (3.60)^2$$
$$= 10.24 + 12.96 = 23.2$$

or

$$s = 4.82 \text{ m}$$

The lateral area can now be found with $r = 3.20$ m and $s = 4.82$ m as follows:

$$L = \pi(3.20)(4.82) = (3.14)(3.20)(4.82) \qquad \text{using Eq. (15-12)}$$
$$= 48.4 \text{ m}^2 \qquad \qquad\qquad\qquad\qquad\qquad\qquad \square$$

EXAMPLE F

A cup is in the shape of part of a cone, as shown in Figure 15-124 on the next page. Using the measurements shown, find the area of the paper needed for this cup.

Since the bottom is a circle with radius 2.20 cm, the area of the bottom is

$$A = \pi r^2 \qquad \text{using Eq. (6-16)}$$
$$= (3.14)(2.20)^2$$
$$= 15.2 \text{ cm}^2$$

To find the area of paper used for the side, we will first find the lateral area of the completed total cone including the smaller cone shown at the bottom which is not part of the cup. With $r = 3.00$ cm and $s = 28.6$ cm, we get

$$L = (3.14)(3.00)(28.6) \qquad \text{using Eq. (15-12)}$$
$$= 269 \text{ cm}^2 \qquad\qquad\qquad\qquad \textit{(Continued on next page)}$$

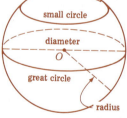

FIGURE 15-124

Now we find the lateral area of the smaller cone at the bottom which is not part of the cup itself. With $r = 2.20$ cm and $s = 21.0$ cm, we get

$$L = (3.14)(2.20)(21.0) \qquad \text{using Eq. (15-12)}$$
$$= 145 \text{ cm}^2$$

The area of the paper needed for the side of the cup is the lateral area of the completed total cone minus the lateral area of the smaller bottom cone. The area of the paper needed for the side is therefore 269 cm² − 145 cm² = 124 cm². Adding that area to the area of the bottom we get a total area of 139 cm². ☐

The final solid figure we shall consider is the **sphere,** which was introduced briefly in Section 6-4. *A sphere is generated by rotating a circle about a diameter. The* **radius** *of the sphere is a line segment joining the center and a point on the sphere. The* **diameter** *of the sphere is a line segment through the center and having its endpoints on the sphere.* The intersection of a plane and a sphere is a *circle.* If the plane of intersection contains the center of the sphere, the circle of intersection is a **great circle.** Other circles in planes not containing the center are **small circles.** These are illustrated in Figure 15-125.

Recall from Section 6-4 that the volume of a sphere can be found from Eq. (6-19).

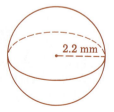

FIGURE 15-125

$$\boxed{V = \frac{4}{3}\pi r^3} \qquad \text{volume of sphere} \qquad (6\text{-}19)$$

Here r is the radius of the sphere.

The total surface area of a sphere is given by

$$\boxed{A = 4\pi r^2} \qquad \text{area of sphere} \qquad (15\text{-}13)$$

EXAMPLE G A spherical ball bearing has a radius of 2.2 mm as shown in Figure 15-126. Its volume is

$$V = \frac{4}{3}\pi r^3 = \frac{4}{3}(3.14)(2.2)^3 \qquad \text{using Eq. (6-19)}$$
$$= 45 \text{ mm}^3$$

Its surface area is

$$A = 4\pi r^2 = 4(3.14)(2.2)^2 \qquad \text{using Eq. (15-13)}$$
$$= 61 \text{ mm}^2$$

FIGURE 15-126

EXAMPLE H Oil with a certain viscosity is propelled from a nozzle in the shape of spherical droplets with diameter $\frac{1}{250}$ in. What is the total volume of 10^{12} of these droplets?

Each individual droplet of oil has volume

$$V = \frac{4}{3}\pi r^3$$

$$= \frac{4}{3}(3.14)\left(\frac{1}{500}\right)^3$$

$$= 3.35 \times 10^{-8} \text{ in.}^3$$

Note that the radius of $\frac{1}{500}$ in. was found by taking one-half of the diameter of $\frac{1}{250}$ in. Since each droplet of oil has a volume of 3.35×10^{-8} in.^{3}, the total volume of 10^{12} droplets is

$$V_t = 10^{12} \times 3.35 \times 10^{-8}$$

$$= 3.35 \times 10^4 \text{ in.}^3$$

$$= 33,500 \text{ in.}^3 \qquad \square$$

EXAMPLE I A grain storage container is in the shape of a cylinder surmounted by a hemisphere (half a sphere). See Figure 15-127. Given that the radius of the cylinder is 40.0 ft and its height is 120 ft, find the volume of the container.

The volume of the container is the volume of the cylinder plus the volume of the hemisphere. By the construction we see that the radius of the hemisphere is the same as the radius of the cylinder. Thus

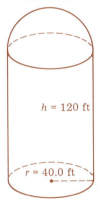

$h = 120$ ft

$r = 40.0$ ft

$$\overset{\text{cylinder}}{} \quad \overset{\text{hemisphere}}{}$$

$$V = \pi r^2 h + \frac{1}{2}\left(\frac{4}{3}\pi r^3\right) = \pi r^2 h + \frac{2}{3}\pi r^3$$

$$= (3.14)(40.0)^2(120) + \frac{2}{3}(3.14)(40.0)^3$$

$$= 603,000 + 134,000 = 737,000 \text{ ft}^3 \qquad \square$$

FIGURE 15-127

Exercises 15-6

In Exercises 1 through 4, find the volumes of the pyramids with the given bases and heights.

1. Base is a polygon with $B = 3600$ ft^2; $h = 45.0$ ft.
2. Base is a polygon with $B = 7850$ cm^2; $h = 38.4$ cm.
3. Base is a square of side 25.0 mm; $h = 4.60$ mm.
4. Base is a square of side 8.50 ft; $h = 4.85$ ft.

In Exercises 5 through 8, find the lateral areas of the pyramids with the given bases, slant heights, or heights.

   5. Base is a polygon with $p = 18.0$ yd; $s = 8.50$ yd.

   6. Base is a polygon with $p = 880$ cm; $s = 350$ cm.

   7. Base is a square of side 12.0 m; $h = 8.00$ m.

   8. Base is a square of side 30.0 in.; $h = 36.0$ in.

In Exercises 9 through 12, find the volumes of the cones with the given base radii (or diameters) and heights.

   9. $r = 10.0$ ft; $h = 14.0$ ft   **10.** $r = 25.0$ cm; $h = 40.0$ cm

   11. $d = 62.8$ cm; $h = 26.3$ cm   **12.** $d = 17.8$ ft; $h = 22.3$ ft

In Exercises 13 through 16, find the lateral areas of the cones with the given base radii (or diameters) and slant heights (or heights).

   13. $r = 7.00$ in.; $s = 9.00$ in.   **14.** $d = 2.80$ m; $s = 1.80$ m

   15. $r = 45.0$ cm; $h = 24.0$ cm   **16.** $d = 17.8$ ft; $h = 10.5$ ft

In Exercises 17 through 20, find the volumes of the spheres with the given radii or diameters.

   17. $r = 3.00$ ft   **18.** $r = 20.0$ cm   **19.** $d = 220$ mm   **20.** $d = 36.2$ ft

In Exercises 21 through 24, find the surface areas of the spheres with the given radii or diameters.

   21. $r = 30.0$ in.   **22.** $r = 60.0$ m   **23.** $d = 346$ mm   **24.** $d = 62.4$ in.

In Exercises 25 through 48, solve the given problems involving pyramids, cones, and spheres.

   25. The Great Pyramid of Egypt has a square base approximately 250 yd on a side. Its height is about 160 yd. What is its volume?

   26. A steel wedge is made in the shape of a pyramid with a square base, 3.50 cm on a side, and a height of 8.25 cm. Given that the density of steel is 7.80 g/cm³, what is the mass of the wedge?

   27. A sheet metal container is in the shape of a pyramid (inverted) with a square base. The side of the square is 12.0 cm and the depth of the container is 8.75 cm. At a cost of 75¢ per square centimeter, what is the cost of the material of the container (no base)? See Figure 15-128.

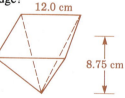

12.0 cm

8.75 cm

FIGURE 15-128

   28. A tent has the shape of a pyramid on a square base. The side of the square is 9.25 ft and the height of the pyramid is 10.5 ft. What is the area of the canvas needed for the four sides and the floor?

   29. A hardness-testing machine has a point in the shape of a cone with radius 2.45 mm and height 9.32 mm. Find the volume of this point.

   30. A fire sprinkler sprays water in the shape of a cone with radius 12 ft and height 8.0 ft. Find the volume of this cone.

   31. A conical cistern 10.0 ft high has a radius at the top of 6.00 ft. Water weighs 62.4 lb/ft³. How many pounds of water does the cistern hold?

   32. A pile of sand is in the shape of a cone. The diameter of the base is 12.0 m and its height is 2.50 m. What is the volume of sand in the pile?

   33. A paper cup is in the shape of a cone with radius 1.80 in. and height 3.50 in. What is the surface area of the cup?

   34. A conical funnel is 8.50 cm deep and 10.0 cm in diameter. What is the surface area of the funnel? (Neglect the opening of the funnel.)

35. What is the weight of a gold sphere 6.00 in. in diameter? The density of gold is 0.697 lb/in.³

36. The circumference of a basketball is about 29.8 in. What is its volume?

37. The circumference of a weather balloon is 6.53 m. Find its volume.

38. What is the difference in surface area between a baseball with diameter 2.75 in. and a soccer ball with a diameter 8.50 in.?

39. The diameter of the moon is about 3480 km. What is the surface area of the moon? Compare the radius of the moon with that of the earth, and then compare the areas.

40. A lampshade is in the shape of a hemisphere of diameter 36.0 cm. What is the surface area of the shade? See Figure 15-129.

41. The side view of a certain rivet is shown in Figure 15-130. It is actually a conical part on a cylindrical part. Find the volume of the rivet.

42. An oil storage tank is in the shape of hemispheres on each end of a cylinder. The length of the cylinder is 45.0 ft, and the diameter of the cylinder (or sphere) is 12.5 ft. What is the volume, in gallons, of the tank? (1.00 ft³ = 7.48 gal) See Figure 15-131.

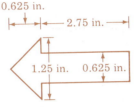

FIGURE 15-129　　　　　**FIGURE 15-130**　　　　　**FIGURE 15-131**

43. A spherical steel bearing has volume 0.7185 cm³. Find its diameter.

44. A tank is in the shape of a cube 2.00 ft on each edge. When the tank is exactly half full of water, a steel sphere is dropped in and the water level rises 0.465 in. Find the diameter of the sphere.

45. Equation (6-19) expresses the volume V of a sphere in terms of its radius r. Express the volume V of a sphere in terms of its diameter d.

46. Equation (15-13) expresses the area A of a sphere in terms of its radius r. Express the area A of a sphere in terms of its diameter d.

47. Derive the formula for the total surface area of a hemispherical volume of radius r (curved surface and flat surface).

48. Derive the formula for the volume formed by placing a hemisphere of radius r on a cone of radius r and height h.

Chapter 15 Formulas

$a^2 + b^2 = c^2$	Pythagorean theorem	(15-1)
$V = lwh$	volume of rectangular solid	(15-2)
$V = Bh$	volume of prism	(15-3)
$A = 2lw + 2wh + 2lh$	area of rectangular solid	(15-4)
$A = 6e^2$	area of cube	(15-5)
$V = \pi r^2 h$	volume of cylinder	(15-6)
$L = 2\pi rh$	lateral area of cylinder	(15-7)
$A = 2\pi r^2 + 2\pi rh = 2\pi r(r + h)$	total area of cylinder	(15-8)
$V = \dfrac{1}{3}Bh$	volume of pyramid	(15-9)
$L = \dfrac{1}{2}ps$	lateral area of pyramid	(15-10)
$V = \dfrac{1}{3}\pi r^2 h$	volume of cone	(15-11)
$L = \pi rs$	lateral area of cone	(15-12)
$A = 4\pi r^2$	area of sphere	(15-13)

15-7 Review Exercises for Chapter 15

In Exercises 1 through 8, answer the given questions.

1. What is the complement of an angle of 29°?
2. What is the supplement of an angle of 29°?
3. What angle is formed when a tangent line touches a circle at a point that is connected to the center?
4. Find the measure of the fourth angle in a quadrilateral with angles of 40°, 120°, and 140°.
5. Find the smallest angle in a right triangle which contains an angle of 63°.
6. Triangles *ABC* and *DEF* are similar. If $\angle A = \angle E = 56°$ and $\angle B = \angle F = 83°$, find the missing angles.
7. If one side of a regular pentagon is 4.30 cm, find its perimeter.
8. If the longest side of an isosceles right triangle is 1.23, find the other sides.

In Exercises 9 through 12, use Figure 15-132 and identify the indicated angles.
(Assume that BD ⊥ AC.)

9. An obtuse angle

10. Two acute angles

11. The complement of ∠ABE

12. The supplement of ∠ABE

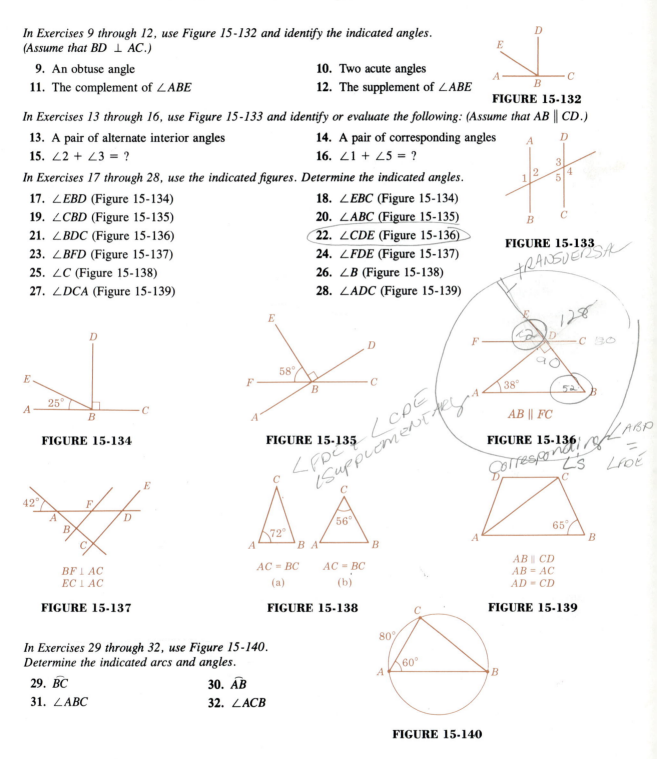

FIGURE 15-132

In Exercises 13 through 16, use Figure 15-133 and identify or evaluate the following: (Assume that AB ∥ CD.)

13. A pair of alternate interior angles

14. A pair of corresponding angles

15. ∠2 + ∠3 = ?

16. ∠1 + ∠5 = ?

In Exercises 17 through 28, use the indicated figures. Determine the indicated angles.

17. ∠EBD (Figure 15-134)

18. ∠EBC (Figure 15-134)

19. ∠CBD (Figure 15-135)

20. ∠ABC (Figure 15-135)

21. ∠BDC (Figure 15-136)

22. ∠CDE (Figure 15-136)

23. ∠BFD (Figure 15-137)

24. ∠FDE (Figure 15-137)

25. ∠C (Figure 15-138)

26. ∠B (Figure 15-138)

27. ∠DCA (Figure 15-139)

28. ∠ADC (Figure 15-139)

FIGURE 15-133

FIGURE 15-134

FIGURE 15-135

FIGURE 15-136

AB ∥ FC

FIGURE 15-137

BF ⊥ AC
EC ⊥ AC

FIGURE 15-138

AC = BC   AC = BC
(a)   (b)

FIGURE 15-139

AB ∥ CD
AB = AC
AD = CD

In Exercises 29 through 32, use Figure 15-140.
Determine the indicated arcs and angles.

29. $\overset{\frown}{BC}$

30. $\overset{\frown}{AB}$

31. ∠ABC

32. ∠ACB

FIGURE 15-140

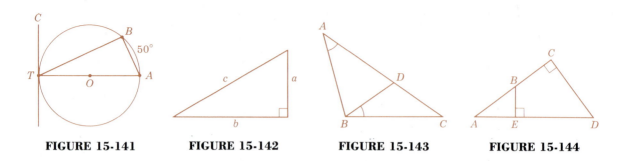

| FIGURE 15-141 | FIGURE 15-142 | FIGURE 15-143 | FIGURE 15-144 |

In Exercises 33 through 36, use Figure 15-141. Line CT is tangent to the circle with center at O. Determine the indicated angles.

33. $\angle BTA$　　　　**34.** $\angle TAB$　　　　**35.** $\angle BTC$　　　　**36.** $\angle ABT$

In Exercises 37 through 44, find the indicated sides of the right triangle shown in Figure 15-142. Where necessary, round off results to three significant digits.

37. $a = 9$, $b = 40$, $c = ?$　　　　　　　**38.** $a = 14$, $b = 48$, $c = ?$
39. $a = 40$, $c = 58$, $b = ?$　　　　　**40.** $b = 56$, $c = 65$, $a = ?$
41. $a = 6.30$, $b = 3.80$, $c = ?$　　　**42.** $a = 126$, $b = 251$, $c = ?$
43. $b = 29.3$, $c = 36.1$, $a = ?$　　　**44.** $a = 0.782$, $c = 0.885$, $b = ?$

In Exercises 45 and 46, solve for the unknown side of the given triangle by use of an appropriate proportion.

45. In Figure 15-143, $AC = 12$, $BC = 8$, $BD = 5$. Find AB.
46. In Figure 15-144, $AB = 4$, $BC = 4$, $CD = 6$. Find BE.

In Exercises 47 and 48, use Figure 15-145 and find the indicated sides. Given: BF is a diameter, CE is tangent at D, AB = 2.0 in., BC = 4.0 in., AE = 18.0 in., BF ∥ CE.

47. $BF = ?$　　　　**48.** $CE = ?$　　　　　　　　**FIGURE 15-145**

In Exercises 49 through 60, find the volumes of the indicated figures for the given values.

49. Prism: base area 40.0 ft², height 15.0 ft　　　**50.** Prism: base area 800 cm², height 50.0 cm
51. Prism: base is right triangle with legs 26.0 cm and 34.0 cm; height 14.0 cm
52. Prism: base is right triangle with legs 6.20 ft and 3.80 ft; height 8.50 ft
53. Cylinder: base radius 36.0 in., altitude 24.0 in.
54. Cylinder: base diameter 120 mm, height 58.0 mm
55. Pyramid: base area 3850 cm², height 125 cm
56. Pyramid: base is square with side 6.44 in., altitude 2.25 in.
57. Cone: base radius 18.2 ft, height 11.5 ft　　　**58.** Cone: base diameter 76.2 mm, height 22.1 mm
59. Sphere: radius 18.0 cm　　　　　　　　　**60.** Sphere: diameter 52.0 ft

In Exercises 61 through 72, find the lateral area or total area of the indicated figure for the given values.

61. Lateral area of prism: base perimeter 60.0 ft, height 15.0 ft
62. Total area of prism of Exercise 51
63. Total area of rectangular solid: length 2.00 m, width 1.50 m, height 1.20 m

64. Total area of cube of edge 3.5 yd

65. Lateral area of cylinder of Exercise 53   66. Total area of cylinder of Exercise 54

67. Lateral area of pyramid: base perimeter 240 cm, slant height 140 cm

68. Lateral area of pyramid of Exercise 56

69. Lateral area of cone of Exercise 57   70. Lateral area of cone of Exercise 58

71. Total area of sphere of Exercise 59   72. Total area of sphere of Exercise 60

In Exercises 73 through 120, solve the given problems.

73. What is the diagonal distance along the floor between corners of a rectangular room 12.5 ft wide and 17.0 ft long?

74. An observer is 550 m from the launch pad of a rocket. After the rocket has ascended vertically to a point 750 m from the observer, what is its height above the launch pad? See Figure 15-146.

75. In a certain pulley system, the center of pulley *A* is 180 cm above and 145 cm to the right of the center of pulley *B*. How far is it between the centers of the pulleys? See Figure 15-147.

76. The base of a 20.0-ft ladder is 6.25 ft from the base of a vertical wall. How far up the wall does the ladder touch?

77. A loop of wire is in the shape shown in Figure 15-148. (The two geometric figures are a semicircle and an isosceles right triangle.) Find the length of the wire loop.

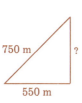

FIGURE 15-146   **FIGURE 15-147**   **FIGURE 15-148**

78. The impedance of a certain alternating-current circuit is 16.5 Ω and the capacitive reactance is 3.75 Ω. Find the resistance in the circuit. (See Exercise 38 of Section 15-3.)

79. Is it possible that sides of 9, 40, and 41 could be the sides of a right triangle?

80. Determine a general formula for the length of the diagonal of a cube (from one corner to the opposite corner).

81. The hypotenuse of a right triangle is 24.0 m and one side is twice the other. Find the perimeter of the triangle.

82. The hypotenuse of a right triangle is 3.00 ft longer than one of the sides, which in turn is 5.00 ft longer than the other side. How long are the sides and the hypotenuse?

83. One of the acute angles of a right triangle is three times the other. How many degrees are there in each angle?

84. In a given triangle, the second angle is three times the first and the third angle is twice the first. How many degrees are there in each angle?

85. A tree casts a shadow of 12 ft and the distance from the end of the shadow to the top of the tree is 13 ft. How high is the tree?

86. A tree and a telephone pole cast shadows as shown in Figure 15-149. Find the height of the telephone pole.

87. The diameter of the sun is 860,000 mi, the diameter of the earth is 7920 mi, and the distance from the earth to the sun (center to center) is 93,000,000 mi. What is the distance from the center of the earth to the end of the shadow due to the rays from the sun?

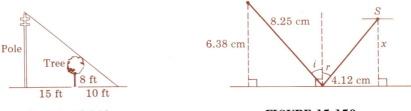

FIGURE 15-149   FIGURE 15-150

88. Light is reflected from a mirror so that the angle of incidence i (see Figure 15-150) equals the angle of reflection r. Suppose that a light source is 6.38 cm from a mirror and a particular ray of light strikes the mirror at the point shown. How far is the screen S from the mirror?

89. A good approximation of the height of a tree can be made by following the procedure suggested in Figure 15-151. By measuring DE, AE, and BC (use a ruler), the length of the tree $DE + EF$ can be found. Find the height of a tree if $AB = 50$ cm, $BC = 30$ cm, $AE = 2400$ cm, and $DE = 150$ cm.

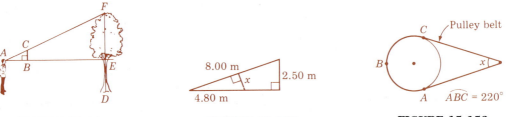

FIGURE 15-151   FIGURE 15-152   FIGURE 15-153

90. What is the length of the steel support in the structure shown in Figure 15-152?

91. A drafting student is making a drawing with a scale of $2\frac{3}{4}$ in. = 5 ft. What distance on her drawing should be used to represent 18 ft 3 in.?

92. On the blueprint of a certain building, a certain hallway is 38.5 in. long. The scale is $1\frac{1}{8}$ in. = 6 ft. How long is the hallway?

93. A square frame is constructed in a circular rim such that the vertices of the square are in contact with the rim. What is the arc intercepted by adjacent sides of the square?

94. The pulley belt shown in Figure 15-153 is in contact with the pulley along an arc of 220°. What is the indicated angle between sections of the belt?

95. A vertical beam has a rectangular cross section 15.4 cm by 10.5 cm. The beam is 240 cm high. What is the lateral area of the beam?

96. A fish tank is 3.50 ft long, 2.75 ft wide, and 2.25 ft high. What is the surface area of the tank? (There is no top.)

97. The circumference of a cylindrical water tank is 152 ft, and its height is 48.5 ft. How many gallons of water can it hold? (1.00 ft^3 = 7.48 gal)

98. How many meters of steel rods with radius 0.500 cm can be made from a steel ingot which is in the shape of a rectangular solid 6.00 cm by 6.00 cm by 24.0 cm?

99. A hollow concrete conduit is 12.0 m long. The inside diameter is 54.0 cm and the outside diameter is 75.0 cm. Find the volume of concrete in the conduit.

100. How many cubic inches of metal are there in a length of pipe 15.5 ft long when the inside diameter is 8.00 in. and the metal is 0.500 in. thick?

101. What is the area of a label (covering the lateral surface) on a cylindrical can 3.00 in. in diameter and 4.25 in. high? Assume the ends just meet and do not overlap.

102. A furnace is built in the shape of a cylinder 1.40 m high and 0.950 m in diameter. How many square meters of insulation are required to cover the sides and top of the furnace?

103. What is the volume of a conical rod with a base diameter of 0.750 cm and a length of 48.0 cm?

104. What is the volume of darkness of the earth's shadow? (Approximate it by a right circular cone with the base through the center of the earth. See Exercise 87.) Compare this with the volume of the earth.

105. An umbrella opens into the shape of a cone 3.00 ft in diameter and 0.500 ft deep. What is the surface area of the umbrella?

106. A sheet metal cover is in the shape of a cone with radius 17.0 cm and height 6.00 cm. What is the surface area of the cover?

107. A marble monument is in the shape of a pyramid with a square base 160 cm on a side. The height of the monument is 140 cm. What is the mass, in kilograms, of the monument? The density of marble is 2.70×10^4 kg/m^3.

108. A concrete base for a piece of machinery is in the shape of a *frustum* of a pyramid. (A frustum is formed by cutting off the top by a plane parallel to the base.) The top of the frustum is a rectangle 18.0 ft by 12.5 ft. The length of the base is 24.0 ft and the depth of the frustum is 3.25 ft. How many cubic feet of concrete were used in making the base? See Figure 15-154.

109. A wedge is in the shape of a prism with a right triangular base. The base has a width of 4.00 in. and a length of 14.5 in. What is the total surface area of the wedge, given that its altitude is 5.38 in.?

110. A hopper containing grain is in the shape of an inverted pyramid (with a square base) surmounted by a cube. The base of the pyramid is 3.25 m on a side and the height of the pyramid is 2.50 m. What is the capacity of the hopper?

111. What is the radius of a 16.0-lb shot? The density of iron used in the shot (spherical) is 450 lb/ft^3.

112. A hollow metal sphere has an outer radius of 1.50 ft and an inner radius of 1.25 ft. What is the volume of the metal in the sphere?

113. How many square centimeters of material are used to make a tennis ball whose circumference is 21.0 cm?

114. A hot water tank is the shape of a right circular cylinder surmounted by a hemisphere. See Figure 15-155. The total height of the tank is 6.75 ft and the diameter of the base is 2.50 ft. How many gallons does the tank contain? (See Exercise 97.)

18.0 ft
12.5 ft

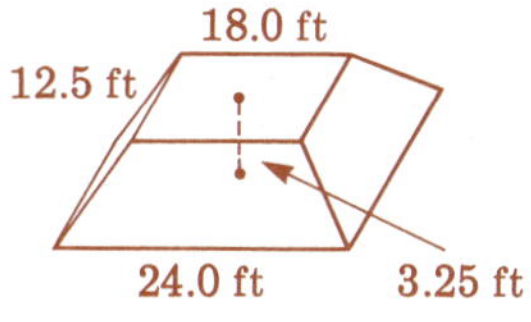

FIGURE 15-154   24.0 ft   3.25 ft

6.75 ft

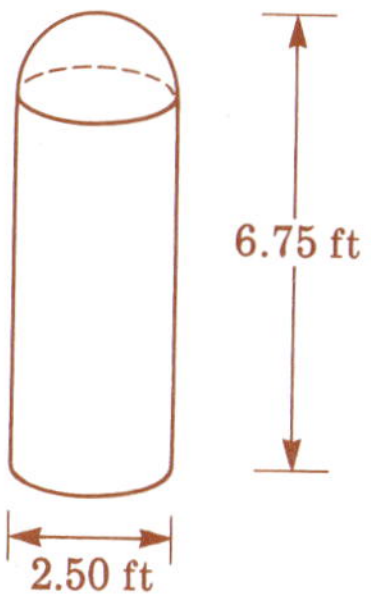

FIGURE 15-155   2.50 ft

115. Derive the formula for the total surface area of a volume formed by placing a hemisphere of radius r on a cylinder of radius r and altitude h. See Figure 15-156.

116. Derive the formula for the volume formed by placing a pyramid of square base (edge e) and altitude h on a cube of edge e.

117. Find the volume of the largest sphere that can be contained within a rectangular box that is 6.00 in. wide, 7.00 in. long, and 8.00 in. high.

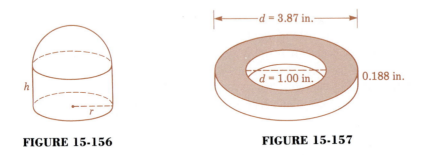

FIGURE 15-156   **FIGURE 15-157**

118. A rubber gasket is 3.87 in. in diameter. It has a hole that is 1.00 in. in diameter, and its thickness is 0.188 in. Find the volume of material needed to fill an order for 2500 such gaskets. See Figure 15-157.

119. The surface area of a sphere is 237 cm². If that same area is used to construct a cylinder with radius 2.00 cm, how high is the cylinder?

120. A lead cube has volume 12.8 cm³. If it is melted and molded into a sphere, by what amount does the surface area change?

CHAPTER SIXTEEN

Trigonometry of Right Triangles

Many applied problems in science and technology require the use of triangles, especially right triangles. Included among the applied problems are those involving forces acting on objects, air navigation, surveying, establishing road slopes, the motion of projectiles, and light refraction in optics.

In **trigonometry** we develop methods for measuring sides and angles of triangles, as well as solving the related applied problems. Because of its many uses, trigonometry is generally recognized as one of the most practical and relevant branches of mathematics.

In Section 16-1 we present the fundamental concept of a trigonometric relation, and in Section 16-2 we develop specific procedures for determining their values. In Section 16-3 we proceed to consider a variety of applications. One of the examples included in that section involves an airplane which loses its engine so that it must glide for a landing. One particular airplane can glide along a path which makes an angle of 18.0° with level ground. If the plane is at 5500 ft when its engine stops, what is the maximum horizontal distance it can go? See Figure 16-1. We will solve this problem in Section 16-3.

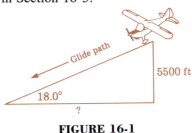

Glide path

18.0°

5500 ft

?

FIGURE 16-1

16-1 The Trigonometric Ratios

The fundamental definitions of trigonometry are based on a property of similar triangles. In Chapter 15 we noted that the corresponding sides of similar triangles are proportional. This property implies that *whenever we have two similar triangles, the ratio of one side to another in the first triangle will be the same as the ratio of the corresponding sides in the second triangle*. To illustrate this very important concept, we will consider the following example.

EXAMPLE A   The two triangles in Figure 16-2 are similar. The corresponding sides are proportional; each side of the second triangle is three times the length of the corresponding side in the first triangle. However, we want to illustrate that ▶ *the ratio of two sides in one of these triangles will be the same as the ratio of the corresponding sides in the other triangle.* To illustrate this, we consider the ratio of the shortest side to the longest side in the first triangle and we get $\frac{3}{5}$. Also, in the second triangle the ratio of the shortest side to the longest side is $\frac{9}{15} = \frac{3}{5}$ so that the same ratio is obtained. Ratios of pairs of other corresponding sides can also be determined and shown to be equal.

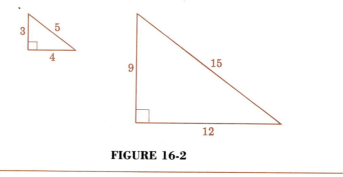

FIGURE 16-2

EXAMPLE B   In Figure 16-3 the two triangles are similar since the corresponding angles are equal. Using the property that the ratio of one side to another in the first triangle is the same as the ratio of the corresponding sides in the second triangle, we get

$$\frac{1.0}{2.0} = \frac{x}{5.0}$$

which implies that $x = 2.5$ cm.

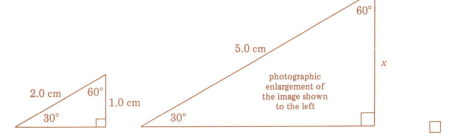

FIGURE 16-3

From Example B we can see that *in any right triangle with angles of 30° and 60°, the ratio of the shortest side to the hypotenuse will always be $\frac{1}{2}$.* It is observations of this type that have led to the formal definition of the **trigonometric ratios** which are based on the right triangle shown in Figure 16-4.

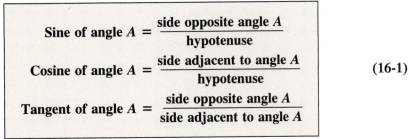

FIGURE 16-4

$$\text{Sine of angle } A = \frac{\text{side opposite angle } A}{\text{hypotenuse}}$$

$$\text{Cosine of angle } A = \frac{\text{side adjacent to angle } A}{\text{hypotenuse}} \qquad \text{(16-1)}$$

$$\text{Tangent of angle } A = \frac{\text{side opposite angle } A}{\text{side adjacent to angle } A}$$

The names of these trigonometric ratios are usually abbreviated as **sin A, cos A,** and **tan A.** The definitions given in Eq. (16-1) can be used to find the trigonometric ratios of any acute angle in any right triangle. Using the definitions in Eq. (16-1) and considering angle *B*, we can also state the following.

$$\sin B = \frac{\text{side opposite angle } B}{\text{hypotenuse}}$$

$$\cos B = \frac{\text{side adjacent angle } B}{\text{hypotenuse}}$$

$$\tan B = \frac{\text{side opposite angle } B}{\text{side adjacent angle } B}$$

Because these definitions are so important, they should be remembered. It is often helpful to use some memory device such as associating "opposite over hypotenuse" with sine by the letters O-H-S which can be remembered from the phrase "Old High School." Cosine can be associated with "adjacent over hypotenuse" by the letters A-H-C ("A Helpful Course"). The letters O-A-T suggest that tangent is associated with "opposite over adjacent." Another common memory device involves the word "SOH-CAH-TOA" which also summarizes the trigonometric ratios. The first three letters suggest that sine is related to the opposite and hypotenuse. "CAH" suggests that cosine is adjacent divided by hypotenuse, while "TOA" represents tangent as opposite divided by adjacent.

It would be helpful to memorize the definitions of sine, cosine, and tangent, then apply those definitions to the triangle given in Figure 16-5. You should be able to cover up the fractions given in Example C and obtain the same values by referring to Figure 16-5 and using the memorized definitions.

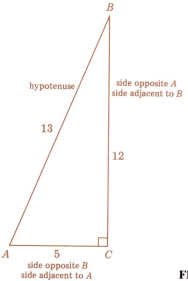

FIGURE 16-5

EXAMPLE C   In the triangle shown in Figure 16-5, we have

$$\sin A = \frac{12}{13} \qquad \sin B = \frac{5}{13} \qquad \frac{\text{side opposite}}{\text{hypotenuse}}$$

$$\cos A = \frac{5}{13} \qquad \cos B = \frac{12}{13} \qquad \frac{\text{side adjacent}}{\text{hypotenuse}}$$

$$\tan A = \frac{12}{5} \qquad \tan B = \frac{5}{12} \qquad \frac{\text{side opposite}}{\text{side adjacent}}$$

In some cases we know only two sides of a right triangle. The third side can be found by using the Pythagorean theorem. The trigonometric ratios can then be found for the angles. This is illustrated in the following example.

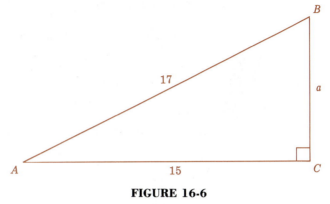

FIGURE 16-6

EXAMPLE D   In Figure 16-6 we know the two sides of the right triangle. We find the third side by using the Pythagorean theorem.

$$a^2 + b^2 = c^2$$
$$a^2 + (15)^2 = (17)^2$$
$$a^2 + 225 = 289$$
$$a^2 = 64$$
$$a = 8$$

Knowing that $a = 8$ we can now find any of the trigonometric ratios. For example,

$$\sin A = \tfrac{8}{17} \text{ and } \tan B = \tfrac{15}{8}. \qquad \square$$

So far, we have defined the three trigonometric ratios of sine, cosine, and tangent. There are three others, called **cotangent**, **secant**, and **cosecant** which are abbreviated and defined as follows:

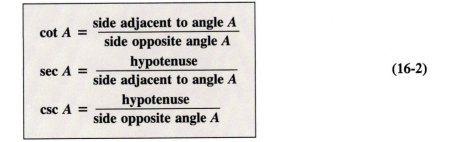

$$\cot A = \frac{\text{side adjacent to angle } A}{\text{side opposite angle } A}$$

$$\sec A = \frac{\text{hypotenuse}}{\text{side adjacent to angle } A} \qquad (16\text{-}2)$$

$$\csc A = \frac{\text{hypotenuse}}{\text{side opposite angle } A}$$

If we compare Eqs. (16-1) to Eqs. (16-2) we can see that cot A is the reciprocal of tan A, sec A is the reciprocal of cos A, and csc A is the reciprocal of sin A. These relationships are expressed in Eqs. (16-3) and they will be especially useful for many calculator computations, which will be discussed in the next section.

$$\cot A = \frac{1}{\tan A} \qquad \sec A = \frac{1}{\cos A} \qquad \csc A = \frac{1}{\sin A} \qquad\qquad (16\text{-}3)$$

EXAMPLE E   Applying Eqs. (16-2) to Figure 16-5, we get

$$\cot A = \frac{5}{12} \qquad \cot B = \frac{12}{5} \qquad \frac{\text{side adjacent}}{\text{side opposite}}$$

$$\sec A = \frac{13}{5} \qquad \sec B = \frac{13}{12} \qquad \frac{\text{hypotenuse}}{\text{side adjacent}}$$

$$\csc A = \frac{13}{12} \qquad \csc B = \frac{13}{5} \qquad \frac{\text{hypotenuse}}{\text{side opposite}} \qquad \square$$

If we have a right triangle and know only one of the trigonometric ratios for a given angle, the values of the other five trigonometric ratios can be found. The following example illustrates how this can be done.

EXAMPLE F   Suppose we have a right triangle and know that sin $A = \frac{9}{10}$. We know that the ratio of the side opposite angle A to the hypotenuse is 9 to 10. For the purposes of finding the other ratios, we may assume that the lengths of these two sides are 9 units and 10 units (see Figure 16-7). We find the third side by using the Pythagorean theorem and we get

$$x = \sqrt{10^2 - 9^2} = \sqrt{19}$$

We can now determine the other five trigonometric ratios of the angle whose sine is given as $\frac{9}{10}$.

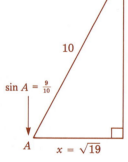

$$\cos A = \frac{\sqrt{19}}{10} \qquad\qquad \cot A = \frac{\sqrt{19}}{9}$$

$$\tan A = \frac{9}{\sqrt{19}} = \frac{9\sqrt{19}}{19} \qquad\qquad \sec A = \frac{10}{\sqrt{19}} = \frac{10\sqrt{19}}{19}$$

$$\csc A = \frac{10}{9}$$

FIGURE 16-7

Using a calculator, we can easily obtain the decimal forms of the above values. $\qquad \square$

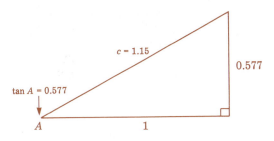

FIGURE 16-8

EXAMPLE G   Given a right triangle in which tan $A = 0.577$, we know that the ratio of the side opposite angle A to the side adjacent to it is 0.577. This means that *we may construct a triangle with sides of* **0.577** *opposite* A *and* 1 *adjacent to* **A,** as shown in Figure 16-8.

The length of the hypotenuse is found by using the Pythagorean theorem as follows:

$$a^2 + b^2 = c^2$$
$$(0.577)^2 + 1^2 = c^2$$
$$1.33 = c^2$$
$$c = 1.15$$

The other five trigonometric ratios for this angle can now be found, as shown below.

$$\sin A = \frac{0.577}{1.15} = 0.502 \qquad \cot A = \frac{1}{0.577} = 1.73$$

$$\cos A = \frac{1}{1.15} = 0.870 \qquad \sec A = \frac{1.15}{1} = 1.15$$

$$\csc A = \frac{1.15}{0.577} = 1.99$$

Exercises 16-1

In Exercises 1 through 4, find the indicated trigonometric ratios in fractional form from Figure 16-9.

1. $\sin A$, $\tan A$, $\cos B$

2. $\cos A$, $\sin B$, $\cot A$

3. $\cot B$, $\sec A$, $\tan B$

4. $\sec B$, $\csc A$, $\csc B$

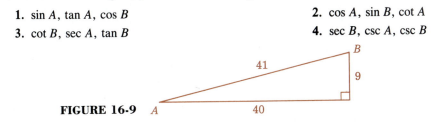

FIGURE 16-9

In Exercises 5 through 8, find the indicated trigonometric ratios in fractional form from Figure 16-10.

5. sin A, sec B, cot A

6. csc A, sin B, cot B

7. tan A, cos B, sec A

8. tan B, csc B, cos A

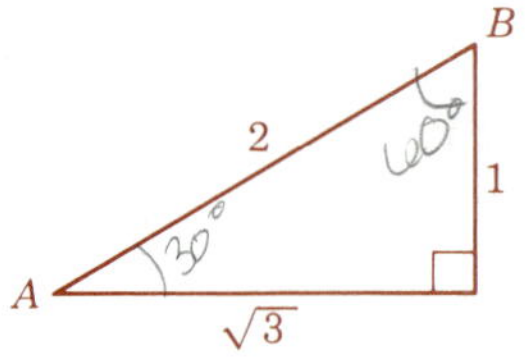

FIGURE 16-10

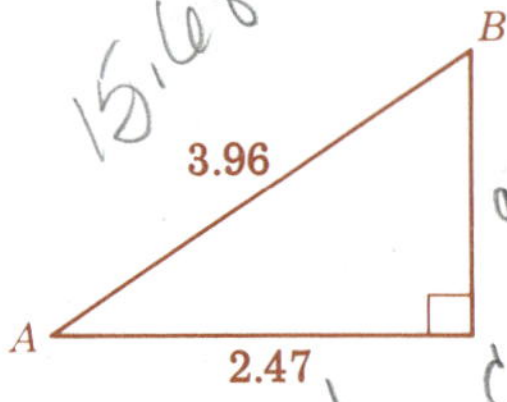

FIGURE 16-11

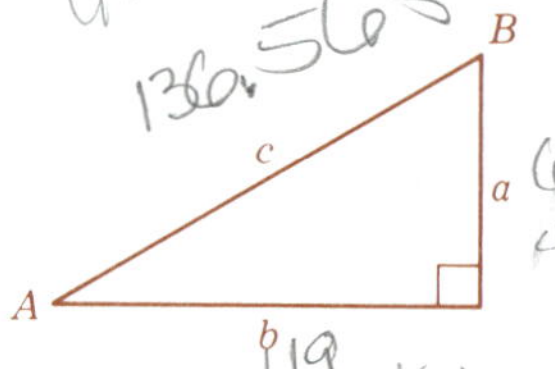

FIGURE 16-12

In Exercises 9 through 12, find the indicated trigonometric ratios in decimal form from Figure 16-11.

9. cos A, tan A, csc B

10. sin B, sec A, cot B

11. sin A, cos B, cot A

12. csc A, tan B, sec B

In Exercises 13 through 24, determine the indicated trigonometric ratios. The listed sides are those shown in Figure 16-12.

13. $a = 4$, $b = 3$. Find sin A and tan B.

14. $a = 8$, $c = 17$. Find cos A and csc B.

15. $b = 7$, $c = 25$. Find cot A and cos B.

16. $a = 9$, $c = 25$. Find sin A and sec B.

17. $a = 1$, $b = 1$. Find cos A and cot B.

18. $a = 3$, $c = 4$. Find sec A and tan B.

19. $b = 15$, $c = 22$. Find csc A and cos B.

20. $a = 67$, $b = 119$. Find cos A and cot B.

21. $a = 1.2$, $b = 1.5$. Find sin A and tan B.

22. $a = 3.44$, $c = 6.82$. Find cot A and csc B.

23. $b = 0.0446$, $c = 0.0608$. Find sin A and sec B.

24. $a = 1673$, $c = 1944$. Find csc A and sin B.

In Exercises 25 through 36, use the given trigonometric ratios to find the indicated trigonometric ratios.

25. If tan $A = 1$, find sin A.

26. If sin $A = \frac{1}{2}$, find cos A.

27. If cos $A = 0.70$, find csc A.

28. If sec $A = 1.6$, find sin A.

29. If cot $A = 0.563$, find cos A.

30. If csc $A = 2.64$, find tan A.

31. If sin $A = 0.720$, find tan A.

32. If tan $A = 0.350$, find cos A.

33. If cos $A = 0.8660$, find sin A.

34. If csc $A = 5.55$, find sec A.

35. If sec $A = \sqrt{3.00}$, find tan A.

36. If cot $A = \sqrt{5.00}$, find sin A.

In Exercises 37 through 44, solve the given problems.

37. From the definitions of the trigonometric ratios, it can be seen that sin A = cos B. What ratio associated with angle B equals tan A? csc A?

38. Construct three right triangles. The first triangle should have sides 3 in., 4 in., and 5 in. The second triangle should have sides 6 cm, 8 cm, and 10 cm. The third triangle should have sides 9 cm, 12 cm, and 15 cm. For each triangle determine the sine and tangent of the smallest angle. What is the relationship between the three triangles? What is true of the trigonometric ratios found?

39. State the definitions of all six trigonometric ratios of angle A in terms of the sides a, b, and c of the triangle shown in Figure 16-13.

40. In Figure 16-13, if $a < b$ is sin A < sin B? Explain.

41. In Figure 16-13, if $a = 5$ and $b = 12$, what is the value of $(\sin A)^2 + (\cos A)^2$?

42. In Figure 16-13, if $a = 8$ and $b = 15$, what is the value of $(\sin A)^2 + (\cos A)^2$?

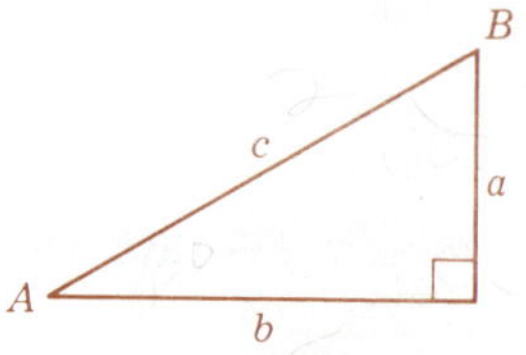

FIGURE 16-13

43. In Figure 16-13, given that $a = 5$ and $b = 12$, calculate the values of sin A, cos A, and tan A. Then show that $(\sin A)/(\cos A) = \tan A$.

44. In Figure 16-13, given that $a = 8$ and $b = 15$, calculate the values of sin A, cos A, and tan A. Then show that $(\sin A)/(\cos A) = \tan A$.

16-2 Values of the Trigonometric Ratios

In the first section of this chapter we defined the trigonometric ratios of an angle, but we did not mention the size of the angle. We did note, however, that the ratio of two sides of a right triangle is the same as the ratio of the two corresponding sides in any similar right triangle. Therefore, *for an angle of a particular size, a given trigonometric ratio has a particular value.* If an angle has a specified number of degrees, there is a specific set of values of trigonometric ratios. The actual method used to determine these values for the purpose of setting up tables requires the use of more advanced mathematics than we shall discuss here. However, by using certain basic geometric properties we can establish the values of the trigonometric ratios for certain angles which are frequently used.

A basic geometric fact is that *in a 30°-60°-90° triangle, the side opposite the 30° angle is one-half the hypotenuse* (see Figure 16-14a). We can easily verify this statement by referring to the equilateral triangle shown in Figure 16-14(b) which has been divided into two smaller congruent triangles by the altitude. Each of the smaller triangles has angles of 30° (at top), 60°, and 90°.

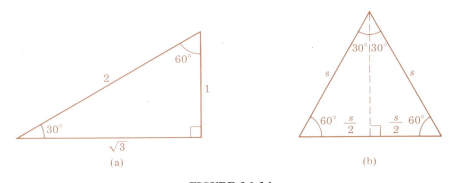

FIGURE 16-14

In Figure 16-14(a), the hypotenuse has been given the value 2, the side opposite the 30° angle has been given the value 1, and from the Pythagorean theorem we determine that the third side has the value $\sqrt{3}$. Using the known angles and sides in Figure 16-14(a), we can now establish all of the trigonometric ratios shown in the following table.

Angle	sin	cos	tan	cot	sec	csc
30°	$\dfrac{1}{2}$	$\dfrac{\sqrt{3}}{2}$	$\dfrac{\sqrt{3}}{3}$	$\sqrt{3}$	$\dfrac{2\sqrt{3}}{3}$	2
60°	$\dfrac{\sqrt{3}}{2}$	$\dfrac{1}{2}$	$\sqrt{3}$	$\dfrac{\sqrt{3}}{3}$	2	$\dfrac{2\sqrt{3}}{3}$

The following example illustrates the use of another geometric property to establish the values of the trigonometric ratios of 45°.

EXAMPLE A

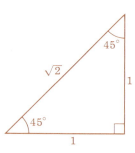

FIGURE 16-15

To find the trigonometric ratios for 45°, we construct an isosceles right triangle, as shown in Figure 16-15. Since the triangle is isosceles, *both acute angles are 45° and the legs are equal.* Each of the equal sides may be given the value 1. From the Pythagorean theorem, we find that the hypotenuse will then be $\sqrt{2}$. Using the values shown in Figure 16-15, we can apply the definitions of the trigonometric ratios to get the following.

$$\sin 45° = \frac{\sqrt{2}}{2} \qquad \cos 45° = \frac{\sqrt{2}}{2} \qquad \tan 45° = 1$$

$$\cot 45° = 1 \qquad \sec 45° = \sqrt{2} \qquad \csc 45° = \sqrt{2} \qquad \square$$

Combining the values from the preceding table for 30° and 60° and the values from Example A for 45°, we can set up the following short table of values of the trigonometric ratios (in decimal form):

Angle	sin	cos	tan	cot	sec	csc
30°	0.500	0.866	0.577	1.732	1.155	2.000
45°	0.707	0.707	1.000	1.000	1.414	1.414
60°	0.866	0.500	1.732	0.577	2.000	1.155

Values of other trigonometric ratios can be found by using either a calculator or the values given in Table 3 in Appendix D at the end of this book. We will describe both methods.

When using a calculator for finding the values of trigonometric ratios, the keys labeled SIN, COS, and TAN are used. Consult the manual written for your particular calculator to *be sure that it is in the* **degree** mode (not radians or grads, which will be described later). Most calculators are automatically in the degree mode as soon as they are turned on. The value of a trigonometric ratio is obtained by pressing the number of degrees and then pressing the appropriate trigonometric key as in the following example.

EXAMPLE B   Using a calculator to find the value of sin 20°, we first verify that the calculator is in the degree mode and we then enter

20 (SIN)   (0.34202014)

to get the displayed value.
The value of tan 52° is found by pressing the keys

52 (TAN)   (1.2799416)

In the absence of special (COT), (SEC), and (CSC) keys, we can use the (SIN), (COS), and (TAN) keys with the reciprocal key labeled $1/x$ to obtain the same result.

$$\text{(COT)} \leftrightarrow \text{(TAN)}(1/x)$$
$$\text{(SEC)} \leftrightarrow \text{(COS)}(1/x)$$
$$\text{(CSC)} \leftrightarrow \text{(SIN)}(1/x)$$

EXAMPLE C   Use a calculator to find cot 65°, sec 65°, and csc 65°.
We enter

65 (TAN)(1/x)   (0.46630766)

to get a result of 0.46630766 for cot 65°.
We get sec 65° by entering

65 (COS)(1/x)   (2.3662016)

and the result is 2.3662016.
Similarly, the value of csc 65° is found by entering

65 (SIN)(1/x)   (1.1033779)

and the result is 1.1033779.

Sometimes we know the value of a trigonometric ratio and we want to find the angle involved. To find the angle when the ratio is known, some calculators have a key labeled INV while others have keys with such labels as SIN⁻¹ or ARCSIN. Again, it is important to consult the manual for your calculator. We assume in the following examples that the (INV) key is available.

EXAMPLE D   Use a calculator to find angle A if sin $A = 0.454$.
The following sequence of keys will yield the desired value:

0.454 [INV] [SIN]

The result is 27.000611, which means that angle A is approximately 27°. ☐

EXAMPLE E   Use a calculator to find angle A if cot $A = 0.105$.
The following sequence of keys will yield the value (in degrees) of angle A.

0.105 [¹⁄ₓ] [INV] [TAN]

the reciprocal of cot A is tan A

The result is 84.005907, which means that angle A is approximately 84°. ☐

Table 3 in Appendix D can also be used for finding angles or trigonometric ratios. This table includes the values of the trigonometric ratios for each degree from 0° to 90°. (If you need more complete tables with greater precision, you can find them in many reference books.) Note that *from 0° to 45° the values of the angle are found in the left column, while the values from 45° to 90° are found in the right column*. This arrangement is convenient because it avoids duplication of the same values. The values of sine, tangent, secant from 0° to 45° (or 45° to 90°) are the same as the values of cosine, cotangent, cosecant from 45° to 90° (or 0° to 45°). This is illustrated in the next example.

EXAMPLE F   In the right triangle shown in Figure 16-16, the two acute angles add up to 90°. Since sin $A = a/c$ and cos $B = a/c$, we see that sin $A = $ cos B. Also, since $A + B = 90°$,

$$B = 90° - A \quad \text{or} \quad \sin A = \cos (90° - A)$$

If $A = 40°$, for example, then

$$\sin 40° = \cos 50° \qquad\qquad ☐$$

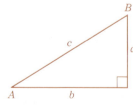

FIGURE 16-16

Example E illustrates why cosine was originally called "complement of sine." In the table, all the values are included for the ratios from 0° to 45° and it is only necessary to relabel the columns (as is done *at the bottom*) to obtain the values for the ratios from 45° to 90°.

The following examples illustrate how the table is used to find values of the ratios and to find the angle when one of the ratios is given.

EXAMPLE G   To find sin 25° from Table 3 in Appendix D, we look in the column labeled "sin" (at the top) to the right of 25, which is in the degrees column. Here we find the number 0.4226. Thus sin 25° = 0.4226.

To find cos 70°, from Table 3 in Appendix D, we must *look* **above** *"cos" which is at the bottom* of the column. Also, we must look to the left of 70 which appears in the right-hand degrees column. Therefore, in the "cos" column (labeled at the bottom) and to the *left* of 70 in the degrees column, we find 0.3420. Thus cos 70° = 0.3420. □

EXAMPLE H   Given that csc α = 3.236, find α. (The Greek letter alpha is often used to designate angles, as are other Greek letters.) First we look for 3.236 in one of the "csc" columns. Finding this value *under* "csc," we then look to the *left*-hand degrees column and determine that α = 18°.

Given that tan α = 2.905, find α to the nearest degree. We look for 2.905 in the columns labeled "tan" and find that the nearest number to 2.905 is 2.904. Since 2.904 is *above* "tan," we must *look to the* **right**-*hand degrees column.* Here we find 71, so α = 71°. □

Examples G and H show how a table of trigonometric values is used where it might be necessary to use a table. Although Table 3 shows values of the trigonometric ratios only for each degree, tables which give such values for at least each 0.1° are available in many reference sources. However, since it is expected that you will be using a calculator for nearly all problems using values of the trigonometric ratios, we restrict our discussion of tables to that just given.

The following example illustrates the use of a trigonometric ratio in an applied problem.

EXAMPLE I   A surveyor needs to determine the angle α for the parcel of land shown in Figure 16-17. Since we already know the two legs of the triangle, we can directly obtain the value of the tangent of α to get

$$\tan \alpha = \frac{62.10 \text{ m}}{136.4 \text{ m}} \qquad \begin{array}{l} \text{side opposite } \alpha \\ \hline \text{side adjacent to } \alpha \end{array}$$

We use the following sequence on a calculator to find the angle α:

62.1 ÷ 136.4 = INV TAN   *24.478758*

which tells us that α = 24.48° (rounded off to hundredths). We note that it is not actually necessary to record tan α, although its value does appear on the display after we press =.

FIGURE 16-17

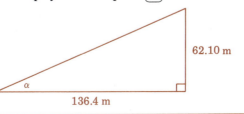

62.10 m

α

136.4 m

□

Exercises 16-2

In Exercises 1 through 16, use a calculator to determine the value of the indicated trigonometric ratio.

1. cos 32.0°   **2.** sin 21.0°   **3.** tan 24.5°   **4.** cot 15.4°

5. sec 48.2°   **6.** csc 56.1°   **7.** sin 66.6°   **8.** cos 52.9°

9. cot 76.6°   **10.** tan 63.7°   **11.** sin 44.8°   **12.** cos 9.2°

13. csc 13.7°   **14.** sec 41.3°   **15.** tan 68.0°   **16.** cot 46.4°

In Exercises 17 through 32, use a calculator to determine the value of the angle α to the nearest 0.1°.

17. sin α = 0.5299   **18.** tan α = 0.2126   **19.** sec α = 1.057   **20.** cos α = 0.7944

21. csc α = 1.149   **22.** cot α = 0.8040   **23.** cos α = 0.4712   **24.** sin α = 0.9888

25. cot α = 0.7620   **26.** sec α = 1.666   **27.** tan α = 0.3250   **28.** csc α = 2.608

29. cos α = 0.09932   **30.** sin α = 0.9464   **31.** cot α = 0.6190   **32.** tan α = 1.900

In Exercises 33 through 36 use a calculator and the given trigonometric ratio to find the indicated trigonometric ratio.

33. sin α = 0.5592; sec α **34.** cos α = 0.8290; tan α

35. tan α = 1.600; csc α **36.** cot α = 0.1584; sin α

In Exercises 37 through 44, find, to the nearest 0.1°, angle α in Figure 16-18 for the given sides of the triangle.

37. $a = 4, c = 5$ **38.** $a = 3, c = 7$

39. $b = 6.2, c = 8.2$ **40.** $a = 3.2, b = 2.0$

41. $a = 15.5, c = 27.3$ **42.** $b = 0.35, c = 0.84$

43. $a = 6580, b = 1230$ **44.** $a = 3.95, c = 45.2$

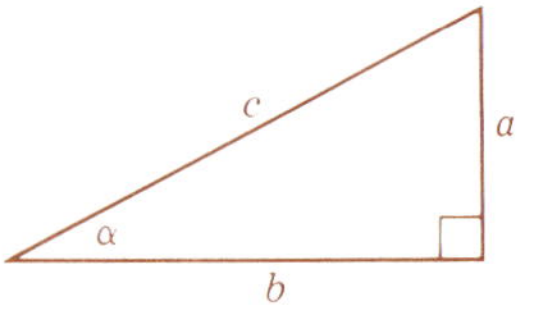

FIGURE 16-18

In Exercises 45 through 48, draw a right triangle, including the indicated angle (use a protractor). Draw the sides adjacent to the angle 10 cm long. By measuring the other sides and using the definitions of the ratios, verify the values of the ratios with those found from a calculator.

45. sin 46°   **46.** cos 20°   **47.** tan 62°   **48.** csc 50°

In Exercises 49 through 60, solve the given problems.

49. In an AC (alternating-current) electric circuit, the instantaneous voltage is found by using the formula

$$v_{\text{inst}} = V_p \sin \theta$$

Find the instantaneous voltage if V_p = 15 V and θ = 63°.

50. Under certain conditions, the height of an overcast cloud layer is found by the formula

$$H = s \tan \theta$$

Find the height H when s = 5750 ft and θ = 27°.

51. In surveying, the horizontal distance between two points is obtained by evaluating $H = s \cos \alpha$. Find H if s = 132.8 m and α = 17.7°.

52. In determining the angle a windshield makes with the dashboard of a car, it is necessary to find A given that tan A = 2.174. Find that angle to the nearest 0.1°.

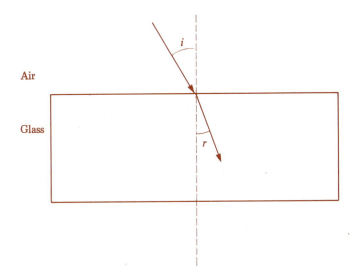

FIGURE 16-19

53. In the study of optics, it is known that a light ray entering glass is bent toward a line perpendicular to the surface, as shown in Figure 16-19. The *index of refraction* of the glass is defined to be

$$n = \frac{\sin i}{\sin r}$$

Find the index of refraction if $i = 72.0°$ and $r = 37.7°$.

54. The work W done by a force F is defined as

$$W = Fd \cos \theta$$

where F is the magnitude of the force, d is the distance through which it acts, and θ is the angle between the direction of the force and the direction of motion. Given that a 25.0-lb force acts through 20.0 ft and the angle between the force and motion is 32.0°, how much work is done by the force?

55. One end of a 25.0-m metal rod lies on a flat surface; the rod makes an angle of 17.0° with the surface. The length of the shadow of the rod due to a light shining vertically down on it is (25.0)(cos 17.0°). Find the length of the shadow.

56. The coefficient of friction for an object on an inclined plane equals the tangent of the angle that the plane makes with the horizontal if the object moves down the plane with a constant speed. The coefficient of friction between a wooden crate and a wooden plank is 0.340 when the crate is moving with a constant speed. What angle does the plank make with the horizontal? See Figure 16-20.

57. Using a calculator, press the key sequence

1.3 [INV] [SIN]

and describe the result.

58. Write the correct key sequence for finding α if csc $\alpha = 3.0000$.

59. Write the correct calculator key sequence for finding sin θ if sec $\theta = 2.05$.

60. Write the correct calculator key sequence for finding sec θ if csc $\theta = 3.0000$.

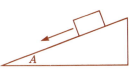

FIGURE 16-20

16-3 Right Triangle Applications

In this section we use examples and exercises to investigate many of the applications of the trigonometric ratios. These include indirect measurement, which was mentioned in the first section of this chapter. First, we consider the general idea of **solving a triangle.**

▶ In every triangle there are three angles and three sides. *If three of these six parts are known, the other three can be found provided that at least one of the known parts is a side. By solving a triangle, we mean determining the values of the six parts.* The strategy for solving right triangles will involve these four steps:

1. If two angles are known, the third angle can be found by using the property that the sum of the three angles is 180°.

2. If two sides are known, find the third side by using the Pythagorean theorem ($c^2 = a^2 + b^2$).

3. If only one side and one of the acute angles are known, find another side by using a trigonometric ratio.

4. If neither of the acute angles is known but two sides are known, both acute angles can be found by using trigonometric ratios.

When using these steps, we should try to avoid using derived values for finding other values. That is, when solving for a particular part of the triangle, try to use only values that were given in the original statement of the problem.

The following examples illustrate solving right triangles.

EXAMPLE A   Given that the hypotenuse of a right triangle is 16.0 and that one of the acute angles is 35.0°, find the other acute angle and the two sides (see Figure 16-21).

Here we know one side ($c = 16.0$) and two angles ($\alpha = 35.0°$ and the right angle is 90°). Using step 1 we can determine that *the third angle is 55.0° since the three angles must add up to* 180°. Since only one side is known, we follow the suggestion of step 3 as we use the sine ratio to find the value of b. Since

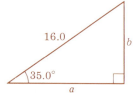

FIGURE 16-21

$$\sin 35.0° = \frac{b}{16.0}$$

given angle ↓ ; ← hypotenuse given

we have

$$b = 16.0 \sin 35.0° = 9.18$$

The calculator steps for this solution are

16 ⊗ 35 [SIN] [=] ⟨ 9.1772230 ⟩ for b

In the calculator steps for b, note that the $\boxed{\text{SIN}}$ key worked only on the number 35 and not on the product of 16 and 35.

Side a can be found by using the cosine relation. Since we have $\cos 35.0° = a/16.0$,

$$a = 16.0 \cos 35.0° = 13.1$$

While we could have found side a using the Pythagorean theorem, it is better to avoid using derived values (such as $b = 9.18$) if possible. We were able to find side a using the original values, not derived values. In addition to the given information, we now know that the other acute angle is $55.0°$ and the other sides are 9.18 and 13.1. *All six parts of the triangle are now known, and the triangle is solved.*　　□

EXAMPLE B　In a right triangle the two legs are 5.00 and 8.00. Find the hypotenuse and the two acute angles (see Figure 16-22).

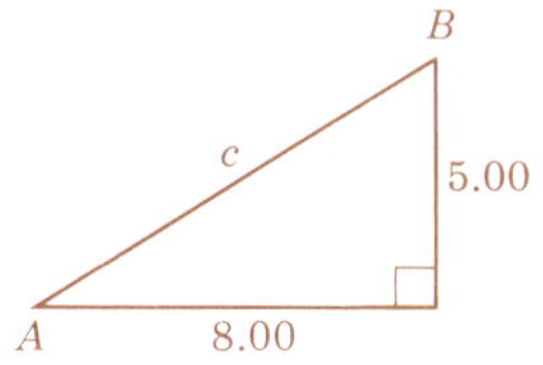

Since we know two sides, we can follow the suggestion of step 2 to use the Pythagorean theorem for finding the third side. We get

$$c = \sqrt{(5.00)^2 + (8.00)^2}$$

— given sides

$$= 9.43$$

FIGURE 16-22

Since neither of the acute angles is known, we will *use a trigonometric ratio to find one of them*. From the figure we see that

$$\tan A = \frac{5.00}{8.00}$$

which means that $\tan A = 0.6250$. From the table or by using a calculator we can now establish that $\angle A = 32.0°$ (to the nearest $0.1°$).

We can now find $\angle B$ by solving

$$\tan B = \frac{8.00}{5.00}$$

With $\tan B = 1.60$ we get $\angle B = 58.0°$. The triangle is now solved since we know all six parts (three angles and three sides). (The use of the tangent ratio is preferred here since both sine and cosine would have used the derived value of $c = 9.43$; if possible, we should not use derived values for other calculations.)　　□

Using the trigonometric ratios in applied problems involves the same approach as solving triangles, although *it is usually one particular part of the triangle that we need to determine*. The following examples illustrate some of the basic applications.

EXAMPLE C   A section of a highway is 4.20 km long and rises along a uniform grade which makes an angle of 3.2° with the horizontal. What is the change in elevation? (See Figure 16-23, which is not drawn to scale.)

We seek the value of the side opposite the known angle and we know the hypotenuse. Since *we are involved with "opposite" and hypotenuse*, the solution is most directly obtained through use of the sine ratio.

given

$$\sin 3.2° = \frac{x}{4.20} \quad \begin{array}{l} \leftarrow \text{required opposite side} \\ \leftarrow \text{given hypotenuse} \end{array}$$

$$x = 4.20 \sin 3.2°$$

$$= 0.234 \text{ km}$$

The highway therefore changes in elevation by a height of 0.234 km.   □

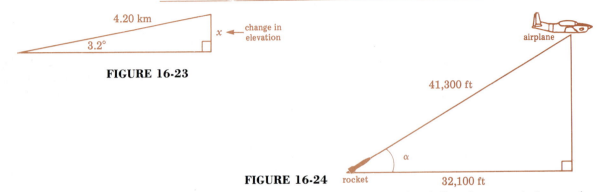

FIGURE 16-23

FIGURE 16-24

EXAMPLE D   In a military training exercise, a rocket is aimed directly at an airplane as in Figure 16-24. If the rocket were to be fired, what angle would its path make with the ground?

Since we know the *hypotenuse and the adjacent side* of the desired angle, we can easily find that angle by using the cosine ratio.

required
angle

$$\cos \alpha = \frac{32,100}{41,300} \quad \begin{array}{l} \leftarrow \text{given adjacent side} \\ \leftarrow \text{given hypotenuse} \end{array}$$

Using the [INV] and [COS] keys on a calculator, we find that the angle (to the nearest 0.1°) is 39.0°.   □

In Example D we expressed the angle by rounding it to the nearest 0.1°. If the sides of the triangle are given with three significant digits, the derived angles should be rounded to the nearest 0.1°. If the sides have four significant digits, the derived angles should be rounded to the nearest 0.01°, and so on.

EXAMPLE E   If a certain airplane loses its only engine, it can glide along a path which makes an angle of 18.0° with the level ground (see Figure 16-25). If the plane is 5500 ft above the ground, what is the maximum horizontal distance it can go?

    In Figure 16-25 we seek the value of the distance x. Since the known side is *opposite* the known angle, and the side to be determined is *adjacent*, the solution can be completed by using the tangent.

$$\tan 18.0° = \frac{5500}{x} \quad \xleftarrow{\text{given opposite side}} \atop \xleftarrow{\text{required adjacent side}}$$

$$x = \frac{5500}{\tan 18.0°} = 17{,}000 \text{ ft}$$

The horizontal distance is 17,000 ft.   □

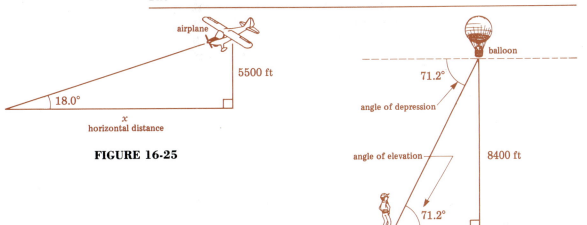

FIGURE 16-25

FIGURE 16-26

EXAMPLE F   While studying the behavior of wind, one weather observer flies a hot air balloon while another observer remains at the point of departure (see Figure 16-26). The **angle of depression** (the angle between the horizontal and the line of sight, *downward with respect to the balloon observer*) is measured by the pilot to be 71.2° as the balloon flies into clouds 8400 ft above ground. Find the horizontal distance between the ground observer and the point directly below the balloon when it entered the clouds.

    In Figure 16-26 the angle of depression is equal to the **angle of elevation** (the angle between the horizontal and the line of sight, *upward with respect to the ground observer*) since alternate interior angles are equal. We therefore know that the angle of elevation is also 71.2° as shown. We now have a right triangle with a known angle and *the opposite and adjacent sides involved*. Using the tangent we get

$$\tan 71.2° = \frac{8400}{x} \quad \xleftarrow{\text{given opposite side}} \atop \xleftarrow{\text{required adjacent side}}$$

$$x = \frac{8400}{\tan 71.2°} = 2900 \text{ ft} \qquad\qquad □$$

EXAMPLE G   A television antenna is on the roof of a building. From a point on the ground 36.0 ft from the building, the angles of elevation of the top and the bottom of the antenna are 51.0° and 42.0°, respectively. How tall is the antenna?

In Figure 16-27 we let x represent the distance from the top of the building to the ground and y represent the distance from the top of the antenna to the ground. *The solution will be the value of* **y − x.** We proceed to find the values of y and x.

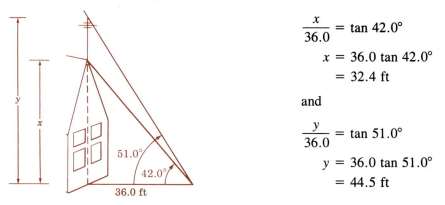

$$\frac{x}{36.0} = \tan 42.0°$$

$$x = 36.0 \tan 42.0°$$

$$= 32.4 \text{ ft}$$

and

$$\frac{y}{36.0} = \tan 51.0°$$

$$y = 36.0 \tan 51.0°$$

$$= 44.5 \text{ ft}$$

FIGURE 16-27

The length of the antenna is $y - x = 12.1$ ft. We were able to determine the length of the antenna without entering the building or climbing onto the roof.   □

Exercises 16-3

In Exercises 1 through 16, solve the triangles with the given parts. The parts are indicated in Figure 16-28. (Angles are indicated only by the appropriate capital letter.)

1. $A = 30.0°$, $a = 12.0$

2. $A = 45.0°$, $b = 16.0$

3. $B = 56.3°$, $c = 22.5$

4. $B = 17.1°$, $a = 15.7$

5. $A = 76.8°$, $c = 31.4$

6. $B = 35.7°$, $b = 1.45$

7. $a = 0.650$, $c = 1.35$

8. $a = 4.70$, $b = 7.40$

9. $b = 5.80$, $c = 45.0$

10. $a = 734$, $b = 129$

11. $a = 9.72$, $c = 10.8$

12. $b = 0.195$, $c = 0.321$

13. $A = 7.0°$, $b = 15.3$

14. $B = 84.5°$, $c = 1730$

15. $a = 65.1$, $b = 98.3$

16. $b = 1.89$, $c = 7.14$

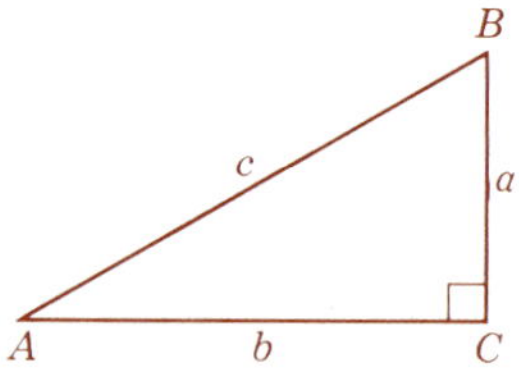

FIGURE 16-28

In Exercises 17 through 48, solve the given problems by finding the appropriate part of a triangle. In each problem, draw a rough sketch.

17. The angle of elevation to the top of a statue is 10.0° from a point 165 ft from the base of the statue. Find the height of the statue. See Figure 16-29.

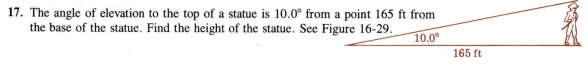

FIGURE 16-29

18. The angle of elevation of the sun is 51.3° at the time a tree casts a shadow 23.7 m long. Find the height of the tree.

19. A robot is on the surface of Mars. The angle of depression from a camera in the robot to a rock on the surface of Mars is 23.7°. The camera is 122 cm above the surface. How far is the camera from the rock?

20. A guy wire whose grounded end is 16.0 ft from the pole it supports makes an angle of 56.0° with the ground. How long is the wire?

21. The angle of inclination of a road is often expressed as *percent grade,* which is the vertical rise divided by the horizontal run (expressed as a percent). See Figure 16-30. A 6% grade corresponds to a road that rises 6 ft for every 100 ft along the horizontal. Find the angle of inclination corresponding to a 6% grade.

22. The angle of inclination of a road is 6.5°. Find the percent grade and the vertical rise that corresponds to a distance of 150 ft along the horizontal. (See Exercise 21.)

23. The bottom of a picture on a wall is on the same level as a person's eye. The picture is 135 cm high, and the angle of elevation to the top of the picture is 23.0°. How far is the person's eye from the bottom of the picture? See Figure 16-31.

24. Along the shore of a river, from a rock at a height of 300 ft above the river, the angle of depression of the closest point on the opposite shore is 12.0°. What is the distance across the river from the base of the height to the closest point on the opposite shore?

25. A point near the top of the Leaning Tower of Pisa is about 50.5 m from a point at the base of the tower (measured along the tower). This top point is also directly above a point on the ground 4.25 m from the same base point. What angle does the tower make with the ground? See Figure 16-32.

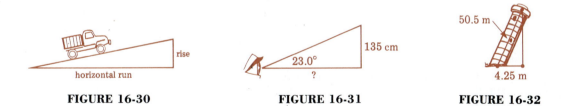

| FIGURE 16-30 | FIGURE 16-31 | FIGURE 16-32 |

26. A roof rafter is 5.25 m long (neglecting the overhang), and its upper end is 1.70 m above the lower end. Find the angle between the rafter and the horizontal.

27. Aviation weather reports include the *ceiling* which is the distance between the ground and the bottom of overcast clouds. A ground observer is 5000 ft from a spotlight which is aimed vertically. The angle of elevation between the ground and the spot of light on the clouds is measured by the observer to be 38.7°. What is the ceiling? See Figure 16-33.

28. A ship is traveling toward a port from the west when it changes course by 18.0° to the north. It then travels 23.0 mi until it is due north of the port. How far was it from port when it turned? See Figure 16-34.

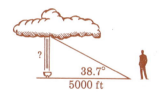
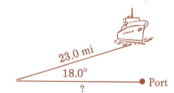

FIGURE 16-33   FIGURE 16-34

29. A draftsman sets the legs of a pair of dividers so that the angle between them is 36.0°. What is the distance between the points if each leg is 4.75 in. long?

30. For an alternating-current circuit, we represent resistance as one leg of a right triangle. The reactance is the other leg, and the impedance is represented by the hypotenuse. In one such triangle, the acute angles are 46.2° and 43.8°. If the smallest side corresponds to a reactance of 29.3 Ω, find the resistance (in ohms) and impedance (also in ohms).

31. A person is building a swimming pool 12.5 m long. The depth at one end is to be 1.00 m and the depth at the other end is to be 2.50 m. Find, to the nearest 0.1°, the angle between the bottom of the pool and the horizontal. See Figure 16-35.

32. A shelf is supported by a straight 65.0-cm support attached to the wall at a point 53.5 cm below the bottom of the shelf. What angle does the support make with the wall? See Figure 16-36.

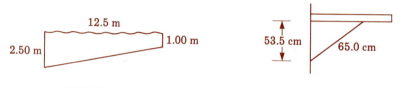

FIGURE 16-35   FIGURE 16-36

33. The total angle through which a pendulum 42.5 in. long swings is 6.2°. Find the horizontal distance between the extreme positions of the pendulum.

34. A television transmitter antenna is 600 ft high. If the angle between the guy wires (attached at the top) and the antenna is 55.0°, how long are the guy wires?

35. An observer in a helicopter 800 ft above the ground notes that the angle of depression of an object is 26.0°. How far from directly below the helicopter is this object? See Figure 16-37.

36. One way of finding the distance from the earth to the moon is indicated in Figure 16-38. From point P the moon is directly overhead, and from point Q the moon is just visible. Both points are on the equator. The angle at E is about 89.0°. Given that the radius of the earth is 6360 km, how far is it to the moon (PM)?

37. An astronaut observes two cities which are 45.0 km apart; the angle between the lines of sight is 10.6°. Given that he is the same distance from each, how high above the surface of the earth is he? See Figure 16-39 and ignore the earth's curvature.

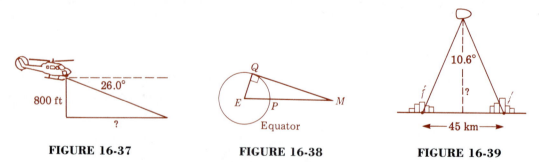

FIGURE 16-37   FIGURE 16-38   FIGURE 16-39

38. A jet cruising at 610 km/h climbs at an angle of 12.0°. What is its gain in altitude in 3 min?

39. What is the distance between two points on a wheel 38.0° apart on the circumference, given that the diameter of the wheel is 28.0 in.? See Figure 16-40.

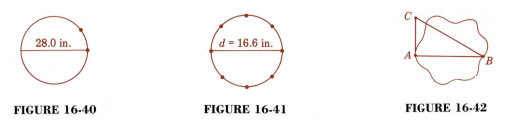

FIGURE 16-40   FIGURE 16-41   FIGURE 16-42

40. Eight bolt holes (spaced equally) are drilled on the circumference of a circle whose diameter is 16.6 in. What is the center-to-center distance between adjacent holes? See Figure 16-41.

41. A surveyor wishes to measure the width of a large lake. He sights a point B on the opposite side of the lake from point A. He then measures off 200 ft from point A to point C, such that $CA \perp AB$. He then determines that $\angle ABC = 8.5°$. How wide is the lake? See Figure 16-42.

42. The angle of depression of a fire noticed directly north of a 24.5-m fire tower is 5.5°. The angle of depression of a stream running east to west, and also north of the tower, is 13.4°. How far is the fire from the stream?

43. From an airplane 5000 ft above the surface of the water, a pilot observes two boats directly ahead. The angles of depression are 20.0° and 12.0°. How far apart are the boats?

44. One observer is directly below a jet flying at an altitude of 2575 ft. At the same time, a second observer measures the angle between the ground and the jet and finds that angle to be 28.68°. What is the distance between the two observers?

45. A surveyor wants to determine the height of a vertical cliff without actually climbing it. She walks 263 ft away from the bottom of the cliff and measures the angle between the ground and the top of the cliff. If this angle is 52.0°, how high is the cliff?

46. An aircraft encoding altimeter and a radar system indicate that a plane is 7350 ft high when it is 17,400 ft from the end of the active runway. If the plane flies directly toward the end of the runway, what angle does its path make with the ground? See Figure 16-43.

47. An observer in a lighthouse uses an instrument to record the angle of depression of ships. If the angle of depression is denoted by A and the observer is 88 ft above sea level, find a formula for the distance d between the ship and the base of the lighthouse.

48. A fire watchtower is built on the top edge of a cliff. From a point on the ground 307 ft from the base of the cliff, the angles of elevation of the top and bottom of the tower are 56.4° and 54.4°, respectively. How tall is the cliff and how tall is the tower? (See Example G.)

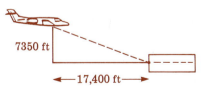

7350 ft

←— 17,400 ft —→

FIGURE 16-43

Chapter 16 Formulas

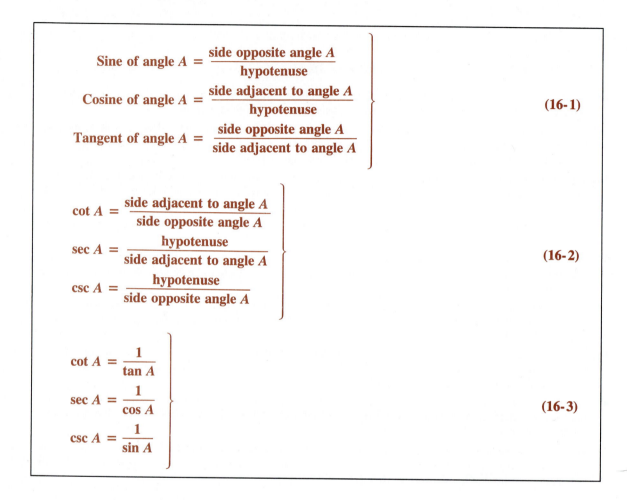

$$\text{Sine of angle } A = \frac{\text{side opposite angle } A}{\text{hypotenuse}}$$

$$\text{Cosine of angle } A = \frac{\text{side adjacent to angle } A}{\text{hypotenuse}}$$

$$\text{Tangent of angle } A = \frac{\text{side opposite angle } A}{\text{side adjacent to angle } A}$$

(16-1)

$$\cot A = \frac{\text{side adjacent to angle } A}{\text{side opposite angle } A}$$

$$\sec A = \frac{\text{hypotenuse}}{\text{side adjacent to angle } A}$$

$$\csc A = \frac{\text{hypotenuse}}{\text{side opposite angle } A}$$

(16-2)

$$\cot A = \frac{1}{\tan A}$$

$$\sec A = \frac{1}{\cos A}$$

$$\csc A = \frac{1}{\sin A}$$

(16-3)

16-4 Review Exercises for Chapter 16

In Exercises 1 through 8, use a calculator to determine the values of the given trigonometric ratios. Round results to three significant digits.

1. sin 47.3°   **2.** cos 61.2°   **3.** tan 7.4°   **4.** sin 52.6°

5. cot 83.4°   **6.** csc 29.8°   **7.** sec 30.5°   **8.** cot 75.9°

In Exercises 9 through 16, use a calculator to determine the value of the angle A to the nearest 0.1°.

9. cos A = 0.8660   **10.** tan A = 3.732   **11.** sin A = 0.7071   **12.** cos A = 0.0175

13. cot A = 19.08   **14.** sec A = 1.766   **15.** csc A = 1.196   **16.** csc A = 1.472

In Exercises 17 through 20, find the indicated trigonometric ratios in fractional form for the angles of the triangle in Figure 16-44.

17. sin *A*, cos *B* **18.** tan *B*, cos *A* **19.** sec *B*, tan *A* **20.** cot *A*, csc *B*

In Exercises 21 through 24, find the indicated trigonometric ratios in decimal form for the angles of the triangle in Figure 16-45.

21. cos *A*, tan *B* **22.** cot *B*, sin *A* **23.** sin *B*, tan *A* **24.** sec *A*, csc *B*

In Exercises 25 through 32, determine the indicated trigonometric ratios. The listed sides are those shown in Figure 16-46. Do not use Table 3 in Appendix D or the $\boxed{\text{SIN}}$ $\boxed{\text{COS}}$ $\boxed{\text{TAN}}$ *keys on a calculator.*

25. $a = 12.0$, $b = 5.00$. Find cos *A* and tan *B*. **26.** $b = 9.00$, $c = 15.0$. Find sec *A* and cot *B*.

27. $a = 8.70$, $b = 2.30$. Find tan *A* and cos *B*. **28.** $a = 125$, $c = 148$. Find sin *A* and sec *B*.

29. $a = 0.0890$, $c = 0.0980$. Find cot *A* and sin *B*. **30.** $b = 1670$, $c = 4200$. Find csc *A* and csc *B*.

31. $b = 0.0943$, $c = 0.105$. Find cos *A* and tan *B*. **32.** $a = 8.60$, $b = 7.90$. Find sin *A* and cot *B*.

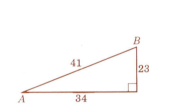

FIGURE 16-44 **FIGURE 16-45** **FIGURE 16-46**

In Exercises 33 through 40, find, to the nearest 0.1°, the indicated angle. The parts of the triangle are those shown in Figure 16-46.

33. $a = 8.00$, $c = 13.0$. Find *A*. **34.** $b = 51.0$, $c = 58.0$. Find *A*.

35. $a = 5.60$, $b = 1.30$. Find *B*. **36.** $a = 0.780$, $c = 2.40$. Find *B*.

37. $b = 3420$, $c = 7200$. Find *B*. **38.** $a = 0.00670$, $b = 0.0156$. Find *A*.

39. $a = 4910$, $b = 3650$. Find *A*. **40.** $b = 0.860$, $c = 0.915$. Find *B*.

In Exercises 41 through 56, use a scientific calculator to solve the triangles with the given parts. The given parts are shown in Figure 16-46.

41. $A = 21.0°$, $c = 6.93$ **42.** $B = 18.0°$, $a = 0.360$

43. $B = 32.7°$, $b = 45.9$ **44.** $B = 7.4°$, $c = 1890$

45. $a = 8.70$, $b = 5.20$ **46.** $a = 0.00760$, $b = 0.0120$

47. $a = 97.0$, $c = 108$ **48.** $b = 17.4$, $c = 54.0$

49. $A = 37.25°$, $c = 13,872$ **50.** $B = 64.333°$, $b = 5280$

51. $A = 0.02°$, $a = 0.00425$ **52.** $B = 89.966°$, $c = 0.03566$

53. $a = 127.35$, $b = 192.60$ **54.** $a = 0.001372$, $c = 0.005605$

55. $b = 2.13520$, $c = 112.503$ **56.** $b = 0.0357$, $c = 112.03$

In Exercises 57 through 60, use the given trigonometric ratios to find the indicated trigonometric ratios.

57. If $\tan A = 0.5774$, find $\sin A$.

58. If $\cos A = 0.4924$, find $\sin A$.

59. If $\sin A = 0.9917$, find $\tan A$.

60. If $\csc A = 3.168$, find $\sec A$.

In Exercises 61 through 88, solve the problems.

61. An approximate equation found in the diffraction of light through a narrow opening is

$$\sin \theta = \frac{\lambda}{d}$$

where λ (the Greek lambda) is the wavelength of light and d is the width of the opening. If $\theta = 1.1°$ and $d = 32.0 \ \mu\text{m}$, what is the wavelength of the light?

62. An equation used for the instantaneous value of electric current in an alternating-current circuit is

$$i = I_m \cos \theta$$

Calculate i for $I_m = 56.0$ mA and $\theta = 10.5°$.

63. In determining the height h of a building which is 220 m distant, a surveyor may use the equation

$$h = 220 \tan \theta$$

where θ is the angle of elevation to the top of the building. What should θ be if $h = 130$ m?

64. Find the amount of work W done by a force F if

$$W = Fd \cos \theta$$

and $F = 57.3$ lb, $d = 23.8$ ft, and $\theta = 27.3°$.

65. A conical pendulum has radius r given by

$$r = L \sin \theta$$

Find r if $L = 1.83$ m and $\theta = 16.8°$.

66. A formula used with a certain type of gear is

$$D = \frac{1}{4}N \sec \theta$$

where D is the pitch diameter of the gear, N is the number of teeth on the gear, and θ is called the spiral angle. If $D = 6.75$ in. and $N = 20$, find θ.

67. A blueprint drawing of a room is a rectangle with sides 4.52 cm and 3.27 cm. What is the angle between the longer wall and a diagonal across the room?

68. When the angle of elevation of the sun is 40.0°, what is the length of the shadow of a tree 65.0 ft tall? See Figure 16-47.

69. The distance from a point on the shore of a lake to an island is 3500 ft. From a point directly above the shore on a cliff the angle of depression to the island is 15.0°. How high is the cliff?

70. The distance from point B to point C along the shore of a river is 500 m. Point A is directly across from point C on the opposite shore. Given that $\angle ABC$ is 21.9°, how wide is the river? See Figure 16-48.

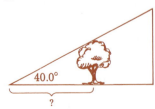

FIGURE 16-47

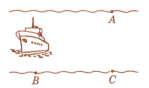

FIGURE 16-48

71. A standard sheet of plywood is rectangular with dimensions 4.00 ft by 8.00 ft. If such a sheet is cut in one straight line from one corner to the opposite corner, find the angles created.

72. A guy wire supporting a television tower is grounded 250 m from the base of the tower. Given that the wire makes an angle of 29.0° with the ground, how high up on the tower is it attached?

73. The span of a roof is 30.0 ft. Its rise is 8.00 ft at the center of the span. What is the angle the roof makes with the horizontal?

74. A roadway rises 85.0 ft for every 1000 ft along the road. Determine the angle of inclination of the roadway.

75. A jet climbs at an angle of 35.0° while traveling at 600 km/h. How long will it take to climb to an altitude of 10,000 m?

76. An observer 3000 ft from the launch pad of a rocket measures the angle of inclination to the rocket soon after lift off to be 65.0°. How high is the rocket, assuming it has moved vertically?

77. An airplane flies a distance of 240 mi in a direction that is 12° east of due north. At that point, the pilot turns to the west. How far must the plane fly in the westerly direction in order to be at a point directly north of the departure point? See Figure 16-49.

78. An astronomer measures the angle between the center of the moon and the edge of the moon and obtains a value of 0°15′. If the distance between the observer and the center of the moon is 237,000 mi, find the radius of the moon.

79. A parachutist exits an airplane at a height of 4250 ft directly over his target. The path of his descent is a straight line which makes an angle of 82.3° with the ground. By what distance does the parachutist miss his target? See Figure 16-50.

FIGURE 16-49   FIGURE 16-50

80. A searchlight is 520 ft from a building. The beam lights up the building at a point 150 ft above the ground. What angle does the beam make with the ground?

81. A parcel of land is as shown in Figure 16-51. What is the angle θ between the road which runs along the front section and the side boundary with length 124 m?

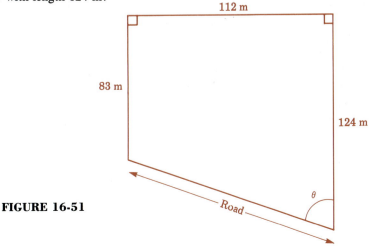

FIGURE 16-51

82. Two laser beams from the same source are transmitted to two points 100 m apart on a flat surface. One beam is perpendicular to the surface and the other makes an angle of 87.5° with the surface. How far from the surface is the source? See Figure 16-52.

83. While on a mountain top 2.0 mi high, the angle of depression to the horizon is measured as 1.8°. Use this information to estimate the radius of the earth.

84. The side of an equilateral triangle is 24.0 in. How long is its altitude?

85. A certain machine part is a regular hexagon (six sides) 0.844 in. on a side. What is the distance across from one side to the opposite side? See Figure 16-53.

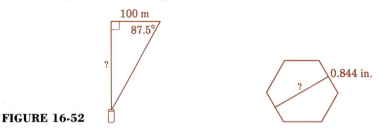

FIGURE 16-52   FIGURE 16-53

86. A storage room 9.0 ft wide, 12.0 ft long, and 8.0 ft high contains a beam which extends from a corner on the floor to the opposite corner on the ceiling. This is the largest beam the room can hold. Find the angle this beam makes with the floor.

87. Given that A and B are acute angles, is $\cos A > \cos B$ if $A > B$? Is $\tan A > \tan B$ if $A > B$?

88. Given that A and B are acute angles, is $\sin A > \sin B$ if $A > B$? Is $\cot A > \cot B$ if $A > B$?

CHAPTER SEVENTEEN

Trigonometry with Any Angle

Chapter 16 introduced the study of trigonometry and showed how right triangles and the trigonometric ratios may be applied to technical problems. It is possible to generalize the definitions of the trigonometric ratios so that we are not restricted to acute angles or right triangles. This chapter presents these general definitions and shows how we may solve problems involving **oblique triangles** *(triangles which do not contain right angles)*. In this chapter we will be able to solve a greater variety of different applied problems because we will not have the restriction of working with right triangles only.

In Chapter 6 we solved a problem related to finding the distance a plane must fly in order to avoid a restricted area. That solution was relatively easy since it involved the circumference of a circle. Suppose that in order to avoid a restricted area, a plane must fly from point A to point C to point B as shown in Figure 17-1. Compare this distance to the distance traveled by a military plane that is cleared to fly directly from point A to point B. This problem, which will be solved later in the chapter, is typical of the type of problem that we can solve using the concepts we will develop.

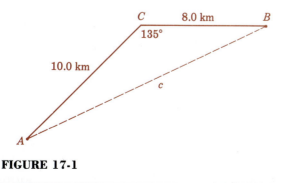

FIGURE 17-1

17-1 Signs of the Trigonometric Functions

We begin our study of oblique triangles by positioning angles so that they can be discussed in terms of coordinate axes. *If the initial side of an angle extends from the origin to the right on the positive x axis in the rectangular coordinate system, the angle is said to be in* **standard position.** See Figure 17-2. The angle is then determined by the position of the terminal side. If the terminal side is in the first quadrant, the angle is called a "first-quadrant angle." Similar terms are used when the terminal side is in the other quadrants. *If the terminal side coincides with one of the axes, the angle is a* **quadrantal angle.**

The measure of an angle in standard position corresponds to the amount of rotation generated when the terminal side is rotated from the initial side. *If the terminal side is rotated in a counterclockwise direction, the angle is said to be* **positive.** *If the rotation is clockwise, the angle is* **negative,** as in Figure 17-2.

EXAMPLE A   An angle of 150° in standard position (see Figure 17-2) has its terminal side in the second quadrant so that 150° *is a second-quadrant angle.* The terminal side is rotated 150° in the counterclockwise direction as shown.   ☐

EXAMPLE B   See Figure 17-2. An angle of −60° in standard position has its terminal side in the fourth quadrant. The terminal side is rotated 60° in the *clockwise* direction as shown.   ☐

When an angle is in standard position, *the terminal side is determined if we know any point, other than the origin, on the terminal side.*

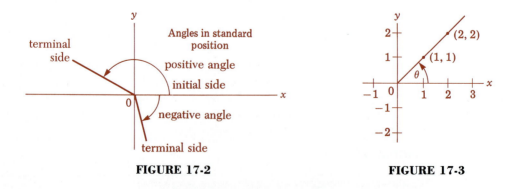

FIGURE 17-2   **FIGURE 17-3**

EXAMPLE C See angle θ in Figure 17-3. The angle θ is in standard position and the terminal side passes through the points (1, 1) and (2, 2) as well as infinitely many other points. Knowing that the terminal side passes through any one of these points makes it possible to determine the terminal side. □

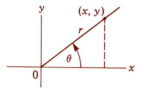

FIGURE 17-4

If we were to stipulate for an angle β that the terminal side passes through (0, 3), we would know that β is one example of a quadrantal angle. Since the terminal side would coincide with one of the axes (the positive y axis in this case), β is a quadrantal angle.

Using angles that are in standard position, we can present more general definitions that are consistent with the definitions of the trigonometric ratios already presented in Chapter 16. Refer to Figure 17-4, where we show three important distances for a given point on the terminal side of the angle. *They are the* **abscissa** *(x coordinate), the* **ordinate** *(y coordinate), and the* **radius vector.** *The radius vector has a distance that extends from the origin to the point.* The radius vector is designated as r, as shown in Figure 17-4.

Using the abscissa (x coordinate), the ordinate (y coordinate), and radius vector $r = \sqrt{x^2 + y^2}$, we define the trigonometric functions as follows.

$$\text{sine } \theta = \frac{y}{r} \qquad \text{cosine } \theta = \frac{x}{r} \qquad \text{tangent } \theta = \frac{y}{x}$$

$$\text{cotangent } \theta = \frac{x}{y} \qquad \text{secant } \theta = \frac{r}{x} \qquad \text{cosecant } \theta = \frac{r}{y}$$

(17-1)

The results given in Eqs. (17-1) will be the same for any point on a given terminal side. The ratios will be different for a different angle having a different terminal side.

The trigonometric *ratios*, defined only in terms of the sides of a right triangle in Chapter 16, are special cases of these more general trigonometric *functions.* **These functions depend on the size of an angle in standard position and may be evaluated for angles of any magnitude** and with a terminal side in any quadrant.

As with the trigonometric ratios, the names of the various functions are abbreviated for convenience. These abbreviations are sin θ, cos θ, tan θ, cot θ, sec θ, and csc θ. Note that a given function is not defined if the denominator is zero. This occurs for tan θ and sec θ when $x = 0$ and for cot θ and csc θ when $y = 0$. In all cases *we assume that $r > 0$*. If $r = 0$, there would be no terminal side and therefore no angle.

EXAMPLE D   Determine the values of the trigonometric functions of the angle with a terminal side passing through the point (3, 4).

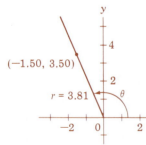

FIGURE 17-5

By placing the angle in standard position, as shown in Figure 17-5, and drawing the terminal side through (3, 4), we see that r is found by applying the Pythagorean theorem. Thus

$$r = \sqrt{3^2 + 4^2} = 5$$

Using the values $x = 3$, $y = 4$, and $r = 5$, we find the following values:

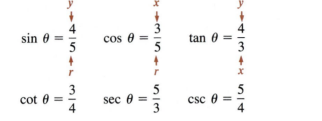

$$\sin \theta = \frac{4}{5} \qquad \cos \theta = \frac{3}{5} \qquad \tan \theta = \frac{4}{3}$$

$$\cot \theta = \frac{3}{4} \qquad \sec \theta = \frac{5}{3} \qquad \csc \theta = \frac{5}{4}$$

EXAMPLE E   Determine the values of the trigonometric functions of the angle with a terminal side which passes through the point (−1.50, 3.50). See Figure 17-6.

From the Pythagorean theorem, we find that

$$r = \sqrt{x^2 + y^2} = \sqrt{(-1.50)^2 + (3.50)^2} = \sqrt{2.25 + 12.25}$$
$$= \sqrt{14.50} = 3.81$$

(−1.50, 3.50)

$r = 3.81$

FIGURE 17-6

Using the values $x = -1.50$, $y = 3.50$, and $r = 3.81$, we have the following results for the trigonometric functions.

$$\sin \theta = \frac{3.50}{3.81} = 0.919 \qquad \cos \theta = \frac{-1.50}{3.81} = -0.394$$

$$\tan \theta = \frac{3.50}{-1.50} = -2.33 \qquad \cot \theta = \frac{-1.50}{3.50} = -0.429$$

$$\sec \theta = \frac{3.81}{-1.50} = -2.54 \qquad \csc \theta = \frac{3.81}{3.50} = 1.09$$

▶   From Example E we see that *the value of a trigonometric function can sometimes be negative.* It is important that we determine the correct sign of each function, and that depends on the location of the terminal side of the angle. We will now determine the sign of each function in each quadrant. In Section 17-2 we will consider the general technique of determining the actual values of the trigonometric functions of any angle.

The sign of sin θ depends on the sign of the y coordinate of the point on the terminal side. This is because sin θ = y/r and r is *positive*. We know that y is positive if the point defining the terminal side is above the x axis and that y is negative if this point is below the x axis. As a result, if the terminal side of the angle is in the first or second quadrant, the value of sin θ will be positive, but if the terminal side is in the third or fourth quadrant, sin θ is *negative*.

Quadrant

	I	II	III	IV
sin θ	+	+	−	−

EXAMPLE F The value of sin 20° is positive, since the terminal side of 20° is in the first quadrant. The value of sin (−200°) is positive, since the terminal side of −200° is in the second quadrant. The values of sin 200° and sin 340° are negative, since the terminal sides of these angles are in the third and fourth quadrants, respectively. □

Since cos θ = x/r, the sign of cos θ depends upon the sign of x. Since x is positive in the first and fourth quadrants, cos θ is *positive* in these quadrants. In the same way, cos θ is *negative* in the second and third quadrants.

Quadrant

	I	II	III	IV
cos θ	+	−	−	+

EXAMPLE G The values of cos 20° and cos (−30°) are positive, since these angles are first- and fourth-quadrant angles, respectively. The values of cos 160° and cos 200° are negative, since these angles are second- and third-quadrant angles, respectively. □

The sign of tan θ depends upon the ratio of y to x. In the first quadrant both x and y are positive, and therefore the ratio y/x is positive. In the third quadrant both x and y are negative, and therefore the ratio y/x is *positive*. In the second and fourth quadrants either x or y is positive and the other is negative, and so the ratio of y/x is *negative*.

Quadrant

	I	II	III	IV
tan θ	+	−	+	−

EXAMPLE H The values of tan 20° and tan 200° are positive, since the terminal sides of these angles are in the first and third quadrants, respectively. The values of tan 160° and tan (−20°) are negative, since the terminal sides of these angles are in the second and fourth quadrants, respectively. □

Since csc θ is defined in terms of r and y, as is sin θ, the sign of csc θ is the same as that of sin θ. For the same reason, cot θ has the same sign as tan θ, and sec θ has the same sign as cos θ. A method for remembering the signs of the functions in the four quadrants is as follows:

Sin Csc	All
Tan Cot	Cos Sec

positive functions

FIGURE 17-7

All functions of first-quadrant angles are positive. The sin θ *and* csc θ *are positive for second-quadrant angles. The* tan θ *and* cot θ *are positive for third-quadrant angles. The* cos θ *and* sec θ *are positive for fourth-quadrant angles. All others are negative.* See Figure 17-7.

This discussion does not include the quadrantal angles—those angles with terminal sides on one of the axes. They are discussed in Section 17-2.

EXAMPLE I   sin 50°, sin 150°, sin (−200°), cos 300°, cos (−40°), tan 220°, tan (−100°), cot 260°, cot (−310°), sec 280°, sec (−37°), csc 140°, and csc (−190°) are all positive.

sin 190°, sin 325°, cos 100°, cos (−95°), tan 172°, tan 295°, cot 105°, cot (−6°), sec 135°, sec (−135°), csc 240°, and csc 355° are all negative.

□

EXAMPLE J   If the angle of 690° is in standard position, use a calculator to find the quadrant in which its terminal side lies.

Using a calculator we enter 690 $\boxed{\text{SIN}}$ to get −0.5, and we then enter 690 $\boxed{\text{COS}}$ to get 0.86602540. Since sin 690° is negative, we know that we have a third- or fourth-quadrant angle. Since cos 690° is positive, we know that we have a first- or fourth-quadrant angle. Both conditions must be met so that 690° is a fourth-quadrant angle.

□

EXAMPLE K   If the index of refraction of a glass plate is found by evaluating n where $n = \sin A/\sin B$, what is known about the sign of that index if angles A and B are both greater than 0° and less than 90°?

Since the sine is positive for first-quadrant angles, sin A and sin B are both positive so that the index of refraction is always positive.

□

Exercises 17-1

In Exercises 1 through 4, draw the given angles in standard position.

1. 30°, 135°   **2.** 100°, 240°   **3.** 200°, −60°   **4.** −30°, 400°

In Exercises 5 through 12, assume that the terminal side of θ passes through the given point and find the values of the given trigonometric ratios in fractional form.

5. (5, 12);   sin θ, tan θ   **6.** ($\sqrt{3}$, 1);   cos θ, cot θ   **7.** (−3, 2);   sec θ, sin θ

8. (−8, 15);   tan θ, csc θ   **9.** (−1, −1);   cot θ, cos θ   **10.** (−3, −4);   sin θ, sec θ

11. (6, −8);   cos θ, sin θ   **12.** (5, −6);   csc θ, cos θ

In Exercises 13 through 20, assume that the terminal side of θ passes through the given point and find the values of the given trigonometric ratios in decimal form.

13. (1.00, 3.00);   cos θ, cot θ   **14.** (2.50, 1.30);   sin θ, tan θ   **15.** (−15.0, 6.20);   tan θ, csc θ

16. (−2.30, 7.40);   sec θ, sin θ   **17.** (−140, −170);   sin θ, sec θ   **18.** (−0.175, −1.05);   cot θ, csc θ

19. (27.3, −17.5);   csc θ, cos θ   **20.** (1.75, −7.50);   cos θ, sin θ

In Exercises 21 through 28, determine the algebraic sign of the given trigonometric functions.

21. sin 45°, cos 130°, tan 350°   **22.** sin (−25°), csc 120°, tan 200°

23. cos 250°, sec 160°, cot (−10°)   **24.** sin 182°, csc (−12°), cos 95°

25. tan 260°, csc 72°, cos 380°   **26.** cot (−93°), sec 295°, tan 110°

27. sin 718°, cot (−570°), sec 520°   **28.** cos 212°, tan 275°, sin (−380°)

In Exercises 29 through 36, determine the quadrant in which the terminal side of θ lies, subject to the given conditions.

29. sin θ is negative, cos θ is positive

30. sin θ is negative, tan θ is negative

31. cos θ is positive, cot θ is positive

32. cos θ is positive, csc θ is negative

33. csc θ is positive, cot θ is negative

34. csc θ is negative, tan θ is negative

35. cot θ is negative, sec θ is negative

36. cot θ is positive, csc θ is negative

In Exercises 37 through 44, use the SIN *and* COS *keys on a calculator to determine the quadrant in which the terminal side of the given angle θ lies. The quadrant can be determined by using the calculator to find the signs of sin θ and cos θ.*

37. θ = 852°　　**38.** θ = 963°　　**39.** θ = −986°　　**40.** θ = −1265°

41. θ = 1537°　　**42.** θ = 1506°　　**43.** θ = −2070°　　**44.** θ = −2280°

17-2 Values of Trigonometric Functions of Any Angle

The trigonometric ratios were introduced in Chapter 16, and in the last section we defined the general trigonometric functions and determined the sign of each function in each of the four quadrants. In this section we shall see how we can find the values of the trigonometric functions of an angle of any magnitude. This will be important when we discuss oblique triangles later in this chapter, and when we discuss graphs of the trigonometric functions in Chapter 19.

Before proceeding, we should realize that it is easy to obtain values of trigonometric functions by pushing the right buttons on a calculator. However, it is critically important to understand the concepts if trigonometry is to be a useful tool which can be applied to technical and scientific analyses. Serious errors and misunderstandings can arise from blind use of calculators.

Two angles in standard position are **coterminal** *if their terminal sides are the same.* It is a property of all angles that *any angle in standard position is coterminal with an angle between* 0° *and* 360°. Since the terminal sides of coterminal angles are the same, *the trigonometric functions of coterminal angles are the same.* The problem of finding the value of the trigonometric function of any angle can therefore be reduced to angles between 0° and 360°.

EXAMPLE A　Coterminal angles

The following pairs of angles are coterminal:

10°	and	370°	**See Fig. 17-8(a)**
90°	and	450°	
350°	and	−10°	
−120°	and	240°	**See Fig. 17-8(b)**

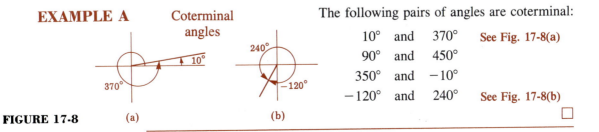

FIGURE 17-8　　(a)　　　　　　　(b)

Since the coterminal angles share the same terminal side, we can conclude that for two coterminal angles, the values of the trigonometric functions are equal. For example, the following equations are valid:

$$\sin 10° = \sin 370°$$
$$\cos 90° = \cos 450°$$
$$\sec 350° = \sec (-10°)$$
$$\tan (-120°) = \tan 240°$$

When attempting to find trigonometric functions of an angle whose terminal side is not in the first quadrant, we will use a **reference angle** denoted by α. In general, *the reference angle α for the given angle θ is the positive acute angle formed by the x axis and the terminal side of θ*. From Figure 17-9 we see that if the given angle θ is in the second quadrant, then $\alpha = 180° - \theta$. Figure 17-10 suggests that for an angle in the third quadrant, the relationship between α and θ is $\alpha = \theta - 180°$. Figure 17-11 shows that when θ is in the fourth quadrant, the reference angle α can be found from $\alpha = 360° - \theta$.

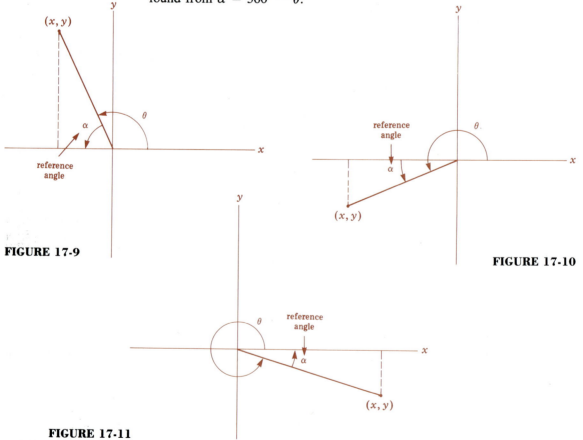

FIGURE 17-9

FIGURE 17-10

FIGURE 17-11

If θ is in the . . .	then the reference angle α can be found from the equation . . .	
second quadrant	$\alpha = 180° - \theta$	**(17-2)**
third quadrant	$\alpha = \theta - 180°$	**(17-3)**
fourth quadrant	$\alpha = 360° - \theta$	**(17-4)**

The reason for using reference angles is that the trigonometric function of an angle θ and its reference angle α are *numerically* the same and might differ only in sign. (But from Section 17-1 we know how to determine the appropriate sign.)

EXAMPLE B   The angle $\theta = 126.9°$ is in the second quadrant and the terminal side passes through the point $(-3, 4)$ as shown in Figure 17-12. The reference angle is $\alpha = 180° - 126.9° = 53.1°$ and we can see from the figure that the x, y, and r values of $(-3, 4)$ and $(3, 4)$ differ only in sign. From Eqs. (17-1) we know that the trigonometric functions depend only on the values of x, y and r. It therefore follows that *the trigonometric functions of* **126.9°** *and* **53.1°** *may differ only in sign.* In the second quadrant, the sine and cosecant are positive while the other four functions are negative. Therefore

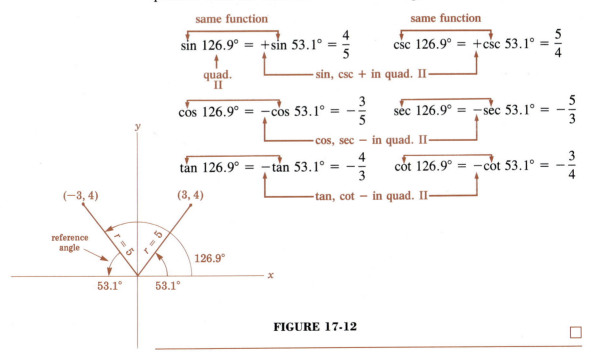

FIGURE 17-12

In Example B we found the values of the trigonometric functions by using Eqs. (17-1) and the fact that the terminal side of 53.1° passes through the point (3, 4) so that $x = 3$, $y = 4$, and $r = 5$. However, it is not necessary to know a point on the terminal side, as illustrated in the next example.

EXAMPLE C　An angle of 250.0° is a third-quadrant angle. Using Eq. (17-3) with $\theta = 250.0°$, we find the reference angle, as shown in Figure 17-13.

$$\alpha = 250.0° - 180° = 70.0°$$

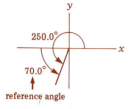

FIGURE 17-13

We can now find the numerical values of the functions of 250.0° *by first finding the values of the same functions of* 70.0°. We then adjust signs to make use of the fact that in the third quadrant the tangent and cotangent are positive while the other four trigonometric functions are negative. We therefore get

$$\sin 250.0° = -\sin 70.0° = -0.940$$
$$\cos 250.0° = -\cos 70.0° = -0.342$$
$$\tan 250.0° = +\tan 70.0° = 2.75$$
$$\cot 250.0° = +\cot 70.0° = 0.364$$
$$\sec 250.0° = -\sec 70.0° = -2.92$$
$$\csc 250.0° = -\csc 70.0° = -1.06$$

same function — reference angle

quadrant III —⌐ └— proper sign for quadrant III

EXAMPLE D　An angle of 318.5° is a fourth-quadrant angle. Using Eq. (17-4) with $\theta = 318.5°$, we find the reference angle, as shown in Figure 17-14.

$$\alpha = 360° - 318.5° = 41.5°$$

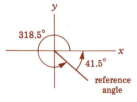

FIGURE 17-14

In the fourth quadrant, the cosine and secant are positive and the other functions are negative. Therefore

$$\sin 318.5° = -\sin 41.5° = -0.663$$
$$\cos 318.5° = +\cos 41.5° = 0.749$$
$$\tan 318.5° = -\tan 41.5° = -0.885$$
$$\cot 318.5° = -\cot 41.5° = -1.13$$
$$\sec 318.5° = +\sec 41.5° = 1.34$$
$$\csc 318.5° = -\csc 41.5° = -1.51$$

quadrant IV —⌐ └— proper sign for quadrant IV

We now demonstrate how we find θ when the value of a function of θ is known. Since the angles of primary importance are those from $0°$ to $360°$, and since each function is positive in two quadrants and negative in the other two quadrants, *we must take care to obtain* both *values of θ for which* $0° < \theta < 360°$.

EXAMPLE E   Given that $\sin \theta = 0.342$, find θ for $0° < \theta < 360°$.

Here we are to find any angles between $0°$ and $360°$ for which $\sin \theta = 0.342$. Since $\sin \theta$ is positive for first- and second-quadrant angles, *there are two such angles.*

Using a calculator, we press the key sequence

0.342 [INV] [SIN] (*19.998772*)

and the display of 19.998772 is rounded to 20.0 so that one angle is $20.0°$. To find the second-quadrant angle for which $\sin \theta = 0.342$, we note that $\sin (180° - 20.0°) = \sin 160°$. Thus the second-quadrant angle is $160.0°$. The values of θ are therefore $20.0°$ and $160.0°$. We note that *the calculator gives only one of the two answers.*   □

In finding an angle when the value of a function is given, we may also know certain additional information. If this information is such that θ is restricted to a particular quadrant, we must be careful to obtain the proper value. This is illustrated in the following example.

EXAMPLE F   Given that $\tan \theta = -1.505$ and $\cos \theta > 0$, find θ for $0° < \theta < 360°$.

Since $\tan \theta$ is negative in the second and fourth quadrants and $\cos \theta$ is positive in the first and fourth quadrants, θ *must be a fourth-quadrant angle.* (That is, where both conditions are met.) Using the calculator sequence

1.505 [+/−] [INV] [TAN] (*−56.397877*)

we get the negative result of $-56.4°$ (rounded). This is a fourth-quadrant angle expressed as a negative angle. The reference angle is therefore $56.4°$, and we can express θ as a fourth-quadrant angle between $0°$ and $360°$ as $\theta = 360° - 56.4° = 303.6°$.   □

With the use of the reference angle α we may find the value of any function so long as the angle lies *in* one of the quadrants. This problem reduces to finding the function of an acute angle. We are left with the case of the terminal side being along one of the axes, a **quadrantal angle.** Using the definitions of the functions and remembering that $r > 0$, we arrive at the values in the following table:

θ	$\sin \theta$	$\cos \theta$	$\tan \theta$	$\cot \theta$	$\sec \theta$	$\csc \theta$
0°	0.0	1.0	0.0	Undef.	1.0	Undef.
90°	1.0	0.0	Undef.	0.0	Undef.	1.0
180°	0.0	−1.0	0.0	Undef.	−1.0	Undef.
270°	−1.0	0.0	Undef.	0.0	Undef.	−1.0
360°	Same as the functions of 0° (same terminal side)					

The values in the table may be verified by referring to the figures in Figure 17-15.

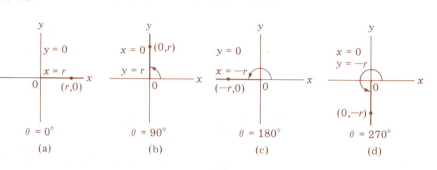

FIGURE 17-15   (a)   (b)   (c)   (d)

EXAMPLE G   Since $\cos \theta = x/r$, we see from Figure 17-15(a) that $\cos 0° = r/r = 1$. Since $\sec \theta = r/x$, we see from Figure 17-15(b) that $\sec 90° = r/0$, which is undefined since division by zero is undefined. Since $\sin \theta = y/r$, we see from Figure 17-15(c) that $\sin 180° = 0/r = 0$. Since $\tan \theta = y/x$, we see from Figure 17-15(d) that $\tan 270° = -r/0$, which is undefined since it has a divisor of zero.   □

The following example further illustrates the use of a calculator in finding values of trigonometric functions.

EXAMPLE H   In this example note that cotangent, secant, and cosecant involve the key labeled $1/x$ with the sine, cosine, and tangent keys.

	Calculator entry	*Typical display*
$\sin 219.3°$	219.3 [SIN]	−0.63338087
$\tan 347.0°$	347.0 [TAN]	−0.23086819
$\cos(-27.0°)$	27.0 [+/−][COS]	0.89100652
$\cot 200.0°$	200.0 [TAN][$1/x$]	2.7474774
$\csc 276.4°$	276.4 [SIN][$1/x$]	−1.0062712
$\sec(-36.0°)$	36.0 [+/−][COS][$1/x$]	1.2360680 □

As we saw in Examples E and F, if we are given the value of a trigonometric function and wish to find the corresponding angle between 0° and 360°, we must be careful when using a calculator. There may be *two solutions* (as in Example E); *the solution may not be in proper form* (as in Example F); or *the solution may actually be incorrect* for the problem being solved (see the fourth and fifth illustrations in Example I which follows). In this last case we can get the proper reference angle from the calculator display, but not the actual solution. Therefore, we must recognize that *blind use of the calculator can lead to an incomplete or incorrect solution.*

EXAMPLE I In this example we illustrate a typical calculator response to problems in which we want to find the values of θ satisfying the given conditions and $0° < \theta < 360°$. The values given as actual solutions are rounded to the nearest tenth of a degree.

	Calculator entry	Calculator display	Actual solution
$\sin \theta = 0.4067$	0.4067 [INV] [SIN]	23.997702	24.0° and 156.0°
$\cos \theta = -0.3746$	0.3746 [+/−] [INV] [COS]	111.99959	112.0° and 248.0°
$\tan \theta = -1.0000$	1 [+/−] [INV] [TAN]	−45	135.0° and 315.0°
$\csc \theta = 1.260,\ \cos \theta < 0$	1.26 [1/x] [INV] [SIN]	52.528005	127.5° ($\theta_{ref} = 52.5°$)
$\sec \theta = -2.000,\ \tan \theta > 0$	2 [+/−] [1/x] [INV] [COS]	120	240° ($\theta_{ref} = 60°$)
$\csc \theta = -2.000$	2 [+/−] [1/x] [INV] [SIN]	−30	210.0° and 330.0°
$\csc \theta = 0$	0 [1/x] [INV] [SIN]	Error	Undef.

EXAMPLE J A pilot flies a triangular route connecting three airports. The distance d (in miles) between two of the airports can be found by solving the equation

$$\frac{134}{\sin 147.0°} = \frac{d}{\sin 12.2°}$$

Multiplying both sides of this equation by $\sin 12.2°$, we get

$$d = \frac{134 \sin 12.2°}{\sin 147.0°} = \frac{134(0.2113)}{0.5446}$$

$$= 52.0$$

We now know that $d = 52.0$ mi.

Exercises 17-2

In Exercises 1 through 8, express the given trigonometric functions in terms of the same function of a positive acute angle.

1. $\sin 165°$, $\cos 230°$ **2.** $\tan 95°$, $\sec 305°$ **3.** $\cos 207°$, $\csc 290°$ **4.** $\sec 100°$, $\cot 218°$

5. tan 342°, sec (−10°)   **6.** cot 104°, sin (−104°)

7. cot 650°, tan (−300°)   **8.** sin 760°, csc (−210°)

In Exercises 9 through 28, determine the values of the given trigonometric functions.

9. sin 213.0°   **10.** cos 307.0°   **11.** tan 275.3°   **12.** cot 97.2°

13. sec 156.5°   **14.** csc 108.4°   **15.** cos 202.1°   **16.** sin 291.7°

17. cot 195.0°   **18.** tan 114.3°   **19.** csc 215.6°   **20.** sec 347.1°

21. sin 102.6°   **22.** cos 171.4°   **23.** tan 250.2°   **24.** cot 322.8°

25. sin (−32.4°)   **26.** cos (−67.8°)   **27.** tan (−112.5°)   **28.** sec (−215.9°)

In Exercises 29 through 44, find θ to the nearest 0.1° for 0° < θ < 360°. (In each case there are two values of θ.)

29. sin θ = −0.8480   **30.** cot θ = −0.2126   **31.** cos θ = 0.4003   **32.** tan θ = −1.830

33. cot θ = 0.5265   **34.** sin θ = 0.6374   **35.** tan θ = 0.2833   **36.** cos θ = −0.9287

37. sin θ = −0.7880   **38.** csc θ = 1.580   **39.** csc θ = −1.580   **40.** sec θ = 6.188

41. sec θ = −1.023   **42.** cot θ = 0.3779   **43.** cos θ = −0.9994   **44.** tan θ = −20.4

In Exercises 45 through 52, find θ to the nearest 0.1° for 0° < θ < 360°, subject to the given conditions.

45. sin θ = −0.4384, cos θ > 0   **46.** cos θ = 0.4083, tan θ < 0

47. tan θ = −1.200, csc θ > 0   **48.** sec θ = 1.526, sin θ < 0

49. csc θ = 1.150, tan θ < 0   **50.** cot θ = −1.540, sec θ > 0

51. cos θ = −0.9870, csc θ < 0   **52.** sin θ = 0.9860, sec θ < 0

In Exercises 53 through 60, solve the given problems.

53. When a weather balloon reaches an altitude of 1250 ft, it is 105 ft downwind from its departure point. The angle θ its path makes with the ground is found by solving the equation

$$\tan \theta = \frac{1250}{105}$$

Find the acute angle θ.

54. A certain alternating-current voltage can be found from the equation

$$V = 170 \cos 725.0°$$

Find the voltage V.

55. Under specified conditions, a force F (in pounds) is determined by solving the equation

$$\frac{F}{\sin 125.0°} = \frac{250}{\sin 35.4°}$$

Find the magnitude of the force.

56. A brace is used to support a frame so that cement can be poured. The angle θ between the brace and the form can be found by solving the equation

$$\frac{4.27}{\sin \theta} = \frac{6.35}{\sin 78.5°}$$

Solve for θ assuming that it is an acute angle.

57. A formula for finding the area of a triangle, knowing sides a and b and angle C, is

$$A = \frac{1}{2}ab \sin C$$

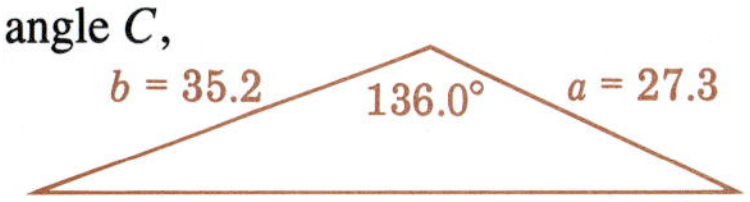

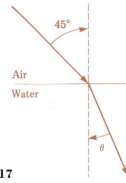

Find the area of a triangle for which $a = 27.3$, $b = 35.2$, and $C = 136.0°$. See Figure 17-16.

FIGURE 17-16

58. In calculating the area of a triangular tract of land, a surveyor uses the formula in Exercise 57. He inserts the values $a = 510$ m, $b = 345$ m, and $C = 125.0°$. Find the area of the tract of land.

59. A highway construction engineer uses the formula

$$\tan \theta = \frac{V^2}{k}$$

to calculate the angle at which a curved section of road should be banked. Find θ if the maximum speed is $V = 60$ mi/h and $k = 86{,}400$ mi^2/h^2.

60. A ray of light enters water and is refracted as shown in Figure 17-17. In this case, the angle of the refracted beam θ can be found by using the index of refraction

$$n = \frac{\sin 45°}{\sin \theta}$$

Find θ if n is 1.33 for water.

FIGURE 17-17

17-3 The Law of Sines

To this point we have limited our study of triangle solution to right triangles. However, many triangles which require solution do not contain a right angle. As we noted earlier, such a triangle is termed an *oblique triangle*. In this and the following sections we learn methods for solving oblique triangles.

When we first introduced the subject of triangle solution, we stated that we need three parts, at least one of them a side, to solve any triangle. With this in mind we may determine that there are four possible combinations of parts from which we may solve a triangle. These combinations are:

Case 1: two angles and one side

Case 2: two sides and the angle opposite one of them

Case 3: two sides and the included angle

Case 4: three sides

There are several ways in which oblique triangles may be solved, but we shall restrict our attention to the two most useful methods, the **Law of Sines** and the **Law of Cosines.** In this section we discuss the Law of Sines and show that it may be used to solve case 1 and case 2.

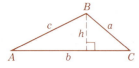

FIGURE 17-18

Let *ABC* be an oblique triangle with sides *a*, *b*, and *c* opposite angles *A*, *B*, and *C*, respectively. By drawing a perpendicular *h* from *B* to side *b*, we see from Figure 17-18 that $\sin A = h/c$ and $\sin C = h/a$. Solving for *h*, we have $h = c \sin A$ and $h = a \sin C$. Therefore $c \sin A = a \sin C$, or

$$\frac{a}{\sin A} = \frac{c}{\sin C}$$

By dropping a perpendicular from *A* to *a* we also find that $c \sin B = b \sin C$, or

$$\frac{b}{\sin B} = \frac{c}{\sin C}$$

Combining these results, we have the **Law of Sines:**

$$\boxed{\frac{a}{\sin A} = \frac{b}{\sin B} = \frac{c}{\sin C}}$$ (17-5)

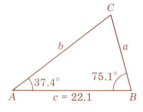

See Appendix B for a computer program that solves triangles by using the Law of Sines.

The Law of Sines is a statement of proportionality between the sides of a triangle and the sines of the angles opposite them. We should note that *there are three actual equations combined in Eq.* (17-5). Also, we might note that the same result will be obtained if we choose another vertex in the derivation so that the Law of Sines is valid for all triangles.

Now we may see how the Law of Sines is applied to the solution of a triangle in which two angles and one side are known (case 1). If two angles are known, the third may be found from the fact that the sum of the angles in a triangle is 180°. At this point we must be able to *determine the ratio between the given side and the sine of the angle opposite it*. Then, by use of the Law of Sines, we may find the other sides.

EXAMPLE A

Given that $c = 22.1$, $A = 37.4°$, and $B = 75.1°$, find *a*, *b*, and *C* (see Figure 17-19).

First, we can see that

$$C = 180° - (37.4° + 75.1°) = 180° - 112.5°$$
$$= 67.5°$$

FIGURE 17-19

We now know side *c* and angle *C*, which allows us to use Eq. (17-5). *Using the equation relating* **a**, **A**, **c**, *and* **C**, we have

$$\frac{a}{\sin 37.4°} = \frac{22.1}{\sin 67.5°}$$

or

$$a = \frac{22.1 \sin 37.4°}{\sin 67.5°} = 14.5 \qquad \text{round to three significant digits}$$

The calculator key sequence for this computation is as follows. (First be sure that the calculator is in the degree mode.)

$$22.1 \; \boxed{\times} \; 37.4 \; \boxed{\text{SIN}} \; \boxed{\div} \; 67.5 \; \boxed{\text{SIN}} \; \boxed{=} \; \boxed{\textit{14.528957}}$$

Now *using the equation relating b, B, c, and C,* we have

$$\frac{b}{\sin 75.1°} = \frac{22.1 \overset{\longleftarrow}{} c}{\sin 67.5°}$$
$$\underset{B}{\Big\uparrow} \qquad \underset{C}{\Big\uparrow}$$

or

$$b = \frac{22.1 \sin 75.1°}{\sin 67.5°} = 23.1$$

The calculator sequence is the same as that for *a* if 37.4 is replaced by 75.1. We now know that *a* = 14.5, *b* = 23.1, and *C* = 67.5°. ☐

In the next example we illustrate use of the Law of Sines in a situation where we know two sides and the angle opposite one of them. This corresponds to case 2 in the previous discussion. We illustrate this case with an applied problem.

EXAMPLE B Electric supply lines must be built so that points *A* and *B* are connected, but a direct path cuts through a marsh. How much longer is the indirect path from point *A* to point *C* and then to point *B*, as shown in Figure 17-20?

From Figure 17-20 we can see that the indirect path from *A* to *C* to *B* is a distance of 197.0 m + 302.9 m = 499.9 m. We must now find the length of side *c* so that we can determine the amount by which 499.9 m exceeds the length of the direct path represented by side *c*.

We will first use the Law of Sines to find angle *B*, which in turn will allow us to find angle *C*. We will then use the Law of Sines a second time so that the length of side *c* can be determined. We begin by finding angle *B* and then angle *C*.

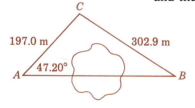

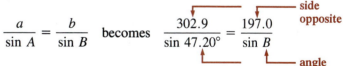

$$\frac{a}{\sin A} = \frac{b}{\sin B} \quad \text{becomes} \quad \frac{302.9}{\sin 47.20°} = \frac{197.0}{\sin B}$$

FIGURE 17-20

which leads to

$$\sin B = \frac{197.0 \sin 47.20°}{302.9} = 0.4772$$

With sin *B* = 0.4772, we determine that *B* = 28.50°. (We might also get *B* = 151.50°, but we discard that possibility since angles *A* and *B* would then exceed 180°.) With *A* = 47.20° and *B* = 28.50°, we conclude that *C* = 104.30° since the sum of angles *A*, *B*, *C* must be 180°.

(Continued on next page)

We now use the Law of Sines a second time so that the length of side c can be determined.

$$\frac{a}{\sin A} = \frac{c}{\sin C} \quad \text{becomes} \quad \frac{302.9}{\sin 47.20°} = \frac{c}{\sin 104.30°}$$

which leads to

$$c = \frac{302.9 \sin 104.30°}{\sin 47.20°} = 400.0 \text{ m}$$

The indirect path is 499.9 m while the direct path is 400.0 m long. The indirect path is therefore 99.9 m longer than the direct path.   □

In Example B we were able to solve the triangle by using the Law of Sines twice. Our ultimate goal was to find the length of side c, but it was necessary to first find angles B and C. The next example illustrates another use of the Law of Sines in solving an applied problem.

EXAMPLE C   A microwave relay tower 10.2 m tall is placed on the top of a building as shown in Figure 17-21. From point D the angles of elevation to the top and bottom of the tower are measured as 50.0° and 45.0°, respectively. Find the height of the building.

Triangle BCD is a right triangle with angles of 45.0°, 45.0°, and 90.0°. We want to find the length of side BC and this can be easily computed if we could determine the length of the hypotenuse BD. To find the length of BD we will consider triangle ABD and use the Law of Sines. It is easy to see that $\angle ADB = 5.0°$. Also, $\angle DBC = 45.0°$ (since the angles of $\triangle BCD$ must yield a total of 180°) which implies that $\angle DBA = 135.0°$ (since $\angle DBA$ and $\angle DBC$ are supplementary). Thus, $\angle DAB = 180° - (135.0° + 5.0°) = 40.0°$ (since the angles of $\triangle ABD$ must yield a total of 180°). Using the Law of Sines in $\triangle ABD$ we get

FIGURE 17-21

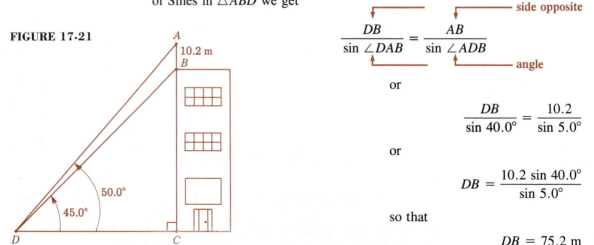

$$\frac{DB}{\sin \angle DAB} = \frac{AB}{\sin \angle ADB}$$

         — side opposite

         — angle

or

$$\frac{DB}{\sin 40.0°} = \frac{10.2}{\sin 5.0°}$$

or

$$DB = \frac{10.2 \sin 40.0°}{\sin 5.0°}$$

so that

$$DB = 75.2 \text{ m}$$

Now that we know the length of side DB, we consider $\triangle BCD$, which is a right triangle, and we get

$$\sin 45.0° = \frac{BC}{75.2}$$

$$BC = (75.2)(\sin 45.0°) = 53.2 \text{ m}$$

The building is 53.2 m tall.   □

▶ Case 2, listed at the beginning of this section (we know two sides and the angle opposite one of them) is called the **ambiguous case** because, depending on the particular values given, *there may be one solution, two solutions, or no solution.* The next example illustrates these possibilities.

EXAMPLE D   (a) In Figure 17-22(a), we are given $a = 1$, $b = 4$, and $\angle A = 30°$. Using the Law of Sines we get

$$\frac{a}{\sin A} = \frac{b}{\sin B}$$

$$\frac{1}{\frac{1}{2}} = \frac{4}{\sin B}$$

$$\sin B = 2$$

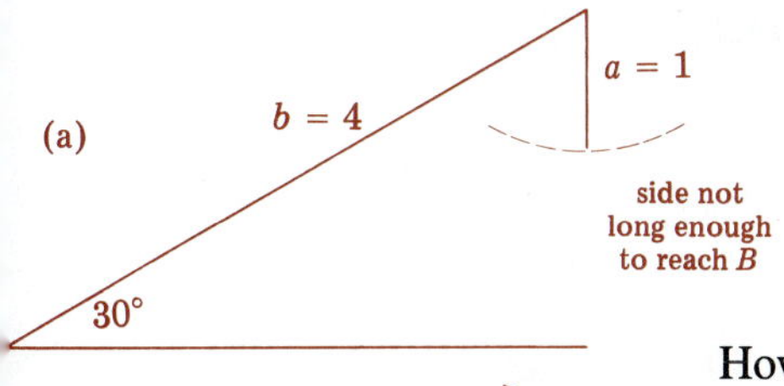

(a)

$b = 4$ $a = 1$

side not long enough to reach B

30°

▶ However, this is impossible because $\sin B$ cannot exceed 1. We conclude that *there is* no solution.

(b) In Figure 17-22(b), we are given $a = 2$, $b = 4$, and $\angle A = 30°$. Using the Law of Sines we get

$$\frac{a}{\sin A} = \frac{b}{\sin B}$$

$$\frac{2}{\frac{1}{2}} = \frac{4}{\sin B}$$

$$\sin B = 1$$

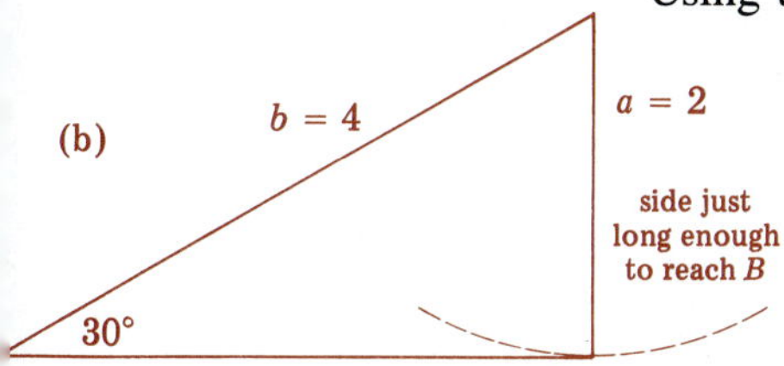

(b)

$b = 4$ $a = 2$

side just long enough to reach B

30°

▶ We conclude that $\angle B = 90°$ and there is *one solution.*

(c) In Figure 17-22(c), we are given $a = 3$, $b = 4$, and $\angle A = 30°$. Using the Law of Sines we get

$$\frac{a}{\sin A} = \frac{b}{\sin B}$$

$$\frac{3}{\frac{1}{2}} = \frac{4}{\sin B}$$

$$\sin B = \frac{2}{3}$$

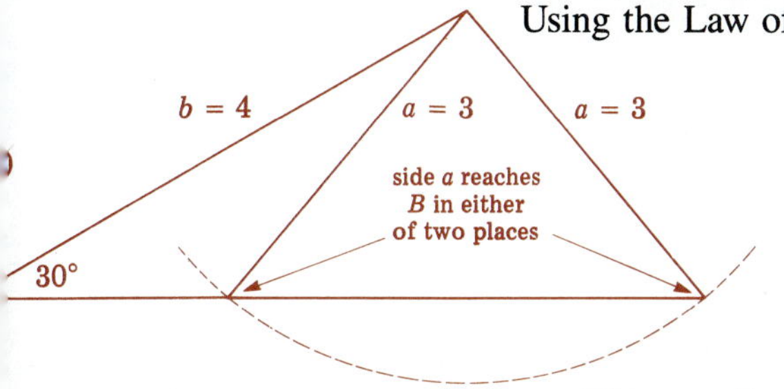

$b = 4$ $a = 3$ $a = 3$

side a reaches B in either of two places

30°

▶ This result means that $\angle B = 41.8°$ or $138.2°$. In this case we have **two solutions.** They are represented by the two possible positions of side a in Figure 17-22(c).   □

FIGURE 17-22

In Example D it would be easy to miss the second solution for the case which led to two solutions. Care must be taken to recognize this ambiguous case so that both solutions are detected. Also, it is possible to have one solution in the ambiguous case with a triangle that is not a right triangle as in Figure 17-22(b). In Example D, for any value of *a* greater than 4, only one solution exists. In general, if the side opposite the given angle is longer than the other given side, then only one solution exists. In Figure 17-23 we illustrate the case for $a = 5$, $b = 4$, and $\angle A = 30°$.

Note that in this case, the side $a = 5$ opposite the given angle of $A = 30°$ is longer than the other given side of $b = 4$, so only one solution exists. That solution is the triangle shown in Figure 17-23. The dashed arc is intended to illustrate that no other angle *B* will produce a triangle.

We have seen that solving triangles in the ambiguous case (case 2) can lead to two solutions, one solution, or no solution. Since it is so easy to miss a second solution when two solutions exist, we might consider the following. *When given two sides and the angle opposite one of them, there are two triangle solutions if these conditions are all met.*

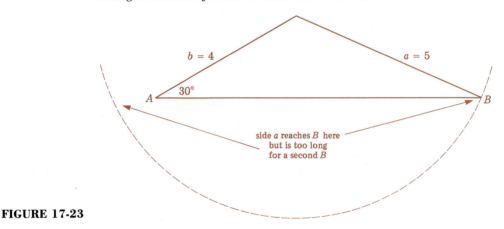

FIGURE 17-23

1. The given angle is acute.
2. The given angle's opposite side is shorter than the adjacent side.
3. The sine of the second angle is calculated to be a value less than one.

A scale drawing of the given data will also help to identify the correct number of solutions.

Exercises 17-3

In Exercises 1 through 24, solve the triangles with the given parts by use of the Law of Sines.

1. $a = 7.16$, $A = 70.8°$, $B = 36.6°$

2. $b = 36.8$, $B = 31.5°$, $C = 43.5°$

3. $c = 2190$, $A = 37.0°$, $B = 34.0°$

4. $a = 4.66$, $B = 17.9°$, $C = 82.6°$

5. $b = 155$, $c = 72.8$, $B = 20.7°$ **6.** $a = 45.0$, $c = 75.0$, $C = 76.4°$

7. $a = 0.926$, $b = 0.228$, $A = 143.2°$ **8.** $a = 232$, $b = 557$, $B = 112.8°$

9. $a = 63.8$, $B = 58.4°$, $C = 22.2°$ **10.** $a = 13.0$, $A = 55.2°$, $B = 67.5°$

11. $b = 438$, $B = 47.4°$, $C = 64.5°$ **12.** $b = 283$, $B = 13.7°$, $C = 76.3°$

13. $a = 26.2$, $b = 22.2$, $B = 48.1°$ **14.** $a = 89.4$, $c = 37.3$, $C = 15.6°$

15. $b = 576$, $c = 730$, $B = 31.4°$ **16.** $a = 0.841$, $b = 0.965$, $A = 57.1°$

17. $a = 94.2$, $c = 68.0$, $A = 69.1°$ **18.** $a = 630$, $b = 670$, $B = 102.0°$

19. $a = 1.43$, $b = 4.21$, $A = 30.4°$ **20.** $b = 15.0$, $c = 55.0$, $B = 75.0°$

21. $a = 100$, $c = 200$, $A = 30.0°$ **22.** $b = 17.2$, $c = 23.7$, $C = 125.2°$

23. $b = 2.14$, $c = 6.73$, $B = 85.2°$ **24.** $b = 16.35$, $c = 10.13$, $C = 74.9°$

In Exercises 25 through 36, use the Law of Sines to solve the given problems.

25. Find the lengths of the supports of the shelf shown in Figure 17-24.

26. The front line of a triangular piece of land is 425 m long and makes angles of 75.0° and 62.5° with the other sides. What are the lengths of the other sides? See Figure 17-25.

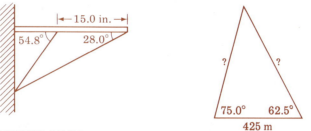

FIGURE 17-24   **FIGURE 17-25**

27. A rocket is fired at an angle of 42.0° with the ground from a point 2500 m directly behind an observer. Soon thereafter it is observed that the rocket is at an angle of elevation of 47.0° directly in front of the observer. How far on a direct line has the rocket moved?

28. The pilot of a light plane notes that the angles of depression of two objects directly ahead are 21.5° and 16.0°. The objects are known to be 1.25 mi apart. How far is the plane on a direct line from the nearer object?

29. A submarine leaves a port and travels due north. At a certain point it turns 45.0° to the left and travels an additional 75.0 km to a point 90.0 km from the port. How far from the port is the point where the submarine turned? See Figure 17-26.

30. Two points A and B are located on opposite sides of a lake. A third point C is positioned so that $\angle B = 52.0°$. If BC is 162 ft and AC is 212 ft, what is the length of side AB? See Figure 17-27.

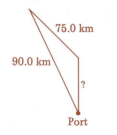

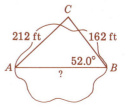

FIGURE 17-27

FIGURE 17-26

31. A communications satellite is directly above the extension of a line between receiving towers A and B. It is determined from radio signals that the angle of elevation from tower A is 88.9° and the angle of elevation from tower B is 87.6°. If A and B are 658 km apart, how far is the satellite from A? (Ignore the curvature of the earth.) See Figure 17-28.

32. A person measures a triangular piece of land and reports the following information: "One side is 96.4 ft long and another side is 65.9 ft long. The angle opposite the shorter side is 48.0°." Could this information be correct?

33. A circular plate with diameter 2.00 m is secured with 10 bolts equally spaced along the outer edge. Find the distance between two adjacent bolts (measured from center to center).

34. A manufacturing plant is designed to be in the shape of a regular pentagon with 88 ft on each side. A security fence surrounds the building to form a circle and each corner of the building is 25 ft from the closest point on the circle. How much fencing is required? See Figure 17-29.

35. A surveyor sets stakes at points A, B, C on a level field. The angles at A and B are measured and found to be 46.7° and 82.9°, respectively. If the distance from point A to point B is 34.9 m, find the distance between the stakes at A and C. See Figure 17-30.

36. An airplane is sighted from points A and B on ground level. Points A and B are 357 ft apart. At point A, the angle of elevation to the airplane is 41.2°; for point B it is 46.9°. How high is the airplane? See Figure 17-31.

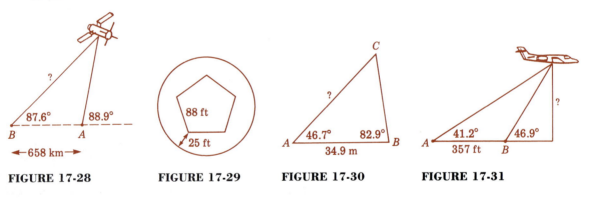

FIGURE 17-28   FIGURE 17-29   FIGURE 17-30   FIGURE 17-31

17-4 The Law of Cosines

In the beginning of Section 17-3, we noted that there were four combinations or cases in which the given information allows a solution of a triangle. In Section 17-3 we used the Law of Sines to solve two of those four cases. In this section we introduce the **Law of Cosines** as a means for solving the remaining two cases.

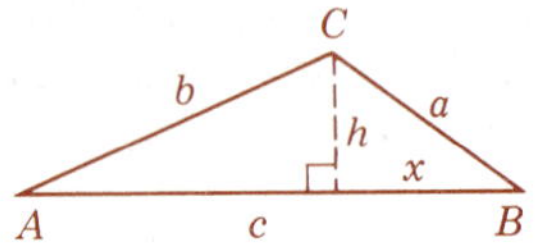

Consider any oblique triangle—for example, the one shown in Figure 17-32. In this triangle we see that $h/b = \sin A$ or that $h = b \sin A$. Also, $c - x = b \cos A$, or $x = c - b \cos A$. From the Pythagorean theorem, we see that $a^2 = h^2 + x^2$, or

$$a^2 = (b \sin A)^2 + (c - b \cos A)^2$$

FIGURE 17-32

Multiplying out, we get

$$\begin{aligned} a^2 &= b^2 \sin^2 A + c^2 - 2bc \cos A + b^2 \cos^2 A \\ &= b^2(\sin^2 A + \cos^2 A) + c^2 - 2bc \cos A \end{aligned}$$

(Here $\sin^2 A = (\sin A)^2$.) Recalling the definitions of the trigonometric functions, $\sin \theta = y/r$ and $\cos \theta = x/r$, which in turn means that $\sin^2 \theta + \cos^2 \theta = (x^2 + y^2)/r^2$. Since $x^2 + y^2 = r^2$, we then conclude that

$$\sin^2 \theta + \cos^2 \theta = 1 \tag{17-6}$$

This equation holds for any angle θ, which means that it holds for angle A.

The above equation for a^2 can now be simplified to

$$a^2 = b^2 + c^2 - 2bc \cos A \qquad \text{Law of Cosines} \tag{17-7}$$

This formula is known as the **Law of Cosines.**

See Appendix B for a computer program that solves a triangle by using the Law of Cosines.

Using the same method, we may also show that

$$b^2 = a^2 + c^2 - 2ac \cos B \tag{17-8}$$

and

$$c^2 = a^2 + b^2 - 2ab \cos C \tag{17-9}$$

These are simply alternative forms of the Law of Cosines. *All three forms express the same basic relationship among the sides of the triangle and the angle opposite one of the sides.*

If we know two sides and the included angle (case 3), we may directly solve for the side opposite the given angle. Then we normally would use the Law of Sines to complete the solution. If we are given all three sides (case 4), we may solve for the angle opposite one of them by the Law of Cosines. Again we usually use the Law of Sines to complete the solution.

EXAMPLE A

Solve the triangle with $a = 4.08$, $b = 4.37$, and $C = 68.0°$ (see Figure 17-33.)

FIGURE 17-33

Using Eq. (17-9) which is the form of the Law of Cosines with side c and angle C, we get

unknown side opposite known angle

$$c^2 = (4.08)^2 + (4.37)^2 - 2(4.08)(4.37) \cos 68.0°$$

known sides

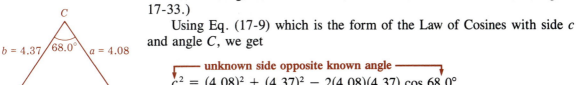

$$c = \sqrt{(4.08)^2 + (4.37)^2 - 2(4.08)(4.37) \cos 68.0°} = 4.73$$

(Continued on next page)

This type of computation can be easily performed on a calculator. We list below the key sequence which can be used to evaluate c. Again, we must verify that our calculator is in the degree mode before we begin.

$4.08 \boxed{x^2} \boxed{+} 4.37 \boxed{x^2} \boxed{-} 2 \boxed{\times} 4.08 \boxed{\times} 4.37 \boxed{\times} 68 \boxed{\text{cos}} \boxed{=} \boxed{\sqrt{x}}$ $\boxed{4.7312925}$

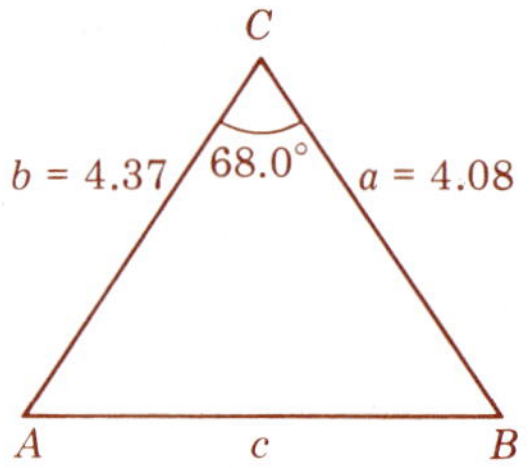

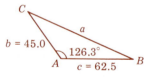

FIGURE 17-33

From the Law of Sines, we now have

$$\frac{4.08}{\sin A} = \frac{4.37}{\sin B} = \frac{4.73}{\sin 68.0°}$$ — sides opposite / angles

Solving for A, we obtain

$$\sin A = \frac{4.08 \sin 68.0°}{4.73}$$

from which we get

$$A = 53.1°$$

by using the calculator sequence

$4.08 \boxed{\times} 68 \boxed{\text{SIN}} \boxed{\div} 4.73 \boxed{=} \boxed{\text{INV}} \boxed{\text{SIN}}$ $\boxed{53.108105}$

Now, since the sum of the three angles is 180°, we have $B = 58.9°$. We conclude that $A = 53.1°$, $B = 58.9°$, and $c = 4.73$. □

EXAMPLE B

Solve the triangle with $b = 45.0$, $c = 62.5$, and $A = 126.3°$ (see Figure 17-34).

Using the form of the Law of Cosines with side a and angle A, we have the following. Since $A = 126.3°$, we note that $\cos A$ is negative.

FIGURE 17-34

— unknown side opposite known angle —

$$a^2 = (45.0)^2 + (62.5)^2 - 2(45.0)(62.5) \cos 126.3°$$
$$a = \sqrt{(45.0)^2 + (62.5)^2 - 2(45.0)(62.5) \cos 126.3°} = 96.2$$

From the Law of Sines, we have

$$\frac{96.2}{\sin 126.3°} = \frac{45.0}{\sin B}$$

Solving for B we get

$$\sin B = \frac{45.0 \sin 126.3°}{96.2}$$

which gives us

$$B = 22.1°$$

Finally, $C = 180° - (126.3° + 22.1°) = 31.6°$. Therefore $a = 96.2$, $B = 22.1°$, and $C = 31.6°$. □

Examples A and B illustrated case 3, where we are given two sides and the included angle. Example C illustrates case 4, where we are given three sides.

EXAMPLE C   A survey team locates three benchmarks. They are equipped with a tape for measuring distances, but they don't have any instruments for measuring angles. Solve the triangle for which $a = 83.8$ ft, $b = 36.7$ ft, and $c = 72.4$ ft. See Figure 17-35.

We may use any of the forms of the Law of Cosines to *find one of the angles*. Choosing the form with a and $\cos A$, we use Eq. (17-7) and solve for $\cos A$ to get

$$\cos A = \frac{b^2 + c^2 - a^2}{2bc} = \frac{(36.7)^2 + (72.4)^2 - (83.8)^2}{2(36.7)(72.4)} = -0.0816$$

which gives us

$A = 94.7°$ **(rounded)**

by using the calculator sequence

$36.7 \;\boxed{x^2}\boxed{+}\; 72.4 \;\boxed{x^2}\boxed{-}\; 83.8 \;\boxed{x^2}\boxed{=}\boxed{\div}\; 2 \boxed{\div}\; 36.7$
$\boxed{\div}\; 72.4 \boxed{=}\boxed{\text{INV}}\boxed{\text{COS}}\;$ (94.682212)

 (Other sequences using parentheses or memory may also be used.) We note that $\cos A$ is negative (the display after the second $=$). **Since $\cos A$ is negative, we know that A is between 90° and 180°.**

Finally, *using the Law of Sines* we determine that

$B = 25.9°$ and $C = 59.4°$

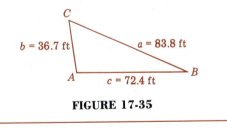

FIGURE 17-35   □

In Example C we used the Law of Cosines to *find the largest angle first*. That is always a good strategy. If the triangle contains an obtuse angle, that fact will then be discovered since the cosine of the obtuse angle will be negative. This will avoid problems when the Law of Sines is used, for the sine of an obtuse angle is positive and **the proper result may not be given directly by the calculator.** Example D is another illustration of the Law of Cosines for solving a triangle with two given sides and the included angle (case 3).

EXAMPLE D   In order to avoid a restricted area, an airplane flies from point A to point C to point B as shown in Figure 17-36. Following this path, the distance traveled is 18.0 km. If a military airplane is allowed to fly directly from A to B, how much shorter is the route?

This problem involves finding the length of side c. Using the Law of Cosines we get

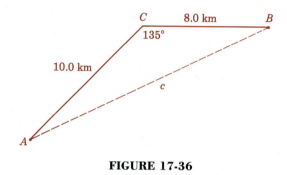

$$\overbrace{c^2 = (8.00)^2 + (10.0)^2 - 2(8.00)(10.0) \cos 135.0°}^{\text{unknown side opposite known angle}}$$

$$c = \sqrt{(8.00)^2 + (10.0)^2 - 2(8.00)(10.0) \cos 135.0°}$$

$$= 16.6 \text{ km}$$

The direct path is therefore $18.0 - 16.6 = 1.4$ km shorter.

FIGURE 17-36

Exercises 17-4

In Exercises 1 through 24, use the Law of Cosines to solve the triangles with the given parts.

1. $a = 22.3$, $b = 16.4$, $C = 87.5°$

2. $b = 16.3$, $c = 30.5$, $A = 42.1°$

3. $a = 7720$, $c = 42,000$, $B = 58.9°$

4. $a = 7.00$, $b = 9.00$, $C = 133.9°$

5. $a = 1510$, $b = 308$, $C = 98.0°$

6. $a = 30.6$, $c = 15.2$, $B = 70.8°$

7. $a = 770$, $b = 934$, $c = 1600$

8. $a = 0.729$, $b = 0.789$, $c = 0.459$

9. $a = 51.2$, $b = 38.4$, $c = 72.6$

10. $a = 72.3$, $b = 65.9$, $c = 98.3$

11. $a = 7.20$, $b = 5.30$, $c = 9.19$

12. $a = 510$, $b = 650$, $c = 221$

13. $a = 238$, $b = 312$, $C = 24.7°$

14. $a = 7.92$, $b = 5.02$, $C = 100.6°$

15. $a = 5.12$, $c = 1.24$, $B = 37.3°$

16. $b = 102$, $c = 84.0$, $A = 56.9°$

17. $a = 637$, $b = 831$, $c = 987$

18. $a = 1.42$, $b = 1.95$, $c = 1.65$

19. $a = 3.47$, $b = 4.52$, $c = 2.25$

20. $a = 825$, $b = 770$, $c = 712$

21. $a = 0.1034$, $b = 0.1287$, $c = 0.1429$

22. $a = 2.160$, $b = 3.095$, $C = 109.10°$

23. $b = 10.21$, $c = 34.26$, $A = 9.65°$

24. $a = 2936$, $b = 1874$, $c = 3468$

In Exercises 25 through 36, use the Law of Cosines to solve the given problems.

25. To measure the distance *AC*, a person walks 620 m from *A* to *B*, then turns 35.0° to face *C*, and walks 730 m to *C*. What is the distance *AC*? See Figure 17-37.

26. An airplane traveling at 900 mi/h leaves the airport at noon going due east. At 2 P.M. the pilot turns 15.3° south of east. How far is the plane from the airport at 3 P.M.?

27. A forest ranger using orienteering walks directly north for a distance of 3.20 km. After making a turn, he then walks 3.82 km. From that location he returns to the starting point by walking 4.67 km. Find the angles of the triangle formed by this route. See Figure 17-38.

          FIGURE 17-37   **FIGURE 17-38**

28. A tract of land is in the shape of a parallelogram. The sides of the tract are 230 m and 475 m, and the longer diagonal is 540 m. What is the larger angle between adjacent sides?

29. A triangular shelf is to be placed in a corner at which the walls meet at an angle of 105.0°. If the edges of the shelf along the walls are 56.0 cm and 65.0 cm, how long is the outer edge of the shelf?

30. One ship 250 mi from a port is on a line 36.5° south of west of the port. A second ship is 315 mi from the port on a line 15.0° south of east of the port. How far apart are the ships?

31. One end of an 8.00 ft pole is 11.55 ft from an observer's eyes and the other end is 17.85 ft from his eyes. Through what angle does the observer see the pole? See Figure 17-39.

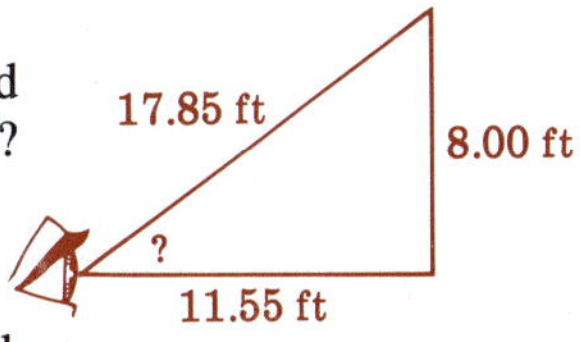

32. A triangular piece of land is bounded by 30.1 m of stone wall, 26.5 m of road frontage, and 50.5 m of fencing. What angle does the fence make with the road?

                                                          FIGURE 17-39

33. A plastic triangle used for drafting has sides of 4.58 in., 9.16 in., and 7.93 in. Find the angles formed by this triangle.

34. A laser experiment involves a prism with a triangular base which has sides of 2.00 cm, 3.00 cm, and 4.00 cm. Find the angles of the triangle.

35. In order to find a single force equivalent to the net effect produced by two other forces acting on an object, it is necessary to find the length of the third side in a triangle. Find that length if the two known sides are 86.0 lb and 110 lb, and the angle between them is 107.0° (See Figure 17-40.)

36. An industrial robot is programmed to cut out a triangular part. A straight cut of 6.223 cm is made. The part is then rotated clockwise 42.50° about the end of that cut, and a second straight cut of 8.255 cm is made. The third cut is made in a straight line that returns to the beginning of the first cut. Find the length of the third cut.

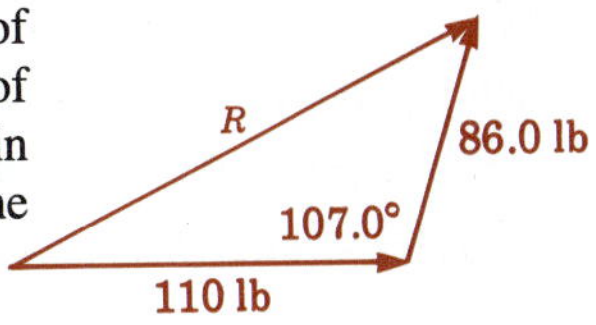

                                                                   FIGURE 17-40

Chapter 17 Formulas

$$\text{sine } \theta = \frac{y}{r} \qquad \text{cosine } \theta = \frac{x}{r} \qquad \text{tangent } \theta = \frac{y}{x}$$

$$\text{cotangent } \theta = \frac{x}{y} \qquad \text{secant } \theta = \frac{r}{x} \qquad \text{cosecant } \theta = \frac{r}{y}$$

(17-1)

If θ is in the . . .	then the reference angle α can be found from the equation . . .	
second quadrant	$\alpha = 180° - \theta$	(17-2)
third quadrant	$\alpha = \theta - 180°$	(17-3)
fourth quadrant	$\alpha = 360° - \theta$	(17-4)

$$\frac{a}{\sin A} = \frac{b}{\sin B} = \frac{c}{\sin C} \qquad \text{Law of Sines} \tag{17-5}$$

$$\sin^2 \theta + \cos^2 \theta = 1 \tag{17-6}$$

$$a^2 = b^2 + c^2 - 2bc \cos A \tag{17-7}$$

$$b^2 = a^2 + c^2 - 2ac \cos B \qquad \text{Law of Cosines} \tag{17-8}$$

$$c^2 = a^2 + b^2 - 2ab \cos C \tag{17-9}$$

17-5 Review Exercises for Chapter 17

In Exercises 1 through 4, find the trigonometric functions of θ given that the terminal side of θ passes through the given point. Express results for Exercises 1 and 2 in fractional form, express results for Exercises 3 and 4 in decimal form.

1. $(4, 3)$ **2.** $(-5, 12)$ **3.** $(7, -2)$ **4.** $(-2, -3)$

In Exercises 5 through 8, determine the quadrant in which the terminal side of θ lies, subject to the given conditions.

5. $\sin \theta$ is positive and $\tan \theta$ is negative

6. $\cos \theta$ is negative and $\sin \theta$ is negative

7. $\sec \theta$ is positive and $\csc \theta$ is negative

8. $\cot \theta$ is negative and $\sec \theta$ is positive

In Exercises 9 through 12, express the given trigonometric functions in terms of the same function of a positive acute angle.

9. $\cos 132°$, $\tan 194°$ **10.** $\sin 243°$, $\cot 318°$ **11.** $\sin 289°$, $\sec (-15°)$ **12.** $\cos 103°$, $\csc (-100°)$

In Exercises 13 through 28, determine the values of the given trigonometric functions.

13. cos 243.0°   **14.** sin 139.0°   **15.** cot 287.0°   **16.** tan 188.0°

17. csc 247.5°   **18.** sec 96.3°   **19.** sin 215.2°   **20.** cos 337.2°

21. tan 291.4°   **22.** cot 109.3°   **23.** tan 256.7°   **24.** cos 172.5°

25. sin 763.0°   **26.** tan 482.0°   **27.** sec 715.4°   **28.** csc 992.0°

In Exercises 29 through 40, find θ to the nearest 0.1° for 0° < θ < 360°. In each case there are two solutions.

29. tan $\theta = 0.7532$ **30.** sin $\theta = -0.5293$ **31.** cos $\theta = -0.4208$ **32.** cot $\theta = -1.829$

33. csc $\theta = 2.175$ **34.** sec $\theta = 1.428$ **35.** sin $\theta = -0.1639$ **36.** cos $\theta = 0.2735$

37. cot $\theta = -2.145$ **38.** tan $\theta = 0.1459$ **39.** sin $\theta = 0.8319$ **40.** sec $\theta = -1.817$

In Exercises 41 through 60, solve the triangles with the given parts.

41. $a = 100$, $A = 47.0°$, $B = 61.3°$

42. $a = 35.0$, $A = 48.2°$, $C = 73.3°$

43. $a = 89.7$, $B = 148.8°$, $C = 10.0°$

44. $A = 36.4°$, $B = 123.6°$, $c = 78.4$

45. $a = 3.20$, $b = 4.50$, $C = 91.7°$

46. $b = 3820$, $c = 2910$, $A = 84.8°$

47. $a = 283$, $c = 278$, $B = 62.0°$

48. $b = 913$, $c = 117$, $A = 57.6°$

49. $b = 16.3$, $c = 18.2$, $B = 54.3°$

50. $a = 902$, $b = 763$, $A = 148.9°$

51. $a = 65.5$, $b = 78.4$, $A = 51.0°$

52. $a = 840$, $b = 1390$, $A = 19.2°$

53. The triangle shown in Figure 17-41(a)

54. The triangle shown in Figure 17-41(b)

55. The triangle shown in Figure 17-41(c)

56. The triangle shown in Figure 17-41(d)

57. The triangle shown in Figure 17-41(e)

58. The triangle shown in Figure 17-41(f)

FIGURE 17-41

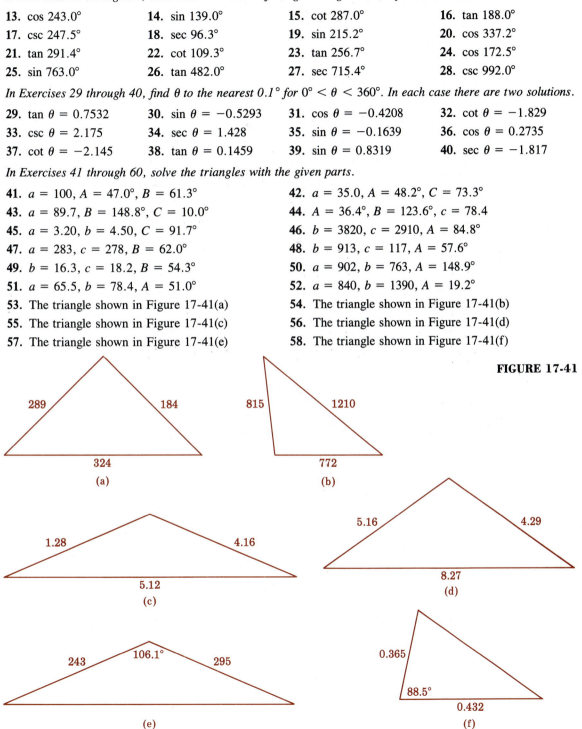

59. The triangle shown in Figure 17-41(g)   **60.** The triangle shown in Figure 17-41(h)

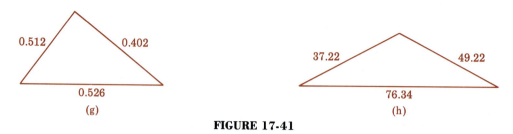

(g)   (h)

FIGURE 17-41

In Exercises 61 through 72, solve the given problems.

61. At a certain instant, the voltage in a certain alternating-current circuit is given by

$$v = 150 \cos 140.0°$$

Find the voltage.

62. At a given instant, the displacement of a particle moving with simple harmonic motion is given by

$$d = 16.0 \sin 250.0°$$

Find the displacement (in centimeters).

63. To find the distance between points A and B on opposite sides of a river, a distance AC is measured as 600 m, where C is on the same side of the river as A. Angle BAC is measured to be 110.0° and angle ACB is 34.5°. What is the distance between A and B? See Figure 17-42.

64. A 28.0-ft pole leans against a slanted wall, making an angle of 25.0° with the wall. If the foot of the pole is 12.5 ft from the foot of the wall, find the angle of inclination of the wall to the ground. See Figure 17-43.

65. A plumb line is dropped from the top peak of a tower which is 80.0 ft tall. If the plumb line reaches a point on the ground 3.10 ft from the center of the base of the tower, by how many degrees is the tower away from being vertical?

66. A triangular patio has sides of 40.0 ft, 32.0 ft, and 28.0 ft. What are the angles between the edges of the patio?

67. Two points on opposite sides of an obstruction are 156 m and 207 m, respectively, from a third point. The lines joining the first two points and the third point intersect at an angle of 100.5° at the third point. How far apart are the two original points? See Figure 17-44.

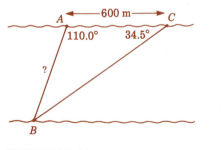

FIGURE 17-42

FIGURE 17-43

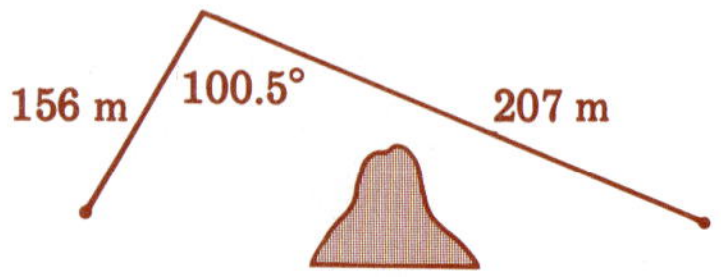

FIGURE 17-44

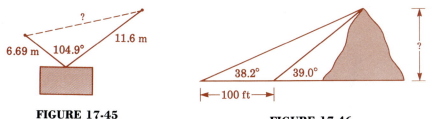

FIGURE 17-45

FIGURE 17-46

68. A crate is being held aloft by two ropes which are tied at the same point on the crate. They are 11.6 m and 6.69 m long, respectively, and the angle between them is 104.9°. Find the distance between the ends of the ropes. See Figure 17-45.

69. An observer measures the angle of elevation to the top of a mountain and obtains a value of 39.0°. After moving 100 ft farther away from the mountain, the angle of elevation is measured as 38.2°. How tall is the mountain? See Figure 17-46.

70. In a military training exercise, two bases are 380 km apart and they are trying to locate an unidentified airplane. The naval base is 40.0° south of west of the army base. Direction-finding instruments indicate that the plane is 74.0° south of east of the army base and 8.3° north of east of the naval base. What is the distance between the plane and the nearest base? See Figure 17-47.

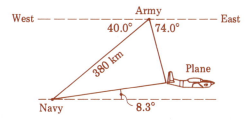

FIGURE 17-47

71. Two cars are at the intersection of two straight roads. One travels 5.36 km on one road, and the other travels 8.91 km on the other road. The drivers contact each other on CB radios and find that they are at points known to be 7.28 km apart. What angle do the roads make at the intersection?

72. The edges of a saw tooth are 1.39 mm and 3.26 mm long. The base of the tooth is 2.05 mm long. At what angle do the edges of the tooth meet?

In Exercises 73 through 80, use a scientific calculator to solve the given problems.

73. If $\theta = 137°15'$, find $\sin \theta$ and $\sec \theta$.

74. If $\sin \theta = 0.6666$ and $\cos \theta$ is negative, find θ.

75. Use the $\boxed{\text{SIN}}$ and $\boxed{\text{COS}}$ keys on a calculator to find the quadrant in which the terminal side of θ lies if $\theta = -1547.335°$.

76. If θ is an angle in a triangle and $\cos \theta = -0.6046001$, find θ.

77. Find the angles in a triangle with sides of $\sqrt{5246.2}$, 75.385, and 80.944.

78. A triangle has sides of 2.465 and 2.837 and the angle between those sides is $52°36'45''$. Solve the triangle.

79. A triangle has angles of $46°30'14''$, $96°18'27''$, and the largest side is 8825.4. Solve the triangle.

80. A side of length 6.258 is opposite an angle of $27°36'$. Another side has length 8.339. Solve the triangle.

Vectors

In previous chapters we have been able to fully describe most quantities by specifying their magnitudes. Generally, we can describe areas, volumes, lengths of objects, time intervals, monetary amounts, temperatures, and many other quantities by specifying a number which represents the magnitude of the quantity. *Such quantities that can be expressed by specifying only their magnitudes are known as* **scalar** *quantities.*

There are many other quantities that can be fully described only when both their magnitude and their direction are specified. Such quantities are known as **vectors.** That is, a vector is a quantity having both a magnitude and a direction. Examples of vectors are velocity, force, and momentum. Vectors are extremely important in many fields of science and technology.

In Figure 18-1 we show a typical vector representation of an applied problem. A pilot flies her plane so that it is pointed in the direction of 60.0° north of west as shown. Her airspeed indicator shows 205 mi/h and the wind is blowing toward the direction of 45.0° north of east. In order to find the true ground speed of the plane and the true direction, we must add the two vectors. We will show how to solve such problems. This particular problem will be solved later in the chapter.

18-1 Introduction to Vectors

We have stated that a vector is a quantity having both magnitude and direction. We begin this section with examples of vectors, and we will then consider procedures for adding and subtracting them.

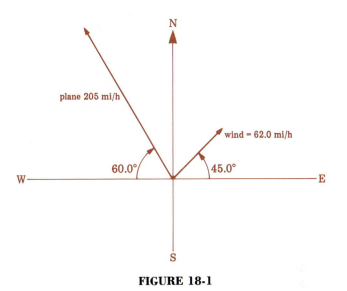

FIGURE 18-1

EXAMPLE A   An example of a vector is found in an aviation weather report describing the wind as coming from the south at 15 mi/h. This quantity has both a *direction* (from the south) and a *magnitude* (15 mi/h) and it is therefore a vector.

□

EXAMPLE B   If we are told that a missile travels 3270 km, we know only the distance traveled. This distance is a scalar quantity. If we later learn that the missile traveled 3270 km on a direct path from San Diego to Moscow, we would be specifying the direction as well as the distance traveled. *Since the direction and magnitude are both known, we now have a vector,* as well as a serious international incident.

□

We know the **displacement** of an object when we know its direction as well as the distance from the starting point. Displacement is therefore a vector quantity.

To add scalar quantities we simply find the sum of the magnitudes. For example, 300 mi + 400 mi = 700 mi illustrates scalar addition. The addition of vector quantities is somewhat more complex, as illustrated by the following example.

EXAMPLE C

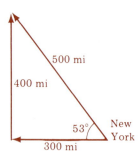

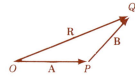

FIGURE 18-2

An air traffic controller at a New York airport instructs the pilot of a jet to fly 300 mi due west, then turn and fly 400 mi due north. Following these instructions will clearly cause the jet to travel 700 mi, but it will not be 700 mi from New York. In fact, we see by using the Pythagorean theorem that the jet is 500 mi from New York (see Figure 18-2). The magnitude of the displacement is 500 mi, which means that the jet is actually 500 mi from New York. The direction of the displacement is indicated by the longest arrow shown in Figure 18-2. Measuring the angle, we find that it is directed at an angle of 53° north of west. Here we have added vectors with magnitudes of 300 and 400 and the result is a vector of magnitude 500. The "discrepancy" is due to the different directions of the two original vectors.   □

From Example C we see that *the addition of vectors must involve consideration of directions as well as magnitudes*. The method used to add the vectors in Example C can be generalized. We begin with some notation: Vector quantities will be represented by **boldface** type while the same letter in *italic* type will represent magnitude only. Using this notation, **A** is a vector of magnitude A. Since handwriting does not lend itself easily to italic and boldface forms, the handwritten notation for a vector will be an arrow placed over the letter as in $\vec{A}$.

FIGURE 18-3

Let **A** and **B** represent vectors directed from O to P and P to Q respectively (see Figure 18-3). *The vector sum **A** + **B** is the vector **R**, from the* **initial point** O *to the* **terminal point** Q. Here vector **R** is called the **resultant**. *In general, a resultant is a single vector which can replace any number of other vectors and still produce the same physical effect*.

In this chapter we will consider graphical and analytical ways of adding vectors. This section will involve only the graphical approach while the remaining sections will consider the analytical approach involving trigonometry and the Pythagorean theorem. With the graphic approach, we add vectors by means of a diagram. There are two basic ways of adding vectors graphically: the **tail-to-head method** and the **parallelogram method.**

The tail-to-head method is illustrated in Figure 18-4. To add **B** to **A**, we shift **B** parallel to itself until its tail touches the head of **A**. In doing so we must *be careful to keep the magnitude and direction of **B** unchanged*. The vector sum **A** + **B** is the vector **R**, which is drawn from the tail of **A** to the head of **B**.

The tail-to-head method can also be used for adding more than two vectors. Starting with one of the vectors, we position the initial point of the second vector at the terminal point of the first. Then the third vector is located so that its initial point is at the terminal point of the second vector, and so on. *The resultant vector is determined by the initial point of the first vector and the terminal point of the last vector*. The order of the vectors does not affect the result.

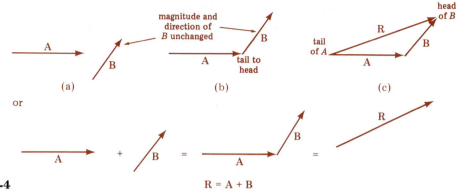

FIGURE 18-4   R = A + B

EXAMPLE D   In Figure 18-5(a) we show the vectors **A**, **B**, and **C** which are to be added. In Figure 18-5(b), the resultant **R** is found from one order, while Figure 18-5(c) shows that the same resultant is obtained with a different order of the same vectors. In addition to the orders of **A** + **B** + **C** and **B** + **A** + **C**, *other orders are also possible and they will produce the same resultant* **R**.

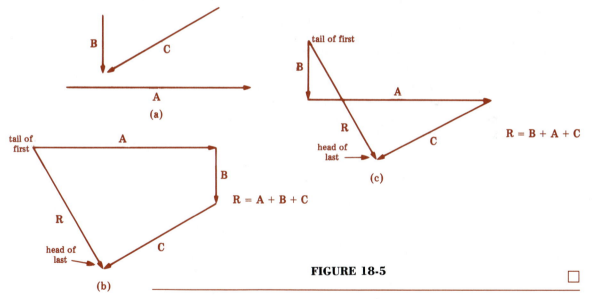

FIGURE 18-5

Example D illustrates the tail-to-head method of adding vectors. Another convenient method of adding two vectors is to *let the two vectors be the sides of a parallelogram. The resultant is then the diagonal of the parallelogram.* The initial point of the resultant is the common initial point of the vectors being added. In this method the vectors are first placed tail-to-tail. See the following example.

EXAMPLE E   Using the parallelogram method, add vectors **A** and **B** given in Figure 18-6(a). Figure 18-6(b) shows the vectors placed tail-to-tail. Figure 18-6(c) shows the construction of the parallelogram while Figure 18-6(d) shows the resultant.

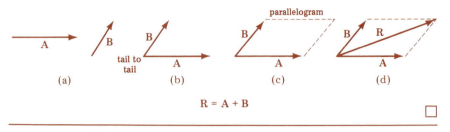

FIGURE 18-6   R = A + B

When we are considering applications of vectors, one method of addition may illustrate the situation better than the other method. For example, if two forces are acting on an object, the parallelogram method provides a good illustration of the actual situation. When we are adding displacements, the tail-to-head method is usually effective in showing the physical situation. Either method may be used to find the resultant for any set of vectors.

EXAMPLE F   A nuclear submarine travels 10 km east and then 5 km to the northeast. The two vectors may be shown as in Figure 18-7(a), for the movement to the northeast started when the eastward movement ceased. Then it is natural to find the resultant displacement as in Figure 18-7(b). By measurement in the figure, the magnitude of the displacement can be found to be 14 km.

A force of 10 lb acts on an object to the right. A second force of 5 lb acts on the same object upward to the right, at an angle of 45° with the horizontal. The two vectors may be shown as in Figure 18-7(c), for the forces act on the same object. Then it is more natural to find the resultant force as in Figure 18-7(d). The magnitude of the resultant force is about 14 lb.

FIGURE 18-7   (a)   (b)   (c)   (d)

*If vector **A** has the same direction as vector **B**, and **A** also has a magnitude n times that of **B**, we may state that **A** = n**B**. Thus 2**A** represents a vector twice as long as **A** but it is in the same direction.*

EXAMPLE G   For the vectors **A** and **B** in Figure 18-4, find 3**A** + 2**B**. See Figure 18-8.

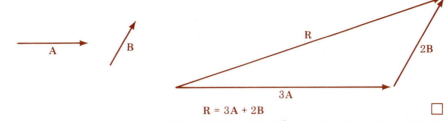

FIGURE 18-8

$R = 3A + 2B$

Vector **B** *may be subtracted from vector* **A** *by reversing the direction of* **B** *and proceeding as in vector addition.* Thus $A - B = A + (-B)$, where the minus sign indicates that vector $-B$ has the opposite direction of vector **B** (but the same magnitude). Vector subtraction is shown in the following example.

EXAMPLE H   For vectors **A** and **B** of Figure 18-4, find $2A - B$. See Figure 18-9.

While we can add, subtract, and find resultants of vectors using these techniques, we should remember that these graphic methods yield results which are approximate, not exact.

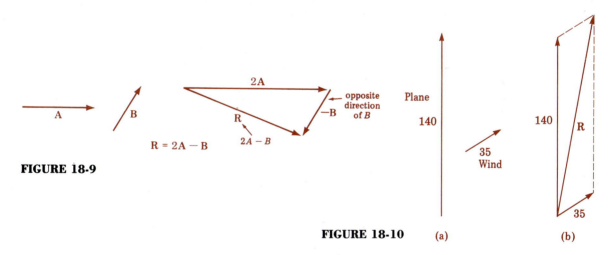

FIGURE 18-9

FIGURE 18-10   (a)   (b)

EXAMPLE I   A pilot flies a small airplane with an air speed of 140 mi/h. The plane is pointed to the north, but is blown off course by a wind of 35 mi/h which is *from* a direction of 45° south of west (toward 45° north of east). See Figure 18-10. Find the true speed of the plane.

From Figure 18-10(b), the resultant **R** has a magnitude of approximately 165 and we conclude that the true speed of the plane is 165 mi/h.

In Example I the solution of 165 mi/h is only an approximation since it depends upon a measurement. The true direction of the plane can be determined by measuring the angle with a protractor, but that is also an approximation. Applied uses of vectors often require more exact results. In the following sections we present methods which will yield more exact results.

Exercises 18-1

In Exercises 1 through 8, determine whether a scalar or a vector is described in part (a) and part (b). Explain your answers.

1. (a) A car travels at 50 mi/h. (b) A car travels at 50 mi/h due south.

2. (a) A person traveled 300 km to the southwest. (b) A person traveled 300 km.

3. (a) A vertical rope supports a 100-g object. (b) A force equivalent to 100 g is applied to an object.

4. (a) A small craft warning reports winds of 25 mi/h. (b) A small craft warning reports winds out of the north at 25 mi/h.

5. (a) An arm of an industrial robot pushes with a 10-lb force downward on a part. (b) A part is being pushed with a 10-lb force by an arm of an industrial robot.

6. (a) Spilled oil is being pushed by the 2-mi/h current in a river. (b) Spilled oil is being pushed south by the 2-mi/h current in a river.

7. (a) A ballistics test shows that a bullet hit a wall at a speed of 400 ft/s. (b) A ballistics test shows that a bullet hit a wall at a speed of 400 ft/s perpendicular to the wall.

8. (a) To denote the *speed* of an object we state how fast it is moving. Thus speed is a ———. (b) To denote the *velocity* of an object we state how fast and in what direction it is moving. Thus velocity is a ———.

*In Exercises 9 through 16, find the sum of the vectors **A** and **B** shown in the indicated figure.*

9. Figure 18-11(a)   **10.** Figure 18-11(b)   **11.** Figure 18-11(c)   **12.** Figure 18-11(d)

13. Figure 18-11(e)   **14.** Figure 18-11(f)   **15.** Figure 18-11(g)   **16.** Figure 18-11(h)

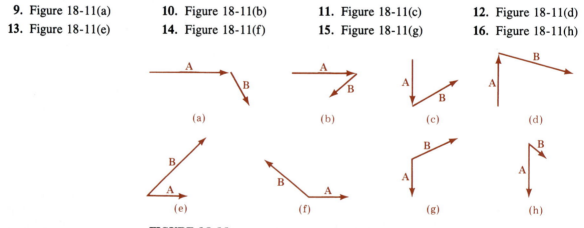

FIGURE 18-11

In Exercises 17 through 36, find the indicated vector sums and differences by means by diagrams, using the vectors in Figure 18-12.

17. A + B   **18. A + C**   **19. A + A**   **20. B + B**

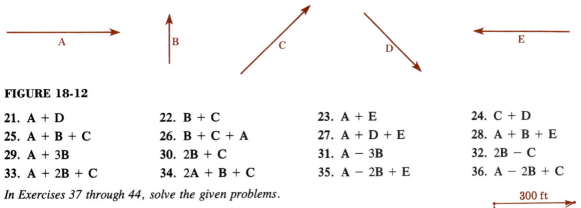

FIGURE 18-12

21. A + D	**22. B + C**	**23. A + E**	**24. C + D**
25. A + B + C	**26. B + C + A**	**27. A + D + E**	**28. A + B + E**
29. A + 3B	**30. 2B + C**	**31. A − 3B**	**32. 2B − C**
33. A + 2B + C	**34. 2A + B + C**	**35. A − 2B + E**	**36. A − 2B + C**

In Exercises 37 through 44, solve the given problems.

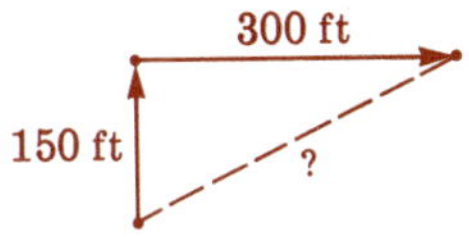

37. A surveyor sets a stake, then travels 150 ft north where a second stake is set. He then travels east 300 ft and sets a third stake. Use a diagram like Figure 18-13 to find the displacement between the first and third stakes.

FIGURE 18-13

38. A motorist travels 20 mi due west and then 6 mi to the northeast. By means of a diagram find where the motorist is relative to the starting position.

39. A test of the behavior of a tire on ice involves two forces. One force of 500 lb is exerted to the side while another force of 1200 lb is exerted to the front. Use a diagram to find the resultant of these two forces.

40. The motor of a boat applies a force that would make the boat travel 10 mi/h in still water. The boat is pointed directly at the opposite shore on a river, and the current is downstream at 6 mi/h. Use a diagram to find the resultant of the two forces caused by the motor and the current.

41. In conducting a stress test on the end of a horizontal beam, two forces are applied. One force of 1800 lb is exerted vertically downward on the beam. Another force of 2700 lb is exerted horizontally at the same point on the beam. By means of a diagram find the resultant of these two forces.

42. A car is stuck in snow. One pedestrian pushes the back of the car with a force having magnitude 50 lb while another pedestrian pushes on the driver's side with a force having magnitude 30 lb. By means of a diagram find the resultant of these two forces.

43. A balloon is rising vertically at the rate of 15 m/s and the wind is blowing from the northwest at 10 m/s. By means of a diagram find the resultant velocity of the balloon.

44. An accident investigator determines that when two cars crashed, one car was traveling on a northbound highway at a speed of 55 mi/h. The second car entered the highway at a speed of 30 mi/h from a side road, and its path was 45° north of west. Use a diagram to find the resultant of the two forces at the point of impact.

18-2 Vector Components

In Section 18-1 we used graphic techniques for adding and subtracting vectors and multiples of vectors. Beside being able to perform these vector operations, there is often a need to represent a given vector as the sum of two other vectors. These two vectors, when added, combine to produce the original vector. *The method is called* **resolving** *the given vector into its*

components. This is usually done within the framework of the rectangular coordinate system. That is, the given vector is resolved into two components; one of the components has a direction along the x axis and the other component has a direction along the y axis. When the components have directions along the axes, they are called **rectangular component vectors.** The following example shows how to resolve a given vector into its rectangular components.

EXAMPLE A Resolve the vector **A**, with $A = 20.0$, which makes an angle of 56.0° with the positive x axis, into two components, one directed along the x axis (the x component) and the other along the y axis (the y component). That is, resolve the given vector into its rectangular components. See Figure 18-14.

In the figure we see that two right triangles are formed. For the right triangle with sides **A** and $\mathbf{A}_x$ we can use the cosine function to get

$$\cos 56.0° = \frac{A_x}{A}$$

(Remember that **A** represents a vector while A is a scalar.) Solving for A_x we get

$$A_x = A \cos 56.0°$$

Since $A = 20.0$ we have

$$A_x = 20.0 \cos 56.0° = 11.2 \qquad \text{(rounded)}$$

using the calculator sequence

$$20 \;\boxed{\times}\; 56 \;\boxed{\cos}\; \boxed{=} \;\boxed{\quad 11.183858\quad}$$

Working with the same right triangle, it can be shown that the dashed side opposite the 56.0° angle is equal to the magnitude A_y. Using the sine function we get

$$\sin 56.0° = \frac{A_y}{A}$$

or

$$A_y = A \sin 56.0° = 16.6 \qquad \text{(rounded)}$$

using the calculator sequence

$$20 \;\boxed{\times}\; 56 \;\boxed{\text{SIN}}\; \boxed{=} \;\boxed{\quad 16.580751\quad}$$

We have now resolved vector **A** into two components. One component has magnitude 11.2 and is in the direction of the positive x axis. The second component has magnitude 16.6 and is in the direction of the positive y axis.

□

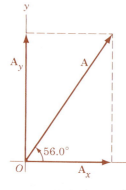

FIGURE 18-14

Some calculators have a key, usually labeled $P \leftrightarrow R$, which allows conversion between "polar" and "rectangular" forms of vectors. This key could be used to resolve a vector into its rectangular components. For example, one calculator allows us to enter $A = 20.0$ and $\theta = 56.0°$ as follows.

$$P \leftrightarrow R$$

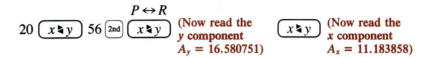

20 $\boxed{x \blacktriangleleft y}$ 56 $\boxed{\text{2nd}}$ $\boxed{x \blacktriangleleft y}$ (Now read the y component $A_y = 16.580751$) $\boxed{x \blacktriangleleft y}$ (Now read the x component $A_x = 11.183858$)

We can generalize the first method used in Example A: When resolving a vector **A** into its rectangular components, *the magnitude of the horizontal component is $A \cos \theta$ and the magnitude of the vertical component is $A \sin \theta$, where θ is the angle of the vector in standard position.* Vector **A** has rectangular components A_x and A_y with magnitudes given by

$$A_x = A \cos \theta \qquad \qquad \text{(18-1)}$$
$$A_y = A \sin \theta \qquad \qquad \text{(18-2)}$$

If the original vector is not in the first quadrant, the rectangular component vectors will be directed along their respective axes so that their sum is the original vector.

Rectangular components in the direction of the positive x axis or positive y axis are represented by positive values, while rectangular components in the direction of the negative x axis or negative y axis will be represented by negative values. This will make it easier to add vectors, as discussed in the following section.

EXAMPLE B

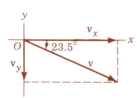

FIGURE 18-15

The velocity of a rocket is 350 km/s, directed to the right, at an angle of 23.5° below the horizontal. What are the horizontal (x component) and vertical (y component) components of the velocity?

We can construct the figure as shown in Figure 18-15. Noting that the angle corresponds to a *clockwise* rotation, we represent it as $-23.5°$ to get

$$v_x = v \cos (-23.5°) = 350 \cos (-23.5°) = 321 \text{ km/s}$$

Also, since v_y is equal to the side opposite the 23.5° angle, we get

$$v_y = v \sin (-23.5°) = 350 \sin (-23.5°) = -140 \text{ km/s}$$

Remember, *we use negative values for rectangular components in the direction of the negative x axis or negative y axis.* ☐

EXAMPLE C In analyzing the stress on a ceiling joist caused by snow on a roof, an architect finds it necessary to resolve the force vector in Figure 18-16(a) into its rectangular components. Find the rectangular component vectors.

From Figure 18-16(b) we can see that one rectangular component vector will be in the direction of the negative x axis while the other is in the direction of the negative y axis. Noting that R_x and R_y are both negative since they are in the direction of the negative x axis and negative y axis, we proceed to use Eqs. (18-1) and (18-2) as follows.

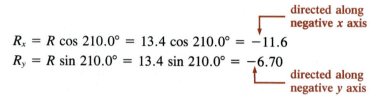

$$R_x = R \cos 210.0° = 13.4 \cos 210.0° = -11.6$$
$$R_y = R \sin 210.0° = 13.4 \sin 210.0° = -6.70$$

directed along negative x axis

directed along negative y axis

The horizontal component vector has magnitude 11.6 lb and is directed along the negative x axis, while the vertical component has magnitude 6.70 lb and is directed along the negative y axis.

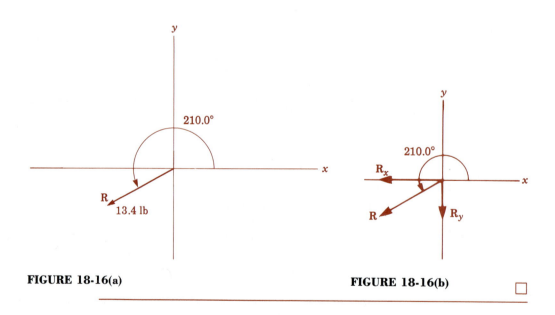

FIGURE 18-16(a)   FIGURE 18-16(b)

EXAMPLE D A cable is anchored at a point as shown in Figure 18-17. The pull on the cable is equivalent to 1360 N and we want to provide weights X and Y so that the three forces at the anchor point offset each other. Note that the weight X actually produces a force in the direction of the negative x axis. Find the forces produced by X and Y.

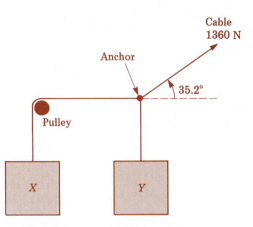

FIGURE 18-17

We will represent by **C** the force vector representing the cable. Resolving **C** into rectangular components we get

$$C_x = C \cos 35.2°$$
$$= 1360 \cos 35.2° = 1110 \text{ N}$$
$$C_y = C \sin 35.2°$$
$$= 1360 \sin 35.2° = 784 \text{ N}$$

From these rectangular components we see that the force of the cable is equivalent to two other forces: one which pulls to the right with 1110 N and one which pulls vertically upward with 784 N. To offset the pull to the right of 1110 N, we should make $X = 1110$ N. To offset the pull upward, we should make $Y = 784$ N. ☐

Exercises 18-2

In Exercises 1 through 4, use a diagram to resolve the vector given in the indicated figure into its x component and y component.

1. Figure 18-18(a)   **2.** Figure 18-18(b)   **3.** Figure 18-18(c)   **4.** Figure 18-18(d)

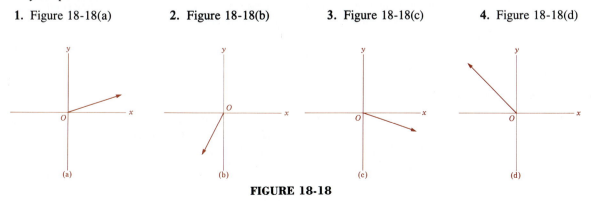

FIGURE 18-18

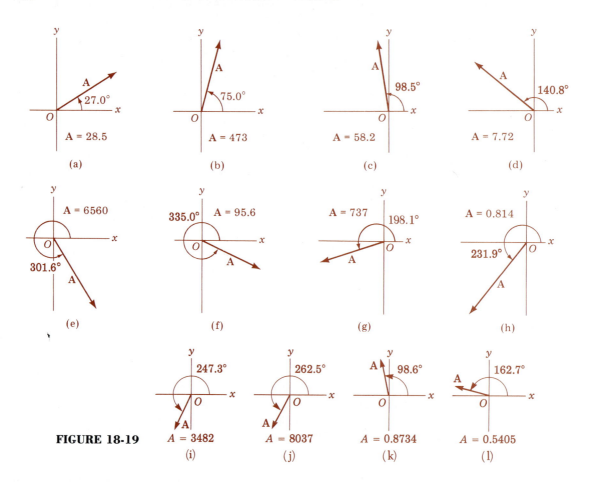

FIGURE 18-19

In Exercises 5 through 16, resolve the vector given in the indicated figure into its x component and y component.

5. Figure 18-19(a)   **6.** Figure 18-19(b)   **7.** Figure 18-19(c)   **8.** Figure 18-19(d)

9. Figure 18-19(e)   **10.** Figure 18-19(f)   **11.** Figure 18-19(g)   **12.** Figure 18-19(h)

13. Figure 18-19(i)   **14.** Figure 18-19(j)   **15.** Figure 18-19(k)   **16.** Figure 18-19(l)

In Exercises 17 through 32, resolve the given vector into its x component and y component. The given angle θ is measured counterclockwise from the positive x axis (in standard position).

17. Magnitude 25.8, $\theta = 45.0°$ **18.** Magnitude 465, $\theta = 30.0°$

19. Magnitude 6.34, $\theta = 57.0°$ **20.** Magnitude 136, $\theta = 18.3°$

21. Magnitude 219, $\theta = 263.0°$ **22.** Magnitude 34.5, $\theta = 201.9°$

23. Magnitude 1370, $\theta = 225.0°$ **24.** Magnitude 29.3, $\theta = 315.0°$

25. Magnitude 87.2, $\theta = 306.9°$ **26.** Magnitude 219, $\theta = 342.0°$

27. Magnitude 0.05436, $\theta = 233.5°$ **28.** Magnitude 0.0204, $\theta = 117.6°$

29. Magnitude 25872, $\theta = 356.20°$ **30.** Magnitude 13724, $\theta = 279.90°$

31. Magnitude 34.666, $\theta = 112.70°$ **32.** Magnitude 58.307, $\theta = 166.40°$

In Exercises 33 through 40, solve the given problems.

33. A plane is 24.0° north of east from a city at a distance of 36.5 mi. Find the horizontal (east) and vertical (north) components of the displacement. See Figure 18-20.

34. Wind is blowing *from* the southeast at 42 mi/h. Resolve this velocity vector into a component directed to the north and a component directed to the west.

35. A force of 780 N acts to the right at an angle of 74.3° above the horizontal on an object. Find the horizontal and vertical components of the force. See Figure 18-21.

36. A bulldozer stuck in mud is being pulled by a cable connected to another bulldozer. The cable makes an angle of 18.0° with an imaginary line representing the forward path of the stuck bulldozer. The magnitude of the force is 8500 lb of pull. What is the magnitude of the component of the force which has the same direction as the stuck bulldozer? See Figure 18-22.

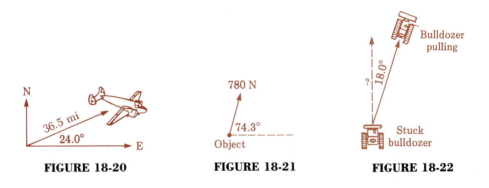

FIGURE 18-20   FIGURE 18-21   FIGURE 18-22

37. In studying the interference of harmonic sound waves, it becomes necessary to find the rectangular components of the vector which has a magnitude of 0.024 m. The vector makes an angle of 58.3° with the positive x axis. Find the rectangular component vectors.

38. A missile makes an angle of 84.8° with a level platform. What is the horizontal force applied by the rocket on its platform if the rocket exerts 155,000 lb of thrust? See Figure 18-23.

39. In a parallel resistance-capacitance (*RC*) electric circuit, the current I_C through the capacitance and the current I_R through the resistance are out of phase by 90°, as depicted in Figure 18-24. If $I_C = 2.35$ A and $I_R = 2.89$ A, find the magnitude of the resultant and the phase angle θ.

40. A 6480-lb truck is parked on a ramp that is at a 10.3° angle with the horizontal. Find the component of the truck's weight that is parallel with the surface of the ramp. See Figure 18-25.

FIGURE 18-23

FIGURE 18-24   FIGURE 18-25

18-3 Vector Addition by Components

See Appendix B for a computer program that will add vectors.

In Section 18-1 we saw how to add vectors graphically, but the results were only approximate. Common problems found in science and technology require the addition of vectors with more exact results. We will use the Pythagorean theorem and the trigonometric ratios to develop a procedure giving the exactness required. The procedure is based on resolving vectors into components, and Section 18-2 provided the foundation for this technique.

We begin with a simple example in which the two given vectors are already at right angles so that, in this particular case, we need not resolve them into x component and y component vectors. The objective is to *add the two vectors to get a resultant vector with a known magnitude and a known direction*.

EXAMPLE A

Add vectors **A** and **B**, with $A = 86.4$ and $B = 67.4$. The vectors are at right angles as shown in Figure 18-26.

Since the vectors **A**, **B**, and their sum **R** form a right triangle, we can *find the magnitude R* of the resultant **R** by using the Pythagorean theorem as follows:

$$R = \sqrt{A^2 + B^2} = \sqrt{(86.4)^2 + (67.4)^2}$$
$$= 110 \quad \text{(rounded)}$$

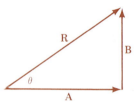

FIGURE 18-26

This computation can be performed with a calculator by using the following sequence of key strokes:

$$86.4 \,\boxed{x^2}\,\boxed{+}\, 67.4 \,\boxed{x^2}\,\boxed{=}\,\boxed{\sqrt{x}}\,\boxed{\textit{109.57974}}$$

Having found the magnitude of the resultant, we can *find its direction* angle θ by using an appropriate trigonometric ratio. We choose tangent because we know the sides opposite and adjacent to θ.

$$\tan \theta = \frac{B}{A} = \frac{67.4}{86.4}$$

so that

$$\theta = 38.0° \quad \text{(rounded)}$$

The value of θ can be found by using a calculator with this key stroke sequence:

$$67.4 \,\boxed{\div}\, 86.4 \,\boxed{=}\,\boxed{\text{INV}}\,\boxed{\text{TAN}}\,\boxed{\textit{37.957529}}$$

The sum of **A** and **B** is the resultant **R** which has magnitude 110 and direction $\theta = 38.0°$, where θ is located as in Figure 18-26. We must remember that *the solution is not complete without determining the direction*. ☐

In Example A the given vectors were directed so that they were at right angles to each other. When adding vectors which are not at right angles, we follow these key steps:

1. Resolve each vector into its x and y components by using

$$A_x = A \cos \theta \quad \text{(18-1)}$$
$$A_y = A \sin \theta \quad \text{(18-2)}$$

2. Combine the x components to determine the x component R_x of the resultant. Also combine the y components to determine the y component R_y of the resultant.

3. Find the magnitude of the resultant by using the Pythagorean theorem with the x and y components of the resultant (as found in step 2).

$$R = \sqrt{R_x^2 + R_y^2} \quad \text{(18-3)}$$

4. Find the direction of the resultant by using tangent with the x and y components of the resultant (as found in step 2). Use

$$\tan \theta = \frac{R_y}{R_x} \quad \text{(18-4)}$$

and adjust θ so that it corresponds to the correct quadrant.

When using Equation (18-4) to find θ, we normally determine the value of R_y/R_x and then use the $\boxed{\text{INV}}\,\boxed{\text{TAN}}$ keys on the calculator. If the resultant is in the first quadrant, the displayed angle will be correct. However, if the resultant is in the second quadrant, R_y/R_x will be negative and the displayed angle will be negative; we can add 180° to this displayed result so that the positive angle corresponds to the standard position of the resultant vector. We can adjust θ in any quadrant as shown in the following table.

Quadrant of resultant	Sign of R_y/R_x	Using a calculator, find θ by entering $R_y \boxed{\div} R_x \boxed{=} \boxed{\text{INV}} \boxed{\text{TAN}}$, then . . .
I	+	displayed angle is correct
II	−	add 180° to displayed angle
III	+	add 180° to displayed angle
IV	−	add 360° to displayed angle

It is important to remember that the resultant vector has not been determined unless both its magnitude *and direction* are known.

The following examples illustrate how to find the magnitude of the resultant of two vectors, and the angle the resultant makes with the x axis, by using steps 1–4.

EXAMPLE B   Add vectors **A** and **B** as shown in Figure 18-27(a).

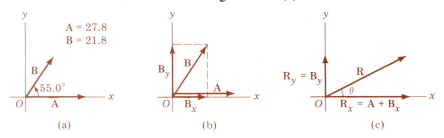

        (a)   (b)   (c)

FIGURE 18-27

step 1   Since vector **A** is directed along the x axis, we see that $A_x = A = 27.8$ and $A_y = 0$. It is necessary now to resolve vector **B** into its components:

$$B_x = B \cos 55.0° = 21.8 \cos 55.0° = 12.5$$
$$B_y = B \sin 55.0° = 21.8 \sin 55.0° = 17.9$$

These components are shown in Figure 18-27(b).

step 2   Now, since the only component in the y direction is B_y, we have $R_y = B_y$. Also, since both **A** and B_x are directed along the positive x axis, we add their magnitudes together to get the magnitude of the x component of the resultant. We get

$$R_x = A + B_x = 27.8 + 12.5 = 40.3$$
$$R_y = B_y = 17.9$$

step 3   We now find the magnitude of the resultant **R** from the Pythagorean theorem:

$$R = \sqrt{R_x^2 + R_y^2} = \sqrt{(40.3)^2 + (17.9)^2}$$
$$= 44.1$$

step 4   The direction of **R** is specified by determining angle θ from

$$\tan \theta = \frac{R_y}{R_x} = \frac{17.9}{40.3} \quad \text{or} \quad \theta = 23.9° \longleftarrow \text{ don't forget the direction}$$

Thus the magnitude of the resultant is 44.1 and it is directed at an angle of 23.9° above the positive x axis (see Figure 18-27(c)).   □

After each vector has been resolved into its components, we must *be very careful to observe the direction of these components.* In Example C, which follows, note that the vector **B** has a positive vertical component and a negative horizontal component.

EXAMPLE C Add vectors **A** and **B** as shown in Figure 18-28(a).

Since neither vector is directed along an axis, it is necessary to resolve both vectors into their x and y components. Following the same procedure as in previous examples, we have (see Figure 18-28(b)):

$$A_x = A \cos 15.0° \quad \text{and} \quad A_y = A \sin 15.0°$$
$$B_x = B \cos 125.0° \quad \text{and} \quad B_y = B \sin 125.0°$$

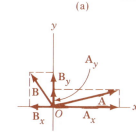

(a)

(b)

Substituting the appropriate values, we get

$$A_x = 570 \cos 15.0° = 551$$
$$A_y = 570 \sin 15.0° = 148$$
$$B_x = 350 \cos 125.0° = -201 \quad \longleftarrow \text{ directed along negative } x \text{ axis}$$
$$B_y = 350 \sin 125.0° = 287$$

We can condense and summarize the above results by using the following table format.

	x components		y components	
A	570 cos 15.0° =	551	570 sin 15.0° = 148	
B	350 cos 125.0° =	−201	350 sin 125.0° = 287	
R	**Resultant**	350 **Add**	435	**Add**

Using Equations (18-3) and (18-4), we have

$$R = \sqrt{R_x^2 + R_y^2} = \sqrt{(350)^2 + (435)^2} = 558$$

$$\tan \theta = \frac{R_y}{R_x} = \frac{435}{350} \quad \text{so that} \quad \theta = 51.2°$$

We conclude that the magnitude of the resultant is 558 and its direction is 51.2° above the positive x axis.

The calculator sequence, using two memories, for this solution is as follows:

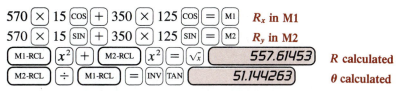

If only one memory is available, then the result for R_y will have to be written down and reentered when necessary. When these results are rounded, we get a magnitude of 558 and a direction of 51.1°. The small discrepancy of 0.1° in the direction is due to the difference in rounding between the calculator solution and the solution summarized in the preceding table. □

FIGURE 18-28

The following examples demonstrate two of the many applications of vectors.

EXAMPLE D

Two forces in the same vertical plane act on an object. One force of 6.85 lb acts to the right at an angle of 18.0° above the horizontal. The other force of 7.44 lb acts to the right at an angle of 51.5° below the horizontal. Find the resultant force.

From the statement of the problem, we show the forces as F_1 and F_2 in Figure 18-29(a). We then resolve each force into its x and y components as follows.

	x components	*y components*
F_1	6.85 cos 18.0° = 6.515	6.85 sin 18.0° = 2.117
	clockwise angle	
F_2	7.44 cos (−51.5°) = 4.632	7.44 sin (−51.5°) = −5.823
R	11.147 Add	−3.706 Add

The resultant force and its components are shown in Figure 18-29(c).

The magnitude of the resultant is found by the Pythagorean theorem:

$$R = \sqrt{R_x^2 + R_y^2} = \sqrt{(11.147)^2 + (-3.706)^2} = 11.7 \text{ lb}$$

The angle θ is found by using Eq. (18-4):

$$\tan \theta = \frac{R_y}{R_x} = \frac{-3.706}{11.147}$$

$$\theta = -18.4°$$

The angle of −18.4° could also be expressed as 341.6°. The resultant force is 11.7 lb and is directed at an angle of 18.4° below the positive x axis.

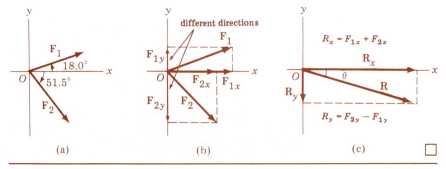

FIGURE 18-29   (a)   (b)   (c)

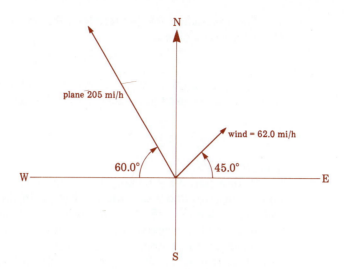

FIGURE 18-30

EXAMPLE E   A pilot flies her plane so that it is pointed in the direction shown in Figure 18-30. Her airspeed indicator shows 205 mi/h. At her altitude there is a 62.0 mi/h wind blowing from the southwest. Find the true speed of the plane relative to the ground, and find the direction the plane is actually headed. (Note that a wind *from* the southwest blows *to* the northeast.)

This problem involves finding the vector which is the sum of the two given vectors. The magnitude of the resultant is the actual speed of the plane; the direction of the resultant is the direction the plane will actually travel. We can treat Figure 18-30 as a regular x-y coordinate system with east corresponding to the positive x axis and north corresponding to the positive y axis. The following table shows the vectors resolved into rectangular components. Note that the angle of 60.0° becomes 120.0° when the plane vector is represented in standard position.

	x component	*y component*
Wind	$62.0 \cos 45.0° =$ 43.8	$62.0 \sin 45.0° =$ 43.8
	┌─── standard position ───┐	
Plane	$205 \cos 120.0° = -102.5$	$205 \sin 120.0° = 177.5$
R	**Resultant** -58.7 **Add**	221.3 **Add**

The plane's horizontal component is negative since it corresponds to the negative x axis.

(Continued on next page)

The magnitude of the resultant **R** can now be found by using the Pythagorean theorem.

$$R = \sqrt{R_x^2 + R_y^2} = \sqrt{(-58.7)^2 + (221.3)^2} = 229$$

We now proceed to find the direction as follows. With $R_x = -58.7$ and $R_y = 221.3$, we use the tangent function to get

$$\tan \theta = \frac{R_y}{R_x} = \frac{221.3}{-58.7} = -3.770$$

Thus $\theta = -75.1°$.

With a positive y rectangular component and a negative x rectangular component, we know that the resultant is in the second quadrant. In this case, the result of $-75.1°$ corresponds to $75.1°$ north of west.

In general, a negative angle indicates that the resultant vector is in the second or fourth quadrant. We can adjust the negative angle by adding $180°$ or $360°$ so that the result is a positive angle for a vector in standard position.

The plane travels at a true speed of 229 mi/h and flies in a direction of $75.1°$ north of west. (Air navigation headings are actually measured in degrees from $0°$ to $360°$ where $0°$ is north and the other angles are measured clockwise from north so that $90°$ is east, $180°$ is south, $270°$ is west, and so on.)   ☐

Exercises 18-3

In Exercises 1 through 8, vectors **A** *and* **B** *are at right angles. Determine the magnitude and the direction of the resultant relative to vector* **A**.

1. $A = 5.00, B = 2.00$

2. $A = 72.0, B = 15.0$

3. $A = 746, B = 1250$

4. $A = 0.962, B = 0.385$

5. $A = 34.7, B = 16.3$

6. $A = 126, B = 703$

7. $A = 12.05, B = 22.46$

8. $A = 0.0348, B = 0.0526$

In Exercises 9 through 16, use the Pythagorean theorem and trigonometric ratios to find the resultant of the vectors in the indicated figure. Determine the magnitude and direction of the resultant.

9. Figure 18-31(a)

10. Figure 18-31(b)

11. Figure 18-31(c)

12. Figure 18-31(d)

13. Figure 18-31(e)

14. Figure 18-31(f)

15. Figure 18-31(g)

16. Figure 18-31(h)

In Exercises 17 through 32, determine the resultant of the vectors with the given magnitudes and directions. Positive angles are measured counterclockwise from the positive x axis, and negative angles are measured clockwise from the positive x axis.

17. A: 8.0, 0°
        B: 6.0, 90°

18. A: 0.766, 44.0°
        B: 0.486, 90.0°

19. A: 47.2, 90.0°
        B: 28.5, 165.0°

20. A: 450, 20.0°
        B: 320, 65.0°

21. A: 566, 155.0°
        B: 1240, 221.0°

22. A: 20.6, 131.1°
        B: 45.1, 318.3°

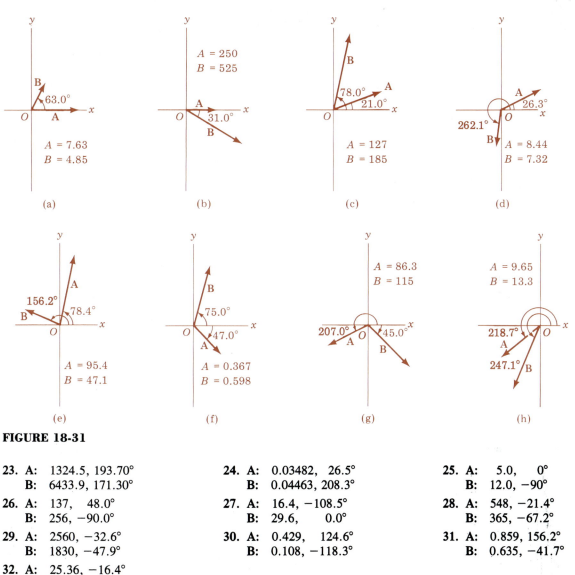

FIGURE 18-31

23. A: 1324.5, 193.70°
 B: 6433.9, 171.30°

24. A: 0.03482, 26.5°
 B: 0.04463, 208.3°

25. A: 5.0, 0°
 B: 12.0, −90°

26. A: 137, 48.0°
 B: 256, −90.0°

27. A: 16.4, −108.5°
 B: 29.6, 0.0°

28. A: 548, −21.4°
 B: 365, −67.2°

29. A: 2560, −32.6°
 B: 1830, −47.9°

30. A: 0.429, 124.6°
 B: 0.108, −118.3°

31. A: 0.859, 156.2°
 B: 0.635, −41.7°

32. A: 25.36, −16.4°
 B: 10.28, 212.8°

In Exercises 33 through 44, solve the given vector problems by using the Pythagorean theorem and the trigonometric ratios.

33. In an automobile, two forces act on the same object and at right angles to each other. One force is 212 lb, and the other is 158 lb. Find the resultant of these two forces.

34. Two forces in the same vertical plane act on an object. One force of 8.35 lb acts to the right at an angle of 16.0° above the horizontal. The other force of 9.83 lb acts to the right at an angle of 63.7° below the horizontal. Find the resultant force.

35. A ship travels 38.0 mi due east from a port and then turns south for 26.0 mi farther. What is the displacement of the ship from the port?

36. A motorboat which travels 7.00 km/h in still water heads directly across a stream which flows at 3.20 km/h. What is the resultant velocity of the boat? See Figure 18-32.

37. A plane is traveling horizontally at 1500 km/h. A rocket is fired horizontally ahead from the plane at an angle of 8.0° from the direction of the plane and with a velocity of 2500 km/h. What is the resultant velocity of the rocket?

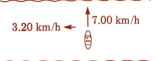

FIGURE 18-32

38. Two forces in the same horizontal plane act on an object. One force of 58.0 lb acts to the right. The second force of 40.8 lb acts to the left at an angle of 23.5° from the line of action of the first force. Find the resultant force.

39. A 50.0-lb object is suspended from the ceiling by a rope. It is then pulled to the side by a horizontal force of 20.0 lb and held in place. What is the tension in the rope? (In order that the object be in equilibrium there should be no net force in any direction. This means that the tension in the rope must equal the resultant of the two forces acting on it. The forces in this case are the 20.0-lb force and the weight of the object.)

40. A plane is headed due north at a speed of 550 mi/h with respect to the air. There is a tail wind blowing from the southeast at 75.0 mi/h. What is the resultant velocity of the plane with respect to the ground? See Figure 18-33.

41. Two voltages V_1 and V_2 are out of phase so that they are represented by two vectors with the same initial point. Find the resultant vector if $V_1 = 120.0$ V, $V_2 = 18.0$ V, and the phase angle between the two vectors is 59.7°.

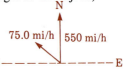

FIGURE 18-33

42. An object is affected by two different substances having magnetic intensities represented by vectors H_1 and H_2. If $H_1 = 106$ A/m, $H_2 = 76.0$ A/m, and the angle between the two vectors is 53.7°, find the resultant.

43. An object with a weight equivalent to 90.0 N is on the floor. Two forces are acting on the object. The first force, of 110 N, acts at an angle of 30.0° upward from the horizontal to the right. The second force, of 75.0 N, acts at an angle of 20.0° upward from the horizontal to the left. Will the object move? (Ignore frictional effects.) See Figure 18-34.

44. A wooden block is on an inclined plane, as shown in Figure 18-35. The block will slip down the plane if the component acting downward along the plane is greater than the frictional force acting upward along the plane. Given that the block weighs 85.0 lb and the plane is inclined at 13.0°, will the block slip if the frictional force is 18.0 lb?

FIGURE 18-34   **FIGURE 18-35**

Chapter 18 Formulas

$A_x = A \cos \theta$	horizontal component	(18-1)
$A_y = A \sin \theta$	vertical component	(18-2)
$R = \sqrt{R_x^2 + R_y^2}$	magnitude of resultant	(18-3)
$\tan \theta = \dfrac{R_y}{R_x}$	direction of resultant	(18-4)

18-4 Review Exercises for Chapter 18

In Exercises 1 through 4, determine whether a scalar or a vector is described in part (a) and part (b). Explain your answers.

1. (a) The airspeed indicator of an airplane shows a speed of 187 mi/h.
   (b) The vertical speed indicator of an airplane shows that the plane is descending at a rate of 500 ft/min.
2. (a) A missile is launched straight up with a speed of 2560 mi/h.
   (b) A missile is cruising at the rate of 2560 mi/h.
3. (a) A 257-lb force is applied to a beam at a particular point.
   (b) A force of 257 lb is applied to a beam in such a way that it is perpendicular down to the beam.
4. (a) Two men are pulling on opposite ends of a rope in opposite directions with forces having magnitudes of 80 lb and 72 lb, respectively.
   (b) Two men are pulling on a rope and their efforts are measured as 80 lb and 72 lb, respectively.

*In Exercises 5 through 8, find the sum of vectors **A** and **B** shown in the indicated figure. Use diagrams.*

5. Figure 18-36(a)   6. Figure 18-36(b)   7. Figure 18-36(c)   8. Figure 18-36(d)

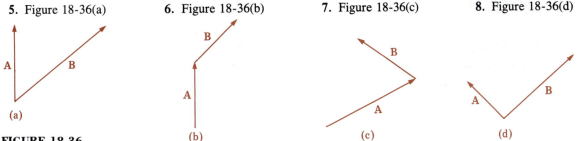

FIGURE 18-36

In Exercises 9 through 16, use appropriate diagrams to find the indicated vector sums and differences for the vectors shown in Figure 18-37.

9. **A** + **B**   10. **A** − **B**
11. **A** + **B** + **C**   12. **A** + **B** + **C** + **D**
13. 2**A** + **B**   14. **A** + $\frac{1}{2}$**B**
15. 2**A** − **B**   16. 2**D** − **C**

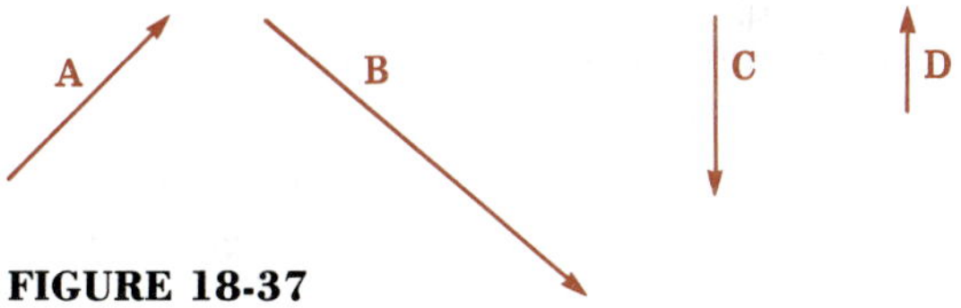

FIGURE 18-37

In Exercises 17 through 20, find the x and y components of the vectors shown in the indicated figures. Use the trigonometric ratios.

17. Figure 18-38(a)   18. Figure 18-38(b)   19. Figure 18-38(c)   20. Figure 18-38(d)

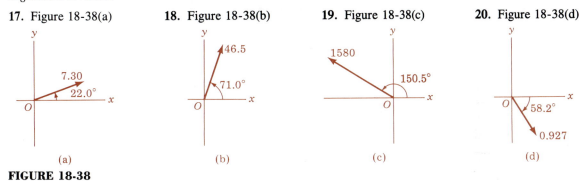

FIGURE 18-38

In Exercises 21 through 24, resolve the given vector into its x component and y component. Each angle is in standard position.

21. Vector A, with $A = 17.2$, $\theta = 22.2°$

22. Vector A, with $A = 127$, $\theta = 132.4°$

23. Vector A, with $A = 16.48$, $\theta = -57.3°$

24. Vector A, with $A = 0.8966$, $\theta = 262.9°$

In Exercises 25 through 28, resolve the described vector into appropriate vertical and horizontal components.

25. A wind of 36 mi/h is blowing in the direction of 15.0° north of west.

26. A Cessna Citation jet is flying at a speed of 270 mi/h in a direction of 20.0° south of west.

27. A cable is anchored on the ground and is connected to the top of a pole. The cable makes an angle of 39.0° with the pole and it pulls on the pole with a force having magnitude 555 lb.

28. A hammer strikes a board at a point. The direction of the hammer makes an angle of 85.0° with the board's surface, and the force of the strike is 27 lb.

In Exercises 29 through 32, add the vectors shown in the indicated figures by using the Pythagorean theorem and the trigonometric ratios.

29. Figure 18-39(a)   **30.** Figure 18-39(b)   **31.** Figure 18-39(c)   **32.** Figure 18-39(d)

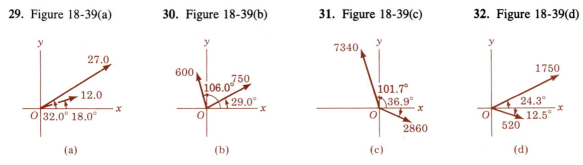

FIGURE 18-39

In Exercises 33 through 40, determine the resultant of the vectors with the given magnitudes and directions.

33. Vector A, with $A = 17.76$, $\theta = 90.00°$
Vector B, with $B = 21.20$, $\theta = 0.00°$

34. Vector A, with $A = 136.4$, $\theta = 50.8°$
Vector B, with $B = 635.0$, $\theta = 17.7°$

35. Vector A, with $A = 0.992$, $\theta = 18.4°$
Vector B, with $B = 0.545$, $\theta = 112.1°$

36. Vector A, with $A = 47.3$, $\theta = -36.2°$
Vector B, with $B = 98.1$, $\theta = 134.2°$

37. Vector A, with $A = 36$, $\theta = 25°$
Vector B, with $B = 49$, $\theta = -90°$

38. Vector A, with $A = 2.39$, $\theta = 51.2°$
Vector B, with $B = 5.08$, $\theta = 85.6°$

39. Vector A, with $A = 75.6$, $\theta = 29.4°$
Vector B, with $B = 48.3$, $\theta = 151.2°$

40. Vector A, with $A = 429$, $\theta = 248.1°$
Vector B, with $B = 301$, $\theta = 301.0°$

In Exercises 41 through 68, solve the given problems.

41. A tree surgeon attempts to drop a tree by making a cut and then pulling with a rope. The angle between the rope and the tree is 47.0° and the rope is pulled with a force which has a magnitude of 155 lb. Only the horizontal component of the force will cause the tree to fall. Find the horizontal component. See Figure 18-40.

42. The force caused by a truck on an inclined loading ramp is represented by a vector which has magnitude 32,400 lb and a direction which is 82.0° below the negative x axis. Resolve this vector into its x and y components.

FIGURE 18-40

43. A missile is launched at an angle of 81.1° with respect to level ground. If it is moving at a speed of 2250 km/h, find the horizontal and vertical components of the velocity.

44. In conducting a ballistics test, it is determined that at one point, a bullet is traveling 800 ft/s along a path which makes an angle of 62.7° with level ground. Find the horizontal and vertical components of the velocity vector.

45. A cable is connected to the front of a boat in a lake. The cable makes an angle of 40.5° with the water surface. If the cable is pulled with a force having magnitude 305 lb, find the force which pulls the boat forward along the surface of the lake. See Figure 18-41.

46. A jet climbs at a speed of 380 mi/h and its path makes an angle of 21.6° with level ground. Find the horizontal and vertical components of the velocity.

47. A Cessna Citation jet is traveling at 315 mi/h in a direction 37.2° north of east. How much farther north is this jet after 1.50 h of flying?

48. In conducting a stress test on an automobile part, two forces are applied at a point. A vertical force of 1200 lb pushes down on the part while a horizontal force of 850 lb pulls the part to the right. Find the resultant of these two forces.

49. A shearing pin is designed to break and disengage gears before damage is done to a machine. In conducting tests on such a pin, two forces are applied. One force pulls vertically upward with a magnitude of 9650 lb while another force pulls horizontally with a magnitude of 9370 lb. Find the resultant of these two forces. See Figure 18-42.

FIGURE 18-41   **FIGURE 18-42**

50. Two forces act on an object at right angles to each other. One force is 87.2 lb and the other is 13.7 lb. Find the resultant of these two forces.

51. Forces of 867 lb and 532 lb act on an object. The angle between these forces is 67.7°. Find the resultant of these two forces.

52. A machine pulls a cart with a rope in the same way that a child pulls on a wagon. The rope makes an angle of 26° with the horizontal, and the rope is pulled with a force of 36 lb. Find the horizontal component of the force. (This is the component that will cause the cart to move horizontally.) See Figure 18-43.

53. A 655-lb cart is on an incline that makes an angle of 34.8° with the horizontal. Find the force parallel to the surface of the incline. (This is the force that must be negated in order to prevent the cart from rolling down the incline.) See Figure 18-44.

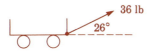

FIGURE 18-43   **FIGURE 18-44**

54. Two cables are connected to a weight and the angle between these cables is 52.7°. The cables are pulled with forces of 176 N and 221 N, respectively. Find the resultant of these two forces. See Figure 18-45.

55. A plane travels 50.0 mi east from an airfield and then travels another 35.0 mi north. What is the plane's displacement from the airfield? See Figure 18-46.

FIGURE 18-45   **FIGURE 18-46**

56. A person walks 3.25 km to the west from an intersection of roads. The person then turns and walks 2.50 km in a direction 20.0° north of west. What is the person's displacement from the intersection?

57. Two voltages V_1 and V_2 are out of phase so that they are represented by two vectors with the same initial point. Find the magnitude of the resultant if $V_1 = 110.0$ V, $V_2 = 80.0$ V, and the phase angle between V_1 and V_2 is 73.4°.

58. A jet travels at the speed of 658 mi/h. If it flies east for 1.25 h, then north for 1.60 h, and northeast for 2.00 h, find its distance from the starting point.

59. A surveyor sets a stake and then proceeds to a point 486.1 m due east of the stake. From that point she walks 390.1 m in a direction 15.2° north of east. What is her displacement from the original stake? See Figure 18-47.

60. A surveyor locates a benchmark 87.2 m to the southeast of his present position. The benchmark is known to be 20.0 m west of a grade stake. What is the surveyor's displacement from the grade stake? See Figure 18-48.

61. A surveyor locates a point at which an upgrade in a road is to begin. That point is 27.3 m northeast of the surveyor and 5.0 m north of a right-of-way marker. What is the surveyor's displacement from the right-of-way marker? See Figure 18-49.

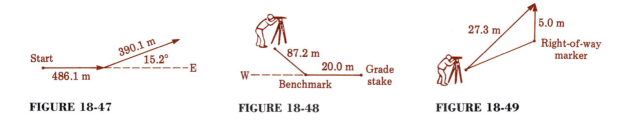

FIGURE 18-47   **FIGURE 18-48**   **FIGURE 18-49**

62. A U.S. Navy ship sails 72.4 km due east. It then turns 30.2° north of east and sails an additional 49.6 km. Find the distance and direction of the ship relative to its departure point.

63. A ship is traveling at 15.0 mi/h. What is the velocity of a person walking across the deck, at right angles to the direction of the ship at 3.00 mi/h, with respect to the sea?

64. A balloon is rising at the rate of 6.50 m/s and at the same time is being blown horizontally by the wind at the rate of 7.50 m/s. Find the resultant velocity.

65. An airplane is pointed in the direction of due south and its airspeed indicator shows 172 mi/h. A wind of 18.0 mi/h is headed toward the direction of 20.0° north of east. Find the true speed and direction of the plane.

66. A national park survey team travels 12 km in a direction 27.0° south of east. They then turn and travel 19 km in a direction 12.0° north of east. Find their displacement from the starting point. See Figure 18-50.

67. A training apparatus for Olympic divers consists of a harness with two cables connected at a point. The angle between the cables is 120.0° and trainers pull on the cables with forces having magnitudes of 32 lb and 28 lb, respectively. Find the resultant of the two forces corresponding to the two cables. See Figure 18-51.

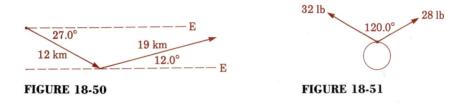

FIGURE 18-50   **FIGURE 18-51**

68. In a test conducted on a seat belt, a force of 5200 lb pulls in the direction toward the front of the car while another force of 4800 lb pulls toward the door on the driver's side. Find the resultant of these two forces.

CHAPTER NINETEEN

Radians and Trigonometric Curves

We have already used the degree as a measure of an amount of rotation. If an angle is 40.0°, then one side must be rotated by an amount of 40.0° in order for it to meet the other side. We saw in Section 6-1 that one complete rotation of a line about a point is defined to be an angle of 360°. It therefore follows that 1° is defined as $\frac{1}{360}$ of one complete rotation of a line about a point.

In the preceding chapters there are many applications in which we used degree measurements of angles. The applied vector problems of Chapter 18, for example, involved degree measurements. However, there are many other situations, both practical and theoretical, in which another type of angle measurement is more appropriate. In this chapter we begin by introducing the **radian**. We then consider applications involving radian measure, and we proceed to use radians in constructing the graphs of trigonometric functions.

As an example of an applied problem using radian measure, consider an analysis of a light airplane that was damaged when the nosewheel collapsed. If the tip of the propeller hit the runway and made an impression equivalent to an object moving 8640 ft/min, we want to find the rate (in revolutions per minute) at which the propeller was turning. The propeller is 5.50 ft from tip to tip. We will solve this problem in Section 19-2. The solution will provide us with information about the status of the aircraft when the accident occurred. We would know if the engine was idling or if it was being powered up for a takeoff.

19-1 Radians

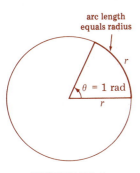

arc length equals radius

FIGURE 19-1

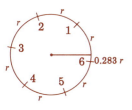

FIGURE 19-2

The **radian** is formally defined as the measure of the angle with these properties:

1. The angle has its vertex at the center of a circle.
2. The angle intercepts an arc on the circle, and the length of that arc equals the length of the radius.

To understand this definition, refer to Figure 19-1. The indicated angle θ has a measure described as 1 radian (rad) since the arc length r equals the radius r.

Because the circumference of any circle is given by $c = 2\pi r$, it follows that the radius may be marked off 2π times (or about 6.283 times) along the circumference, as shown in Figure 19-2. *One complete rotation therefore corresponds to 2π rad.* Since one complete rotation also corresponds to 360°, we conclude that 2π rad = 360°. Dividing by 2 we get

$$\pi \text{ rad} = 180° \tag{19-1}$$

We can divide both sides of Eq. (19-1) by 180 to get

$$1° = \frac{\pi}{180} \text{ rad} = 0.01745 \text{ rad} \tag{19-2}$$

If we divide both sides of Eq. (19-1) by π we get

$$1 \text{ rad} = \frac{180°}{\pi} = 57.3° \tag{19-3}$$

Since Eqs. (19-1) through (19-3) express relationships between degrees and radians, these equations suggest ways of converting between radians and degrees. We summarize these procedures as follows:

See Appendix B for a computer program that converts degrees to radians.

degrees → radians	radians → degrees
Multiply the given number of degrees by $\frac{\pi}{180}$ rad	Multiply the given number of radians by $\frac{180°}{\pi}$

EXAMPLE A We convert 180° to radians as follows:

$$180° = (180)\left(\frac{\pi}{180}\right) = \pi = 3.14 \text{ rad}$$

We convert 240° to radians as follows:

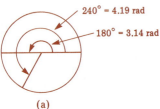

$$240° = (240)\left(\frac{\pi}{180}\right) = \frac{4\pi}{3} = 4.19 \text{ rad}$$

See Figure 19-3(a). To convert 1.6 rad to degrees, we get

$$1.6 \text{ rad} = (1.6)\left(\frac{180}{\pi}\right) = \frac{288°}{\pi} = 91.7°$$

To convert 0.50 rad to degrees, we get

$$0.50 \text{ rad} = (0.50)\left(\frac{180}{\pi}\right) = \frac{90°}{\pi} = 28.6°$$

FIGURE 19-3   See Figure 19-3(b). ☐

One advantage of using radians is that it often simplifies formulas and calculations involving π. Because of this, radian measures are often expressed in terms of π. The next example illustrates this.

EXAMPLE B Convert 30°, 45°, and 60° to radian measure and leave the answers in terms of π.

converting degrees to radians

$$30° = (30)\left(\frac{\pi}{180}\right) = \frac{\pi}{6} \text{ rad}$$

$$45° = (45)\left(\frac{\pi}{180}\right) = \frac{\pi}{4} \text{ rad}$$

$$60° = (60)\left(\frac{\pi}{180}\right) = \frac{\pi}{3} \text{ rad}$$ ☐

If we have a radian measure already expressed in terms of π, we can easily find the corresponding measure in degrees since the conversion will cause two values of π to cancel out through division. See the following example.

EXAMPLE C   Convert $\pi/2$ and $2\pi/3$ to degree measure.

converting radians to degrees

$$\frac{\pi}{2} \text{ rad} = \left(\frac{\pi}{2}\right)\left(\frac{180}{\pi}\right) = 90°$$

$$\frac{2\pi}{3} \text{ rad} = \left(\frac{2\pi}{3}\right)\left(\frac{180}{\pi}\right) = 120°$$ ☐

Many modern uses of trigonometric ratios involve treating them as functions of real numbers instead of angle measurements. For example, we might consider $y = \sin x$ where x is a real number and not an angle measured in degrees or radians. This becomes important as we proceed to consider many applications in which angle measurements are inappropriate. In electronics, for example, we might describe the relationship between current i and time t by $i = 5 \sin 60\pi t$. Here, t is not an angle and it would be inappropriate to express t in degrees or radians. This brings us to a very important point. Since π is the *ratio* of the circumference of a circle to its diameter, it is one distance divided by another. As a result, radians really have no units and *radian measure amounts to measuring angles in terms of numbers*. As a result of this useful property of radians, **angle measurements lacking indicated units are understood to be in radian units.**

EXAMPLE D   To convert 1.75 to degrees, we get

no units indicates radian measure

$$1.75 = (1.75)\left(\frac{180}{\pi}\right) = \frac{315°}{\pi} = 100°$$

To convert 270° to an equivalent number without units, we get

$$270° = (270)\left(\frac{\pi}{180}\right) = \frac{3\pi}{2} = 4.71$$ no units indicates radian measure

In this example 1.75 and 4.71 are expressed without units, so they are understood to be radian measure.   ☐

When using the trigonometric functions with degree measures, confusion tends to be minimized because these functions operate on the degree measures to yield results which are numbers without units. For example, it should be clear that $\sin 30° = 0.5000$ is correct while $\sin 0.5000 = 30°$ is fundamentally wrong. With radian measure, however, we may be dealing with two numbers, both of which lack units. We should therefore *be careful to correctly interpret expressions such as sin 1 and sin θ = 1.* The expression **sin 1 *is equivalent to* sin 57.3°** (since 1 rad = 57.3°). But *if* **sin θ = 1,** *then θ is the angle for which the sine is* **1.** Since sin 90° = 1, we conclude that θ = 90°, or θ = $\pi/2$. The following examples provide additional illustrations whereby we evaluate expressions involving radians.

EXAMPLE E   1. Evaluate cos $\pi/3$. ◄—no units indicates radian measure
Since

converting radians to degrees

$$\frac{\pi}{3} = \left(\frac{\pi}{3}\right)\left(\frac{180}{\pi}\right) = 60° \quad \text{we get}$$

$$\cos\frac{\pi}{3} = \cos 60° = 0.5000.$$

2. Evaluate tan 1.37. ◄—no units indicates radian measure
Since

$$1.37 = (1.37)\left(\frac{180}{\pi}\right) = 78.5° \quad \text{we get}$$

$$\tan 1.37 = \tan 78.5° = 4.915. \qquad \square$$

EXAMPLE F   Find a number t (not in degrees) for which sin t = 0.9178.
     Since sin t = 0.9178, we could get t = 66.6°, but we want t as a number not in degree units. We therefore convert 66.6° to radians to get

$$t = (66.6)\left(\frac{\pi}{180}\right) = 1.162$$

so that sin 1.162 = 0.9178.

no units indicates radian measure $\square$

     When evaluating trigonometric expressions involving radians, it is not necessary to convert to degrees so that the calculations can be performed. That is, we can evaluate expressions like sin 2 or cos 4.5 directly without first converting 2 radians and 4.5 radians to degree measure.
     Most scientific calculators have a capability for selecting the degree mode or radian mode. If the radian mode is selected, then all entered angles are treated as radians. While *in the radian mode,* an entry of

2 [SIN] [*0.90929743*]

will yield 0.90929743 since 2 radians = 114.6° and sin 114.6° = 0.90929743. While *in the degree mode,* the same entry would yield 0.03489950 since sin 2° = 0.03489950.
     Many scientific calculators provide keys for converting between radians and degrees. It is common to identify these keys with labels such as D-R or DRG where D represents degrees, R represents radians, and G represents

grads. (*A* **grad** *is $\frac{1}{400}$ of one complete rotation so that* $360° = 400$ *grads*.) Consult the manual for your calculator to see how to select the mode. Also study the examples given for converting between degrees and radians. In any event, these conversions can be made by multiplying by the appropriate conversion factor as indicated in the following two examples.

EXAMPLE G Convert 39.4° to radians by using a scientific calculator.

We have already converted degrees to radians by multiplying the given number of degrees by $\pi/180$. The correct key sequence is therefore

$$39.4 \boxed{\times} \boxed{\pi} \boxed{\div} 180 \boxed{=} \boxed{\textit{0.68765973}}$$

The key represented by $\boxed{\pi}$ may involve a key whose secondary purpose is to give the value of π; if that is the case, precede the key by the key labeled INV or 2nd. If such a key is not available, use 3.1416 instead of π. This whole process can be streamlined by noting that $\pi/180 = 0.01745$ so that 39.4×0.01745 will yield the radian equivalent of 39.4°. That is, to convert an angle to radians, we can simply multiply the angle by 0.01745. ☐

EXAMPLE H Convert 2.37 to degrees.

We can convert radians to degrees by multiplying by $180°/\pi$, or 57.296°. The entry of

$$2.37 \boxed{\times} 57.296 \boxed{=} \boxed{\textit{135.79152}}$$

shows that 2.37 is equivalent to 135.8° (rounded to tenths). ☐

We should be sure that we understand the underlying theory so that errors will not arise from thoughtless use of calculators. The next example leads to two solutions even though the calculator might seem to suggest that there is only one solution.

EXAMPLE I Find t in radians if $\sin t = 0.8415$ and $0 < t < 2\pi$.

Using a calculator *in the radian mode* we can easily determine that $\sin 1.0001 = 0.8415$ so that $t = 1.0001$. But in addition to this first-quadrant angle, *there is also a second-quadrant angle* for which $\sin t = 0.8415$. Using the idea of reference angles summarized by Eqs. (17-2), (17-3), and (17-4), we find the second-quadrant angle in radians as follows (see Eq. 17-2).

For reference:
θ is standard position angle
α is reference angle

Quadrant
II $\alpha = 180° - \theta$
(17-2)
III $\alpha = \theta - 180°$
(17-3)
IV $\alpha = 360° - \theta$
(17-4)

equivalent to 180°
$$\theta = \pi - 1.0001 = 2.1415$$
reference angle

Note that we used Eq. (17-2) and solved for θ, with the angles expressed in radians instead of degrees. The two solutions are $t = 1.0001$ and $t = 2.1415$. ☐

If we wish to find the value of a function of an angle greater than $\pi/2$, we must first determine which quadrant the angle is in and then find the reference angle. In this process we should note the radian measure equivalents of 90°, 180°, 270°, and 360°.

For 90° we have $\dfrac{\pi}{2} = 1.571$;

for 180° we have $\pi = 3.142$;

for 270° we have $\dfrac{3\pi}{2} = 4.712$;

for 360° we have $2\pi = 6.283$.

EXAMPLE J Find sin 3.402 using the proper reference angle.

Since 3.402 is greater than 3.142 but less than 4.712, we know that this angle is in the third quadrant and has a reference angle of $3.402 - 3.142 = 0.260$. Thus

$$\overbrace{}^{\text{sin is } - \text{ in third quadrant}}$$
$$\sin 3.402 = -\sin 0.260 = -0.2575$$

Now find cos 5.210. Since 5.210 is between 4.712 and 6.283, we know that this angle is in the fourth quadrant and its reference angle is $6.283 - 5.210 = 1.073$. Thus

$$\overbrace{}^{\text{cos is } + \text{ in fourth quadrant}}$$
$$\cos 5.210 = \cos 1.073 = 0.4773$$
□

EXAMPLE K The voltage (in volts) in a circuit is described by the equation

$$v = 112 \sin 0.540t$$

Find the voltage at $t = 1.35$ s.

With $t = 1.35$ s, we evaluate v as follows.

$$\overbrace{}^{\text{substitute for } t}$$
$$v = 112 \sin 0.540(1.35)$$
$$= 112 \sin 0.729 = 112(0.666)$$
$$= 74.6 \text{ V}$$
□

We have discussed radians and their use with calculators. Computer programs also use radian measure. The BASIC programming language normally includes the functions SIN(X), COS(X), TAN(X), where X must be given in radian measure. In Appendix B, the programs for solving triangles and adding vectors incorporate steps that convert radians to degrees and degrees to radians.

Exercises 19-1

In Exercises 1 through 8, express the given angle measurements in radians. (Leave answers in terms of π.)

1. 40°, 16°   **2.** 36°, 240°   **3.** 55°, 330°   **4.** 20°, 335°

5. 30°, 135°   **6.** 60°, 300°   **7.** 175°, 210°   **8.** 52°, 280°

In Exercises 9 through 16, the given numbers express angle measure. Express the measure of each angle in terms of degrees.

9. $\dfrac{2\pi}{3}, \dfrac{\pi}{5}$ **10.** $\dfrac{\pi}{10}, \dfrac{4\pi}{5}$ **11.** $\dfrac{\pi}{12}, \dfrac{5\pi}{4}$ **12.** $\dfrac{2\pi}{9}, \dfrac{3\pi}{2}$

13. $\dfrac{14\pi}{15}, \dfrac{7\pi}{9}$ **14.** $\dfrac{11\pi}{12}, \dfrac{8\pi}{5}$ **15.** $\dfrac{7\pi}{36}, \dfrac{3\pi}{4}$ **16.** $\dfrac{51\pi}{30}, \dfrac{23\pi}{12}$

In Exercises 17 through 24, express the given angles in radian measure. (Round to three significant digits and use 3.14 as an approximation for π.)

17. 46.0°   **18.** 72.0°   **19.** 190.0°   **20.** 253.0°

21. 278.6°   **22.** 98.6°   **23.** 182.4°   **24.** 327.9°

In Exercises 25 through 32, the numbers express angle measure. Express the measure of each angle in terms of degrees.

25. 0.80   **26.** 0.25   **27.** 2.50   **28.** 1.75

29. 3.25   **30.** 5.00   **31.** 12.4   **32.** 75

In Exercises 33 through 44, evaluate the given trigonometric functions by first changing the radian measure to degree measure.

33. $\cos \dfrac{\pi}{3}$ **34.** $\sin \dfrac{\pi}{4}$ **35.** $\tan \dfrac{2\pi}{3}$ **36.** $\cos \dfrac{7\pi}{12}$

37. $\sin 1.05$ **38.** $\cos 0.785$ **39.** $\tan 0.875$ **40.** $\cot 1.43$

41. $\sec \dfrac{5\pi}{18}$ **42.** $\csc \dfrac{3\pi}{8}$ **43.** $\csc 3.00$ **44.** $\sec 5.66$

In Exercises 45 through 52, evaluate the given trigonometric functions directly, without first changing the radian measure to degree measure.

45. $\sin 1.0300$ **46.** $\tan 0.1450$ **47.** $\sec 3.650$ **48.** $\csc 4.766$

49. $\cot 6.180$ **50.** $\cos 5.050$ **51.** $\csc 2.875$ **52.** $\sec 3.135$

In Exercises 53 through 60, find the values of θ for $0 < \theta < 2\pi$. In each case there are two solutions.

53. $\cos \theta = 0.5000$ **54.** $\sin \theta = 0.9975$ **55.** $\tan \theta = -2.8198$ **56.** $\cos \theta = -0.7900$

57. $\sin \theta = 0.4350$ **58.** $\cot \theta = 3.916$ **59.** $\csc \theta = -1.030$ **60.** $\sec \theta = 2.625$

In Exercises 61 through 68, solve the given problems.

61. The *end slope* of the free end of a deflected beam is given as $\theta = 0.00256$ rad. Convert this value to degree measure.

62. In studying the trajectory of a projectile, the horizontal component of velocity is given by $v_0 \cos \theta$. Find that component if $v_0 = 66$ m/s and $\theta = 0.668$.

63. Under certain conditions, the instantaneous voltage in a power line is given by

$$v = 170 \sin 377t$$

where t is time in seconds. Find v if $t = 0.0050$ s.

64. One pulley rotates through 340.2° while another pulley rotates through an angle of 6.00 radians. Which pulley rotates more?

65. In surveying, the horizontal distance between two points is calculated from

$$H = s \cos \alpha$$

Find H if $s = 37.4$ m and $\alpha = 0.465$.

66. In analyzing the behavior of an object attached to a spring, the time t (in seconds) required for the object to move halfway in to the center from its initial position is found from

$$\cos kt = \frac{1}{2}$$

If $k = 10$, find the time t.

67. A flywheel rotates through 4.5 revolutions. What angle corresponds to this amount of rotation? Express the answer in degrees and in radians.

68. At cruising speed, the crankshaft of a light aircraft engine rotates 2300 r/min. Express 2300 complete revolutions as an angle in degree measure and in radian measure.

In Exercises 69 through 76, use a scientific calculator to convert the given angle measure to radian measure.

69. 56.344°   **70.** 426.558°   **71.** 37.987°   **72.** 16.125°

73. 27°35′   **74.** 327°17′   **75.** 8°40′50″   **76.** 6°30′45″

In Exercises 77 through 80, use a scientific calculator to convert the given angle measure from radians to degrees.

77. 2.55   **78.** 1.375   **79.** $\dfrac{\pi}{13}$ **80.** $\dfrac{7.3\pi}{9.2}$

In Exercises 81 through 84, use a scientific calculator to evaluate the given trigonometric functions.

81. $\sin 3.46227$ **82.** $\cos \dfrac{\pi}{4.9365}$ **83.** $\sin 0.000134\pi$ **84.** $\sin 0.000786$

19-2 Applications of Radian Measure

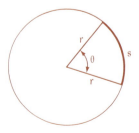

FIGURE 19-4

Radian measure has many applications in mathematics and technology. In this section we examine some of the common uses of radian measure in geometric and technical applications.

In Section 15-2 we saw that a central angle of a circle is an angle with its vertex at the center of the circle. If the central angle is measured in radians, we can develop a simple and useful relationship between the *central angle* θ, *the length s of the corresponding arc,* and the radius r of the circle (see Figure 19-4). *The relationship between s, r, and θ is*

$$s = r\theta \qquad \text{(where } \theta \text{ is in radians)} \qquad \text{(19-4)}$$

Equation (19-4) is justified by these observations:

1. *The length of an arc on a circle is proportional to its central angle* so that $s = k\theta$, where k is the constant of proportionality.

2. For a complete circle, the arc length s is the circumference $2\pi r$ and the central angle θ is 2π. In this case $s = k\theta$ becomes $2\pi r = k(2\pi)$ so that $k = r$ and Eq. (19-4) is established.

EXAMPLE A

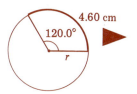

FIGURE 19-5

For a circle with radius $r = 15.0$ ft, find the length of the arc with central angle $\theta = \pi/3$. See Figure 19-5.

Using Eq. (19-4) we get

θ in radians

$$s = (15.0)\left(\frac{\pi}{3}\right)$$

$$= 15.7 \text{ ft}$$

EXAMPLE B

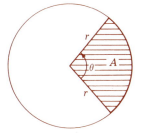

FIGURE 19-6

Given that a circle has an arc with length 4.60 cm and central angle 120.0°, find the radius. See Figure 19-6.

We again use Eq. (19-4) with $s = 4.60$ cm and $\theta = (120.0)(\pi/180)$ so that *the central angle is in radian measure.* Equation (19-4) becomes

$$4.60 = r(120.0)\left(\frac{\pi}{180}\right) = \frac{120\pi r}{180}$$

θ in radians

or

$$r = \frac{(180)(4.60)}{120\pi} = 2.20 \text{ cm}$$

FIGURE 19-7

In Figure 19-7 we give an example of a **sector** of a circle. A *sector of a circle is the area enclosed by two radii and an arc of the circle.* Just as the arc length is proportional to the central angle, *the area of a sector is also proportional to the central angle so that $A = k\theta$*, where θ is in radian measure. Knowing that a complete circle has central angle $\theta = 2\pi$ and area $A = \pi r^2$, we can solve for the constant of proportionality k to get Eq. (19-5).

$$A = \frac{1}{2}\theta r^2 \qquad \text{(where } \theta \text{ is in radians)} \qquad (19\text{-}5)$$

Here A is the area of the sector, r is the radius, and θ is the central angle in radians.

EXAMPLE C

A circle has radius 36.0 cm. Find the area of a sector with a central angle of 24.0°. See Figure 19-8.

We must **be sure to express θ in radian measure.** Using Eq. (19-5) we get

$$A = \frac{1}{2}(24.0)\left(\frac{\pi}{180}\right)(36.0)^2$$

────────── **θ in radians**

$$= 271 \text{ cm}^2$$

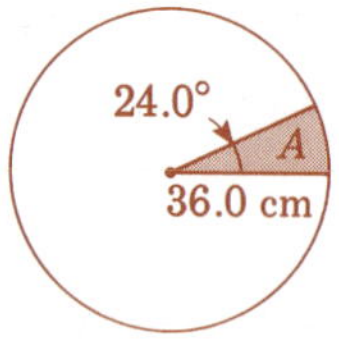

24.0°

A

36.0 cm

FIGURE 19-8

EXAMPLE D

The area of a sector of a circle with radius 2.50 m is known to be 3.28 m². Find the central angle corresponding to this sector. See Figure 19-9.

Using Eq. (19-5) with $A = 3.28$ m² and $r = 2.50$ m, we get

$$3.28 = \frac{1}{2}\theta(2.50)^2$$

Solving for θ we get

$$\theta = \frac{2(3.28)}{(2.50)^2} = 1.05 \quad \text{— no units indicates radian measure}$$

This means that the central angle is 1.05 rad, or 60.2°.

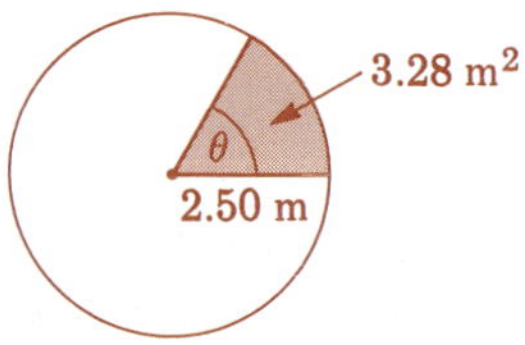

3.28 m²

θ

2.50 m

FIGURE 19-9

Our next application involves velocity. Recall that velocity is a vector with both magnitude and direction, whereas speed is a scalar which corresponds to the magnitude of velocity. When an object travels in a circular path at a constant speed, the distance traveled is the length of the arc which is its path. Referring to Eq. (19-4) which already involves arc length, we divide both sides by *t* to get

$$\frac{s}{t} = \frac{r\theta}{t}$$

In this expression s/t is a speed since it represents distance per unit of time. Also, θ/t represents the change in radians per unit of time. We therefore express the above equation as follows.

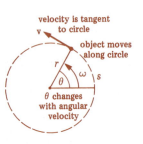

velocity is tangent
to circle
object moves
along circle
θ changes
with angular
velocity

FIGURE 19-10

$$v = \omega r \qquad \text{(where } \omega \text{ is in radians per unit time)} \qquad \text{(19-6)}$$

In Eq. (19-6), *v is the magnitude of the* **linear velocity** *and ω (the Greek omega) is the* **angular velocity** *of an object moving around a circle of radius r.* See Figure 19-10. The units of ω are radians per unit of time. In practice, the angular velocity is often given in revolutions per unit of time, and for such cases it is necessary to convert ω to radians per unit of time before substitution in Eq. (19-6). This conversion is not difficult if we remember that one revolution corresponds to 2π radians.

EXAMPLE E

An object is moving in a circular path with radius 10.0 cm, with an angular velocity of 3.80 rad/s. Find the linear velocity of this object. See Figure 19-11.

Using Eq. (19-6) we get

$$\overbrace{\hspace{2cm}}^{\text{\small ω in rad/s}}$$

$$v = (3.80)(10.0)$$
$$= 38.0 \text{ cm/s}$$

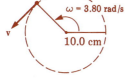

FIGURE 19-11

Therefore, the linear velocity of the object is 38.0 cm/s in the circular path.

□

EXAMPLE F

A floppy disk commonly used in computer systems has a radius of 2.625 in. and it rotates with an angular velocity of 360 r/min. (We use r/min to designate revolutions per minute.) Find the linear velocity of a point on the outer edge. See Figure 19-12.

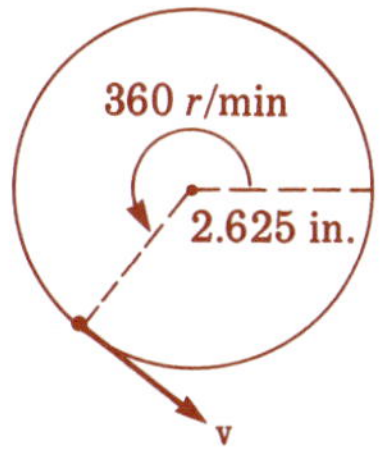

We can use Eq. (19-6) here, but we must be sure to express ω as radians per unit time, not revolutions per unit time. Since each complete revolution is equivalent to 2π rad, we have

$$\omega = (360)(2\pi) = 720\pi \text{ rad/min}$$

Now we use Eq. (19-6) and get

$$v = (720\pi)(2.625)$$

FIGURE 19-12

$$= 5940 \text{ in./min}$$

The linear velocity of a point on the outer edge is therefore 5940 in./min. We can see that a point nearer the middle has a lesser linear velocity since r is less for such a point.

□

EXAMPLE G

A connecting rod serves as a link between a piston and a rotating disk. The disk rotates at 2800 r/min. If the rod is connected to the disk at a point 1.55 in. from the center, find the linear velocity of the point of connection. See Figure 19-13.

We can again use Eq. (19-6) provided that we express the angular velocity ω in radians per unit time, not revolutions per unit time. Since 2800 r/min is equivalent to $(2800)(2\pi)$ rad/min, we get

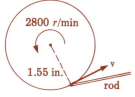

$$\overbrace{\hspace{5cm}}^{\text{\small ω in rad/min}}$$

$$v = (2800)(2\pi)(1.55) = 27{,}300 \text{ in./min}$$

The linear velocity is 27,300 in./min.

□

FIGURE 19-13

EXAMPLE H   A light aircraft was damaged when the nosewheel collapsed. If the tip of the propeller hit the runway and made an impression equivalent to an object moving 8640 ft/min, find the rate (in revolutions per minute) at which the propeller was turning. The propeller is 5.50 ft from tip to tip.

The velocity of 8640 ft/min is the linear velocity v. Noting that $v = 8640$ ft/min and $r = 5.50$ ft/2 $= 2.75$ ft, we use Eq. (19-6) to get

$$v = \omega r$$
$$8640 = \omega(2.75)$$

which means that $\omega = 8640/2.75 = 3140$ rad/min. We convert the angular velocity to revolutions per minute by dividing by 2π and we get $3140/2\pi = 500$ r/min. The propeller was turning at the rate of 500 r/min, indicating that the engine was idling.   □

Exercises 19-2

In Exercises 1 through 8, the radius of a circle and the central angle are given. (a) Find the length of the arc subtended by the central angle. (b) Find the area of the sector.

1. $r = 3.65$ cm, $\theta = 2.01$ **2.** $r = 15.78$ in., $\theta = \dfrac{\pi}{6}$ **3.** $r = 412$ mm, $\theta = 49.3°$

4. $r = 0.275$ m, $\theta = 98.6°$ **5.** $r = 2.37$ ft, $\theta = \dfrac{\pi}{12}$ **6.** $r = 9.27$ cm, $\theta = 120.0°$

7. $r = 6.55$ in., $\theta = 235.0°$ **8.** $r = 3960$ mi, $\theta = 2.38$

In Exercises 9 through 12, an object is moving in a circular path with the given radius and angular velocity. Find the linear velocity.

9. $r = 12.5$ cm, 5.65 rad/s   **10.** $r = 42.39$ mm, 362.0 rad/min

11. $r = 1.28$ ft, 861 r/min   **12.** $r = 3.75$ in., 453 r/s

In Exercises 13 through 40, solve the given problems.

13. In a circle of radius 36.0 cm, find the length of the arc subtended on the circumference by a central angle of 60.0°.

14. In a circle of diameter 12.4 m, find the length of the arc subtended on the circumference by a central angle of 36.0°.

15. Find the area of the circular sector indicated in Exercise 13.

16. Find the area of a sector of a circle given that the central angle is 150.0° and the diameter is 39.50 m.

17. Find the radian measure of an angle at the center of a circle of radius 73.0 cm which intercepts an arc length of 118 cm.

18. Find the central angle of a circle which intercepts an arc length of 928 mm when the radius of the circle is 985 mm.

19. Two concentric (same center) circles have radii of 12.5 ft and 18.0 ft. Find the portion of the area of the sector of the larger circle which is outside the smaller circle when the central angle is 60.0°. See Figure 19-14.

20. In a circle of radius 2.40 ft, the arc length of a sector is 4.00 ft. What is the area of the sector?

21. A pendulum 1.05 m long oscillates through an angle of 6.2°. Find the distance through which the end of the pendulum swings in going from one extreme position to the other.

22. The radius of the earth is about 3960 mi. What is the length, in miles, of an arc of the earth's equator for a central angle of 1.0°?

23. In turning, an airplane traveling at 320 mi/h moves through a circular arc for 2 min. What is the radius of the circle, given that the central angle of the arc is 300.0°?

24. An ammeter needle is deflected 37.0° by a current of 0.150 A. The needle is 3.50 cm long and a circular scale is used. How long is the scale for a maximum current of 1.00 A? See Figure 19-15.

25. A record whose radius is 15.1 cm rotates at 33 r/min. Through what total distance does a point on the rim travel in 1 min?

26. For the record in Exercise 25, how far does a point halfway out, along a radius, move in 15 s?

27. Two streets meet at an angle of 84.0°. What is the length of the piece of curved curbing at the intersection if it is constructed along the arc of a circle 18.0 ft in radius? See Figure 19-16.

28. In traveling three-fourths of the way along a traffic circle a car travels 0.203 mi. What is the radius of the traffic circle? See Figure 19-17.

FIGURE 19-15

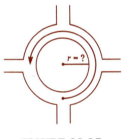

FIGURE 19-16   **FIGURE 19-17**

29. A truck whose tires are 33.0 in. in diameter is traveling at 55 mi/h. What is the angular velocity of the tires in radians per second?

30. A satellite circles the earth four times each 24.0 h. If its altitude is constant at 22,040 mi, what is its velocity? (The earth's radius is about 3960 mi.)

31. The moon is about 237,000 mi from the earth. What is its angular velocity about the earth in radians per second? It takes the moon about 28 days to make one revolution.

32. A 50-kW wind turbine rotates at 36 r/min. What is the linear velocity of a point on the end of a blade, if the blade is 12.0 m long (from the center of rotation)?

33. The armature of a dynamo is 54.0 cm in diameter and is rotating at 900 r/min. What is the linear velocity of a point on the rim of the dynamo?

34. A pulley belt 16.0 in. long takes 3.25 s to make one complete revolution. The diameter of the pulley is 4.57 in. What is the pulley's angular velocity in revolutions per minute?

35. An IBM memory recording disk has a diameter of 14.0 in. and rotates at 3600 r/min. What is the linear velocity of a point on the rim in feet per minute?

36. The propeller of an airplane is 5.50 ft from tip to tip, and it rotates at 2350 r/min while cruising. What is the linear velocity of a point on the tip of the propeller?

37. A conical tent is made from a circular piece of canvas 16.0 ft in diameter, with a sector of central angle 120.0° removed. What is the surface area of the tent? See Figure 19-18.

38. A circular sector whose central angle is 210.0° is cut from a circular piece of sheet metal of diameter 18.0 cm. A cone is then formed by bringing the two radii of the sector together. What is the lateral surface area of the cone?

39. A telephone dial has a diameter of 2.875 in. After dialing a 9, it takes 1.481 s for the dial to rotate 292.74°. Find the linear velocity of a point on the circumference of the dial.

40. The height of a distant object can be approximated by considering the object to be an arc of a circle. Refer to Figure 19-19 and find the height of the building by using Eq. (19-4). Also find the height by using the tangent ratio in a right triangle, and then compare the two results.

In Exercises 41 through 44, another use of radians is illustrated. Solve the given problems. Use Eq. (19-7) in each solution.

41. It can be shown through advanced mathematics that an excellent approximate method of evaluating sin θ or tan θ is given by

$$\sin \theta = \tan \theta = \theta \qquad\qquad\qquad (19\text{-}7)$$

provided that the values of θ are small (the equivalent of a few degrees or less) and θ is expressed in radians. Equation (19-7) is particularly useful for very small values of θ—even some scientific calculators cannot adequately handle angles of 1″ or 0.001° or less. Verify Eq. (19-7) by changing 1″ to radians and then evaluating sin 1″ and tan 1″.

42. Using Eq. (19-7), evaluate tan 0.0005°. (See Exercise 41.)

43. An astronomer observes that a star 14.5 light years away moves through an angle of 0.000054° in one year. Assuming it moves in a straight line perpendicular to the initial line of observation, how many miles does the star move? (One light year = 5.88 × 10¹² mi.)

44. In calculating the back line of a lot, a surveyor discovers an error of 0.05° in angle measurement. If the lot is 480.0 ft deep, by how much is the back line calculation in error? (See Figure 19-20.)

FIGURE 19-18

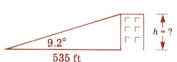

FIGURE 19-19

FIGURE 19-20

19-3 Graphs of *y = a* sin *x* and *y = a* cos *x*

We have seen that the trigonometric functions can be used in many technical applications. Angles measured in degrees are not appropriate for many of those applications and radians are used instead. In this and the following two sections we develop procedures for graphing trigonometric functions using only radian measures. Also, since the sine and cosine functions are of the greatest importance, we shall restrict our attention to them.

We will construct our graphs on the rectangular coordinate system and *angles will be expressed as* **numbers,** and these numbers may have any appropriate unit of measurement. We begin with the function $y = \sin x$ and we construct a table of values of x and y for that function.

x	0	$\frac{\pi}{6}$	$\frac{\pi}{3}$	$\frac{\pi}{2}$	$\frac{2\pi}{3}$	$\frac{5\pi}{6}$	π	$\frac{7\pi}{6}$	$\frac{4\pi}{3}$	$\frac{3\pi}{2}$	$\frac{5\pi}{3}$	$\frac{11\pi}{6}$	2π
y	0	0.5	0.87	1	0.87	0.5	0	−0.5	−0.87	−1	−0.87	−0.5	0

If we plot the points corresponding to the pairs of x and y values in the table, we obtain the graph shown in Figure 19-21.

Proceeding in the same way, we develop the table of values and graph which correspond to $y = \cos x$ (see Figure 19-22).

x	0	$\frac{\pi}{6}$	$\frac{\pi}{3}$	$\frac{\pi}{2}$	$\frac{2\pi}{3}$	$\frac{5\pi}{6}$	π	$\frac{7\pi}{6}$	$\frac{4\pi}{3}$	$\frac{3\pi}{2}$	$\frac{5\pi}{3}$	$\frac{11\pi}{6}$	2π
y	1	0.87	0.5	0	−0.5	−0.87	−1	−0.87	−0.5	0	0.5	0.87	1

The graphs shown in Figure 19-21 and Figure 19-22 actually continue indefinitely to the right and to the left as they repeat the same basic pattern shown. This repetitive nature of both graphs is often referred to as the property of being **periodic.** It is this property that makes the trigonometric functions especially relevant in the study of phenomena which are cyclic in nature, such as electric current, sound waves, radio waves, orbital motion, fluid motion, vibration of a spring, and so on.

Comparison of the graphs in Figures 19-21 and 19-22 should reveal that the two graphs have the same shape. The cosine curve is actually the sine curve moved $\pi/2$ units to the left.

Having graphed $y = \sin x$ and $y = \cos x$, we will now consider a variation of these two basic equations. If we want to graph $y = a \sin x$, we sim-

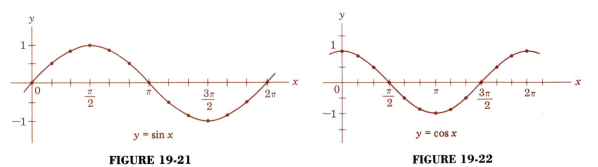

$y = \sin x$

FIGURE 19-21

$y = \cos x$

FIGURE 19-22

ply note that each height y obtained for the graph of $y = \sin x$ is multiplied by the number a. Instead of reaching up to 1 and down to -1, the curve will now rise and fall between a and $-a$. The largest value of y is therefore $|a|$ instead of 1. We use the absolute value since a may be a negative number. *The number $|a|$ is called the* **amplitude** *of the curve and it represents the maximum y value of the curve.* These statements also apply to $y = a \cos x$.

EXAMPLE A   Plot the curve of $y = 4 \sin x$.

   We can develop the following table of values. Note that each value of y is four times the corresponding value of y in the table for $y = \sin x$. The graph is shown in Figure 19-23.

x	0	$\dfrac{\pi}{6}$	$\dfrac{\pi}{3}$	$\dfrac{\pi}{2}$	$\dfrac{2\pi}{3}$	$\dfrac{5\pi}{6}$	π	$\dfrac{7\pi}{6}$	$\dfrac{4\pi}{3}$	$\dfrac{3\pi}{2}$	$\dfrac{5\pi}{3}$	$\dfrac{11\pi}{6}$	2π
y	0	2	3.5	4	3.5	2	0	-2	-3.5	-4	-3.5	-2	0

   From the equation

$$\overset{\displaystyle \text{amplitude}}{y = 4 \sin x}$$

we determine that the amplitude is 4, as indicated. By comparing Figure 19-23 to Figure 19-21, we see that the shapes are the same except that the graph for $y = 4 \sin x$ reaches high points and low points which are four times as large as the graph for $y = \sin x$.

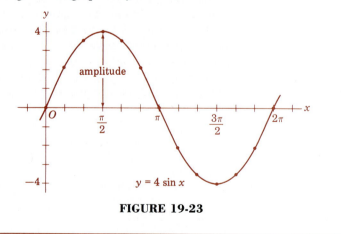

FIGURE 19-23

EXAMPLE B   Plot the curve of $y = -2 \cos x$.

   The following table of values can be used for constructing the graph shown in Figure 19-24.

x	0	$\dfrac{\pi}{6}$	$\dfrac{\pi}{3}$	$\dfrac{\pi}{2}$	$\dfrac{2\pi}{3}$	$\dfrac{5\pi}{6}$	π	$\dfrac{7\pi}{6}$	$\dfrac{4\pi}{3}$	$\dfrac{3\pi}{2}$	$\dfrac{5\pi}{3}$	$\dfrac{11\pi}{6}$	2π
y	-2	-1.7	-1	0	1	1.7	2	1.7	1	0	-1	-1.7	-2

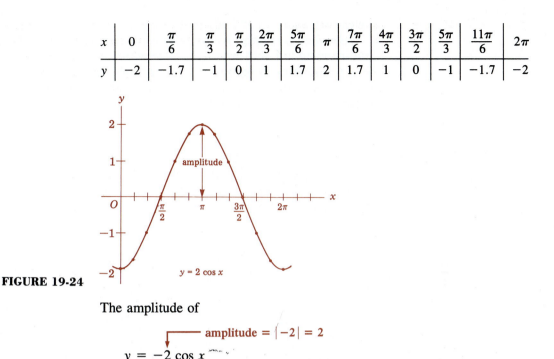

FIGURE 19-24

The amplitude of

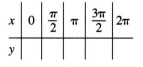

amplitude = $|-2| = 2$

$$y = -2 \cos x$$

is 2, as shown. _**The negative sign has the effect of inverting the curve.**_

$\square$

It is not always necessary to develop a table of values like the tables included in the preceding examples. Given any equation of the form $y = a \sin x$ or $y = a \cos x$, we can quickly construct a sketch of the curve that includes the intercepts and amplitude. In the next example we will sketch the curve instead of plotting the curve from a table of values. We will use only a few important values along with our knowledge of the basic shape of the curve and the effect of the amplitude.

For graphs of the form $y = a \sin x$ or $y = a \cos x$, the values of x for which the curve either crosses the x axis or reaches its lowest or highest points are $x = 0$, $\pi/2$, π, $3\pi/2$, 2π. It is these particular values of x which we should definitely include in our abbreviated table. We can begin by setting up the table as follows:

x	0	$\dfrac{\pi}{2}$	π	$\dfrac{3\pi}{2}$	2π
y					

The corresponding values of y are then found by substituting each value for x in the original equation. The pairs of values included in the completed table are then plotted and connected by a smooth curve.

EXAMPLE C   Sketch the graph of $y = \frac{1}{2} \cos x$.

Instead of developing a large table of values, we set up a table of values corresponding to points where the curve either crosses the x axis or where the curve achieves its highest and lowest heights. Starting with those values of x shown in the above incomplete table, we proceed to find the corresponding values of y as follows:

$$y = \frac{1}{2} \cos 0 = \frac{1}{2}(1) = \frac{1}{2} \qquad \text{one of highest points}$$

$$y = \frac{1}{2} \cos \frac{\pi}{2} = \frac{1}{2}(0) = 0 \qquad \text{crosses axis}$$

$$y = \frac{1}{2} \cos \pi = \frac{1}{2}(-1) = -\frac{1}{2} \qquad \text{one of lowest points}$$

$$y = \frac{1}{2} \cos \frac{3\pi}{2} = \frac{1}{2}(0) = 0 \qquad \text{crosses axis}$$

$$y = \frac{1}{2} \cos 2\pi = \frac{1}{2}(1) = \frac{1}{2} \qquad \text{one of highest points}$$

The table can now be completed as

x	0	$\frac{\pi}{2}$	π	$\frac{3\pi}{2}$	2π
y	$\frac{1}{2}$	0	$-\frac{1}{2}$	0	$\frac{1}{2}$

We now plot the points $\left(0, \frac{1}{2}\right)$, $\left(\frac{\pi}{2}, 0\right)$, $\left(\pi, -\frac{1}{2}\right)$, $\left(\frac{3\pi}{2}, 0\right)$, and $\left(2\pi, \frac{1}{2}\right)$ and connect them as shown in Figure 19-25.

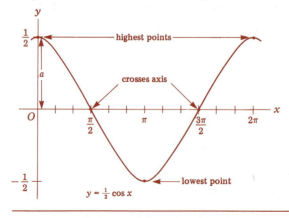

FIGURE 19-25

EXAMPLE D In analyzing the current in an alternating-current circuit, a technician uses the equation $y = -110 \sin x$, where x represents time (in seconds) and y represents current (in amperes). Sketch the graph of $y = -110 \sin x$.

In the accompanying table we list the important values associated with this curve.

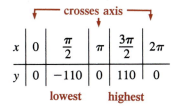

We now plot these five points and connect them by following the general shape of the sine curve. The sketch is shown in Figure 19-26.

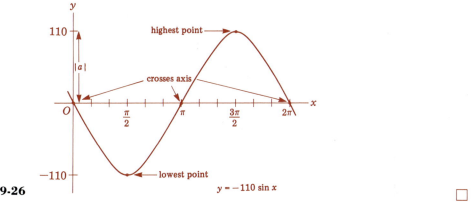

FIGURE 19-26

There are calculators and computers that are capable of displaying the graphs of functions similar to those discussed in this section, as well as the functions found in the following sections of this chapter.

Exercises 19-3

In Exercises 1 through 4, complete the following table for the given functions and then plot the resulting graph.

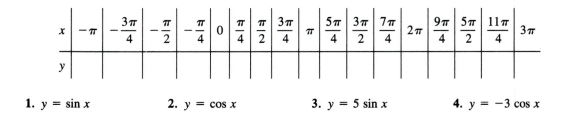

x	$-\pi$	$-\frac{3\pi}{4}$	$-\frac{\pi}{2}$	$-\frac{\pi}{4}$	0	$\frac{\pi}{4}$	$\frac{\pi}{2}$	$\frac{3\pi}{4}$	π	$\frac{5\pi}{4}$	$\frac{3\pi}{2}$	$\frac{7\pi}{4}$	2π	$\frac{9\pi}{4}$	$\frac{5\pi}{2}$	$\frac{11\pi}{4}$	3π
y																	

1. $y = \sin x$ **2.** $y = \cos x$ **3.** $y = 5 \sin x$ **4.** $y = -3 \cos x$

In Exercises 5 through 8, refer to the computer-generated graphs and identify the equation of the function displayed. The hash marks on the axes represent one unit each, so that the positive x axis runs from 0 to about 6.3 (or 2π). In each case, the equation is of the form y = a sin x or y = a cos x.

5. Figure 19-27   **6.** Figure 19-28   **7.** Figure 19-29   **8.** Figure 19-30

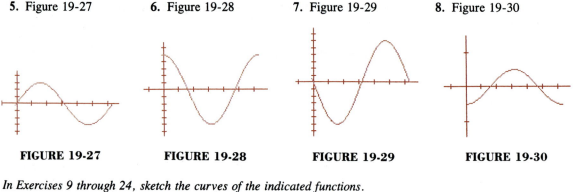

    FIGURE 19-27   **FIGURE 19-28**   **FIGURE 19-29**   **FIGURE 19-30**

In Exercises 9 through 24, sketch the curves of the indicated functions.

9. $y = 2 \cos x$ **10.** $y = 4 \cos x$ **11.** $y = \frac{3}{2} \cos x$ **12.** $y = \frac{9}{4} \cos x$

13. $y = 2 \sin x$ **14.** $y = 3 \sin x$ **15.** $y = 0.6 \sin x$ **16.** $y = \frac{5}{2} \sin x$

17. $y = -\cos x$ **18.** $y = -4 \cos x$ **19.** $y = -1.2 \cos x$ **20.** $y = -0.8 \cos x$

21. $y = -\sin x$ **22.** $y = -6 \sin x$ **23.** $y = -\frac{5}{4} \sin x$ **24.** $y = -0.4 \sin x$

Although units of π are often convenient, we must remember that π is simply a number. Numbers which are not multiples of π may be used as well. In Exercises 25 through 36, plot the indicated graphs by finding the values of y corresponding to the values of 0, 1, 2, 3, 4, 5, 6, and 7 for x. (Remember: The numbers 0, 1, 2, and so forth represent radian measure.) Values from a calculator may be used.

25. $y = \cos x$ **26.** $y = 2 \cos x$ **27.** $y = \sin x$ **28.** $y = 3 \sin x$

29. $y = \frac{2}{3} \sin x$ **30.** $y = \frac{3}{2} \sin x$ **31.** $y = -2 \cos x$ **32.** $y = -4.5 \cos x$

33. $y = -2.6 \cos x$ **34.** $y = \frac{1}{3} \cos x$ **35.** $y = -3.5 \sin x$ **36.** $y = -0.8 \sin x$

19-4 Graphs of $y = a \sin bx$ and $y = a \cos bx$

In Section 19-3 we considered the basic graphs of $y = \sin x$ and $y = \cos x$. We also discussed one variation in these basic graphs as we saw the effects of changing amplitude to graph equations of the form $y = a \sin x$ or $y = a \cos x$. In this section we will consider another variation.

We have noted that the sine and cosine functions are periodic in the sense that they repeat a basic pattern. In fact, all of the equations considered in Section 19-3 have graphs which repeat the pattern every 2π units of x. For such periodic functions, we now find it useful to define the **period** of a function F as the number P where $F(x) = F(x + P)$. Simply stated, *the period P refers to the x distance between any point and the next point at which the same pattern of y values starts repeating*. We might think of the period as the distance (along the x axis) required to get one complete cycle of a repeating pattern.

In Section 19-3 we saw that the amplitude can be changed by changing the coefficient preceding the function. For example, $y = \sin x$ and $y = 4 \sin x$ have amplitudes of 1 and 4, respectively. We will now show that *we can change the period of the sine and cosine functions by altering the coefficient of the angle*. To do this, we will now graph $y = \sin 4x$. For this graph we choose suitable values for x, then multiply those values by 4, and then find the sine of the result.

x	0	$\dfrac{\pi}{16}$	$\dfrac{\pi}{8}$	$\dfrac{3\pi}{16}$	$\dfrac{\pi}{4}$	$\dfrac{5\pi}{16}$	$\dfrac{3\pi}{8}$	$\dfrac{7\pi}{16}$	$\dfrac{\pi}{2}$	$\dfrac{9\pi}{16}$	$\dfrac{5\pi}{8}$
$4x$	0	$\dfrac{\pi}{4}$	$\dfrac{\pi}{2}$	$\dfrac{3\pi}{4}$	π	$\dfrac{5\pi}{4}$	$\dfrac{3\pi}{2}$	$\dfrac{7\pi}{4}$	2π	$\dfrac{9\pi}{4}$	$\dfrac{5\pi}{2}$
$y = \sin 4x$	0	0.7	1	0.7	0	-0.7	-1	-0.7	0	0.7	1

Plotting the values of x and y in this table, we get the graph shown in Figure 19-31.

Upon examination of the table and the graph in Figure 19-31, we can see that $y = \sin 4x$ completes one full cycle between $x = 0$ and $x = \pi/2$ so that the period is $\pi/2$. Since $y = \sin x$ has a period of 2π while $y = \sin 4x$ has a period of $\pi/2$, we can see that the 4 had the effect of reducing the period to one-fourth of the period of $y = \sin x$. This suggests the following generalization which is valid: *Both* $\sin x$ *and* $\cos x$ *have a period of* 2π, *whereas* $\sin bx$ *and* $\cos bx$ *both have a period of* $2\pi/b$.

We have just seen that $y = \sin x$ and $y = \sin 4x$ have different periods. Note very carefully that *the period is changed by altering the coefficient of the angle*. Both $y = \sin x$ and $y = 4 \sin x$ (see Figure 19-32) have the same period but *different amplitudes,* whereas $y = \sin x$ and $y = \sin 4x$ have the same amplitude but *different periods*.

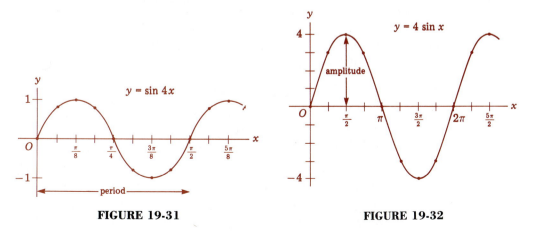

FIGURE 19-31   **FIGURE 19-32**

EXAMPLE A   Find the period of $y = \sin 5x$.

The period of $y = \sin 5x$ is $\dfrac{2\pi}{5}$. This means that the graph of $y = \sin 5x$ will repeat the same cycle every $2\pi/5$ (approximately 1.26) units of x.   □

EXAMPLE B   Find the periods of $y = \cos 3x$, $y = \cos \frac{1}{2}x$, and $y = \cos \pi x$.

The period of $y = \cos 3x$ is $\dfrac{2\pi}{3}$.

The period of $y = \cos \frac{1}{2}x$ is $\dfrac{2\pi}{\frac{1}{2}} = 4\pi$.

The period of $y = \cos \pi x$ is $2\pi/\pi = 2$.

For each of these three cases, the amplitude is 1.   □

The concepts of amplitude and period can be summarized as follows: *The functions $y = a \sin bx$ and $y = a \cos bx$ both have amplitude $|a|$ and period $2\pi/b$.*

For $y = a \sin bx$ or $y = a \cos bx$:		
amplitude	$|a|$	**Curve goes as high as $|a|$ and as low as $-|a|$.**
period	$\dfrac{2\pi}{b}$	**One full cycle is completed in** $\dfrac{2\pi}{b}$ **units of x.**

These properties are extremely useful in developing sketches, as the following examples illustrate.

EXAMPLE C   A piston is designed to oscillate so that its vertical displacement is described by the equation $y = 4 \sin 2x$, where x is the time (in seconds) and y is the displacement (in centimeters). Sketch the graph of $y = 4 \sin 2x$ for $0 \le x \le 2\pi$.

For the given function we know that the amplitude is 4 and the period is $2\pi/2 = \pi$. The amplitude of 4 indicates that the curve can go up to 4 and down to -4. The period of π indicates that a complete cycle will be repeated for every π units of x. The graph shown in Figure 19-33 is a result of *using the amplitude and period plus our knowledge of the basic sine curve.*

These lead us to the following values:

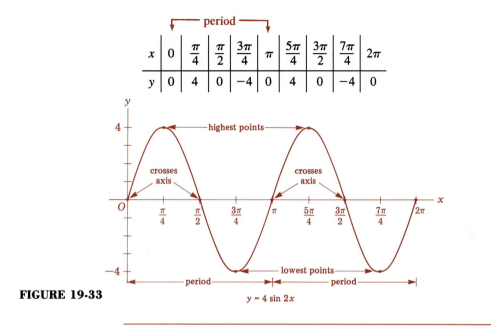

	period								
x	0	$\dfrac{\pi}{4}$	$\dfrac{\pi}{2}$	$\dfrac{3\pi}{4}$	π	$\dfrac{5\pi}{4}$	$\dfrac{3\pi}{2}$	$\dfrac{7\pi}{4}$	2π
y	0	4	0	-4	0	4	0	-4	0

FIGURE 19-33

$y = 4 \sin 2x$

EXAMPLE D   Sketch the graph of $y = -3 \cos 4x$ for $0 \le x \le 2\pi$.

a ⌐   ⌐ b

We might begin by recalling the basic pattern of the cosine function. Considering amplitude next, we see that this function will go up to 3 and down to -3. Also, the negative sign will cause the basic curve to be inverted. Finally, the period is $2\pi/4 = \pi/2$, so that a full cycle will be completed every $\pi/2$ units of x. The sketch can therefore be developed by transforming the basic cosine curve as follows: *Invert it, stretch it out vertically and complete the first cycle between $x = 0$ and $x = \pi/2$, then repeat the pattern.* The table of important values which identifies key points on the graph for the first cycle is as follows. The graph is shown in Figure 19-34.

x	0	$\dfrac{\pi}{8}$	$\dfrac{\pi}{4}$	$\dfrac{3\pi}{8}$	$\dfrac{\pi}{2}$
y	-3	0	3	0	-3

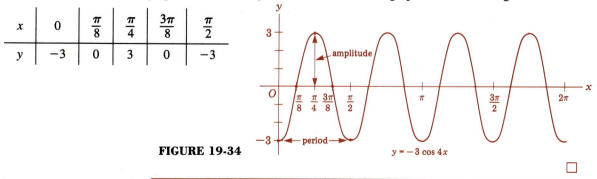

FIGURE 19-34

$y = -3 \cos 4x$

EXAMPLE E   Sketch the function $y = \sin \pi x$ for $0 \le x \le \pi$.

For this function the amplitude is 1 and the period is $2\pi/\pi = 2$. Since the value of the period is not expressed in terms of π, *it will be more convenient to use regular decimal units for* x. The following table lists important values and the graph is shown in Figure 19-35.

x	0	0.5	1	1.5	2	2.5	3
y	0	1	0	−1	0	1	0

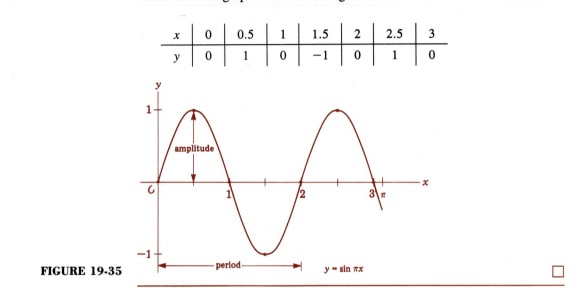

FIGURE 19-35

$y = \sin \pi x$

Exercises 19-4

In Exercises 1 through 20, find the period of each given function.

1. $y = \sin 3x$ **2.** $y = 2 \sin 3x$ **3.** $y = 2 \sin 4x$ **4.** $y = 3 \sin 10x$

5. $y = 4 \cos 3x$ **6.** $y = 4 \cos 10x$ **7.** $y = -2 \sin 6x$ **8.** $y = -3 \sin 8x$

9. $y = -3 \cos 5x$ **10.** $y = -4 \cos 6x$ **11.** $y = 5 \cos 2\pi x$ **12.** $y = 3 \cos 10\pi x$

13. $y = 5 \sin \frac{1}{2}x$ **14.** $y = -4 \sin \frac{1}{3}x$ **15.** $y = -4 \cos \frac{2}{3}x$ **16.** $y = \frac{1}{3} \cos \frac{1}{4}x$

17. $y = 1.5 \sin \frac{3}{2}\pi x$ **18.** $y = 0.8 \cos \frac{1}{3}\pi x$ **19.** $y = 4.5 \cos \pi^2 x$ **20.** $y = \frac{1}{2} \sin \frac{x}{\pi}$

In Exercises 21 through 40, sketch the graphs of the functions given for Exercises 1 through 20.

In Exercises 41 through 48, the period is given for a function of the form $y = \sin bx$. Write the equation corresponding to the given values of the period.

41. π **42.** $\pi/3$ **43.** 2   **44.** 3

45. 1   **46.** 6   **47.** $2\pi/5$ **48.** $3\pi/7$

In Exercises 49 through 52, refer to the computer-generated graph and identify the equation of the function displayed. The hash marks on the axes represent one unit each. In each case, the equation is of the form $y = a \sin bx$ or $y = a \cos bx$.

49. Figure 19-36   **50.** Figure 19-37   **51.** Figure 19-38   **52.** Figure 19-39

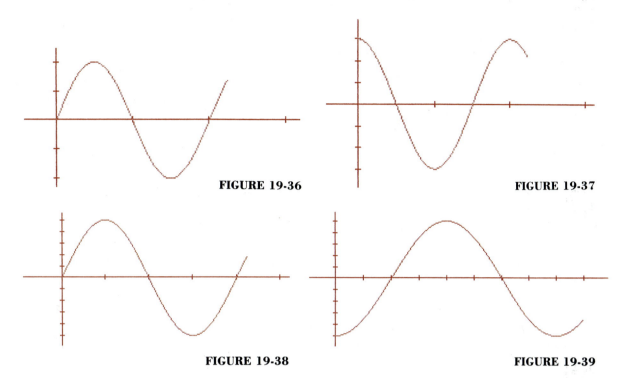

FIGURE 19-36

FIGURE 19-37

FIGURE 19-38

FIGURE 19-39

In Exercises 53 through 60, sketch the indicated graphs.

53. The electric current in a certain alternating-current circuit is given by

$$i = 4 \sin 80\pi t$$

where i is the current in amperes and t is the time in seconds. Sketch the graph for $0 \le t \le 0.1$ s.

54. A generator produces a voltage given by

$$V = 120 \cos 30\pi t$$

where t is the time in seconds. Sketch the graph of V versus t for $0 \le t \le 1$ s.

55. The vertical displacement x of a certain object oscillating at the end of a spring is given by

$$x = 7.5 \cos 3\pi t$$

where x is measured in inches and t in seconds. Sketch the graph of x versus t for $0 \le t \le 1$ s.

56. The velocity of a piston in a certain engine is given by

$$v = 5.0 \cos 1600\pi t$$

where v is the velocity in centimeters per second and t is the time in seconds. Sketch the graph of v versus t for $0 \le t \le 0.01$ s.

57. A buoy floats in water. Its vertical motion caused by waves is described by the equation

$$y = 2.37 \sin 0.598t$$

The height y is in feet and the time t is in seconds. Sketch the graph of y versus t and include at least one full cycle of this periodic function.

58. In analyzing the pressure variations caused by the sound wave of a singer hitting the note A, the following equation is determined.

$$y = 0.00001 \sin \frac{8\pi}{3}x$$

The variable y is measured in atmospheres, and x is in meters. Sketch the graph of y versus x and include at least one full cycle of this periodic function.

59. When a tuning fork is activated with a frequency of 60 cycles per second, the pressure p at time t seconds is described by

$$p = \sin 120\pi t$$

Sketch the graph of p versus t and include at least one full cycle of this periodic function.

60. The blade of a saber saw moves vertically up and down at the rate of 18 strokes per second. The vertical displacement of the tip of the blade is described by the equation

$$y = 1.21 \sin 36\pi t$$

where y is in centimeters and t is in seconds. Sketch the graph of y versus t and include at least one full cycle of this periodic function.

19-5 Graphs of $y = a \sin (bx + c)$ and $y = a \cos (bx + c)$

Beginning with the basic graphs of $y = \sin x$ and $y = \cos x$, we have shown the effects caused by changes in amplitude and period. In this section we consider one more variation. In the function $y = a \sin (bx + c)$, c is the **phase angle** and its effect can be seen in the next example.

EXAMPLE A   Sketch the graph of $y = \sin \left(x + \dfrac{\pi}{2} \right)$.

For this function the amplitude is 1, the period is 2π, but the phase angle is $c = \pi/2$. We first present selected important values in the following table and develop the graph shown in Figure 19-40.

x	$-\dfrac{\pi}{2}$	0	$\dfrac{\pi}{2}$	π	$\dfrac{3\pi}{2}$	2π
y	0	1	0	-1	0	1

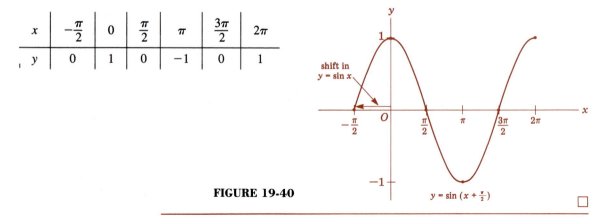

FIGURE 19-40

Examining Figure 19-40, we see that the graph of $y = \sin \left(x + \dfrac{\pi}{2} \right)$ is identical to the graph of $y = \sin x$ with the exception that it is shifted $\pi/2$ units to the left. Since the phase angle $c = \pi/2$ is the only variation from $y = \sin x$, it follows that *the phase angle causes the curve to shift horizontally*. From Example A it might seem that the amount of shift is $-c$, but it is actually $-\dfrac{c}{b}$, as illustrated in the next example. *The quantity* $-\dfrac{c}{b}$ *is called the* **displacement** *(or* **phase shift***)*. Remember, the curve shifts to the left if $-\dfrac{c}{b} < 0$ and to the right if $-\dfrac{c}{b} > 0$.

EXAMPLE B   Sketch the graph of $y = \cos \left(2x + \dfrac{\pi}{4} \right)$.

For this function, $a = 1$, $b = 2$, $c = \pi/4$ so that the amplitude is 1, the period is $2\pi/b = 2\pi/2 = \pi$, and the displacement is $-c/b = -(\pi/4)/2 = -\pi/8$.

This is a shift of $\pi/8$ to the *left* since $-c/b < 0$. If we begin a cycle of the curve at $x = -\pi/8$, it will be completed π units to the right of that starting point, namely $(-\pi/8) + \pi = 7\pi/8$; ***the next cycle then begins at*** $\boldsymbol{7\pi/8}$***,*** and so on. The following table lists some important values and the graph is shown in Figure 19-41.

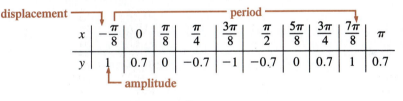

displacement					period						
x	$-\dfrac{\pi}{8}$	0	$\dfrac{\pi}{8}$	$\dfrac{\pi}{4}$	$\dfrac{3\pi}{8}$	$\dfrac{\pi}{2}$	$\dfrac{5\pi}{8}$	$\dfrac{3\pi}{4}$	$\dfrac{7\pi}{8}$	π	
y	1	0.7	0	-0.7	-1	-0.7	0	0.7	1	0.7	

amplitude

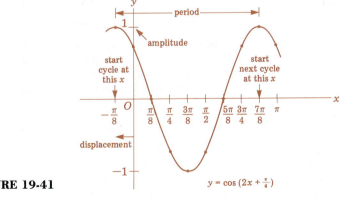

FIGURE 19-41

After studying Example B we can see that we have three important quantities to determine. They are summarized as follows:

For $y = a \sin (bx + c)$ or $y = a \cos (bx + c)$:		
amplitude	$\lvert a \rvert$	Curve goes as high as $\lvert a \rvert$ and as low as $-\lvert a \rvert$.
period	$\dfrac{2\pi}{b}$	One full cycle is completed in $\dfrac{2\pi}{b}$ units of x.
displacement	$-\dfrac{c}{b}$	Curve is displaced $-\dfrac{c}{b}$ units of x

See Appendix B for a computer program that will generate coordinates for graphs of functions of the form $y = a \sin (bx + c)$.

Here is a helpful hint in identifying which values of x are of enough significance to include in the tables we have been developing: First begin with the displacement value of $-c/b$ and then add one-fourth of the period to get the second x value. To this result add one-fourth of the period to get the third x value, and continue adding one-fourth of the period until at least five points are obtained. This procedure should determine for *one full cycle of the curve those points which are on the x axis or at the highest or lowest points of the curve*. These are important values.

$$a \longrightarrow \qquad b \qquad \qquad c = -\frac{\pi}{6}$$

EXAMPLE C   Sketch the graph of $y = -3 \sin \left(4x - \dfrac{\pi}{6} \right)$.

The amplitude is 3 and the negative sign preceding the 3 has the effect of inverting the curve. The period is $2\pi/4 = \pi/2$. The displacement of $-\dfrac{(-\pi/6)}{4} = \dfrac{\pi}{24}$ means that the graph is shifted to the *right* by an amount of $\pi/24$ units of x. We can consider $x = \pi/24$ to be the start of one cycle of the curve. Since the period is $\pi/2$, one-fourth of the period is

$$\frac{1}{4}\left(\frac{\pi}{2}\right) = \frac{\pi}{8}$$

Adding this to $\pi/24$, we get

$$\frac{\pi}{24} + \frac{\pi}{8} = \frac{\pi}{24} + \frac{3\pi}{24} = \frac{4\pi}{24} = \frac{\pi}{6}$$

which is the next important value of x. Continuing to add $\pi/8$, we get x values of $7\pi/24$, $5\pi/12$ and $13\pi/24$. These values are included in the following table, and the corresponding points are plotted in the graph of Figure 19-42.

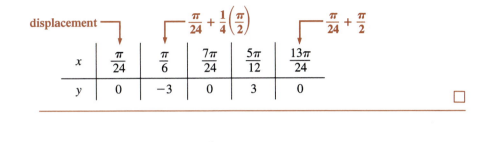

displacement		$-\dfrac{\pi}{24} + \dfrac{1}{4}\left(\dfrac{\pi}{2}\right)$		$-\dfrac{\pi}{24} + \dfrac{\pi}{2}$
x	$\dfrac{\pi}{24}$	$\dfrac{\pi}{6}$	$\dfrac{7\pi}{24}$	$\dfrac{5\pi}{12}$	$\dfrac{13\pi}{24}$
y	0	-3	0	3	0

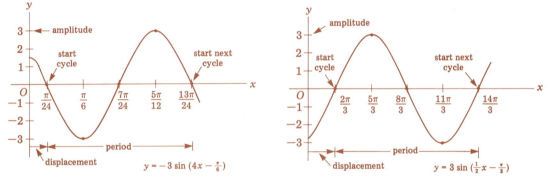

FIGURE 19-42   **FIGURE 19-43**

$$y = -3 \sin \left(4x - \tfrac{\pi}{6}\right) \qquad y = 3 \sin \left(\tfrac{1}{2}x - \tfrac{\pi}{3}\right)$$

EXAMPLE D   Sketch the graph of $y = 3 \sin \left(\dfrac{1}{2}x - \dfrac{\pi}{3}\right)$.

where $a = 3$, $b = \dfrac{1}{2}$, $c = -\dfrac{\pi}{3}$.

The amplitude is 3 and the period is $\dfrac{2\pi}{1/2} = 4\pi$. The displacement of $-\dfrac{-\pi/3}{1/2} = \dfrac{2\pi}{3}$ indicates that there is a shift of $\dfrac{2\pi}{3}$ to the right. Using this information and the following table of key values, we get the graph shown in Figure 19-43.

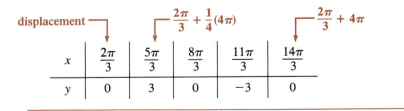

displacement		$-\dfrac{2\pi}{3} + \dfrac{1}{4}(4\pi)$		$-\dfrac{2\pi}{3} + 4\pi$
x	$\dfrac{2\pi}{3}$	$\dfrac{5\pi}{3}$	$\dfrac{8\pi}{3}$	$\dfrac{11\pi}{3}$	$\dfrac{14\pi}{3}$
y	0	3	0	-3	0

EXAMPLE E The current in a certain alternating-current circuit is described by the equation $y = 15 \sin \left(60\pi x + \dfrac{\pi}{6} \right)$. Sketch the curve.

The amplitude is 15, the period is $\dfrac{2\pi}{60\pi} = \dfrac{1}{30}$, and the displacement is $-\dfrac{(\pi/6)}{60\pi} = -\dfrac{1}{360}$. This represents a shift of 1/360 to the *left*, since the displacement is negative. Because the period and displacement are not expressed in terms of π, we choose values of x that do not involve π. Following the preceding hint, we get the values in the following table:

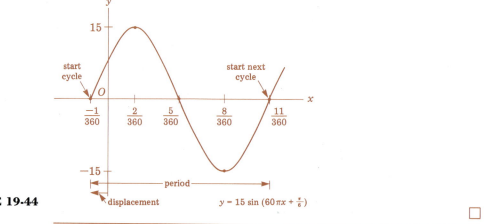

displacement ——→ ↱ $-\dfrac{1}{360} + \dfrac{1}{4}\left(\dfrac{1}{30}\right)$ ↱ $-\dfrac{1}{360} + \dfrac{1}{30} = \dfrac{11}{360}$

x	$-\dfrac{1}{360}$	$\dfrac{2}{360}$	$\dfrac{5}{360}$	$\dfrac{8}{360}$	$\dfrac{11}{360}$
y	0	15	0	-15	0

Using the values in this table and knowledge of the amplitude, period, and displacement, we get the graph shown in Figure 19-44.

FIGURE 19-44

$y = 15 \sin \left(60\pi x + \dfrac{\pi}{6}\right)$

Exercises 19-5

In Exercises 1 through 24, determine the amplitude, period, and displacement for each function. Then sketch the graphs of the functions.

1. $y = \sin \left(x + \dfrac{\pi}{3} \right)$ **2.** $y = 2 \sin \left(x - \dfrac{\pi}{3} \right)$ **3.** $y = \cos \left(x - \dfrac{\pi}{3} \right)$

4. $y = 5 \cos \left(x + \dfrac{\pi}{4} \right)$ **5.** $y = 3 \sin \left(2x + \dfrac{\pi}{4} \right)$ **6.** $y = -\sin \left(2x - \dfrac{\pi}{4} \right)$

7. $y = -4 \cos(3x + \pi)$

8. $y = 6 \cos\left(2x - \dfrac{\pi}{2}\right)$

9. $y = 2 \sin\left(\dfrac{1}{3}x + \dfrac{\pi}{2}\right)$

10. $y = -\dfrac{1}{5} \sin\left(\dfrac{1}{4}x - \dfrac{\pi}{2}\right)$

11. $y = 6 \cos\left(\dfrac{2}{3}x + \dfrac{\pi}{3}\right)$

12. $y = \dfrac{1}{6} \cos\left(\dfrac{1}{2}x - \pi\right)$

13. $y = 10 \sin(\pi x - 1)$

14. $y = -8 \sin(2\pi x + \pi)$

15. $y = \dfrac{3}{2} \cos\left(3\pi x - \dfrac{1}{3}\right)$

16. $y = \dfrac{3}{5} \cos\left(4\pi x + \dfrac{\pi}{5}\right)$

17. $y = -1.2 \sin\left(\pi x - \dfrac{\pi}{6}\right)$

18. $y = 0.4 \sin\left(2\pi x + \dfrac{1}{4}\right)$

19. $y = 6 \cos\left(\pi x - \dfrac{1}{2}\right)$

20. $y = 8 \cos(5\pi x + 0.1)$

21. $y = \dfrac{3}{4} \sin(\pi^2 x + \pi)$

22. $y = -\sin\left(4x - \dfrac{2}{\pi}\right)$

23. $y = -\dfrac{5}{2} \cos\left(\pi x - \dfrac{\pi^2}{3}\right)$

24. $y = \pi \cos\left(\dfrac{1}{\pi}x - \dfrac{3}{2}\right)$

In Exercises 25 through 32, sketch the indicated curves.

25. A wave traveling in a string may be represented by the equation

$$y = A \sin 2\pi\left(\dfrac{t}{T} - \dfrac{x}{\lambda}\right)$$

Here A is the amplitude, t is the time the wave has traveled, x is the distance from the origin, T is the time required for the wave to travel one *wavelength* λ (the Greek lambda). Sketch three cycles of the wave for which $A = 3$ cm, $T = 0.2$ s, $\lambda = 36$ cm, and $x = 9$ cm. (Sketch y versus t.)

26. The cross section of a particular water wave is

$$y = 1.8 \sin\left(\dfrac{\pi x}{4} - \dfrac{\pi}{6}\right)$$

where x and y are measured in feet. Sketch two cycles of the graph of y versus x.

27. A particular electromagnetic wave is described by

$$y = a \cos\left(6\pi \times 10^{12}t - \dfrac{\pi}{3}\right)$$

Sketch two cycles of the graph of y (in centimeters) versus t (in seconds).

28. The voltage in a certain alternating-current circuit is given by

$$y = 240 \sin\left(120\pi t + \dfrac{\pi}{4}\right)$$

where t represents the time in seconds. Sketch three cycles of the curve.

29. An automobile flywheel has a timing mark on its outer edge. The height of the timing mark on the rotating flywheel is given by

$$y = 3.55 \sin\left(x - \dfrac{\pi}{4}\right)$$

where y is in inches and x is the angle of rotation. Sketch y versus x and include at least one full cycle of this function.

30. For a certain swinging pendulum, if s is the arc length (in meters) measured from the bottom of the path and t is the time in seconds, the motion is described by

$$s = 4 \cos \left(0.9t + \frac{\pi}{2} \right)$$

Sketch s versus t and include at least one full cycle of this function.

31. The sound created by one machine is slightly out of phase with that of another machine so that the pressure variations caused by the second machine are described by

$$y = 0.00001 \sin \left(\frac{8\pi}{3}x - \frac{3}{16} \right)$$

where y is measured in atmospheres and x is in meters. Sketch the graph of y versus x and include at least one full cycle of this function.

32. A circular disk suspended by a thin wire attached to the center at one of its flat faces is twisted through an angle θ. See Figure 19-45. Torsion in the wire tends to turn the disk back in the opposite direction so that the name "torsion pendulum" is given to this device. The angular displacement as a function of time is given by

$$\theta = \theta_0 \cos (\omega t + \alpha)$$

where θ_0 is the maximum angular displacement, ω is a constant which depends on the properties of the disk and wire, and α is the phase angle. Plot the graph of θ versus t if $\theta_0 = 0.22$ rad, $\omega = 2.0$ rad/s, and $\alpha = \pi/4$.

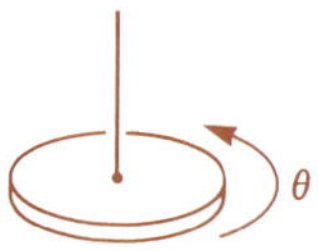

FIGURE 19-45

Chapter 19 Formulas

π rad $= 180°$	(19-1)
$1° = \dfrac{\pi}{180}$ rad $= 0.01745$ rad	(19-2)
1 rad $= \dfrac{180°}{\pi} = 57.3°$	(19-3)
$s = r\theta$ (where θ is in radians) arc length	(19-4)
$A = \dfrac{1}{2}\theta r^2$ (where θ is in radians) area of sector	(19-5)
$v = \omega r$ (where ω is in radians per unit time) linear velocity	(19-6)

19-6 Review Exercises for Chapter 19

In Exercises 1 through 4, express the given angles in terms of π.

1. 63.0°, 137.0°　　　　**2.** 67.5°, 202.5°　　　**3.** 46.0°, 10.0°　　　　**4.** 318.0°, 284.0°

In Exercises 5 through 12, the given numbers represent angle measure. Express the measure of each angle in degrees.

5. $\dfrac{\pi}{9}, \dfrac{5\pi}{6}$　　　　　**6.** $\dfrac{6\pi}{5}, \dfrac{5\pi}{8}$　　　　　**7.** $\dfrac{7\pi}{18}, \dfrac{9\pi}{20}$　　　　**8.** $\dfrac{19\pi}{10}, \dfrac{7\pi}{5}$

9. 0.625　　　　　**10.** 1.33　　　　　　**11.** 3.45　　　　　　**12.** 12.38

In Exercises 13 through 20, express the given angles in radians (not in terms of π).

13. 75.0°　　　　　**14.** 110.0°　　　　**15.** 340.0°　　　　**16.** 15.5°

17. 152.5°　　　　**18.** 215.4°　　　　**19.** 9.3°　　　　　**20.** 422.0°

In Exercises 21 through 24, find θ for $0 < \theta < 2\pi$.

21. $\sin \theta = 0.6361$　　**22.** $\cos \theta = 0.9925$　　**23.** $\cos \theta = -0.4147$　　**24.** $\tan \theta = 2.087$

In Exercises 25 through 28, the radius and central angle are given for a circle. (a) Find the length of the arc subtended by the central angle. (b) Find the area of the sector.

25. $r = 12.4$ cm, $\theta = 35.7°$　　　　　　**26.** $r = 507$ ft, $\theta = 95.6°$

27. $r = 7.24$ in., $\theta = \dfrac{\pi}{10}$　　　　　**28.** $r = 0.683$ m, $\theta = \dfrac{4\pi}{3}$

In Exercises 29 through 32, an object is moving in a circular path with the given radius and angular velocity. Find the linear velocity.

29. $r = 0.365$ m, 12.6 rad/min　　　　**30.** $r = 237$ in., 9.59 rad/min

31. $r = 1.25$ m, 45.0 r/min　　　　　**32.** $r = 82.6$ cm, 360 r/min

In Exercises 33 through 52, sketch the curves of the given trigonometric functions.

33. $y = \dfrac{5}{2} \cos x$　　　　　**34.** $y = -6 \cos x$　　　　　**35.** $y = -2.5 \sin x$

36. $y = 3.6 \sin x$　　　　　**37.** $y = 3 \cos 4x$　　　　　**38.** $y = 5 \cos \dfrac{1}{2} x$

39. $y = 2 \sin 6x$　　　　　**40.** $y = 5 \sin 2x$　　　　　**41.** $y = \cos \pi x$

42. $y = 5 \cos 4\pi x$　　　　**43.** $y = 5 \sin 2\pi x$　　　　**44.** $y = -\sin 3\pi x$

45. $y = -3 \cos \left(4x + \dfrac{\pi}{6}\right)$　　**46.** $y = 2 \cos \left(\dfrac{1}{2} x + \dfrac{\pi}{4}\right)$　　**47.** $y = 3 \sin \left(4x - \dfrac{\pi}{2}\right)$

48. $y = 2 \sin \left(\dfrac{1}{2} x - \dfrac{\pi}{4}\right)$　　**49.** $y = -\cos \left(\pi x + \dfrac{\pi}{3}\right)$　　**50.** $y = 2 \cos (4\pi x - \pi)$

51. $y = 6 \sin \left(4\pi x - \dfrac{\pi}{2}\right)$　　**52.** $y = 5 \sin (2\pi x + \pi)$

In Exercises 53 through 56, refer to the computer-generated graph and identify the equation of the function displayed. The hash marks on the axes represent one unit each. In each case, the equation is of the form $y = a \sin bx$ *or* $y = a \cos bx$.

53. Figure 19-46   **54.** Figure 19-47   **55.** Figure 19-48   **56.** Figure 19-49

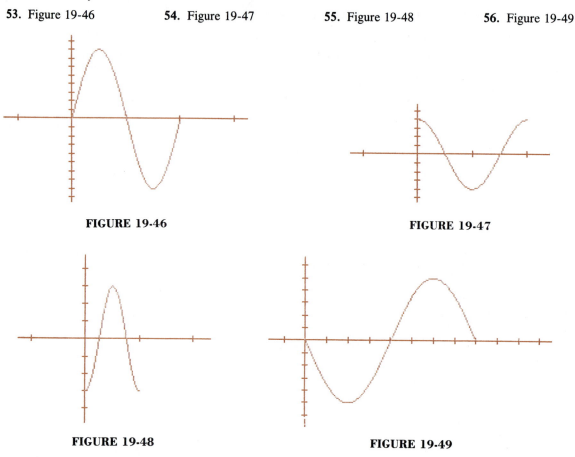

FIGURE 19-46   **FIGURE 19-47**

FIGURE 19-48   **FIGURE 19-49**

In Exercises 57 through 80, solve the given problems.

57. The displacement of a particle moving with simple harmonic motion is given by

$$d = A \cos 5t$$

where A is the maximum displacement and t is the time. Find d given that $A = 12.0$ cm and $t = 0.150$ s.

58. The current in a certain alternating-current circuit is given by

$$i = I \sin 25t$$

where I is the maximum possible current and t is the time. Find i for $t = 0.100$ s and $I = 150$ mA.

59. St. Louis has a latitude of $38.7°$ north. How far is it from the equator? (The radius of the earth is about 3960 mi.)

60. A thermometer needle passes through $50.0°$ when the temperature changes $36.0°$C. See Figure 19-50. If the needle is 2.80 cm long and a circular scale is used, how long is the scale for a maximum temperature change of $140.0°$C?

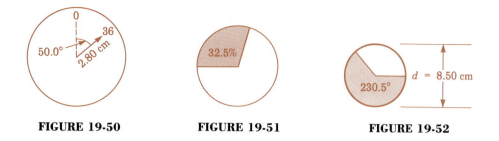

FIGURE 19-50   **FIGURE 19-51**   **FIGURE 19-52**

61. In constructing a pie chart, a sector of a circle must have an area that is 32.5% of the total area. Find the central angle in degrees. See Figure 19-51.

62. A satellite is in orbit at an altitude of 18,600 mi above the surface of the earth, and it makes 1.02 revolutions about the earth each 24.5 h. Find the linear velocity of this satellite. (The radius of the earth is 3960 mi.)

63. A piece of circular filter paper 8.50 cm in diameter is folded so that its effective filtering area is the same as that of a sector with central angle of 230.5°. What is the filtering area? See Figure 19-52.

64. A funnel is made from a circular piece of sheet metal 25.0 cm in radius from which two pieces are removed. The first piece removed is a circle of radius 2.50 cm at the center, and the second piece removed is a sector of central angle 200.0°. What is the surface area of the funnel? See Figure 19-53.

65. A computer cooling fan has a blade with radius 5.58 cm and it rotates at 3600 r/min. What is the linear velocity of a point at the end of the blade?

66. The armature of a dynamo is 1.25 ft in diameter and is rotating at the rate of 1250 r/min. What is the linear velocity of a point on the outside of the armature?

67. A pulley of radius 3.60 in. is belted to another pulley of radius 5.75 in. The smaller pulley rotates at 35.0 r/s. What is the angular velocity of the larger pulley? See Figure 19-54.

FIGURE 19-53

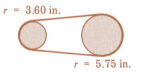

$r = 3.60$ in.

$r = 5.75$ in.

FIGURE 19-54

68. A lathe is to cut material at the rate of 115 m/min. Calculate the radius of a cylindrical piece that is turned at the rate of 120 r/min.

69. An emery wheel 45.0 cm in diameter has a cutting speed of 1280 m/min. Find the number of revolutions per minute that the wheel makes.

70. A Black & Decker circular saw has a blade 7.25 in. in diameter which rotates at 4900 r/min. What is the linear velocity of a point at the end of one of the teeth?

71. A jet is flying westward along the equator with the sun directly overhead. How fast must the jet fly in order to keep the sun directly overhead? (Assume that the earth's radius is 3960 mi and that the earth rotates about its axis in exactly 24 h.)

72. A person calculates the distance traveled by a point on the equator as the point travels for exactly 24 h about the earth's center. That person assumed a radius of 3960 mi and that the point makes exactly one complete revolution in 24 h. If it actually requires 24 h 3 min and 55.909 s (mean solar day) to make one complete revolution, find the amount of error in the first result.

73. A simple pendulum is started by giving it a velocity from its equilibrium position. The angle θ between the vertical and the pendulum is given by

$$\theta = a \sin \left(\sqrt{\frac{g}{l}} t \right)$$

where a is the amplitude in radians, g ($= 32$ ft/s^2) is the acceleration due to gravity, l is the length of the pendulum in feet, and t is the length of time of the motion. Sketch two cycles of θ as a function of t for the pendulum whose length is 2.0 ft and $a = 0.1$ rad.

74. The displacement of a certain water wave (height above the calm water level) is given by

$$d = 3.20 \sin 2.50t$$

where d is measured in meters and t in seconds. Find d, given that $t = 1.25$ s.

75. The voltage in a certain alternating-current circuit is given by

$$v = v_{max} \sin 36t$$

where v_{max} is the maximum possible voltage and t is the time in seconds. Find v for $t = 0.0050$ s and $v_{max} = 120$ V.

76. The electric current in a certain circuit is given by

$$i = i_0 \sin \left(\frac{t}{\sqrt{LC}} \right)$$

where i_0 is the initial current, L is an inductance, and C is a capacitance. Sketch two cycles of i as a function of t (in seconds) for the case where $i_0 = 15.0$ A, $L = 1$ H, and $C = 25$ μF.

77. A certain object is oscillating at the end of a spring. See Figure 19-55. The displacement as a function of time is given by the relation

$$y = 2.4 \cos \left(12t + \frac{\pi}{6} \right)$$

where y is measured in meters and t in seconds. Sketch the graph of y versus t.

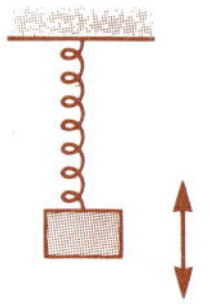

78. The current in an alternating-current circuit is described by the equation

FIGURE 19-55

$$y = 110 \sin \left(60\pi x - \frac{\pi}{4} \right)$$

Sketch the graph of y versus x.

79. A laser beam transmitted with a "width" of 0.0005° makes a circular spot of radius 1.63 km on a distant object. How far is the object from the source of the laser beam? (See Exercise 41 from Section 19-2.)

80. An observer on earth records an angle of 15″ between the top and bottom of Venus. If the distance between Venus and earth is 1.04×10^8 mi, what is the diameter of Venus? (See Exercise 41 from Section 19-2.)

CHAPTER TWENTY

Complex Numbers

In Section 10-3 we introduced the concept of the imaginary number j. In Chapter 11 we used imaginary numbers in describing the solutions to certain quadratic equations. In this chapter we extend the previous brief discussions of imaginary numbers as we define **complex numbers.** We then examine the basic arithmetic operations as they apply to complex numbers, and we also discuss graphical representations of complex numbers. We conclude the chapter by representing complex numbers in a special form known as the *polar form.*

Complex numbers are used in many applications, including the development of the shape of the cross section of an airplane wing. See Figure 20-1. In constructing the curve shown in Figure 20-1, it is necessary to evaluate expressions such as

$$\frac{1}{2}\left[(1 + 2j) + \frac{1}{1 + 2j}\right].$$

where the expression $1 + 2j$ is typical of the complex numbers discussed

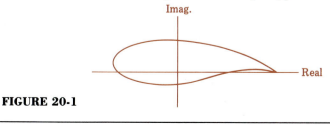

FIGURE 20-1

in this chapter. We will learn how to evaluate such expressions, and the particular expression given above will be evaluated later in the chapter.

20-1 Introduction to Complex Numbers

In Section 10-3 we saw that problems involving the square root of a negative number were resolved by defining the imaginary unit j, where

$$j^2 = -1 \quad \text{or} \quad j = \sqrt{-1} \tag{10-4}$$

A number in the form of bj (b is a real number) is called an *imaginary number*. We used imaginary numbers again in Chapter 11 when we solved certain quadratic equations. For example, the quadratic equation $x^2 + 4 = 0$ has $2j$ and $-2j$ as solutions. We begin with Example A that reviews the earlier work done with imaginary numbers.

EXAMPLE A

$$\sqrt{-4} = \sqrt{4j^2} = 2j$$
$$\sqrt{-25} = \sqrt{25j^2} = 5j$$

Also,

$$(8j)^2 = 64j^2 = -64$$
$$(-5j)^2 = 25j^2 = -25$$
$$(-9j)^2 = 81j^2 = -81$$

☐

EXAMPLE B   In an alternating-current circuit, the inductive reactance (in ohms) is represented by $\sqrt{-9}$. We can express this value as

$$\sqrt{-9} = \sqrt{9j^2} = 3j$$

The inductive reactance is $3j\ \Omega$.

☐

EXAMPLE C

$$(\sqrt{-16})^2 = (\sqrt{16j^2})^2 = (j\sqrt{16})^2$$
$$= 16j^2 = -16$$

☐

In Example C, it is easy to make the mistake of concluding that

$$(\sqrt{-16})^2 = \sqrt{-16}\ \sqrt{-16} = \sqrt{256} = 16 \ \text{(wrong!)}$$

The correct result is -16, not 16. In general, Eq. (10-8) states that

$$\boxed{\sqrt{ab} = \sqrt{a}\,\sqrt{b}} \qquad a \text{ and } b \text{ are positive} \tag{10-8}$$

but this is valid only for positive values of a and b; it is *not* valid if a or b is a negative number. We must remember that we should never use Eq. (10-8) with negative numbers as in $\sqrt{-16}\,\sqrt{-16} = \sqrt{256} = 16$, which is wrong. Much of this difficulty can be avoided if we follow this simple rule: *when working with square roots of negative numbers, express each in terms of j immediately and before proceeding.* This can be accomplished by applying the following.

$$\boxed{\sqrt{-a} = j\sqrt{a}} \qquad \text{(where } a > 0\text{)} \tag{10-6}$$

EXAMPLE D Express $\sqrt{-5}$, $\sqrt{-12}$, and $-\sqrt{-18}$ in terms of j.

$$\sqrt{-5} = \sqrt{5j^2} = j\sqrt{5}$$
$$\sqrt{-12} = \sqrt{(4)(3)j^2} = 2j\sqrt{3}$$
$$-\sqrt{-18} = -\sqrt{(9)(2)j^2} = -3j\sqrt{2}$$

When the result contains a factor of j, we usually write the j *before* any radical so that it is clear that j is not under the radical.

Using real numbers and imaginary numbers, we can define a **complex number** *to be any number that can be expressed in the form of $a + bj$,* where the numbers a and b are both real. The standard form of $a + bj$ is referred to as the **rectangular form** of a complex number. The number a is called the **real part** while b is called the **imaginary part**. If $a = 0$, then the number bj is called a **pure imaginary** number. If $b = 0$, then the number a is a **real number**. With these definitions, the real numbers become a subset of the collection of all complex numbers.

EXAMPLE E To express $-7 + \sqrt{-12}$ in the rectangular form of $a + bj$, we proceed as follows.

$$-7 + \sqrt{-12} = -7 + \sqrt{(4)(3)j^2}$$
$$= -7 + 2j\sqrt{3}$$

A complex number has the form $a + bj$, whereas an imaginary number can be written as bj.

EXAMPLE F $3 + 4j$ is a complex number with 3 as the real part and 4 as the imaginary part.

   $-5 - 6j$ is a complex number with -5 as the real part and -6 as the imaginary part.

   $6 - j$ is a complex number with 6 as the real part and -1 as the imaginary part.

   $8j$ is a complex number with 0 as the real part and 8 as the imaginary part; $8j$ is also an imaginary number.

   7 is a complex number with 7 as the real part and 0 as the imaginary part; 7 is also a real number.   □

It is often necessary to raise j to different powers. The value of j^n is either 1, j, -1, or $-j$ provided that n is zero or a positive integer. We can evaluate j^n as shown in the following example.

EXAMPLE G
$$j^0 = 1$$
$$j^1 = j$$
$$j^2 = -1$$
$$j^3 = (j^2)(j) = -j$$
$$j^4 = j^2 j^2 = (-1)(-1) = 1$$
$$j^5 = j^4 j = (1)j = j$$

pattern of 1, j, -1, $-j$, repeats   □

If we were to continue Example G with the successive powers of j, we would find that the pattern of 1, j, -1, $-j$ repeats. We can summarize the results as follows.

$$j^n = \begin{cases} 1 & \text{for } n = 0, 4, 8, \ldots \\ j & \text{for } n = 1, 5, 9, \ldots \\ -1 & \text{for } n = 2, 6, 10, \ldots \\ -j & \text{for } n = 3, 7, 11, \ldots \end{cases}$$

We conclude this section with a basic definition that will be used later. *The* **conjugate** *of the complex number a + bj is the complex number a − bj.* To find the conjugate of any complex number, simply change the sign of the imaginary part.

EXAMPLE H The conjugate of $2 - 7j$ is $2 + 7j$.
The conjugate of $-5 + 2j$ is $-5 - 2j$.
The conjugate of $-8 - 3j$ is $-8 + 3j$.   □

Exercises 20-1

In Exercises 1 through 12, express each number in terms of j.

1. $\sqrt{-16}$ **2.** $\sqrt{-25}$ **3.** $-\sqrt{-9}$ **4.** $-\sqrt{-36}$

5. $\sqrt{-0.25}$ **6.** $-\sqrt{-0.16}$ **7.** $\sqrt{-27}$ **8.** $\sqrt{-45}$

9. $-\sqrt{-48}$ **10.** $-\sqrt{-75}$ **11.** $\sqrt{-0.0004}$ **12.** $-\sqrt{-0.0009}$

In Exercises 13 through 20, express each number in the rectangular form $a + bj$.

13. $5 + \sqrt{-1}$ **14.** $-3 + \sqrt{-1}$ **15.** $6 - \sqrt{-4}$ **16.** $-9 - \sqrt{-9}$

17. $-4 + \sqrt{-8}$ **18.** $-8 - \sqrt{-54}$ **19.** $14 - \sqrt{-63}$ **20.** $-7 + \sqrt{-28}$

In Exercises 21 through 32, simplify each expression. (Be sure to introduce j immediately before proceeding.)

21. $(\sqrt{-9})^2$ **22.** $(\sqrt{-4})^2$ **23.** $(\sqrt{-6})^2$

24. $(\sqrt{-0.5})^2$ **25.** $\sqrt{(-3)^2}$ **26.** $\sqrt{(-7)^2}$

27. $\sqrt{-3}\sqrt{-4}$ **28.** $\sqrt{-2}\sqrt{-8}$ **29.** $\sqrt{-3}\sqrt{-12}$

30. $\sqrt{-5}\sqrt{-7}$ **31.** $\sqrt{-3}\sqrt{-11}$ **32.** $\sqrt{-13}\sqrt{-5}$

In Exercises 33 through 44, simplify each expression.

33. j^6 **34.** j^8 **35.** j^9 **36.** j^{10}

37. j^{65} **38.** j^{103} **39.** j^{402} **40.** j^{604}

41. $(-j)^{22}$ **42.** $-j^{43}$ **43.** $-j^{81}$ **44.** $(-j)^{64}$

In Exercises 45 through 52, find the conjugate of each complex number.

45. $3 - 8j$ **46.** 7 **47.** -4 **48.** $-5 + 4j$

49. $9j$ **50.** $-3j$ **51.** $-2 - 7j$ **52.** $3 - 8j$

20-2 Arithmetic of Complex Numbers

In this section we consider the basic operations of addition, subtraction, multiplication, and division as they apply to complex numbers. Before defining those operations, we must first establish the rule that is used to determine whether two complex numbers are equal. We stipulate that $a + bj = x + yj$ *if and only if $a = x$ and $b = y$.* That is, two complex numbers are equal if and only if the real parts are equal and the imaginary parts are equal.

EXAMPLE A   If $a + bj = 6 + 3j$, then $a = 6$ and $b = 3$.

If $a + bj = 2 - 5j$, then $a = 2$ and $b = -5$.

If $a + bj = 7$, then $a = 7$ and $b = 0$.

If $a + bj = -8j$, then $a = 0$ and $b = -8$.   ☐

We can now proceed to consider the basic operations of arithmetic. We can add (or subtract) two complex numbers by combining the real parts and combining the imaginary parts.

$$(a + bj) + (c + dj) = (a + c) + (b + d)j \qquad \text{(20-1)}$$

EXAMPLE B

$$(2 + 3j) + (6 + 4j) = (2 + 6) + (3 + 4)j = 8 + 7j$$
$$(6 + j) + (-2 - 8j) = (6 - 2) + (1 - 8)j = 4 - 7j$$
$$(7 + 5j) - (3 + 2j) = (7 - 3) + (5 - 2)j = 4 + 3j$$
$$(-2 + 4j) - (6 - 9j) = (-2 - 6) + (4 - (-9))j = -8 + 13j \quad \square$$

When multiplying two complex numbers, we can treat them as binomials. In Section 7-3 we saw that multiplication of two binomials involves multiplying each term of one binomial by each term of the other. When multiplying $a + bj$ and $c + dj$, we get

$$(a + bj)(c + dj) = a(c) + a(dj) + bj(c) + bj(dj)$$
$$= ac + adj + bcj + bdj^2$$
$$= (ac - bd) + (ad + bc)j$$

We summarize this result in Eq. (20-2).

$$(a + bj)(c + dj) = (ac - bd) + (ad + bc)j \qquad \text{(20-2)}$$

It is not necessary to memorize the exact form of Eq. (20-2) since that expression shows that we multiply two complex numbers by following the same procedure used in the algebraic multiplication of any two binomials.

EXAMPLE C

$$(2 + 3j)(6 + 4j) = 12 + 8j + 18j + 12j^2$$
$$= 12 + 8j + 18j + 12(-1)$$
$$= (12 - 12) + (8 + 18)j$$
$$= 26j \qquad \qquad \square$$

EXAMPLE D

$$(6 + j)(-2 - 8j) = -12 - 48j - 2j - 8j^2$$
$$= -12 - 48j - 2j - 8(-1) \qquad -12 + 8 = -4$$
$$= -4 - 50j \qquad \qquad \square$$

EXAMPLE E   Multiply the complex number $c + dj$ by its conjugate $c - dj$. We get

$$(c + dj)(c - dj) = c(c) + c(-dj) + dj(c) + dj(-dj)$$

zero

$$= c^2 - cdj + cdj - d^2 j^2$$

$$= c^2 - d^2(-1)$$

$$= c^2 + d^2 \qquad \square$$

The most important principle that must be remembered for multiplication is that *each term* of the first complex number must be multiplied by *each term* of the second complex number. In Example C, $(2 + 3j)(6 + 4j)$ is *not* equal to $2(6) + 3j(4j) = 12 + 12j^2 = 12 - 12 = 0$. When adding or subtracting complex numbers, we add or subtract the real parts and then add or subtract the imaginary parts, but multiplication of complex numbers involves more than multiplying the real parts and multiplying the imaginary parts.

When dividing a number (real or complex) by a *complex* number, we multiply the numerator and denominator by the conjugate of the denominator. In Example E we saw that when the complex number $c + dj$ is multiplied by its conjugate $c - dj$, the result is the real number $c^2 + d^2$. Therefore, multiplication of the numerator and denominator by the conjugate of the denominator will eliminate j from the denominator. We can summarize division in Eq. (20-3).

$$\frac{a + bj}{c + dj} = \frac{(a + bj)(c - dj)}{(c + dj)(c - dj)} = \frac{(ac + bd) + (bc - ad)j}{c^2 + d^2} \qquad (20\text{-}3)$$

EXAMPLE F

$$\frac{2 + 3j}{3 + 4j} = \frac{(2 + 3j)(3 - 4j)}{(3 + 4j)(3 - 4j)}$$

conjugate

$$= \frac{6 - 8j + 9j - 12j^2}{9 - 12j + 12j - 16j^2}$$

$$= \frac{6 + j - 12(-1)}{9 - 16(-1)}$$

$$= \frac{18 + j}{25} \qquad \square$$

The result in Example F could be expressed in the standard rectangular form of $a + bj$ as $\frac{18}{25} + \frac{1}{25}j$, but it is usually left as a single fraction.

EXAMPLE G

$$\frac{5j}{2 - 7j} = \frac{5j(2 + 7j)}{(2 - 7j)(2 + 7j)} = \frac{10j + 35j^2}{4 + 14j - 14j - 49j^2}$$

$$= \frac{10j + 35(-1)}{4 - 49(-1)}$$

$$= \frac{-35 + 10j}{53}$$

☐

EXAMPLE H

$$\frac{3 + 5j}{-4j} = \frac{(3 + 5j)(4j)}{(-4j)(4j)} = \frac{12j + 20j^2}{-16j^2}$$

conjugate

$$= \frac{12j + 20(-1)}{-16(-1)}$$

$$= \frac{-20 + 12j}{16} = \frac{-5 + 3j}{4}$$

reduce

☐

EXAMPLE I   Express the number

$$\frac{1}{2}\left[(1 + 2j) + \frac{1}{1 + 2j}\right]$$

in the standard rectangular form $a + bj$. Such expressions are used in determining the cross section of an airplane wing.

We can begin by expressing $\dfrac{1}{1 + 2j}$ in standard rectangular form as follows.

$$\frac{1}{1 + 2j} = \frac{1(1 - 2j)}{(1 + 2j)(1 - 2j)} = \frac{1 - 2j}{1 - 2j + 2j - 4j^2}$$

$$= \frac{1 - 2j}{1 - 4(-1)} = \frac{1 - 2j}{5} \quad \text{or} \quad \frac{1}{5} - \frac{2}{5}j$$

The original expression can now be evaluated as

$$\frac{1}{2}\left[(1 + 2j) + \left(\frac{1}{5} - \frac{2}{5}j\right)\right]$$

$$= \frac{1}{2}\left[\left(1 + \frac{1}{5}\right) + \left(2 - \frac{2}{5}\right)j\right] \quad \text{using Eq. (20-1)}$$

$$= \frac{1}{2}\left[\frac{6}{5} + \frac{8}{5}j\right]$$

$$= \frac{3}{5} + \frac{4}{5}j$$

☐

Exercises 20-2

In Exercises 1 through 48, perform the indicated operations and express all answers in the rectangular form $a + bj$.

1. $(4 - 6j) + (1 - j)$

2. $(3 - 2j) + (5 + j)$

3. $(7 + j) - (3 - 4j)$

4. $(-9 - 2j) - (3 - 5j)$

5. $(1 + \sqrt{-9}) + (2 - \sqrt{4})$

6. $(3 - \sqrt{-16}) - (-5 + \sqrt{-25})$

7. $(2 - \sqrt{-36}) - (2 + \sqrt{-36})$

8. $(-3 - \sqrt{-64}) + (5 - \sqrt{-49})$

9. $6j + (3 - 8j) + (-1 - 4j)$

10. $(7 - 2j) + (3 - j) + 4j$

11. $\sqrt{-25} + \sqrt{-36} - (3 - 9j)$

12. $(6 - 7j) - \sqrt{-16} - (-5 - 8j)$

13. $5j(6j)$

14. $-3j(2j)$

15. $(2 - j)(2 + j)$

16. $(-3 + 4j)(-3 - 4j)$

17. $(2 - 7j)(3 + 6j)$

18. $(-3 - 9j)(-2 - 8j)$

19. $6j(2j)(-3j)$

20. $-j(-3j)(8j)$

21. $\sqrt{-4}(2 - \sqrt{-9})$

22. $\sqrt{-9}(-3 + \sqrt{-16})$

23. $(2 + \sqrt{-25})(3 - \sqrt{-36})$

24. $(7 - \sqrt{16})(-2 - \sqrt{-9})$

25. $\dfrac{2j}{3 - 4j}$

26. $\dfrac{-3j}{-2 + 5j}$

27. $\dfrac{3 + 8j}{2j}$

28. $\dfrac{5 - 6j}{-3j}$

29. $\dfrac{7 + 3j}{5 + 6j}$

30. $\dfrac{6 - 4j}{9 - 3j}$

31. $\dfrac{-4 + 7j}{1 - 3j}$

32. $\dfrac{-2 + 9j}{-2 - 3j}$

33. $\sqrt{-9} \div (2 + \sqrt{-4})$

34. $\sqrt{-25} \div (4 - \sqrt{-9})$

35. $(2 + \sqrt{-25}) \div (3 - \sqrt{-9})$

36. $(5 + \sqrt{-81}) \div (-2 + \sqrt{-64})$

37. $(1 - j)^2$

38. $(-2 + 3j)^2$

39. $(1 + j)^2 + (2 - 5j)$

40. $(-6 + 2j) - (2 - j)^2$

41. $5j(1 + 3j)^2$

42. $-3j(2 - 7j)^2$

43. $(2 + j)(3 - j) \div (1 - j)$

44. $(3 - 2j)(8 - 3j) \div (2 + 5j)$

45. $6j - (1 + j)(1 + 2j)(1 + 3j)$

46. $-2j + (2 - 3j)(2 + 3j)(3 + 2j)$

47. $(4 - 3j)^2 - (4 + 3j)^2$

48. $(2 - 5j)^2 + (-3 + 4j)^2$

In Exercises 49 through 52, multiply the given complex number by its conjugate.

49. $6 + 8j$

50. $5 - 3j$

51. $-2 + 7j$

52. $-3 - j$

In Exercises 53 through 56, perform the indicated operations and express all answers in the rectangular form of $a + bj$.

53. In an alternating-current circuit, the impedance Z is given by

$$Z = R + j(X_L - X_C)$$

where R is the resistance (in ohms), X_L is the inductive reactance (in ohms), and X_C is the capacitive reactance (in ohms). Find the impedance if $R = 14.7 \ \Omega$, $X_L = 10.2 \ \Omega$, and $X_C = 12.3 \ \Omega$.

54. In an alternating-current circuit, the impedance Z, the current I, and the voltage V are related by

$$Z = \frac{V}{I}$$ Find the impedance (in ohms) if $V = 26 + 3j$ volts and $I = 6 - 2j$ amperes.

55. In Exercise 54, find the voltage (in volts) if $Z = -0.20 + 1.3j$ ohms and $I = 4.0 - 3.0j$ amperes.

56. In an alternating-current circuit, two impedances Z_1 and Z_2 are included so that the total impedance Z_T is given by

$$Z_T = \frac{Z_1 Z_2}{Z_1 + Z_2}$$

Find the total impedance if $Z_1 = 2 + 3j$ ohms and $Z_2 = 3 - 4j$ ohms.

20-3 Graphical Representation of Complex Numbers

We represent complex numbers in rectangular form as $a + bj$, where a is the real part and b is the imaginary part. Since each complex number is associated with the pair of numbers a and b, we can represent complex numbers graphically in a rectangular coordinate system. The usual format involves a *horizontal axis called the* **real axis** *and a vertical axis called the* **imaginary axis**. This combination of axes allows us to form the **complex plane**.

EXAMPLE A   In Figure 20-2 we depict the complex number $3 + 5j$ as it is graphed in the complex plane.   ☐

When constructing the graph of a complex number, we can draw a line from the origin to the point. In this way, we can represent complex numbers in the same way that we represent vectors. In general, *a complex number can be considered to be a vector from the origin to its point in the complex plane*.

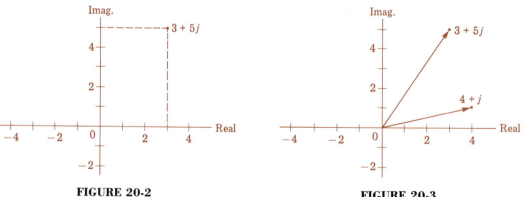

FIGURE 20-2   FIGURE 20-3

EXAMPLE B   In Figure 20-3 we show the two complex numbers $3 + 5j$ and $4 + j$. These two complex numbers are shown as vectors.   ☐

In Example B we graphed the complex numbers $3 + 5j$ and $4 + j$. From Section 20-2 we know that the sum of $3 + 5j$ and $4 + j$ is equal to $7 + 6j$. In Figure 20-4 we show that the vectors $3 + 5j$ and $4 + j$ can be added graphically by following the same parallelogram method used for adding vectors in Section 18-1. After representing both complex numbers as vectors, we complete a parallelogram as shown in Figure 20-4. The sum of the two original vectors is represented by a diagonal of the parallelogram. (We always use the diagonal that extends from the origin.)

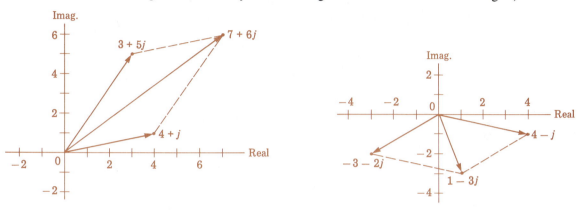

FIGURE 20-4   FIGURE 20-5

EXAMPLE C   Add the complex numbers $4 - j$ and $-3 - 2j$ graphically.

In Figure 20-5 we show the original two vectors with the completed parallelogram. The sum $1 - 3j$ is shown as the diagonal of the parallelogram. □

EXAMPLE D   Subtract $2 - 3j$ from $4j$ graphically.

Subtracting $2 - 3j$ is equivalent to adding $-2 + 3j$. In Figure 20-6 we show the addition of $-2 + 3j$ and $4j$. The sum is $-2 + 7j$. Note that when graphing the complex number $4j$, we consider that number to be in rectangular form as $0 + 4j$ so that the point $(0, 4)$ is plotted in the complex plane. □

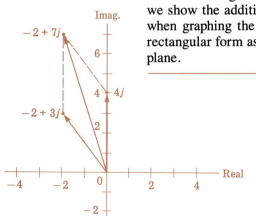

FIGURE 20-6

There are many advantages in representing complex numbers as vectors. This association is helpful in applications. In electronics, for example, the complex plane can be used to represent resistance as the horizontal axis and reactance as the vertical axis so that their sum represents impedance.

EXAMPLE E One common application of complex numbers involves alternating-current circuits. The impedance Z (in ohms) is the sum of the resistance R (in ohms), the capacitive reactance X_C (in ohms), and the inductive capacitance X_L (in ohms). It is standard to represent R along the positive real axis, X_C along the negative real axis, and X_L along the positive imaginary axis. In Figure 20-7, find the impedance Z by adding the given values of R, X_C, and X_L.

We first combine $2.5j$ and $-5.5j$ to get $-3.0j$ as the sum of the inductive reactance and capacitive reactance. The impedance is now represented by $8.3 - 3.0j$ ohms.

FIGURE 20-7

Exercises 20-3

In Exercises 1 through 8, graph each complex number as a vector in the complex plane.

1. $3 + 2j$ 2. $5 - 3j$ 3. $3j$
4. 4 5. $-2.5 + 3.1j$ 6. $6.2 - 0.8j$
7. $0.3 - 0.2j$ 8. $-1.6 + 2.2j$

In Exercises 9 through 28, perform the indicated operations graphically and check the results algebraically.

9. $(4 + 3j) + (1 + j)$ 10. $(2 + j) + (3 + 9j)$ 11. $5j + (2 + j)$
12. $-2j + (3 + 5j)$ 13. $(6 + 2j) - j$ 14. $(8 + 3j) - 4j$
15. $(9 + 7j) - (2 + 6j)$ 16. $(-5 + 4j) - (3 + 4j)$ 17. $(-2 + 3j) + (6 - 4j)$
18. $(-1 - 7j) - (2 - 3j)$ 19. $(5 - 6j) - (-4 + 9j)$ 20. $(8 + 2j) + (-3 + 5j)$
21. $(2 - j) + (2 + j)$ 22. $(1 + 9j) + (8 - 7j)$ 23. $(-8 - 4j) - (5 - 6j)$
24. $(3 + 4j) - (3 - 4j)$ 25. $(-8 - 7j) - (7 + 4j)$ 26. $(-1 + 8j) + (4 + 5j)$
27. $(7 + 4j) + (5 - 8j)$ 28. $(9 - 3j) - (6 - 8j)$

In Exercises 29 through 32, graph the given complex number, its negative, and its conjugate on the same set of axes in the complex plane.

29. $2 + 3j$ 30. $5 - 4j$ 31. $-3 + 7j$ 32. $-6 - j$

20-4 Polar Form of Complex Numbers

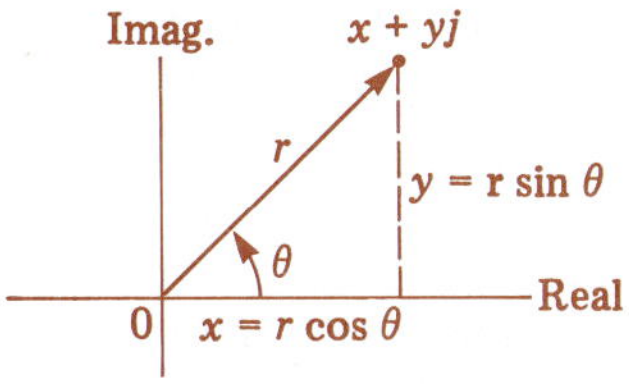

In Section 20-3 we saw that there is a relationship between complex numbers and vectors. We will now use that relationship to develop another way of expressing complex numbers. Refer to Figure 20-8, where the general complex number $x + yj$ is depicted in the same way that we depict a vector. That complex number is a distance r from the origin, and the angle θ is in standard position. For any complex number, if we know the magnitude r and the direction θ, then we can locate the correct point in the complex plane.

From Figure 20-8 we can use trigonometry and the Pythagorean theorem to get

FIGURE 20-8

$$x = r \cos \theta \qquad (20\text{-}4)$$

$$y = r \sin \theta \qquad (20\text{-}5)$$

$$r^2 = x^2 + y^2 \qquad (20\text{-}6)$$

$$\tan \theta = \frac{y}{x} \qquad (20\text{-}7)$$

We can begin with the rectangular form $x + yj$ of a complex number and then substitute using Eqs. (20-4) and (20-5) to get

$$x + yj = r \cos \theta + jr \sin \theta$$

or

$$x + yj = r(\cos \theta + j \sin \theta) \qquad (20\text{-}8)$$

The form $r(\cos \theta + j \sin \theta)$ is called the **polar form** *(or* **trigonometric form***) of a complex number. The length r is called the* **absolute value** *(or the* **modulus***), and the angle θ is called the* **argument** *of the complex number.* The polar form of $r(\cos \theta + j \sin \theta)$ is sometimes abbreviated as $r\angle\theta$ or r cis θ. That is,

$$r \text{ cis } \theta = r(\cos \theta + j \sin \theta) \qquad (20\text{-}9)$$

$$r \angle \theta = r(\cos \theta + j \sin \theta) \qquad (20\text{-}10)$$

EXAMPLE A $4(\cos 30° + j \sin 30°) = 4\angle 30°$

$5\angle 210° = 5(\cos 210° + j \sin 210°)$

$8 \text{ cis } 93° = 8(\cos 93° + j \sin 93°)$ ☐

EXAMPLE B   Represent the complex number $3 + 4j$ graphically; then express it in the polar form $r(\cos \theta + j \sin \theta)$.

From the rectangular form of $3 + 4j$ we see that $x = 3$ and $y = 4$. We plot the point $(3, 4)$ in Figure 20-9. Using Eq. (20-6) we find that the length of r is

$$r = \sqrt{x^2 + y^2} = \sqrt{3^2 + 4^2} = 5$$

Also, we can find the direction θ by using Eq. (20-7) as follows.

$$\tan \theta = \frac{y}{x} = \frac{4}{3} = 1.333$$

With $\tan \theta = 1.333$, we can use the $\boxed{\text{INV}}\boxed{\text{TAN}}$ keys on a calculator to establish that $\theta = 53.1°$. Knowing both r and θ, we can express $3 + 4j$ in polar form as

$$5(\cos 53.1° + j \sin 53.1°)$$

In Figure 20-10 we include the values of r and θ.   ☐

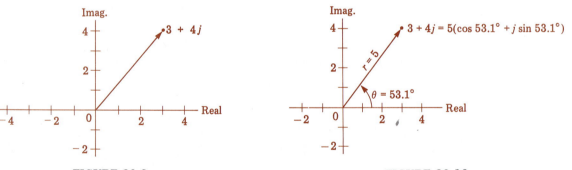

FIGURE 20-9   FIGURE 20-10

In Example B we showed how to convert a complex number from the rectangular form $x + yj$ to the polar form $r(\cos \theta + j \sin \theta)$. Some calculators are designed to allow direct conversions between the rectangular and polar forms. On one calculator, we can convert $3 + 4j$ to polar form by entering the sequence of keys

$$P \leftrightarrow R$$

3 $\boxed{x \blacktriangleleft y}$ 4 $\boxed{\text{INV}}\boxed{\text{2nd}}$ $\boxed{x \blacktriangleleft y}$ $\boxed{\mathit{53.130102}}$

to get the display 53.130102, which is the value of θ in degrees. We could then press the key labeled $\boxed{x \geqslant y}$ to get the display 5, which is the value of r.

EXAMPLE C By combining resistance, inductive capacitance, and reactive capacitance, we find that we can represent the impedance in a circuit as $3.14 - 2.07j$ ohms. We will represent that number graphically and then express it in the polar form $r(\cos \theta + j \sin \theta)$.

We first plot the point $x = 3.14$ and $y = -2.07$ in the complex plane. See Figure 20-11. We must now find the values of r and θ so that we can determine the polar form of the given complex number. Using Eq. (20-6) we get

$$r = \sqrt{x^2 + y^2} = \sqrt{3.14^2 + (-2.07)^2}$$
$$= 3.76$$

Using Eq. (20-7) we get

$$\tan \theta = \frac{y}{x} = \frac{-2.07}{3.14} = -0.6592$$

With $\tan \theta = -0.6592$, we use a calculator to determine that $\theta = -33.4°$, and that result agrees with the fourth-quadrant location of the number as it is shown in Figure 20-11.

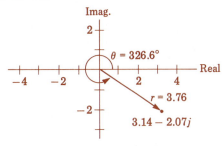

FIGURE 20-11 □

EXAMPLE D Express the complex number

$$5(\cos 150.0° + j \sin 150.0°)$$

in rectangular form.

Evaluating $\cos 150.0°$ and $\sin 150.0°$, the number can be expressed in rectangular form as follows.

$$5(\cos 150.0° + j \sin 150.0°) = 5(-0.866 + 0.500j)$$
$$= -4.33 + 2.50j \qquad □$$

EXAMPLE E   Express 12.3 ∠239.4° in rectangular form.

We know that the polar form 12.3∠239.4° is equivalent to the polar form 12.3 (cos 239.4° + j sin 239.4°). Evaluating cos 239.4° = −0.509 and sin 239.4° = −0.861, we get

$$12.3\angle239.4° = 12.3(\cos 239.4° + j \sin 239.4°)$$
$$= 12.3(-0.509 - 0.861\,j)$$
$$= -6.26 - 10.6\,j \qquad \square$$

EXAMPLE F   Express −6.26 − 10.6 j in polar form.

Since x = −6.26 and y = −10.6, we use Eq. (20-6) to find r.

$$r = \sqrt{x^2 + y^2}$$
$$= \sqrt{(-6.26)^2 + (-10.6)^2} = 12.3$$

Using Eq. (20-7) we find that

$$\tan \theta = \frac{y}{x} = \frac{-10.6}{-6.26} = 1.693$$

Using a calculator, we would find that θ = 59.4°, but Example E and Figure 20-12 both show that 59.4° is *wrong!*

Because −6.26 − 10.6 j is in the *third* quadrant, θ must fall between 180° and 270°, so 59.4 is incorrect. Here, the graph of Figure 20-12 helps us to avoid the mistake of simply accepting the value of 59.4° that the calculator would provide. Again, blind use of the calculator will lead to an error; we must *understand* what's happening so that we can adjust the calculator results. However, 59.4° is the reference angle. From Figure 20-12 we can see that the correct value for θ is 239.4°. With r = 12.3 and θ = 239.4°, we can express the number in polar form as

$$12.3(\cos 239.4° + j \sin 239.4°)$$

which agrees with the result of Example E and Figure 20-12.

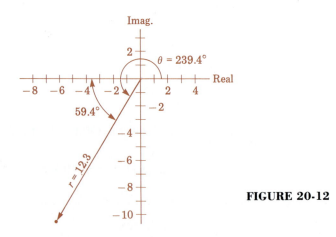

FIGURE 20-12

□

Exercises 20-4

In Exercises 1 through 16, convert each complex number from the given rectangular form to the polar form of $r(\cos \theta + j \sin \theta)$. *Use a graph to be sure that* θ *corresponds to the correct quadrant.*

1. $4 + 3j$

2. $-4 + 3j$

3. $4 - 3j$

4. $-4 - 3j$

5. $2.45 - 3.78j$

6. $-4.18 + 1.56j$

7. $3.66 + 5.39j$

8. $-852 - 631j$

9. $-\sqrt{2} - j\sqrt{2}$

10. $\sqrt{3} - j$

11. $-1 + j\sqrt{3}$

12. $1 + j\sqrt{3}$

13. -10

14. 7

15. $6j$

16. $-3j$

In Exercises 17 through 24, express each given complex number in the polar form $r(\cos \theta + j \sin \theta)$.

17. $6.15\angle 43.0°$

18. $256\angle 184.6°$

19. 10.3 cis $335.2°$

20. 73.5 cis $109.6°$

21. $0.348\angle 76.4°$

22. 8.19 cis $27.4°$

23. 56.0 cis $212.5°$

24. 403 cis $8.3°$

In Exercises 25 through 40, convert each complex number from the given polar form to the rectangular form $x + yj$.

25. $3.00(\cos 70.0° + j \sin 70.0°)$

26. $5.00(\cos 245.0° + j \sin 245.0°)$

27. $1.50(\cos 321.2° + j \sin 321.2°)$

28. $4.50(\cos 255.7° + j \sin 255.7°)$

29. $10(\cos 270° + j \sin 270°)$

30. $20(\cos 0° + j \sin 0°)$

31. $25(\cos 180° + j \sin 180°)$

32. $65(\cos 90° + j \sin 90°)$

33. $50\angle 35°$

34. $75\angle 240°$

35. 12.4 cis $300°$

36. 2.93 cis $210.0°$

37. $12.48\angle 36.25°$

38. $758.9\angle 163.7°$

39. 2.194 cis $235.00°$

40. 5.724 cis $123.64°$

In Exercises 41 through 44, solve the given problems.

41. The voltage (in volts) measured on a wire is given as $E = 57.3\angle 7.2°$. Express this number in rectangular form.

42. The current (in microamperes) in a microprocessor circuit is given as $4.25\angle 21.5°$. Express this number in rectangular form.

43. Since complex numbers can be associated with vectors, find the magnitude and direction of a force that is represented by $64.8 - 49.0j$ pounds.

44. Since complex numbers can be associated with vectors, find the magnitude and direction of a displacement vector represented by $0.427 + 0.158j$ millimeters.

Chapter 20 Formulas

$(a + bj) + (c + dj) = (a + c) + (b + d)j$	addition and subtraction	(20-1)
$(a + bj)(c + dj) = (ac - bd) + (ad + bc)j$	multiplication	(20-2)
$\dfrac{a + bj}{c + dj} = \dfrac{(a + bj)(c - dj)}{(c + dj)(c - dj)} = \dfrac{(ac + bd) + (bc - ad)j}{c^2 + d^2}$	division	(20-3)
$x = r \cos \theta$		(20-4)
$y = r \sin \theta$		(20-5)
$r^2 = x^2 + y^2$		(20-6)
$\tan \theta = \dfrac{y}{x}$		(20-7)
$x + yj = r(\cos \theta + j \sin \theta)$	rectangular and polar form	(20-8)
$r \operatorname{cis} \theta = r(\cos \theta + j \sin \theta)$		(20-9)
$r \angle \theta = r(\cos \theta + j \sin \theta)$		(20-10)

20-5 Review Exercises for Chapter 20

In Exercises 1 through 8, express each number in terms of j.

1. $\sqrt{-64}$ **2.** $-\sqrt{-100}$ **3.** $-\sqrt{-400}$ **4.** $\sqrt{-81}$

5. $\sqrt{-54}$ **6.** $\sqrt{-40}$ **7.** $-\sqrt{-56}$ **8.** $-\sqrt{-63}$

In Exercises 9 through 12, express each number in the rectangular form x + yj.

9. $-6 + \sqrt{-100}$ **10.** $5 - \sqrt{-144}$ **11.** $3 - \sqrt{-48}$ **12.** $-2 + \sqrt{-27}$

In Exercises 13 through 16, simplify each expression.

13. $(\sqrt{-25})^2$ **14.** $\sqrt{(-8)^2}$ **15.** $\sqrt{-7}\sqrt{-2}$ **16.** $\sqrt{-2}\sqrt{-32}$

In Exercises 17 through 20, simplify each expression.

17. j^{14} **18.** $-j^{21}$ **19.** $(-j)^{40}$ **20.** $-j^{15}$

In Exercises 21 through 32, perform the indicated operations and express all answers in the rectangular form x + yj.

21. $(12 + 3j) + (2 + j)$ **22.** $(6 - 7j) - (5 + 4j)$

23. $(-10 - 3j) - (-9 + 4j)$ **24.** $(-7 + 5j) + (-8 - 4j)$

25. $(3 + 6j)(2 + j)$

26. $(10 - 2j)(3 + j)$

27. $(-3 - 4j)(-3 - 5j)$

28. $(-8 + 9j)(-5 - 7j)$

29. $(2 - 10j) \div (3 + j)$

30. $(-5 + j) \div (4 - j)$

31. $(6 + 2j) \div (-7 - 2j)$

32. $(10 + 15j) \div (-5j)$

In Exercises 33 through 36, find the sum of the given number and its conjugate, and find the product of the given number and its conjugate.

33. $6 + 8j$

34. $-2 + 5j$

35. $7 - 3j$

36. $-4 - 7j$

In Exercises 37 through 40, perform the indicated operations graphically and check the results algebraically.

37. $(1 - j) + (2 + 4j)$ **38.** $3j - (5 - 4j)$ **39.** $-4j + (-9 - 3j)$ **40.** $(2 + 8j) - (8 - 7j)$

In Exercises 41 through 44, convert each complex number from the given rectangular form to the polar form $r(\cos \theta + j \sin \theta)$. Use a graph to be sure that θ corresponds to the correct quadrant.

41. $5 + 12j$ **42.** $7.38 - 4.16j$ **43.** $-18.0 - 12.3j$ **44.** $-62.4 + 87.3j$

In Exercises 45 through 48, convert each complex number from the given polar form to the rectangular form $x + yj$.

45. $15.2(\cos 70.2° + j \sin 70.2°)$

46. $0.912(\cos 216.4° + j \sin 216.4°)$

47. $16.7\angle 327.5°$

48. $3.45\angle 170.1°$

In Exercises 49 through 52, answer the given questions or perform the indicated operations.

49. Is $1 + 2j$ a solution to the quadratic equation

$$x^2 - 2x + 5 = 0?$$

Is its conjugate a solution?

50. Show that when the complex number $a + bj$ is multiplied by its conjugate, the product is a real number.

51. In an alternating-current circuit, the voltage V, the impedance Z, and the current I are related by the equation $V = IZ$. Find V (in volts) if $I = 5 - 4j$ amperes and $Z = -3 + 6j$ ohms.

52. In Exercise 51, find the current (in amperes) if $V = 10 + 12j$ volts and $Z = -5 + 8j$ ohms.

The Scientific Calculator

Since the development of the first electronic calculators in the early 1970s, there has been a wide variety of different calculators with different features and methods of operation. Some of the calculators now available allow us to display the graphs of functions. These same calculators are also capable of performing the computations we normally expect from a "scientific calculator." This appendix is intended to be a brief reference for the use of scientific calculators. More extensive discussions involving calculators are included throughout the text where they are appropriate. Since the use and labeling of keys will vary among different models, it should be stressed that you should consult the manual for your particular calculator. Also, do not expect the calculator to rectify or compensate for a lack of understanding of the related mathematics—gross errors could be created.

In making entries or calculations, it is possible that a number with too many digits or one which is too large or too small cannot be entered. Also, certain operations do not have defined results. If such an entry or calculation is attempted, the calculator will make an error indication such as E, ⊓, or a flashing display. Operations that can result in an error indication include division by zero, square root of a negative number, logarithm of a negative number, and the inverse trigonometric function of a value outside of the appropriate interval.

Also, some calculators will give a special display when batteries need charging or replacement. The user should become acquainted with any special displays the calculator may use.

We assume in this appendix that the calculator uses *algebraic logic* so that operations are done according to the following *hierarchy*. (See Section 1-11.)

1. All expressions enclosed within parentheses are evaluated.

2. All of the following are performed as they occur: *square, square root, reciprocal, powers, logarithms, percents, sine, cosine, tangent*.

3. All *multiplications* and *divisions* are performed.

4. All *additions* and *subtractions* are performed.

We now list important keys along with brief descriptions and examples. Some of these keys will not be meaningful until the appropriate text material has been covered.

Keys	Examples	Text reference (if appropriate)
[0],[1],...,[9] **Digit Keys** These keys are used to enter the digits 0 through 9 in the display, or to enter an exponent of 10 when scientific notation is used.	To enter: 37514 Sequence: [3],[7],[5],[1],[4] Display: 37514.	
[.] **Decimal Point Key** This key is used to enter a decimal point.	To enter: 375.14 Sequence: [3],[7],[5],[.],[1],[4] Display: 375.14	
[+/-] **Change Sign Key** This key is used to change the sign of the number on the display, or to change the sign of the exponent when scientific notation is used. (May be designated as [CHS].)	To enter: −375.14 Sequence: [3],[7],[5],[.],[1],[4],[+/-] Display: −375.14 (From here on the entry of a number will be shown as one operation.)	3-1
[π] **Pi Key** This key is used to enter π to the number of digits of the display. On some calculators a key will have a secondary purpose of providing the value of π. See the second function key.	To enter: π Sequence: [π] Display: 3.1415927 (For calculators with dual purpose keys, see the second function key.)	6-2

(Continued on next page)

Keys	Examples	Text reference (if appropriate)
EE **Enter Exponent Key** This key is used to enter an exponent when scientific notation is used. After the key is pressed, the exponent is entered. For a negative exponent, the +/− key is pressed after the exponent is entered. (May be designated as EEX.)	To enter: 2.936×10^8 Sequence: 2.936, EE, 8 Display: 2.936 08 To enter: -2.936×10^{-8} Sequence: 2.936, +/−, EE, 8, +/− Display: −2.936 −08	10-2
= **Equals Key** This key is used to complete a calculation to give the required result.	See the following examples for illustrations of the use of this key.	
+ **Add Key** This key is used to add the next entry to the previous entry or result.	Evaluate: 37.56 + 241.9 Sequence: 37.56, +, 241.9, = Display: 279.46	1-11
− **Subtract Key** This key is used to subtract the next entry from the previous entry or result.	Evaluate: 37.56 − 241.9 Sequence: 37.56, −, 241.9, = Display: −204.34	1-11
× **Multiply Key** This key is used to multiply the previous entry or result by the next entry.	Evaluate: 8.75 × 30.92 Sequence: 8.75, ×, 30.92, = Display: 270.55	1-11
÷ **Divide Key** This key is used to divide the previous entry or result by the next entry.	Evaluate: 8.75 ÷ 30.92 Sequence: 8.75, ÷, 30.92, = Display: 0.28298836 (truncated or rounded off)	1-11
CE **Clear Entry Key** This key is used to clear the last entry. Its use will not affect any other part of a calculation. On some calculators one press of the C/CE or CL key is used for this purpose.	Evaluate: 37.56 + 241.9, with an improper entry of 242.9 Sequence: 37.56, +, 242.9, CE, 241.9, = Display: 279.46	
C **Clear Key** This key is used to clear the display and information being calculated (not including memory), so that a new calculation may be started. For calculators with a C/CE or C key, and no CE key, a second press on these keys is used for this purpose.	To clear previous calculation: Sequence: C Display: 0.	

Keys	Examples	Text reference (if appropriate)
[2nd] **Second Function Key** This key is used on calculators on which many of the keys serve dual purposes. It is pressed before the second key functions are activated. (This key may be labeled as F.)	Evaluate: $(37.4)^2$ on a calculator where x^2 is a second use of a key Sequence: 37.4, [2nd], $[x^2]$ Display: 1398.76	
$[x^2]$ **Square Key** This key is used to square the number on the display.	Evaluate: $(37.4)^2$ Sequence: 37.4, $[x^2]$ Display: 1398.76	
$[\sqrt{x}]$ **Square Root Key** This key is used to find the square root of the number on the display.	Evaluate: $\sqrt{37.4}$ Sequence: 37.4, $[\sqrt{x}]$ Display: 6.1155539	1-10
$[1/x]$ **Reciprocal Key** This key is used to find the reciprocal of the number on the display.	Evaluate: $\dfrac{1}{37.4}$ Sequence: 37.4, $[1/x]$ Display: 0.02673797	1-11
$[y^x]$ **y to the x Power Key** This key is used to raise y, the first entry, to the x power, the second entry. (This key may be labeled as x^y.)	Evaluate: $(3.73)^{1.5}$ Sequence: 3.73, $[y^x]$, 1.5, $[=]$ Display: 7.2038266	1-11
$[\%]$ **Percent Key** This key converts the number displayed to a percent.	Convert 16.5 to a percent. Sequence: 16.5, $[\%]$ Display: 0.165	1-11
[DRG] **Degree-Radian-Grad Key** This key is used to designate a displayed angle as being measured in degrees, radians, or grads. (On many calculators, several keys may be involved.)		19-1
[SIN] **Sine Key** This key is used to find the sine of the angle on the display.	Evaluate: $\sin 37.4°$ Sequence: [DRG], 37.4, [SIN] Display: 0.60737584	16-2
[COS] **Cosine key** This key is used to find the cosine of the angle on the display.	Evaluate: $\cos 2.475$ (rad) Sequence: [DRG], 2.475, [COS] Display: −0.78593303	16-2

(Continued on next page)

Keys	*Examples*	*Text reference (if appropriate)*
[TAN] **Tangent Key** This key is used to find the tangent of the angle on the display.	Evaluate: tan (−24.9°) Sequence: [DRG], 24.9, [+/−], [TAN] Display: −0.46418455	16-2
Cotangent, Secant, Cosecant These functions are found through their reciprocal relation with the tangent, cosine, and sine functions, respectively.	Evaluate: cot 2.841 (rad) Sequence: [DRG], 2.841, [TAN], [1/x] Display: −3.2259549	16-2
[INV] **Inverse Trigonometric Function Key** This key is used (prior to the appropriate trigonometric function key) to find the angle whose trigonometric function is on the display. Some calculators have [ARCSIN] or equivalent keys for this purpose. Also, some calculators use the [INV] key for other inverse functions.	Evaluate: Arcsin 0.1758 (the angle whose sine is 0.1758) in degrees. Sequence: [DRG], 0.1758, [INV], [SIN] Display: 10.125217	16-2
[LOG] **Common Logarithm Key** This key is used to find the logarithm to the base 10 of the number on the display.	Evaluate: log 37.45 Sequence: 37.45, [LOG] Display: 1.5734518	12-1
[LN] **Natural Logarithm Key** This key is used to find the logarithm to the base e of the number on the display.	Evaluate: ln 0.8421 Sequence: 0.8421, [LN] Display: −0.17185651	12-3
[10^x] **10 to the x Power Key** This key is used to find antilogarithms (base 10) of the number on the display. (The [y^x] key can be used for this purpose, if the calculator does not have this key.)	Evaluate: Antilog 0.7265 (or $10^{0.7265}$) Sequence: 0.7265, [10^x] Display: 5.3272122	12-1
[e^x] **e to the x Power Key** This key is used to raise e to the power on the display.	Evaluate: $e^{-4.05}$ Sequence: 4.05, [+/−], [e^x] Display: 0.01742237	12-3
[(][)] **Parentheses Keys** These keys are used to specify the order in which expressions are evaluated.	Evaluate: 2(3 + 4) Sequence: 2, [×], [(], 3, [+], 4, [)], [=] Display: 14.	1-11
[STO] **Store in Memory Key** This key is used to store the displayed number in the memory.	Store in memory: 56.02 Sequence: 56.02, [STO]	

Keys	Examples	*Text reference (if appropriate)*

RCL **Recall from Memory Key**
This key is used to recall the number in the memory to the display. (May be designated as MR.)

Recall from memory: 56.02
Sequence: RCL
Display: 56.02

M **Other Memory Keys**
Some calculators use an M key to store a number in the memory. It also may add the entry to the number in the memory. On such calculators CM (Clear Memory) is used to clear the memory. There are also keys for other operations on the number in the memory.

BASIC Programming

This appendix serves as a brief introduction to the programming language BASIC. The name BASIC is an acronym for Beginners All-Purpose Symbolic Instruction Code, and that phrase correctly suggests that this language is designed for beginners.

Two major advantages of the BASIC language are its simplicity and widespread availability on computer systems of all sizes. Instead of attempting a comprehensive study of the BASIC programming language, we will introduce the important fundamentals. We will then present examples of BASIC programs which can be used to solve problems of the type presented in this book. The programs will illustrate the fundamental features of BASIC which we present. In some cases, the programs could be made more efficient by using more advanced programming concepts, but it is our intent to provide simple programs which can be understood by beginners.

A computer program is a step-by-step list of instructions which the computer must follow. These instructions are usually designed so that they will lead to the solution of some problem. Before actually writing the program, there must first be a clear understanding of the process that leads to the solution. A BASIC program consists of a sequence of numbered lines. While the lines could be numbered 1, 2, 3, 4, and so on, it is better to use a numbering sequence such as 10, 20, 30, 40, and so on. The latter numbering system has the advantage that additional lines can be inserted without retyping the entire program. Also, a line can be corrected by simply entering the corrected line.

EXAMPLE A

Entering these lines . . .   results in this program.

```
10 LET A=10          10 LET A=10
20 LET B=20          15 LET R=15
30 LET C=30          20 LET B=20
15 LET R=156789      30 LET C=30
15 LET R=15
```

☐

In BASIC, the arithmetic operations are represented as follows:

BASIC symbol	Operation	Example	Result of example
+	addition	3+4	7
−	subtraction	7−5	2
*	multiplication	6*3	18
/	division	12/4	3
˄	exponentiation	2˄3	8

There may be variations in different BASIC versions. With the operation of exponentiation, for example, some systems will use ** or ↑ instead of ˄. The arithmetic operations listed in the preceding table do follow the same *hierarchy* used in scientific calculators. First, all exponents are evaluated. Second, all multiplications and divisions are performed. Third, all additions and subtractions are performed. Expressions enclosed within parentheses are evaluated first.

In addition to the listed arithmetic operations, there are several mathematical functions. The following table illustrates some important mathematical functions available in BASIC.

BASIC expression	Function	Example	Result of example	Comment
SIN (X)	$\sin x$	SIN (1.2)	0.932039086	x is in radians
COS (X)	$\cos x$	COS (3.14)	−0.9999987317	x is in radians
TAN (X)	$\tan x$	TAN (2.8)	−0.3555298317	x is in radians
ATN (X)	arctangent x	ATN (0.75)	0.6435011088	Result is in radians
LOG (X)	$\ln x$	LOG (2.1)	0.7419373447	Natural logarithm
SQR (X)	$\sqrt{x}$	SQR (9)	3	Square root
ABS (X)	$|x|$	ABS (−8)	8	Absolute value

When writing BASIC programs, in addition to using the arithmetic operations and mathematical functions, we also use the following special statements:

1. The LET instruction directs the computer to assign a given value to a variable. For example, LET D = 6 causes the letter D to be assigned the value 6 in all subsequent instances. The expression LET C = 0 causes the letter C to represent the number 0, while LET C = C + 1 causes C to increase its previous value by 1. Many computers will permit omission of the word LET so that LET C = 0 can be entered as C = 0.

2. The INPUT instruction directs the computer to pause while you enter your data. For example, INPUT D causes the computer to wait until a number is entered; the letter D assumes the value that is entered.

3. The PRINT instruction directs the computer to print or display the value of some variable. For example, PRINT R causes the value of R to be displayed. PRINT "HELLO" causes HELLO to be displayed.

4. The GOTO command causes the computer to proceed directly to a specified line of a program instead of continuing with the normal sequence. GOTO 50 causes a jump directly to the line numbered 50 in the program.

5. The IF . . . THEN . . . command causes a jump to an indicated line of a program only if some condition is met. For example, IF A = B THEN 50 directs the computer to proceed directly to line 50 of the program if the value of A equals that of B; if A does not equal B, the jump to line 50 is ignored and the normal sequence continues. The condition following IF can involve only comparisons (equal =, less than <, greater than >, not equal <>, greater than or equal to >=, or less than or equal to <=).

6. The FOR-NEXT statement causes a sequence of lines to be executed a specific number of times. The brief program below shows the format for using the FOR-NEXT statement. In this program, lines 20, 30, 40 are executed; then the program "loops" back to repeat lines 20, 30, 40 a second time. This repetition continues until the lines have been executed 500 times. The program will then continue with any additional lines that may follow.

```
10 FOR J=1 TO 500
20 ...
30 ...
40 ...
50 NEXT J
```

7. The END instruction terminates a program.

8. The REM statement allows remarks which can be used to describe the nature of the program. For example,

```
10 REM  COMPUTES AREA OF CIRCLE
```

might be the first line of a program. The REM statement is used for documentation purposes and does not affect the actual operation of the program.

The arithmetic operations, mathematical functions, and statements are used within a BASIC program. There are five important system commands which are not used within the programs themselves. These system commands tell the computer what to do in the course of its operations.

1. After a program is entered, the RUN command causes the program to be run.

2. The LIST command causes the program to be listed or displayed.

3. The NEW command erases the existing program and provides a clean "slate" for a new program.

4. The SAVE command is used to save or store a program. A common arrangement is to enter an identifying name after SAVE so that the stored program can be referenced at some future time. For example, SAVE QUADRATIC could be entered as a command which causes a particular program to be stored in a location identified by the name QUADRATIC.

5. The LOAD command is used to retrieve a program which has already been stored. For example, LOAD QUADRATIC could be used to gain access to the program stored in the location identified by the name QUADRATIC.

To illustrate the preceding fundamentals of the BASIC programming language, we begin with a simple program. An important process used frequently in science and industry involves conversions or reductions, as discussed in Section 2-2.

EXAMPLE B Develop a BASIC program for converting inches to centimeters.

```
10 REM CONVERTS INCHES TO CENTIMETERS
20 PRINT "ENTER THE NUMBER OF INCHES:"
30 INPUT INCH
40 CM=INCH*2.54
50 PRINT INCH;" IN. = ";CM;" CM"
60 END
```

The first line of this program consists of the remark that briefly describes the general nature of the program. In the line numbered 20, the computer is instructed to print or display the given list of characters enclosed within the quotes. Line 30 causes the program to stop until we enter or "input" a number of inches which is then assigned to INCH. Line 40 causes the corresponding number of centimeters to be computed. Line 50 causes the results to be printed or displayed. We now list sample runs of this program. In these sample runs and in the remaining examples, the colored text refers to entries required by the user. These sample runs follow.

```
RUN

ENTER THE NUMBER OF INCHES:
? 12
12 IN. = 30.48 CM


RUN

ENTER THE NUMBER OF INCHES:
? 36
36 IN. = 91.44 CM
```

EXAMPLE C   Develop a BASIC program which can be used to convert angles from degree measure to radian measure.

A program listing is provided here. The sample runs which follow correspond to the values used in Example A of Section 19-1.

```
10 REM CONVERTS ANGLE FROM DEGREES TO RADIANS
20 PRINT "ENTER THE NUMBER OF DEGREES: "
30 INPUT D
40 R=D*3.1415927/180
50 PRINT D;" DEGREES = ";R;" RADIANS"
60 END

RUN

ENTER THE NUMBER OF DEGREES:
? 180
180 DEGREES = 3.14159 RADIANS

RUN

ENTER THE NUMBER OF DEGREES:
? 240
240 DEGREES = 4.18879 RADIANS
```

EXAMPLE D   Develop a BASIC program which can be used to convert temperature readings from Fahrenheit degrees to Celsius degrees.

The program listing and sample runs follow.

```
10 REM CONVERTS FAHRENHEIT TO CELSIUS
20 PRINT "ENTER THE TEMPERATURE IN DEGREES FAHRENHEIT:"
30 INPUT F
40 C=(F-32)*(5/9)
50 PRINT "********************************************
       ********"
60 PRINT F; " DEGREES FAHRENHEIT EQUALS ";C;" DEGREES
       CELSIUS"
70 PRINT "********************************************
       ********"
80 END
```

```
RUN

ENTER THE TEMPERATURE IN DEGREES FAHRENHEIT:
? 32
*************************************************************
 32 DEGREES FAHRENHEIT EQUALS 0 DEGREES CELSIUS
*************************************************************


RUN

ENTER THE TEMPERATURE IN DEGREES FAHRENHEIT:
? 212
*************************************************************
 212 DEGREES FAHRENHEIT EQUALS 100 DEGREES CELSIUS
*************************************************************


RUN

ENTER THE TEMPERATURE IN DEGREES FAHRENHEIT:
? 70
*************************************************************
 70 DEGREES FAHRENHEIT EQUALS 21.1111 DEGREES CELSIUS
*************************************************************
```

□

EXAMPLE E   Develop a BASIC program for solving quadratic equations.

The program listing and sample runs follow. This program uses Eq. (11-2) and will work only for quadratic equations having real roots. If the roots are imaginary, an error message will be given by the computer. It is possible to expand this program so that it would accommodate imaginary roots. The sample runs correspond to Examples A and B given in Section 11-3.

```
10 REM SOLVES QUAD EQ VIA FORMULA 11-2 IN TEXT
20 PRINT "ENTER THE COEFFICIENT OF THE SQUARE OF X:"
30 INPUT A
40 PRINT "NOW ENTER THE COEFFICIENT OF X:"
50 INPUT B
60 PRINT "NOW ENTER THE CONSTANT:"
70 INPUT C
80 ROOT=SQR(B^2-4*A*C)
90 X1=(-B+ROOT)/(2*A)
100 X2=(-B-ROOT)/(2*A)
110 PRINT "********************************"
120 PRINT "X = ";X1;" OR ";X2
130 PRINT "********************************"
140 END
```

RUN

```
ENTER THE COEFFICIENT OF THE SQUARE OF X:
? 1
NOW ENTER THE COEFFICIENT OF X:
? -5
NOW ENTER THE CONSTANT:
? 6
****************************
X = 3 OR 2
****************************
```

RUN

```
ENTER THE COEFFICENT OF THE SQUARE OF X:
? 2
NOW ENTER THE COEFFICIENT OF X:
? 7
NOW ENTER THE CONSTANT:
? -3
****************************
X = .386001 OR -3.886
****************************
```

EXAMPLE F Develop a BASIC program which can be used to find the area and the circumference of a circle.

In the program which follows, line 20 establishes P as the approximate value of π. Line 50 then corresponds to Eq. (6-8) and line 60 corresponds to Eq. (6-16). The sample runs correspond to examples from Sections 6-2 and 6-3.

```
10 REM FINDS AREA AND CIRCUMFERENCE OF CIRCLE
20 P=3.1415927
30 PRINT "ENTER RADIUS:"
40 INPUT R
50 C=2*P*R
60 AREA=P*(R^2)
70 PRINT "CIRCUMFERENCE = ";C
80 PRINT "AREA = ";AREA
90 END
```

```
RUN

ENTER RADIUS:
? 2.00
CIRCUMFERENCE = 12.5664
AREA = 12.5664
```

```
RUN

ENTER RADIUS:
? 2.73
CIRCUMFERENCE = 17.1531
AREA = 23.414
```

EXAMPLE G Develop a BASIC program for finding the perimeter and area of a triangle for which the lengths of the three sides are all known.

In line 50 of the program display which follows, Eq. (6-1) is used to find the perimeter of a triangle. The area of the triangle is found in lines 60 and 70 which correspond to Eqs. (6-13) and (6-14).

```
10 REM FINDS PERIMETER AND AREA OF TRIANGLE
20 REM REQUIRES 3 KNOWN SIDES
30 PRINT "ENTER THE 3 SIDES OF THE TRIANGLE, SEPARATED
        BY COMMAS:"
40 INPUT A,B,C
50 PER=A+B+C
60 S=PER/2
70 AREA=SQR(S*(S-A)*(S-B)*(S-C))
80 PRINT "PERIMETER = ";PER
90 PRINT "AREA = ";AREA
100 END
```

(Continued on next page)

```
RUN

ENTER THE 3 SIDES OF THE TRIANGLE, SEPARATED BY COMMAS:
? 6,7,8
PERIMETER = 21
AREA = 20.3332
```
□

EXAMPLE H Develop a BASIC program which can be used to solve a system of two equations with two variables.

The program that follows can be used to solve systems of equations such as those found in Chapter 14. This program uses the method of determinants (Cramer's rule) as discussed in Section 14-4. This program will not work if the system is inconsistent or dependent. Following the program listing is a sample run which corresponds to Example A of Section 14-2. The two sets of entered coefficients represent the following system of equations:

$$2x + y = 4$$
$$3x - y = 1$$

```
10 REM SOLVES SYSTEM OF 2 LINEAR EQUATIONS
20 PRINT "ENTER COEFFICIENTS FOR 1ST EQUATION, SEPARATED
         BY COMMAS:"
30 INPUT A,B,C
40 PRINT "ENTER COEFFICIENTS FOR 2ND EQUATION, SEPARATED
         BY COMMAS:"
50 INPUT D,E,F
60 X=(C*E-B*F)/(A*E-B*D)
70 Y=(A*F-C*D)/(A*E-B*D)
80 PRINT "**********"
90 PRINT "X = ";X
100 PRINT "Y = ";Y
110 PRINT "**********"
120 END
```

```
RUN

ENTER COEFFICIENTS FOR 1ST EQUATION, SEPARATED BY COMMAS:
? 2,1,4
ENTER COEFFICIENTS FOR 2ND EQUATION, SEPARATED BY COMMAS:
? 3,-1,1
**********
X = 1
Y = 2
**********
```
□

EXAMPLE I   Develop a BASIC program which will use the Pythagorean theorem to find the third side of a right triangle.

In the program which follows, we must supply the lengths of two sides of a right triangle. The third side can be found directly from Eq. (15-1) provided that neither of the two sides is the hypotenuse. If one of the two known sides is the hypotenuse, then a variation of Eq. (15-1) is used. The sample runs which follow the program listing correspond to Examples A and B of Section 15-3.

```
10 REM USES PYTHAGOREAN THM TO FIND 3RD SIDE OF A RIGHT
      TRIANGLE
20 REM REQUIRES 2 KNOWN SIDES
30 PRINT "ENTER THE NUMBER 1 IF THE HYPOTENUSE IS
      KNOWN:"
40 PRINT "(OTHERWISE, ENTER 0 OR ANY OTHER NUMBER)"
50 INPUT TYPE
60 IF TYPE=1 THEN 120
70 PRINT "ENTER THE LENGTHS OF THE 2 SIDES, SEPARATED BY
      COMMAS:"
80 INPUT A,B
90 C=SQR(A^2+B^2)
100 PRINT "THE LENGTH OF THE HYPOTENUSE IS ";C
110 GOTO 180
120 PRINT "ENTER THE LENGTH OF THE HYPOTENUSE:"
130 INPUT C
140 PRINT "ENTER THE LENGTH OF THE OTHER KNOWN SIDE:"
150 INPUT A
160 B=SQR(C^2-A^2)
170 PRINT "THE LENGTH OF THE 3RD SIDE IS ";B
180 END
```

RUN

```
ENTER THE NUMBER 1 IF THE HYPOTENUSE IS KNOWN:
(OTHERWISE, ENTER 0 OR ANY OTHER NUMBER)
? 0
ENTER THE LENGTHS OF THE 2 SIDES, SEPARATED BY COMMAS:
? 5,12
THE LENGTH OF THE HYPOTENUSE IS 13
```

(Continued on next page)

```
RUN

ENTER THE NUMBER 1 IF THE HYPOTENUSE IS KNOWN:
(OTHERWISE, ENTER 0 OR ANY OTHER NUMBER)
? 1
ENTER THE LENGTH OF THE HYPOTENUSE:
? 25
ENTER THE LENGTH OF THE OTHER KNOWN SIDE:
? 20
THE LENGTH OF THE 3RD SIDE IS 15                          □
```

In the following four examples, we consider the different methods for solving triangles. These four examples illustrate uses of the Law of Sines and Law of Cosines. We stated in the beginning of Section 17-3 that there are four possible combinations of parts from which we may solve a triangle. The following four examples include programs related to each of the four cases listed in Section 17-3. These programs are especially appropriate since they relate to some of the more complex calculations included in the text.

EXAMPLE J   Develop a BASIC program which uses the Law of Sines to solve a triangle in which we know the values of two angles and a side. The following program does incorporate the Law of Sines in lines 450 and 460. Lines 420, 430, and 440 convert the angles from degree measure to radian measure. This conversion is necessary because the BASIC SIN function requires that the angle must be in radians. Lines 380, 390, 400, and 410 are designed to arrange the angles and sides in correct configuration. The sample run uses the data from Example A of Section 17-3.

```
300 REM LAW OF SINES USED TO SOLVE TRIANGLE GIVEN 2
        ANGLES AND A SIDE
310 P=3.1415927
320 PRINT "ENTER TWO KNOWN ANGLES, SEPARATED BY A
        COMMA:"
330 INPUT A,B
340 C=180-A-B
350 PRINT "ENTER THE KNOWN SIDE:"
360 INPUT SC
370 PRINT "THE THREE ANGLES ARE ";A,B,C
380 PRINT "NOW ENTER THE ANGLE WHICH IS OPPOSITE THE
        KNOWN SIDE:"
390 INPUT C
400 PRINT "NOW ENTER THE OTHER TWO ANGLES SEPARATED BY A
        COMMA:"
```

```
410 INPUT A,B
420 A=P*A/180
430 B=P*B/180
440 C=P*C/180
450 SB=SC*SIN(B)/SIN(C)
460 SA=SC*SIN(A)/SIN(C)
470 PRINT "THE OTHER SIDES ARE ";SA;SB
480 END

RUN

ENTER TWO KNOWN ANGLES, SEPARATED BY A COMMA:
? 37.4,75.1
ENTER THE KNOWN SIDE:
? 22.1
THE THREE ANGLES ARE 37.4 75.1 67.5
NOW ENTER THE ANGLE WHICH IS OPPOSITE THE KNOWN SIDE:
? 67.5
NOW ENTER THE OTHER TWO ANGLES SEPARATED BY A COMMA:
? 37.4,75.1
THE OTHER SIDES ARE 14.529 23.1166                          □
```

EXAMPLE K Write a BASIC program which uses the Law of Cosines to find the third side of a triangle when two sides and their included angle are known.

The following program uses the Law of Cosines on line 200 which is the BASIC expression for

$$c = \sqrt{a^2 + b^2 - 2ab \cos C}$$

In this program we represent sides a, b, and c by SA, SB, and SC. The program listing is followed by a sample run which uses the data from Example A of Section 17-4.

```
100 REM LAW OF COSINES USED TO FIND 3RD SIDE OF TRIANGLE
110 REM REQUIRES TWO SIDES AND INCLUDED ANGLE
120 P=3.1415927
130 PRINT "ENTER LENGTH OF THE FIRST SIDE:"
140 INPUT SA
150 PRINT "ENTER LENGTH OF THE SECOND SIDE:"
160 INPUT SB
170 PRINT "ENTER ANGLE (IN DEGREES) INCLUDED BETWEEN THE
          2 SIDES:"
180 INPUT C
190 C=C*P/180
200 SC=SQR(SA^2+SB^2-2*SA*SB*COS(C))
210 PRINT "THE THIRD SIDE HAS LENGTH ";SC
220 END                              (Continued on next page)
```

```
RUN

ENTER LENGTH OF THE FIRST SIDE:
? 4.08
ENTER LENGTH OF THE SECOND SIDE:
? 4.37
ENTER ANGLE (IN DEGREES) INCLUDED BETWEEN THE 2 SIDES:
? 68.0
THE THIRD SIDE HAS LENGTH 4.73129
```

EXAMPLE L　Develop a BASIC program which can be used to find the three angles of a triangle for which the lengths of the three sides are known.

　　The following program uses the Law of Cosines. Lines 540, 550, and 560 compute the cosines of angles A, B, and C. Lines 570, 580, and 590 use the ATN function to find the angles in radians, while lines 600, 610, and 620 convert the angles to degree measure. (The arctangent function ATN is used since there is no $\cos^{-1} x$ or arccos x function available. Also, we can use the formula

$$A = \arctan \sqrt{-1 + \frac{1}{(\cos A)^2}}$$

to find angle A when its cosine is known.) Lines 630, 640, and 650 test for the presence of a negative value for cosine; if an angle has a negative cosine, then it is obtuse and we therefore use the supplement on line 670, 690, or 710. This program has a flaw: If any angle is $90°$, then one of the lines 570, 580, or 590 would result in division by zero. This can be corrected by checking the three sides to see if they satisfy the Pythagorean theorem or by checking the values of cosine to see if a zero results. Appropriate corrections can then be made.

　　Following the program listing is a sample run which uses the data from Example C of Section 17-4.

```
500 REM LAW OF COSINES USED TO FIND 3 ANGLES OF
        TRIANGLE, GIVEN 3 SIDES
510 P=3.1415927
520 PRINT "ENTER 3 SIDES OF TRIANGLE, SEPARATED
        BY COMMAS;"
530 INPUT SA,SB,SC
540 COSA=-(SA^2-SB^2-SC^2)/(2*SB*SC)
550 COSB=-(SB^2-SA^2-SC^2)/(2*SA*SC)
560 COSC=-(SC^2-SA^2-SB^2)/(2*SA*SB)
570 A=ATN(SQR(-1+1/COSA^2))
580 B=ATN(SQR(-1+1/COSB^2))
590 C=ATN(SQR(-1+1/COSC^2))
```

```
600 A=A*180/P
610 B=B*180/P
620 C=C*180/P
630 IF COSA<0 THEN 670
640 IF COSB<0 THEN 690
650 IF COSC<0 THEN 710
660 GOTO 720
670 A=180-A
680 GOTO 720
690 B=180-B
700 GOTO 720
710 C=180-C
720 PRINT "THE THREE ANGLES IN DEGREES ARE ";A;B;C
730 END

RUN

ENTER 3 SIDES OF TRIANGLE, SEPARATED BY COMMAS:
? 83.8,36.7,72.4
THE THREE ANGLES IN DEGREES ARE 94.6822 25.8799 59.4379
```

□

EXAMPLE M Develop a BASIC program which can be used to solve a triangle for which we know two sides and the angle opposite one of them.

This program is somewhat complex since it incorporates the Law of Sines and the Law of Cosines to consider situations which may result in no solution, one solution, or two solutions. This is the ambiguous case discussed in Section 17-3. Following the program listing are three sample runs which correspond to the three possibilities illustrated in Example D of Section 17-3.

Line 790 uses the Law of Sines to compute sin A. In line 800 we are referred to the impossible case if sin A is greater than 1. (We used 1.0001 to correct for values which incorrectly appear to exceed 1 because of rounding errors.) On line 810, we are referred to the case which could result in two solutions if sin A is less than 1. (We use 0.9999 to correct for values which incorrectly appear to be less than 1 because of rounding errors.) Line 820 begins treatment of the case for which sin $A = 1$.

```
700 REM LAWS OF SINES AND COSINES USED TO SOLVE TRIANGLE
710 REM REQUIRES ANGLE WITH OPPOSITE SIDE AND ADJACENT
        SIDE
720 P=3.1415927
730 PRINT "ENTER KNOWN ANGLE IN DEGREES:"
740 INPUT B
```

(Continued on next page)

```
750  PRINT "ENTER SIDE ADJACENT TO KNOWN ANGLE:"
760  INPUT SA
770  PRINT "ENTER LENGTH OF SIDE OPPOSITE THE KNOWN
            ANGLE:"
780  INPUT SB
790  SINEA=SA*SIN(B*P/180)/SB
800  IF SINEA>1.0001 THEN 1010
810  IF SINEA<0.9999 THEN 880
820  A=90
830  C=180-A-B
840  SC=SB*SIN(C*P/180)/SIN(B*P/180)
850  PRINT "THE OTHER ANGLES ARE ";A;C
860  PRINT "THE OTHER SIDE HAS LENGTH ";SC
870  GOTO 1020
880  TANG=SINEA/SQR(1-SINEA^2)
890  A=ATN(TANG)*180/P
900  C=180-A-B
910  SC=SB*SIN(C*P/180)/SIN(B*P/180)
920  PRINT "CASE 1: THE OTHER ANGLES ARE ";A;C
930  PRINT "        THE OTHER SIDE HAS LENGTH ";SC
940  A=180-A
950  C=180-A-B
960  IF C<0.05 THEN 1020
970  SC=SB*SIN(C*P/180)/SIN(B*P/180)
980  PRINT "CASE 2: THE OTHER ANGLES ARE ";A;C
990  PRINT "        THE OTHER SIDE HAS LENGTH ";SC
1000 GOTO 1020
1010 PRINT "NO SUCH TRIANGLE"
1020 END
```

RUN

```
ENTER KNOWN ANGLE IN DEGREES:
? 30
ENTER SIDE ADJACENT TO KNOWN ANGLE:
? 4
ENTER LENGTH OF SIDE OPPOSITE THE KNOWN ANGLE:
? 1
NO SUCH TRIANGLE
```

RUN

```
ENTER KNOWN ANGLE IN DEGREES:
? 30
ENTER SIDE ADJACENT TO KNOWN ANGLE:
? 4
```

```
ENTER LENGTH OF SIDE OPPOSITE THE KNOWN ANGLE:
? 2
THE OTHER ANGLES ARE 90 60
THE OTHER SIDE HAS LENGTH 3.4641

RUN

ENTER KNOWN ANGLE IN DEGREES:
? 30
ENTER SIDE ADJACENT TO KNOWN ANGLE:
? 4
ENTER LENGTH OF SIDE OPPOSITE THE KNOWN ANGLE:
? 3
CASE 1: THE OTHER ANGLES ARE 41.8103 108.19
        THE OTHER SIDE HAS LENGTH 5.70017
CASE 2: THE OTHER ANGLES ARE 138.19 11.8103
        THE OTHER SIDE HAS LENGTH 1.22803          □
```

Since Examples J, K, L, and M consider the four cases for solving triangles, the programs they list could be combined into one large program for solving any triangle. Such a comprehensive program would require some minor modifications which would direct the computer to the appropriate segment of the program. We used line numbers which would facilitate such a project.

In the next program we consider graphs of the type disucssed in Section 19-5.

EXAMPLE N Develop a BASIC program which will give the coordinates of points on graphs of functions in the form of $y = a \sin (bx + c)$.

The listed program is designed to provide the coordinates of twenty-six points between $x = 0$ and $x = 10$. To get more points which are closer together, we should modify line 130 so that 0.4 is replaced by a smaller number such as 0.1. To get points to the right of $x = 10$, we should change line 140 by replacing 10 with a larger number. Also, we can make this program work for functions of the type $y = a \cos (bx + c)$ by modifying line 110 so that COS replaces SIN. This program does allow functions of the type $y = a \sin bx$ or $y = a \sin x$. To get coordinates for $y = 3 \sin 4x$, for example, enter the values of 3, 4, and 0 for A, B, and C, respectively. Following the program listing, a sample run is shown which uses the function $y = -3 \sin [4x - (\pi/6)]$ from Example C of Section 19-5. The value of $\pi/6$ is entered as 0.52359878.

```
10 REM GENERATES COORDINATES OF POINTS FOR GRAPH OF
20 REM ******* Y=A SIN(BX+C) *******
30 X=0
40 PRINT "ENTER THE VALUE OF A IN Y=A SIN(BX+C):"
```

(Continued on next page)

```
50  INPUT A
60  PRINT "ENTER THE VALUE OF B IN Y=A SIN(BX+C):"
70  INPUT B
80  PRINT "ENTER THE VALUE OF C IN Y=A SIN(BX+C):"
90  INPUT C
100 PRINT "X","Y"
110 Y=A*SIN(B*X+C)
120 PRINT X,Y
130 X=X+0.4
140 IF X<10 THEN 110
150 END
```

RUN

```
ENTER THE VALUE OF A IN Y=A SIN(BX+C):
? -3
ENTER THE VALUE OF B IN Y=A SIN(BX+C):
? 4
ENTER THE VALUE OF C IN Y=A SIN(BX+C):
? -0.52359878
X               Y
 0               1.5
 .4             -2.64077
 .8             -1.34578
 1.2             2.71936
 1.6             1.18697
 2              -2.78868
 2.4            -1.02412
 2.8             2.84849
 3.2             .857765
 3.6            -2.89858
 4              -.688488
 4.4             2.93879
 4.8             .516867
 5.2            -2.96897
 5.6            -.34348
 6               2.98903
 6.4             .168921
 6.8            -2.9989
 7.2             .621419E-2
 7.6             2.99853
 8              -.181322
 8.4            -2.98794
 8.8             .355812
 9.2             2.96717
 9.6            -.529083
 10             -2.93627
```

EXAMPLE O Develop a BASIC program which can be used to add vectors.

We follow the approach outlined in Section 18-3. The program begins on lines 20 and 30 where the resultant vector is initially given an x component (RX) and a y component (RY) with zero magnitudes. For each vector we enter the magnitude and the angle. Each vector is broken down into its X and Y components. Those components are then added to RX and RY so that after all vectors are supplied, RX and RY are the final components of the resultant vector. The magnitude of the resultant vector is then computed on line 160, and the angle is computed on lines 170 through 250. Following the program listing is a sample run which corresponds to Example C in Section 18-3.

```
10  REM ***ADDS VECTORS***
20  RX=0
30  RY=0
40  PRINT "ENTER MAGNITUDE OF VECTOR:"
50  PRINT "(ENTER -9 WHEN DONE)"
60  INPUT M
70  IF M=-9 THEN 160
80  PRINT "ENTER ANGLE FOR VECTOR:"
90  INPUT A
100 A=A*3.1415927/180
110 X=M*COS(A)
120 Y=M*SIN(A)
130 RX=RX+X
140 RY=RY+Y
150 GOTO 40
160 MAG=SQR ((RX^2)+(RY^2))
170 IF RX=0 AND RY>0 THEN ANGLE=90
180 IF RX=0 AND RY<0 THEN ANGLE=270
190 IF RX=0 AND RY=0 THEN ANGLE=0
200 IF RX=0 THEN 260
210 ANGLE=ATN(RY/RX)
220 ANGLE=ANGLE*180/3.1415927
230 IF RX<0 AND RY>0 THEN ANGLE=ANGLE+180
240 IF RX<0 AND RY<0 THEN ANGLE=ANGLE+180
250 IF RX>0 AND RY<0 THEN ANGLE=ANGLE+360
260 PRINT "MAGNITUDE OF RESULTANT = ";MAG
270 PRINT "ANGLE OF RESULTANT     = "ANGLE;"
         DEGREES"
280 PRINT "X COMPONENT OF RESULTANT IS ";RX
290 PRINT "Y COMPONENT OF RESULTANT IS ";RY
300 END
```

(Continued on next page)

```
RUN

ENTER MAGNITUDE OF VECTOR:
(ENTER -9 WHEN DONE)
? 570
ENTER ANGLE FOR VECTOR:
? 15.0
ENTER MAGNITUDE OF VECTOR:
(ENTER -9 WHEN DONE)
? 350
ENTER ANGLE FOR VECTOR:
? 125.0
ENTER MAGNITUDE OF VECTOR:
(ENTER -9 WHEN DONE)
? -9
MAGNITUDE OF RESULTANT =  557.6145
ANGLE OF RESULTANT      =  51.14427  DEGREES
X COMPONENT OF RESULTANT IS  349.8259
Y COMPONENT OF RESULTANT IS  434.2301
```

APPENDIX C

Numbers in Base Two

With computers in such wide use today, it is of value to consider briefly how numbers are written for use on computers. We shall see that although only two symbols are used to write the numbers, the same idea of positional notation is employed. We shall also discover that the same methods are employed in the operations of addition and multiplication.

In Section 1-1 we noted that ten symbols (0, 1, 2, 3, 4, 5, 6, 7, 8, 9) are used to write numbers. We also saw the use of positional notation in writing numbers. Then, in Section 1-9, we noted how positional notation and powers of ten are related and found that ten is the base of our number system. Numbers used in a computer system are written in **base two**. *Only two symbols, 0 and 1, are necessary, and the various positions denote various powers of two.*

EXAMPLE A

The number 1101 in base two can be interpreted as

$$1101 = 1(2^3) + 1(2^2) + 0(2) + 1 \qquad \text{see diagram at left}$$

By this we see that the numeral on the right indicates the number of 1's, the next numeral indicates the number of 2's, the next numeral the number of 4's ($2^2 = 4$ in the same way that $10^2 = 100$), and the next numeral the number of 8's ($2^3 = 8$). Therefore

$$1101_2 = 13_{10} \qquad 1(2^3) + 1(2^2) + 0(2) + 1 = 1(10) + 3$$

Note the manner in which the base is indicated. It should be emphasized that 1101_2 and 13_{10} are both the number 13; the way in which 13 is written is the only difference.

1 1 0 1

$1(2^3) + 1(2^2) + 0(2) + 1$

The probable reason for ten being the base of our number system is that we have ten fingers, and many ancient peoples used them for counting. The reason for base two being used in computers is that it simply requires that either there is *no* electric current (0) in a given circuit or there *is* a current (1).

Example A shows the basis for changing a number in base two into a number in base ten. That is, we evaluate each position to determine whether that power of two is counted. In the following example we convert a number from base two into base ten. The base two number system is also referred to as the **binary** number system.

EXAMPLE B   Change 1011011_2 into a number in base ten.

Interpreting 1011011_2 in terms of powers of two, we have

$$1011011_2 = 1(2^6) + 0(2^5) + 1(2^4) + 1(2^3) + 0(2^2) + 1(2) + 1$$
$$= 1(64) + 0(32) + 1(16) + 1(8) + 0(4) + 1(2) + 1$$
$$= 64 + 16 + 8 + 2 + 1$$
$$= 91_{10}$$

When we change a number from base ten to base two we determine the largest power of two that will divide into the number and successive remainders. The following example illustrates the method.

EXAMPLE C   Change 106_{10} into a number in base two.

▶   *We look for the largest power of two that will divide into* **106.** Since $2^7 = 128$, we know that 2^7 is too large. Trying $2^6 = 64$, we see that 64 will divide into 106 once, with a remainder of 42. Therefore we know that the left-hand 1 in the result will represent the 2^6 position. We now try the next smallest power of 2 to determine whether or not it will divide into 42, the remainder. We do this, for we have so far determined that

$$106_{10} = 2^6 + 42$$

Noting that $2^5 = 32$ will divide into 42 once with a remainder of 10, we now have

$$106_{10} = 2^6 + 2^5 + 10$$

We now see that $2^4 = 16$ will not divide into 10. Next we try $2^3 = 8$ and see that it divides into 10 with a remainder of 2. This tells us that

$$106_{10} = 2^6 + 2^5 + 2^3 + 2$$

We therefore have 1's in the 2^6, 2^5, 2^3, and 2 positions and 0's in the 2^4, 2^2, and 1 positions, which means

$$106_{10} = 1(2^6) + 1(2^5) + 0(2^4) + 1(2^3) + 0(2^2) + 1(2) + 0(1)$$

or

$$106_{10} = 1101010_2 \qquad \Box$$

We shall now briefly discuss the addition and multiplication of numbers in base two. Only three basic additions are necessary. They are $0 + 0 = 0$, $1 + 0 = 1$, and $1 + 1 = 10$. This last one undoubtedly looks strange, but all it says is that "one plus one equals two." We must remember, however, that two is written as 10 (one 2 and no 1's) in base two. Therefore, if we are adding two numbers whose sum is two (10) in any column, it is necessary to carry the 1. Consider the following example.

EXAMPLE D   Add 1100101_2 and 1001101_2.

Setting up the addition in the ordinary fashion, we have

```
1100101
1001101
```

Starting at the right, as usual, we have $1 + 1 = 10$. Thus we place a 0 under the 1's and carry a 1 into the next column. At this point we have

$$\begin{array}{r} \overset{1}{1100101} \\ \underline{1001101} \\ 0 \end{array} \quad 1 + 1 = 10$$

where the small 1 indicates the amount carried. In the 2's column, we then have $1 + 0 + 0 = 1$, which necessitates no carrying. Continuing in the same fashion, we have

$$\begin{array}{r} \overset{1 \quad 11 \ 1}{1100101} \\ \underline{1001101} \\ 10110010 \end{array}$$

If these numbers were converted to base ten, we would see that we just added 101 and 77 to arrive at 178. $\Box$

In multiplication, the basic process is the same. Three multiplication facts are needed. They are $0 \times 0 = 0$, $0 \times 1 = 0$, and $1 \times 1 = 1$. Consider the following example.

EXAMPLE E   Multiply 1101_2 by 101_2.

Setting up the multiplication in the usual fashion and following normal multiplication procedures with the multiplication facts given above we get

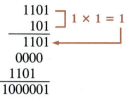

$$\begin{array}{r} 1101 \\ 101 \\ \hline 1101 \\ 0000 \\ 1101 \\ \hline 1000001 \end{array}$$

Note the use of addition in the process. We have just multiplied 13 by 5 to get 65 in base ten.   □

Exercises

In Exercises 1 through 8, change the given numbers from base two to base ten.

1. 11
2. 110
3. 1001
4. 10110
5. 1101100
6. 1000011
7. 100110001
8. 111000011

In Exercises 9 through 16, change the given numbers from base ten to base two.

9. 6
10. 10
11. 17
12. 19
13. 46
14. 51
15. 79
16. 145

In Exercises 17 through 24, perform the indicated additions in base two. Check your results by converting all numbers to base ten.

17. 10 + 11
18. 11 + 101
19. 110 + 1001
20. 1010 + 1110
21. 10011 + 11010
22. 11110 + 10110
23. 1100011 + 1110110
24. 1011010 + 1001111

In Exercises 25 through 32, perform the indicated multiplications in base two. Check your results by converting all numbers to base ten.

25. 11 × 10
26. 101 × 10
27. 101 × 11
28. 110 × 11
29. 1101 × 1001
30. 1011 × 1101
31. 1100111 × 10110
32. 1001110 × 11011

In Exercises 33 through 40, subtract the given numbers in base two. Check your results by converting all numbers to base ten. Borrowing is done in just the same manner as with numbers in base ten.

33. 1011 − 101
34. 1010 − 110
35. 110110 − 11001
36. 1001011 − 101101
37. 101111 − 10101
38. 100000 − 10011
39. 1100111 − 1001
40. 1111001 − 110110

Tables

TABLE 1. Squares and Square Roots

N	N^2	$\sqrt{N}$	$\sqrt{10N}$	N	N^2	$\sqrt{N}$	$\sqrt{10N}$
1.0	1.00	1.000	3.162	5.5	30.25	2.345	7.416
1.1	1.21	1.049	3.317	5.6	31.36	2.366	7.483
1.2	1.44	1.095	3.464	5.7	32.49	2.387	7.550
1.3	1.69	1.140	3.606	5.8	33.64	2.408	7.616
1.4	1.96	1.183	3.742	5.9	34.81	2.429	7.681
1.5	2.25	1.225	3.873	6.0	36.00	2.449	7.746
1.6	2.56	1.265	4.000	6.1	37.21	2.470	7.810
1.7	2.89	1.304	4.123	6.2	38.44	2.490	7.874
1.8	3.24	1.342	4.243	6.3	39.69	2.510	7.937
1.9	3.61	1.378	4.359	6.4	40.96	2.530	8.000
2.0	4.00	1.414	4.472	6.5	42.25	2.550	8.062
2.1	4.41	1.449	4.583	6.6	43.56	2.569	8.124
2.2	4.84	1.483	4.690	6.7	44.89	2.588	8.185
2.3	5.29	1.517	4.796	6.8	46.24	2.608	8.246
2.4	5.76	1.549	4.899	6.9	47.61	2.627	8.307
2.5	6.25	1.581	5.000	7.0	49.00	2.646	8.367
2.6	6.76	1.612	5.099	7.1	50.41	2.665	8.426
2.7	7.29	1.643	5.196	7.2	51.84	2.683	8.485
2.8	7.84	1.673	5.292	7.3	53.29	2.702	8.544
2.9	8.41	1.703	5.385	7.4	54.76	2.720	8.602
3.0	9.00	1.732	5.477	7.5	56.25	2.739	8.660
3.1	9.61	1.761	5.568	7.6	57.76	2.757	8.718
3.2	10.24	1.789	5.657	7.7	59.29	2.775	8.775
3.3	10.89	1.817	5.745	7.8	60.84	2.793	8.832
3.4	11.56	1.844	5.831	7.9	62.41	2.811	8.888
3.5	12.25	1.871	5.916	8.0	64.00	2.828	8.944
3.6	12.96	1.897	6.000	8.1	65.61	2.846	9.000
3.7	13.69	1.924	6.083	8.2	67.24	2.864	9.055
3.8	14.44	1.949	6.164	8.3	68.89	2.881	9.110
3.9	15.21	1.975	6.245	8.4	70.56	2.898	9.165
4.0	16.00	2.000	6.325	8.5	72.25	2.915	9.220
4.1	16.81	2.025	6.403	8.6	73.96	2.933	9.274
4.2	17.64	2.049	6.481	8.7	75.69	2.950	9.327
4.3	18.49	2.074	6.557	8.8	77.44	2.966	9.381
4.4	19.36	2.098	6.633	8.9	79.21	2.983	9.434
4.5	20.25	2.121	6.708	9.0	81.00	3.000	9.487
4.6	21.16	2.145	6.782	9.1	82.81	3.017	9.539
4.7	22.09	2.168	6.856	9.2	84.64	3.033	9.592
4.8	23.04	2.191	6.928	9.3	86.49	3.050	9.644
4.9	24.01	2.214	7.000	9.4	88.36	3.066	9.695
5.0	25.00	2.236	7.071	9.5	90.25	3.082	9.747
5.1	26.01	2.258	7.141	9.6	92.16	3.098	9.798
5.2	27.04	2.280	7.211	9.7	94.09	3.114	9.849
5.3	28.09	2.302	7.280	9.8	96.04	3.131	9.899
5.4	29.16	2.324	7.348	9.9	98.01	3.146	9.950

TABLE 2. Four-Place Logarithms of Numbers

N	0	1	2	3	4	5	6	7	8	9
10	0000	0043	0086	0128	0170	0212	0253	0294	0334	0374
11	0414	0453	0492	0531	0569	0607	0645	0682	0719	0755
12	0792	0828	0864	0899	0934	0969	1004	1038	1072	1106
13	1139	1173	1206	1239	1271	1303	1335	1367	1399	1430
14	1461	1492	1523	1553	1584	1614	1644	1673	1703	1732
15	1761	1790	1818	1847	1875	1903	1931	1959	1987	2014
16	2041	2068	2095	2122	2148	2175	2201	2227	2253	2279
17	2304	2330	2355	2380	2405	2430	2455	2480	2504	2529
18	2553	2577	2601	2625	2648	2672	2695	2718	2742	2765
19	2788	2810	2833	2856	2878	2900	2923	2945	2967	2989
20	3010	3032	3054	3075	3096	3118	3139	3160	3181	3201
21	3222	3243	3263	3284	3304	3324	3345	3365	3385	3404
22	3424	3444	3464	3483	3502	3522	3541	3560	3579	3598
23	3617	3636	3655	3674	3692	3711	3729	3747	3766	3784
24	3802	3820	3838	3856	3874	3892	3909	3927	3945	3962
25	3979	3997	4014	4031	4048	4065	4082	4099	4116	4133
26	4150	4166	4183	4200	4216	4232	4249	4265	4281	4298
27	4314	4330	4346	4362	4378	4393	4409	4425	4440	4456
28	4472	4487	4502	4518	4533	4548	4564	4579	4594	4609
29	4624	4639	4654	4669	4683	4698	4713	4728	4742	4757
30	4771	4786	4800	4814	4829	4843	4857	4871	4886	4900
31	4914	4928	4942	4955	4969	4983	4997	5011	5024	5038
32	5051	5065	5079	5092	5105	5119	5132	5145	5159	5172
33	5185	5198	5211	5224	5237	5250	5263	5276	5289	5302
34	5315	5328	5340	5353	5366	5378	5391	5403	5416	5428
35	5441	5453	5465	5478	5490	5502	5514	5527	5539	5551
36	5563	5575	5587	5599	5611	5623	5635	5647	5658	5670
37	5682	5694	5705	5717	5729	5740	5752	5763	5775	5786
38	5798	5809	5821	5832	5843	5855	5866	5877	5888	5899
39	5911	5922	5933	5944	5955	5966	5977	5988	5999	6010
40	6021	6031	6042	6053	6064	6075	6085	6096	6107	6117
41	6128	6138	6149	6160	6170	6180	6191	6201	6212	6222
42	6232	6243	6253	6263	6274	6284	6294	6304	6314	6325
43	6335	6345	6355	6365	6375	6385	6395	6405	6414	6425
44	6435	6444	6454	6464	6474	6484	6493	6503	6513	6522
45	6532	6542	6551	6561	6571	6580	6590	6599	6609	6618
46	6628	6637	6646	6656	6665	6675	6684	6693	6702	6712
47	6721	6730	6739	6749	6758	6767	6776	6785	6794	6803
48	6812	6821	6830	6839	6848	6857	6866	6875	6884	6893
49	6902	6911	6920	6928	6937	6946	6955	6964	6972	6981
50	6990	6998	7007	7016	7024	7033	7042	7050	7059	7067
51	7076	7084	7093	7101	7110	7118	7126	7135	7143	7152
52	7160	7168	7177	7185	7193	7202	7210	7218	7226	7235
53	7243	7251	7259	7267	7275	7284	7292	7300	7308	7316
54	7324	7332	7340	7348	7356	7364	7372	7380	7388	7396

TABLE 2. (Continued)

N	0	1	2	3	4	5	6	7	8	9
55	7404	7412	7419	7427	7435	7443	7451	7459	7466	7474
56	7482	7490	7497	7505	7513	7520	7528	7536	7543	7551
57	7559	7566	7574	7582	7589	7597	7604	7612	7619	7627
58	7634	7642	7649	7657	7664	7672	7679	7686	7694	7701
59	7709	7716	7723	7731	7738	7745	7752	7760	7767	7774
60	7782	7789	7796	7803	7810	7818	7825	7832	7839	7846
61	7853	7860	7868	7875	7882	7889	7896	7903	7910	7917
62	7924	7931	7938	7945	7952	7959	7966	7973	7980	7987
63	7993	8000	8007	8014	8021	8028	8035	8041	8048	8055
64	8062	8069	8075	8082	8089	8096	8102	8109	8116	8122
65	8129	8136	8142	8149	8156	8162	8169	8176	8182	8189
66	8195	8202	8209	8215	8222	8228	8235	8241	8248	8254
67	8261	8267	8274	8280	8287	8293	8299	8306	8312	8319
68	8325	8331	8338	8344	8351	8357	8363	8370	8376	8382
69	8388	8395	8401	8407	8414	8420	8426	8432	8439	8445
70	8451	8457	8463	8470	8476	8482	8488	8494	8500	8506
71	8513	8519	8525	8531	8537	8543	8549	8555	8561	8567
72	8573	8579	8585	8591	8597	8603	8609	8615	8621	8627
73	8633	8639	8645	8651	8657	8663	8669	8675	8681	8686
74	8692	8698	8704	8710	8716	8722	8727	8733	8739	8745
75	8751	8756	8762	8768	8774	8779	8785	8791	8797	8802
76	8808	8814	8820	8825	8831	8837	8842	8848	8854	8859
77	8865	8871	8876	8882	8887	8893	8899	8904	8910	8915
78	8921	8927	8932	8938	8943	8949	8954	8960	8965	8971
79	8976	8982	8987	8993	8998	9004	9009	9015	9020	9025
80	9031	9036	9042	9047	9053	9058	9063	9069	9074	9079
81	9085	9090	9096	9101	9106	9112	9117	9122	9128	9133
82	9138	9143	9149	9154	9159	9165	9170	9175	9180	9186
83	9191	9196	9201	9206	9212	9217	9222	9227	9232	9238
84	9243	9248	9253	9258	9263	9269	9274	9279	9284	9289
85	9294	9299	9304	9309	9315	9320	9325	9330	9335	9340
86	9345	9350	9355	9360	9365	9370	9375	9380	9385	9390
87	9395	9400	9405	9410	9415	9420	9425	9430	9435	9440
88	9445	9450	9455	9460	9465	9469	9474	9479	9484	9489
89	9494	9499	9504	9509	9513	9518	9523	9528	9533	9538
90	9542	9547	9552	9557	9562	9566	9571	9576	9581	9586
91	9590	9595	9600	9605	9609	9614	9619	9624	9628	9633
92	9638	9643	9647	9652	9657	9661	9666	9671	9675	9680
93	9685	9689	9694	9699	9703	9708	9713	9717	9722	9727
94	9731	9736	9741	9745	9750	9754	9759	9763	9768	9773
95	9777	9782	9786	9791	9795	9800	9805	9809	9814	9818
96	9823	9827	9832	9836	9841	9845	9850	9854	9859	9863
97	9868	9872	9877	9881	9886	9890	9894	9899	9903	9908
98	9912	9917	9921	9926	9930	9934	9939	9943	9948	9952
99	9956	9961	9965	9969	9974	9978	9983	9987	9991	9996

TABLE 3. Trigonometric Functions: Degrees, Radians

Degrees	Radians	Sin	Cos	Tan	Cot	Sec	Csc		
0	0.0000	0.0000	1.0000	0.0000	—	1.000	—	1.5708	90
1	0.0175	0.0175	0.9998	0.0175	57.29	1.000	57.30	1.5533	89
2	0.0349	0.0349	0.9994	0.0349	28.64	1.001	28.65	1.5359	88
3	0.0524	0.0523	0.9986	0.0524	19.08	1.001	19.11	1.5184	87
4	0.0698	0.0698	0.9976	0.0699	14.30	1.002	14.34	1.5010	86
5	0.0873	0.0872	0.9962	0.0875	11.43	1.004	11.47	1.4835	85
6	0.1047	0.1045	0.9945	0.1051	9.514	1.006	9.567	1.4661	84
7	0.1222	0.1219	0.9925	0.1228	8.144	1.008	8.206	1.4486	83
8	0.1396	0.1392	0.9903	0.1405	7.115	1.010	7.185	1.4312	82
9	0.1571	0.1564	0.9877	0.1584	6.314	1.012	6.392	1.4137	81
10	0.1745	0.1736	0.9848	0.1763	5.671	1.015	5.759	1.3963	80
11	0.1920	0.1908	0.9816	0.1944	5.145	1.019	5.241	1.3788	79
12	0.2094	0.2079	0.9781	0.2126	4.705	1.022	4.810	1.3614	78
13	0.2269	0.2250	0.9744	0.2309	4.331	1.026	4.445	1.3439	77
14	0.2443	0.2419	0.9703	0.2493	4.011	1.031	4.134	1.3265	76
15	0.2618	0.2588	0.9659	0.2679	3.732	1.035	3.864	1.3090	75
16	0.2793	0.2756	0.9613	0.2867	3.487	1.040	3.628	1.2915	74
17	0.2967	0.2924	0.9563	0.3057	3.271	1.046	3.420	1.2741	73
18	0.3142	0.3090	0.9511	0.3249	3.078	1.051	3.236	1.2566	72
19	0.3316	0.3256	0.9455	0.3443	2.904	1.058	3.072	1.2392	71
20	0.3491	0.3420	0.9397	0.3640	2.747	1.064	2.924	1.2217	70
21	0.3665	0.3584	0.9336	0.3839	2.605	1.071	2.790	1.2043	69
22	0.3840	0.3746	0.9272	0.4040	2.475	1.079	2.669	1.1868	68
23	0.4014	0.3907	0.9205	0.4245	2.356	1.086	2.559	1.1694	67
24	0.4189	0.4067	0.9135	0.4452	2.246	1.095	2.459	1.1519	66
25	0.4363	0.4226	0.9063	0.4663	2.145	1.103	2.366	1.1345	65
26	0.4538	0.4384	0.8988	0.4877	2.050	1.113	2.281	1.1170	64
27	0.4712	0.4540	0.8910	0.5095	1.963	1.122	2.205	1.0996	63
28	0.4887	0.4695	0.8829	0.5317	1.881	1.133	2.130	1.0821	62
29	0.5061	0.4848	0.8746	0.5543	1.804	1.143	2.063	1.0647	61
30	0.5236	0.5000	0.8660	0.5774	1.732	1.155	2.000	1.0472	60
31	0.5411	0.5150	0.8572	0.6009	1.664	1.167	1.942	1.0297	59
32	0.5585	0.5299	0.8480	0.6249	1.600	1.179	1.887	1.0123	58
33	0.5760	0.5446	0.8387	0.6494	1.540	1.192	1.836	0.9948	57
34	0.5934	0.5592	0.8290	0.6745	1.483	1.206	1.788	0.9774	56
35	0.6109	0.5736	0.8192	0.7002	1.428	1.221	1.743	0.9599	55
36	0.6283	0.5878	0.8090	0.7265	1.376	1.236	1.701	0.9425	54
37	0.6458	0.6018	0.7986	0.7536	1.327	1.252	1.662	0.9250	53
38	0.6632	0.6157	0.7880	0.7813	1.280	1.269	1.624	0.9076	52
39	0.6807	0.6293	0.7771	0.8098	1.235	1.287	1.589	0.8901	51
40	0.6981	0.6428	0.7660	0.8391	1.192	1.305	1.556	0.8727	50
41	0.7156	0.6561	0.7547	0.8693	1.150	1.325	1.524	0.8552	49
42	0.7330	0.6691	0.7431	0.9004	1.111	1.346	1.494	0.8378	48
43	0.7505	0.6820	0.7314	0.9325	1.072	1.367	1.466	0.8203	47
44	0.7679	0.6947	0.7193	0.9657	1.036	1.390	1.440	0.8029	46
45	0.7854	0.7071	0.7071	1.0000	1.000	1.414	1.414	0.7854	45
		Cos	Sin	Cot	Tan	Csc	Sec	Radians	Degrees

TABLE 4. Natural Logarithms of Numbers

N	$\ln N$	N	$\ln N$	N	$\ln N$
0.0		4.5	1.5041	9.0	2.1972
0.1	$7.6974 - 10$	4.6	1.5261	9.1	2.2083
0.2	$8.3906 - 10$	4.7	1.5476	9.2	2.2192
0.3	$8.7960 - 10$	4.8	1.5686	9.3	2.2300
0.4	$9.0837 - 10$	4.9	1.5892	9.4	2.2407
0.5	$9.3069 - 10$	5.0	1.6094	9.5	2.2513
0.6	$9.4892 - 10$	5.1	1.6292	9.6	2.2618
0.7	$9.6433 - 10$	5.2	1.6487	9.7	2.2721
0.8	$9.7769 - 10$	5.3	1.6677	9.8	2.2824
0.9	$9.8946 - 10$	5.4	1.6864	9.9	2.2925
1.0	0.0000	5.5	1.7047	10	2.3026
1.1	0.0953	5.6	1.7228	11	2.3979
1.2	0.1823	5.7	1.7405	12	2.4849
1.3	0.2624	5.8	1.7579	13	2.5649
1.4	0.3365	5.9	1.7750	14	2.6391
1.5	0.4055	6.0	1.7918	15	2.7081
1.6	0.4700	6.1	1.8083	16	2.7726
1.7	0.5306	6.2	1.8245	17	2.8332
1.8	0.5878	6.3	1.8405	18	2.8904
1.9	0.6419	6.4	1.8563	19	2.9444
2.0	0.6931	6.5	1.8718	20	2.9957
2.1	0.7419	6.6	1.8871	25	3.2189
2.2	0.7885	6.7	1.9021	30	3.4012
2.3	0.8329	6.8	1.9169	35	3.5553
2.4	0.8755	6.9	1.9315	40	3.6889
2.5	0.9163	7.0	1.9459	45	3.8067
2.6	0.9555	7.1	1.9601	50	3.9120
2.7	0.9933	7.2	1.9741	55	4.0073
2.8	1.0296	7.3	1.9879	60	4.0943
2.9	1.0647	7.4	2.0015	65	4.1744
3.0	1.0986	7.5	2.0149	70	4.2485
3.1	1.1314	7.6	2.0281	75	4.3175
3.2	1.1632	7.7	2.0412	80	4.3820
3.3	1.1939	7.8	2.0541	85	4.4427
3.4	1.2238	7.9	2.0669	90	4.4998
3.5	1.2528	8.0	2.0794	95	4.5539
3.6	1.2809	8.1	2.0919	100	4.6052
3.7	1.3083	8.2	2.1041		
3.8	1.3350	8.3	2.1163		
3.9	1.3610	8.4	2.1282		
4.0	1.3863	8.5	2.1401		
4.1	1.4110	8.6	2.1518		
4.2	1.4351	8.7	2.1633		
4.3	1.4586	8.8	2.1748		
4.4	1.4816	8.9	2.1861		

TABLE 5. Quantities and Their Associated Units

Quantity	Unit with symbols U.S. Customary	Metric (SI)
Length	foot (ft)	meter (m)
Mass	slug	kilogram (kg)
Force	pound (lb)	newton (N)
Time	second (s)	second (s)
Area	ft^2	m^2
Volume	ft^3	m^3
Capacity	gallon (gal)	liter (L)
Velocity	ft/s	m/s
Acceleration	ft/s^2	m/s^2
Density	lb/ft^3	kg/m^3
Pressure	lb/ft^2	Pascal (Pa)
Energy, work	ft-lb	joule (J)
Power	horsepower (hp)	watt (W)
Period	s	s
Frequency	1/s	hertz (Hz)
Angle	radian (rad)	radian (rad)
Electric current	ampere (A)	ampere (A)
Electric charge	coulomb (C)	coulomb (C)
Electric potential	volt (V)	volt (V)
Capacitance (electricity)	farad (F)	farad (F)
Inductance (electricity)	henry (H)	henry (H)
Resistance (electricity)	ohm (Ω)	ohm (Ω)
Temperature	Fahrenheit degree (°F)	Celsius degree (°C)
Quantity of heat	British thermal unit (Btu)	joule (J)
Amount of substance		mole (mol)
Luminous intensity	candlepower (cp)	candela (cd)

Notes: A discussion of units, symbols, and prefixes is found in Chapter 2. The symbols shown above are those used in the text. Some of them were adopted along with the SI system. This means that we use s rather than sec for seconds, A rather than amp for amperes, and C rather than coul for coulombs. Other units such as volt and farad are generally not spelled out, a common practice in the past.

The symbol for liter has several variations. L is recognized for use in the United States and Canada, l is the symbol recognized by the International Committee of Weights and Measures, and ℓ is also recognized in several countries.

The Celsius degree (°C) is not actually an SI unit. The *kelvin* (K) is the base unit of thermodynamic temperature. For *temperature intervals*, 1°C = 1K, whereas the temperature in kelvins equals the temperature in degrees Celsius plus 273.15.

Other units of time along with their symbols which are recognized for use with the SI system and are used in this text are: minute (min) and hour (h).

The symbol which is used for revolutions is r.

TABLE 6. Metric Prefixes

Prefix	Factor	Symbol	Prefix	Factor	Symbol
exa	10^{18}	E	deci	10^{-1}	d
peta	10^{15}	P	centi	10^{-2}	c
tera	10^{12}	T	milli	10^{-3}	m
giga	10^{9}	G	micro	10^{-6}	μ
mega	10^{6}	M	nano	10^{-9}	n
kilo	10^{3}	k	pico	10^{-12}	p
hecto	10^{2}	h	femto	10^{-15}	f
deca	10^{1}	da	atto	10^{-18}	a

Answers to Odd-Numbered Exercises

Exercises 1-1, page 6

1. 152   **3.** 1234   **5.** 2420   **7.** 21,410   **9.** 25,295   **11.** 139,639   **13.** 5602   **15.** 581
17. 949   **19.** 9798   **21.** 30,579   **23.** 95,682   **25.** 8 ft 2 in.   **27.** 18 lb   **29.** 5 ft 5 in.
31. 4 lb 11 oz   **33.** 10 ft 3 in.   **35.** 11 ft 10 in.   **37.** 29 yd 1 ft   **39.** 25 ft 4 in.   **41.** 11 in.
43. 4 ft 8 in.   **45.** 8 yd 2 ft   **47.** 10 in.   **49.** 530 ft   **51.** $1691   **53.** 4 h 41 min 30 s
55. 278 gal   **57.** 156 ft   **59.** 10 h 43 min   **61.** 188 kg   **63.** 16,250 mi^2

Exercises 1-2, page 12

1. 10,534   **3.** 1,314,976   **5.** 1,048,576   **7.** 236,894,403   **9.** 244   **11.** 324   **13.** 2048
15. 1981, rem 104   **17.** 47,804   **19.** 3,183,390   **21.** 150   **23.** 72,220   **25.** 160 mi^2
27. 238 in^2 **29.** 10,360 cm^2 **31.** 432 in^2 **33.** 10,115 mi   **35.** $1280   **37.** 330 ft
39. 18 mi/gal   **41.** 40 mi   **43.** $5690   **45.** 48   **47.** No

Exercises 1-3, page 16

1. $\frac{5}{9}$ **3.** $\frac{1}{7}$ **5.** $\frac{7}{13}$; $\frac{3}{16}$ **7.** $\frac{9}{8}$; $\frac{1}{12}$ **9.** 1; 6   **11.** 32; 1   **13.** $1\frac{2}{3}$ **15.** $4\frac{12}{13}$ **17.** $3\frac{53}{75}$
19. $53\frac{4}{25}$ **21.** $\frac{17}{5}$ **23.** $\frac{79}{8}$ **25.** $\frac{223}{13}$ **27.** $\frac{423}{4}$ **29.** $\frac{17}{24}$ **31.** $\frac{10}{23}$ **33.** $8\frac{1}{2}$ in.   **35.** $3\frac{3}{4}$ gal
37. $\frac{37}{10}$ in.   **39.** $\frac{3}{8}$

Exercises 1-4, page 23

1. $\frac{6}{14}$ **3.** $\frac{4}{5}$ **5.** $\frac{24}{78}$ **7.** $\frac{5}{13}$ **9.** $\frac{91}{175}$ **11.** $\frac{32}{2}$ **13.** $\frac{1}{2}$ **15.** $\frac{3}{2}$ **17.** $\frac{4}{5}$ **19.** $\frac{3}{5}$ **21.** $\frac{2}{5}$
23. $\frac{2}{5}$ **25.** 4   **27.** 15   **29.** 3   **31.** 4   **33.** 40   **35.** 16   **37.** $2 \times 2 \times 5$
39. $2 \times 2 \times 2 \times 2$ **41.** $2 \times 2 \times 3 \times 3$ **43.** $2 \times 2 \times 2 \times 2 \times 3$ **45.** 3×19
47. $3 \times 5 \times 7$ **49.** $\frac{6}{7}$ **51.** $\frac{4}{3}$ **53.** $\frac{7}{3}$ **55.** $\frac{2}{3}$ **57.** $\frac{1}{4}$ **59.** $\frac{3}{5}$ **61.** $\frac{3}{4}$ in.   **63.** $\frac{7}{16}$
65. $\frac{3}{10}$ **67.** $\frac{2}{5}$ **69.** $\frac{1}{5}$ **71.** $\frac{16}{31}$

Exercises 1-5, page 29

1. 4   **3.** 24   **5.** 72   **7.** 300   **9.** $\frac{4}{5}$ **11.** $\frac{2}{7}$ **13.** $\frac{3}{4}$ **15.** $\frac{1}{8}$ **17.** $\frac{13}{12}$ **19.** $\frac{1}{10}$ **21.** $\frac{7}{3}$
23. $\frac{47}{40}$ **25.** $\frac{28}{39}$ **27.** $\frac{83}{24}$ **29.** $\frac{29}{42}$ **31.** $\frac{73}{72}$ **33.** $\frac{73}{42}$ **35.** $\frac{13}{5}$ **37.** $\frac{1}{2}$ in.   **39.** $9\frac{1}{2}\Omega$ **41.** $26\frac{2}{3}$ ft
43. $1\frac{17}{24}$ in.   **45.** $\frac{5}{24}$ **47.** $1\frac{23}{32}$ in.

Exercises 1-6, page 36

1. $\frac{6}{77}$ **3.** $\frac{6}{5}$ **5.** $\frac{21}{20}$ **7.** $\frac{5}{27}$ **9.** $\frac{5}{18}$ **11.** $\frac{3}{2}$ **13.** $\frac{14}{9}$ **15.** $\frac{2}{17}$ **17.** 2 **19.** $\frac{2}{15}$ **21.** $\frac{11}{4}$
23. $\frac{27}{64}$ **25.** $\frac{168}{23}$ **27.** $\frac{5}{4}$ **29.** 7 **31.** $\frac{188}{351}$ **33.** $\frac{57}{4}$ **35.** $\frac{60}{209}$ **37.** $\frac{1}{5}$; $\frac{1}{13}$ **39.** 2; 5 **41.** $\frac{3}{16}$; $\frac{2}{7}$
43. $\frac{16}{81}$; $\frac{10}{71}$ **45.** $8\frac{1}{2}$ in. **47.** 12 **49.** $\frac{9}{26}$ acre/h **51.** $8\frac{4}{5}$ qt **53.** $\frac{2}{5}$ in. **55.** $87\frac{3}{11}$ mi/h

Exercises 1-7, page 41

1. $4(10) + 7(1) + \frac{3}{10}$ **3.** $4(100) + 2(10) + 9(1) + \frac{4}{10} + \frac{8}{100} + \frac{6}{1000}$ **5.** 27.3 **7.** 57.54 **9.** 8.03
11. 17.4 **13.** 0.4 **15.** 0.21 **17.** 1.7 **19.** 0.499 **21.** $\frac{4}{5}$ **23.** $\frac{9}{20}$ **25.** $\frac{267}{50}$ **27.** $\frac{63}{2500}$
29. 31.295 **31.** 2817.256 **33.** 8.763 **35.** 0.02628 **37.** 13.99952 **39.** 9.882
41. 124.992 **43.** 0.1215 **45.** 12.5 **47.** 295.4 **49.** 5.20 **51.** 0.46 **53.** 0.6 V
55. $177.49 **57.** 28.5 Ω **59.** 235.88 L **61.** $8\frac{1}{4}$ ft; 8 ft 3 in. **63.** $298.20 **65.** $213.18
67. 244 m

Exercises 1-8, page 47

1. 0.08 **3.** 2.36 **5.** 0.003 **7.** 0.056 **9.** 27% **11.** 321% **13.** 0.64% **15.** 700%
17. $\frac{3}{10}$ **19.** $\frac{1}{40}$ **21.** $\frac{6}{5}$ **23.** $\frac{57}{10,000}$ **25.** 60% **27.** 57.1% **29.** 22.9% **31.** 266.7%
33. 13 **35.** 5.304 **37.** 25% **39.** 7.5% **41.** 50 **43.** 400 **45.** $40.80 **47.** 32%
49. $17.64 **51.** $3395.75 **53.** 35% **55.** 0.3% **57.** 97% **59.** $512 **61.** 66%
63. 4.48%

Exercises 1-9, page 51

1. 8^3 **3.** 2^4 **5.** 3^5 **7.** 10^5 **9.** 8×8 **11.** $3 \times 3 \times 3 \times 3 \times 3 \times 3$
13. $7 \times 7 \times 7 \times 7 \times 7 \times 7 \times 7 \times 7$ **15.** $5 \times 5 \times 5 \times 5 \times 5 \times 5$
17. $(8 \times 10^3) + (5 \times 10^2) + (4 \times 10) + 3$ **19.** $5 + (7 \times \frac{1}{10}) + (3 \times \frac{1}{10^2}) + (9 \times \frac{1}{10^3})$ **21.** 243
23. 64 **25.** 0.09 **27.** 42.875 **29.** 4 **31.** 11 **33.** 4 **35.** 2 **37.** 0.4 **39.** 0.3
41. 108 **43.** 4500 **45.** 891 **47.** 0.637 **49.** 70,000 J **51.** 25 **53.** 30 ft **55.** 30 Ω
57. 600,000,000 **59.** 676,000

Exercises 1-10, page 55

1. 4.41 **3.** 15.21 **5.** 1.449 **7.** 1.975 **9.** 5.657 **11.** 7.416 **13.** 13.78 **15.** 65.57
17. 0.2098 **19.** 0.08062 **21.** 4.24 **23.** 2.1 s **25.** 2.53 s **27.** 75.9 kPa **29.** 5.8 in.
31. 12.2 ft

Exercises 1-11, page 60

1. 3.07 **3.** 6.44 **5.** 56 **7.** 6 **9.** 3 **11.** 8 **13.** 8 **15.** 32 **17.** 1.4166667
19. 0.51461988 **21.** 11.916667 **23.** 6.6071429 **25.** 4.4 **27.** 0.9 **29.** 120 **31.** 132
33. 4.3743571 **35.** 0.01 **37.** 0.33333333 **39.** 0.02222222 **41.** 0.34 **43.** 3.456
45. 39.76 in^3 **47.** 0.002 Ω **49.** 403.2 **51.** 124.8 g

Exercises 1-12, page 61

1. 13,330 **3.** 2779 **5.** 121.112 **7.** 1876.96 **9.** 4,365,872 **11.** 428 **13.** 72.312
15. 0.014 **17.** 37.11312 **19.** $\frac{5}{13}$ **21.** $\frac{19}{10}$ **23.** $\frac{71}{60}$ **25.** $\frac{401}{525}$ **27.** $\frac{22}{5}$ **29.** $\frac{11}{6}$ **31.** $\frac{4}{3}$
33. $\frac{215}{33}$ **35.** 7 **37.** 0.04 **39.** 10 **41.** 48 **43.** 49 **45.** $\frac{24}{7}$; $3\frac{2}{5}$ **47.** $\frac{257}{8}$; $17\frac{4}{7}$ **49.** $\frac{3}{5}$
51. $\frac{2}{5}$ **53.** $\frac{9}{2}$; $\frac{7}{22}$ **55.** $\frac{9}{10}$; $\frac{43}{34}$ **57.** 0.28125 **59.** $\frac{14}{25}$ **61.** 0.82; $\frac{41}{50}$ **63.** 0.0055; $\frac{11}{2000}$

65. 93.4%   **67.** 8%   **69.** 2.098   **71.** 14.44   **73.** 6.557   **75.** 126.5   **77.** 0.734   **79.** 3.06
81. 4.028   **83.** 5.59   **85.** 17.98 in.   **87.** 29.54 in.   **89.** 21.5 ft   **91.** 20.6 Ω **93.** $\frac{7}{32}$ in.
95. $\frac{3}{10}$ qt   **97.** 84.06 m   **99.** 0.008 lb   **101.** $58.65   **103.** 2.5%   **105.** 9.5 Ω **107.** $20\frac{9}{16}$ in.

Exercises 2-1, page 70

1. mA; 1 mA = 0.001 A   **3.** kV; 1 kV = 1000 V   **5.** kW; 1kW = 1000W
7. ML; 1ML = 1,000,000 L   **9.** Megavolt; 1 MV = 1,000,000 V
11. Microsecond; 1 μs = 0.000001 s   **13.** Centivolt; 1 cV = 0.01 V
15. Nanoampere; 1 nA = 0.000000001 A   **17.** (a) m^2; (b) ft^2 **19.** (a) m/s; (b) ft/s   **21.** 4 s
23. 8 m   **25.** 4 cs   **27.** 3 μs   **29.** 8 cg   **31.** 4 kg   **33.** mm^2 **35.** mi/gal   **37.** 1/s
39. lb/ft^2 **41.** kg $\cdot$ m/s^2 **43.** A$\cdot$s

Exercises 2-2, page 74

1. 10   **3.** 63,360   **5.** 288 in^2 **7.** 30.48 cm   **9.** 220 qt   **11.** 23.7 L   **13.** 52,000 cm^2
15. 9.05 ft^3 **17.** 3.09 L   **19.** 69.13 kg   **21.** 8.29 lb   **23.** 1829 mm   **25.** 90 t
27. 50 mi/h   **29.** 3.8 L   **31.** 32.2 ft/s^2 **33.** 1000 kg/m^3 **35.** 3270 ft/s   **37.** 0.47 L/s
39. 60.9 kcal

Excercises 2-3, page 80

1. Exact   **3.** Approximate   **5.** Approximate   **7.** Both 1's are approximate, but 2.80 is exact
9. 3; 4   **11.** 4; 4   **13.** 3; 3   **15.** 4; 5   **17.** (a) 3.764; (b) 3.764   **19.** (a) 0.01; (b) 30.8
21. (a) Same; (b) 78.0   **23.** (a) 0.004; (b) Same   **25.** 5.71; 5.7   **27.** 6.93; 6.9   **29.** 4100; 4100
31. 46800; 47000   **33.** 501; 500   **35.** 0.215; 0.22   **37.** 128.25 ft; 128.35 ft   **39.** 81.5 L; 82.5 L
41. 0.1733 qt   **43.** 91.4 m   **45.** Answer varies with calculators
47. Most calculators will yield 24 so that the accuracy is reduced to 2

Exercises 2-4, page 86

1. 39.0   **3.** 1.78   **5.** 754.0   **7.** 2.59   **9.** 113   **11.** 430   **13.** 110   **15.** 790   **17.** 6.9
19. 9.35   **21.** 10.2   **23.** 22   **25.** 17.62   **27.** 18.85   **29.** 13.2   **31.** 8.1   **33.** 23.33
35. 0.632   **37.** 62.23   **39.** 0.367   **41.** 196 ft   **43.** 262,144 bytes   **45.** 64.3 ft/s
47. 154 m   **49.** 5440 ft   **51.** 56.76 L   **53.** 433 V
55. Too many significant digits; time has only two significant digits

Exercises 2-5, page 88

1. 4; 4   **3.** 3; 4   **5.** (a) 7.32; (b) Same   **7.** (a) Same; (b) 207.31   **9.** 98.5; 98
11. 60500; 61000   **13.** 673; 670   **15.** 0.700; 0.70   **17.** 438.7   **19.** 10.89   **21.** 0.12
23. 0.057   **25.** 6.9   **27.** 7.6   **29.** 31.44   **31.** 29   **33.** Microgram; 1 μg = 0.000001 g
35. ks; 1 ks = 1000 s   **37.** 0.385 cm^3 **39.** 13 cm   **41.** 35.62 pt   **43.** 760 L   **45.** 740 in^3
47. 1.10 m/s   **49.** 1.061 t/ft^2 **51.** 30,630,000 cm^3/min   **53.** 0.1855 in.; 0.1865 in.
55. $537.50   **57.** 21.84 g   **59.** 0.021 m   **61.** 320.2 calc/s   **63.** 0.67 kg/L   **65.** 0.000287 in^2
67. Cannot obtain the difference to tenths, since the initial time was not recorded to tenths

Exercises 3-1, page 94

1. Add 3 to 6   **3.** Add 7 to +2   **5.** Add −2 to −8   **7.** Subtract +6 from −3

9.   11.   13.   15.

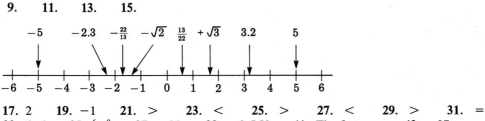

17. 2   **19.** −1   **21.** >   **23.** <   **25.** >   **27.** <   **29.** >   **31.** =
33. 6, 6   **35.** $\frac{6}{7}, \frac{8}{5}$ **37.** −30   **39.** −0.5 V   **41.** The first one   **43.** −37
45. −30°C < −5°C   **47.** −2 V > −5 V   **49.** Answer varies with calculators
51. Enter $\boxed{0}\ \boxed{-}\ \boxed{3}\ \boxed{=}$

Exercises 3-2, page 100

1. +11   **3.** −15   **5.** +3   **7.** −8   **9.** +7   **11.** −3   **13.** −10   **15.** +16   **17.** +1
19. −6   **21.** −8   **23.** +1   **25.** +6   **27.** −17   **29.** −2   **31.** −6
33. (a) −5°C; (b) −15°C   **35.** 63 cm   **37.** 17 A   **39.** +5000   **41.** −7   **43.** 9   **45.** −3
47. 7

Exercises 3-3, page 105

1. +80   **3.** −63   **5.** −84   **7.** +30   **9.** −56   **11.** +240   **13.** +360   **15.** 0
17. −168   **19.** +36; +36   **21.** +49; −49   **23.** −1024   **25.** +1; −1   **27.** −128; −128
29. −50 mm   **31.** 7000 ft   **33.** 994 lb   **35.** 10 ft   **37.** 18   **39.** −56   **41.** −8   **43.** −81

Exercises 3-4, page 108

1. +4   **3.** −9   **5.** −11   **7.** +15   **9.** 0   **11.** Undefined   **13.** −16   **15.** +3   **17.** +3
19. −5   **21.** −$130   **23.** −250 ft   **25.** 5 days   **27.** 5250 ft   **29.** −557   **31.** 8
33. −21   **35.** 0

Exercises 3-5, page 114

1. 20   **3.** −14   **5.** −49   **7.** −1   **9.** −69   **11.** −5   **13.** −106   **15.** −20   **17.** 14
19. −101   **21.** 19   **23.** 3   **25.** −12   **27.** −2   **29.** −1000 ft   **31.** −78.0 m
33. 46.2 gal   **35.** 2500 gal   **37.** 16   **39.** −9   **41.** 22.5   **43.** −0.75

Exercises 3-6, page 115

1. <   **3.** >   **5.**

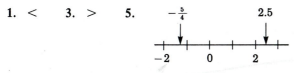

7. $\frac{2}{3}, \frac{1}{4}$ **9.** −2   **11.** −13   **13.** −63   **15.** +9   **17.** +10
19. +19   **21.** +96   **23.** −60   **25.** +10   **27.** −6   **29.** +256   **31.** +5   **33.** 0
35. −7   **37.** +10   **39.** −16   **41.** 5 qt   **43.** 26 ft   **45.** 396°C   **47.** −4   **49.** $440
51. −7 m/s   **53.** 1.04 A   **55.** −0.8°F   **57.** 5980   **59.** −2   **61.** −5   **63.** 2

Exercises 4-1, page 122

1. ab 3. x^2 5. $2w^2$ 7. a^3b^2 9. b, c 11. $7, p, q, r$ 13. i, i, R 15. a, b, c, c, c
17. 6   19. 2π 21. $4\pi e$ 23. mw^2 25. $x = 4y$ 27. $m = 10c$ 29. $V = 7.48lwd$

31. $d = \dfrac{1}{2}gt^2$ 33. $N = 5280x$ 35. $N = 9xy$ 37. $N = 24n$ 39. $C = 4cs$ 41. 1500 mi

43. 1.5 V   45. 720 gal   47. 81.5 ft   49. 13,200 ft   51. 3456 bolts   53. 6970 gal
55. 172 ft   57. $40.80   59. 4536 m²

Exercises 4-2, page 128

1. $x^2, 4xy, -7x$ 3. $12, -5xy, 7x, -\dfrac{x}{8}$ 5. $3x$ and $2x$ 7. x and $5x$ 9. $-8\,mn$ and $-mn$

11. $6(x - y)$ and $-3(x - y)$ 13. $6a(a - x)$ 15. $x^2(a - x)(a + x)$ 17. $\dfrac{2}{5a}$ 19. $\dfrac{6}{a - b}$

21. $K = 328$ 23. $t = 2$ 25. $L = 50.0 - x$ 27. $A = 2x^2 + 4lx$ 29. $V = I(R + r)$

31. $A = 2\pi r(r + h)$ 33. $A = \dfrac{a + b + c + d + e}{5}$ 35. $T = 6.28\sqrt{\dfrac{l}{g}}$ 37. 22.6 ft

39. 77.8 ft²   41. 7   43. 2.54 s   45. 9.741 m²   47. 3.121 V   49. 2.24 s   51. 5.349 m²

Exercises 4-3, page 134

1. $6x + y$ 3. $7a - 4b^2$ 5. $5s + 3t$ 7. $5x + 5y$ 9. $6a - 3b$ 11. $\dfrac{7x}{2}$ 13. $\dfrac{1}{2xy}$

15. $7a - 15b$ 17. $12x + y$ 19. $9R + 12S + 5RS$ 21. $-5x + 5y - 4xy$ 23. $4x$
25. $5rs - 15r$ 27. $5abc - c^2 - b$ 29. $5ax - 3$ 31. $5x - 10y$ 33. $8x - 2y + b$
35. $5x^2 + 6xy$ 37. $4x + 5a$ ft   39. $20a + 10b + 5$ cents   41. $2xy + 5y + 15$ 43. $3ay$ min
45. 40   47. 78   49. $1.75   51. 456 cm²   53. 846   55. -576 57. 9921.45
59. -9.907

Exercises 4-4, page 136

1. $-7a$ and $5a$ 3. $5(a - b)$ and $-7(a - b)$ 5. $7a - 4b$ 7. $12ax - 18bx$ 9. $3x + y$
11. $c - ac$ 13. $2a - b$ 15. $3a + 11b - ab$ 17. $2ax + 2a$ 19. $2a - ax$
21. $5x - xy$ 23. $3ab + a - 2b$ 25. $3 + 5a + ax$ 27. $13a + 3b$ 29. $5a + 4b$
31. $4ax - 5xy + 2ay$ 33. $6a^2$ 35. $12x$ 37. 42   39. 98   41. -26 43. 218,448
45. $x = x^2 - 7$ 47. $C = \dfrac{Q}{V}$ 49. $p = 3s$ 51. $I = 1000t^2$ 53. $I = 4xr + 0.06x$

55. $C = 40n + 145$ 57. $N = 33xt + 20x + 48t$ 59. $I = 14 - 4t$ 61. 21 in.   63. $12,000
65. (a) $4.65; (b) $6.25   67. 2,312,736   69. 52   71. 190.09

Exercises 5-1, page 145

1. 2   3. 3   5. 7   7. 35   9. -18 11. -4 13. 9   15. -2 17. 3   19. -6
21. -1 23. 8   25. 1   27. $\frac{1}{2}$ 29. $\frac{1}{3}$ 31. 2   33. 9   35. 32   37. 3   39 68

41. -2 43. $\dfrac{25}{21}$ 45. First has no solution; second is an identity   47. a, d 49. 0.906

51. 1.45

Exercises 5-2, page 149

1. $\dfrac{N}{A-s}$ **3.** $\dfrac{R_1L_2}{L_1}$ **5.** v_2-at **7.** $\dfrac{PV}{R}$ **9.** $\dfrac{Id^2}{5300E}$ **11.** $\dfrac{yd}{ml}$ **13.** $180-A-C$

15. $\dfrac{RM}{CV}$ **17.** $\dfrac{MD_m}{D_p}$ **19.** $\dfrac{L-3.14r_1-2d}{3.14}$ **21.** $\dfrac{L-L_0}{L_0t}$ **23.** $\dfrac{p-p_a+dgy_1}{dg}$

25. $\dfrac{n_1A+n_2A-n_2p_2}{n_1}$ **27.** $\dfrac{f_su-fu}{f}$ **29.** $a-bc$ **31.** $\dfrac{f-3y}{a}$ **33.** $\dfrac{3y-2ay}{2a}$ **35.** $2b+4$

37. $\dfrac{x_1-x_2-3a}{a}$ **39.** $2R_3-R_1$ **41.** $\dfrac{3y+6-7ay}{7a}$ **43.** $\dfrac{x-3a-ax}{a}$ **45.** $3A-a-b$

47. $\dfrac{E-Ir}{I}$ **49.** $\dfrac{I-xr_1}{x+1000}$ **51.** $\dfrac{C-x}{7}$

Exercises 5-3, page 155

1. (a) $8>6$; (b) $-5<1$ **3.** (a) $3>-6$; (b) $1<2$ **5.** (a) $2>0$; (b) $-2>-3$
7. (a) $15>10$; (b) $48>-32$ **9.** $-1>-6$ **11.** $-1\ge-2$ **13.** $x>12$ **15.** $x<5$
17. $x>18$ **19.** $x<-6$ **21.** $x\le4$ **23.** $x\le-1$ **25.** $x<-4$ **27.** $x<-2$
29. $x>-1$ **31.** $x\ge6$ **33.** Less than \$87,500 **35.** More than 50 mi/h
37. $12.9\le x\le15.5$ **39.** Greater than 35 h **41.** $x\le2.33$ **43.** $x\ge14.69$

Exercises 5-4, page 160

1. $220\ \Omega$ **3.** \$1.20 **5.** 85 mA **7.** 16 gal/min, 12 gal/min, 22 gal/min
9. 320 lines/min, 800 lines/min **11.** 13 at 5¢, 8 at 25¢ **13.** \$21 **15.** 800 gal, 1200 gal, 2400 gal
17. 4 h **19.** 3600 ft/s **21.** 75 km/h, 85 km/h **23.** 16, 64 **25.** 150 mA, 200 mA, 300 mA
27. \$1800, \$2500 **29.** 1,900,000 first year; 2,600,000 second year **31.** $166\frac{2}{3}$ g
33. 50 acres at \$200; 90 acres at \$300 **35.** 48 mi/h **37.** 344 m **39.** \$11,000

Exercises 5-5, page 165

1. $\frac{12}{5};\frac{3}{23}$ **3.** $\frac{7}{1};\frac{1}{6}$ **5.** $\frac{2}{3};\frac{2}{3}$ **7.** $\frac{2}{11};\frac{2}{7}$ **9.** $\frac{15}{4}$ **11.** $\frac{3}{10}$ **13.** $\frac{1}{6}$ **15.** $\frac{4}{9}$ **17.** 2 m/s
19. $\frac{2}{9}$ lb/ft^3 **21.** $\frac{5}{4}$ **23.** $\frac{6}{7}$ **25.** 5 **27.** 9 **29.** 14 **31.** $\frac{35}{2}$ **33.** $\frac{1}{4}$ **35.** 4540 g
37. $\frac{100}{3}$ in. **39.** 6 m, 9 m **41.** 4,000,000 gal reg; 10,000,000 gal lead-free **43.** 1.44 m
45. 1.2 **47.** \$21,700 **49.** 6 h **51.** 37 h **53.** 9.3 mm **55.** $\frac{9}{5}$

Exercises 5-6, page 172

1. $y=kt$ **3.** $y=ks^2$ **5.** $t=\dfrac{k}{y}$ **7.** $y=kst$ **9.** $y=\dfrac{ks}{t}$ **11.** $x=\dfrac{kyz}{t^2}$ **13.** $y=5s$

15. $s=2t^3$ **17.** $u=\dfrac{272}{d^2}$ **19.** $y=\dfrac{9x}{t}$ **21.** 16 **23.** $\dfrac{32}{5}$ **25.** 36 **27.** $\dfrac{36}{49}$ **29.** $\dfrac{72}{5}$

31. 81 **33.** $v=32t$ **35.** $p=30{,}000t$ **37.** $F=\dfrac{kQ_1Q_2}{s^2}$ **39.** 35.4 hp **41.** $0.116\ \Omega$

43. 240 r/min **45.** $F=0.005231\,Av^2$ **47.** 27.79 s

Exercises 5-7, page 174

1. 8 **3.** 9 **5.** -4 **7.** 4 **9.** -3 **11.** $\dfrac{1}{2}$ **13.** $R-R_1-R_2$ **15.** $\dfrac{ms_1}{r}$ **17.** $\dfrac{d_m+A}{A}$

19. $\dfrac{PT - M_2V_2}{V_1}$ **21.** $\dfrac{wL}{R(w + L)}$ **23.** $\dfrac{W + H_1 - H_2}{S_1 - S_2}$ **25.** $\dfrac{2a + ax}{3}$ **27.** $\dfrac{ax + ab - bx}{b}$

29. $\dfrac{6 - x - a^2}{a}$ **31.** $\dfrac{2 - 3a}{6}$ **33.** $x < 9$ **35.** $x \geq 5$ **37.** $x < -3$ **39.** $x < -4$

41. $x < -2$ **43.** $x \leq 3$ **45.** $\frac{2}{3}$ **47.** $\frac{6}{1}$ **49.** $\frac{8}{3}$ **51.** $\frac{9}{2}$ **53.** 4   **55.** 4   **57.** $\frac{7}{3}$

59. $\frac{21}{2}$ **61.** $y = 6x$ **63.** $m = \dfrac{15}{\sqrt{r}}$ **65.** 5   **67.** $\frac{81}{5}$ **69.** 36 in., 54 in.   **71.** \$150, \$50

73. 37.2 ft by 24.8 ft   **75.** 48 ft   **77.** 480 m   **79.** 8 at 14¢; 13 at 22¢   **81.** 2.5 h
83. 1850 km/h, 2150 km/h   **85.** 10 L   **87.** \$9000   **89.** 20 lb   **91.** 3.51 s   **93.** 2.45
95. 71.5820 mi/h

Exercises 6-1, page 182

1. 26°   **3.** 104°   **5.** 56°24′   **7.** 136°27′   **9.** 156.25°   **11.** 67.1°   **13.** $\angle CBD$, $\angle DBA$
15. $\triangle CDB$ **17.** $\triangle ADC$ **19.** $AC = 4$ in., $BC = 3$ in.   **21.** $AB \parallel DC$, $AD \parallel BC$ **23.** 135°
25. AB and DC **27.** AO or OB **29.**   **31.**

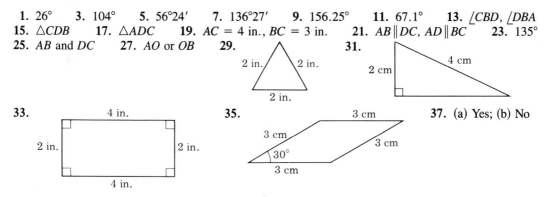

33.   **35.**   **37.** (a) Yes; (b) No

39. Two isosceles triangles   **41.** 10 cm   **43.** 54°   **45.** 50°   **47.** 94 km   **49.** 66°   **51.** West
53. 21°   **55.** 46°3′36″

Exercises 6-2, page 188

1. 12 ft   **3.** 123 mm   **5.** 896 m   **7.** 84.8 in.   **9.** 193.8 mm   **11.** 57.1 in.   **13.** 2.6 m
15. 0.60 mi   **17.** 168.0 cm   **19.** 230.6 ft   **21.** 94.8 cm   **23.** 21.2 in.   **25.** 36 ft
27. 23.00 cm   **29.** 45.36 in.   **31.** 24.22 m   **33.** $p = 2s + a$ **35.** $p = 2r + \pi r$ **37.** $p = 5s$
39. $p = 2a + b_1 + b_2$ **41.** \$33   **43.** 24,900 mi   **45.** 890 ft   **47.** \$1312   **49.** 25.1 ft
51. 1428 ft   **53.** 26.00 in.   **55.** 1.58 mm   **57.** 6.28 in.   **59.** 4.58 cm   **61.** 234.3 cm
63. Answer varies

Exercises 6-3, page 197

1. 2700 cm²   **3.** 170 in²   **5.** 57.8 in²   **7.** 6.45 in²   **9.** 2450 mm²   **11.** 24.4 cm²
13. 106 in²   **15.** 40 ft²   **17.** 0.240 m²   **19.** 340 cm²   **21.** 238 ft²   **23.** 15,700 cm²
25. 130 cm²   **27.** 12 ft²   **29.** 0.000258 km²   **31.** 31.0 m²   **33.** 0.717 ft²   **35.** 1692 ft²
37. 21.65 in²   **39.** 1.27 ft²   **41.** (a) 36 in.; (b) 60 in²   **43.** (a) 60 cm; (b) 225 cm²
45. (a) 30 in.; (b) 30 in²   **47.** (a) 26.6 in.; (b) 35.0 in²   **49.** $A = bh_1 + \frac{1}{2}bh_2$
51. $A = r^2 + 2rh + \frac{1}{2}\pi r^2$ **53.** \$179.20   **55.** 30 ft²   **57.** 10.5 ft²   **59.** 122,000 m²
61. 1070 cm²   **63.** 196 cm²   **65.** Multiplied by 4   **67.** Circle   **69.** 46,010,000 ft²
71. 0.4729 m²

Exercises 6-4, page 202

1. 20,100 mm^3 **3.** 0.080 cm^3 **5.** 4600 m^3 **7.** 1,500,000 cm^3 **9.** 3400 cm^3 **11.** 0.512 in^3
13. 0.011 m^3 **15.** 7156 cm^3 **17.** 85,200 cm^3 **19.** 14,700 m^3 **21.** 21,000 in^3
23. 0.735 mm^3 **25.** 1728 in^3 **27.** 1,000,000 cm^3 **29.** 1300 ft^3 **31.** 90 ft^3 **33.** 15.9
35. No **37.** 4.71 m^3 **39.** 311 in^3 **41.** 11.9 lb **43.** 1,390,000,000,000,000 km^3
45. 21,200 gal **47.** 1.271 mm^3

Exercises 6-5, page 205

1. 37°30′ **3.** 12°33′ **5.** 63.5° **7.** 105.9° **9.** 40.2 in. **11.** 25.5 mm **13.** 27.2 m
15. 1798 ft **17.** 278 cm **19.** 26.7 ft **21.** 153 mm **23.** 15.8 mm **25.** 224 cm^2
27. 3.08 yd^2 **29.** 4.80 cm^2 **31.** 3.06 in^2 **33.** 0.00148 km^2 **35.** 35.4 in^2 **37.** 140,000 mm^2
39. 3.30 m^2 **41.** 1200 mm^2 **43.** 9.30 mm^2 **45.** 3.60 m^3 **47.** 42.9 yd^3 **49.** 678 cm^3
51. 27.4 yd^3 **53.** $p = 2b + 2c$ **55.** $p = 3a + \pi a$ **57.** $A = \frac{1}{2}ac$ **59.** $A = \frac{1}{2}\pi a^2 + \frac{3}{2}ah$
61. $19,440 **63.** 204 m^2 **65.** 30,700 lb **67.** 15 cm **69.** 26,200 mi **71.** 6.30 ft^2
73. 8.77 ft^2 **75.** 47.6 m^3 **77.** 9.26 cm^3 **79.** 1.77 m^3 **81.** 150 gal **83.** 63,000 ft^3
85. 54 in^3 **87.** 1.41 ft^2 **89.** 49.4 in^2 **91.** 28,300 cm^3 **93.** 10,200 ft^3 **95.** 51.396111°

Exercises 7-1, page 214

1. $3s - xy - a$ **3.** $9y - 3x - 4a - 9xy$ **5.** $3x + 2xy$ **7.** $-3x^2 + 3xy - 2s$ **9.** $2a + 3$
11. $-4 - 4x$ **13.** $2s - 1$ **15.** $-9 + 4y$ **17.** $10x + 1$ **19.** $2s + 2$ **21.** $6x - 12$
23. $7t - 5x - 5p^2 + 9$ **25.** $2a^2 + 3x + 1$ **27.** $7 + 2t$ **29.** 1 **31.** $\frac{3}{2}$ **33.** $5 - 4a$
35. $\frac{6}{5}$ **37.** $2x + 28$ **39.** $2R + 30$ ohms **41.** $2M - 40$ **43.** $16t_1^2 - 16t_2^2 + 32$ **45.** $\frac{I + mv}{m}$
47. $C + P_1 - L$ **49.** -0.2840 **51.** 1.21

Exercises 7-2, page 222

1. x^{10} **3.** y^8 **5.** x^3 **7.** a **9.** t^{10} **11.** n^{14} **13.** $\frac{1}{p^{10}}$ **15.** $4n^3$ **17.** y^{15} **19.** p^{25}
21. $-ax$ **23.** $\frac{2at^4}{c^2}$ **25.** a^{40} **27.** r^{150} **29.** $-\frac{x}{4r^2}$ **31.** $\frac{7t^2u}{6}$ **33.** $a^2x^4b^2$ **35.** $-a^5t^{10}$
37. $-\frac{x}{3y^2}$ **39.** $\frac{t^3}{3s^3}$ **41.** $-84r^2s^2t^4$ **43.** $-8s^3t^9x^3$ **45.** True: e only **47.** True: a, b, c
49. $2x^3$ **51.** $I = \frac{E}{R}$

Exercises 7-3, page 225

1. $2a^2 + 6ax$ **3.** $6a^2x - 3a^4$ **5.** $-2s^2tx + 2st^3y$ **7.** $-3x^3y^2 + 9ax^2y^7$ **9.** $x^2 - 4x + 3$
11. $s^2 + s - 6$ **13.** $2x^2 + x - 1$ **15.** $10v^2 + 17v + 3$ **17.** $a^2 - 3ax + 2x^2$
19. $6a^2 + ac - 2c^2$ **21.** $2x^2 + 18x - 5tx - 45t$ **23.** $4a^2 - 81p^2y^2$ **25.** $2a^3 - 5a^2 - 13a - 5$
27. $a^2 + 2axy - 4ax - 2x^2y + 3x^2$ **29.** $x^2 - 4x + 4$ **31.** $x^2 + 4xy + 4y^2$
33. $x^3 - 3x^2 - 10x + 24$ **35.** $x^3 + 3x^2 + 3x + 1$ **37.** True: c only **39.** True: b, c
41. $x^2 + x - 6$ ft **43.** $200x^2 + 1600x + 3000$ **45.** $(13)(7) = 100 - 9$ **47.** $16t^2 - 64t + 48$
49. $kL_1T_1 - kL_1T_2 - kL_2T_1 + kL_2T_2$ **51.** 8Ω, 3Ω

Exercises 7-4, page 231

1. $6b + 5$ **3.** $3m - 1$ **5.** $-a^2x^2 + ax$ **7.** $y^2 - 2xy^3$ **9.** $bc - ab^3c^4 - 2a$

11. $-ab^2 + 2a^2b^3 + 1 + b$ **13.** $x - 3$ **15.** $2x + 1$ **17.** $4x - 3 + \dfrac{2}{2x + 3}$

19. $2x^2 - 3x - 4$ **21.** $2x^2 - x + 3$ **23.** $x^3 + x^2 + x + 1$ **25.** $5(x - 1)$

27. $2x^3 - x^2 - 2x + 1 - \dfrac{2}{3x - 1}$ **29.** $a - 3b$ **31.** $x^2 - 4xy - 2y^2$ **33.** True: b, c

35. True: a, b   **37.** $a + 5$ **39.** $a^2 - 5$ **41.** $2x + 3$ kg   **43.** $6r - 4 + \dfrac{18}{r + 2}$

45. $5x - 6 + \dfrac{12}{x + 2}$ **47.** $10r - 2 + \dfrac{2.4}{6r + 1.2}$ **49.** $2.01x^4 - 0.691x^3 + 2.98$

51. $21.61x^2 - 2.643x - 81.73$

Exercises 7-5, page 233

1. $3a - x$ **3.** $9x - 13y$ **5.** $2n - 4$ **7.** $8y$ **9.** $-6a^3b^6$ **11.** $56x^4y^9z^8$ **13.** $27a^3b^6$

15. $16x^8y^4z^{12}$ **17.** $-4ax^3$ **19.** $\dfrac{5x}{y^4z^3}$ **21.** $8 - 5x$ **23.** $5x - 2a$ **25.** $4x - 7$

27. $5b - 10$ **29.** $2x^5 - 6x^3$ **31.** $-2a^3x + 2a^2t$ **33.** $2x^2 + x - 21$ **35.** $6a^2 - 11ab - 10b^2$
37. $x^3 + 1$ **39.** $-2x^3 + 6x^2 + 4x - 16$ **41.** $-2y^3 + 3x^3$ **43.** $h - 3j^2 - 6h^3j^3$ **45.** $x + 4$

47. $x^2 - x + 1 - \dfrac{11}{2x + 3}$ **49.** $x^2 - x + 1$ **51.** $2x^2 - x - 4$ **53.** $x + 2y$ **55.** $x^2 + x + 1$

57. 7   **59.** $a^2 + 1$ **61.** $1.71C + 33.6$ **63.** $\dfrac{1 - C - 3I_t}{3}$ **65.** $-4x^2 + 8x$ **67.** $36x - 4x^2$

69. $mv_2{}^2 - mv_1{}^2$ **71.** $lh + 2alht + a^2lht^2$ **73.** $r_1 - ar_1 + r_2 - 2ar_2 + a^2r_2$ **75.** $p_0 - kx$
77. $2a + 5$ **79.** $2x^2 + 20x + 50$; 450 lb   **81.** $2841x^8$ **83.** $1024x^{10} - 256x^8 - 1$

Exercises 8-1, page 241

1. $5(x + y)$ **3.** $7(a^2 - 2bc)$ **5.** $a(a + 2)$ **7.** $2x(x - 2)$ **9.** $3(ab - c)$ **11.** $2p(2 - 3q)$
13. $3y^2(1 - 3z)$ **15.** $abx(1 - xy)$ **17.** $6x(1 - 3y)$ **19.** $3ab(a + 3)$ **21.** $acf(abc - 4)$
23. $ax^2y^2(x + y)$ **25.** $2(x + y - z)$ **27.** $5(x^2 + 3xy - 4y^3)$ **29.** $2x(3x + 2y - 4)$
31. $4pq(3q - 2 - 7q^2)$ **33.** $7a^2b^2(5ab^2c^2 + 2b^3c^3 - 3a)$ **35.** $3a(2ab - 1 + 3b^2 - 4ab^2)$
37. $nR\left(\dfrac{T_2}{T_1} - 1\right)$ **39.** $I(R_1 + R_2 + R_3)$ **41.** $2wh(7 + 2w)$ **43.** $2(lw + lh + hw)$
45. $2x(x^2 - 3x + 5)$ **47.** $wx(x^3 - 2Lx^2 + L^3)$ **49.** (a) Yes; (b) No, $2x(x - 4)$
51. (a) Yes; (b) No   **53.** (a) Yes; (b) No; (c) No   **55.** (a) No; (b) Yes; (c) No

Exercises 8-2, page 247

1. $\sqrt{9} = 3$; $\sqrt{16} = 4$ **3.** $\sqrt{121} = 11$ **5.** $\sqrt{x^6} = x^3$; $\sqrt{16x^6} = 4x^3$ **7.** $\sqrt{a^2b^4} = ab^2$
9. $x^2 - y^2$ **11.** $4a^2 - b^2$ **13.** $49a^2x^4 - p^6$ **15.** $x^2 + 4x + 4$ **17.** $(a + 1)(a - 1)$
19. $(t + 3)(t - 3)$ **21.** $(4 + x)(4 - x)$ **23.** $(2x^2 + y)(2x^2 - y)$
25. $(10 + a^2b)(10 - a^2b)$ **27.** $(ab + y)(ab - y)$ **29.** $(9x^2 + 2y^3)(9x^2 - 2y^3)$
31. $5(x^2 + 3)(x^2 - 3)$ **33.** $4(x + 5y)(x - 5y)$ **35.** $(x^2 + 1)(x + 1)(x - 1)$
37. $4(x^2 + 9y^2)$ **39.** $3s^2(t^4 + 4)$ **41.** $8(2x^2y^2 + 3a^2b^2)$ **43.** $5ax(x - 8a)$
45. $(x + y + 1)(x + y - 1)$ **47.** $(5 + x + y)(5 - x - y)$ **49.** $4xy$
51. $a(x + y + 1)(x + y - 1)$ **53.** 399   **55.** 39,900   **57.** $(c + 8)(c - 8)$
59. $c\pi(r_1 + r_2)(r_1 - r_2)$ **61.** $(C + kp)(C - kp)$ **63.** $kw(h_2 + h_1)(h_2 - h_1)$
65. $k(T_2{}^2 + 1)(T_2 + 1)(T_2 - 1)$ **67.** $2\pi d(h_1a_1 + h_2a_2)(h_1a_1 - h_2a_2)$

Exercises 8-3, page 254

1. $(x + 1)(x + 2)$ **3.** $(x + 4)(x - 3)$ **5.** $(y - 5)(y + 1)$ **7.** $(x + 5)^2$ **9.** $(2q + 1)(q + 5)$
11. $(3x + 1)(x - 3)$ **13.** $(5c - 1)(c + 7)$ **15.** Prime   **17.** $(2s - 3t)(s - 5t)$
19. $(5x + 2)(x + 3)$ **21.** $(2x - 3)(2x - 1)$ **23.** $(6q + 1)(2q + 3)$ **25.** $(6t - 5u)(t + 2u)$
27. $(4x - 3)(2x + 3)$ **29.** $(4x - 3)(x + 6)$ **31.** $(4n - 5)(2n + 3)$ **33.** $2(x - 3)(x - 8)$
35. $2(2x - 3z)(x + 2z)$ **37.** $2x(x + 1)(x + 2)$ **39.** $a(5x - y)(2x + 5y)$ **41.** $3a(x + 5)(x - 3)$
43. $7a^3(2x + 1)(x - 1)$ **45.** $(x + 1)^2$ **47.** $(x - 4)^2$ **49.** $(2x + 1)^2$ **51.** $(3x - 1)^2$
53. $N(r + 1)^2$ **55.** $2(p - 4)(p - 50)$ **57.** $2(x - 4)(x - 8)$ **59.** $4x(x - 6)^2$
61. $3x(2x - 1)(x - 2)$ **63.** $4(4t + 1)(t - 8)$

Exercises 8-4, page 259

1. $8 = 2^3$ **3.** $64 = 4^3; 27x^3 = (3x)^3$ **5.** $x^6 = (x^2)^3; 8x^6 = (2x^2)^3$ **7.** $a^3x^9 = (ax^3)^3$
9. $x^3 + y^3$ **11.** $x^3 + 8$ **13.** $8a^3 - b^3$ **15.** $x^6 - 8$ **17.** $(a - 1)(a^2 + a + 1)$
19. $(t + 2)(t^2 - 2t + 4)$ **21.** $(1 - x)(1 + x + x^2)$ **23.** $(2x + 3a)(4x^2 - 6ax + 9a^2)$
25. $(2x^2 - y)(4x^4 + 2x^2y + y^2)$ **27.** $(ax - y^2)(a^2x^2 + axy^2 + y^4)$ **29.** $8x(x + 1)(x^2 - x + 1)$
31. $ax^2(1 - y)(1 + y + y^2)$ **33.** $2k(R_1 + 2R_2)(R_1^2 - 2R_1R_2 + 4R_2^2)$
35. $2(3t_1^2 - t_2)(9t_1^4 + 3t_1^2t_2 + t_2^2)$ **37.** $(5s^2 - 4t^3)(25s^4 + 20s^2t^3 + 16t^6)$
39. $(ab + c^5)(a^2b^2 - abc^5 + c^{10})$ **41.** $(1 - ax)(1 + ax)(1 + ax + a^2x^2)(1 - ax + a^2x^2)$
43. $(1 - x - y)[1 + x + y + (x + y)^2]$ **45.** $N(x - y)(x^2 + xy + y^2)$
47. $\frac{4}{3}\pi(r_1 + r_2)(r_1^2 - r_1r_2 + r_2^2)$ **49.** $k(r_1 - r_2)(r_1^2 + r_1r_2 + r_2^2)$
51. $k\pi(r_1 - r_2)(r_1^2 + r_1r_2 + r_2^2)$ **53.** $V(T_1 - T_2)(c_1 - c_2)(c_1^2 + c_1c_2 + c_2^2)$
55. $\frac{nw^2}{24p^2}(L_1 - L_2)(L_1^2 + L_1L_2 + L_2^2)$

Exercises 8-5, page 261

1. $5(a - c)$ **3.** $3a(a + 2)$ **5.** $4ab(3a + 1)$ **7.** $8stu^2(1 - 3s^2)$ **9.** $(2x + y)(2x - y)$
11. $(4y^2 + x)(4y^2 - x)$ **13.** $(x + 1)^2$ **15.** $(x - 1)(x - 6)$ **17.** $a(x^2 + 3ax - a^2)$
19. $2nm(m^2 - 2nm + 3n^2)$ **21.** $4t^2(p^3 - 3t^2 - 1 + a)$ **23.** $2xy^3(1 - 7x + 8y - 3x^2y^2)$
25. $(4rs + 3y)(4rs - 3y)$ **27.** Prime   **29.** $(2x + 7)(x + 1)$ **31.** $(5s + 2)(s - 1)$
33. $(7t + 1)(2t - 3)$ **35.** $(3x + 1)^2$ **37.** $(x + y)(x + 2y)$ **39.** $(5c - d)(2c + 5d)$
41. $(8x + 7)(11x - 12)$ **43.** $2(x + 3y)(x - 3y)$ **45.** $8x^4y^2(xy + 2)(xy - 2)$
47. $3a(x - 3)(x + 4)$ **49.** $3r(6r - 13s)(3r + 4s)$ **51.** $16y^3(y - 1)(y - 3)$
53. $5(x^2 + 5)(x^2 - 5)$ **55.** $(4x^2 + 1)(2x + 1)(2x - 1)$ **57.** $(x + 3)(x^2 - 3x + 9)$
59. $(2x + 1)(4x^2 - 2x + 1)$ **61.** $axy(x - y)(x^2 + xy + y^2)$
63. $(5R_1 + 2aR_2)(25R_1^2 - 10aR_1R_2 + 4a^2R_2^2)$ **65.** $i(R_1 + R_2 + R_3)$ **67.** $P(N + 2)$
69. $k(D + 2r)(D - 2r)$ **71.** $(v_2 - 3v_1)(v_2 - v_1)$ **73.** $\dfrac{(u - 1)^2}{(u + 1)^2}$ **75.** $(T - 10)(T + 530)$
77. $V(T_1 - T_2)(k_1 - k_2)(k_1^2 + k_1k_2 + k_2^2)$ **79.** $\frac{4}{3}\pi(2r_1 + 3r_2)(4r_1^2 - 6r_1r_2 + 9r_2^2)$

Exercises 9-1, page 270

1. $\dfrac{4}{6}; \dfrac{10}{15}$ **3.** $-\dfrac{15}{20}; -\dfrac{-15}{-20}$ **5.** $\dfrac{6a^3x}{2a^2b}; \dfrac{3a^2bx}{ab^2}$ **7.** $\dfrac{-10ax^3}{-2bx}; \dfrac{10a^2x^3}{2abx}$ **9.** $\dfrac{3x}{x^2 - 2x}; \dfrac{3x + 6}{x^2 - 4}$
11. $\dfrac{x^2 - 2xy + y^2}{x^2 - y^2}; \dfrac{x^2 - y^2}{x^2 + 2xy + y^2}$ **13.** $\dfrac{4}{7}$ **15.** $\dfrac{2a}{3a^2}$ **17.** $\dfrac{2}{x + 1}$ **19.** $\dfrac{2x - 1}{x + 1}$ **21.** $\dfrac{3(x - 2)}{x + 2}$
23. $\dfrac{-(2 + x)}{x - 3}$ **25.** $\dfrac{2}{9}$ **27.** $\dfrac{1}{3}$ **29.** $3x$ **31.** $\dfrac{ab}{4}$ **33.** $\dfrac{8}{9}$ **35.** $\dfrac{2t}{7r^2s}$ **37.** $\dfrac{2x - 1}{x - 2}$

39. $\dfrac{(x+1)(x+2)}{2(x+3)}$　　**41.** $\dfrac{x+1}{x-1}$　　**43.** $\dfrac{x}{x+2}$　　**45.** $\dfrac{3x-2}{4x+3}$　　**47.** $\dfrac{x+3y}{3y}$　　**49.** $\dfrac{1-3x}{3x+1}$

51. $\dfrac{5-x}{2+x}$　　**53.** $-3x$　　**55.** $\dfrac{2-x}{3+x}$　　**57.** $\dfrac{3}{7}$　　**59.** All of them

Exercises 9-2, page 277

1. $\dfrac{1}{8n}$; $13s$　　**3.** $\dfrac{3b}{a}$; $\dfrac{a}{3b}$　　**5.** $\dfrac{x-y}{x+y}$; $\dfrac{x^2}{x^2+y^2}$　　**7.** $\dfrac{a+b}{a}$; $-\dfrac{V}{IR}$　　**9.** $\dfrac{4}{9t}$　　**11.** $\dfrac{2a}{15}$　　**13.** $\dfrac{rt}{12}$

15. $\dfrac{81}{256}$　　**17.** $\dfrac{a^{10}}{32x^5}$　　**19.** $\dfrac{a^3x^9}{b^6}$　　**21.** $\dfrac{26}{35cx}$　　**23.** $\dfrac{24mx}{7}$　　**25.** $\dfrac{y}{45x}$　　**27.** $\dfrac{1}{2a^2b^5}$　　**29.** $\dfrac{a+3b}{a+b}$

31. $\dfrac{5x(x-y)}{6}$　　**33.** $\dfrac{x+1}{x+2}$　　**35.** $\dfrac{(x+3)(x-3)}{(x-2)(x-4)}$　　**37.** $\dfrac{5b-2}{10}$　　**39.** $\dfrac{3(a-b)}{(a-2b)(a+b)}$

41. $\dfrac{(s+2)(s+7)}{(s-12)(s+11)}$　　**43.** $\dfrac{(x-1)^2}{3x+2}$　　**45.** $\dfrac{2a(a+b)}{(a-b)(2a+b)}$　　**47.** $\dfrac{81a(2x-3y)(2x+3y)}{(x-y)(x-y)}$

49. $\dfrac{pq}{p+q}$　　**51.** $\dfrac{1}{79f^2c}$　　**53.** $\dfrac{4000}{x}$　　**55.** $I = \dfrac{2t(t+1)}{3(2t+1)}$

Exercises 9-3, page 283

1. 18　　**3.** 36　　**5.** $12a$　　**7.** $40t$　　**9.** $90y$　　**11.** $9x^2$　　**13.** $8x^2$　　**15.** $420ax$　　**17.** $375ax^2$
19. $75a^3$　　**21.** $96a^3b^3$　　**23.** $15a^2$　　**25.** $60a^2cx^3$　　**27.** $8x(x-1)$　　**29.** $3a(a+3)$
31. $6ax(a-3)$　　**33.** $6x(x-y)(x+y)$　　**35.** $(a-2b)(a+2b)(a+b)$　　**37.** $12(x-3)(x+3)^2$
39. $2(x-3y)(x+3y)(3x+2y)$

Exercises 9-4, page 287

1. $\dfrac{20}{36a} - \dfrac{21}{36a}$　　**3.** $\dfrac{5b}{abx} + \dfrac{a}{abx} - \dfrac{4bx}{abx}$　　**5.** $\dfrac{8x}{2x^2(x-1)} - \dfrac{3}{2x^2(x-1)}$

7. $\dfrac{x(x+2)^2}{2(x-2)(x+2)^2} + \dfrac{10(x+2)}{2(x-2)(x+2)^2} - \dfrac{6x(x-2)}{2(x-2)(x+2)^2}$　　**9.** $\dfrac{19}{10x}$　　**11.** $\dfrac{2b-5a}{3ab}$　　**13.** $\dfrac{6x+3}{x^2}$

15. $\dfrac{26b-25}{40b}$　　**17.** $\dfrac{8y-b}{by^2}$　　**19.** $\dfrac{9+2x^2}{3x^3y}$　　**21.** $\dfrac{2xy+5x^2-3}{x^2y}$　　**23.** $\dfrac{42yz-15xz+2xy}{12xyz}$

25. $\dfrac{4(2a-1)}{(a-2)(a+2)}$　　**27.** $\dfrac{2x^2+3x+9}{4(x-3)(x+3)}$　　**29.** $\dfrac{y^2-6y-3}{3(y-3)(y+3)}$　　**31.** $\dfrac{-(x+6)}{2(2x+3)}$

33. $\dfrac{3x^2-17x+14}{(2-3x)(2+3x)}$　　**35.** $\dfrac{4+15x-5x^2}{(x-2)(x+2)}$　　**37.** $\dfrac{2x^2+3x-125}{3(x+5)(x-5)}$　　**39.** $\dfrac{7p^2+4p+8q-175q^2}{8(p-5q)^2}$

41. $\dfrac{R_1+R_2}{R_1R_2}$　　**43.** $\dfrac{41}{24}$ in^2　　**45.** $\dfrac{120-60x^2+5x^4-x^6}{120}$　　**47.** $\dfrac{R_2-R_1}{R_1}$　　**49.** $\dfrac{II_0-I_r-I_t}{I_0}$

51. $\dfrac{g_m+8}{g_m^2}$　　**53.** $\dfrac{p^2-2gm^2rM}{2mr^2}$　　**55.** $\dfrac{h_1P_1^2-h_2P_2^2}{(h_1+h_2)(h_1-h_2)}$

Exercises 9-5, page 296

1. 4　　**3.** -3　　**5.** 10　　**7.** $-\dfrac{3}{4}$　　**9.** $1-3b$　　**11.** $4a+2$　　**13.** $\dfrac{2a-4}{3a^2}$　　**15.** $-\dfrac{16}{7b}$　　**17.** 2

19. $\dfrac{7}{3}$　　**21.** $\dfrac{3}{4}$　　**23.** No solution　　**25.** $\dfrac{-b}{3b-1}$　　**27.** $\dfrac{3}{2n-4}$　　**29.** $\dfrac{52}{11}$　　**31.** $\dfrac{pf}{p-f}$

33. $\dfrac{2D_p}{D_0-D_p}$　　**35.** $\dfrac{2\pi^2r^2(P-p)}{m^2}$　　**37.** \$2000　　**39.** 1.8 mi　　**41.** 1.7 h　　**43.** 240 h

Exercises 9-6, page 298

1. $\dfrac{3rt^4}{s^3}$ **3.** $\dfrac{a}{3bc^2}$ **5.** $\dfrac{4}{x-2y}$ **7.** $\dfrac{p+q}{3+2p^2}$ **9.** $\dfrac{a}{2b}$ **11.** $\dfrac{3x+y}{2x-y}$ **13.** 18 **15.** $12t^2$

17. $48b^2t$ **19.** $20x^2(x-2)$ **21.** $\dfrac{10a}{3x^2}$ **23.** $\dfrac{15y}{4x}$ **25.** $\dfrac{6}{a}$ **27.** $\dfrac{2bu}{av}$ **29.** $\dfrac{10b-3a}{5a^2b}$

31. $\dfrac{10cd+c-6}{2c^2d}$ **33.** $\dfrac{8}{27}$ **35.** $\dfrac{a^4x^4}{81y^8}$ **37.** $\dfrac{2}{x(x+1)}$ **39.** $\dfrac{x-5}{4}$ **41.** $a(a-1)$

43. $(3x+2y)(y-2x)$ **45.** $\dfrac{5x+9}{x^2(x+3)}$ **47.** $\dfrac{(x-1)(x-3)}{(x-2)^2}$ **49.** $\dfrac{-(2x+3)^2}{(2x+5)(x-3)(x+3)}$

51. $\dfrac{-2x^3+9x^2-43x+15}{x(x+5)(x-5)(2x-1)}$ **53.** $\dfrac{2}{5}$ **55.** $\dfrac{(x+1)(x-3)}{(x-2)(x+3)}$ **57.** 9 **59.** $\dfrac{a+6}{4(b-a)}$

61. No solution **63.** $-\dfrac{8}{21}$ **65.** $\dfrac{nle^2E}{2mv}$ **67.** $\dfrac{3r-h}{12r^3}$ **69.** $\dfrac{g_2^2-2g_1g_2+g_1^2}{g_2^2}$

71. $\dfrac{C^2L^2w^4-2LCw^2+1}{C^2w^2}$ **73.** $\dfrac{\mu R}{r+\mu R+R}$ **75.** $Prt+P$ **77.** $\dfrac{q_2D-df}{d+D}$ **79.** $\dfrac{2akmM}{amv^2+kmM}$

81. 3.4 min **83.** \$150,000

Exercises 10-1, page 306

1. $\dfrac{1}{t^5}$ **3.** $\dfrac{1}{x^4}$ **5.** x^3 **7.** R_1^3 **9.** $\dfrac{3}{c^2}$ **11.** $\dfrac{c}{3}$ **13.** 1 **15.** 1 **17.** 5 **19.** $9y^2$

21. 3^6 **23.** 6 **25.** $\dfrac{1}{ax}$ **27.** $\dfrac{2}{c^8}$ **29.** $\dfrac{y^2}{x^6}$ **31.** $\dfrac{x^2}{125}$ **33.** $\dfrac{s}{t^2}$ **35.** $\dfrac{x^4}{8y^4}$ **37.** $\dfrac{b^7}{9a}$

39. $\dfrac{4}{25a^2b^2}$ **41.** $\dfrac{y^5}{x^4}$ **43.** $\dfrac{a^5c^2}{18}$ **45.** $\dfrac{1}{R_1}+\dfrac{1}{R_2}$ **47.** $N=\dfrac{N_0}{e^{kt}}$ **49.** $\dfrac{R}{hR+1}$

51. $g\cdot cm^{-3}; m\cdot s^{-2}$ **53.** $362\ Btu/h\cdot ft^2$ **55.** $\dfrac{rR}{R+r}$

Exercises 10-2, page 312

1. 5 **3.** -4 **5.** 4,000,000 **7.** 0.08 **9.** 2.17 **11.** 0.00365 **13.** 3×10^3
15. 7.6×10^{-2} **17.** 7.04×10^{-1} **19.** 9.21×10^0 **21.** 5.3×10^{-5} **23.** 2.01×10^9
25. 1.55×10^8 **27.** 9.30×10^{-3} **29.** 4.65×10^{-5} **31.** 6.740×10^{-9} **33.** 1.30×10^1
35. 1.26×10^7 **37.** $6.5\times10^6\ g$ **39.** $4.5\times10^{-2}\ m$ **41.** $3.92\times10^{-2}\ L$ **43.** $8.06\times10^{-5}\ s$
45. $2\times10^9\ Hz$ **47.** $9.1\times10^{-28}\ g$ **49.** 4,000,000,000,000,000,000,000,000,000,000 lb
51. 0.000001 in. **53.** $3.6\times10^8\ km^2$ **55.** 0.0000000000016 W **57.** $3.6\times10^7\ mi$
59. 0.000000000001 W/cm^2 **61.** $6\times10^{-19}\ J$ **63.** 30,000,000,000 cm/s **65.** $2.59\times10^{10}\ cm^2$
67. $6.06\times10^7\ Hz$ **69.** 4.90×10^{19} **71.** 6.9×10^{10}

Exercises 10-3, page 317

1. 7 **3.** -12 **5.** 0.4 **7.** -0.2 **9.** 20 **11.** -40 **13.** 2 **15.** -2 **17.** -5
19. 0.5 **21.** 2 **23.** 3 **25.** 2 **27.** $2j$ **29.** $-20j$ **31.** $0.7j$ **33.** -25 **35.** -20
37. -12 **39.** 64 **41.** Rational, irrational, rational, rational, imaginary
43. Irrational, rational, imaginary, irrational, rational **45.** 6.00 ft **47.** 4.71 s **49.** $\frac{4}{3}$ **51.** 0.1

Exercises 10-4, page 322

1. $\dfrac{\sqrt{2}}{2}$ **3.** $\dfrac{2\sqrt{5}}{5}$ **5.** $\dfrac{\sqrt{a}}{a}$ **7.** $\dfrac{\sqrt{ab}}{b}$ **9.** $\dfrac{\sqrt{15}}{5}$ **11.** $\dfrac{a\sqrt{3a}}{3}$ **13.** $2\sqrt{3}$ **15.** $2\sqrt{7}$

17. $3\sqrt{5}$ **19.** $5\sqrt{6}$ **21.** $7\sqrt{3}$ **23.** $9\sqrt{3}$ **25.** $c\sqrt{a}$ **27.** $ab\sqrt{a}$ **29.** $2ac\sqrt{bc}$

31. $4x^2z^2\sqrt{5yz}$ **33.** $\dfrac{b\sqrt{3a}}{6}$ **35.** $\dfrac{x\sqrt{10y}}{5a^4}$ **37.** $3\sqrt[3]{2}$ **39.** $2a\sqrt[3]{a}$ **41.** $2a^2\sqrt[4]{a}$

43. $3a^2\sqrt[4]{3a^3}$ **45.** $3a^2x^3\sqrt[4]{2a^2}$ **47.** $2rs^2t^2\sqrt[7]{2t^2}$ **49.** $\dfrac{V\sqrt{gl}}{gl}$ **51.** $\dfrac{2\sqrt{3.14A}}{3.14}$

53. 80 ft, 160 ft   **55.** 107.5, 112.5   **57.** $V = \dfrac{k\sqrt[3]{PW^2}}{W}$ **59.** $d = \dfrac{k\sqrt[3]{16JC^2}}{C}$

Exercises 10-5, page 328

1. $2\sqrt{7}$ **3.** $4\sqrt{7} + \sqrt{5}$ **5.** $5\sqrt{3}$ **7.** $9\sqrt{10}$ **9.** 10   **11.** $5\sqrt{3}$ **13.** $9\sqrt{2}$ **15.** $\sqrt{7}$

17. $-\sqrt{2} - 4\sqrt{3}$ **19.** $4\sqrt{a}$ **21.** $(3 + 4a)\sqrt{2a}$ **23.** $7a\sqrt{2} - 2\sqrt{3a}$ **25.** $\sqrt{21} - 9\sqrt{2}$

27. 26   **29.** $a\sqrt{b} + 3a\sqrt{c}$ **31.** $1 + \sqrt{6}$ **33.** $-17 - 3\sqrt{15}$ **35.** $2a - \sqrt{ac} - 15c$

37. $7 + 4\sqrt{3}$ **39.** $\dfrac{\sqrt{6} + 2}{2}$ **41.** 7   **43.** $\dfrac{7 - \sqrt{21}}{4}$ **45.** $\dfrac{4 + 3\sqrt{2}}{2}$ **47.** $\dfrac{a - 2\sqrt{ab}}{a - 4b}$

49. $\dfrac{R_1\sqrt{R_2} - R_2\sqrt{R_1}}{R_1 - R_2}$ **51.** $2a$ **53.** $\dfrac{13 - 2\sqrt{30}}{7}$ **55.** $75\sqrt{2}$ ft   **57.** 8.46   **59.** 0.473

Exercises 10-6, page 334

1. $\sqrt{5}$ **3.** $\sqrt[4]{a}$ **5.** $\sqrt[5]{x^3}$ **7.** $\sqrt[3]{R^7}$ **9.** $a^{1/3}$ **11.** $x^{1/2}$ **13.** $x^{2/3}$ **15.** $b^{8/5}$ **17.** 3

19. 4   **21.** 2   **23.** -2 **25.** 8   **27.** 16   **29.** 27   **31.** 4   **33.** $\dfrac{1}{6}$ **35.** $\dfrac{1}{2}$ **37.** $\dfrac{1}{2}$

39. 18   **41.** 2   **43.** 2   **45.** $\dfrac{1}{48}$ **47.** $-\dfrac{1}{14}$ **49.** $a^{3/2}$ **51.** $a^{3/4}b$ **53.** $x^{29/15}$ **55.** $x^{5/6}$

57. $x^{1/5} + y^{1/5} = k$ **59.** $x = A^{1/2}$ **61.** $\dfrac{(c^2 - v^2)^{1/2}}{c}$ **63.** $r_p = \dfrac{2}{3DK^{2/3}I^{1/3}}$ **65.** 4.72×10^{22}

67. $B = \dfrac{2\pi kI}{R}$ **69.** 2.1   **71.** 6.7   **73.** 24.5   **75.** 10.3

Exercises 10-7, page 336

1. $\frac{1}{10}$ **3.** 9   **5.** $\frac{1}{6}$ **7.** 13   **9.** 5   **11.** -16 **13.** $\frac{1}{4}$ **15.** $-\frac{3}{11}$ **17.** 10   **19.** 10

21. 343   **23.** 1331   **25.** $\frac{8}{9}$ **27.** 324   **29.** $9j$ **31.** $-0.8j$ **33.** 5.70×10^2

35. 3.25×10^0 **37.** 7.69×10^{-4} **39.** 8.695×10^1 **41.** 3000   **43.** 314,000   **45.** $\dfrac{3b}{a^2}$

47. $\dfrac{m^4}{n^2}$ **49.** $\dfrac{2x^5}{3y^3}$ **51.** $-\dfrac{x^2y^3}{a^2}$ **53.** $a^{7/12}$ **55.** $a^{7/6}$ **57.** $\dfrac{x^{1/2}}{y}$ **59.** $s^{2/3}t^{7/3}$ **61.** $2\sqrt{11}$

63. $6\sqrt{2}$ **65.** $8\sqrt{2}$ **67.** $2\sqrt[3]{5}$ **69.** $\dfrac{\sqrt{11}}{11}$ **71.** $\dfrac{2\sqrt{7}}{7}$ **73.** $2a$ **75.** $5b\sqrt{5c}$ **77.** $\dfrac{\sqrt{6a}}{a}$

79. $\dfrac{2\sqrt{21a}}{3a}$ **81.** $2a\sqrt[3]{2}$ **83.** $2a\sqrt[5]{2a^3}$ **85.** $-\sqrt{7}$ **87.** $5\sqrt{7} - 2\sqrt{6}$ **89.** $\sqrt{70} - 5\sqrt{10}$

91. $(6 - 3a)\sqrt{2}$ **93.** $-6\sqrt{3}$ **95.** $2\sqrt{15} - 30$ **97.** $a\sqrt{b} - 3\sqrt{5ab}$ **99.** $27 - 5\sqrt{30}$

101. $81 - 17\sqrt{42}$ **103.** $2a - 3b - 5\sqrt{ab}$ **105.** $\dfrac{2 + \sqrt{10}}{3}$ **107.** $5\sqrt{2} - 7$

109. 2.75×10^{10} Hz   **111.** 2×10^{-7} in.   **113.** 0.00000000000000000016 C

115. 0.0000000000005 m   **117.** 9800 J   **119.** $\dfrac{r_1 r_2}{(\mu - 1)(r_2 - r_1)}$ **121.** $v = \dfrac{\sqrt{2emV}}{m}$ **123.** $\dfrac{21}{8}$ in.

125. 6.4 V **127.** 288 **129.** $\dfrac{\sqrt{3}}{3}$ **131.** 0.182 ft **133.** 7.478×10^{18}
135. 1.0045×10^2 **137.** 2.67 **139.** 9.37 **141.** 28.77 **143.** 0.000842

Exercises 11-1, page 343

1. $4x^2 - 3x + 2 = 0$ **3.** $-x^2 + 8x = 0$ **5.** $x^2 = 0$ **7.** $x^2 - x - 1 = 0$
9. $x^2 - 7x - 4 = 0$; $a = 1, b = -7, c = -4$ **11.** Not quadratic
13. $x^2 + 4x + 4 = 0$; $a = 1, b = 4, c = 4$ **15.** $7x^2 - x = 0$; $a = 7, b = -1, c = 0$
17. Not quadratic **19.** $-6x^2 + 3x - 1 = 0$; $a = -6, b = 3, c = -1$ **21.** 2, 3 **23.** 2
25. $-1, 2$ **27.** $\dfrac{1}{2}, 1$ **29.** $3, -4$ **31.** $3x^2 - 10x + 8 = 0$; none of the listed values
33. $s^2 - 4s + 4 = 0$; 2 **35.** $n^2 - 9 = 0$; $-3, 3$ **37.** $-x^2 + 6x - 5 = 0$
39. $16t^2 - 96t + 128 = 0$ **41.** $w(w + 2) = 48$ becomes $w^2 + 2w - 48 = 0$
43. $4\pi r^2 - \pi r^2 = 4$ becomes $3\pi r^2 - 4 = 0$

Exercises 11-2, page 349

1. $3, -3$ **3.** $-2, 1$ **5.** $-2, 5$ **7.** $-2, \frac{1}{3}$ **9.** $-2, \frac{1}{2}$ **11.** $-\frac{5}{2}, \frac{1}{3}$ **13.** $-\frac{1}{2}, \frac{2}{3}$ **15.** $\frac{1}{5}, 4$
17. $-1, \frac{9}{2}$ **19.** $-2, -2$ **21.** $0, 8$ **23.** $-\frac{7}{2}, \frac{1}{2}$ **25.** $\frac{7}{2}, \frac{7}{2}$ **27.** $0, 7$ **29.** $-\frac{1}{3}, 3$
31. $-2a, 2a$ **33.** $2, 17$ **35.** $-6, -2$ **37.** $192°F$ **39.** $e = 5$ cm **41.** 1000 mi/h
43. 3 ft, 4 ft, 7 ft

Exercises 11-3, page 357

1. $3, -1$ **3.** $-\dfrac{1}{2}, -3$ **5.** $\dfrac{-5 \pm \sqrt{13}}{2}$ **7.** $2 \pm \sqrt{6}$ **9.** $\dfrac{1}{2}, \dfrac{3}{2}$ **11.** $\dfrac{4}{3}, -\dfrac{4}{3}$ **13.** $-\dfrac{1}{2}, \dfrac{7}{2}$
15. $-1 + 2j, -1 - 2j$ **17.** $\dfrac{1 \pm \sqrt{33}}{2}$ **19.** $\dfrac{-1 \pm \sqrt{33}}{4}$ **21.** $0, 7$ **23.** $\dfrac{3 \pm \sqrt{55}\,j}{4}$
25. $-\dfrac{2}{3}, 4$ **27.** $-\dfrac{1}{3a}, -\dfrac{3}{2a}$ **29.** $-3, 1$ **31.** $-1, -\dfrac{1}{2}$ **33.** 16, 17 **35.** 12 cm by 17 cm
37. 8.50 s, 0.441 s **39.** 0.382 atm **41.** ± 4 A **43.** 64.9 m **45.** $-0.34, -1.94$
47. $4.20, -0.05$

Exercises 11-4, page 358

1. $-3, -4$ **3.** $6, -\dfrac{1}{2}$ **5.** $-\dfrac{1}{6}, 6$ **7.** $\dfrac{3}{4}, \dfrac{3}{4}$ **9.** $0, -\dfrac{7}{5}$ **11.** $-11, 10$ **13.** $3, -8$
15. $-3, -3$ **17.** $\dfrac{5}{6}, 1$ **19.** $1, 1$ **21.** $-2 \pm \sqrt{2}$ **23.** $\dfrac{-5 \pm \sqrt{33}}{4}$ **25.** $3 \pm \sqrt{15}$
27. $\dfrac{-1 \pm j\sqrt{39}}{4}$ **29.** $-2j, 2j$ **31.** $-3, 3$ **33.** $1, -9$ **35.** $\dfrac{1 \pm \sqrt{11}}{2}$ **37.** $1 \pm 2\sqrt{2}$
39. $\dfrac{1 \pm \sqrt{41}}{5}$ **41.** $2 \pm \sqrt{3}$ **43.** $\dfrac{1 \pm j\sqrt{7}}{4}$ **45.** $-1 \pm j$ **47.** 27 **49.** 4 in., 5 in.
51. $-r \pm \sqrt{r^2 - k^2}$ **53.** 6 s **55.** $\dfrac{-v \pm \sqrt{v^2 + 2as}}{a}$ **57.** 5, 20 **59.** 7 mm **61.** 33 poles
63. 300 mi/h **65.** $1, -0.35$ **67.** $0.98, -0.30$ **69.** 6.1199 cm **71.** 5.16 ft

Exercises 12-1, page 365

1. $\log 100 = 2$ **3.** $\log 0.01 = -2$ **5.** $\log 2884 = 3.4600$ **7.** $\log 0.0003594 = -3.4444$
9. $\log_2 1024 = 10$ **11.** $\log_6 216 = 3$ **13.** $10^1 = 10$ **15.** $10^3 = 1000$ **17.** $10^{-2} = 0.01$
19. $10^{2.7536} = 567$ **21.** $2^3 = 8$ **23.** $3^5 = 243$ **25.** 4   **27.** -3 **29.** 6   **31.** -3
33. 4   **35.** 3   **37.** 0.8609   **39.** 1.9713   **41.** 4.9191   **43.** 4.6990   **45.** $8.7267 - 10$
47. $8.4843 - 10$ **49.** 6   **51.** 2   **53.** 625   **55.** 10   **57.** 2   **59.** 100,000   **61.** 6.1271
63. 4.3010   **65.** 9.9868   **67.** -7.7959

Exercises 12-2, page 372

1. 46.3   **3.** 54.7   **5.** 12,600   **7.** 115,000   **9.** 2.19   **11.** 1.36   **13.** 0.0296   **15.** 12.9
17. 3.74   **19.** 4.12   **21.** 20.8   **23.** 0.243   **25.** 5.18   **27.** 2.38   **29.** 1.65   **31.** 20.1
33. 29.2 in^2 **35.** 8.4   **37.** 85.4 ft^3 **39.** 2.13 V   **41.** 26.7 bu/acre   **43.** 332 m/s
45. 3,010,000 ft   **47.** 1.29   **49.** 4.56×10^{192} **51.** 3.27×10^{150} **53.** 1.34×10^{154}
55. 3.76×10^{414}

Exercises 12-3, page 378

1. 2.1282   **3.** -0.1054 **5.** 5.2983   **7.** -0.9808 **9.** -6.9078 **11.** 8.2161   **13.** 9.90
15. 304.2   **17.** 28.1   **19.** 5.46×10^{18} **21.** 2.7080   **23.** 4.8282   **25.** 4.3944   **27.** 7.6132
29. 3.0910   **31.** 15,000 yr   **33.** 3.04 yr   **35.** 674 yr   **37.** 0.0927 s   **39.** 8.50%
41. 6.3363   **43.** 248.4547   **45.** 317   **47.** 0.6709

Exercises 12-4, page 382

1. 0.8528   **3.** -0.8147 **5.** 1.4050   **7.** 4.4970   **9.** 3.025   **11.** 70.34   **13.** 3357
15. 2.862×10^{-5} **17.** 1443 m/s   **19.** 1481 m/s   **21.** $4°$ **23.** $46°$ **25.** 445 mg
27. 228 mg   **29.** 45 days   **31.** 131 days   **33.** 46.77   **35.** 3.483×10^7 **37.** 106.2 lb
39. 1.970×10^8 mi^2

Exercises 12-5, page 384

1. $\log 10,000 = 4$ **3.** $\log_6 1296 = 4$ **5.** $\log 10 = 1$ **7.** $\log 0.00288 = -2.54$ **9.** $2^7 = 128$
11. $10^5 = 100,000$ **13.** $e^{3.0} = 20$ **15.** $4^5 = 1024$ **17.** 7.14   **19.** 1.79   **21.** 1.77×10^{-5}
23. 153.9   **25.** 3.43   **27.** 30.9   **29.** 7.67×10^4 **31.** 9.716   **33.** 1260   **35.** 3.35×10^7
37. 1462   **39.** 4.81   **41.** 1.05   **43.** 36.17   **45.** 692   **47.** 35.2   **49.** 1.4110
51. $9.4892 - 10$ **53.** 1.7917   **55.** 0.4055   **57.** 3.7478   **59.** 4.5949   **61.** 2.59
63. 140 dB   **65.** 50%   **67.** 1.0×10^{-7} **69.** 4.60   **71.** 152 mi   **73.** \$13,900   **75.** 162 ft/s
77. 736.0 m^2 **79.** 0.0200 s   **81.** 1.5750   **83.** 2.7840   **85.** 221.651   **87.** 3.83604
89. 1.26×10^{120} **91.** 1.04×10^{159}

Exercises 13-1, page 392

1. y is the dependent variable; x is the independent variable
3. p is the dependent variable; V is the independent variable
5. Multiply the value of the independent variable by 3
7. Subtract the square of the independent variable from 2   **9.** $f(x) = 5 - x$ **11.** $f(t) = t^2 - 3t$
13. 0, 3   **15.** 7, -5 **17.** -2, $-\frac{5}{4}$ **19.** -1, 2.91   **21.** -1, 8   **23.** 2, 2   **25.** $\frac{1}{2}$, undefined
27. $a^2 - 2a^4;\dfrac{a - 2}{a^2}$ **29.** Function   **31.** Not a function   **33.** $c = 2\pi r$ **35.** $A = \dfrac{p^2}{16}$

37. $t = 0.02n$ **39.** $s = 3v$ **41.** $R_1 = R_2 + 1500$ **43.** $i = \dfrac{V}{5}$

Exercises 13-2, page 395

1. $A(2, 1)$, $B(-2, 3)$ **3.** $E(4, 0)$, $F(-2, 1)$ **5.** $I(1, 5.5)$, $J(3, 5.5)$ **7.** $M(-9.5, 0)$, $N(-9.5, 2)$

9.   **11.**   **13.**   **15.**

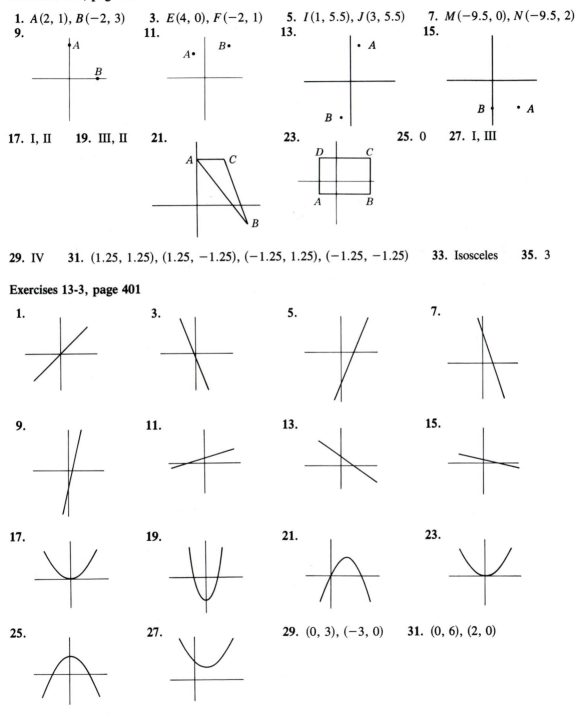

17. I, II   **19.** III, II   **21.**   **23.**   **25.** 0   **27.** I, III

29. IV   **31.** $(1.25, 1.25)$, $(1.25, -1.25)$, $(-1.25, 1.25)$, $(-1.25, -1.25)$ **33.** Isosceles   **35.** 3

Exercises 13-3, page 401

1.   **3.**   **5.**   **7.**

9.   **11.**   **13.**   **15.**

17.   **19.**   **21.**   **23.**

25.   **27.**   **29.** $(0, 3)$, $(-3, 0)$ **31.** $(0, 6)$, $(2, 0)$

33.

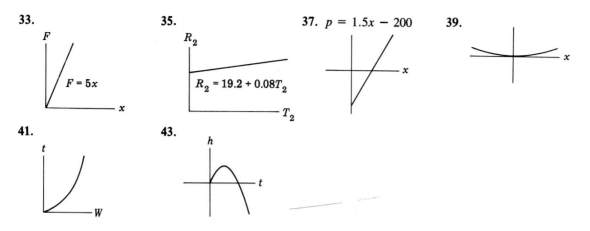

$F = 5x$

35.

$R_2 = 19.2 + 0.08T_2$

37. $p = 1.5x - 200$

39.

41.

43.

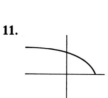

Exercises 13-4, page 405

1.

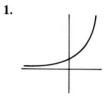

3.

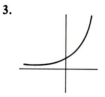

5.

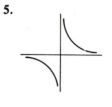

7.

9.

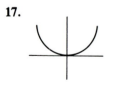

11.

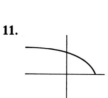

13.

15.

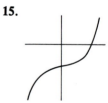

17.

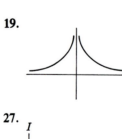

19.

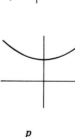

21.

23.

25.

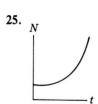

27.

29.

31.

33.

35.

Exercises 13-5, page 410

1. 1.4, −1.2   **3.** 9, −8   **5.** 6.5, 11.6   **7.** 9.2, 1.7   **9.** 8.9, 2.1   **11.** −2.1, 5.0   **13.** 0.7
15. −1.2, 1.7   **17.** 1.7   **19.** 4.2   **21.** 3.5   **23.** 1.5   **25.** 0.0065 C, 0.0001 C
27. 35 g, 70 g   **29.** 340 m/s, 346 m/s, 354 m/s, 360 m/s   **31.** 800 Ω, 78 Ω, 15 Ω, 11 Ω
33. 28 lb   **35.** 2.1 in.

Exercises 13-6, page 416

1.

3.

5.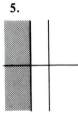

7.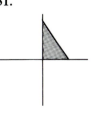

9.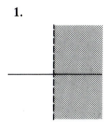

11.

13.

15.

17.

19.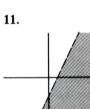

21.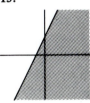

23.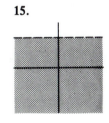

25.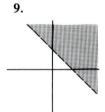

27.

29.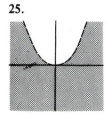

31.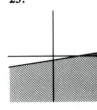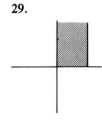

33.

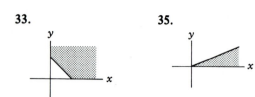

35.

Exercises 13-7, page 419

1.

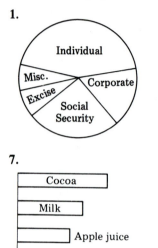

3.

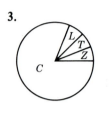

5.

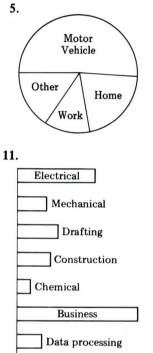

7.

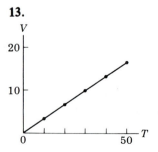

9.

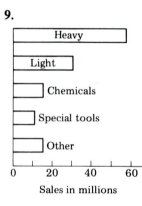

11.

13.

15.

17.

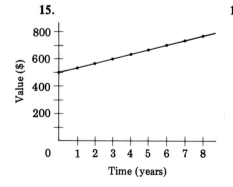

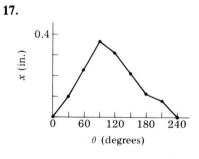

19.

21.

23.

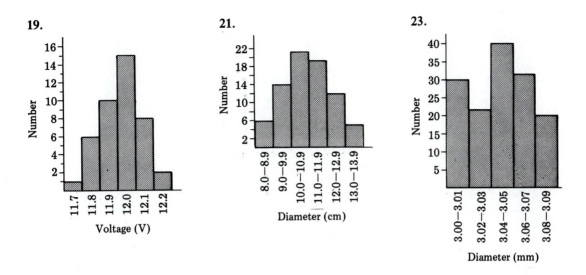

Exercises 13-8, page 422

1. 3, 2   **3.** 4, $\frac{5}{3}$ **5.** 3, $-\frac{1}{2}$ **7.** 12, -3.36 **9.** 0, 8   **11.** 1, 5

13.

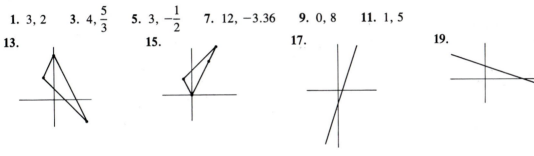

15.

17.

19.

21.

23.

25.

27.

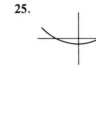

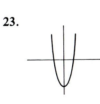

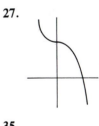

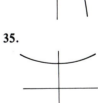

29.

31.

33.

35.

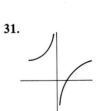

37.   **39.**   **41.**   **43.**

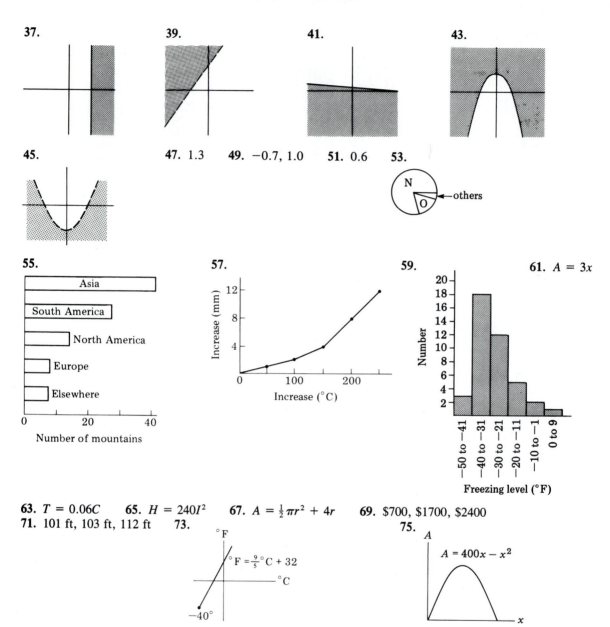

45.   **47.** 1.3   **49.** −0.7, 1.0   **51.** 0.6   **53.**

N
O ←others

55.   **57.**   **59.**   **61.** $A = 3x$

55.

Asia

South America

North America

Europe

Elsewhere

0   20   40

Number of mountains

57.

Increase (mm)

12

8

4

0   100   200

Increase (°C)

59.

Number

20
18
16
14
12
10
8
6
4
2

−50 to −41   −40 to −31   −30 to −21   −20 to −11   −10 to −1   0 to 9

Freezing level (°F)

63. $T = 0.06C$ **65.** $H = 240I^2$ **67.** $A = \frac{1}{2}\pi r^2 + 4r$ **69.** \$700, \$1700, \$2400
71. 101 ft, 103 ft, 112 ft   **73.**   **75.**

°F

$F = \frac{9}{5}°C + 32$

°C

−40°

A

$A = 400x - x^2$

x

Exercises 14-1, page 431

1. $x = 3.0, y = 0.0$ **3.** $x = 2.0, y = 1.0$ **5.** $r = 4.0, x = -3.0$ **7.** $x = 2.0, y = 2.0$
9. $x = -1.0, y = 2.0$ **11.** $x = -2.0, y = 1.5$ **13.** $a = -1.9, b = -0.5$ **15.** Inconsistent
17. $p = 2.4, q = 1.1$ **19.** $x = 1.9, y = -2.2$ **21.** $x = 5.9, y = -0.2$ **23.** $x = 0.8, y = 3.6$
25. Inconsistent   **27.** Dependent   **29.** Dependent   **31.** Inconsistent   **33.** $x = 95, y = 25$
35. $x = 33$ lb, $y = 67$ lb   **37.** $r_1 = 3.5$ mi/h, $r_2 = 1.5$ mi/h   **39.** $R_1 = 2.5\ \Omega, R_2 = 1.0\ \Omega$

Exercises 14-2, page 436

1. $x = 5, y = 8$　　**3.** $x = 6, y = 4$　　**5.** $x = \frac{2}{5}, y = \frac{3}{5}$　　**7.** $x = -1, y = 2$
9. $x = 2, y = 1$　　**11.** $x = 0, y = 0$　　**13.** $x = 6, y = 0$　　**15.** $u = -\frac{6}{17}, k = \frac{20}{17}$
17. $x = \frac{81}{10}, y = \frac{63}{10}$　　**19.** $x = -2, y = -1$　　**21.** $r = 4, s = -2$　　**23.** $x = \frac{32}{31}, k = -\frac{69}{31}$
25. $x = 1000, y = 200$　　**27.** $r_1 = 1.25, r_2 = 0.75$　　**29.** $V_1 = 40$ V, $V_2 = 20$ V
31. $b = 24$ ft, $r = 13$ ft　　**33.** $x = -0.995, y = 1.36$　　**35.** $x = -0.2268, y = 1.873$

Exercises 14-3, page 441

1. $x = 5, y = 2$　　**3.** $x = 2, y = 1$　　**5.** $m = 4, n = 8$　　**7.** $d = -1, t = 4$　　**9.** $x = 1, n = 5$
11. $x = 9, y = 2$　　**13.** Inconsistent　　**15.** $a = 1, b = 2$　　**17.** $p = \frac{3}{2}, q = -\frac{1}{12}$
19. $x = -14, y = -3$　　**21.** $m = -4, n = 6$　　**23.** $x = 1, y = 1$　　**25.** $V_1 = 6.0$ V, $V_2 = 1.5$ V
27. $x = \frac{1}{16}, y = 1$　　**29.** $c_1 = 10$ ft, $c_2 = 4$ ft　　**31.** 650 math, 500 verbal　　**33.** $m = 15$ L, $c = 5$ L
35. $x = 18$ lb, $y = 10$ lb　　**37.** $x = -17.0, y = -22.5$　　**39.** $x = 3.666, y = -0.4703$

Exercises 14-4, page 447

1. -2　　**3.** 31　　**5.** 0　　**7.** -2　　**9.** 6　　**11.** 52　　**13.** $x = 1, y = 5$　　**15.** $x = 2, y = 3$
17. $s = 4, t = 3$　　**19.** $v_1 = 5, v_2 = 3$　　**21.** $x = -6, y = 7$　　**23.** $x = 12, y = 9$
25. $x = -12, y = -10$　　**27.** $x = 3, y = \frac{1}{2}$　　**29.** $x = 0.2, y = 0.3$　　**31.** $x = \frac{1}{2}, y = \frac{1}{4}$
33. $R_1 = 24,000\ \Omega, R_2 = 8000\ \Omega$　　**35.** $x = 5.3, y = 10.7$　　**37.** $f = \$4.50, b = \$0.75$
39. $R_1 = 40.2\ \Omega, R_2 = 15.5\ \Omega$　　**41.** $x = 1.43, y = -0.653$　　**43.** $x = 0.006088, y = 0.001498$

Exercises 14-5, page 452

1. $x = 1, y = 2, z = 3$　　**3.** $x = -1, y = 2, z = 2$　　**5.** $x = 1, y = -1, z = 2$
7. $x = 5, y = 4, z = -1$　　**9.** $x = 4, y = -1, z = 3$　　**11.** $x = 1, y = 1, z = 1$
13. $x = 8, y = 5, z = -2$　　**15.** $x = 7, y = -1, z = -2$　　**17.** $x = 10, y = 12, z = -8$
19. $x = 1, y = 2, z = -8$　　**21.** $x = 4, y = 8, z = 12$　　**23.** $x = 4, y = 6, z = -3$
25. $I_1 = 1$ A, $I_2 = 5$ A, $I_3 = 6$ A　　**27.** $x = 10$ cm³, $y = 50$ cm³, $z = 20$ cm³
29. $r_1 = 12$ mi/gal, $r_2 = 16$ mi/gal, $r_3 = 24$ mi/gal　　**31.** $x = 0.05, y = 0.08, z = 0.06$

Exercises 14-6, page 460

1. 8 A, -3 A　　**3.** 10.5 ft from one end　　**5.** 75 h, 30 h　　**7.** 8 days, 14 days
9. \$35,000 at 8%; \$10,000 at 6%　　**11.** 400 ft by 500 ft　　**13.** 90,000; 60,000　　**15.** 1 yr; 1.5 yr
17. 42%　　**19.** 5 Ω, 9 Ω　　**21.** 8 h, 10 h　　**23.** 16, 48　　**25.** 80 g, 40 g　　**27.** 15.8 lb
29. 30 A, 10 A, 80 A　　**31.** 20 gal/min, 60 gal/min, 10 gal/min

Exercises 14-7, page 462

1. $x = \frac{12}{5}, y = \frac{6}{5}$　　**3.** $x = 4, y = 4$　　**5.** $x = 4, y = 4$　　**7.** $p = \frac{18}{5}, q = \frac{2}{5}$　　**9.** $u = 1, v = -1$
11. $a = 12, b = -3$　　**13.** $x = -9, y = 11$　　**15.** $y = 1, z = -\frac{1}{2}$　　**17.** Dependent
19. $x = -\frac{7}{3}, y = 2$　　**21.** $x = \frac{43}{19}, y = -\frac{22}{19}$　　**23.** $s = \frac{94}{107}, t = -\frac{22}{107}$　　**25.** $x = 100, y = -1$
27. $x = \frac{119}{201}, y = -\frac{59}{201}$　　**29.** $x = 12, y = 24$　　**31.** $r = \frac{1}{3}, s = \frac{1}{5}$　　**33.** $x = 1.0, y = 0.3$
35. $x = 1.6, y = -1.2$　　**37.** $u = 2.0, v = -6.0$　　**39.** Inconsistent　　**41.** -2　　**43.** -7
45. $x = -7, y = 2$　　**47.** $x = \frac{1}{2}, y = 4$　　**49.** $x = 2, y = -2, z = 3$　　**51.** $x = 6, y = 4, z = -3$
53. $i_1 = -\frac{3}{22}$ A, $i_2 = -\frac{39}{110}$ A　　**55.** $m = 416$ mol/h, $n = 331$ mol/h
57. $I_1 = 5$ A, $I_2 = 2$ A, $I_3 = -7$ A　　**59.** $x = 600$ g, $y = 400$ g, $z = 80$ g　　**61.** 45 spots/h, 30 spots/h
63. $d = 40.0$ ft, $s = 30.0$ ft　　**65.** 72 refunds, 112 additional　　**67.** \$300 fixed cost, 25¢ per booklet
69. 45 mi/h, 55 mi/h　　**71.** 750 mL of 5% solution; 250 mL of 25% solution　　**73.** 15 Ω, 20 Ω, 40 Ω
75. 12 nuts, 12 bolts, 24 washers　　**77.** $x = -30.3, -28.5$　　**79.** $x = 0.6348, 0.03160$

Exercises 15-1, page 469

1. 53° **3.** 21° **5.** ∠*BEC* or ∠*CED*
7. ∠*AEB* and ∠*BEC*; ∠*BEC* and ∠*CED*; ∠*AEC* and ∠*CED*; ∠*AEB* and ∠*BED* (any two pairs)
9. ∠*CBE* and ∠*EBA* **11.** ∠*CBD* **13.** 25° **15.** 115° **17.** 40° **19.** 62°
21. ∠1 and ∠5; ∠3 and ∠4 **23.** ∠1 and ∠3 or ∠4 and ∠5 **25.** 50° **27.** 130° **29.** 58°
31. 148° **33.** 40° **35.** 50° **37.** 40° **39.** 100° **41.** 37°22′18″ **43.** 7°13″

Exercises 15-2, page 476

1. 56° **3.** 48° **5.** 68° **7.** 80° **9.** 60° **11.** 120° **13.** 32° **15.** 76° **17.** *AE, GD*
19. *GC, BC* **21.** ∠*BOG* **23.** $\widehat{BG}, \widehat{GC}$ **25.** 60° **27.** 30° **29.** 110° **31.** 35° **33.** 3
35. 5 **37.** 72°, 72° **39.** 60°, 60°, 60° **41.** 135° each **43.** 138° **45.** 117°
47. Horizontally

Exercises 15-3, page 482

1. 5 **3.** 17 **5.** 8 **7.** 4.90 **9.** 10.6 **11.** 28.3 **13.** 59.9 **15.** 40.9 **17.** 2.67
19. 39.1 **21.** 5.66 cm **23.** $p = 37.9$ in, $A = 55.4$ in² **25.** 10,400 ft **27.** 12.5 ft **29.** 206
31. 521 ft **33.** 4.73 m **35.** 17.3 ft **37.** 47.0 cm **39.** 30.4 kV/m **41.** 30.7 cm

43. 990 ft **45.** 29.7 km **47.** 28.3 ft **49.** $h = \dfrac{s\sqrt{3}}{2}$ **51.** 562 mi

Exercises 15-4, page 489

1. 20°, 100°, 60° **3.** 16.5, 19.5 **5.** 40°, 65°, 75° **7.** 12.7, 70°, 70° **9.** Similar
11. Neither **13.** ∠*E*, side *EF* **15.** ∠*U*, side *ST* **17.** 10 **19.** 12 **21.** 21 **23.** 20
25. 16 **27.** 8 **29.** **31.** 3.3 cm, 4.0 cm

33. ∠*XKY* = ∠*NKF*, ∠*KXY* = ∠*KNF*, ∠*XYK* = ∠*NFK* **35.** 8.0 in. **37.** 6.7 ft **39.** 350 cm
41. $2\frac{1}{4}$ in. **43.** 1200 km **45.** 1530 km, 1800 km, 760 km **47.** 20.7 m **49.** 5.3 ft
51. 104 m **53.** 23.3 ft **55.** 17 cm

Exercises 15-5, page 498

1. 6900 ft³ **3.** 99,000 cm³ **5.** 8.85 ft³ **7.** 22,300 cm³ **9.** (a) 795 ft³; (b) 1170 ft³
11. (a) 2.69 m²; (b) 4.03 m² **13.** (a) 12,300 in²; (b) 18,900 in² **15.** (a) 57.4 cm²; (b) 85.0 cm²
17. 18,800 cm³ **19.** 2540 ft³ **21.** 14,700,000 mm³ **23.** 14,800 in³
25. (a) 226,000 mm²; (b) 791,000 mm² **27.** (a) 124 in²; (b) 546 in² **29.** (a) 641 cm²; (b) 1550 cm²
31. (a) 3360 ft²; (b) 15,100 ft² **33.** 5.50 m³ **35.** 594 in² **37.** 56.6 cm³ **39.** 123,000 bu
41. 18,100 cm² **43.** 600 ft³

Exercises 15-6, page 505

1. 54,000 ft³ **3.** 958 mm³ **5.** 76.5 yd² **7.** 240 m² **9.** 1470 ft³ **11.** 27,100 cm³
13. 198 in² **15.** 7210 cm² **17.** 113 ft³ **19.** 5,570,000 mm³ **21.** 11,300 in²

23. 376,000 mm² **25.** 3,330,000 yd³ **27.** $191 **29.** 58.6 mm³ **31.** 23,500 lb **33.** 22.3 in²
35. 78.8 lb **37.** 4.70 m³ **39.** 38,000,000 km² **41.** 1.10 in³ **43.** 1.111 cm **45.** $V = \dfrac{\pi d^2}{6}$
47. $A = 3\pi r^2$

Exercises 15-7, page 508

1. 61° **3.** 90° **5.** 27° **7.** 21.5 cm **9.** ∠CBE **11.** ∠EBD **13.** ∠2 and ∠5
15. 180° **17.** 65° **19.** 32° **21.** 52° **23.** 132° **25.** 36° **27.** 50° **29.** 120° **31.** 40°
33. 25° **35.** 65° **37.** 41 **39.** 42 **41.** 7.36 **43.** 21.1 **45.** 7.5 **47.** 6.3 in.
49. 600 ft³ **51.** 6190 cm³ **53.** 97,700 in³ **55.** 160,000 cm³ **57.** 3990 ft³ **59.** 24,400 cm³
61. 900 ft² **63.** 14.4 m² **65.** 5430 in² **67.** 16,800 cm² **69.** 1230 ft² **71.** 4070 cm²
73. 21.1 ft **75.** 231 cm **77.** 32.8 cm **79.** Yes **81.** 56.1 m **83.** 22.5°, 67.5° **85.** 5.0 ft
87. 864,000 mi **89.** 1600 cm **91.** 10.0 in. **93.** 180° **95.** 12,400 cm² **97.** 667,000 gal
99. 2,550,000 cm³ **101.** 40.0 in² **103.** 7.07 cm³ **105.** 7.44 ft² **107.** 32,300 kg
109. 238 in² **111.** 0.204 ft **113.** 140 cm² **115.** $A = 3\pi r^2 + 2\pi rh$ **117.** 113 in³
119. 16.9 cm

Exercises 16-1, page 521

1. $\frac{9}{41}, \frac{9}{41}, \frac{9}{41}$ **3.** $\frac{9}{40}, \frac{41}{40}, \frac{40}{9}$ **5.** $\frac{1}{2}, 2, \sqrt{3}$ **7.** $\frac{\sqrt{3}}{3}, \frac{1}{2}, \frac{2\sqrt{3}}{3}$ **9.** 0.624, 1.25, 1.60
11. 0.782, 0.782, 0.798 **13.** $\frac{4}{5}, \frac{3}{4}$ **15.** $\frac{7}{24}, \frac{24}{25}$ **17.** $\frac{\sqrt{2}}{2}, 1$ **19.** 1.4, 0.73 **21.** 0.63, 1.3
23. 0.679, 1.47 **25.** $\frac{\sqrt{2}}{2}$ **27.** 1.4 **29.** 0.491 **31.** 1.04 **33.** 0.5000 **35.** 1.41
37. $\cot B, \sec B$ **39.** $\sin A = \dfrac{a}{c}$; $\cos A = \dfrac{b}{c}$; $\tan A = \dfrac{a}{b}$; $\sec A = \dfrac{c}{b}$; $\csc A = \dfrac{c}{a}$; $\cot A = \dfrac{b}{a}$ **41.** 1
43. $\frac{5}{13}, \frac{12}{13}, \frac{5}{12}$; $\frac{5}{13} \div \frac{12}{13} = \frac{5}{12}$

Exercises 16-2, page 528

1. 0.8480 **3.** 0.4557 **5.** 1.500 **7.** 0.9178 **9.** 0.2382 **11.** 0.7046 **13.** 4.222
15. 2.475 **17.** 32.0° **19.** 18.9° **21.** 60.5° **23.** 61.9° **25.** 52.7° **27.** 18.0° **29.** 84.3°
31. 58.2° **33.** 1.206 **35.** 1.179 **37.** 53.1° **39.** 40.9° **41.** 34.6° **43.** 79.4° **45.** 0.72
47. 1.9 **49.** 13 V **51.** 126.5 m **53.** 1.56 **55.** 23.9 m **57.** Error display
59. 2.05 $\boxed{1/x}$ $\boxed{\text{INV}}$ $\boxed{\text{COS}}$ $\boxed{\text{SIN}}$

Exercises 16-3, page 534

1. $B = 60.0°, b = 20.8, c = 24.0$ **3.** $A = 33.7°, a = 12.5, b = 18.7$
5. $B = 13.2°, a = 30.6, b = 7.17$ **7.** $A = 28.8°, B = 61.2°, b = 1.18$
9. $A = 82.6°, B = 7.4°, a = 44.6$ **11.** $A = 64.2°, B = 25.8°, b = 4.71$
13. $B = 83.0°, a = 1.88, c = 15.4$ **15.** $A = 33.5°, B = 56.5°, c = 118$ **17.** 29.1 ft **19.** 304 cm
21. 3.4° **23.** 318 cm **25.** 85.2° **27.** 4000 ft **29.** 2.94 in. **31.** 6.8° **33.** 4.60 in.
35. 1640 ft **37.** 243 km **39.** 9.12 in. **41.** 1340 ft **43.** 9790 ft **45.** 337 ft **47.** $d = \dfrac{88}{\tan A}$

Exercises 16-4, page 538

1. 0.735 **3.** 0.130 **5.** 0.116 **7.** 1.16 **9.** 30.0° **11.** 45.0° **13.** 3.0° **15.** 56.7°
17. $\frac{23}{41}, \frac{23}{41}$ **19.** $\frac{41}{23}, \frac{23}{34}$ **21.** 0.549, 0.656 **23.** 0.549, 1.52 **25.** 0.385, 0.417 **27.** 3.78, 0.967
29. 0.461, 0.418 **31.** 0.898, 2.04 **33.** 38.0° **35.** 13.1° **37.** 28.4° **39.** 53.4°
41. $B = 69.0°, a = 2.48, b = 6.47$ **43.** $A = 57.3°, a = 71.5, c = 85.0$

45. $A = 59.1°$, $B = 30.9°$, $c = 10.1$ **47.** $A = 63.9°$, $B = 26.1°$, $b = 47.5$
49. $B = 52.75°$, $a = 8397$, $b = 11,040$ **51.** $B = 89.98°$, $b = 12.18$, $c = 12.18$
53. $A = 33.473°$, $B = 56.527°$, $c = 230.90$ **55.** $A = 88.9125°$, $B = 1.08748°$, $a = 112.483$
57. 0.5000   **59.** 7.713   **61.** 0.614 μm   **63.** 30.6°   **65.** 0.529 m   **67.** 35.9°   **69.** 940 ft
71. 26.6°, 63.4°   **73.** 28.1°   **75.** 0.0291 h   **77.** 50 mi   **79.** 575 ft   **81.** 70°   **83.** 4000 mi
85. 1.46 in.   **87.** No; yes

Exercises 17-1, page 548

1. **3.**

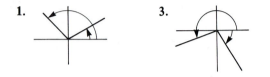

5. $\dfrac{12}{13}, \dfrac{12}{5}$ **7.** $-\dfrac{\sqrt{13}}{3}, \dfrac{2}{\sqrt{13}}$ **9.** $1, -\dfrac{1}{\sqrt{2}}$ **11.** $\dfrac{3}{5}, -\dfrac{4}{5}$ **13.** 0.316, 0.333
15. -0.413, 2.62   **17.** -0.772, -1.57 **19.** -1.85, 0.842   **21.** $+\ -\ -$ **23.** $-\ -\ -$
25. $+\ +\ +$ **27.** $-\ -\ -$ **29.** IV   **31.** I   **33.** II   **35.** II   **37.** II   **39.** II   **41.** II
43. Quadrantal angle (positive y axis)

Exercises 17-2, page 555

1. sin 15°, $-$cos 50°   **3.** $-$cos 27°, $-$csc 70°   **5.** $-$tan 18°, sec 10°   **7.** $-$cot 70°, tan 60°
9. -0.5446 **11.** -10.78 **13.** -1.090 **15.** -0.9265 **17.** 3.732   **19.** -1.718
21. 0.9759   **23.** 2.778   **25.** -0.5358 **27.** 2.414   **29.** 238.0°, 302.0°   **31.** 66.4°, 293.6°
33. 62.2°, 242.2°   **35.** 15.8°, 195.8°   **37.** 232.0°, 308.0°   **39.** 219.3°, 320.7°
41. 167.8°, 192.2°   **43.** 178.0°, 182.0°   **45.** 334.0°   **47.** 129.8°   **49.** 119.6°   **51.** 189.2°
53. 85.2°   **55.** 354 lb   **57.** 334   **59.** 2.4°

Exercises 17-3, page 562

1. $C = 72.6°$, $b = 4.52$, $c = 7.23$ **3.** $C = 109.0°$, $a = 1390$, $b = 1300$
5. $A = 149.7°$, $C = 9.6°$, $a = 221$ **7.** $B = 8.5°$, $C = 28.3°$, $c = 0.733$
9. $A = 99.4°$, $b = 55.1$, $c = 24.4$ **11.** $A = 68.1°$, $a = 552$, $c = 537$
13. $A_1 = 61.5°$, $C_1 = 70.4°$, $c_1 = 28.1$; $A_2 = 118.5°$, $C_2 = 13.4°$, $c_2 = 6.89$
15. $A_1 = 107.3°$, $C_1 = 41.3°$, $a_1 = 1060$; $A_2 = 9.9°$, $C_2 = 138.7°$, $a_2 = 191$
17. $B = 68.5°$, $C = 42.4°$, $b = 93.8$ **19.** No solution   **21.** $B = 60.0°$, $C = 90.0°$, $b = 173$
23. No solution   **25.** 15.6 in., 27.2 in.   **27.** 21,000 m   **29.** 19.7 km   **31.** 29,000 km
33. 0.618 m   **35.** 44.9 m

Exercises 17-4, page 568

1. $A = 55.3°$, $B = 37.2°$, $c = 27.1$ **3.** $A = 9.9°$, $C = 111.2°$, $b = 38,600$
5. $A = 70.9°$, $B = 11.1°$, $c = 1580$ **7.** $A = 18.2°$, $B = 22.2°$, $C = 139.6°$
9. $A = 42.3°$, $B = 30.3°$, $C = 107.4°$ **11.** $A = 51.5°$, $B = 35.1°$, $C = 93.4°$
13. $A = 46.1°$, $B = 109.2°$, $c = 138$ **15.** $A = 132.4°$, $C = 10.3°$, $b = 4.20$
17. $A = 39.9°$, $B = 56.8°$, $C = 83.4°$ **19.** $A = 48.6°$, $B = 102.3°$, $C = 29.1°$
21. $A = 44.37°$, $B = 60.51°$, $C = 75.12°$ **23.** $B = 4.05°$, $C = 166.30°$, $a = 24.25$ **25.** 1290 m
27. 42.8°, 54.3°, 82.9°   **29.** 96.2 cm   **31.** 19.8°   **33.** 30.0°, 60.0°, 90.0°   **35.** 158 lb

Exercises 17-5, page 570

1. $\sin \theta = \frac{3}{5}$, $\cos \theta = \frac{4}{5}$, $\tan \theta = \frac{3}{4}$, $\cot \theta = \frac{4}{3}$, $\sec \theta = \frac{5}{4}$, $\csc \theta = \frac{5}{3}$
3. $\sin \theta = -0.275$, $\cos \theta = 0.962$, $\tan \theta = -0.286$, $\cot \theta = -3.50$, $\sec \theta = 1.04$, $\csc \theta = -3.64$
5. II 7. IV 9. $-\cos 48°$, $\tan 14°$ 11. $-\sin 71°$, $\sec 15°$ 13. -0.4540
15. -0.3057 17. -1.082 19. -0.5764 21. -2.552 23. 4.230 25. 0.6820
27. 1.003 29. $37.0°, 217.0°$ 31. $114.9°, 245.1°$ 33. $27.4°, 152.6°$ 35. $189.4°, 350.6°$
37. $155.0°, 335.0°$ 39. $56.3°, 123.7°$ 41. $C = 71.7°$, $b = 120$, $c = 130$
43. $A = 21.2°$, $b = 128$, $c = 43.1$ 45. $A = 34.8°$, $B = 53.5°$, $c = 5.60$
47. $A = 59.8°$, $C = 58.2°$, $b = 289$
49. $A_1 = 60.6°$, $C_1 = 65.1°$, $a_1 = 17.5$; $A_2 = 10.8°$, $C_2 = 114.9°$, $a_2 = 3.75$
51. $B_1 = 68.5°$, $C_1 = 60.5°$, $c_1 = 73.4$; $B_2 = 111.5°$, $C_2 = 17.5°$, $c_2 = 25.3$ 53. $62.4°, 83.3°, 34.3°$
55. $10.5°, 36.4°, 133.1°$ 57. $41.1°, 32.8°, 431$ 59. $65.4°, 45.5°, 69.1°$ 61. -115 V
63. 585 m 65. $2.2°$ 67. 281 m 69. 2787 ft 71. $54.8°$ 73. $0.6788, -1.362$ 75. III
77. $55.064°, 58.565°, 66.370°$ 79. $6441.1, 5366.9, 37°11'19''$

Exercises 18-1, page 580

1. (a) Scalar: only magnitude is specified; (b) Vector: magnitude and direction are specified
3. (a) Vector: magnitude and direction are specified; (b) Scalar: only magnitude is specified
5. (a) Vector: magnitude and direction are specified; (b) Scalar: only magnitude is specified
7. (a) Scalar: only magnitude is specified; (b) Vector: magnitude and direction are specified

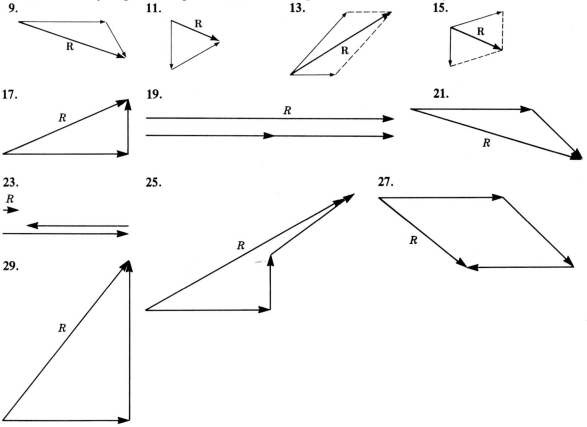

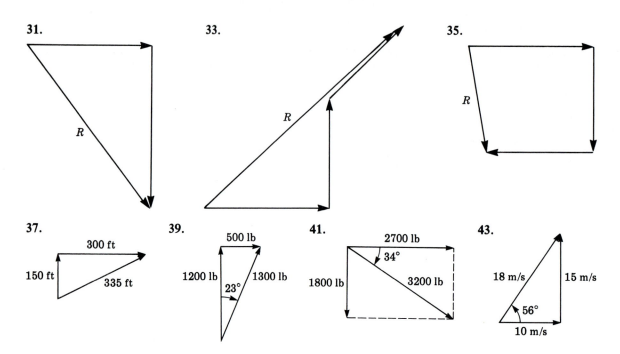

31.   **33.**   **35.**

37.   **39.**   **41.**   **43.**

Exercises 18-2, page 585

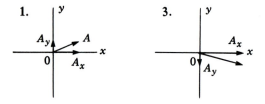

1.   **3.**

5. $A_x = 25.4, A_y = 12.9$ **7.** $A_x = -8.60, A_y = 57.6$ **9.** $A_x = 3340, A_y = -5590$
11. $A_x = -701, A_y = -229$ **13.** $A_x = -1344, A_y = -3212$ **15.** $A_x = -0.1306, A_y = 0.8636$
17. $A_x = 18.2, A_y = 18.2$ **19.** $A_x = 3.45, A_y = 5.32$ **21.** $A_x = -26.7, A_y = -217$
23. $A_x = -969, A_y = -969$ **25.** $A_x = 52.4, A_y = -69.7$ **27.** $A_x = -0.03233, A_y = -0.04370$
29. $A_x = 25815, A_y = -1714.6$ **31.** $A_x = -13.378, A_y = 31.981$
33. 33.3 mi (east); 14.8 mi (north)   **35.** 211 N (left); 751 N (vertical)
37. 0.013 (x axis); 0.020 (y axis)   **39.** 3.72 A, 39.1°

Exercises 18-3, page 594

1. $R = 5.39, \theta = 21.8°$ **3.** $R = 1460, \theta = 59.2°$ **5.** $R = 38.3, \theta = 25.2°$
7. $R = 25.49, \theta = 61.8°$ **9.** $R = 10.7, \theta = 23.7°$ **11.** $R = 276, \theta = 55.3°$
13. $R = 115, \theta = 102.0°$ **15.** $R = 121, \theta = 272.1°$ **17.** $R = 10, \theta = 36.9°$
19. $R = 61.1, \theta = 116.8°$ **21.** $R = 1560, \theta = 201.6°$ **23.** $R = 7675.1, \theta = 175.07°$
25. $R = 13, \theta = 293°$ **27.** $R = 28.9, \theta = 327.5°$ **29.** $R = 4352, \theta = 321.0°$
31. $R = 0.321, \theta = 193.7°$ **33.** 264 lb at an angle that is 36.7° from the 212 lb force
35. 46.0 mi, 34.4° S of E   **37.** 3990 km/h, 5.0° from direction of plane   **39.** 53.9 lb, $\theta = 68.2°$
41. 130 V at an angle of 6.9° from V_1 **43.** Yes, to the right

Exercises 18-4, page 597

1. (a) Scalar: only magnitude is specified; (b) Vector: magnitude and direction are specified
3. (a) Scalar: only magnitude is specified; (b) Vector: magnitude and direction are specified
5.   **7.**   **9.**

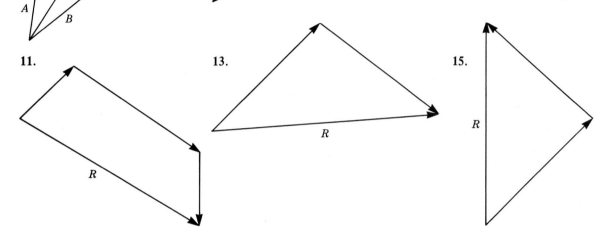

11.   **13.**   **15.**

17. $A_x = 6.77$, $A_y = 2.73$ **19.** $A_x = -1380$, $A_y = 778$
21. $A_x = 15.9$, $A_y = 6.50$ **23.** $A_x = 8.903$, $A_y = -13.87$ **25.** 35 mi/h (west), 9.3 mi/h (north)
27. 431 lb (down), 349 lb (horizontally)   **29.** $R = 38.8$, $\theta = 27.7°$ **31.** $R = 5530$, $\theta = 81.7°$
33. $R = 27.66$, $\theta = 40.0°$ **35.** $R = 1.10$, $\theta = 48.0°$ **37.** $R = 47$, $\theta = 314°$
39. $R = 64.8$, $\theta = 68.7°$ **41.** 113 lb   **43.** 348 km/h (horizontally), 2220 km/h (vertically)
45. 232 lb   **47.** 286 mi   **49.** $R = 13{,}500$ lb, $\theta = 45.8°$ (above the horizontal)
51. $R = 1180$ lb, $\theta = 43.0°$ (away from the force of 532 lb)   **53.** 374 lb   **55.** 61.0 mi, 35.0° N of E
57. 153 V   **59.** $R = 868.6$, $\theta = 6.8°$ N of E   **61.** $R = 24$ m, $\theta = 36.5°$ N of E
63. 15.3 mi/h, 11.3° off ship's path   **65.** 167 mi/h, 84.2° S of E   **67.** 30 lb, 53.4° from the 32 lb force

Exercises 19-1, page 609

1. $\dfrac{2\pi}{9}, \dfrac{4\pi}{45}$ **3.** $\dfrac{11\pi}{36}, \dfrac{11\pi}{6}$ **5.** $\dfrac{\pi}{6}, \dfrac{3\pi}{4}$ **7.** $\dfrac{35\pi}{36}, \dfrac{7\pi}{6}$ **9.** 120°, 36°   **11.** 15°, 225°
13. 168°, 140°   **15.** 35°, 135°   **17.** 0.802   **19.** 3.31   **21.** 4.86   **23.** 3.18   **25.** 46°
27. 143°   **29.** 186°   **31.** 710°   **33.** 0.5000   **35.** −1.732   **37.** 0.8674   **39.** 1.197
41. 1.556   **43.** 7.086   **45.** 0.85730   **47.** −1.145   **49.** −9.657   **51.** 3.796
53. 1.047, 5.236   **55.** 1.9116, 5.0532   **57.** 0.4500, 2.692   **59.** 4.470, 4.954   **61.** 0.147°
63. 160 V   **65.** 33.4 m   **67.** 1620°, 9π rad   **69.** 0.98339   **71.** 0.66300   **73.** 0.4814
75. 0.151504   **77.** 146°   **79.** 13.8°   **81.** −0.315209   **83.** 0.000421

Exercises 19-2, page 614

1. (a) 7.34 cm; (b) 13.4 cm² 　　**3.** (a) 355 mm; (b) 73,000 mm² 　　**5.** (a) 0.620 ft; (b) 0.735 ft²
7. (a) 26.9 in.; (b) 88.0 in² 　　**9.** 70.6 cm/s 　　**11.** 6920 ft/min 　　**13.** 37.7 cm 　　**15.** 679 cm²
17. 1.62 　　**19.** 87.8 ft² 　　**21.** 0.114 m 　　**23.** 2.0 mi 　　**25.** 3130 cm 　　**27.** 26.4 ft 　　**29.** 59 rad/s
31. 2.6×10^{-6} rad/s 　　**33.** 153,000 cm/min 　　**35.** 13,200 ft/min 　　**37.** 134 ft² 　　**39.** 4.959 in./s
41. 4.85×10^{-6} 　　**43.** 80,400,000 mi

Exercises 19-3, page 621

1. 0, −0.7, −1, −0.7, 0, 0.7, 1, 0.7, 0, −0.7, −1, −0.7, 0, 0.7, 1, 0.7, 0

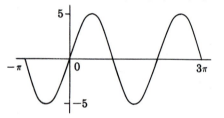

3. 0, −3.5, −5, −3.5, 0, 3.5, 5, 3.5, 0, −3.5, −5, −3.5, 0, 3.5, 5, 3.5, 0 　　**5.** $y = 3 \sin x$

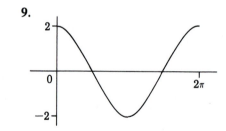

7. $y = -6 \sin x$ 　　**9.**

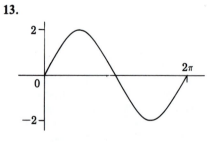

11.

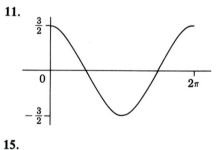

13.

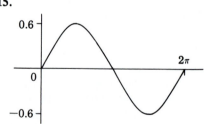

15.

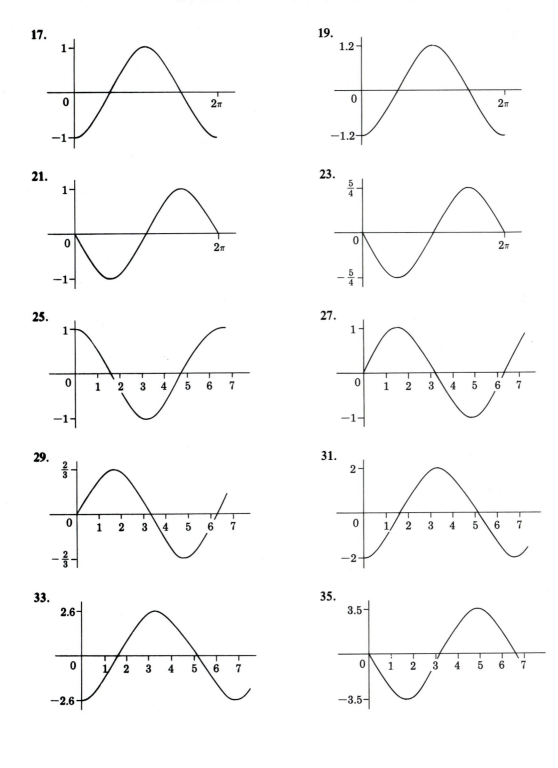

Exercises 19-4, page 626

1. $\dfrac{2\pi}{3}$ 3. $\dfrac{\pi}{2}$ 5. $\dfrac{2\pi}{3}$ 7. $\dfrac{\pi}{3}$ 9. $\dfrac{2\pi}{5}$ 11. 1   13. 4π 15. 3π 17. $\dfrac{4}{3}$ 19. $\dfrac{2}{\pi}$

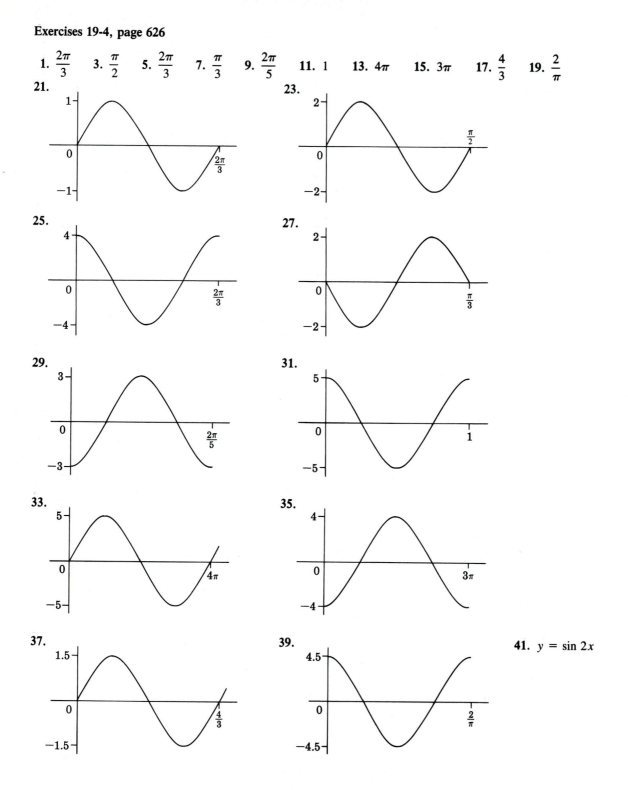

21.

23.

25.

27.

29.

31.

33.

35.

37.

39.

41. $y = \sin 2x$

43. $y = \sin \pi x$ **45.** $y = \sin 2\pi x$ **47.** $y = \sin 5x$ **49.** $y = 2 \sin \pi x$ **51.** $y = 5 \sin \dfrac{\pi}{2}x$

53. **55.**

57. **59.**

Exercises 19-5, page 632

1. $1, 2\pi, -\dfrac{\pi}{3}$ **3.** $1, 2\pi, \dfrac{\pi}{3}$

5. $3, \pi, -\dfrac{\pi}{8}$ **7.** $4, \dfrac{2\pi}{3}, -\dfrac{\pi}{3}$

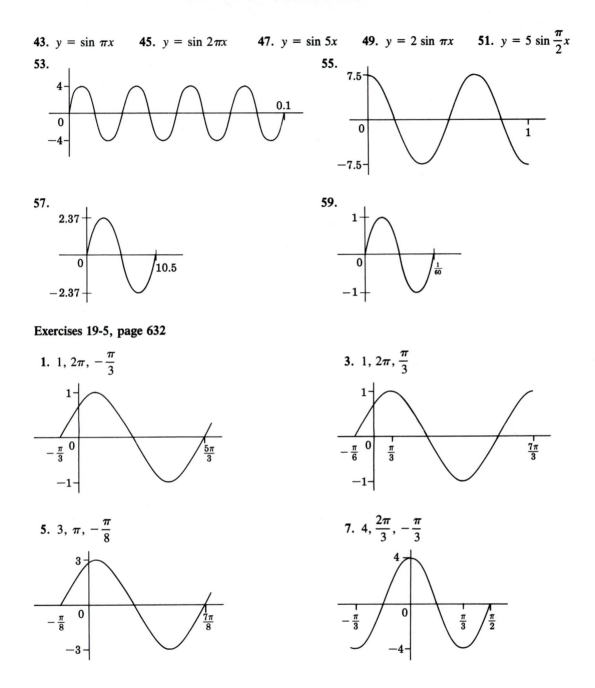

9. $2, 6\pi, -\dfrac{3\pi}{2}$

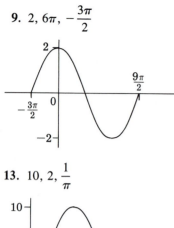

11. $6, 3\pi, -\dfrac{\pi}{2}$

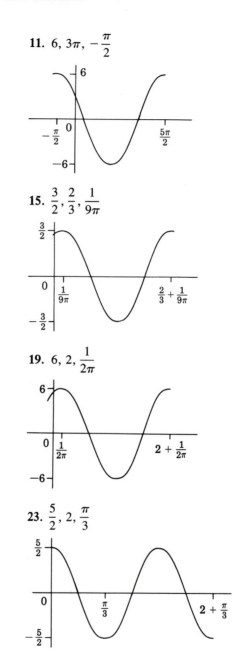

13. $10, 2, \dfrac{1}{\pi}$

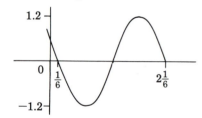

15. $\dfrac{3}{2}, \dfrac{2}{3}, \dfrac{1}{9\pi}$

17. $1.2, 2, \dfrac{1}{6}$

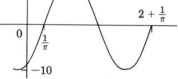

19. $6, 2, \dfrac{1}{2\pi}$

21. $\dfrac{3}{4}, \dfrac{2}{\pi}, -\dfrac{1}{\pi}$

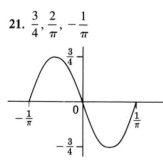

23. $\dfrac{5}{2}, 2, \dfrac{\pi}{3}$

25.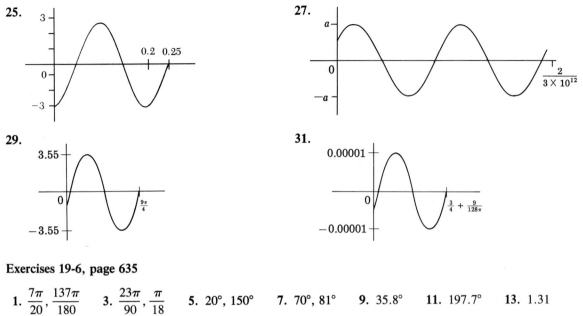

27.

29.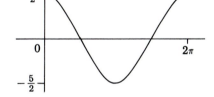

31.

Exercises 19-6, page 635

1. $\dfrac{7\pi}{20}, \dfrac{137\pi}{180}$ **3.** $\dfrac{23\pi}{90}, \dfrac{\pi}{18}$ **5.** 20°, 150° **7.** 70°, 81° **9.** 35.8° **11.** 197.7° **13.** 1.31

15. 5.934 **17.** 2.662 **19.** 0.16 **21.** 0.6894, 2.452 **23.** 1.998, 4.285

25. (a) 7.73 cm; (b) 47.9 cm² **27.** (a) 2.27 in.; (b) 8.23 in² **29.** 4.60 m/min **31.** 353 m/min

33.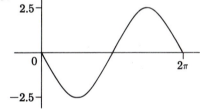

35.

37.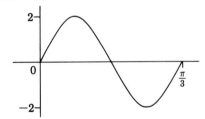

39.

41.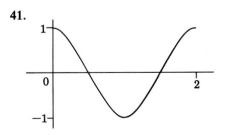

43.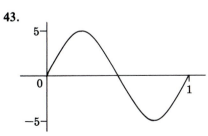

45.
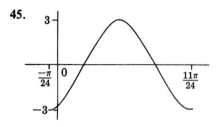

47.
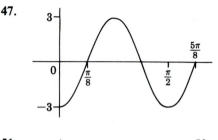

49.
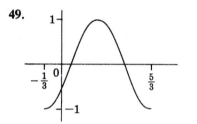

51.

53. $y = 8 \sin \pi x$

55. $y = -3 \cos 2\pi x$ **57.** 8.78 cm   **59.** 2670 mi   **61.** 117.0° or 2.04 rad   **63.** 36.3 cm²
65. 126,000 cm/min   **67.** 138 rad/s or 21.9 r/s   **69.** 905 r/min   **71.** 1040 mi/h
73.   **75.** 21 V   **77.**   **79.** 374,000 km

Exercises 20-1, page 643

1. $4j$ **3.** $-3j$ **5.** $0.5j$ **7.** $3j\sqrt{3}$ **9.** $-4j\sqrt{3}$ **11.** $0.02j$ **13.** $5 + j$ **15.** $6 - 2j$
17. $-4 + 2j\sqrt{2}$ **19.** $14 - 3j\sqrt{7}$ **21.** -9 **23.** -6 **25.** 3   **27.** $-2\sqrt{3}$ **29.** -6
31. $-\sqrt{33}$ **33.** -1 **35.** j **37.** j **39.** -1 **41.** -1 **43.** $-j$ **45.** $3 + 8j$
47. -4 **49.** $-9j$ **51.** $-2 + 7j$

Exercises 20-2, page 647

1. $5 - 7j$ **3.** $4 + 5j$ **5.** $1 + 3j$ **7.** $-12j$ **9.** $2 - 6j$ **11.** $-3 + 20j$ **13.** -30
15. 5   **17.** $48 - 9j$ **19.** $36j$ **21.** $-6 + 4j$ **23.** $36 + 3j$ **25.** $\dfrac{-8 + 6j}{25}$ **27.** $\dfrac{8 - 3j}{2}$
29. $\dfrac{53 - 27j}{61}$ **31.** $\dfrac{-5 - j}{2}$ **33.** $\dfrac{3 + 3j}{4}$ **35.** $\dfrac{-3 + 7j}{6}$ **37.** $-2j$ **39.** $2 - 3j$
41. $-30 - 40j$ **43.** $3 + 4j$ **45.** $-10 + 6j$ **47.** $-48j$ **49.** 100   **51.** 53
53. $14.7 - 2.1j\,\Omega$ **55.** $3.1 + 5.8j$ V

Exercises 20-3, page 650

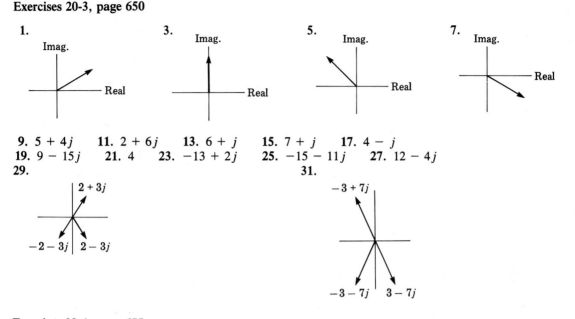

9. $5 + 4j$ **11.** $2 + 6j$ **13.** $6 + j$ **15.** $7 + j$ **17.** $4 - j$
19. $9 - 15j$ **21.** 4 **23.** $-13 + 2j$ **25.** $-15 - 11j$ **27.** $12 - 4j$
29. **31.**

Exercises 20-4, page 655

1. $5(\cos 37° + j \sin 37°)$ **3.** $5(\cos 323° + j \sin 323°)$ **5.** $4.50(\cos 302.9° + j \sin 302.9°)$
7. $6.52(\cos 55.8° + j \sin 55.8°)$ **9.** $2(\cos 225° + j \sin 225°)$ **11.** $2(\cos 120° + j \sin 120°)$
13. $10(\cos 180° + j \sin 180°)$ **15.** $6(\cos 90° + j \sin 90°)$ **17.** $6.15(\cos 43.0° + j \sin 43.0°)$
19. $10.3(\cos 335.2° + j \sin 335.2°)$ **21.** $0.348(\cos 76.4° + j \sin 76.4°)$
23. $56.0(\cos 212.5° + j \sin 212.5°)$ **25.** $1.03 + 2.82j$ **27.** $1.17 - 0.940j$ **29.** $-10j$
31. -25 **33.** $41 + 29j$ **35.** $6.2 - 11j$ **37.** $10.06 + 7.380j$ **39.** $-1.258 - 1.797j$
41. $56.8 + 7.18j$ V **43.** Magnitude: 81.2 lb; direction: 322.9°

Exercises 20-5, page 656

1. $8j$ **3.** $-20j$ **5.** $3j\sqrt{6}$ **7.** $-2j\sqrt{14}$ **9.** $-6 + 10j$ **11.** $3 - 4j\sqrt{3}$ **13.** -25
15. $-\sqrt{14}$ **17.** -1 **19.** 1 **21.** $14 + 4j$ **23.** $-1 - 7j$ **25.** $15j$ **27.** $-11 + 27j$
29. $\dfrac{-2 - 16j}{5}$ **31.** $\dfrac{-46 - 2j}{53}$ **33.** Sum: 12; product: 100 **35.** Sum: 14; product: 58
37. $3 + 3j$ **39.** $-9 - 7j$ **41.** $13(\cos 67° + j \sin 67°)$ **43.** $21.8 (\cos 214.3° + j \sin 214.3°)$
45. $5.15 + 14.3j$ **47.** $14.1 - 8.97j$ **49.** Yes; yes **51.** $9 + 42j$ V

Exercises from Appendix C, page 686

1. 3 **3.** 9 **5.** 108 **7.** 305 **9.** 110 **11.** 10001 **13.** 101110 **15.** 1001111 **17.** 101
19. 1111 **21.** 101101 **23.** 11011001 **25.** 110 **27.** 1111 **29.** 1110101
31. 100011011010 **33.** 110 **35.** 11101 **37.** 11010 **39.** 1011110

Index